Springer Handbooks of Computational Statistics

Series Editors

James E. Gentle
Wolfgang K. Härdle
Yuichi Mori

For further volumes:
http://www.springer.com/series/7286

Vincenzo Esposito Vinzi
Wynne W. Chin
Jörg Henseler
Huiwen Wang

Editors

Handbook of Partial Least Squares

Concepts, Methods and Applications

Springer

Editor-in-Chief
Vincenzo Esposito Vinzi
ESSEC Business School of Paris and Singapore
Department of Information Systems & Decision Sciences
Avenue Bernard Hirsch - B.P. 50105
95021 Cergy-Pontoise Cedex France
vinzi@essec.fr

Jörg Henseler
Nijmegen School of Management
Institute for Management Research
Radboud Universiteit Nijmegen
6500 HK Nijmegen
The Netherlands
j.henseler@fm.ru.nl

Editors
Wynne W. Chin
Bauer Faculty Fellow
Department of Decision and Information Sciences
C.T. Bauer College of Business
334 Melcher Hall, room 280D
University of Houston
Houston, Texas 77204-6282
wchin@uh.edu

Huiwen Wang
School of Economic Management
BeiHang University
37, XueYuan Road, HaiDian District
Beijing 100191
P. R. China
wanghw@vip.sina.com

ISBN 978-3-540-32825-4 e-ISBN 978-3-540-32827-8
DOI 10.1007/978-3-540-32827-8
Springer Heidelberg Dordrecht London New York

Library of Congress Control Number: 2009943435

Cover design: deblik, Berlin, Germany

Printed on acid-free paper

Springer is part of Springer Science+Business Media (www.springer.com)

Contents

PLS Regression

List of Contributors

Sönke Albers Institute of Innovation Research, Christian-Albrechts-University at Kiel, Westring 425, 24098 Kiel, Germany, albers@bwl.uni-kiel.de

Maria H. Almeida Faculty of Economics, New University of Lisbon, Campus de Campolide, 1099-032 Lisbon, Portugal, mhalmeida@fe.unl.pt

Silvano Amato Dipartimento di Matematica e Statistica, Università degli Studi di Napoli "Federico II", Via Cintia 26, Complesso Monte S. Angelo, 80126 Napoli, Italy, silvano.amato@gmail.com

Francisco Arteaga Department of Statistics, Universidad Católica de Valencia San Vicente Martir, Guillem de Castro, 175, Valencia 46008, Spain, francisco.arteaga@ucv.es

Carmen Barroso Management and Marketing Department, University of Seville, Ramón y Cajal, 1, 41018 Sevilla, Spain, barroso@us.es

Silvia Boßow-Thies Capgemini Telecom Media and Networks Deutschland GmbH, Neues Kanzler Eck 21, 10719 Berlin, Germany, silvia.bossow-thies@capgemini.com

Gabriel Cepeda Carrión Departamento de Administración de Empresas y Marketing, Universidad de Sevilla, Ramón y Cajal, 1, 41018 Sevilla, Spain, gabi@us.es

Wynne W. Chin Department of Decision and Information Sciences, Bauer College of Business, University of Houston, TX, USA, wchin@uh.edu

Pedro S. Coelho ISEGI – New University of Lisbon, Campus de Campolide, 1070-032 Lisbon, Portugal, psc@isegi.unl.pt

Clara Cordeiro Department of Mathematics, FCT, University of Algarve, Campus de Gambelas, 8005-139 Faro, Portugal, ccordei@ualg.pt

Sébastien Crochemore Materials Engineering Department, Technocentre Renault 1, avenue du Golf, API TCR LAB 252, 78 288 Guyancourt Cedex, France, sebastien.crochemore@renault.com

Ahmad Daryanto Department of Business Analysis, Systems and Information Management, Newcastle Business School, City Campus East, Northumbria University, Newcastle upon Tyne, NE1 8ST, UK, ahmad.daryanto@northumbria.ac.uk

Ko de Ruyter Department of Marketing and Marketing Research, Maastricht University, P.O. Box 616 6200 MD, The Netherlands, k.deruyter@mw.unimaas.nl

Jens Dibbern Department of Information Engineering, Institute of Information Systems, University of Bern, Engehaldenstr. 8, Room 204, 3012 Bern, Switzerland, jens.dibbern@iwi.unibe.ch

Theo K. Dijkstra SNS Asset Management, Research and Development, Pettelaarpark 120, P.O. Box 70053, 5201 DZ 's-Hertogenbosch, The Netherlands, theo.dijkstra@snsam.nl
and
University of Groningen, Economics and Econometrics, Zernike Complex, P.O. Box 800, 9700 AV, Groningen, The Netherlands, t.k.dijkstra@rug.nl

Paulo Alexandre de Oliveira Duarte Departamento de Gestão e Economia, Universidade da Beira Interior, Estrada do Sineiro, 6200-209 Covilhã, Portugal, pduarte@ubi.pt

Markus Eberl Senior Consultant Models and Methods, TNS Infratest Forschung GmbH, Landsberger Straße 338, 80687 München, Germany, markus.eberl@tns-infratest.com

Andreas Eggert University of Paderborn, Warburger Str. 100, 33098 Paderborn, Germany, andreas.eggert@notes.upb.de

Lennart Eriksson Umetrics Inc, 17 Kiel Ave, Kinnelon, NJ 07405, USA, Lennart.eriksson@umetrics.com

Jacob Eskildsen School of Business, University of Aarhus, Haslegaardsvej 10, 8210 Aarhus V, Denmark, Eskildsen@asb.dk

Vincenzo Esposito Vinzi Dept. of Information Systems and Decision Sciences ESSEC Business School of Paris,
Avenue Bernard Hirsch – B.P. 50105, 95021 Cergy-Pontoise, Cedex, France, vinzi@essec.fr

Thierry Fahmy Addinsoft, 40 rue Damrémont, 75018 Paris, France, fahmy@xlstat.com

Georg Fassott Department of Marketing, University of Kaiserslautern, Postfach 30 49, 67653 Kaiserslautern, Germany, fassott@wiwi.uni-kl.de

Martina González Gallarza Faculty of Economics, Department of Marketing, Universitat de Valencia, Avenida de los Naranjos s/n, Valencia 46022, Spain, martina.gallarza@uv.es

Ina Garnefeld University of Paderborn, Warburger Str. 100, 33098 Paderborn, Germany, ina.garnefeld@notes.upb.de

Irene Gil Department of Marketing, Universitat de València, Avenida de los Naranjos s/n, Valencia 46022, Spain, irene.gil@uv.es

Oliver Götz University of Münster, Marketing Centrum Münster, Institute of Marketing, Am Stadtgraben 13-15, 48143 Münster, Germany, o.goetz@uni-muenster.de

Dwayne D. Gremler Department of Marketing, College of Business Administration, Bowling Green State University, Bowling Green, OH 43403, USA, gremler@cba.bgsu.edu

Christiane Guinot Biometrics and Epidemiology unit, C.E.R.I.E.S, 20 rue Victor Noir, 92521 Neuilly sur Seine, France, christiane.guinot@ceries-lab.com
and
Computer Science Laboratory, Ecole Polytechnique, University of Tours, France

Mohamed Hanafi Unité Mixte de Recherche (ENITIAA-INRA) en Sensométrie et Chimiométrie, ENITIAA, Rue de la Géraudière – BP 82225, Nantes 44322, Cedex 3, France, hanafi@enitiaa-nantes.fr

Sabrina Helm University of Arizona, John and Doris Norton School of Family and Consumer Sciences, 650 N. Park Ave, P.O. Box 210078, Tucson, AZ 85721-0078, USA, helm@email.arizona.edu

Jörg Henseler Nijmegen School of Management, Radboud University Nijmegen, P.O. Box 9108, 6500 HK Nijmegen, The Netherlands, j.henseler@fm.ru.nl

Lutz Hildebrandt Institute of Marketing, Humboldt University Berlin, Unter den Linden 6, 10099 Berlin, Germany, hildebr@wiwi.hu-berlin.de

John Hulland Katz Business School, University of Pittsburgh, Pittsburgh, PA 15260, USA, jhulland@katz.pitt.edu

Nouna Kettaneh NNS Consulting, 42 Pine Hill Rd, Hollis, NH 03049, USA, nouna@nnsconsulting.com

Manfred Krafft University of Münster, Marketing Centrum Münster, Institute of Marketing, Am Stadtgraben 13-15, 48143 Münster, Germany, mkrafft@uni-muenster.de

Henning Kreis Marketing-Department, Freie Universität Berlin, School of Business and Economics, Otto-von-Simson-Str. 19, 14195 Berlin, Germany, henning.kreis@fu-berlin.de

Kai Kristensen School of Business, University of Aarhus, Haslegaardsvej 10, 8210 Aarhus V, Denmark, kak@asb.dk

Dapeng Li Agricultural Bank of China, Beijing 100036, China, zh.ldp@intl.abocn.com

Kerstin Liehr-Gobbers Hering Schuppener Consulting, Kreuzstraße 60, 40210 Düsseldorf, Germany, kliehr@heringschuppener.com

Alexandra Machás Polytechnic Institute of Lisbon, Escola Superior de Comunicação Social Campus de Benfica do IPL, 1549-014 Lisboa, Portugal, amachas@escs.ipl.pt

Harald Martens Norwegian Food Research Institute, Matforsk, 1430 Ås, Norway, harald.martens@matforsk.no
and
Faculty of Life Sciences, Department of Food Science, University of Copenhagen, Rolighedsvej 30, 1958 Frederiksberg C, Denmark
and
Norwegian University of Life Sciences, IKBM/CIGENE, P.O. Box 5003, 1432 Ås, Norway

Magni Martens Norwegian Food Research Institute, Matforsk, 1430 Ås, Norway, magni.martens@matforsk.no
and
Faculty of Life Sciences, Department of Food Science, University of Copenhagen, Rolighedsvej 30, 1958 Frederiksberg C, Denmark

Emmanuelle Mauger Biometrics and Epidemiology unit, C.E.R.I.E.S, 20 rue Victor Noir, 92521 Neuilly sur Seine, France, emmanuelle.mauger@ceries-lab.com

Meng Jie School of Statistics, Central University of Finance and Economics, Beijing 100081, China, mengjie517@126.com

Maria Manuela Neves Department of Mathematics, Instituto Superior de Agronomia, Technical University of Lisbon (TULisbon), Tapada da Ajuda, 1349-017 Lisboa, Portugal, manela@isa.utl.pt

Laure Nokels Materials Engineering Department, Technocentre Renault, 1, avenue du Golf, API TCR LAB 2 52, 78 288 Guyancourt Cedex, France, laure.nokels@renault.com

Coskun Ozkan Department of Industrial Engineering, Kocaeli University, Veziroglu Yerleskesi, 41040 Kocaeli, Turkey, coskun_ozkan@yahoo.com

Mário Lino Barata Raposo Departamento de Gestão e Economia, Universidade da Beira Interior, Estrada do Sineiro, 6200-209 Covilhã, Portugal, mraposo@ubi.pt

Robert K. Rayner Market Strategies International, 20255 Victor Parkway, Suite 400, Livonia, MI 48152, USA, bob_rayner@marketstrategies.com

Christian M. Ringle University of Hamburg, Institute for Industrial Management and Organizations, Von-Melle-Park 5, 20146 Hamburg, Germany, c.ringle@smartpls.de
and
University of Technology Sydney, Centre for Management and Organization Studies, P.O. Box 123, Broadway NSW 2007, Australia

José L. Roldán Management and Marketing Department, University of Seville, Ramón y Cajal, 1, 41018 Sevilla, Spain, jlroldan@us.es

David Martín Ruiz Escuela Universitaria de Estudios Empresariales, 41012 Sevilla, Spain, dmartin@us.es

Michael J. Ryan Ross School of Business, University of Michigan, P.O. Box 105, Bass Harbor, ME 04653, USA, mikeryan@umich.edu

Solve Sæbø Department of Chemistry, Biotechnology and Food Science (IKBM), Norwegian University of Life Sciences, P.O. Box 5003, 1432 Ås, Norway, solve.sabo@umb.no

Gilbert Saporta Conservatoire National des Arts et Métiers, Chaire de Statistique Appliquée, case 441, 292 rue Saint Martin, 75141 Paris, Cedex 03, France, gilbert.saporta@cnam.fr

Silvia Squillacciotti EDF R&D, Département ICAME, 1 avenue du Général de Gaulle, 92141 Clamart, France, silvia.fagnoni@edf.fr

Valentina Stan Groupe ESSCA Angers, 1 Rue Lakanal – BP 40348, 49003 Angers, Cedex 01, France, valentina_titu@yahoo.fr

Sandra Streukens Department of Marketing and Strategy, Universiteit Hasselt, Campus Diepenbeek, Agoralaan Gebouw DBE 3590 Diepenbeek, Belgium, sandra.streukens@uhasselt.be

Ekrem Tatoglu Faculty of Economics and Administrative Sciences, Chair of International Trade and Business, Bahcesehir University, Besiktas, Istanbul, Turkey, ekremt@bahcesehir.edu.tr

Dirk Temme Chair of Retailing and Service Management, Schumpeter School of Business and Economics, Bergische Universität Wuppertal, Gaußstr. 20, 42097 Wuppertal, Germany, Temme@wiwi.uni-wuppertal.de

Michel Tenenhaus Department of SIAD, HEC School of Management, 1 rue de la Libération, 78351 Jouy-en-Josas, France, tenenhaus@hec.fr

Laura Trinchera Dipartimento di Matematica e Statistica, Università degli Studi di Napoli "Federico II", Via Cintia, 26 – Complesso Monte S. Angelo, 80126 Napoli, Italy, ltrinche@unina.it

Ali Turkyilmaz Department of Industrial Engineering, Fatih University, Buyukcekmece, 34500 Istanbul, Turkey, aturkyilmaz@fatih.edu.tr

Manuel J. Vilares ISEGI – New University of Lisbon, Campus de Campolide, 1070-032 Lisbon, Portugal, mjv@isegi.unl.pt

Wang Huiwen School of Economics and Management, Beihang University, 37 Xueyuan Road, Haidian District, Beijing 100083, China, wanghw@vip.sina.com

Judith H. Washburn John H. Sykes College of Business, University of Tampa, 401 W. Kennedy Blvd., UT Box 48F, Tampa, FL 33606, USA, jwashburn@ut.edu

Sven Wende University of Hamburg, Institute for Industrial Management and Organizations, Von-Melle-Park 5, 20146 Hamburg, Germany, s.wende@smartpls.de

Martin Wetzels Department of Marketing and Marketing Research, Maastricht University, P.O. Box 616 6200 MD, The Netherlands, m.wetzels@mw.unimaas.nl

Alexander Will University of Hamburg, Institute for Industrial Management and Organizations, Von-Melle-Park 5, 20146 Hamburg, Germany, mail@alexwill.de

Bradley Wilson School of Media and Communication, RMIT University, 124 LaTrobe Street, GPO Box 2476V, Melbourne, Victoria 3000, Australia, brad.wilson@rmit.edu.au

Svante Wold NNS Consulting, 42 Pine Hill Rd, Hollis, NH 03049, USA, sbwold@gmail.com

Wu Zaibin School of Economics and Management, Beihang University, 37 Xueyuan Road, Haidian District, Beijing 100083, China, binship@126.com

Selim Zaim Department of Management, Fatih University, Buyukcekmece, Istanbul 34500, Turkey, szaim@fatih.edu.tr

Editorial: Perspectives on Partial Least Squares

Vincenzo Esposito Vinzi, Wynne W. Chin, Jörg Henseler, and Huiwen Wang

1 Partial Least Squares: A Success Story

This Handbook on Partial Least Squares (PLS) represents a comprehensive presentation of the current, original and most advanced research in the domain of PLS methods with specific reference to their use in Marketing-related areas and with a discussion of the forthcoming and most challenging directions of research and perspectives. The Handbook covers the broad area of PLS Methods from Regression to Structural Equation Modeling, from methods to applications, from software to interpretation of results. This work features papers on the use and the analysis of latent variables and indicators by means of the PLS Path Modeling approach from the design of the causal network to the model assessment and improvement. Moreover, within the PLS framework, the Handbook addresses, among others, special and advanced topics such as the analysis of multi-block, multi-group and multi-structured data, the use of categorical indicators, the study of interaction effects, the integration of classification issues, the validation aspects and the comparison between the PLS approach and the covariance-based Structural Equation Modeling.

V. Esposito Vinzi
ESSEC Business School of Paris, Avenue Bernard Hirsch – B.P. 50105, 95021 Cergy-Pontoise, France
e-mail: vinzi@essec.fr

W.W. Chin
Department of Decision and Information Sciences, Bauer College of Business, University of Houston, TX, USA
e-mail: wchin@uh.edu

J. Henseler
Nijmegen School of Management, Radboud University Nijmegen, P.O. Box 9108, 6500 HK Nijmegen, The Netherlands
e-mail: j.henseler@fm.ru.nl

H. Wang
School of Economics and Management, Beihang University, 37 Xueyuan Road, Haidian District, Beijing 100083, China
e-mail: wanghw@vip.sina.com

V. Esposito Vinzi et al. (eds.), *Handbook of Partial Least Squares*, Springer Handbooks of Computational Statistics, DOI 10.1007/978-3-540-32827-8_1,

Most chapters comprise a thorough discussion of applications to problems from Marketing and related areas. Furthermore, a few tutorials focus on some key aspects of PLS analysis with a didactic approach. This Handbook serves as both an introduction for those without prior knowledge of PLS but also as a comprehensive reference for researchers and practitioners interested in the most recent advances in PLS methodology.

Different Partial Least Squares (PLS) cultures seem to have arisen following the original work by Herman Wold (1982): PLS regression models (PLS-R, Wold et al. 1983; Tenenhaus 1998) and PLS Path Modeling (PLS-PM, Lohmöller 1989; Tenenhaus et al. 2005). As a matter of fact, up to now, the two cultures are somehow oriented to different application fields: chemometrics and related fields for PLS-R; econometrics and social sciences for PLS-PM. While experiencing this internal diversity, most often the PLS community has to cope also with external diversities due to other communities that, grown up under the classical culture of statistical inference, seem to be quite reluctant in accepting the PLS approach to data analysis as a well-grounded statistical approach.

Generally speaking, PLS-PM is a statistical approach for modeling complex multivariable relationships among observed and latent variables. In the past few years, this approach has been enjoying increasing popularity in several sciences. Structural Equation Models include a number of statistical methodologies allowing the estimation of a causal theoretical network of relationships linking latent complex concepts, each measured by means of a number of observable indicators. From the standpoint of structural equation modeling, PLS-PM is a component-based approach where the concept of causality is formulated in terms of linear conditional expectation. Herman Wold (1969, 1973, 1975b, 1980, 1982, 1985, 1988) developed PLS as an alternative to covariance-based structural equation modeling as represented by LISREL-type models (Jöreskog, 1978) with, preferably, maximum likelihood estimation. He introduced PLS as a *soft modeling* technique in order to emphasize the difference in methodology for estimating structural equation models (Fornell and Bookstein, 1982; Schneeweiß, 1991). Soft modeling refers to the ability of PLS to exhibit greater flexibility in handling various modeling problems in situations where it is difficult or impossible to meet the hard assumptions of more traditional multivariate statistics. Within this context, "soft" is only attributed to distributional assumptions and not to the concepts, the models or the estimation techniques (Lohmöller, 1989). As an alternative to the classical covariance-based approach, PLS-PM is claimed to seek for optimal linear predictive relationships rather than for causal mechanisms thus privileging a prediction-relevance oriented discovery process to the statistical testing of causal hypotheses. From the standpoint of data analysis, PLS-PM may be also viewed as a very flexible approach to multi-block (or multiple table) analysis. Multi-block situations arise when a few sets of variables are available for the same set of samples. Tenenhaus and Hanafi (2007) show direct relationships between PLS-PM and several techniques for multi-block analysis obtained by properly specifying relationships in the structural model and by mixing the different estimation options available in PLS-PM. This approach clearly shows how the *data-driven* tradition of multiple table analysis can be merged in the *theory-driven* tradition of structural equation modeling to allow running analysis of multi-block data in light of current knowledge on conceptual relationships between tables. In both structural equation modeling and multi-block data analysis, PLS-PM may enhance even further its potentialities, and provide effective added value, when exploited in the case of formative epistemic relationships between manifest variables and their respective latent variables. In PLS-PM latent variables are estimated as linear combinations of the manifest variables and thus they are more naturally defined as emergent constructs (with formative indicators) rather than latent constructs (with reflective indicators). As a matter of fact, formative relationships are more and more commonly used in the applications, especially in the marketing domain, but pose a few problems for the statistical estimation. This mode is based on multiple OLS regressions between each latent variable and its own formative indicators. As known, OLS regression may yield unstable results in presence of important correlations between explanatory variables, it is not feasible when the number of statistical units is smaller than the number of variables nor when missing data affect the dataset. Thus, it seems quite natural to introduce a PLS-R external estimation mode inside the PLS-PM algorithm so as to overcome the mentioned problems, preserve the formative relationships and remain coherent with the component-based and prediction-oriented nature of PLS-PM. Apart from the external estimation module, the implementation of PLS-R within PLS-PM may be extended also to the internal estimation module (as an alternative OLS regression) and to the estimation of path coefficients for the structural model upon convergence of the PLS-PM algorithm and estimation of the latent variable scores. Such an extensive implementation, that might well represent a playground towards the merging of the two PLS cultures, opens a wide range of new possibilities and further developments: different dimensions can be chosen for each block of latent variables; the number of retained components can be chosen by referring to the PLS-R criteria; the well established PLS-R validation and interpretation tools can be finally imported in PLS-PM; new optimizing criteria are envisaged for multi-block analyses; mutual causality with the so-called feedback relationships may be more naturally estimated and so on so forth.

Each chapter of this Handbook focuses on statistical methodology but also on selected applications from real world problems that highlight the usefulness of PLS Methods in Marketing-related areas and their feasibility to different situations. Beside presenting the most recent developments related to the statistical methodology of the PLS-PM approach, this Handbook addresses quite a few open issues that, also due to their relevance in several applications, are of major importance for improving and assessing models estimated by PLS-PM. This work finally wishes to convey the idea that, when exploring and modeling complex data structures, PLS-PM has the promising role of being the basis for merging the two PLS cultures while also benefiting those external cultures traditionally grounded on either data-driven or theory-driven approaches. There are several reasons for the increasing popularity of PLS Path Modeling. They are mainly related to the flexible methodological framework provided by this approach that well adapts

Fig. 1 The PLS handbook's editors in Beijing (April 2006). From left to right: Jörg Henseler as the Prince, Vincenzo Esposito Vinzi (Editor-in-Chief) as the Emperor, Huiwen Wang as the Empress, and Wynne W. Chin as the Minister

to several application fields. For instance, national customer satisfaction indices (e.g. the Swedish Barometer of Satisfaction by Fornell (1992), the American Customer Satisfaction Index by Fornell et al. (1996)) have become the application *par excellence* of PLS Path Modeling. Many other applications are found in Strategic Management (Birkinshaw et al., 1995; Hulland, 1999), Knowledge Management (Gray and Meister, 2004), Information Technology Management (Gefen and Straub, 1997; Yi and Davis, 2003; Venkatesh and Agarwal, 2006) as well as within various disciplines of Marketing, such as Relationship Marketing (Reinartz et al., 2004), Business-to-Business Marketing (Ulaga and Eggert, 2006) and International Marketing (Singh et al., 2006), just to mention a short, and by no means exhaustive, list of references.

2 The Handbook in a Nutshell

This Handbook consists of three parts featuring 33 papers selected after three rounds of a peer reviewing process. In the first part, contemporary methodological developments of PLS analysis are the focus. The second part contains a set of applications of PLS in the field of Marketing as well as in related fields. The pedagogical contributions in the third part reflect tutorials on key aspects of PLS analysis.

2.1 *Part I: Methods of Partial Least Squares*

2.1.1 PLS Path Modeling: Concepts, Model Estimation, and Assessment

Theo K. Dijkstra: Latent Variables and Indices – Herman Wold's Basic Design and Partial Least Squares

In this chapter it is shown that the PLS-algorithms typically converge if the covariance matrix of the indicators satisfies (approximately) the 'basic design', a factor analysis type of model. The algorithms produce solutions to fixed point equations; the solutions are smooth functions of the sample covariance matrix of the indicators. If the latter matrix is asymptotically normal, the PLS estimators will share this property. The probability limits, under the basic design, of the PLS-estimators for loadings, correlations, multiple R^2's, coefficients of structural equations et cetera will differ from the true values. But the difference is decreasing, tending to zero, in the 'quality' of the PLS estimators for the latent variables. It is indicated how to correct for the discrepancy between true values and the probability limits. The contribution deemphasizes the 'normality'-issue in discussions about PLS versus ML: in employing either method one is not required to subscribe to normality; they are 'just' different ways of extracting information from second-order moments.

Dijkstra also proposes a new 'back-to-basics' research program, moving away from factor analysis models and returning to the original object of constructing indices that extract information from high-dimensional data in a predictive, useful way. For the generic case one would construct informative linear compounds, whose constituent indicators have non-negative weights as well as non-negative loadings, satisfying constraints implied by the path diagram. Cross-validation could settle the choice between various competing specifications. In short: it is argued for an upgrade of principal components and canonical variables analysis.

Vincenzo Esposito Vinzi, Laura Trinchera, and Silvano Amato: PLS Path Modeling: From Foundations to Recent Developments and Open Issues for Model Assessment and Improvement

In this chapter the Authors first present the basic algorithm of PLS Path Modeling by discussing some recently proposed estimation options. Namely they introduce the development of new estimation modes and schemes for multidimensional (formative) constructs, i.e. the use of PLS Regression for formative indicators, and the use of path analysis on latent variable scores to estimate path coefficients Furthermore, they focus on the quality indexes classically used to assess the performance of the model in terms of explained variances. They also present some recent developments in PLS Path Modeling framework for model assessment and improvement, including a non-parametric GoF-based procedure for assessing the statistical significance of path coefficients. Finally, they discuss the REBUS-PLS algorithm that enables to improve the prediction performance of the model by capturing unobserved heterogeneity. The chapter ends with a brief sketch of open issues in the area that, in the Authors' opinion, currently represent major research challenges.

Wynne W. Chin: Bootstrap Cross-validation Indices for PLS Path Model Assessment

The goal of PLS path modeling is primarily to estimate the variance of endogenous constructs and in turn their respective manifest variables (if reflective). Models with significant jackknife or bootstrap parameter estimates may still be considered invalid in a predictive sense. In this paper, Chin attempts to reorient researchers from the current emphasis of assessing the significance of parameter estimates (e.g., loadings and structural paths) to that of predictive validity. Specifically, his paper examines how predictive indicator weights estimated for a particular PLS structural model are when applied on new data from the same population. Bootstrap resampling is used to create new data sets where new R-square measures are obtained for each endogenous construct in a model. Chin introduces the weighted summed (WSD) R-square representing how predictive the original sample weights are in a new data context (i.e., a new bootstrap sample). In contrast, the Simple Summed (SSD) R-square examines the predictiveness using the simpler approach of unit weights. From this, Chin develops his Relative Performance Index (RPI) representing the degree to which the PLS weights yield better predictiveness for endogenous constructs than the simpler procedure of performing regression after simple summing of indicators. Chin also introduces a Performance from Optimized Summed Index (PFO) to contrast the WSD R-squares to the R-squares obtained when the PLS algorithm is used on each new bootstrap data set. Results from 2 simulation studies are presented. Overall, in contrast to Q-square which examines predictive relevance at the indicator level, the RPI and PFO indices are shown to provide additional information to assess predictive relevance of PLS estimates at the construct level. Moreover, it is argued that this approach can be applied to other same set data indices such as AVE (Fornell and Larcker, 1981) and GoF (Tenenhaus, Amato, and Esposito Vinzi, 2004) to yield RPI-AVE, PFO-AVE. RPI-GoF, and PFO-GoF indices.

2.1.2 PLS Path Modeling: Extensions

Michel Tenenhaus and Mohamed Hanafi: A Bridge Between PLS Path Modeling and Multiblock Data Analysis

A situation where J blocks of variables $X_1, \dots, X_J$ are observed on the same set of individuals is considered in this paper. A factor analysis approach is applied to blocks instead of variables. The latent variables (LV's) of each block should well explain their own block and at the same time the latent variables of same order should be as highly correlated as possible (positively or in absolute value). Two path models can be used in order to obtain the first order latent variables. The first one is related to confirmatory factor analysis: each LV related to one block is connected to all the LV's related to the other blocks. Then, PLS Path Modeling is used with mode A and centroid scheme. Use of mode B with centroid and factorial schemes is also discussed. The second model is related to hierarchical factor analysis. A causal model is built by relating the LV's of each block X_j to the LV of the super-block X_{J+1} obtained by concatenation of $X_1, \ldots, X_J$. Using PLS estimation of this model with mode A and path-weighting scheme gives an adequate solution for finding the first order latent variables. The use of mode B with centroid and factorial schemes is also discussed. The higher order latent variables are found by using the same algorithms on the deflated blocks. The first approach is compared with the MAXDIFF/MAXBET Van de Geer's algorithm (1984) and the second one with the ACOM algorithm (Chessel and Hanafi, 1996). Sensory data describing Loire wines are used to illustrate these methods.

Michel Tenenhaus, Emmanuelle Mauger, and Christiane Guinot: Use of ULS-SEM and PLS-SEM to Measure a Group Effect in a Regression Model Relating Two Blocks of Binary Variables

The objective of this constribution is to describe the use of unweighted least squares structural equation modelling and partial least squares path modelling in a regression model relating two blocks of binary variables when a group effect can influence the relationship. These methods were applied on the data of a questionnaire investigating sun exposure behaviour addressed to a cohort of French adult in the context of the SU.VI.MAX epidemiological study. Sun protection and exposure behaviours were described according to gender and class of age (less than 50 at inclusion in the study versus more than 49). Significant statistical differences were found between men and women, and between classes of age. This paper illustrates the various stages in the construction of latent variables or scores, based on qualitative data. These kind of scores is widely used in marketing to provide a quantitative measure of the phenomenon studied before proceeding to a more detailed analysis.

Arteaga Francisco, Martina G. Gallarza, and Irene Gil: A New Multiblock PLS Based Method to Estimate Causal Models. Application to the Post-consumption Behavior in Tourism

This chapter presents a new method to estimate causal models based on the Multi-block PLS method (MBPLS) from Wangen and Kowalski (1988). The new method is compared with the classical LVPLS algorithm from Lohmöller (1989), using an academic investigation on the post-consumption behaviour of a particular profile of university students. The results for both methods are quite similar, but the explained percentage of variance for the endogenous latent variables is slightly higher for the MBPLS based method. Bootstrap analysis shows that confidence intervals are slightly smaller for the MBPLS based method.

Wynne W. Chin and Jens Dibbern: A Permutation Based Procedure for Multi-Group PLS Analysis – Results of Tests of Differences on Simulated Data and a Cross Cultural Analysis of the Sourcing of Information System Services Between Germany and the USA

This paper presents a distribution free procedure for performing multi-group PLS analysis. To date, multi-group comparison of PLS models where differences in path estimates for different sampled populations have been relatively naive. Often, researchers simply examine and discuss the difference in magnitude of particular model path estimates for two or more data sets. Problems can occur if the assumption of normal population distribution or similar sample size is not tenable. This paper by Chin and Dibbern presents an alternative distribution free approach via an approximate randomization test - where a subset of all possible data permutations between sample groups is made. The performance of this permutation procedure is applied on both simulated data and a study exploring the differences of factors that impact outsourcing between the countries of US and Germany.

2.1.3 PLS Path Modeling with Classification Issues

Christian M. Ringle, Sven Wende, and Alexander Will: Finite Mixture Partial Least Squares Analysis: Methodology and Numerical Examples

In a wide range of applications for empirical data analysis, the assumption that data is collected from a single homogeneous population is often unrealistic. In particular, the identification of different groups of consumers and their appropriate consideration in partial least squares (PLS) path modeling constitutes a critical issue in marketing. The authors introduce a finite mixture PLS software implementation, which separates data on the basis of the estimates' heterogeneity in the inner path model. Numerical examples using experimental as well as empirical data allow the verification of the methodology's effectiveness and usefulness. The approach permits a reliable identification of distinctive customer segments along with characteristic estimates for relationships between latent variables. Researchers and practitioners can employ this method as a model evaluation technique and thereby assure that results on the aggregate data level are not affected by unobserved heterogeneity in the inner path model estimates. Otherwise, the analysis provides further indications on how to treat that problem by forming groups of data in order to perform a multi-group path analysis.

Silvia Squillacciotti: Prediction oriented classification in PLS Path Modeling

Structural Equation Modeling methods traditionally assume the homogeneity of all the units on which a model is estimated. In many cases, however, this assumption may turn to be false; the presence of latent classes not accounted for by the global model may lead to biased or erroneous results in terms of model parameters and model quality. The traditional multi-group approach to classification is often unsatisfying for several reasons; above all because it leads to classes homogeneous only with respect to external criteria and not to the theoretical model itself.

In this paper, a prediction-oriented classification method in PLS Path Modelling is proposed. Following PLS Typological Regression, the proposed methodology aims at identifying classes of units showing the lowest distance from the models in the space of the dependent variables, according to PLS predictive oriented logic. Hence, the obtained groups are homogeneous with respect to the defined path model. An application to real data in the study of customers' satisfaction and loyalty will be shown.

Valentina Stan and Gilbert Saporta: Conjoint use of variables clustering and PLS structural equations modeling

In the PLS approach, it is frequently assumed that the blocks of variables satisfy the assumption of unidimensionality. In order to fulfill at best this assumption, this contribution uses clustering methods of variables. illustrate the conjoint use of variables clustering and PLS path modeling on data provided by PSA Company (Peugeot Citroën) on customer satisfaction. The data are satisfaction scores on 32 manifest variables given by 2922 customers.

2.1.4 PLS Path Modeling for Customer Satisfaction Studies

Kai Kristensen and Jacob K. Eskildsen: Design of PLS-based Satisfaction Studies

This chapter focuses on the design of PLS structural equation models with respect to satisfaction studies. The authors summarize the findings of previous studies, which have found the PLS technique to be affected by aspects as the skewness of manifest variables, multicollinearity between latent variables, misspecification, question order, sample size as well as the size of the path coefficients. Moreover, the authors give recommendations based on an empirical PLS project conducted at the Aarhus School of Business. Within this project five different studies have been conducted, covering a variety of aspects of designing PLS-based satisfaction studies.

Clara Cordeiro, Alexandra Machás, and Maria Manuela Neves: A Case Study of a Customer Satisfaction Problem – Bootstrap and Imputation Techniques

Bootstrap is a resampling technique proposed by Efron. It has been used in many fields, but in case of missing data studies one can find only a few references. Most studies in marketing research are based in questionnaires, that, for several reasons present missing responses. The missing data problem is a common issue in market research. Here, a customer satisfaction model following the ACSI barometer from

Fornell will be considered. Sometimes, not all customer experience all services or products. Therefore, one may have to deal with missing data, taking the risk of reaching non-significant impacts of these drivers on CS and resulting in inaccurate inferences. To estimate the main drivers of Customer Satisfaction, Structural Equation Models methodology is applied. For a case study in mobile telecommunications several missing data imputation techniques were reviewed and used to complete the data set. Bootstrap methodology was also considered jointly with imputation techniques to complete the data set. Finally, using Partial Least Squares (PLS) algorithm, the authors could compare the above procedures. It suggests that bootstrapping before imputation can be a promising idea.

Manuel J. Vilares, Maria H. Almeida, and Pedro Simões Coelho: Comparison of Likelihood and PLS Estimators for Structural Equation Modeling – A Simulation with Customer Satisfaction Data

Although PLS is a well established tool to estimate structural equation models, more work is still needed in order to better understand its relative merits when compared to likelihood methods. This paper aims to contribute to a better understanding of PLS and likelihood estimators' properties, through the comparison and evaluation of these estimation methods for structural equation models based on customer satisfaction data. A Monte Carlo simulation is used to compare the two estimation methods. The model used in the simulation is the ECSI (European Customer Satisfaction Index) model, constituted by 6 latent variables (image, expectations, perceived quality, perceived value, customer satisfaction and customer loyalty). The simulation is conducted in the context of symmetric and skewed response data and formative blocks, which constitute the typical framework of customer satisfaction measurement. In the simulation we analyze the ability of each method to adequately estimate the inner model coefficients and the indicator loadings. The estimators are analyzed both in terms of bias and precision. Results have shown that globally PLS estimates are generally better than covariance-based estimates both in terms of bias and precision. This is particularly true when estimating the model with skewed response data or a formative block, since for the model based on symmetric data the two methods have shown a similar performance.

John Hulland, M.J. Ryan, and R.K. Rayner: Modeling Customer Satisfaction: A Comparative Performance Evaluation of Covariance Structure Analysis versus Partial Least Squares

Partial least squares (PLS) estimates of structural equation model path coefficients are believed to produce more accurate estimates than those obtained with covariance structure analysis (CVA) using maximum likelihood estimation (MLE) when one or more of the MLE assumptions are not met. However, there exists no empirical support for this belief or for the specific conditions under which it will occur.

MLE-based CVA will also break down or produce improper solutions whereas PLS will not. This study uses simulated data to estimate parameters for a model with 5 independent latent variables and 1 dependent latent variable under various assumption conditions. Data from customer satisfaction studies were used to identify the form of typical field-based survey distributions. Our results show that PLS produces more accurate path coefficients estimates when sample sizes are less than 500, independent latent variables are correlated, and measures per latent variable are less than 4. Method accuracy does not vary when the MLE multinormal distribution assumption is violated or when the data do not fit the theoretical structure very well. Both procedures are more accurate when the independent variables are uncorrelated, but MLE estimations break down more frequently under this condition, especially when combined with sample sizes of less than 100 and only two measures per latent variable.

2.1.5 PLS Regression

Swante Wold, Lennart Eriksson, and Nouna Kettaneh-Wold: PLS in Data Mining and Data Integration

Data mining by means of projection methods such as PLS (projection to latent structures), and their extensions is discussed. The most common data analytical questions in data mining are covered, and illustrated with examples.

1. Clustering, i. e., finding and interpreting "natural" groups in the data,
2. Classification and identification, e. g., biologically active compounds vs. inactive,
3. Quantitative relationships between different sets of variables, e. g., finding variables related to quality of a product, or related to time, seasonal or/and geographical change.

Sub-problems occurring in both (1) to (3) are discussed.

1. Identification of outliers and their aberrant data profiles,
2. Finding the dominating variables and their joint relationships, and
3. Making predictions for new samples.

The use of graphics for the contextual interpretation of results is emphasized. With many variables and few observations – a common situation in data mining – the risk to obtain spurious models is substantial. Spurious models look great for the training set data, but give miserable predictions for new samples. Hence, the validation of the data analytical results is essential, and approaches for that are discussed.

Solve Sæbø, Harald Martens, and Magni Martens: Three-block Data Modeling by Endo- and Exo-LPLS Regression

In consumer science it is common to study how various products are liked or ranked by various consumers. In this context, it is important to check if there are different consumer groups with different product preference patterns. If systematic consumer grouping is detected, it is necessary to determine the person characteristics, which differentiate between these consumer segments, so that they can be reached selectively. Likewise it is important to determine the product characteristics that consumer segments seem to respond differently to.

Consumer preference data are usually rather noisy. The products×persons data table (X_1) usually produced in consumer preference studies may therefore be supplemented with two types of background information: a products×product-property data table (X_2) and a person×person-property data table (X_3). These additional data may be used for stabilizing the data modelling of the preference data X_1 statistically. Moreover, they can reveal the product-properties that are responded to differently by the different consumer segment, and the person-properties that characterize these different segments. The present chapter outlines a recent approach to analyzing the three types of data tables in an integrated fashion and presents new modelling methods in this context.

Huiwen Wang, Jie Meng, and Michel Tenenhaus: Regression Modelling Analysis on Compositional Data

In data analysis of social, economic and technical fields, compositional data is widely used in problems of proportions to the whole. This paper develops regression modelling methods of compositional data, discussing the relationships of one compositional data to one or more than one compositional data and the interrelationship of multiple compositional data. By combining centered logratio transformation proposed by Aitchison (1986) with Partial Least Squares (PLS) related techniques, that is PLS regression, hierarchical PLS and PLS path modelling, respectively, particular difficulties in compositional data regression modelling such as sum to unit constraint, high multicollinearity of the transformed compositional data and hierarchical relationships of multiple compositional data, are all successfully resolved; moreover, the modelling results rightly satisfies the theoretical requirement of log-contrast. Accordingly, case studies of employment structure analysis of Beijing's three industries also illustrate high goodness-of-fit and powerful explainability of the models.

2.2 Part II: Applications to Marketing and Related Areas

Sönke Albers: PLS and Success Factor Studies in Marketing

While in consumer research the "Cronbachs α - LISREL"-paradigm has emerged for a better separation of measurement errors and structural relationships, it is shown in this chapter that studies which involve an evaluation of the effectiveness of marketing instruments require the application of PLS. This is because one no longer distinguishes between constructs and their reflecting measures but rather between abstract marketing policies (constructs) and their forming detailed marketing instruments (indicators). It is shown with the help of examples from literature that many studies of this type applying LISREL have been misspecified and had better made use of the PLS approach. The author also demonstrates the appropriate use of PLS in a study of success factors for e-businesses. He concludes with recommendations on the appropriate design of success factor studies including the use of higher-order constructs and the validation of such studies.

Carmen Barroso, Gabriel Cepeda Carrión, and José L. Roldán: Applying Maximum Likelihood and PLS on Different Sample Sizes – Studies on Servqual Model and Emloyee Behaviour Model

Structural equation modeling (SEM) has been increasingly utilized in marketing and management areas. This rising deployment of SEM suggests addressing comparisons between different SEM approaches. This would help researchers to choose which SEM approach is more appropriate for their studies. After a brief review of the SEM theoretical background, this study analyzes two models with different sample sizes by employing two different SEM techniques to the same set of data. The two SEM techniques compared are: Covariance-based SEM (CBSEM), specifically maximum likelihood (ML) estimation, and Partial Least Square (PLS). After the study findings, the paper provides insights in order to suggest to the researchers when to analyze models with CBSEM or PLS. Finally, practical suggestions about PLS use are added and we discuss whether they are considered by researchers.

Paulo Alexandre O. Duarte and Mario Lino B. Raposo: A PLS Model to Study Brand Preference – An Application to the Mobile Phone Market

Brands play an important role in consumers' daily life and can represent a big asset for companies owning them. Due to the very close relationship between brands and consumers, and the specific nature of branded products as an element of consumer life style, the branded goods industry needs to extend its knowledge of the process of brand preference formation in order to enhance brand equity. This chapter shows how Partial Least Squares (PLS) path modeling can be used to successfully test complex models where other approaches would fail due to the high number of relationships, constructs and indicators, here with an application to brand preference formation for mobile phones. With a wider set of explanatory factors than prior studies, this one explores the factors that contribute to the formation of brand preference using a PLS model to understand the relationship between those and consumer preference on mobile phone brands. The results reveal that brand identity, personality, and image, together with self-image congruence have the highest impact on brand preference. Some other factors linked to the consumer and the situation also affect preference, but in a lower degree.

Markus Eberl: An Application of PLS in Multi-group Analysis – The Need for Differentiated Corporate-level Marketing in the Mobile Communications Industry

The paper focuses on the application of a very common research issue in marketing: the analysis of the differences between groups' structural relations. Although PLS path modeling has some advantages over covariance-based structural equation modeling (CBSEM) regarding this type of research issue – especially in the presence of formative indicators – few publications employ this method. This paper therefore presents an exemplary model that examines the effects of corporate-level marketing activities on corporate reputation as a mediating construct and, finally, on customer loyalty. PLS multi-group analysis is used to empirically test for differences between stakeholder groups in a sample from Germany's mobile communications industry.

Sabrina Helm, Andreas Eggert, and Ina Garnefeld: Modelling the Impact of Corporate Reputation on Customer Satisfaction and Loyalty Using PLS

Reputation is one of the most important intangible assets of a firm. For the most part, recent articles have investigated its impact on firm profitability whereas its effects on individual customers have been neglected. Using data from consumers of an international consumer goods producer, this paper (1) focuses on measuring and discussing the relationships between corporate reputation, consumer satisfaction, and consumer loyalty and (2) examines possible moderating and mediating effects among the constructs. We find that reputation is an antecedent of satisfaction and loyalty that has hitherto been neglected by management. Furthermore, we find that more than half of the effect of reputation onto loyalty is mediated by satisfaction. This means that reputation can only partially be considered a substitute for a consumer's own experiences with a firm. In order to achieve consumer loyalty, organizations need to create both, a good reputation and high satisfaction.

David Martín Ruíz, Dwayne D. Gremler, Judith H. Washburn, and Gabriel Cepeda Carrión: Reframing Customer Value in a Service-based Paradigm: An Evaluation of a Formative Measure in a Multi-industry, Cross-cultural Context

Customer value has received much attention in the recent marketing literature, but relatively little research has specifically focused on inclusion of service components when defining and operationalizing customer value. The purpose of this study is to gain a deeper understanding of customer value by examining several service elements, namely service quality, service equity, and relational benefits, as well as perceived sacrifice, in customers' assessments of value. A multiple industry, cross-cultural setting is used to substantiate our inclusion of service components and to examine whether customer value is best modeled using formative or reflective measures. Our results suggest conceptualizing customer value with service components can be supported empirically, the use of formative components of service value can be supported both theoretically and empirically and is superior to a reflective operationalization of the construct, and that our measure is a robust one that works well across multiple service contexts and cultures.

Sandra Streukens, Martin Wetzels, Ahmad Daryanto, and Ko de Ruyter: Analyzing Factorial Data Using PLS: Application in an Online Complaining Context

Structural equation modeling (SEM) can be employed to emulate more traditional analysis techniques, such as MANOVA, discriminant analysis, and canonical correlation analysis. Recently, it has been realized that this emulation is not restricted to covariance-based SEM, but can easily be extended to components-based SEM, or partials least squares (PLS) path analysis. This chapter presents a PLS path modeling apllication to a fixed-effects, between-subjects factorial design in an online complaint context.

Silvia Thies and Sönke Albers: Application of PLS in Marketing: Content Strategies in the Internet

In an empirical study the strategies are investigated that content providers follow in their compensation policy with respect to their customers. The choice of the policy can be explained by the resource-based view and may serve as recommendations. The authors illustrate how a strategy study in marketing can be analyzed with the help of PLS thereby providing more detailed and actionable results. Firstly, complex measures have to be operationalized by more specific indicators, marketing instruments in this case, which proved to be formative in the most cases. Only by using PLS it was possible to extract the influence of every single formative indicator on the final constructs, i. e. the monetary form of the partnerships. Secondly, PLS allows for more degrees of freedom so that a complex model could be estimated with a number of cases that would not be sufficient for ML-LISREL. Thirdly, PLS does not work with distributional assumptions while significance tests can still be carried out with the help of bootstrapping. The use of PLS is recommended for future strategy studies in marketing because it is possible to extract the drivers at the indicator level so that detailed recommendations can be given for managing marketing instruments.

Ali Türkyilmaz, Ekrem Tatoğlu, Selim Zaim, and Coşkun Özkan: Use of PLS in TQM Research – TQM Practices and Business Performance in SMEs

Advances in structural equation modeling (SEM) techniques have made it possible for management researchers to simultaneously examine theory and measures. When using sophisticated SEM techniques such as covariance based structural equation modeling (CBSEM) and partial least squares (PLS), researchers must be aware of their underlying assumptions and limitations. SEM models such as PLS can help total quality management (TQM) researchers to achieve new insights. Researchers in the area of TQM need to apply this technique properly in order to better understand the complex relationships proposed in their models. This paper makes an attempt to apply PLS in the area of TQM research. In doing that special emphasis was placed on identifying the relationships between the most prominent TQM constructs and business performance based on a sample of SMEs operating in Turkish textile industry. The analysis of PLS results indicated that a good deal of support has been found for the proposed model where a satisfactory percentage of the variance in the dependent constructs is explained by the independent constructs.

Bradley Wilson: Using PLS to Investigate Interaction Effects Between Higher Order Branding Constructs

This chapter illustrates how PLS can be used when investigating causal models with moderators at a higher level of abstraction. This is accomplished with the presentation of a marketing example. This example specifically investigates the influence of brand personality on brand relationship quality with involvement being a moderator. The literature is reviewed on how to analyse moderational hypotheses with PLS. Considerable work is devoted to the process undertaken to analyse higher order structures. The results indicate that involvement does moderate the main effects relationship between brand personality and brand relationship quality.

2.3 Part III: Tutorials

Wynne W. Chin: How to Write Up and Report PLS analyses

The objective of this paper is to provide a basic framework for researchers interested in reporting the results of their PLS analyses. Since the dominant paradigm in reporting Structural Equation Modeling results is covariance based, this paper begins by providing a discussion of key differences and rationale that researchers can use to support their use of PLS. This is followed by two examples from the discipline of Information Systems. The first consists of constructs with reflective indicators (mode A). This is followed up with a model that includes a construct with formative indicators (mode B).

Oliver Götz, Kerstin Liehr-Gobbers, and Manfred Krafft: Evaluation of Structural Equation Models using the Partial Least Squares Approach

This paper gives a basic comprehension of the partial least squares approach. In this context, the aim of this paper is to develop a guide for the evaluation of structural equation models, using the current statistical methods methodological knowledge by specifically considering the Partial-Least-Squares (PLS) approach's requirements. As an advantage, the PLS method demands significantly fewer requirements compared to that of covariance structure analyses, but nevertheless delivers consistent estimation results. This makes PLS a valuable tool for testing theories. Another asset of the PLS approach is its ability to deal with formative as well as reflective indicators, even within one structural equation model. This indicates that the PLS approach is appropriate for explorative analysis of structural equation models, too, thus offering a significant contribution to theory development. However, little knowledge is available regarding the evaluating of PLS structural equation models. To overcome this research gap a broad and detailed guideline for the assessment of reflective and formative measurement models as well as of the structural model had been developed. Moreover, to illustrate the guideline, a detailed application of the evaluation criteria had been conducted to an empirical model explaining repeat purchasing behaviour.

Jörg Henseler and Georg Fassott: Testing Moderating Effects in PLS Path Models: An Illustration of Available Procedures

Along with the development of scientific disciplines, namely social sciences, hypothesized relationships become more and more complex. Besides the examination of direct effects, researchers are more and more interested in moderating effects. Moderating effects are evoked by variables, whose variation influences the strength or the direction of a relationship between an exogenous and an endogenous variable. Investigators using partial least squares path modeling need appropriate means to test their models for such moderating effects. Henseler and Fassott illustrate the identification and quantification of moderating effects in complex causal structures by means of Partial Least Squares Path Modeling. They also show that group comparisons, i.e. comparisons of model estimates for different groups of observations, represent a special case of moderating effects, having the grouping variable as a categorical moderator variable. In their contribution, Henseler and Fassott provide profound answers to typical questions related to testing moderating effects within PLS path models:

1. How can a moderating effect be drawn in a PLS path model, taking into account that available software only permits direct effects?
2. How does the type of measurement model of the independent and the moderator variables influence the detection of moderating effects?
3. Before the model estimation, should the data be prepared in a particular manner? Should the indicators be centered (having a mean of zero), standardized (having a mean of zero and a standard deviation of one), or manipulated in any other way?
4. How can the coefficients of moderating effects be estimated and interpreted? And, finally,
5. How can the significance of moderating effects be determined?

Borrowing from the body of knowledge on modeling interaction effect within multiple regression, Henseler and Fassott develop a guideline on how to test moderating effects in PLS path models. In particular, they create a graphical representation of the necessary steps and decisions to make in form of a flow chart. Starting with the analysis of the type of data available, via the measurement model specification, the flow chart leads the researcher through the decisions on how to prepare the data and how to model the moderating effect. The flow chart ends with the bootstrapping, as the preferred means to test significance, and the final interpretation of the model outcomes which are to be made by the researcher. In addition to this tutorial-like contribution on the modelation of moderating effects by means of Partial Least Squares Path Modeling, readers interested in modeling interaction effects can find many modelling examples in this volume, particularly in the contributions by Chin & Dibbern; Eberl; Guinot, Mauger, Malvy, Latreille, Ambroisine, Ezzedine, Galan, Hercberg & Tenenhaus; Streukens, Wetzels, Daryanto & de Ruyter; and Wilson.

Dirk Temme, Henning Kreis, and Lutz Hildebrandt: Comparison of Current PLS Path Modeling Software – Features, Ease-of-Use, and Performance

After years of stagnancy, PLS path modeling has recently attracted renewed interest from applied researchers in marketing. At the same time, the availability of software alternatives to Lohmöller's LVPLS package has considerably increased (PLS-Graph, PLS-GUI, SPAD-PLS, SmartPLS). To help the user to make an informed decision, the existing programs are reviewed with regard to requirements, methodological options, and ease-of-use; their strengths and weaknesses are identified. Furthermore, estimation results for different simulated data sets, each focusing on a specific issue (sign changes and bootstrapping, missing data, and multi-collinearity), are compared.

Zaibin Wu, Jie Meng, and Huiwen Wang: Introduction to SIMCA-P and Its Application

SIMCA-P is a kind of user-friendly software developed by Umetrics, which is mainly used for the methods of principle component analysis (PCA) and partial least square (PLS) regression. This paper introduces the main glossaries, analysis cycle and basic operations in SIMCA-P via a practical example. In the application section, this paper adopts SIMCA-P to estimate the PLS model with qualitative variables in independent variables set and applies it in the sand storm prevention in Beijing. Furthermore, this paper demonstrates the advantage of lowering the wind erosion by Conservation Tillage method and shows that Conservation Tillage is worth promotion in Beijing sand storm prevention.

Laure Nokels, Thierry Fahmy, and Sebastien Crochemore: Interpretation of the Preferences of Automotive Customers Applied to Air Conditioning Supports by Combining GPA and PLS Regression

A change in the behavior of the automotive customers has been noticed throughout the last years. Customers feel a renewed interest in the intangible assets of perceived quality and comfort of environment. A concrete case of study has been set up to analyze the preferences for 15 air conditioning supports. Descriptive data obtained by flash profiling with 5 experts on the photographs of 15 air conditioning supports are treated by Generalized Procrustes Analysis (GPA). The preferences of 61 customers are then explained by Partial Least Squares (PLS) regression applied to the factors selected from the GPA. The results provided by the XLSTAT GPA and PLS regression functions help to quickly identify the items that have a positive or negative impact on the customers' preferences, and to define products that fit the customers' expectations.

Acknowledgements The PLS Handbook Editors are very grateful to Rosaria Romano and Laura Trinchera from the University of Naples Federico II (Italy) for their endeavor and enthusiasm as editorial assistants and to the more than 50 referees for their highly professional contribution to the three rounds of the peer reviewing process. A special thank goes also to the owners and the staff of the Hotel San Michele in Anacapri (Island of Capri, Italy) for offering a very peaceful and inspiring summer environment during the completion of the editorial work.

References

Birkinshaw, J., Morrison, A., and Hulland, J. (1995). Structural and Competitive Determinants of a Global Integration Strategy. *Strategic Management Journal*, 16(8):637–655.

Fornell, C. (1992). A National Customer Satisfaction Barometer. The Swedish Experience. *Journal of Marketing*, 56(1):6–21.

Fornell, C. and Bookstein, F. L. (1982). Two structural equation models: LISREL and PLS applied to consumer exit-voice theory. *Journal of Marketing Research*, 19(4):440–452.

Fornell, C., Johnson, M., Anderson, E., Cha, J., and Bryant, B. (1996). The American Customer Satisfaction Index: Nature, Purpose, and Findings. *Journal of Marketing*, 60(4):7–18.

Gefen, D. and Straub, D. (1997). Gender Differences in the Perception and Use of E-Mail: An Extension to the Technology Acceptance Model. *MIS Quarterly*, 21(4):389–400.

Gray, P. H. and Meister, D. B. (2004). Knowledge sourcing effectiveness. *Management Science*, 50(6):821–834.

Hulland, J. (1999). Use of Partial Least Squares (PLS) in Strategic Management Research: A Review of Four Recent Studies. *Strategic Management Journal*, 20(2):195–204.

Jöreskog, K. G. (1978). Structural analysis of covariance and correlation matrices. *Psychometrika*, 43(4):443–477.

Lohmöller, J.-B. (1989). *Latent Variable Path Modeling with Partial Least Squares*. Physica, Heidelberg.

Reinartz, W. J., Krafft, M., and Hoyer, W. D. (2004). The Customer Relationship Management Process: Its Measurement and Impact on Performance. *Journal of Marketing Research*, 41(3):293–305.

Schneeweiß, H. (1991). Models with latent variables: LISREL versus PLS. *Statistica Neerlandica*, 45(2):145–157.

Singh, N., Fassott, G., Chao, M., and Hoffmann, J. (2006). Understanding international web site usage. *International Marketing Review*, 23(1):83–97.

Tenenhaus, M. (1998). *La Régression PLS: théorie et pratique*. Technip, Paris.

Tenenhaus, Esposito Vinzi, V., Chatelin, Y.-M., and Lauro, C. (2005). PLS path modeling. *Computational Statistics & Data Analysis*, 48(1):159–205.

Ulaga, W. and Eggert, A. (2006). Value-Based Differentiation in Business Relationships: Gaining and Sustaining Key Supplier Status. *Journal of Marketing*, 70(1):119–136.

Venkatesh, V. and Agarwal, R. (2006). Turning visitors into customers: a usability-centric perspective on purchase behavior in electronic channels. *Management Science*, 52(3):367–382.

Wangen, L. and Kowalski, B. (1988). A multiblock partial least squares algorithm for investigating complex chemical systems. *Journal of Chemometrics*, 3:3–20.

Wold, H. (1973). Nonlinear iterative partial least squares (NIPALS) modelling. some current developments. In Krishnaiah, P. R., editor, *Proceedings of the 3rd International Symposium on Multivariate Analysis*, pages 383–407, Dayton, OH.

Wold, H. (1975a). Modeling in complex situations with soft information. In *Third World Congress of Econometric Society*.

Wold, H. (1975b). Soft modelling by latent variables: The non-linear iterative partial least squares (NIPALS) approach. In *Perspectives on Probability and Statistics, Festschrift (65th Birthday) for M. S. Bartlett*, pages 117–142.

Wold, H. (1980). Model construction and evaluation when theoretical knowledge is scarce: Theory and application of PLS. In Kmenta, J. and Ramsey, J. B., editors, *Evaluation of Econometric Models*, pages 47–74. Academic Press, New York.

Wold, H. (1982). Soft modeling: The basic design and some extensions. In Jöreskog, K. G. and Wold, H., editors, *Systems Under Indirect Observations: Part I*, pages 1–54. North-Holland, Amsterdam.

Wold, H. (1985). *Partial Least Squares*, volume 6, pages 581–591. John Wiley & Sons, New York et al.

Wold, H. (1988). Specification, predictor. In Kotz, S. and Johnson, N. L., editors, *Encyclopedia of Statistical Sciences*, volume 8, pages 587–599. Wiley, New York.

Wold, H. and Lyttkens, E. (1969). Nonlinear iterative partial least squares (NIPALS) estimation procedures. *Bulletin of the International Statistical Institute*, 43:29–51.

Wold, S., Martens, H., and Wold, H. (1983). The multivariate calibration problem in chemistry solved by the PLS method. In *In: Ruhe, A., Kagstrom, B. (Eds.), Proceedings of the Conference on Matrix Pencils. Lectures Notes in Mathematics*, Heidelberg. Springer.

Yi, M. and Davis, F. (2003). Developing and Validating an Observational Learning Model of Computer Software Training and Skill Acquisition. *Information Systems Research*, 14(2): 146–169.

Part I
Methods

PLS Path Modeling: Concepts, Model Estimation and Assessment
Chapters 1–3

PLS Path Modeling: Extensions
Chapters 4–7

PLS Path Modeling with Classification Issues
Chapters 8–10

PLS Path Modeling for Customer Satisfaction Studies
Chapters 11–14

PLS Regression
Chapters 15–17

Chapter 1
Latent Variables and Indices: Herman Wold's Basic Design and Partial Least Squares

Theo K. Dijkstra

Abstract In this chapter it is shown that the PLS-algorithms typically converge if the covariance matrix of the indicators satisfies (approximately) the "basic design", a factor analysis type of model. The algorithms produce solutions to fixed point equations; the solutions are smooth functions of the sample covariance matrix of the indicators. If the latter matrix is asymptotically normal, the PLS-estimators will share this property. The probability limits, under the basic design, of the PLS-estimators for loadings, correlations, multiple R's, coefficients of structural equations et cetera will differ from the true values. But the difference is decreasing, tending to zero, in the "quality" of the PLS estimators for the latent variables. It is indicated how to correct for the discrepancy between true values and the probability limits. We deemphasize the "normality"-issue in discussions about PLS versus ML: in employing either method one is not required to subscribe to normality; they are "just" different ways of extracting information from second-order moments.

We also propose a new "back-to-basics" research program, moving away from factor analysis models and returning to the original object of constructing indices that extract information from high-dimensional data in a predictive, useful way. For the generic case we would construct informative linear compounds, whose constituent indicators have non-negative weights as well as non-negative loadings, satisfying constraints implied by the path diagram. Cross-validation could settle the choice between various competing specifications. In short: we argue for an upgrade of principal components and canonical variables analysis.

T.K. Dijkstra
SNS Asset Management, Research and Development, Pettelaarpark 120, P. O. Box 70053, 5201 DZ 's-Hertogenbosch, The Netherlands
e-mail: theo.dijkstra@snsam.nl
and
University of Groningen, Economics and Econometrics, Zernike Complex, P.O. Box 800, 9700 AV, Groningen, The Netherlands
e-mail: t.k.dijkstra@rug.nl

V. Esposito Vinzi et al. (eds.), *Handbook of Partial Least Squares*, Springer Handbooks of Computational Statistics, DOI 10.1007/978-3-540-32827-8_2,

1.1 Introduction

Partial Least Squares is a family of regression based methods designed for the analysis of high dimensional data in a low-structure environment. Its origin lies in the sixties, seventies and eighties of the previous century, when Herman O.A. Wold vigorously pursued the creation and construction of models and methods for the social sciences, where "soft models and soft data" were the rule rather than the exception, and where approaches strongly oriented at prediction would be of great value. The author was fortunate to witness the development firsthand for a few years. Herman Wold suggested (in 1977) to write a PhD-thesis on LISREL versus PLS in the context of latent variable models, more specifically of "*the basic design*". I was invited to his research team at the Wharton School, Philadelphia, in the fall of 1977. Herman Wold also honoured me by serving on my PhD-committee as a distinguished and decisive member. The thesis was finished in 1981. While I moved into another direction (specification, estimation and statistical inference in the context of model uncertainty) PLS sprouted very fruitfully in many directions, not only as regards theoretical extensions and innovations (multilevel, nonlinear extensions et cetera) but also as regards applications, notably in chemometrics, marketing, and political sciences. The PLS regression oriented methodology became part of main stream statistical analysis, as can be gathered from references and discussions in important books and journals. See e. g. Hastie et al. (2001), or Stone and Brooks (1990), Frank and Friedman (1993), Tenenhaus et al. (2005), there are many others. This chapter will not cover these later developments, others are much more knowledgeable and are more up-to-date than I am. Instead we will go back in time and return to one of the real starting points of PLS: *the basic design*. We will look at PLS here as a method for structural equation modelling and estimation, as in Tenenhaus et al. (2005). Although I cover ground common to the latter's review I also offer additional insights, in particular into the distributional assumptions behind the basic design, the convergence of the algorithms and the properties of their outcomes. In addition, ways are suggested to modify the outcomes for the tendency to over- or underestimate loadings and correlations. Although I draw from my work from the period 1977–1981, which, as the editor graciously suggested is still of some value and at any rate is not particularly well-known, but I also suggest new developments, by stepping away from the latent variable paradigm and returning to the formative years of PLS, where principal components and canonical variables were the main source of inspiration.

In the next section we will introduce the basic design, somewhat extended beyond its archetype. It is basically a second order factor model where each indicator is directly linked to one latent variable only. Although the model is presented as "distribution free" the very fact that conditional expectations are always assumed to be linear does suggest that multinormality is lurking somewhere in the background. We will discuss this in Sect. 1.3, where we will also address the question whether normality is important, and to what extent, for the old "adversary" LISREL. Please note that as I use the term LISREL it does not stand for a specific well-known statistical software package, but for the maximum likelihood estimation and testing approach for latent variable models, under the *working hypothesis* of multivariate normality. There is no implied value judgement about other approaches or packages that have entered the market in the mean time. In Sect. 1.3 we also recall some relevant estimation theory for the case where the structural specification is incorrect or the distributional assumptions are invalid.

The next section, number 4, appears to be the least well-known. I sketch a proof there, convincingly as I like to believe, that the PLS algorithms will converge from arbitrary starting points to unique solutions, fixed points, with a probability tending to one when the sample size increases and the sample covariance matrix has a probability limit that is compatible with the basic design, or is sufficiently close to it.

In Sect. 1.5 we look at the values that PLS attains at the limit, in case of the basic design. We find that correlations between the latent variables will be *under*estimated, that this is also true for the squared multiple correlation coefficients for regressions among latent variables, and the consequences for the estimation of the structural form parameters are indicated; we note that loadings counterbalance the tendency of correlations to be underestimated, by *over*estimation. I suggest ways to correct for this lack of consistency, in the probabilistic sense.

In the Sect. 1.6, we return to what I believe is the origin of PLS: the construction of indices by means of linear compounds, in the spirit of principal components and canonical variables. This section is really new, as far as I can tell. It is shown that for any set of indicators there always exist *proper indices*, i. e. linear compounds with non-negative coefficients that have non-negative correlations with their indicators. I hint at the way constraints, implied by the path diagram, can be formulated as side conditions for the construction of indices. The idea is to take the indices as the fundamental objects, as the carriers or conveyers of information, and to treat path diagrams as relationships between the indices in their own right. Basically, this approach calls for the replacement of fullblown unrestricted principal component or generalized canonical variable analyses by the construction of proper indices, satisfying modest, "theory poor" restrictions on their correlation matrix. This section calls for further exploration of these ideas, acknowledging that in the process PLS's simplicity will be substantially reduced.

The concluding Sect. 1.7 offers some comments on McDonald's (1996) thought provoking paper on PLS; the author gratefully acknowledges an unknown referee's suggestion to discuss some of the issues raised in this paper.

1.2 A Second Order Factor Model, the "Basic Design"

Manifest variables, or indicators, are observable variables who are supposed to convey information about the behavior of *latent* variables, theoretical concepts, who are not directly observable but who are fundamental to the scientific enterprise in almost any field, see Kaplan (1946). In the social sciences *factor models* are the vehicle most commonly used for the analysis of the interplay between latent and manifest variables. Model construction and estimation used to be focussed mainly on the specification, validation and interpretation of factor loadings and underlying factors (latent variables), but in the seventies of the previous century the relationships between the factors themselves became a central object of study. The advent of optimization methods for high-dimensional problems, like the Fletcher-Powell algorithm, see Ortega and Rheinboldt (1970) e. g., allowed research teams to develop highly flexible and user-friendly software for the analysis, estimation and testing of second order factor models, in which relationships between the factors themselves are explicitly incorporated. First Karl G. Jöreskog from Uppsala, Sweden, and his associates developed LISREL, then later, in the eighties, Peter M. Bentler from UCLA designed EQS, and others followed. However, approaches like LISREL appeared to put high demands on the specification of the theoretical relationships: one was supposed to supply a lot of structural information on the theoretical covariance matrix of the indicators. And also it seemed that, ideally, one needed plenty of independent observations on these indicators from a multinormal distribution! Herman O. A. Wold clearly saw the potential of these methods for the social sciences but objected to their informational and distributional demands, which he regarded as unrealistic for many fields of inquiry, especially in the social sciences. Moreover, he felt that estimation and description had been put into focus, at the expense of prediction. Herman Wold had a lifelong interest in the development of predictive and robust statistical methods. In econometrics he pleaded forcefully for "recursive modelling" where every single equation could be used for prediction and every parameter had a predictive interpretation, against the current of mainstream "simultaneous equation modelling". For the latter type of models he developed the Fix-Point estimation method, based on a predictive reinterpretation and rewriting of the models, in which the parameters were estimated iteratively by means of simple regressions. In 1966 this approach was extended to principal components, canonical variables and factor analysis models: using least squares as overall predictive criterion, parameters were divided into subsets in such a way that with any one of the subsets kept fixed at previously determined values, the remaining set of parameters would solve a straightforward regression problem; roles would be reversed and the regressions were to be continued until consecutive values for the parameters differed less then a preassigned value, see Wold (1966) but also Wold (1975). The finalizations of the ideas, culminating into PLS, took place in 1977, when Herman Wold was at the Wharton School, Philadelphia. Incidentally, since the present author was a member of Herman Wold's research team at the Wharton School in Philadelphia in the fall of 1977, one could be tempted to believe that he claims some of the credit for this development. In fact, if anything, my attempts to incorporate structural information into the estimation process, which complicated it substantially, urged Herman Wold to intensify his search for further simplification. I will try to revive my attempts in the penultimate section...

For analytical purposes and for comparisons with LISREL-type of alternatives Herman Wold put up a second order factor model, called the "basic design". In the remainder of this section we will present this model, somewhat extended, i.e. with fewer assumptions. The next section then takes up the discussion concerning the "multivariate normality of the vector of indicators", the hard or "heroic" assumption of LISREL as Herman Wold liked to call it. Anticipating the drift of the argument: the difference between multinormality and the distributional assumptions in PLS is small or large depending on whether the distance between independence and zero correlation is deemed small or large. Conceptually, the difference *is* large, since two random vectors X and Y are independent if and only if "every" real function of X is uncorrelated with "every" real function of Y, not just the linear functions. But any one who has ever given a Stat1 course knows that the psychological distance is close to negligible...

More important perhaps is the fact that multinormality and independence of the observational vectors is *not* required for consistency of LISREL-estimators, all that is needed is that the sample covariance matrix S is a consistent estimator for the theoretical covariance matrix Σ. The existence of Σ and independence of the observational vectors is more than sufficient, there is in fact quite some tolerance for dependence as well. Also, asymptotic normality of the estimators is assured without the assumption of multinormality. All that is needed is asymptotic normality of S, and that is quite generally the case. Asymptotic optimality, and a proper interpretation of calculated standard errors as standard errors, as well as the correct use of test-statistics however does indeed impose heavy restrictions on the distribution, which make the distance to multinormality, again psychologically spoken, rather small, and therefore to PLS rather large...

There is however very little disagreement about the difference in structural information, PLS is much more modest and therefore more realistic in this regard than LISREL. See Dijkstra (1983, 1988, 1992) where further restrictions, relevant for *both* approaches, for valid use of frequentist inference statistics are discussed, like the requirement that the model was *not* specified interactively, using the data at hand.

Now for the "basic design".

We will take all variables to be centered at their mean, so the expected values are zero, and we assume the existence of all second order moments. Let η be a vector of latent variables which can be partitioned in a subvector $\eta_{\mathbf{n}}$ of e**n**dogenous latent variables and a subvector $\eta_{\mathbf{x}}$ of e**x**ogenous latent variables. These vectors obey the following set of structural equations with conformable matrices B and Γ and a (residual) vector ζ with the property that $E(\zeta \mid \eta_{\mathbf{x}}) = 0$:

$$\eta_{\mathbf{n}} = B\eta_{\mathbf{n}} + \Gamma\eta_{\mathbf{x}} + \zeta \tag{1.1}$$

The inverse of $(I - B)$ is assumed to exist, and the (zero-) restrictions on B, Γ and the covariance matrices of η_x and ζ are sufficient for identification of the structural parameters. An easy consequence is that

$$E(\eta_{\mathbf{n}} \mid \eta_{\mathbf{x}}) = (I - B)^{-1}\,\Gamma\eta_{\mathbf{x}} \equiv \Pi\eta_{\mathbf{x}} \tag{1.2}$$

which expresses the intended use of the reduced form, prediction, since no function of $\eta_{\mathbf{x}}$ will predict $\eta_{\mathbf{n}}$ better than $\Pi\eta_{\mathbf{x}}$ in terms of mean squared error. Note that the original basic design is less general, in the sense that B is sub-diagonal there and that for each i larger than 1 the conditional expectation of ζ_i given $\eta_\mathbf{x}$ and the first $i - 1$ elements of $\eta_\mathbf{n}$ is zero. In other words, originally the model for the latent variables was assumed to be a *causal chain*, where every equation, whether from the reduced or the structural form, has a predictive use and interpretation.

Now assume we have a vector of indicators y which can be divided into subvectors, one subvector for each latent variable, such that for the i-th subvector y_i the following holds:

$$y_i = \lambda_i \eta_i + \epsilon_i \tag{1.3}$$

where λ_i is a vector of loadings, with as many components as there are indicators for η_i, and the vector ϵ_i is a random vector of measurement errors. It is assumed that $E(y_i \mid \eta_i) = \lambda_i \eta_i$ so that the errors are uncorrelated with the latent variable of the same equation. Wold assumes that measurement errors relating to different latent variables are uncorrelated as well. In the original basic design he assumes that the elements of each ϵ_i are mutually uncorrelated, so that their covariance matrix is diagonal. We will postulate instead that $V_i \equiv E\epsilon_i\epsilon_i^\top$ has at least one zero element (or equivalently, with more than one indicator, because of the symmetry and the fact that is a covariance matrix, at least two zero elements). To summarize:

$$\Sigma_{ij} \equiv E y_i y_j^\top = \rho_{ij}\lambda_i\lambda_j \text{ for } i \neq j \tag{1.4}$$

where ρ_{ij} stands for the correlation between η_i and η_j, adopting the convention that latent variables have unit variance, and

$$\Sigma_{ii} = \lambda_i\lambda_i^\top + V_i. \tag{1.5}$$

So the ρ_{ij}'s and the loading vectors describe the correlations at the first level, of the indicators, and the structural equations yield the correlations at the second level, of the latent variables. It is easily seen that all parameters are identified: equation (4) determines the direction of λ_i apart from a sign factor and (5) fixes its length, therefore the ρ_{ij}'s are identified (as well as the V_i's), and they on their turn allow determination of the structural form parameters, given Σ of course.

1.3 Distributional Assumptions: Multinormality or "Distribution Free"?

The (extended) basic design does not appear to impose heavy constraints on the distribution of the indicators: the existence of second order moments, some zero conditional expectations and a linear structure, that's about it. Multinormality seems conceptually way off. But let us take an arbitrary measurement equation

$$y_i = \lambda_i \eta_i + \epsilon_i \tag{1.6}$$

and instead of assuming that $E(\epsilon_i \mid \eta_i) = 0$, we let ϵ_i and η_i be stochastically independent, which *implies* a zero conditional expectation. As Wold assumes the elements of ϵ_i to be uncorrelated, let us take them here mutually independent. For $E(\eta_i \mid y_i)$ we take it to be linear as well, so assuming here and in the sequel invertibility of matrices whenever this is needed

$$E(\eta_i \mid y_i) = \lambda_i^\top (\Sigma_{ii})^{-1} y_i \propto \lambda_i^\top V_i^{-1} y_i \tag{1.7}$$

If now all loadings, all elements of λ_i, differ from zero, we *must* have multinormality of the vector $(y_i; \eta_i; \epsilon_i)$ as follows from a characterization theorem in Kagan et al. (1973), see in particular theorem 10.5.3. Let us modify and extend each measurement equation as just described, and let all measurement errors be mutually independent. Then for one thing each element of η will be normal and ϵ, the vector obtained by stacking the ϵ_i's, will be multinormal.

If we now turn to the structural equations, we will take for simplicity the special case of a *complete causal chain*, where B is square and lower diagonal and the elements of the residual vector ζ are mutually independent. A characterization due to Cramér states that when the sum of independent variables is normal, all constituents of this sum are normal, and Cramér and Wold have shown that a vector is multinormal if and only if every linear function of this vector is normal. Combining these characterizations one is easily led to the conclusion that $(y; \eta; \zeta; \epsilon)$ is multinormal. See Dijkstra (1981) for a more elaborate discussion and other results.

So, roughly, if one strengthens zero conditional expectations to independence and takes all conditional expectations to be linear, one gets multinormality. It appears that psychologically PLS and multinormality are not far apart. But the appreciation of these conditions is not just a matter of taste, or of mathematical/statistical maturity. Fundamentally it is an empirical matter and the question of their (approximate) validity ought to be settled by a thorough analysis of the data. If one has to reject them, how sad is that? The linear functions we use for prediction are then no longer least squares optimal in the set of *all* functions, but best linear approximations only to these objects of desire (in the population,that is). If we are happy with linear approximations, i.e. we understand them *and* can use them to good effect, then who cares about multinormality, or for that matter about linearity of conditional expectations? In the author's opinion, normality has a pragmatic justification only. Using it as a working hypothesis in combination with well worn "principles", like least squares or, yes, maximum likelihood, often leads to useful results, which as a bonus usually satisfy appealing consistency conditions.

It has been stated and is often repeated, seemingly thoughtlessly, that LISREL is *based* on normality, in the sense that its use *requires* the data to be normally distributed. This is a prejudice that ought to be cancelled. One can use the maximum entropy principle, the existence of second order moments, and the likelihood principle to motivate the choice of the fitting function that LISREL employs. But at the end of the day this function is just one way of fitting a theoretical covariance matrix $\Sigma(\theta)$ to a sample covariance matrix S, where the fit is determined by the difference between the eigenvalues of $S\Sigma^{-1}$ and the eigenvalues of the identity matrix. To elaborate just a bit:

If we denote the p eigenvalues of $S\Sigma^{-1}$ by $\gamma_1, \gamma_2, \ldots, \gamma_p$ the LISREL fitting function can be written as $\sum_{i=1}^{i=p} (\gamma_i - \log \gamma_i - 1)$. Recall that for real positive numbers $0 \leq x - \log x - 1$ everywhere with equality only for $x = 1$. Therefore the LISREL criterion is always nonnegative and zero only when *all* eigenvalues are equal to 1. The absolute minimum is reached if and only if a θ can be found such that $S = \Sigma(\theta)$. So if $S = \Sigma(\theta_*)$ for some θ_* and identifiability holds, LISREL will find it. Clearly, other functions of the eigenvalues will do the trick, GLS is one of them. See Dijkstra (1990) for an analysis of the class of Swain functions. The "maximum likelihood" estimator $\widehat{\theta}$ is a well-behaved, many times differentiable function of S, which yields θ when evaluated at $S = \Sigma(\theta)$. In other words, if S is close to $\Sigma(\theta)$ the estimator is close to θ and it is locally a linear function of S. It follows that when S tends in probability to its "true value", $\Sigma(\theta)$, then $\widehat{\theta}$ will do the same and moreover, if S is asymptotically normal, then $\widehat{\theta}$ is.

Things become more involved when the probability limit of S, *plim*(S), does *not* satisfy the structural constraints as implied by the second order factor model at hand, so there is *no* θ for which $\Sigma(\theta)$ equals *plim*(S). We will summarize in a stylized way what can be said about the behavior of estimators in the case of Weighted Least Squares, which with proper weighting matrices include LISREL, i. e. maximum likelihood under normality, and related fitting functions as well. The result will be relevant also for the analysis of reduced form estimators using PLS.

To simplify notation we will let $\sigma(\theta)$ stand for the vector of non-redundant elements of the smooth matrix function $\Sigma(\theta)$ and s does the same for S. We will let $\overline{s}$ stand for *plim*(S). Define a fitting function $F(s, \sigma(\theta) \mid W)$ by

$$F(s, \sigma(\theta) \mid W) \equiv (s - \sigma(\theta))^\top W (s - \sigma(\theta)) \qquad (1.8)$$

where W is some symmetric random matrix of appropriate order whose *plim*, $\overline{W}$, exists as a positive definite matrix (non-random matrices can be handled as well). The vector θ varies across a suitable set, non-empty and compact or such that F has a compact level set. We postulate that the minimum of $F\left(\overline{s}, \sigma(\theta) \mid \overline{W}\right)$ is attained in a unique point $\theta\left(\overline{s}, \overline{W}\right)$, depending on the probability limits of S and W. One can show that F tends in probability to $F\left(\overline{s}, \sigma(\theta) \mid \overline{W}\right)$ *uniformly* with respect to θ. This implies that the estimator $\widehat{\theta}(s, W) \equiv \arg\min(F)$ will tend to $\theta\left(\overline{s}, \overline{W}\right)$ in probability. Different fitting functions will produce different probability limits, if the model is incorrect. With sufficient differentiability and asymptotic normality we can say more (see Dijkstra 1981 e. g.), using the implicit function theorem on the first-order conditions of the minimization problem. In fact, when

$$\sqrt{n}\begin{bmatrix} (s - \overline{s}) \\ vec\left(W - \overline{W}\right) \end{bmatrix} \longrightarrow \mathbf{N}\left(0, \begin{bmatrix} V_{ss} & V_{sw} \\ V_{ws} & V_{ww} \end{bmatrix}\right) \qquad (1.9)$$

where n is the number of observations, *vec* stacks the elements columnwise and the convergence is in distribution to the normal distribution, indicated by **N**, and we define:

$$\Delta \equiv \partial\sigma/\partial\theta^{\top} \tag{1.10}$$

evaluated at $\theta\left(\overline{s},\overline{W}\right)$, and M is a matrix with typical element M_{ij}:

$$M_{ij} \equiv \left[\partial^2\sigma^{\top}/\partial\theta_i\partial\theta_j\right] W\left[\sigma-\overline{s}\right] \tag{1.11}$$

and $\widetilde{V}$ equals by definition

$$\left[\Delta^{\top}\overline{W}, \left[\overline{s}-\sigma\right]^{\top}\otimes\Delta^{\top}\right]\begin{bmatrix} V_{ss} & V_{sw} \\ V_{ws} & V_{ww}\end{bmatrix}\begin{bmatrix} \overline{W}\Delta \\ \left[\overline{s}-\sigma\right]\otimes\Delta\end{bmatrix} \tag{1.12}$$

with σ and its partial derivatives in M and $\widetilde{V}$ also evaluated at the same point $\theta\left(\overline{s},\overline{W}\right)$, then we can say that $\sqrt{n}\left(\widehat{\theta}\left(s,W\right)-\theta\left(\overline{s},\overline{W}\right)\right)$ will tend to the normal distribution with zero mean and covariance matrix Ω, say, with

$$\Omega \equiv \left(\Delta^{\top}\overline{W}\Delta+M\right)^{-1}\widetilde{V}\left(\Delta^{\top}\overline{W}\Delta+M\right)^{-1}. \tag{1.13}$$

This may appear to be a somewhat daunting expression, but it has a pretty clear structure. In particular, observe that if $\overline{s}=\sigma\left(\theta\left(\overline{s},\overline{W}\right)\right)$, in other words, *if* the structural information contained in Σ is correct, then M becomes 0 and $\widetilde{V}$ which sums 4 matrices looses 3 of them, and so the asymptotic covariance of the estimator $\widehat{\theta}\left(s,W\right)$ reduces to:

$$\left(\Delta^{\top}\overline{W}\Delta\right)^{-1}\Delta^{\top}\overline{W}V_{ss}\overline{W}\Delta\left(\Delta^{\top}\overline{W}\Delta\right)^{-1} \tag{1.14}$$

which simplifies even further to

$$\left(\Delta^{\top}V_{ss}^{-1}\Delta\right)^{-1} \tag{1.15}$$

when $\overline{W}=V_{ss}^{-1}$. In the latter case we have asymptotic efficiency: no other fitting function will produce a smaller asymptotic covariance matrix. LISREL belongs to this class, provided the structure it implicitly assumes in V_{ss} is correct. More precisely, it is sufficient when the element in V_{ss} corresponding with the asymptotic covariance between s_{ij} and s_{kl} equals $\sigma_{ik}\sigma_{jl}+\sigma_{il}\sigma_{jk}$. This is the case when the underlying distribution is multinormal. Elliptical distributions in general will yield an asymptotic covariance matrix that is proportional to the normal V_{ss}, so they are efficient as well. The author is unaware of other suitable distributions. So LISREL rests for inference purposes on a *major* assumption, that is in the opinion of the author not easily met. If one wants LISREL to produce reliable standard errors, one would perhaps be well advised to use the bootstrap. By the way, there are many versions of the theorem stated above in the literature, the case of a correct model is particularly well covered. In fact, we expect the results on asymptotic efficiency to be so well known that references are redundant.

To summarize, if the model is correct in the sense that the structural constraints on Σ are met, and S is consistent and W has a positive definite probability limit then the classical fitting functions will produce estimators that tend in probability to the true value. If the model is not correct, they will tend to the best fitting value as determined by the particular fitting function chosen. The estimators are normal, asymptotically, when S and W are (jointly), whether the structural constraints are met or not. Asymptotic efficiency is the most demanding property and is not to be taken for granted. A truly major problem that we do not discuss is model uncertainty, where the model itself is random due to the interaction between specification, estimation and validation on the same data set, with hunches taken from the data to improve the model. This wreaks havoc on the standard approach. No statistics school really knows how to deal with this. See for discussions e. g. Leamer (1978), Dijkstra (1988) or Hastie et al. (2001).

In the next sections we will see that under the very conditions that make LISREL consistent, PLS is not consistent, but that the error will tend to zero when the quality of the estimated latent variables, as measured by their correlation with the true values, tends to 1 by increasing the number of indicators per latent variable.

1.4 On the PLS-Algorithms: Convergence Issues and Functional Properties of Fixed Points

The basic approach in PLS is to construct *proxies* for the latent variables, in the form of linear compounds, by means of a sequence of alternating least squares algorithms, each time solving a local, linear problem, with the aim to extract the predictive information in the sample. Once the compounds are constructed, the parameters of the structural and reduced form are estimated with the proxies replacing the latent variables. The particular information embodied in the structural form is not used explicitly in the determination of the proxies. The information actually used takes the presence or absence of variables in the equations into account, but not the implied zero constraints and multiplicative constraints on the reduced form (:the classical rank constraints on submatrices of the reduced form as implied by the structural form).

There are two basic types of algorithms, called *mode A* and *mode B*, and a third type, *mode C*, that mixes these two. Each mode generates an estimated weight vector $\widehat{w}$, with typical subvector $\widehat{w}_i$ of the same order as y_i. These weight vectors are fixed points of mappings defined algorithmically. If we let S_{ij} stand for the sample equivalent of Σ_{ij}, and sign_{ij} for the sign of the sample correlation between the estimated proxies $\widehat{\eta}_i \equiv \widehat{w}_i^\top y_i$ and $\widehat{\eta}_j \equiv \widehat{w}_j^\top y_j$, and C_i is the index set that collects the labels of latent variables which appear at least once on different sides of the structural equations in which η_i appears, we have for *mode A:*

$$\widehat{w}_i \propto \sum_{j \in C_i} \text{sign}_{ij} \cdot S_{ij} \widehat{w}_j \text{ and } \widehat{w}_i^\top S_{ii} \widehat{w}_i = 1. \tag{1.16}$$

As is easily seen the i-th weight vector is obtainable by a regression of the i-th subvector of indicators y_i on the scalar $\widehat{a}_i \equiv \sum_{j\epsilon C_i} \text{sign}_{ij} \cdot \widehat{\eta}_j$, so the weights are determined by the ability of $\widehat{a}_i$ to predict y_i. It is immediate that when the basic design matrix Σ replaces S the corresponding fixed point $\overline{w}_i$, say, is proportional to λ_i. But note that this requires at least two latent variables. In a stand-alone situation mode A produces the first principal component, and there is no simple relationship with the loading vector. See Hans Schneeweiss and Harald Mathes (1995) for a thorough comparison of factor analysis and principal components. Mode A and principal components share a lack of scale-invariance, they are both sensitive to linear scale transformations. McDonald (1996) has shown essentially that mode A corresponds to maximization of the sum of absolute values of the covariances of the proxies, where the sum excludes the terms corresponding to latent variables which are not directly related. The author gratefully acknowledges reference to McDonald (1996) by an unknown referee.

For *mode B* we have:

$$\widehat{w}_i \propto S_{ii}^{-1} \sum_{j\epsilon C_i} \text{sign}_{ij} \cdot S_{ij} \widehat{w}_j \text{ and } \widehat{w}_i^{\top} S_{ii} \widehat{w}_i = 1. \tag{1.17}$$

Clearly, $\widehat{w}_i$ is obtained by a regression that reverses the order compared to mode A: here $\widehat{a}_i$, defined similarly, is regressed on y_i. So the indicators are used to predict the sign-weighted sum of proxies. With only two latent variables mode B will produce the first canonical variables of their respective indicators, see Wold (1966, 1982) e. g. Mode B is a genuine generalization of canonical variables: it is equivalent to the maximization of the sum of absolute values of the correlations between the proxies, $\widehat{w}_i^{\top} S_{ij} \widehat{w}_j$, taking only those i and j into account that correspond to latent variables which appear at least once on different sides of a structural equation. A Lagrangian analysis will quickly reveal this. The author noted this, in 1977, while he was a member of Herman Wold's research team at the Wharton School, Philadelphia. It is spelled out in his thesis (1981). Kettenring (1971) has introduced other generalizations, we will return to this in the penultimate section. Replacing S by Σ yields a weight vector $\overline{w}_i$ proportional to $\Sigma_{ii}^{-1} \lambda_i$, so that the "population proxy" $\overline{\eta}_i \equiv \overline{w}_i^{\top} y_i$ has unit correlation with the best linear least squares predictor for η_i in terms of y_i. This will be true as well for those generalizations of canonical variables that were analyzed by Kettenring (1971). Mode B is scale-invariant, in the sense that linear scale transformations of the indicators leave $\widehat{\eta}_i$ and $\overline{\eta}_i$ undisturbed.

Mode C mixes the previous approaches: some weight vectors satisfy mode A, others satisfy mode B type of equations. As a consequence the products of mode C mix the properties of the other modes as well. In the sequel we not dwell upon this case. Suffice it to say that with two sets of indicators, two latent variables, mode C produces a variant of the well-known MIMIC-model.

Sofar we have simply assumed that the equations as stated have solutions, that they actually *have* fixed points, and the iterative procedure to obtain them has been merely hinted at. To clarify this, let us discuss a simple case first. Suppose we have three latent variables connected by just one relation $\eta_3 = \beta_{31}\eta_1 + \beta_{32}\eta_2$ plus a least squares residual, and let us use mode B. The fixed point equations specialize to:

$$\widehat{w}_1 = \widehat{c}_1 S_{11}^{-1} \cdot [\mathrm{sign}_{13} \cdot S_{13}\widehat{w}_3] \tag{1.18}$$
$$\widehat{w}_2 = \widehat{c}_2 S_{22}^{-1} \cdot [\mathrm{sign}_{23} \cdot S_{23}\widehat{w}_3] \tag{1.19}$$
$$\widehat{w}_3 = \widehat{c}_3 S_{33}^{-1} \cdot [\mathrm{sign}_{13} \cdot S_{31}\widehat{w}_1 + \mathrm{sign}_{23} \cdot S_{32}\widehat{w}_2]\,. \tag{1.20}$$

The scalar $\widehat{c}_i$ forces $\widehat{w}_i$ to have unit length in the metric of S_{ii}.The iterations start with arbitrary nonzero choices for the $\widehat{w}_i$'s, which are normalized as required, the *sign*-factors are determined, and a cycle of updates commences: inserting $\widehat{w}_3$ into (18) and (19) gives updated values for $\widehat{w}_1$ and $\widehat{w}_2$, which on their turn are inserted into (20), yielding an update for $\widehat{w}_3$, then new sign-factors are calculated, and we return to (18) et cetera. This is continued until the difference between consecutive updates is insignificant. Obviously, this procedure allows of small variations, but they have no impact on the results. Now define a function G, say by

$$G\,(w_3, S) \equiv c_3 S_{33}^{-1} \cdot \left[c_1 S_{31} S_{11}^{-1} S_{13} + c_2 S_{32} S_{22}^{-1} S_{23}\right] \cdot w_3 \tag{1.21}$$

where c_1 is such that $c_1 S_{11}^{-1} S_{13} w_3$ has unit length in the metric of S_{11}, c_2 is defined similarly, and c_3 gives G unit length in the metric of S_{33}. Clearly G is obtained by consecutive substitutions of (18) and (19) into (20). Observe that:

$$G\,(w_3, \Sigma) = \overline{w}_3 \tag{1.22}$$

for every value of w_3 (recall that $\overline{w}_3 \propto \Sigma_{33}^{-1}\lambda_3$). A very useful consequence is that the derivative of G with respect to w_3, evaluated at $(\overline{w}_3, \Sigma)$ equals *zero*. Intuitively, this means that for S not too far away from Σ, $G\,(w_3, S)$ maps two different vectors w_3, which are not too far away from $\overline{w}_3$, on points which are *closer* together than the original vectors. In other words, as a function of w_3, $G\,(w_3, S)$ will be a local contraction mapping. With some care and an appropriate mean value theorem one may verify that our function does indeed satisfy the conditions of Copson's *Fixed point theorem with a parameter*, see Copson (1979), Sects. 80–82. Consequently, G *has* a unique fixed point $\widehat{w}_3\,(S)$ in a neighborhood of $\overline{w}_3$ for every value of S in a neighborhood of Σ, and it *can* be found by successive substitutions: for an arbitrary starting value sufficiently close to $\overline{w}_3$ the ensuing sequence of points converges to $\widehat{w}_3\,(S)$ which satisfies $\widehat{w}_3\,(S) = G\,(\widehat{w}_3\,(S), S)$. Also note that if $plim(S) = \Sigma$ then the first iterate from an arbitrary starting point will tend to $\overline{w}_3$ in probability, so if the sample is sufficiently large the conditions for a local contraction mapping will be satisfied with an arbitrarily high probability. Essentially, any choice of starting vector will do. The mapping $\widehat{w}_3\,(S)$ is continuous, in fact it is continuously differentiable, as follows quickly along familiar lines of reasoning in proofs of implicit function theorems. So asymptotic normality is shared with S. The other weight vectors are smooth transformations of $\widehat{w}_3\,(S)$, so they will be well-behaved as well.

It is appropriate now to point out that what we have done with mode B for three latent variables can also be done for the other modes, and the number of latent variables is irrelevant: reshuffle (16) and (17), if necessary, so that the weights corresponding to the exogenous latent variables are listed first; we can express them in terms of the endogenous weight vectors, $w_{\mathbf{n}}$, say, so that after insertion in the equations for the latter a function $G(w_{\mathbf{n}}, S)$ can be defined with the property that $G(w_{\mathbf{n}}, \Sigma) = \overline{w}_{\mathbf{n}}$ and we proceed as before. We obtain again a well-defined fixed point $\widehat{w}(S)$ by means of successive substitutions. Let us collect this in a theorem (Dijkstra, 1981; we ignore trivial regularity assumptions that preclude loading vectors like λ_i to consist of zeros only; and similarly, we ignore the case where Σ_{ij} is identically zero for every $j \in C_i$):

Theorem 1.1. *If* $\text{plim}(S) = \Sigma$ *where* Σ *obeys the restrictions of the basic design, then the PLS algorithms will converge for every choice of starting values to unique fixed points of (16) and (17) with a probability tending to one when the number of sample observations tends to* ∞. *These fixed points are continuously differentiable functions of* S, *their probability limits satisfy the fixed point equations with* S *replaced by* Σ. *They are asymptotically normal when* S *is.*

As a final observation in this section: if $plim(S) = \Sigma_*$ which is *not* a basic design matrix but comes sufficiently close to it, then the PLS-algorithms will converge in probability to the fixed point defined by $\widehat{w}(\Sigma_*)$. We will again have good numerical behavior and local linearity.

1.5 Correlations, Structural Parameters, Loadings

In this section we will assume without repeatedly saying so that $plim(S) = \Sigma$ for a Σ satisfying the requirements of the extended basic design except for one problem, indicated below in the text. Recall the definition of the *population proxy* $\overline{\eta}_i \equiv \overline{w}_i^\top y_i$ where $\overline{w}_i \equiv plim(\widehat{w}_i)$ depends on the mode chosen; for mode A $\overline{w}_i$ is proportional to λ_i and for mode B it is proportional to $\Sigma_{ii}^{-1}\lambda_i$. Its sample counterpart, the *sample proxy*, is denoted by $\widehat{\eta}_i \equiv \widehat{w}_i^\top y_i$. In PLS the sample proxies replace the latent variables. Within the basic design, however, this replacement can never be exhaustive unless there are no measurement errors. We can measure the quality of the proxies by means of the squared correlation between η_i and $\overline{\eta}_i$: $R^2(\eta_i, \overline{\eta}_i) = \left(\overline{w}_i^\top \lambda_i\right)^2$. In particular, for mode A we have

$$R_A^2(\eta_i, \overline{\eta}_i) = \frac{\left(\lambda_i^\top \lambda_i\right)^2}{\lambda_i^\top \Sigma_{ii} \lambda_i} \tag{1.23}$$

and for mode B:

$$R_B^2(\eta_i, \overline{\eta}_i) = \lambda_i^\top \Sigma_{ii}^{-1} \lambda_i \tag{1.24}$$

as is easily checked. It is worth recalling that the mode B population proxy is proportional to the best linear predictor of η_i in terms of y_i, which is not true for mode A. Also note that the Cauchy-Schwarz inequality immediately entails that R_A^2 is always less than R_B^2 unless λ_i is proportional to $\Sigma_{ii}^{-1}\lambda_i$ or equivalently, to $V_i^{-1}\lambda_i$; for diagonal V_i this can only happen when all measurement error variances are equal. For every mode we have that

$$R^2\left(\overline{\eta}_i,\overline{\eta}_j\right) = \left(\overline{w}_i^\top \Sigma_{ij}\overline{w}_j\right)^2 = \rho_{ij}^2 \cdot R^2\left(\eta_i,\overline{\eta}_i\right)\cdot R^2\left(\eta_j,\overline{\eta}_j\right) \tag{1.25}$$

and we observe that in the limit the PLS-proxies will *under*estimate the squared correlations between the latent variables. This is also true of course for two-block canonical variables: they *under*estimate the correlation between the underlying latent variables eventhough they maximize the correlation between linear compounds. It is not typical for PLS of course. Methods like Kettenring's share this property. The error depends in a simple way on the quality of the proxies, with mode B performing best.

The structural bias does have consequences for the estimation of structural form and reduced form parameters as well. If we let R stand for the correlation matrix of the latent variables, $\overline{R}$ does the same for the population proxies, and K is the diagonal matrix with typical element $R\left(\eta_i,\overline{\eta}_i\right)$, we can write

$$\overline{R} = KRK + I - K^2. \tag{1.26}$$

So conditions of the Simon-Blalock type, like zero partial correlation coefficients, even if satisfied by R will typically not be satisfied by $\overline{R}$. Another consequence is that *squared multiple correlations* will be *under*estimated as well: the value that PLS obtains in the limit, using proxies, for the regression of η_i on other latent variables never exceeds the fraction $R^2\left(\eta_i,\overline{\eta}_i\right)$ of the "true" squared multiple correlation coefficient. This is easily deduced from a well-known characterization of the squared multiple correlation: it is the maximum value of $1-\beta^\top R\beta$ with respect to β where R is the relevant correlation matrix of the variables, and β is a conformable vector whose i-th component is forced to equal 1 (substitution of the expression for $\overline{R}$ quickly yields the upper bound as stated). The upper bound can be attained only when the latent variables other than η_i are measured without flaw.

In general we have that the regression matrix for the population proxies equals $\overline{\Pi}$, say, with

$$\overline{\Pi} = \overline{R}_{\mathbf{nx}}\overline{R}_{\mathbf{xx}}^{-1} = K_{\mathbf{n}}\Pi R_{\mathbf{xx}}K_{\mathbf{x}}\overline{R}_{\mathbf{xx}}^{-1} \tag{1.27}$$

where subscripts indicate appropriate submatrices, the definitions will be clear. Now we assumed that B and Γ could be identified from Π. It is common knowledge in econometrics that this is equivalent to the existence of rank restrictions on submatrices of Π. But since $\overline{R}$ differs from R these relations will be disturbed and $\overline{\Pi}$ will *not* satisfy them, except on sets of measure zero in the parameterspace. This makes the theory hinted at in Sect. 1.3 relevant. With p replacing s, and π replacing σ for maximum similarity, if so desired, we can state that classical estimators for the structural form parameters will asymptotically center around (B_*,Γ_*) say, which are such that $(I-B_*)^{-1}\,\Gamma_*$ fits $\overline{\Pi}$ "best". "Best" will depend on the estimation procedure chosen and $\overline{\Pi}$ varies with the mode. In principle, the well-known delta method can be used to get standard errors, but we doubt whether that is really feasible (which is something of an understatement). The author, Dijkstra (1982, 1983), suggested to use the bootstrap as a general tool. Later developments, such as the stationary bootstrap for time series data, has increased the value of the method even more, but care must be used for a proper application; in particular, one should resample the observations on the indicators, not on the sample proxies, for a decent analysis of sampling uncertainty.

Turning now to the loadings, some straightforward algebra easily yields that both modes will tend to *over*estimate them in absolute value, mode B again behaving better than mode A, in the limit that is. The loadings are in fact estimated by

$$\widehat{\lambda}_i \equiv S_{ii}\widehat{w}_i. \tag{1.28}$$

and the error covariance matrices can be calculated as

$$\widehat{V}_i \equiv S_{ii} - \widehat{\lambda}_i\widehat{\lambda}_i^{\top}. \tag{1.29}$$

(Note that $\widehat{V}_i\widehat{w}_i = 0$, so the estimated errors are linearly dependent, which will have some consequences for second level analyses, not covered here). Inserting population values for sampling values we get for mode A that $\overline{\lambda}_i$, the probability limit of $\widehat{\lambda}_i$, is proportional to $\Sigma_{ii}\lambda_i$. For mode B we note that $\overline{\lambda}_i$ is proportional to λ_i with a proportionality factor equal to the square root of 1 over $R^2(\eta_i, \overline{\eta}_i)$. Mode B, but not mode A, will reproduce Σ_{ij} exactly in the limit. For other results, all based on straightforward algebraic manipulations we refer to Dijkstra (1981).

So in general, not all parameters will be estimated consistently. Wold, in a report that was published as Chap. 1 in Jöreskog and Wold (1982), introduced the auxiliary concept of 'consistency at large' which captures the idea that the inconsistency will tend to zero if more indicators of sufficient quality can be introduced for the latent variables. The condition as formulated originally was

$$\frac{\left[E\left(\overline{w}_i^{\top}\epsilon_i\right)^2\right]^{\frac{1}{2}}}{\overline{w}_i^{\top}\lambda_i} \to 0. \tag{1.30}$$

This is equivalent to $R^2(\eta_i, \overline{\eta}_i) \to 1$. Clearly, if these correlations are large, PLS will combine numerical expediency with consistency. If the proviso is not met in a sufficient degree the author (Dijkstra, 1981) has suggested to use some simple "corrections". E. g. in the case of mode B one could first determine the scalar $\widehat{f}_i$ say that minimizes, *assuming uncorrelated measurement errors*,

$$\operatorname{trace}\left(\left[S_{ii} - \operatorname{diag}(S_{ii}) - \left[f_i^2 \cdot \widehat{\lambda}_i\widehat{\lambda}_i^{\top} - \operatorname{diag}\left(f_i^2 \cdot \widehat{\lambda}_i\widehat{\lambda}_i^{\top}\right)\right]\right]^2\right) \tag{1.31}$$

for all real f_i and which serves to rescale $\widehat{\lambda}_i$. We get

$$\widehat{f}_i^2 = \frac{\widehat{\lambda}_i^{\top}\left[S_{ii} - \operatorname{diag}(S_{ii})\right]\widehat{\lambda}_i}{\widehat{\lambda}_i^{\top}\left[\widehat{\lambda}_i\widehat{\lambda}_i^{\top} - \operatorname{diag}\left(\widehat{\lambda}_i\widehat{\lambda}_i^{\top}\right)\right]\widehat{\lambda}_i}. \tag{1.32}$$

One can check that $\widehat{\widehat{f_i}\widehat{\lambda}_i}$ tends in probability to λ_i. In addition we have that $p\lim\left(\widehat{f}_i^2\right)$ equals $R_B^2\left(\eta_i, \overline{\eta}_i\right)$. So one could in principle get consistent estimators for R, the correlation matrix of the latent variables by reversing (25) so to speak. But a more direct approach can also be taken by minimization of

$$\text{trace}\left\{\left[S_{ij} - r_{ij}\widehat{f}_i\widehat{f}_j \cdot \widehat{\lambda}_i\widehat{\lambda}_j^\top\right]^\top \cdot \left[S_{ij} - r_{ij}\widehat{f}_i\widehat{f}_j \cdot \widehat{\lambda}_i\widehat{\lambda}_j^\top\right]\right\} \tag{1.33}$$

for r_{ij}. This produces the consistent estimator

$$\widehat{r}_{ij} \equiv \frac{\widehat{\lambda}_i^\top S_{ij}\widehat{\lambda}_j}{\widehat{f}_i\widehat{f}_j \cdot \widehat{\lambda}_i^\top\widehat{\lambda}_i \cdot \widehat{\lambda}_j^\top\widehat{\lambda}_j}. \tag{1.34}$$

With a consistent estimator for R we can also estimate B and Γ consistently. We leave it to the reader to develop alternatives. The author is not aware of attempts in the PLS-literature to implement this idea or related approaches. Perhaps the development of second and higher order levels has taken precedence over refinements to the basic design because that just comes naturally to an approach which mimics principal components and canonical variables so strongly. But clearly, the bias can be substantial if not dramatic, whether it relates to regression coefficients, correlations, structural form parameters or loadings as the reader easily convinces himself by choosing arbitrary values for the $R^2\left(\eta_i, \overline{\eta}_i\right)$'s; even for high quality proxies the disruption can be significant, and it is parameter dependent. So if one adheres to the latent variable paradigm, bias correction as suggested here or more sophisticated approaches seems certainly to be called for.

1.6 Two Suggestions for Further Research

In this section we depart from the basic design with its adherence to classical factor analysis modelling, and return so to speak to the original idea of constructing indices by means of linear compounds. We take the linear indices as the fundamental objects and we read path diagrams as representing relationships between the indices in their own right. What we try to do here is to delineate a research program that should lead to the construction of *proper indices*, more about them below, that satisfy the restrictions implied by a path diagram. In the process PLS will loose a lot of its simplicity: proper indices impose inequality restrictions on the indices, and we will no longer do regressions with sums of sign weighted indices, if we do regressions at all, but with sums that somehow reflect the pattern of relationships. The approach is highly provisional and rather unfinished.

As a general principle indicators are selected on the basis of a presumed monotonous relationship with the underlying concept: they are supposed to reflect increases or decreases in the latent variable on an empirically relevant range (without loss of generality we assume that indicators and latent variable are supposed to vary in the same direction). The ensuing *index* should mirror this: not only the *weights* (the coefficients of the indicators in the index) but also the *correlations* between the indicators and the index ought to be positive, or at least non-negative. In practice, a popular first choice for the index is the first principal component of the indicators, the linear compound that best explains total variation in the data. If the correlations between the indicators happen to be positive, Perron-Frobenius' theorem tells us that the first principal component will have positive weights, and of course it has positive correlations with the indicators as well. If the proviso is not met we cannot be certain of these appealing properties. In fact, it often happens that the first principal component is not acceptable as an index, and people resort to other weighting schemes, usually rather simple ones, like sums or equally weighted averages of the indicators. It is not always checked whether this simple construct is positively correlated with its indicators.

Here we will establish that with every non-degenerate vector of indicators is associated a set of *admissible indices*: linear compounds of the indicators with non-negative coefficients whose correlations with the indicators are non-negative. The set of admissible or *proper* weighting vectors is a convex polytope, generated by a finite set of extreme points. In a stand-alone situation, where the vector of indicators is not linked to other indicator-vectors one could project the first principal component on this convex polytope in the appropriate metric, or choose another point in the set,e.g. the point whose average squared correlation with the indicators is maximal. In the regular situation, with more than one block of manifest variables, we propose to choose weighting vectors from each of the admissible sets, such that the ensuing correlation matrix of the indices optimizes one of the distance functions suggested by Kettenring (1971), like: GENVAR (the generalized variance or the determinant of the correlation matrix), MINVAR, its minimal eigenvalue or MAXVAR, its maximal eigenvalue. GENVAR and MINVAR have to be minimized, MAXVAR maximized. The latter approach yields weights such that the total variation of the corresponding indices is explained as well as possible by one factor. The MINVAR-indices will move more tightly together than any other set of indices, in the sense that the variance of the minimum variance combination of the indices will be smaller, at any rate not larger, than the corresponding variance of any other set of indices. GENVAR is the author's favorite, it can be motivated in terms of total variation, or in terms of the volume of (confidence) ellipsoids; see Anderson (1984, in particular Chap. 7.5), or Gantmacher (1977, reprint of 1959, in particular Chap. 9, Sect. 5). Alternatively, GENVAR can be linked to *entropy*. The latent variables which the indices represent are supposed to be mutually informative, in fact they are analyzed together for this very reason. If we want indices that are mutually as informative as possible, we should minimize the entropy of their distribution. This is equivalent to the minimization of the determinant of their covariance or correlation matrix, if we adopt the "most neutral" distribution for the indicators that is consistent with the existence of the second order moments: the normal distribution. (The expression "most neutral" is a non-neutral translation of "maximum entropy"...). Also, as pointed out by Kettenring (1971), the GENVAR indices satisfy an appealing *consistency* property:

the index of every block, given the indices of the other blocks, is the first canonical variable of the block in question relative to the other indices; so every index has maximum multiple correlation with the vector of the other indices.

For the situation where the latent variables are arranged in a path diagram, that embodies a number of zero constraints on the structural form matrices (the matrix linking the exogenous latent variables to the endogenous latent variables, and the matrix linking the latter to each other), we suggest to optimize one of Kettenring's distance functions subject to these constraints. Using Bekker and Dijkstra (1990) and Bekker et al. (1994) the zero constraints can be transformed by symbolic calculations into zero constraints and multiplicative constraints on the regression equations linking the endogenous variables to the exogenous latent variables. In this way we can construct admissible, mutually informative indices, embedded in a theory-based web of relationships.

Now for some detail.

1.6.1 Proper Indices

Let Σ be an arbitrary positive definite covariance or correlation matrix of a random vector X of order p by 1, where p is any natural number. We will prove that there is always a p by 1 vector w with non-negative elements, adding up to 1, such that the vector Σw that contains the covariances between X and the "index" $w^\top X$,has no negative elements as well (note that at least one element must be positive, since the positive definiteness of Σ and the fact that the weights add up to one preclude the solution consisting of zeros only). Intuitively, one might perhaps expect such a property since the angle between any w and its image Σw is acute due to Σ's positive definiteness.

Consider the set:

$$\left\{x \in \mathbb{R}^p : x \geq 0, \iota^\top x = 1, \Sigma x \geq 0\right\} \tag{1.35}$$

where ι is a column vector containing p ones. The defining conditions can also be written in the form $Ax \leq b$ with

$$A \equiv \begin{bmatrix} +\iota^\top \\ -\iota^\top \\ -I \\ \Sigma \end{bmatrix} \text{ and } b \equiv \begin{bmatrix} +1 \\ -1 \\ 0 \\ 0 \end{bmatrix} \tag{1.36}$$

where I is the p by p identity matrix, and the zero vectors in b each have p components. Farkas' lemma (see e. g. Alexander Schrijver 2004, in particular corollary 2.5a in Sect. 2.3.) implies that the set

$$\{x \in \mathbb{R}^p : Ax \leq b\} \tag{1.37}$$

is not empty if and only if the set

$$\left\{y \in \mathbb{R}^{2p+2} : y \geq 0, y^\top A = 0, y^\top b < 0\right\} \tag{1.38}$$

is empty. If we write $y^\top$ as $\left(y_1, y_2, u^\top, v^\top\right)$ where u and v are both of order p by 1, we can express $y^\top A = 0$ as

$$v^\top \Sigma + u^\top + (y_2 - y_1) \cdot \iota^\top = 0 \tag{1.39}$$

and the inequalities in (1.38) require that u and v must be non-negative and that $y_2 - y_1$ is positive. If we postmultiply (1.39) by v we get:

$$v^\top \Sigma v + u^\top v + (y_2 - y_1) \cdot \iota^\top v = 0 \tag{1.40}$$

which entails that v is zero and therefore from (1.39) that u as well as $y_2 - y_1$ are zero. (Note that this is true even when Σ is just positive semi-definite). We conclude that the second set is empty, so the first set is nonempty indeed! Therefore there are always admissible indices for any set of indicators. We can describe this set in some more detail if we write the conditions in "standard form" as in a linear programming setting. Define the matrix $\boldsymbol{A}$ as:

$$\boldsymbol{A} \equiv \begin{bmatrix} \iota^\top & 0^\top \\ \Sigma & -I \end{bmatrix} \tag{1.41}$$

where ι is again of order p by 1, and the dimensions of the other entries follow from this. Note that $\boldsymbol{A}$ has $2p$ columns. It is easily verified that the matrix $\boldsymbol{A}$ has full rowrank $p+1$ if Σ is positive definite. Also define a $p+1$ by 1 vector $\boldsymbol{b}$ as $[1;0]$, a 1 stacked on top of p zeros, and let s be a p by 1 vector of "slack variables". The original set can now be reframed as:

$$\left\{x \in \mathbb{R}^p, s \in \mathbb{R}^p : \boldsymbol{A} \cdot \begin{bmatrix} x \\ s \end{bmatrix} = \boldsymbol{b},\ x \geq 0, s \geq 0\right\} \tag{1.42}$$

Clearly this is a convex set, a convex polytope in fact, that can be generated by its extreme points. The latter can be found by selecting $p+1$ independent columns from $\boldsymbol{A}$, resulting in a matrix $\boldsymbol{A}_B$, say, with B for "basis", and checking whether the product of the inverse of $\boldsymbol{A}_B$ times $\boldsymbol{b}$ has nonnegative elements only (note that $\boldsymbol{A}_B^{-1}\boldsymbol{b}$ is the first column of the inverse of $\boldsymbol{A}_B$). If so, the vector $[x; s]$ containing zeros corresponding to the columns of $\boldsymbol{A}$ which were not selected, is an extreme point of the enlarged space (x, s). Since the set is bounded, the corresponding subvector x is an extreme point of the original (x)-space. In principle we have to evaluate $\binom{2p}{p+1}$ possible candidates. A special and trivial case is where the elements of Σ are all non-negative: all weighting vectors are acceptable, and, as pointed out before, the first principal component (suitably normalized) is one of them.

1.6.2 Potentially Useful Constraints

As indicated before we propose to determine for every block of indicators its set of admissible proper indices, and then choose from each of these sets an index such that some suitable function of the correlation matrix of the selected indices is optimized; we suggested the determinant (minimize) or the first eigenvalue (maximize), and others. A useful refinement may be the incorporation of a priori constraints on the relationships between the indices. Typically one employs a pathdiagram that embodies zero or multiplicative constraints on regression coefficients. It may happen e.g. that two indices are believed to be correlated only because of their linear dependence on a third index, so that the conditional correlation between the two given the third is zero: $\rho_{23.1}$, say, equals 0. This is equivalent to postulating that the entry in the second row and third column of the inverse of the correlation matrix of the three indices is zero (see Cox and Nanny Wermuth (1988), in particular the Sects. 3.1–3.4). More complicated constraints are generated by zero constraints on structural form matrices. E. g. the matrix that links three endogenous latent variables to each other might have the following structure:

$$B = \begin{bmatrix} \beta_{11} & 0 & 0 \\ \beta_{21} & \beta_{22} & \beta_{23} \\ 0 & \beta_{32} & \beta_{33} \end{bmatrix} \tag{1.43}$$

and the effect of the remaining exogenous latent variables on the first set is captured by

$$\Gamma = \begin{bmatrix} 0 & \gamma_{12} \\ \gamma_{21} & 0 \\ 0 & 0 \end{bmatrix} \tag{1.44}$$

Observe that not all parameters are identifiable, not even after normalization (β_{23} will be unidentifiable). But the matrix of regression coefficients, of the regressions of the three endogenous latent variables on the two endogenous latent variables, taking the given structure into account, satisfies both zero constraints as well as multiplicative constraints. In fact, this matrix, Π, say, with $\Pi \equiv B^{-1}\Gamma$, can be parameterized in a minimal way as follows (see Bekker et al. (1994), Sect. 5.6):

$$\Pi = \begin{bmatrix} 0 & \theta_3 \\ \theta_1 & \theta_1\theta_4 \\ \theta_2 & \theta_2\theta_4 \end{bmatrix} \tag{1.45}$$

So $\Pi_{11} = 0$ and $\Pi_{21}\Pi_{32} - \Pi_{22}\Pi_{31} = 0$. These restrictions should perhaps not be wasted when constructing indices. They can be translated into restrictions on the inverses of appropriate submatrices of the correlation matrix of the latent variables. Bekker et al. (1994) have developed software for the automatic generation of minimal parameterizations.

Some small scale experiments by the author, using the constraints of properness and those implied by a path diagram, were encouraging (to the author), and only a few lines of MATLAB-code were required. But clearly a lot of development work and testing remains to be done. For constructing and testing indices a strong case can be made for *cross-validation*, which naturally honoures one of *the* purposes of the entire exercise: prediction of observables. It fits rather naturally with the low-structure environment for which PLS was invented, with its soft or fuzzy relationships between (composite) variables. See e. g. Geisser (1993) and Hastie et al. (2002) for cross-validation techniques and analyses. Cross-validation was embraced early by Herman Wold. He also saw clearly the potential of the related *Jackknife*-method, see Wold (1975).

1.7 Conclusion

I have described and analyzed some of PLS' properties in the context of a latent variable model. It was established that one may expect the algorithms to converge, from essentially arbitary starting values, to unique fixed-points. As a function of the sample size these points do not necessarily converge to the parameters of the latent variable model, in fact their limits or theoretical values may differ substantially from the "true" value if the quality of the proxies is not (very) high. But in principle it is possible to adjust the PLS-estimators in a simple way to cancel the induced distortions, within the context of the (extended) basic design. I also outlined an approach where the indices are treated as the fundamental objects, and where the path diagrams serve to construct meaningful, proper indices, satisfying constraints that are relatively modest.

There are other approaches construed as alternatives to PLS. One such approach, as pointed out by a referee, is due to McDonald (1996) who designed six methods for the estimation of latent variable models as the basic design. These methods all share a least squares type of fitting function and a deliberate distortion of the underlying latent variable model. His method I e. g. minimizes the sum of squares of the difference between S and $\Sigma(\theta)$ as a function of θ, where θ contains the loadings as well as the structural parameters of the relationships between the latent variables, and where all measurement error variances are *a priori* taken to be zero. Once the optimal value for θ is obtained, weighting vectors for the composites are chosen proportional to the estimated loading vectors. McDonald applies his methods as well as PLS to a particular, simple population correlation matrix, with known parameters. Method I is the favorite of the referee who referred me to McDonald (1996), but McDonald himself carefully avoids to state his overall preferences. Clearly, one set of parameters is no basis for a well-established preference, as McDonald takes care to point out on page 254, and again on page 262: the results will typically be rather parameter dependent. I think it is relevant to note the fact, which is not difficult to show, that Method I's loading vectors based on true parameters, their probability limits, are typically *not* proportional to the true loadings, as opposed to PLS mode B.

Table 2 of McDonald (1996) confirms this. Moreover, the ensuing proxies are *not* proportional to the best linear predictors of the latent variables (in terms of their direct indicators), again unlike PLS mode B. A necessary and sufficient condition for proportionality in the context of the basic design with unrestricted correlations between the latent variables, is that the loading vectors are eigenvectors of the corresponding error covariance matrices; if the latter are diagonal the unique factors of each block should have identical variances.

One reviewer of McDonald's paper, apparently a member of the "PLS-camp", suggested that among users of PLS there is an emerging consensus that PLS represents a philosophy rather different from the standard philosophy of what quantitative behavioral science is doing: PLS is mainly prediction-oriented whereas the traditional approach is mainly inference-oriented. I tend to agree with this reviewer, if only for the fact that in each and every one of Wold's contributions to statistics "prediction" and "predictive specifications" are central, key terms. And there is also the embryonic PLS-model of principal components, which served as one of the starting points of PLS (or NIPALS as it was called then in 1966): loadings *as well as* "latent" variables are viewed and treated as parameters to be estimated with a least squares "prediction" criterion leading to linear compounds as estimates for the latent variables. So in this context at least, the approach appears to be entirely natural. But I would maintain that it is still in need of serious development and explication. Somehow the latent variable model, the basic design, seems to have interfered in a pernicious way by posturing as *the* unique and proper way to analyze and model high-dimensional data; this may have (as far as I can see) impeded further developments. Without wanting to sound presumptuous, my contribution contained in Sect. 1.6 can be seen as an attempt to revive what I believe to be the original program. Perhaps PLS could re-orient itself by focussing on (proper) index building through prediction-based cross-validation. McDonald clearly disagrees with the reviewer of his paper about the prediction versus inference issue, and counters by claiming that, if it were true, since "we cannot do better than to use multivariate regressions or canonical variate analysis", one would expect to see a preference among PLS users for multivariate regressions, or if they *must* use a path model they should prefer mode B to mode A. Since this does not seem to happen in practice he infers the invalidity of the reviewer's statement. McDonald has a point when the true parameters are known, but not when they are subject to estimation. If the goal is prediction, this goal is as a rule served best by simplifying the maintained model even more than we would do if description were just what we were after. In fact, predictors based on a moderately incorrect version of the "true model" usually outperform those constructed on the basis of a more elaborate, more correct version, see Dijkstra (1989) or Hastie et al. (2002). In other words, one can certainly not dismiss path models and indices if prediction is called for.

The final issue raised by McDonald at the very end of his paper concerns the use and appropriateness of latent variable models (in what follows the emphasis is mine). He contends that because of factor score indeterminacy, a small number of indicators makes a latent variable model quite inappropriate; indeed, we need lots of them if we want to do any *serious* work using the model (this is an "inescapable fact"). But if we have a large number of indicators per latent variable, a simple average of the former will do an adequate job in replacing the latter, so we then no longer need the model (in other words, the model is either inappropriate or redundant). In my opinion this point of view is completely at odds with the notion of an acceptable model being a useful approximation to part of reality, latent variable modelling is no exception. If a model is to be any good for empirical explanation, prediction or otherwise, it should *not* be a complete and correct specification. See among many e. g. Kaplan (1946, 1964), or Hastie et al. (2002). A suitable methaphor is a map, that by its very nature *must* yield a more or less distorted picture of "angles and distances"; maps that are one-to-one can't get us anywhere. The technical merits of McDonald's paper are not disputed here, but the philosophical and methodological content I find hard to understand and accept.

The reviewer of the present chapter concludes from McDonalds results that "PLS was a mistake, and Method I should have been invented instead. PLS should simply be abandoned". I disagree. I contend that PLS' philosophy potentially has a lot to offer. In my view there is considerable scope in the social sciences, especially in high-dimensional, low-structure, fuzzy environments, for statistical approaches that specify and construct rather simple "index-models" through serious predictive testing. PLS in one version or the other still appears to have untapped sources, waiting to be exploited.

References

Anderson, T. W. (1984). *An introduction to multivariate statistical analysis*. New York: Wiley.

Bekker, P. A., & Dijkstra, T. K. (1990). On the nature and number of the constraints on the reduced form as implied by the structural form. *Econometrica, 58* , 507–514.

Bekker, P. A., Merckens, A. , & Wansbeek, T. J. (1994). *Identification, Equivalent models, and computer algebra*. Boston: Academic.

Copson, E. T. (1968). *Metric spaces*. Cambridge: Cambridge University Press.

Cox, D. R., & Wermuth, N. (1998). *Multivariate dependencies- models, analysis and interpretation*. Boca Raton: Chapman & Hall.

Dijkstra, T. K. (1981). *Latent variables in linear stochastic models,* PhD-thesis, (second edition (1985), Amsterdam: Sociometric Research Foundation).

Dijkstra, T. K. (1982). Some comments on maximum likelihood and partial least squares methods, *Research Report UCLA, Dept. Psychology,* a shortened version was published in 1983.

Dijkstra, T. K. (1983). Some comments on maximum likelihood and partial least squares methods. *Journal of Econometrics, 22* , 67–90.

Dijkstra, T. K. (1988). *On model uncertainty and its statistical implications*. Heidelberg: Springer.

Dijkstra, T. K. (1989). Reduced form estimation, hedging against possible misspecification. *International Economic Review, 30*(2), 373–390.

Dijkstra, T. K. (1990). Some properties of estimated scale invariant covariance structures. *Psychometrika, 55* , 327–336.

Dijkstra, T. K. (1992). On statistical inference with parameter estimates on the boundary of the parameter space. *British Journal of Mathematical and Statistical Psychology, 45*, 289–309.

Frank, I. E., & Friedman, J. H. (1993). A statistical view of some chemometric regression tools. *Technometrics, 35*, 109–135.

Gantmacher, F. R. (1977). *The theory of matrices,* Vol. 1. New York: Chelsea.

Geisser, S. (1993). *Predictive inference: an introduction.*New York: Chapman&Hall.
Hastie, T., Tibshirani, R., & Friedman, J. (2002). *The elements of statistical learning.* New York: Springer.
Jöreskog, K. G., & Wold, H. O. A. (Eds.), (1982). *Systems under indirect observation, Part II.* Amsterdam: North-Holland.
Kagan, A. M., Linnik, Y. V., & Rao, C. R. (1973). *Characterization problems in mathematical statistics.* New York: Wiley.
Kaplan, A. (1946). Definition and specification of meaning. *The Journal of Philosophy, 43*, 281–288.
Kaplan, A. (1964). *The conduct of inquiry.* New York: Chandler.
Kettenring, J. R. (1971). Canonical analysis of several sets of variables. *Biometrika, 58*, 433–451.
Leamer, E. E. (1978). *Specification searches.* New York: Wiley.
McDonald, R. P. (1996). Path analysis with composite variables. *Multivariate Behavioral Research, 31*(2), 239–270.
Ortega, J. M., & Rheinboldt, W. C. (1970). *Iterative solution of nonlinear equations in several variables.* New York: Academic.
Schrijver, A. (2004). *A course in combinatorial optimization.* Berlin: Springer.
Schneeweiss, H., & Mathes, H. (1995). Factor analysis and principal components. *Journal of Multivariate Analysis, 55*, 105–124.
Stone, M., & Brooks, R. J. (1990). Continuum regression: cross-validated sequentially constructed prediction embracing ordinary least squares, partial least squares and principal components regression. *Journal of the Royal Statistical Society, Series B (Methodological), 52*, 237–269.
Tenenhaus, M., Esposito Vinzi, V., Chatelin, Y.-M., & Lauro, C. (2005). PLS path modelling. *Computational Statistics & Data Analysis, 48*, 159–205.
Wold, H. O. A. (1966). Nonlinear estimation by iterative least squares procedures. In David, F. N. (Ed.), *Research Papers in statistics, Festschrift for J. Neyman,* (pp. 411–414). New York: New York.
Wold, H. O. A. (1975). Path models with latent variables: the NIPALS approach. In H. M. Blalock, A., Aganbegian, A., F. M.Borodkin, R. Boudon, V. Capecchi, (Eds.), *Quantitative Sociology,* (pp. pp. 307–359). New York: Academic.
Wold, H. O. A. (1982). Soft modelling: the basic design and some extensions. in Jöreskog, K. G., & Wold, H. O. A. (eds), *Systems under indirect observation, Part II,* pp. 1–5. Amsterdam: Northholland.

Chapter 2
PLS Path Modeling: From Foundations to Recent Developments and Open Issues for Model Assessment and Improvement

Vincenzo Esposito Vinzi, Laura Trinchera, and Silvano Amato

Abstract In this chapter the authors first present the basic algorithm of PLS Path Modeling by discussing some recently proposed estimation options. Namely, they introduce the development of new estimation modes and schemes for multidimensional (formative) constructs, i.e. the use of PLS Regression for formative indicators, and the use of path analysis on latent variable scores to estimate path coefficients Furthermore, they focus on the quality indexes classically used to assess the performance of the model in terms of explained variances. They also present some recent developments in PLS Path Modeling framework for model assessment and improvement, including a non-parametric GoF-based procedure for assessing the statistical significance of path coefficients. Finally, they discuss the REBUS-PLS algorithm that enables to improve the prediction performance of the model by capturing unobserved heterogeneity. The chapter ends with a brief sketch of open issues in the area that, in the Authors' opinion, currently represent major research challenges.

2.1 Introduction

Structural Equation Models (SEM) (Bollen 1989; Kaplan 2000) include a number of statistical methodologies meant to estimate a network of causal relationships, defined according to a theoretical model, linking two or more latent complex concepts, each measured through a number of observable indicators. The basic idea is that complexity inside a system can be studied taking into account a causality network among latent concepts, called Latent Variables (LV), each measured by

V. Esposito Vinzi
ESSEC Business School of Paris, Department of Information Systems and Decision Sciences, Avenue Bernard Hirsch - B.P. 50105, 95021 Cergy-Pontoise, Cedex, France
e-mail: vinzi@essec.fr

L. Trinchera and S. Amato
Dipartimento di Matematica e Statistica, Università degli Studi di Napoli "Federico II", Via Cintia, 26 - Complesso Monte S. Angelo, 80126 Napoli, Italy
e-mail: ltrinche@unina.it, silvano.amato@gmail.com

V. Esposito Vinzi et al. (eds.), *Handbook of Partial Least Squares*, Springer Handbooks of Computational Statistics, DOI 10.1007/978-3-540-32827-8_3,

several observed indicators usually defined as Manifest Variables (MV). It is in this sense that Structural Equation Models represent a joint-point between Path Analysis (Tukey 1964; Alwin and Hauser 1975) and Confirmatory Factor Analysis (CFA) (Thurstone 1931).

The PLS (Partial Least Squares) approach to Structural Equation Models, also known as PLS Path Modeling (PLS-PM) has been proposed as a component-based estimation procedure different from the classical covariance-based LISREL-type approach. In Wold's (1975a) seminal paper, the main principles of *partial least squares* for *principal component analysis* (Wold 1966) were extended to situations with more than one block of variables. Other presentations of PLS Path Modeling given by Wold appeared in the same year (Wold 1975b, c). Wold (1980) provides a discussion on the theory and the application of Partial Least Squares for path models in econometrics. The specific stages of the algorithm are well described in Wold (1982) and in Wold (1985). Extensive reviews on the PLS approach to Structural Equation Models with further developments are given in Chin (1998) and in Tenenhaus et al. (2005).

PLS Path Modeling is a component-based estimation method (Tenenhaus 2008a). It is an iterative algorithm that separately solves out the blocks of the measurement model and then, in a second step, estimates the path coefficients in the structural model. Therefore, PLS-PM is claimed to explain at best the residual variance of the latent variables and, potentially, also of the manifest variables in any regression run in the model (Fornell and Bookstein 1982). That is why PLS Path Modeling is considered more as an exploratory approach than as a confirmatory one. Unlike the classical covariance-based approach, PLS-PM does not aim at reproducing the sample covariance matrix. PLS-PM is considered as a *soft modeling* approach where no strong assumptions (with respect to the distributions, the sample size and the measurement scale) are required. This is a very interesting feature especially in those application fields where such assumptions are not tenable, at least in full. On the other side, this implies a lack of the classical parametric inferential framework that is replaced by empirical confidence intervals and hypothesis testing procedures based on resampling methods (Chin 1998; Tenenhaus et al. 2005) such as jackknife and bootstrap. It also leads to less ambitious statistical properties for the estimates, e.g. coefficients are known to be biased but consistent at large (Cassel et al. 1999, 2000). Finally, PLS-PM is more oriented to optimizing predictions (explained variances) than statistical accuracy of the estimates.

In the following, we will first present the basic algorithm of PLS-PM by discussing some recently proposed estimation options and by focusing on the quality indexes classically used to assess the performance (usually in terms of explained variances) of the model (Sect. 2.2). Then, we will present a non-parametric GoF-based procedure for assessing the statistical significance of path coefficients (Sect. 2.3.1). Finally, we will present the REBUS-PLS algorithm that enables to improve the prediction performance of the model in presence of unobserved heterogeneity (Sect. 2.4). This chapter ends with a brief sketch of open issues in the area that, in our opinion, currently represent major research challenges (Sect. 2.5).

2.2 PLS Path Modeling: Basic Algorithm and Quality Indexes

2.2.1 The Algorithm

PLS Path Modeling aims to estimate the relationships among Q ($q = 1, \ldots, Q$) blocks of variables, which are expression of unobservable constructs. Essentially, PLS-PM is made of a system of interdependent equations based on simple and multiple regressions. Such a system estimates the network of relations among the latent variables as well as the links between the manifest variables and their own latent variables.

Formally, let us assume P variables ($p = 1, \ldots, P$) observed on N units ($n = 1, \ldots, N$). The resulting data (x_{npq}) are collected in a partitioned data table $\boldsymbol{X}$:

$$\boldsymbol{X} = \left[\boldsymbol{X}_1, \ldots, \boldsymbol{X}_q, \ldots, \boldsymbol{X}_Q\right]$$

where $\boldsymbol{X}_q$ is the generic q-th block made of P_q variables.

As well known, each Structural Equation Model is composed by two sub-models: the measurement model and the structural model. The first one takes into account the relationships between each latent variable and the corresponding manifest variables, while the structural model takes into account the relationships among the latent variables.

In the PLS Path Modeling framework, the structural model can be written as:

$$\boldsymbol{\xi}_j = \beta_{0j} + \sum_{q:\boldsymbol{\xi}_q \to \boldsymbol{\xi}_j} \beta_{qj} \boldsymbol{\xi}_q + \boldsymbol{\zeta}_j \tag{2.1}$$

where $\boldsymbol{\xi}_j (j = 1, \ldots, J)$ is the generic endogenous latent variable, β_{qj} is the generic path coefficient interrelating the q-th exogenous latent variable to the j-th endogenous one, and $\boldsymbol{\zeta}_j$ is the error in the inner relation (i.e. disturbance term in the prediction of the j-th endogenous latent variable from its explanatory latent variables).

The measurement model formulation depends on the direction of the relationships between the latent variables and the corresponding manifest variables (Fornell and Bookstein 1982). As a matter of fact, different types of measurement model are available: the *reflective model* (or outwards directed model), the *formative model* (or inwards directed model) and the *MIMIC model* (a mixture of the two previous models).

In a *reflective model* the block of manifest variables related to a latent variable is assumed to measure a unique underlying concept. Each manifest variable reflects (is an effect of) the corresponding latent variable and plays a role of endogenous variable in the block specific measurement model. In the reflective measurement model, indicators linked to the same latent variable should covary: changes in one indicator imply changes in the others. Moreover, internal consistency has to be checked, i.e. each block is assumed to be homogeneous and unidimensional. It is important to notice that for the *reflective models*, the measurement model reproduces the factor analysis model, in which each variable is a function of the underlying factor. In more formal terms, in a *reflective model* each manifest variable is related to the corresponding latent variable by a simple regression model, i.e.:

$$\boldsymbol{x}_{pq} = \lambda_{p0} + \lambda_{pq}\boldsymbol{\xi}_q + \boldsymbol{\epsilon}_{pq} \tag{2.2}$$

where λ_{pq} is the loading associated to the p-th manifest variable in the q-th block and the error term $\boldsymbol{\epsilon}_{pq}$ represents the imprecision in the measurement process. Standardized loadings are often preferred for interpretation purposes as they represent correlations between each manifest variable and the corresponding latent variable.

An assumption behind this model is that the error $\boldsymbol{\epsilon}_{pq}$ has a zero mean and is uncorrelated with the latent variable of the same block:

$$E(\boldsymbol{x}_{pq}|\boldsymbol{\xi}_q) = \lambda_{p0} + \lambda_{pq}\boldsymbol{\xi}_q. \tag{2.3}$$

This assumption, defined as *predictor specification*, assures desirable estimation properties in classical Ordinary Least Squares (OLS) modeling.

As the *reflective* block reflects the (unique) latent construct, it should be homogeneous and *unidimensional*. Hence, the manifest variables in a block are assumed to measure the same unique underlying concept. There exist several tools for checking the block homogeneity and unidimensionality:

(a) *Cronbach's alpha*: this is a classical index in reliability analysis and represents a strong tradition in the SEM community as a measure of internal consistency. A block is considered homogenous if this index is larger than 0.7 for confirmatory studies. Among several alternative and equivalent formulas, this index can be expressed as:

$$\alpha = \frac{\sum_{p \neq p'} cor(\boldsymbol{x}_{pq}, \boldsymbol{x}_{p'q})}{P_q + \sum_{p \neq p'} cor(\boldsymbol{x}_{pq}, \boldsymbol{x}_{p'q})} \times \frac{P_q}{P_q - 1} \tag{2.4}$$

where P_q is the number of manifest variables in the q-th block.

(b) *Dillon-Goldstein's* (or *Jöreskog's*) *rho* (Wertz et al. 1974) better known as *composite reliability*: a block is considered homogenous if this index is larger than 0.7

$$\rho = \frac{(\sum_{p=1}^{P_q} \lambda_{pq})^2}{(\sum_{p=1}^{P_q} \lambda_{pq})^2 + \sum_{p=1}^{P_q}(1 - \lambda_{pq}^2)}. \tag{2.5}$$

(c) *Principal component analysis of a block*: a block may be considered unidimensional if the first eigenvalue of its correlation matrix is higher than 1, while the others are smaller (Kaiser's rule). A bootstrap procedure can be implemented to assess whether the eigenvalue structure is significant or rather due to sampling fluctuations. In case unidimensionality is rejected, eventual groups of unidimensional sub-blocks can be identified by referring to patterns of variable-factor correlations displayed on the loading plots.

According to Chin (1998), *Dillon-Goldstein's rho* is considered to be a better indicator than *Cronbach's alpha*. Indeed, the latter assumes the so-called tau equivalence (or parallelity) of the manifest variables, i.e. each manifest variable is assumed to be equally important in defining the latent variable. *Dillon-Goldstein's rho* does not make this assumption as it is based on the results from the model (i.e. the loadings) rather than the correlations observed between the manifest variables in the dataset. *Cronbach's alpha* actually provides a lower bound estimate of reliability.

In the *formative model* , each manifest variable or each sub-block of manifest variables represents a different dimension of the underlying concept. Therefore, unlike the reflective model, the formative model does not assume homogeneity nor unidimensionality of the block. The latent variable is defined as a linear combination of the corresponding manifest variables, thus each manifest variable is an exogenous variable in the measurement model. These indicators need not to covary: changes in one indicator do not imply changes in the others and internal consistency is no more an issue. Thus the measurement model could be expressed as:

$$\boldsymbol{\xi}_q = \sum_{p=1}^{P_q} \omega_{pq} \boldsymbol{x}_{pq} + \boldsymbol{\delta}_q \tag{2.6}$$

where ω_{pq} is the coefficient linking each manifest variable to the corresponding latent variable and the error term $\boldsymbol{\delta}_q$ represents the fraction of the corresponding latent variable not accounted for by the block of manifest variables. The assumption behind this model is the following *predictor specification*:

$$E(\boldsymbol{\xi}_q | \boldsymbol{x}_{pq}) = \sum_{p=1}^{P_q} \omega_{pq} \boldsymbol{x}_{pq}. \tag{2.7}$$

Finally, the *MIMIC model* is a mixture of both the reflective and the formative models within the same block of manifest variables.

Independently from the type of measurement model, upon convergence of the algorithm, the standardized latent variable scores ($\hat{\boldsymbol{\xi}}_q$) associated to the q-th latent variable ($\boldsymbol{\xi}_q$) are computed as a linear combination of its own block of manifest variables by means of the so-called *weight relation* defined as:

$$\hat{\boldsymbol{\xi}}_q = \sum_{p=1}^{P_q} w_{pq} \boldsymbol{x}_{pq} \tag{2.8}$$

where the variables $\boldsymbol{x}_{pq}$ are centred and w_{pq} are the outer weights. These weights are yielded upon convergence of the algorithm and then transformed so as to produce standardized latent variable scores. However, when all manifest variables are observed on the same measurement scale and all outer weights are positive, it is interesting and feasible to express these scores in the original scale (Fornel 1992). This is achieved by using normalized weights $\tilde{w}_{pq}$ defined as:

$$\tilde{w}_{pq} = \frac{w_{pq}}{\sum_{p=1}^{P_q} w_{pq}} \text{ with } \sum_{p=1}^{P_q} \tilde{w}_{pq} = 1 \; \forall q : P_q > 1. \tag{2.9}$$

It is very important not to confound the *weight relation* defined in (2.8) with a *formative model*. The *weight relation* only implies that, in PLS Path Modeling, any latent variable is defined as a weighted sum of its own manifest variables. It does not affect the direction of the relationship between the latent variable and its own manifest variables in the outer model. Such a direction (inwards or outwards) determines how the weights used in (2.8) are estimated.

In PLS Path Modeling an iterative procedure permits to estimate the outer weights (w_{pq}) and the latent variable scores ($\widehat{\xi}_q$). The estimation procedure is named *partial* since it solves blocks one at a time by means of alternating single and multiple linear regressions. The path coefficients (β_{qj}) are estimated afterwards by means of a regular regression between the estimated latent variable scores in accordance with the specified network of structural relations. Taking into account the regression framework of PLS Path Modeling, we prefer to think of such a network as defining a predictive path model for the endogenous latent variables rather than a causality network. Indeed, the emphasis is more on the accuracy of predictions than on the accuracy of estimation.

The estimation of the outer weights is achieved through the alternation of the *outer* and the *inner* estimation steps, iterated till convergence. It is important to underline that no formal proof of convergence of this algorithm has been provided until now for models with more than two blocks. Nevertheless, empirical convergence is usually observed in practice.

The procedure works on centred (or standardized) manifest variables and starts by choosing arbitrary initial weights w_{pq}. Then, in the outer estimation stage, each latent variable is estimated as a linear combination of its own manifest variables:

$$\boldsymbol{v}_q \propto \pm \sum_{p=1}^{P_q} w_{pq} \boldsymbol{x}_{pq} = \pm \boldsymbol{X}_q \boldsymbol{w}_q \tag{2.10}$$

where $\boldsymbol{v}_q$ is the standardized (zero mean and unitary standard deviation) outer estimate of the q-th latent variable $\boldsymbol{\xi}_q$, the symbol $\propto$ means that the left side of the equation corresponds to the standardized right side and the "$\pm$" sign shows the sign ambiguity. This ambiguity is usually solved by choosing the sign making the outer estimate positively correlated to a majority of its manifest variables. Anyhow, the user is allowed to invert the signs of the weights for a whole block in order to make them coherent with the definition of the latent variable.

In the inner estimation stage, each latent variable is estimated by considering its links with the other Q' adjacent latent variables:

$$\boldsymbol{\vartheta}_q \propto \sum_{q'=1}^{Q'} e_{qq'} \boldsymbol{\nu}_{q'} \tag{2.11}$$

where $\boldsymbol{\vartheta}_q$ is the standardized inner estimate of the q-th latent variable $\boldsymbol{\xi}_q$ and each inner weight ($e_{qq'}$) is equal (in the so-called *centroid scheme*) to the sign of the correlation between the outer estimate $\boldsymbol{\nu}_q$ of the q-th latent variable and the outer estimate of the q' latent variable $\boldsymbol{\nu}_{q'}$ connected with $\boldsymbol{\nu}_q$. Inner weights can be obtained also by means of other schemes than the centroid one. Namely, the three following schemes are available:

1. *Centroid scheme* (the Wold's original scheme): take the sign of the correlation between the outer estimate $\boldsymbol{\nu}_q$ of the q-th latent variable and the outer estimate $\boldsymbol{\nu}_{q'}$ connected with $\boldsymbol{\nu}_q$.
2. *Factorial scheme* (proposed by Lohmöller): take the correlation between the outer estimate $\boldsymbol{\nu}_q$ of the q-th latent variable and the outer estimate $\boldsymbol{\nu}_{q'}$ connected with $\boldsymbol{\nu}_q$.
3. *Structural or path weighting scheme*: take the regression coefficient between $\boldsymbol{\nu}_q$ and the $\boldsymbol{\nu}_{q'}$ connected with $\boldsymbol{\nu}_q$ if $\boldsymbol{\nu}_q$ plays the role of dependent variable in the specific structural equation, or take the correlation coefficient in case it is a predictor.

Even though the path weighting scheme seems the most coherent with the direction of the structural relations between latent variables, the centroid scheme is very often used as it adapts well to cases where the manifest variables in a block are strongly correlated to each other. The factorial scheme, instead, is better suited to cases where such correlations are weaker. In spite of different common practices, we strongly advice to use the path weighting scheme. Indeed, this is the only estimation scheme that explicitly considers the direction of relationships as specified in the predictive path model.

Once a first estimate of the latent variables is obtained, the algorithm goes on by updating the outer weights w_{pq}.

Two different *modes* are available to update the outer weights. They are closely related to, but do not coincide with, the *formative* and the *reflective* modes:

- *Mode A* : each outer weight w_{pq} is updated as the regression coefficient in the simple regression of the p-th manifest variable of the q-th block ($\boldsymbol{x}_{pq}$) on the inner estimate of the q-th latent variable $\boldsymbol{\vartheta}_q$. As a matter of fact, since $\boldsymbol{\vartheta}_q$ is standardized, the generic outer weight w_{pq} is obtained as:

$$w_{pq} = cov\left(\boldsymbol{x}_{pq}, \boldsymbol{\vartheta}_q\right) \tag{2.12}$$

i.e. the regression coefficient reduces to the covariance between each manifest variable and the corresponding inner estimate of the latent variable. In case the manifest variables have been also standardized, such a covariance becomes a correlation.

- *Mode B* : the vector $\boldsymbol{w}_q$ of the weights w_{pq} associated to the manifest variables of the q-th block is updated as the vector of the regression coefficients in the multiple regression of the inner estimate of the q-th latent variable $\boldsymbol{\vartheta}_q$ on the manifest variables in $\boldsymbol{X}_q$:

$$\boldsymbol{w}_q = \left(\boldsymbol{X}_q'\boldsymbol{X}_q\right)^{-1} \boldsymbol{X}_q'\boldsymbol{\vartheta}_q \tag{2.13}$$

where $\boldsymbol{X}_q$ comprises the P_q manifest variables $\boldsymbol{x}_{pq}$ previously centred and scaled by $\sqrt{1/N}$.

As already said, the choice of the outer weight estimation mode is strictly related to the nature of the measurement model. For a *reflective (outwards directed) model* the *Mode A* is more appropriate, while *Mode B* is better for a *formative (inwards directed) model*. Furthermore, *Mode A* is suggested for endogenous latent variables, while *Mode B* for the exogenous ones.

In case of a one-block PLS model, *Mode A* leads to the same results (i.e. outer weights, loadings and latent variable scores) as for the first standardized principal component in a Principal Component Analysis (PCA). This reveals the reflective nature of PCA that is known to look for components (weighted sums) explaining the corresponding manifest variables at best. Instead, *Mode B* coherently provides an indeterminate solution when applied to a one-block PLS model. Indeed, without an inner model, any linear combination of the manifest variables is perfectly explained by the manifest variables themselves.

It is worth noticing that *Mode B* may be affected by multicollinearity between manifest variables belonging to the same block. If this happens, PLS regression (Tenenhaus 1998; Wold et al. 1983) may be used as a more stable and better interpretable alternative to OLS regression to estimate outer weights in a formative model, thus defining a *Mode PLS* (Esposito Vinzi 2008, 2009; Esposito Vinzi and Russolillo 2010). This mode is available in the PLSPM module of the XLSTAT software [1] (Addinsoft 2009). As a matter of fact, it may be noticed that Mode A consists in taking the first component from a PLS regression, while Mode B takes all PLS regression components (and thus coincides with OLS multiple regression). Therefore, running a PLS regression and retaining a certain number (that may be different for each block) of significant PLS components is meant as an intermediate

[1] XLSTAT-PLSPM is the ultimate PLS Path Modeling software implemented in XLSTAT (http://www.xlstat.com/en/products/xlstat-plspm/), a data analysis and statistical solution for Microsoft Excel. XLSTAT allows using the PLS approach (both PLS Path modeling and PLS regression) without leaving Microsoft Excel. Thanks to an intuitive and flexible interface, XLSTAT-PLSPM permits to build the graphical representation of the model, then to fit the model, to display the results in Excel either as tables or graphical views. As XLSTAT-PLSPM is totally integrated with the XLSTAT suite, it is possible to further analyze the results with the other XLSTAT features. Apart from the classical and fundamental options of PLS Path Modeling, XLSTAT-PLSPM comprises several advanced features and implements the most recent methodological developments.

mode between Mode A and Mode B. This new Mode PLS adapts well to formative models where the blocks are multidimensional but with fewer dimensions than the number of manifest variables.

The PLS Path Modeling algorithm alternates the outer and the inner estimation stages by iterating till convergence. Up to now convergence has been proved only for path diagrams with one or two blocks (Lyttkens et al. 1975). However, for multi-block models, convergence is practically always encountered in practice.

Upon convergence, the estimates of the latent variable scores are obtained according to 2.8. Thus, PLS Path Modeling provides a direct estimate of the latent variable individual scores as aggregates of manifest variables that naturally involve measurement error. The price of obtaining these scores is the inconsistency of the estimates.

Finally, structural (or path) coefficients are estimated through OLS multiple/simple regressions among the estimated latent variable scores. PLS regression can nicely replace OLS regression for estimating path coefficients whenever one or more of the following problems occur: missing latent variable scores, strongly correlated latent variables, a limited number of units as compared to the number of predictors in the most complex structural equation. A PLS regression option for path coefficients is implemented in the PLSPM module of the XLSTAT software (Addinsoft 2009). This option permits to choose a specific number of PLS components for each endogenous latent variable.

A schematic description of the PLS Path Modeling algorithm by Löhmoller (with specific options for the sake of brevity) is provided in Algorithm 1. This is the best known procedure for the computation of latent variable scores and it is the one implemented in the PLSPM module of the XLSTAT software. There exists a second and less known procedure initially proposed in Wold (1985). The Löhmoller's procedure is more advantageous and easier to implement. However, the Wold's procedure seems to be more interesting for proving convergence properties of the PLS algorithm as it is monotonically convergent (Hanafi 2007). Indeed, at present PLS Path Modeling is often blamed not to optimize a well identified global scalar function. However, very promising researches on this topic are on going and interesting results are expected soon (Tenenhaus 2008b; Tenenhaus and Tenenhaus 2009).

In Lohmöller (1987) and in Lohmöller (1989) Wold's original algorithm was further developed in terms of options and mathematical proprieties. Moreover, in Tenenhaus and Esposito Vinzi (2005) new options for computing both inner and outer estimates were implemented together with a specific treatment for missing data and multicollinearity while enhancing the data analysis flavour of the PLS approach and its presentation as a general framework to the analysis of multiple tables.

A comprehensive application of the PLS Path Modeling algorithm to real data will be presented in Sect. 2.4.2 after dealing with the problem of capturing unobserved heterogeneity for improving the model prediction performance.

Algorithm 1 : PLS Path Modeling based on Löhmoller's algorithm with the following options: centroid scheme, standardized latent variable scores, OLS regressions

Input: $\boldsymbol{X} = [\boldsymbol{X}_1, \ldots, \boldsymbol{X}_q, \ldots, \boldsymbol{X}_Q]$, i.e. Q blocks of centred manifest variables;
Output: $\boldsymbol{w}_q$, $\hat{\boldsymbol{\xi}}_q$, $\boldsymbol{\beta}_j$;

1: **for all** $q = 1, \ldots, Q$ **do**
2: initialize $\boldsymbol{w}_q$
3: $\boldsymbol{\nu}_q \propto \pm \sum_{p=1}^{P_q} w_{pq} \boldsymbol{x}_{pq} = \pm \boldsymbol{X}_q \boldsymbol{w}_q$
4: $e_{qq'} = sign\left[cor\left(\boldsymbol{\nu}_q, \boldsymbol{\nu}_{q'}\right)\right]$ following the centroid scheme
5: $\boldsymbol{\vartheta}_q \propto \sum_{q'=1}^{Q'} e_{qq'} \boldsymbol{\nu}_{q'}$
6: update $\boldsymbol{w}_q$:

(a) $w_{pq} = cov(\boldsymbol{x}_{pq}, \boldsymbol{\vartheta}_q)$ for Mode A (outwards directed model)

(b) $\boldsymbol{w}_q = \left(\frac{X_q' X_q}{N}\right)^{-1} \left(\frac{X_q' \boldsymbol{\vartheta}_q}{N}\right)$ for Mode B (inwards directed model)

7: **end for**
8: **Steps 1–7 are repeated until convergence** on the outer weights is achieved, i.e. until:

$$max\{w_{pq,\text{current iteration}} - w_{pq,\text{previous iteration}}\} < \Delta$$

where Δ is a convergence tolerance usually set at 0.0001 or less
9: **Upon convergence**:

(1) for each block the standardized latent variable scores are computed as weighted aggregates of manifest variables:

$$\hat{\boldsymbol{\xi}}_q \propto \boldsymbol{X}_q \boldsymbol{w}_q,$$

(2) for each endogenous latent variable $\boldsymbol{\xi}_j$ ($j = 1, \ldots, J$), the vector of path coefficients is estimated by means of OLS regression as:

$$\boldsymbol{\beta}_j = \left(\hat{\Xi}' \hat{\Xi}\right)^{-1} \hat{\Xi}' \hat{\boldsymbol{\xi}}_j,$$

where $\hat{\Xi}$ includes the scores of the latent variables that explain the j-th endogenous latent variable $\boldsymbol{\xi}_j$, and $\hat{\boldsymbol{\xi}}_j$ is the latent variable score of the j-th endogenous latent variable

2.2.2 The Quality Indexes

PLS Path Modeling lacks a well identified global optimization criterion so that there is no *global fitting function* to assess the goodness of the model. Furthermore, it is a variance-based model strongly oriented to prediction. Thus, model validation mainly focuses on the model predictive capability. According to PLS-PM structure, each part of the model needs to be validated: the *measurement model*, the *structural model* and the overall model. That is why, PLS Path Modeling provides three different fit indexes: the *communality* index, the *redundancy* index and the *Goodness of Fit* (*GoF*) index.

For each q-th block in the model with more than one manifest variable (i.e. for each block with $P_q > 1$) the quality of the measurement model is assessed by means of the *communality* index:

$$Com_q = \frac{1}{P_q} \sum_{p=1}^{P_q} cor^2 \left(\boldsymbol{x}_{pq}, \hat{\boldsymbol{\xi}}_q \right) \forall q : P_q > 1. \tag{2.14}$$

This index measures how much of the manifest variables variability in the q-th block is explained by their own latent variable scores $\hat{\boldsymbol{\xi}}_q$. Moreover, the communality index for the q-th block is nothing but the average of the squared correlations (squared loadings in case of standardized manifest variables) between each manifest variable in the q-th block and the corresponding latent variable scores.

It is possible to assess the quality of the whole measurement model by means of the *average communality* index, i.e:

$$\overline{Com} = \frac{1}{\sum_{q:P_q>1} P_q} \sum_{q:P_q>1} P_q Com_q. \tag{2.15}$$

This is a weighted average of all the Q block-specific *communality* indexes (see (2.14)) with weights equal to the number of manifest variables in each block. Moreover, since the *communality* index for the q-th block is nothing but the average of the squared correlation in the block, then the *average communality* is the average of all the squared correlations between each manifest variable and the corresponding latent variable scores in the model, i.e.:

$$\overline{Com} = \frac{1}{\sum_{q:P_q>1} P_q} \sum_{q:P_q>1} \sum_{p=1}^{P_q} cor^2 \left(\boldsymbol{x}_{pq}, \hat{\boldsymbol{\xi}}_q \right). \tag{2.16}$$

Let us focus now on the structural model. Although the quality of each structural equation is measured by a simple evaluation of the R^2 fit index, this is not sufficient to evaluate the whole structural model. Specifically, since the structural equations are estimated once the convergence is achieved and he latent variable scores are estimated, then the R^2 values only take into account the fit of each regression equation in the structural model.

It would be a wise choice to replace this current practice by a path analysis on the latent variable scores considering all structural equations simultaneously rather than as independent regressions. We see two advantages in this proposal: the path coefficients would be estimated by optimizing a single discrepancy function based on the difference between the observed covariance matrix of the latent variable scores and the same covariance matrix implied by the model; the structural model could be assessed as a whole in terms of a chi-square test related to the optimized discrepancy function. We have noticed, through several applications, that such a procedure does not actually change the prediction performance of the model in terms of explained variances for the endogenous latent variables. Up to now, no available software has implemented the path analysis option in a PLS-PM framework.

In view of linking the prediction performance of the measurement model to the structural one, the *redundancy* index computed for the j-th endogenous block, measures the portion of variability of the manifest variables connected to the j-th endogenous latent variable explained by the latent variables directly connected to the block, i.e.:

$$Red_j = Com_j \times R^2\left(\hat{\xi}_j, \hat{\xi}_{q:\xi_q \to \xi_j}\right). \tag{2.17}$$

A global quality measure of the structural model is also provided by the *average redundancy* index, computed as:

$$\overline{Red} = \frac{1}{J}\sum_{j=1}^{J} Red_j \tag{2.18}$$

where J is the total number of endogenous latent variables in the model.

As aforementioned, there is no overall fit index in PLS Path Modeling. Nevertheless, a global criterion of goodness of fit has been proposed by Tenenhaus et al. (2004): the GoF index. Such an index has been developed in order to take into account the model performance in both the measurement and the structural model and thus provide a single measure for the overall prediction performance of the model. For this reason the GoF index is obtained as the geometric mean of the *average communality* index and the average R^2 value:

$$GoF = \sqrt{\overline{Com} \times \overline{R^2}} \tag{2.19}$$

where the average R^2 value is obtained as:

$$\overline{R^2} = \frac{1}{J} R^2\left(\hat{\xi}_j, \hat{\xi}_{q:\xi_q \to \xi_j}\right). \tag{2.20}$$

As it is partly based on average communality, the GoF index is conceptually appropriate whenever measurement models are reflective. However, communalities may be also computed and interpreted in case of formative models knowing that, in such a case, we expect lower communalities but higher R^2 as compared to reflective models. Therefore, for practical purposes, the GoF index can be interpreted also with formative models as it still provides a measure of overall fit.

According to (2.16) and (2.20) the GoF index can be rewritten as:

$$GoF = \sqrt{\frac{\sum_{q:P_q>1}\sum_{p=1}^{P_q} Cor^2\left(\boldsymbol{x}_{pq}, \hat{\xi}_q\right)}{\sum_{q:P_q>1} P_q} \times \frac{\sum_{j=1}^{J} R^2\left(\hat{\xi}_j, \hat{\xi}_{q:\xi_q \to \xi_j}\right)}{J}}. \tag{2.21}$$

A normalized version is obtained by relating each term in (2.21) to the corresponding maximum value. In particular, it is well known that in principal component analysis the best rank one approximation of a set of variables $\boldsymbol{X}$ is given by the eigenvector associated to the largest eigenvalue of the $\boldsymbol{X}'\boldsymbol{X}$ matrix. Furthermore, the sum of the squared correlations between each variable and the first principal component of $\boldsymbol{X}$ is a maximum.

Therefore, if data are mean centred and with unit variance, the left term under the square root in (2.21) is such that $\sum_{p=1}^{P_q} cor^2\left(\boldsymbol{x}_{pq}, \hat{\xi}_q\right) \leq \lambda_{(q)}^1$, where $\lambda_{(q)}^1$ is the first eigenvalue obtained by performing a Principal Component Analysis on the q-th block of manifest variables. Thus, the normalized version of the first term of the GoF is obtained as:

$$T_1 = \frac{1}{\sum_{q:P_q>1} P_q} \sum_{q:P_q>1} \frac{\sum_{p=1}^{P_q} cor^2\left(\boldsymbol{x}_{pq}, \hat{\xi}_q\right)}{\lambda_{(q)}^1}. \tag{2.22}$$

In other words, here the sum of the communalities in each block is divided by the first eigenvalue of the block itself.

As concerning the right term under the square root in (2.19), the normalized version is obtained as:

$$T_2 = \frac{1}{J} \sum_{j=1}^{J} \frac{R^2\left(\hat{\xi}_j, \hat{\xi}_{q:\xi_q \rightarrow \xi_j}\right)}{\rho_j^2} \tag{2.23}$$

where ρ_j is the first canonical correlation of the canonical analysis between $\boldsymbol{X}_j$ containing the manifest variables associated to the j-th endogenous latent variable, and a matrix containing the manifest variables associated to all the latent variables explaining $\boldsymbol{\xi}_j$.

Thus, according to (2.21), (2.22) and (2.23), the relative GoF index is:

$$GoF_{rel} = \sqrt{\frac{1}{\sum_{q:P_q>1} P_q} \sum_{q:P_q>1} \frac{\sum_{p=1}^{P_q} Cor^2\left(\boldsymbol{x}_{pq}, \hat{\xi}_q\right)}{\lambda_{(q)}^1} \times \frac{1}{J} \sum_{j=1}^{J} \frac{R^2\left(\hat{\xi}_j, \hat{\xi}_{q:\xi_q \rightarrow \xi_j}\right)}{\rho_j^2}}. \tag{2.24}$$

This index is bounded between 0 and 1. Both the GoF and the relative GoF are descriptive indexes, i.e. there is no inference-based threshold to judge the statistical significance of their values. As a rule of thumb, a value of the relative GoF equal to or higher than 0.90 clearly speaks in favour of the model.

As PLS Path Modeling is a *soft modeling* approach with no distributional assumptions, it is possible to estimate the significance of the parameters trough cross-validation methods like jack-knife and bootstrap (Efron and Tibshirani 1993). Moreover, it is possible to build a cross-validated version of all the quality indexes (i.e. of the *communality* index, of the *redundancy* index, and of the GoF index) by means of a *blindfolding* procedure (Chin 1998; Lohmöller 1989).

Bootstrap confidence intervals for both the absolute and the relative Goodness of Fit Indexes can be computed. In both cases the inverse cumulative distribution function (*cdf*) of the GoF ($\boldsymbol{\Phi}_{GoF}$) is approximated using a bootstrap-based procedure. B (usually $>$ 100) re-samples are drawn from the initial dataset of N units defining the bootstrap population. For each of the B re-samples, the GoF^b index is computed, with $b = 1 \cdots B$. The values of GoF^b are then used for computing the Monte Carlo approximation of the inverse *cdf*, $\boldsymbol{\Phi}^B_{GoF}$. Thus, it is possible to compute the bounds of the empirical confidence interval from the bootstrap distribution at the $(1-\alpha)$ confidence level by using the percentiles as:

$$\left[\boldsymbol{\Phi}^B_{GoF}\left(\alpha/2\right), \boldsymbol{\Phi}^B_{GoF}\left(1-\alpha/2\right)\right]. \tag{2.25}$$

Several applications have shown that the variability of the GoF values is mainly due to the inner model while the outer model contribution to GoF is very stable across the different bootstrap re-samples.

2.3 Prediction-Based Model Assessment

In this section we present a non-parametric GoF-based bootstrap validation procedure for assessing the statistical significance of path coefficients (individually or by sub-sets).

In order to simplify the discussion we will refer to a very simple model with only three latent variables: $\boldsymbol{\xi}_1$, $\boldsymbol{\xi}_2$ and $\boldsymbol{\xi}_3$ (see Fig. 2.1). The structural relations defined in Fig. 2.1 are formalized by the following equations:

$$\begin{aligned} \boldsymbol{\xi}_2 &= \beta_{02} + \beta_{12}\boldsymbol{\xi}_1 + \zeta_2 \\ \boldsymbol{\xi}_3 &= \beta_{03} + \beta_{13}\boldsymbol{\xi}_1 + \beta_{23}\boldsymbol{\xi}_2 + \zeta_3 \end{aligned} \tag{2.26}$$

where β_{qj} ($q = 1,2$ and $j = 2,3$) stands for the path coefficient linking the q-th latent variable to the j-th endogenous latent variable, and ζ_j is the error term associated to each endogenous latent variable in the model.

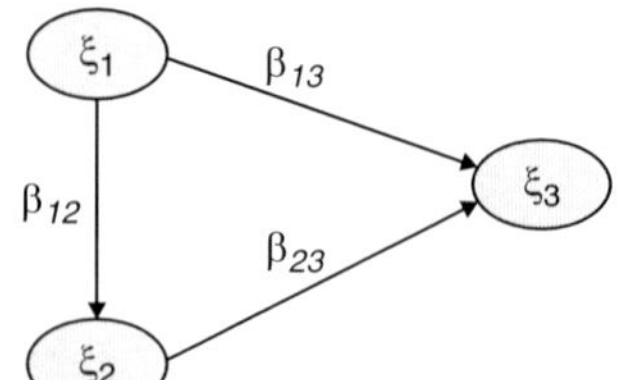

Fig. 2.1 Path diagram of the structural model specified in (2.26)

Equation (2.26) defines a structural model with only three latent variables and with three structural paths. In the following, first we present a non-parametric inferential procedure based on the GoF index to assess the statistical significance of a single path coefficient (Sect. 2.3.1). Then, we discuss the case of an omnibus test on all the path coefficients or on sub-sets of theirs (Sect. 2.3.2).

2.3.1 Hypothesis Testing on One Path Coefficient

Here we want to test if a generic path coefficient β_{qj} is different from 0, i.e.

$$\begin{aligned} \mathrm{H}_0 &: \beta_{qj} = 0 \\ \mathrm{H}_1 &: \beta_{qj} \neq 0 \end{aligned} \tag{2.27}$$

The null hypothesis of $\beta_{qj} = 0$ is tested against the alternative hypothesis that $\beta_{qj} \neq 0$, thus a two-tailed test is performed.

In order to perform this hypotesis testing procedure, we need to define a proper test statistic and the corresponding distribution under the null hypothesis. In particular, the GoF index will be used to test the hypotheses set in (2.27), while the corresponding distribution under the null hypothesis will be obtained by using a bootstrap procedure.

Let GoF_{H_0} be the GoF value under the null hypothesis, $\boldsymbol{\Phi}$ be the inverse cumulative distribution function (cdf) of the GoF_{H_0}, F be the cdf of $\boldsymbol{X}$, and $\boldsymbol{\Phi}^{(B)}$ be the B-sample bootstrap approximation of $\boldsymbol{\Phi}$. In order to approximate $\boldsymbol{\Phi}$ by means of $\boldsymbol{\Phi}^{(B)}$ we need to define a B-sample bootstrap estimate of F under the null hypothesis ($\hat{F}_{H_0{}^{(b)}}$), i.e. such that the null hypothesis is true. Remembering that $\boldsymbol{X}$ is the partitioned matrix of the manifest variables, the sample estimates of F are defined on the basis of $p(\boldsymbol{x}'_n) = \frac{1}{N}$, where $n = 1, 2, \ldots, N$ and $p(\boldsymbol{x}'_n)$ is the probability to extract the n-th observation from the matrix $\boldsymbol{X}$.

Suppose we want to test the null hypothesis that no linear relationship exists between $\boldsymbol{\xi}_2$ and $\boldsymbol{\xi}_3$. In other words, we want to test the null hypothesis that the coefficient β_{23} linking $\boldsymbol{\xi}_2$ to $\boldsymbol{\xi}_3$ is equal to 0:

$$\begin{aligned} \mathrm{H}_0 &: \beta_{23} = 0 \\ \mathrm{H}_1 &: \beta_{23} \neq 0 \end{aligned} \tag{2.28}$$

In order to reproduce the model under H_0 the matrix of the manifest variables associated to $\boldsymbol{\xi}_3$, i.e. $\boldsymbol{X}_3$, can be deflated by removing the linear effect of $\boldsymbol{X}_2$, where $\boldsymbol{X}_2$ is the block of manifest variables associated to $\boldsymbol{\xi}_2$. In particular, the deflated matrix $\boldsymbol{X}_{3(2)}$ is obtained as:

$$\boldsymbol{X}_{3(2)} = \boldsymbol{X}_3 - \boldsymbol{X}_2 \left(\boldsymbol{X}'_2 \boldsymbol{X}_2\right)^{-1} \boldsymbol{X}'_2 \boldsymbol{X}_3. \tag{2.29}$$

Thus, the estimate of F under the null hypothesis is $\hat{F}_{[\boldsymbol{X}_1, \boldsymbol{X}_2, \boldsymbol{X}_{3(2)}]}$.

Once the estimate of cdf of $\boldsymbol{X}$ under the null hypothesis is defined, the B-sample bootstrap approximation $\boldsymbol{\Phi}^{(B)}$ of $\boldsymbol{\Phi}$ is obtained by repeating B times the following procedure.

For each b: $b = 1, 2 \ldots, B$:

1. Draw a random sample from $\hat{F}_{[X_1, X_2, X_{3(2)}]}$.
2. Estimate the model under the null hypothesis for the sample obtained at the previous step.
3. Compute the GoF value, $GoF_{H_0}^{(b)}$.

The choice of B depends on several aspects such as: the sample size, the number of manifest variables and the complexity of the structural model. Usually, we prefer to choose $B \geq 1000$.

The decision on the null hypothesis is taken by referring to the inverse cdf of GoF_{H_0}. In particular, the test is performed at a nominal size α, by comparing the GoF value for the model defined in (2.26), computed on the original data, to the $(1-\alpha)^{th}$ percentile of $\boldsymbol{\Phi}^{(B)}$. If $GoF > \boldsymbol{\Phi}_{(1-\alpha)}^{(B)}$, then we reject the null hypothesis.

A schematic representation of the procedure to perform a non-parametric Bootstrap GoF-based test on a single path-coefficient is given in Algorithm 2.

Algorithm 2 : Non-parametric Bootstrap GoF-based test of a path-coefficient

Hypotheses on the coefficient β_{qj}:

$$\begin{aligned} \mathrm{H}_0 &: \beta_{qj} = 0 \\ \mathrm{H}_1 &: \beta_{qj} \neq 0 \end{aligned} \tag{2.30}$$

Estimate the specified structural model on the original dataset (bootstrap population) and compute the GoF index.

Deflate the endogenous block of manifest variable $\boldsymbol{X}_j$: $\boldsymbol{X}_{j(q)} = \boldsymbol{X}_j - \boldsymbol{X}_q \left(\boldsymbol{X}_q' \boldsymbol{X}_q\right)^{-1} \boldsymbol{X}_q' \boldsymbol{X}_j$.

Define B large enough.

for all $b = 1, \ldots, B$ **do**

  Draw a sample from $\hat{F}_{[X_1, X_2, X_{3(2)}]}$.

  Estimate the model under the null hypothesis.

  Compute the GoF value named $GoF_{H_0^b}$.

end for

By comparing the original GoF index to the inverse cdf of GoF_{H_0} accept or reject H_0.

2.3.2 Hypothesis Testing on the Whole Set of Path Coefficients

The procedure described in Sect. 2.3.1 can be easily generalized in order to test a sub-set of path coefficients or all of them at the same time. If the path coefficients are tested simultaneously, then this omnibus test can be used for an overall assessment of the model. This test is performed by comparing the default model specified by the user to the so-called baseline models, i.e the *saturated* model and the *independence* or *null* model. The *saturated* model is the least restrictive model where all the structural relations are allowed (i.e. all path coefficients are free parameters). The *null* model is the most restrictive model with no relations among latent variables (i.e. all path coefficients are constrained to be 0). Following the structure of the model defined in figure 2.1, the null model is the model where : $\beta_{12} = \beta_{13} = \beta_{23} = 0$, while the saturated model coincides with the one in figure 2.1. More formally:

$$\begin{aligned} &\mathrm{H}_0 : \beta_{12} = \beta_{13} = \beta_{23} = 0 \\ &\mathrm{H}_1 : \text{At least one } \beta_{qj} \neq 0 \end{aligned} \tag{2.31}$$

As for the simple case described in Sect. 2.3.1 we need to properly deflate $\boldsymbol{X}$ in order to estimate $\boldsymbol{\Phi}^{(B)}$. In particular, each endogenous block $\boldsymbol{X}_j$ has to be deflated according to the specified structural relations by means of orthogonal projection operators. In the model defined by (2.26), the block of manifest variables linked to $\boldsymbol{\xi}_2$ ($\boldsymbol{X}_2$) has to be deflated by removing the linear effect of $\boldsymbol{\xi}_1$ on $\boldsymbol{\xi}_2$, while the block of the manifest variables linked to $\boldsymbol{\xi}_3$ ($\boldsymbol{X}_3$) has to be deflated by removing the linear effect of both $\boldsymbol{\xi}_1$ and $\boldsymbol{\xi}_2$. However, since $\boldsymbol{\xi}_2$ is an endogenous latent variable, the deflated block $\boldsymbol{X}_{2(1)}$ has to be taken into account when deflating $\boldsymbol{X}_3$. In other words, the deflation of the block $\boldsymbol{X}_2$ is obtained as:

$$\boldsymbol{X}_{2(1)} = \boldsymbol{X}_2 - \boldsymbol{X}_1 \left(\boldsymbol{X}_1'\boldsymbol{X}_1\right)^{-1} \boldsymbol{X}_1'\boldsymbol{X}_2$$

while, the deflation of the block $\boldsymbol{X}_3$ is obtained as:

$$\boldsymbol{X}_{3(1,2)} = \boldsymbol{X}_3 - \left[\boldsymbol{X}_1, \boldsymbol{X}_{2(1)}\right] \left(\left[\boldsymbol{X}_1, \boldsymbol{X}_{2(1)}\right]' \left[\boldsymbol{X}_1, \boldsymbol{X}_{2(1)}\right]\right)^{-1} \left[\boldsymbol{X}_1, \boldsymbol{X}_{2(1)}\right]' \boldsymbol{X}_3.$$

As we deal with a recursive model, it is always possible to build blocks that verify the null hypothesis by means of a proper sequence of deflations.

The algorithm described in Sect. 2.3.1 and in Algorithm 2 can be applied to $\hat{F}_{[\boldsymbol{X}_1, \boldsymbol{X}_{2(1)}, \boldsymbol{X}_{3(1,2)}]}$ in order to construct an inverse cdf of $\boldsymbol{\Phi}^{(B)}$ such that H_0 is true. The test is performed at a nominal confidence level α, by comparing the GoF value for the model defined in (2.26) to the $(1-\alpha)^{th}$ percentile of $\boldsymbol{\Phi}^{(B)}$ built upon $\hat{F}_{[\boldsymbol{X}_1, \boldsymbol{X}_{2(1)}, \boldsymbol{X}_{3(1,2)}]}$. If $GoF > \boldsymbol{\Phi}^{(B)}_{(1-\alpha)}$, then the null hypothesis is rejected. By comparing the GoF value obtained for the default model on the bootstrap population with the $GoF^{(b)}_{\mathrm{H}_0}$ obtained from bootstrap samples ($b - 1, 2, \ldots, B$), an empirical p-value can be computed as:

$$\text{p-value} = \frac{\sum_{b=1}^{B} I_b}{B} \tag{2.32}$$

where

$$I_b = \begin{cases} 1 \text{ if } GoF^{(b)}_{\mathrm{H}_0} \geq GoF \\ 0 \text{ otherwise} \end{cases} \tag{2.33}$$

and B is the number of Bootstrap re-samples.

As stated in (2.31), the above procedure tests the null hypothesis that all path coefficients are equal to zero against the alternative hypothesis that at least one of the coefficients is different from zero. By defining a proper deflation strategy, tests on any sub-set of path coefficients can be performed. Stepwise procedures can also be defined in order to identify a set of significant coefficients.

2.3.3 *Application to Simulated Data*

In this subsection we apply the procedures for testing path coefficients to simulated data.

Data have been generated according to the basic model defined in Fig. 2.2. This model is a simplified version of the one defined in Fig. 2.1.

According to Fig. 2.2, the structural model is specified by the equation:

$$\xi_3 = \beta_{03} + \beta_{13}\xi_1 + \beta_{23}\xi_2 + \zeta_3 \tag{2.34}$$

Three different tests have been performed on the simulated data-set. In particular, we perform a test:

1. On the whole model:

$$\begin{aligned} &\mathrm{H_0}: \beta_{13} = \beta_{23} = 0 \\ &\mathrm{H_1}: \text{At least one } \beta_{qj} \neq 0 \end{aligned} \tag{2.35}$$

2. On the coefficient β_{13}

$$\begin{aligned} &\mathrm{H_0}: \beta_{13} = 0 \\ &\mathrm{H_1}: \beta_{13} \neq 0 \end{aligned} \tag{2.36}$$

3. On the coefficient β_{23}

$$\begin{aligned} &\mathrm{H_0}: \beta_{23} = 0 \\ &\mathrm{H_1}: \beta_{23} \neq 0 \end{aligned} \tag{2.37}$$

2.3.3.1 Simulation Scheme

The following procedure has been used in order to simulate the manifest variables for the model in Fig. 2.2 with a sample size of 50 units:

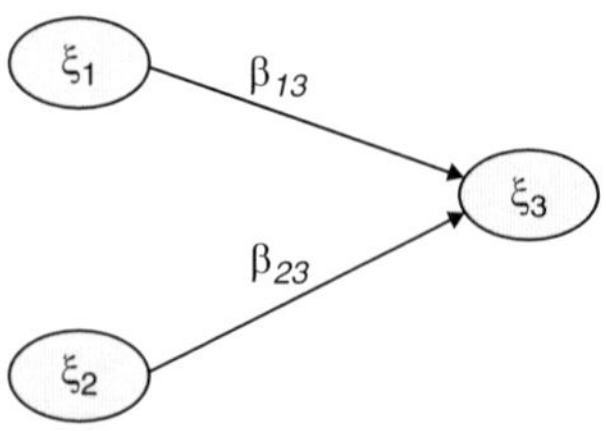

Fig. 2.2 Path diagram of the structural model specified by (2.34)

1. For each exogenous block, three manifest variables have been randomly generated according to a multivariate normal distribution. In particular, the manifest variables linked to the latent variable $\boldsymbol{\xi}_1$ come from a multivariate normal distribution with means equal to 2 and standard deviations equal to 1.5 for every manifest variable. The manifest variables of block 2 come from a multivariate normal distribution with means equal to 0 and standard deviations equal to 1 for every manifest variable.
2. The exogenous latent variables $\boldsymbol{\xi}_1$ and $\boldsymbol{\xi}_2$ have been computed as a standardized aggregate of the manifest variables obtained in the first step. An error term (from a normal distribution with zero mean and standard deviation equal to $1/4$ of the manifest variables' standard deviation) has been added to both exogenous latent variables.
3. The manifest variables corresponding to the endogenous latent variable $\boldsymbol{\xi}_3$ have been generated as a standardized aggregate of $\boldsymbol{\xi}_1$ and $\boldsymbol{\xi}_2$ plus an error term (from a normal distribution with zero mean and standard deviation equal to 0.25).

2.3.3.2 Results

Table 2.1 reports the path coefficients and the GoF values obtained by running the PLS-PM algorithm on the simulated dataset.

According to the procedure described in Sect. 2.3.2 we need to deflate the data in different ways in order to perform the three different types of tests. Namely, in order to perform the first test ($H_0 : \beta_{13} = \beta_{23} = 0$) we need to deflate the block $\boldsymbol{X}_3$ with regards to $\boldsymbol{X}_2$ and $\boldsymbol{X}_1$ (Test 1), while the second test ($H_0 : \beta_{13} = 0$) is performed by deflating the block $\boldsymbol{X}_3$ only with regards to $\boldsymbol{X}_1$ (Test 2) and the last test ($H_0 : \beta_{23} = 0$) is performed by deflating the block $\boldsymbol{X}_3$ with regards to $\boldsymbol{X}_2$ (Test 3).

Under each null hypothesis, bootstrap resampling has been performed to obtain the bootstrap approximation $\boldsymbol{\Phi}^{(B)}$ of $\boldsymbol{\Phi}$. Bootstrap distributions have been approximated by 1,000 pseudo-random samples.

The histograms of the bootstrap approximations of the GoF distributions under the null hypotheses for Test 1, Test 2 and Test 3 are shown in Figs. 2.3–2.5, respectively. These histograms seem to reveal fairly normal distributions.

Table 2.2 reports the values of the critical thresholds computed for test sizes $\alpha = 0.10$ and $\alpha = 0.05$ on the bootstrap distribution for the three different tests. The $p - values$, computed according to the formula in (2.32), are also shown. On this basis, the null hypotheses for Test 1 and Test 2 have been correctly rejected by the proposed procedure. Nevertheless, the proposed test accepts the null hypothesis for Test 3 even if this hypothesis is false. This is due to the very weak value for the corresponding path coefficient, i.e. $\beta_{23} = 0.05$.

Table 2.1 Results from the simulated data-set

β_{13}	0.94
β_{23}	0.05
GoF (Absolute)	0.69

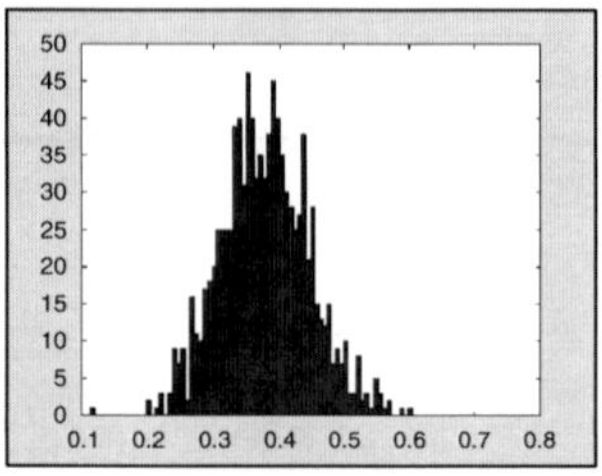

Fig. 2.3 Histogram of the bootstrap approximation of the GoF distribution under the null hypothesis in Test 1

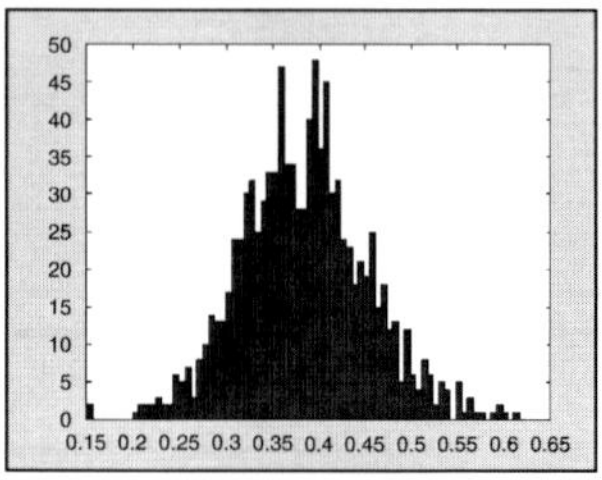

Fig. 2.4 Histogram of the bootstrap approximation of the GoF distribution under the null hypothesis in Test 2

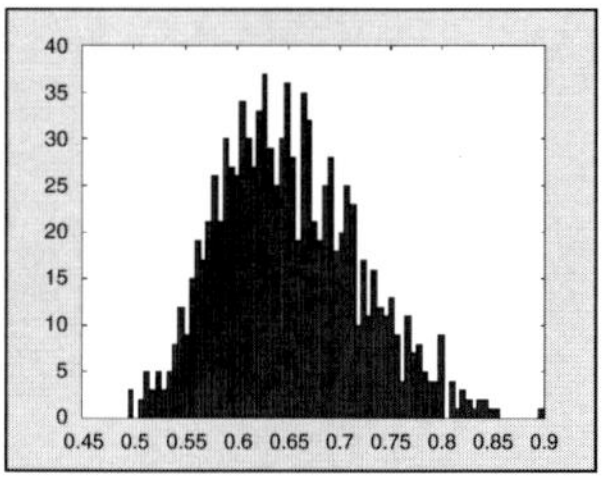

Fig. 2.5 Histogram of the bootstrap approximation of the GoF distribution under the null hypothesis in Test 3

Table 2.2 Thresholds and p-values from bootstrap distributions (1,000 re-samples)

	$\alpha = 0.10$	$\alpha = 0.05$	p-value
Test 1	0.46	0.49	0
Test 2	0.47	0.50	0
Test 3	0.74	0.77	0.27

Further researches are needed to investigate features of the GoF distribution as well as the statistical power of the proposed tests and their sensitivity with respect to the size of the coefficients, the sample size and the complexity of the structural model.

2.4 Heterogeneity in PLS Path Modeling

In this section we discuss how to improve the prediction performance and the interpretability of the model by allowing for unobserved heterogeneity.

Indeed, heterogeneity among units is an important issue in statistical analysis. Treating the sample as homogeneous, when it is not, may seriously affect the quality of the results and lead to biased interpretation. Since human behaviors are complex, looking at groups or classes of units having similar behaviors will be particularly hard. Heterogeneity can hardly be detected using external information, i.e. using *a priori* clustering approach, especially in social, economic and marketing areas. Moreover, in several application fields (e.g. marketing) more attention is being given to clustering methods able to detect groups that are homogeneous in terms of their responses (Wedel and Kamakura 2000). Therefore, *response-based* clustering techniques are becoming more and more important in statistical literature.

Two types of heterogeneity could be affecting the data: observed and unobserved heterogeneity (Tenenhaus et al. 2010; Hensler and Fassott 2010; Chin and Dibbern 2007). In the first case the composition of classes is known *a priori*, while in the second case information on the number of classes or on their composition is not available.

So far in this paper we have assumed homogeneity over the observed set of units. In other words, all units are supposed to be well represented by a unique model estimated on the whole sample, i.e. *the global model*.

In a Structural Equation Model, the two cases of observed and unobserved heterogeneity match with the presence of a discrete moderating factor that, in the first case is manifest, i.e. an observed variable, while in the second case is latent, i.e. an unobserved variable (Chin and Dibbern 2007).

Usually heterogeneity in Structural Equation Models is handled by first forming classes on the basis of external variables or on the basis of standard clustering techniques applied to manifest and/or latent variables, and then by using the multi-group analysis introduced by Jöreskog (1971) and Sörbom (1974). However, heterogeneity in the models may not be necessarily captured by well-known observed variables playing the role of moderating variables (Hahn et al. 2002). Moreover, *post-hoc* clustering techniques on manifest variables, or on latent variable scores, do not take at all into account the model itself. Hence, while the local models obtained by cluster analysis on the latent variable scores will lead to differences in the group averages of the latent variables but not necessarily to different models, the same method performed on the manifest variables is unlikely to lead to different and well-separated models. This is true for both the model parameters and the means of latent variable scores. In addition, *a priori* unit clustering in Structural Equation Models is not conceptually acceptable since no structural relationship among the variables is postulated: when information concerning the relationships among variables is available (as it is in the theoretical causality network), classes should be looked for while taking into account this important piece of information. Finally, even in Structural Equation Models, the need is pre-eminent for a *response-based* clustering method, where the obtained classes are homogeneous with respect to the postulated model. Dealing with heterogeneity in PLS Path Models implies looking for *local models* characterized by class-specific model parameters.

Recently, several methods have been proposed to deal with unobserved heterogeneity in PLS-PM framework (Hahn et al. 2002; Ringle et al. 2005; Squillacciotti 2005; Trinchera and Esposito Vinzi 2006; Trinchera et al. 2006; Sanchez and Aluja 2006, 2007; Esposito Vinzi et al. 2008; Trinchera 2007). To our best knowledge, five approaches exist to handle heterogeneity in PLS Path Modeling: the Finite Mixture PLS, proposed by Hahn et al. (2002) and modified by Ringle et al.

(2010) (see Chap. 8 of this Handbook), the PLS Typological Path Model presented by Squillacciotti (2005) (see Chap. 10 of this Handbook) and modified by Trinchera and Esposito Vinzi (2006) and Trinchera et al. (2006), the PATHMOX by Sanchez and Aluja (2006), the PLS-PM based Clustering (PLS-PMC) by Ringle and Schlittgen (2007) and the Response Based Unit Segmentation in PLS Path Modeling (REBUS-PLS) proposed by Trinchera (2007) and Esposito Vinzi et al. (2008).

In the following we will discuss the REBUS-PLS approach in detail.

2.4.1 The REBUS-PLS Algorithm

A new method for unobserved heterogeneity detection in PLS-PM framework was recently presented by Trinchera (2007) and Esposito Vinzi et al. (2008). REBUS-PLS is an iterative algorithm that permits to estimate at the same time both the unit membership to latent classes and the class specific parameters of the local models. The core of the algorithm is a so-called *closeness measure* (CM) between units and models based on residuals (2.38). The idea behind the definition of this new measure is that if latent classes exist, units belonging to the same latent class will have similar local models. Moreover, if a unit is assigned to the correct latent class, its performance in the local model computed for that specific class will be better than the performance of the same unit considered as supplementary in the other local models.

The CM used in the REBUS-PLS algorithm represents an extension of the distance used in PLS-TPM by Trinchera et al. (2006), aiming at taking into account both the measurement and the structural models in the clustering procedure. In order to obtain local models that fit better than the global model, the chosen *closeness measure* is defined according to the structure of the Goodness of Fit (GoF) index, the only available measure of global fit for a PLS Path Model. According to the $DmodY$ distance used in PLS Regression (Tenenhaus 1998) and the distance used by Esposito Vinzi and Lauro (2003) in PLS Typological Regression all the computed residuals are weighted by quality indexes: the importance of residuals increases while the quality index decreases. That is why the communality index and the R^2 values are included in the CM computation.

In a more formal terms, the *closeness measure* (CM) of the n-th unit to the k-th local model, i.e. to the latent model corresponding to the k-th latent class, is defined as:

$$CM_{nk} = \sqrt{\frac{\sum_{q=1}^{Q}\sum_{p=1}^{P_q}\left[\frac{e_{npqk}^2}{Com(\hat{\xi}_{qk}, x_{pq})}\right]}{\frac{\sum_n^N\sum_{q=1}^{Q}\sum_{p=1}^{P_q}\left[\frac{e_{npqk}^2}{Com(\hat{\xi}_{qk}, x_{pq})}\right]}{(N-t_k-1)}} \times \frac{\sum_{j=1}^{J}\left[\frac{f_{njk}^2}{R^2\left(\hat{\xi}_j, \hat{\xi}_{q:\xi_q \to \xi_j}\right)}\right]}{\frac{\sum_n^N\sum_{j=1}^{J}\left[\frac{f_{njk}^2}{R^2\left(\hat{\xi}_j, \hat{\xi}_{q:\xi_q \to \xi_j}\right)}\right]}{(N-t_k-1)}}} \tag{2.38}$$

where:
$Com\left(\mathbf{x}_{pq}, \xi_{qk}\right)$ is the communality index for the p-th manifest variable of the q-th block in the k-th latent class;
e_{npqk} is the measurement model residual for the n-th unit in the k-th latent class, corresponding to the p-th manifest variable in the q-th block, i.e. the communality residuals;
f_{njk} is the structural model residual for the n-th unit in the k-th latent class, corresponding to the j-th endogenous block;
N is the total number of units;
t_k is the number of extracted components. Since all blocks are supposed to be reflective, the value of t_k will always be equal to 1.

As for the GoF index, the left-side term of the product in (2.38) refers to the measurement models for all the Q blocks in the model, while the right-side term refers to the structural model. It is important to notice that both the measurement and the structural residuals are computed for each unit with respect to each local model regardless of the membership of the units to the specific latent class. In computing the residual from the k-th latent model, we expect that units belonging to the k-th latent class show smaller residuals than units belonging to the other $(K - 1)$ latent classes.

As already said, two kinds of residuals are used to evaluate the closeness between a unit and a model: the measurement or communality residuals and the structural residuals. For a thorough description of the REBUS-PLS algorithm and the computation of the communality and the structural residuals, refer to the original REBUS-PLS papers (Trinchera 2007; Esposito Vinzi et al. 2008).

The choice of the *closeness measure* in (2.38) as a criterion for assigning units to classes has two major advantages. First, unobserved heterogeneity can now be detected in both the measurement and the structural models. If two models show identical structural coefficients, but differ with respect to one or more outer weights in the exogenous blocks, REBUS-PLS is able to identify this source of heterogeneity, which might be of major importance in practical applications. Moreover, since the *closeness measure* is defined according to the structure of the Goodness of Fit (GoF) index, the identified local models will show a better prediction performance.

The CM expressed by (2.38) is only the core of an iterative algorithm allowing us to obtain a *response-based* clustering of the units.

As a matter of fact, REBUS-PLS is an iterative algorithm (see Fig. 2.6). The first step of the REBUS-PLS algorithm involves estimating the global model on all the observed units, by performing a simple PLS Path Modeling analysis. In the second step, the communality and the structural residuals of each unit from the global model are obtained. The number of classes (K) to be taken into account during the successive iterations and the initial composition of the classes are obtained by performing a hierarchical cluster analysis on the computed residuals (both from the measurement and the structural models). Once the number of classes and their initial composition are obtained, a PLS Path Modeling analysis is performed on each class and K provisional local models are estimated. The group-specific parameters computed at the previous step are used to compute the communality and the structural

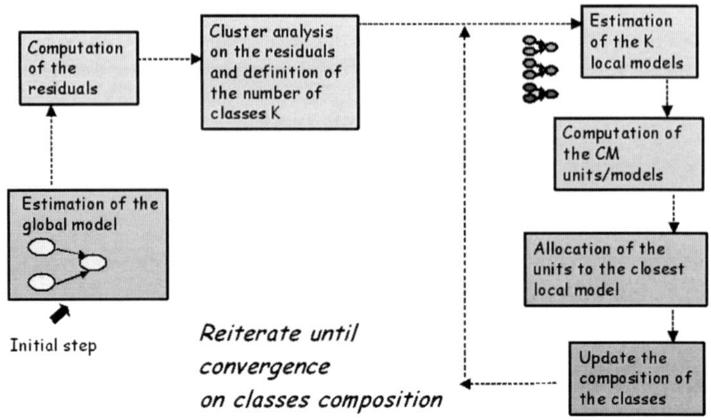

Fig. 2.6 A schematic representation of the REBUS-PLS algorithm

residuals of each unit from each local model. Then the CM of each unit from each local model is obtained according to (2.38). Each unit is, therefore, assigned to the closest local model, i.e. to the model from which it shows the smallest CM value. Once the composition of the classes is updated, K new local models are estimated. The algorithm goes on until the threshold of a stopping rule is achieved.

Stability on class composition from one iteration to the other is considered as a stopping rule. The authors suggest using the threshold of less than 5% of units changing class from one iteration to the other as a stopping rule. Indeed, REBUS-PLS usually assures convergence in a small number of iterations (i.e. less than 15). It is also possible not to define a threshold as a stopping rule and run the algorithm until the same groups are formed in successive iterations. In fact, if no stopping rule is imposed once the "best" model is obtained in the REBUS-PLS viewpoint, i.e. once each unit is correctly assigned to the closest local model, the algorithm provides the same partition of the units at successive iterations.

If the sample size is large, it is possible to have such boundary units that change classes time after time at successive iterations. This leads to obtaining a series of partitions (i.e. of local model estimates) that repeat themselves in successive iterations. In order to avoid the "boundary" unit problem the authors suggest always defining a stopping rule.

Once the stability on class composition is reached, the final local models are estimated. The class-specific coefficients and indexes are then compared in order to explain differences between detected latent classes. Moreover the quality of the obtained partition can be evaluated through a new index (i.e. the *Group Quality Index* - GQI) developed by Trinchera (2007). This index is a reformulation of the *Goodness of Fit* index in a multi-group perspective, and it is also based on residuals. A detailed presentation of the GQI, as well as a simulation study aiming at assessing GQI properties, can be found in Trinchera (2007). The GQI index is equal to the GoF in the case of a unique class, i.e. when $K = 1$ and $n_1 = N$. In other words, the *Group Quality Index* computed for the whole sample as a unique class is equal to the GoF index computed for the global model. Instead, if local models performing better than the global one are detected, the GQI index will be higher than the GoF value computed for the global model.

Trinchera (2007) performed a simulation study to assess GQI features. In particular, it is suggested that a relative improvement of the GQI index from the global model to the detected local models higher than 25% can be considered as a satisfactory threshold to prefer the detected unit partition to the aggregate data solution. Finally, the quality of the detected partition can be assessed by a permutation test (Edgington 1987) involving T random replications of the unit partition (keeping constant the group proportions as detected by REBUS-PLS) so as to yield an empirical distribution of the GQI index.

The GQI obtained for the REBUS-PLS partition is compared to the percentiles of the empirical distribution to decide whether local models are performing significantly better than the global one. Trinchera (2007) has shown that, in case of unobserved heterogeneity and apart from the outlier solutions, the GQI index computed for the aggregate level is the minimum value obtained for the empirical distribution of the GQI.

If external concomitant variables are available, an *ex-post* analysis on the detected classes can be performed so as to characterize the detected latent classes and improve interpretability of their composition.

So far, REBUS-PLS is limited to reflective measurement models because the measurement residuals come from the simple regressions between each manifest variable in a block and the corresponding latent variable. Developments of the REBUS-PLS algorithm to the formative measurement models are on going.

2.4.2 Application to Real Data

Here, we present a simple and clear example to show the REBUS-PLS ability to capture unobserved heterogeneity on empirical data. We use the same data as in Ringle et al. (2010). This dataset comes from the Gruner&Jahr's Brigitte Communication Analysis performed in 2002 that specifically concerns the Benetton fashion brand. REBUS-PLS has been performed using a SAS-IML macro developed by Trinchera (2007).

The Benetton dataset is composed of ten manifest variables observed on 444 German women. Each manifest variable is a question in the Gruner&Jahr's Brigitte Communication Analysis of 2002. The women had to answer each question using a four-point scale from "low" to "high".

The structural model for Benetton's brand preference, as used by Ringle et al. (2010), consists of one latent endogenous *Brand Preference* variable, and two latent exogenous variables, *Image* and *Character*. All manifest variables are linked to the corresponding latent variable via a reflective measurement model. Figure 2.7 illustrates the path diagram with the latent variables and the employed manifest variables. A list of the used manifest variables with the corresponding meanings is shown in Table 2.3.

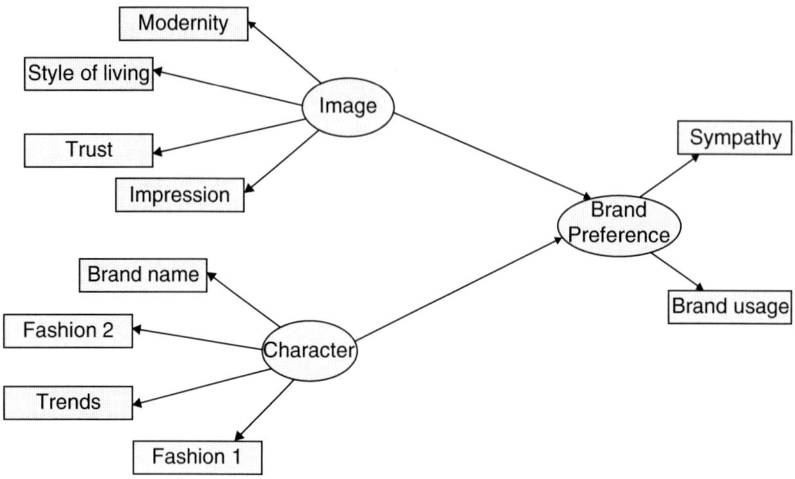

Fig. 2.7 Path diagram for Benetton data

Table 2.3 Manifest (MV) and latent variables (LV) definition for Benetton data

LV Name	MV Name	Concepts
Image	Modernity	It is modern and up to date
	Style of living	Represents a great style of life
	Trust	This brand can be trusted
	Impression	I have a clear impression of this brand
Character	Brand name	A brand name is very important to me
	Fashion 2	I often talk about fashion
	Trends	I am interested in the latest trends
	Fashion 1	Fashion is a way to express who I am
Brand	Sympathy	Sympathy
Preference	Brand usage	Brand usage

A PLS Path Modeling analysis on the whole sample has been performed with standardized manifest variables. As it is obvious, the global model estimates are consistent with the ones obtained by Ringle *et al.* in their study (see Chap. 8). Since all the blocks in the model are supposed to be reflective, then they should be homogeneous and *unidimensional*. Hence, first of all we have to check for block homogeneity and unidimensionality. Table 2.4 shows values of the tools presented in Sect. 2.2.1 for checking the block homogeneity and unidimensionality. According to Chin (1998), all the blocks are considered homogenous, i.e. the *Dillon-Goldstein's rho* is always larger than 0.7. Moreover, the three blocks are unidimensional as only the first eigenvalues for each block are greater than one. Therefore, the reflective model is appropriate.

A simple overview of the global model results is proposed in Fig. 2.8. According to the global model results *Image* seems to be the most important driver for *Brand Preference*, with a path coefficient equal to 0.423. The influence of the

Table 2.4 Homogeneity and unidimensionality of MVs blocks

LV Name	# of MVs	Cronbach's α	D.G.'s ρ	PCA eigenvalues
Image	4	0.869	0.911	2.873
				0.509
				0.349
				0.269
Character	4	0.874	0.914	2.906
				0.479
				0.372
				0.243
Brand preference	2	0.865	0.937	1.763
				0.237

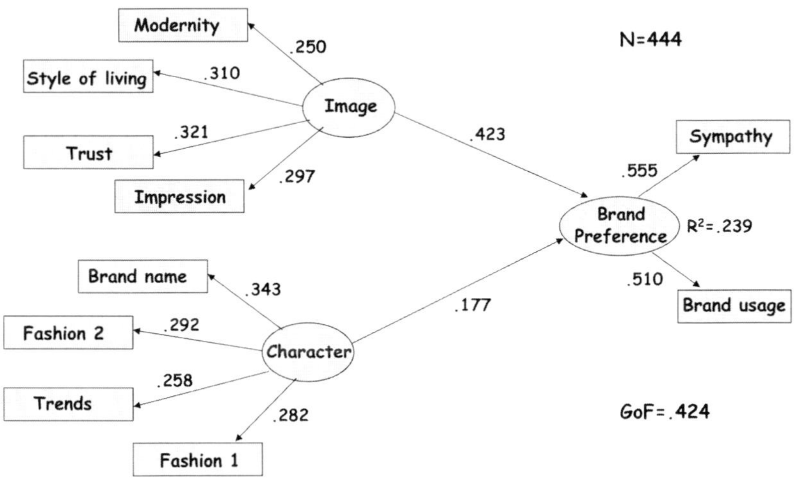

Fig. 2.8 Global model results from Benetton data obtained by using a SAS-IML macro

exogenous latent variable *Character* is considerably weaker (path coefficient of 0.177). Nevertheless, the R^2 value associated with the endogenous latent variable *Brand Preference* is quite low, being equal to 0.239. Ringle et al. (2010) consider this value as a moderate level for a PLS Path Model. In our opinion, an R^2 value of 0.239 has to be considered as unsatisfactory, and could be used as a first sign of possible unobserved heterogeneity in the data. Looking at the measurement models, all the relationships in the reflective measurement models have high factor loadings (the smallest loading has a value of 0.795, see Table 2.5). In Fig. 2.8 the outer weights used for yielding standardized latent variable scores are shown. In the *Brand Preference* block, *Sympathy* and *Brand Usage* have similar weights. Instead, differences arise in both exogenous blocks. Finally, the global model on Benetton data shows a value for the absolute GoF equal to 0.424 (see Table 2.6). The quite low value of the GoF index might also suggest that we have to look for more homogeneous segments among the units.

Table 2.5 Measurement model results for the global and the local models obtained by REBUS-PLS

		Global	Class 1	Class 2	Class 3
Number of units		444	105	141	198
Outer weights	Modernity	0.250	0.328	0.278	0.291
Image	Style of living	0.310	0.264	0.314	0.270
	Trust	0.321	0.284	0.315	0.375
	Impression	0.297	0.292	0.267	0.273
Outer weights	Brand name	0.343	0.342	0.262	0.298
Character	Fashion2	0.292	0.276	0.345	0.314
	Trends	0.258	0.266	0.323	0.335
	Fashion1	0.282	0.314	0.213	0.231
Outer weights	Sympathy	0.555	0.549	0.852	0.682
Brand preference	Brand Usage	0.510	0.637	0.575	0.547
Standardized loadings	Modernity	0.795	0.827	0.810	0.818
Image	Style of living	0.832	0.834	0.860	0.735
	Trust	0.899	0.898	0.890	0.895
	Impression	0.860	0.865	0.840	0.834
Standardized loadings	Brand name	0.850	0.832	0.842	0.822
Character	Fashion2	0.894	0.846	0.929	0.908
	Trends	0.859	0.850	0.902	0.878
	Fashion1	0.801	0.819	0.788	0.762
Standardized loadings	Sympathy	0.944	0.816	0.819	0.855
Brand preference	Brand Usage	0.933	0.867	0.526	0.762
Communality	Modernity	0.632	0.685	0.657	0.668
Image	Style of living	0.693	0.695	0.740	0.541
	Trust	0.808	0.806	0.792	0.801
	Impression	0.739	0.748	0.706	0.696
Communality	Brand name	0.722	0.692	0.709	0.676
Character	Fashion2	0.799	0.715	0.864	0.825
	Trends	0.738	0.722	0.814	0.770
	Fashion1	0.642	0.670	0.620	0.581
Communality	Sympathy	0.891	0.666	0.671	0.730
Brand preference	Brand Usage	0.871	0.752	0.277	0.581

A more complete outline of the global model results is provided in Table 2.5 for the outer model and in Table 2.6 for the inner model. These tables contain also the class-specific results in order to make it easier to compare the segments.

Performing REBUS-PLS on Benetton data leads to detecting three different classes of units showing homogeneous behaviors. As a matter of fact, the cluster analysis on the residuals from the global model (see Fig. 2.9) suggests that we should look for two or three latent classes. Both partitions have been investigated. The three classes partition is preferred as it shows a higher *Group Quality Index*. Moreover, the GQI index computed for the two classes solution ($GQI = 0.454$) is close to the GoF value computed for the global model (i.e. the GQI index in the case of only one global class, $GoF = 0.424$). Therefore, the 25% improvement

Table 2.6 Structural model results for the global model and the local models obtained by REBUS-PLS

		Global	Class 1	Class 2	Class 3
Number of units		444	105	141	198
Path	Image	0.423	0.420	0.703	0.488
Coefficients		[0.331; 0.523]	[0.225; 0.565]	[0.611; 0.769]	[0.314; 0.606]
on brand	Character	0.177	0.274	0.319	0.138
preference		[0.100; 0.257]	[0.078; 0.411]	[0.201; 0.408]	[0.003; 0.311]
Redundancy	Brand preference	0.210	0.207	0.322	0.180
R^2		0.239	0.292	0.680	0.275
Brand preference		[0.166; 0.343]	[0.162; 0.490]	[0.588; 0.775]	[0.195; 0.457]
R^2	Image	0.81	0.67	0.79	0.90
contributions	Character	0.19	0.33	0.21	0.10
GoF value		0.424	0.457	0.682	0.435
		[0.354; 0.508]	[0.325; 0.596]	[0.618; 0.745]	[0.366; 0.577]

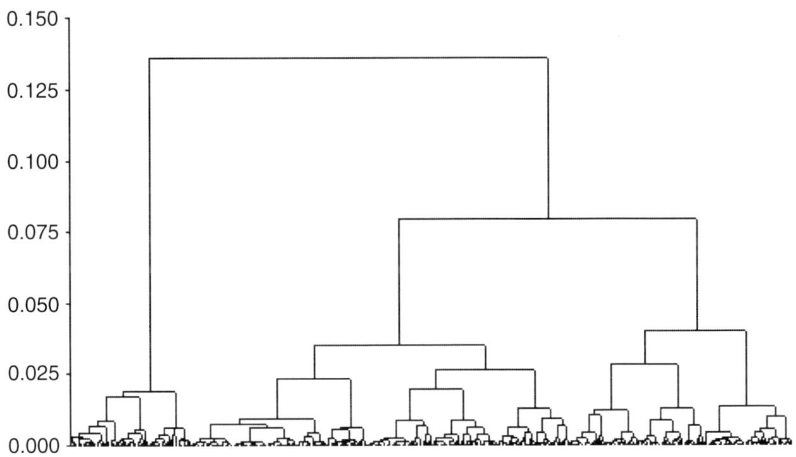

Fig. 2.9 Dendrogramme obtained by a cluster analysis on the residuals from the global model (Step 3 of the REBUS-PLS algorithm)

foreseen for preferring the partition in two classes is not achieved. Here, only the results for the three classes partition are presented.

The first class is composed of 105 units, i.e around 24% of the whole sample. This class is characterized by a path coefficient linking the latent variable *Character* to the endogenous latent variable *Brand Preference* higher than the one obtained for the global model. Moreover, differences in unit behaviors arise also with respect to the outer weights in the *Brand Preference* block, i.e. *Brand Usage* shows a higher weight than *Sympathy*. The GoF value for this class (0.457) is similar to the one for the global model (0.424). Figure 2.10 shows the estimates obtained for this class.

The second class, instead, shows a definitely higher GoF value of 0.682 (see Table 2.6). This class is composed of around 32% of the whole sample, and

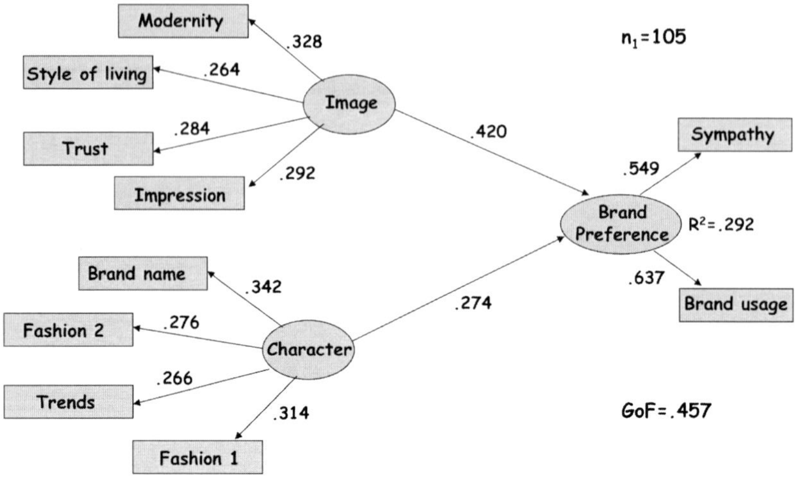

Fig. 2.10 Local model results for the first class detected by the REBUS-PLS algorithm on Benetton data

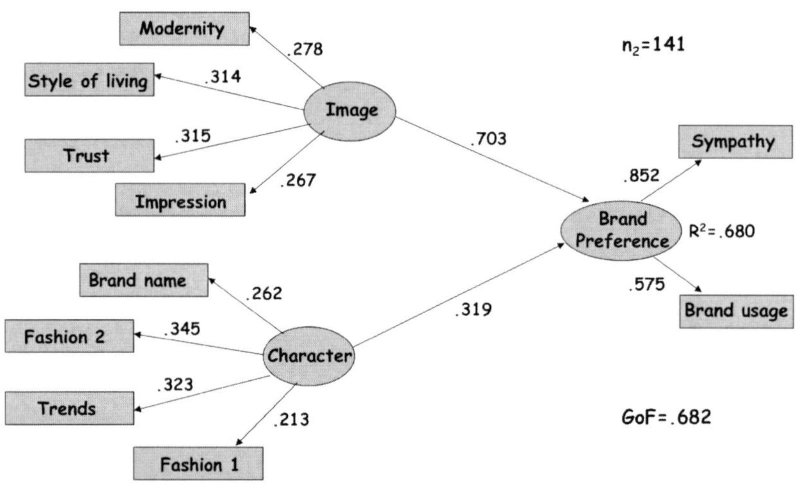

Fig. 2.11 Local model results for the second class detected by the REBUS-PLS algorithm on Benetton data

is characterized by a much higher path coefficient associated to the relationship between the *Image* and the *Brand Preference*. Looking at the measurement model (see Table 2.5), differences arise in the *Brand Preference* block and in the *Character* block. As a matter of fact, the communality index (i.e. the square of the correlation) between the manifest variable *Brand Usage* and the corresponding latent variable *Brand Preference* is really lower than the one obtained for the global model as well as for the first local model described above. Other differences for this second class may be detected by looking at the results provided in Fig. 2.11.

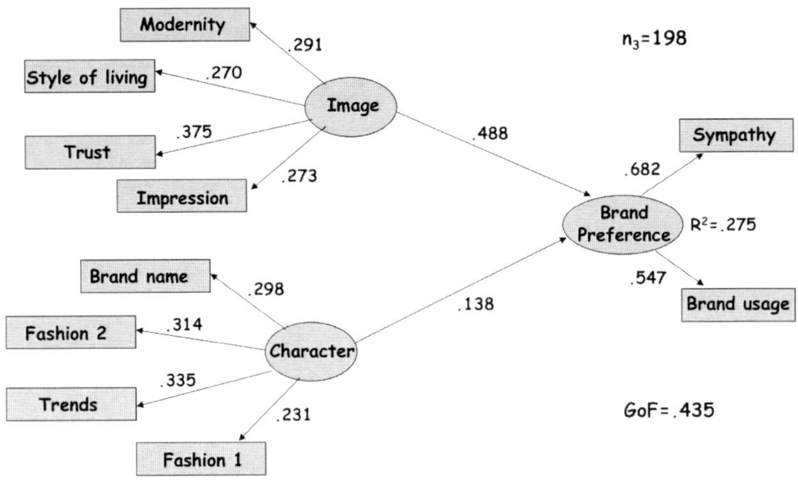

Fig. 2.12 Local model results for the third class by the REBUS-PLS algorithm on Benetton data

Finally, the results for the third class are presented in Fig. 2.12. This class is composed of 198 units., i.e. more than 44% of the whole sample. It is characterized by a very weak relationship between the latent variable *Character* and the endogenous latent variable *Brand Preference*. Moreover, the 95% bootstrap confidence interval shows that this link is close to be non significant as the lower bound is very close to 0 (see Table 2.6). Differences arise also with respect to the measurement model, notably in the *Image* block. As a matter of fact, in this class the manifest variable *Style of living* shows a very low correlation compared with the other models (both local and global).
Nonetheless, the quality index values computed for this third local model are only slightly different from the ones in the global model ($R^2 = 0.275$ and $GoF = 0.435$).

The three classes solution shows a *Group Quality Index* equal to 0.531. In order to validate the REBUS-PLS based partition, an empirical distribution of the GQI values is yielded by means of permutations. The whole sample has been randomly divided 300 times into three classes of the same size as the ones detected by REBUS-PLS. The GQI has been computed for each of the random partitions of the units. The empirical distribution of the GQI values for a three classes partition is then obtained (see Fig. 2.13). As expected, the GQI value from the REBUS-PLS partition is definitely an extremely high value of the distribution thus showing that the REBUS-PLS based partition is better than a random assignment of the units into three classes.

Moreover, in Fig. 2.14, it is possible to notice that the GQI computed for the global model (i.e. the GoF value) is a very small value in the GQI distribution. Therefore, the global model has to be definitely considered as being affected by heterogeneity.

Ringle et al. (2010) apply FIMIX-PLS to Benetton data (see Chap. 8) and identify only two classes. The first one (80.9% of the whole sample) is very similar to

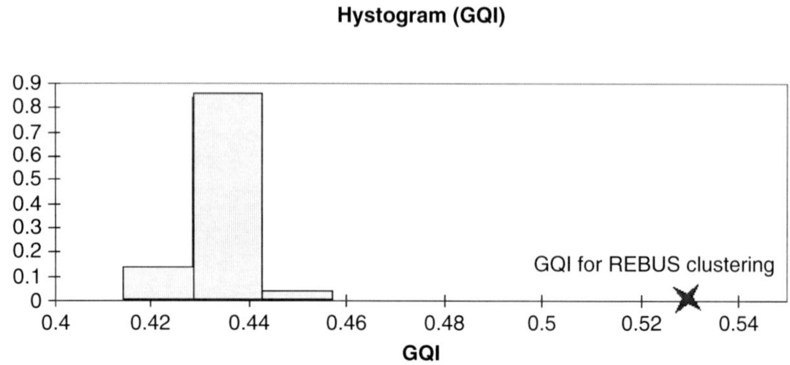

Fig. 2.13 Empirical distribution of the GQI computed on 300 random partitions of the original sample in three classes

Simple Statistics	GQI
No. of observations	301
Minimum	0.421
Maximum	0.531
1° Quartile	0.430
Median	0.433
3° Quartile	0.436
Mean	0.434
Variance (n-1)	0.000
Standard Deviation (n-1)	0.008
Lower bound on mean (95%)	0.424
Upper bound on mean (95%)	0.441

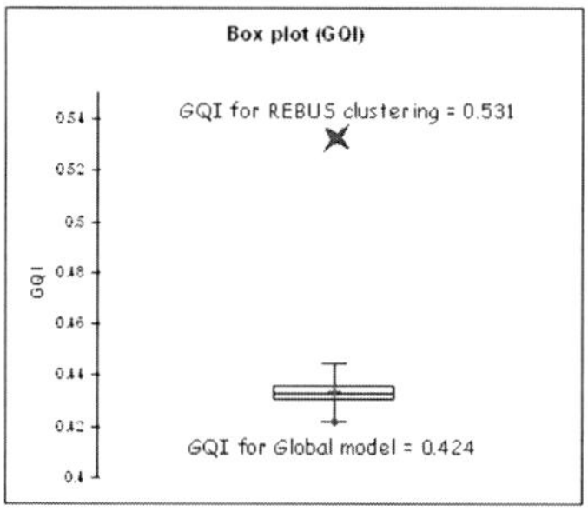

Fig. 2.14 Descriptive statistics for the GQI empirical distribution

the global model results in terms of path coefficients. Nevertheless, the R^2 value associated to the endogenous latent variable *Brand Preference* is equal to 0.108. This value is even smaller than for the global model ($R^2 = 0.239$). The second detected class, instead, is similar to the second class obtained by REBUS-PLS. As a matter of fact, also in this case the exogenous latent variable *Image* seems be the most important driver for *Brand Preference*, showing an R^2 close to 1.

In order to obtain local models that are different also for the measurement model, Ringle et al. (2010) apply a two-step strategy. In the first step they simply apply FIMIX-PLS. Successively they use external/concomitant variables to look for groups overlapping the FIMIX-based ones. Nevertheless, also in this two-step procedure the obtained results are not better than the ones provided by the REBUS-PLS-based partition. As a matter of fact, the R^2 value and the GoF value for the first local model are smaller than for the global model. The local model for the largest class (80% of the whole sample) performs worse than the global model, and worse than all the REBUS-PLS based local models.

The REBUS-PLS algorithm turned out to be a powerful tool to detect unobserved heterogeneity in both experimental and empirical data.

2.5 Conclusion and Perspectives

In the previous sections, where needed, we have already enhanced some of the on going research related to the topics of interest for this chapter. Namely, the development of new estimation modes and schemes for multidimensional (formative) constructs, a path analysis on latent variable scores to estimate path coefficients, the use of GoF-based non parametric tests for the overall model assessment, a sensitivity analysis for these tests, the generalization of REBUS-PLS to capturing heterogeneity in formative models.

We like to conclude this chapter by proposing a short list of further open issues that, in our opinion, currently represent the most important and promising research challenges in PLS Path Modeling:

- Definition of optimizing criteria and unifying functions related to classical or modified versions of the PLS-PM algorithm both for the predictive path model between latent variables and for the analysis of multiple tables.
- Possibility of imposing constraints on the model coefficients (outer weights, loadings, path coefficients) so as to include any information available a priori as well as any hypothesis (e.g. equality of coefficients across different groups, conjectures on model parameters) in the model estimation phase.
- Specific treatment of categorical (nominal and ordinal) manifest variables.
- Specific treatment of non-linearity both in the measurement and the structural model.
- Outliers identification, i.e. assessment of the influence of each statistical unit on the estimates of the outer weights for each block of manifest variables.
- Development of robust alternatives to the current OLS-based PLS Path Modeling algorithm.
- Development of a model estimation procedure based on optimizing the GoF index, i.e. on minimizing a well defined fit function.
- Possibility of specifying feedback relationships between latent variables so as to investigate mutual causality.

The above mentioned issues represent fascinating topics for researchers from both Statistics and applied disciplines.

There is nothing vague or fuzzy about soft modeling;
the technical argument is entirely rigorous
Herman Wold

Acknowledgements The participation of V. Esposito Vinzi to this research was supported by the Research Center of the ESSEC Business School of Paris. The participation of L. Trinchera to this research was supported by the MIUR (Italian Ministry of Education, University and Research) grant "Multivariate statistical models for the ex-ante and the ex-post analysis of regulatory impact", coordinated by C. Lauro (2006).

References

Addinsoft (2009). *XLSTAT 2009*. France: Addinsoft. http://www.xlstat.com/en/products/xlstat-plspm/

Alwin, D. F., and Hauser, R. M. (1975). The decomposition of effects in path. *American Sociological Review, 40*, 36–47.

Bollen, K. A. (1989). *Structural equations with latent variables*. New York: Wiley.

Cassel, C., Hackl, P., and Westlund, A. (1999). Robustness of partial least-squares method for estimating latent variable quality structures. *Journal of Applied Statistics, 26*, 435–446.

Cassel, C., Hackl, P., and Westlund, A. (2000). On measurement of intangible assets: a study of robustness of partial least squares. *Total Quality Management, 11*, 897–907.

Chin, W. W. (1998). The partial least squares approach for structural equation modeling. *in* G. A. Marcoulides (Ed.), Modern methods for business research (pp. 295–236). London: Lawrence Erlbaum Associates.

Chin, W. W., and Dibbern, J. (2007). A permutation based procedure for multigroup PLS analysis: results of tests of differences on simulated data and a cross cultural analysis of the sourcing of information system services between germany and the USA. In: V. Esposito Vinzi, W. Chin, J. Hensler, and H. Wold (Eds.,) *Handbook PLS and Marketing*. Berlin, Heidelberg, New York: Springer.

Edgington, E. (1987). *Randomization test*. New York: Marcel Dekker.

Efron, B., and Tibshirani, R. J. (1993). *An introduction to the bootstrap*. New York: Chapman&Hall.

Esposito Vinzi, V. (2008). The contribution of PLS regression to PLS path modelling: formative measurement model and causality network in the structural model. In: *Joint Statistical Meetings (JSM) 2008*, American Statistical Association, Denver, Colorado, United States of America, August 7th 2008.

Esposito Vinzi, V. (2009). PLS path modeling and PLS regression: a joint partial least squares component-based approach to structural equation modeling. In: *IFCS@GFKL – Classification as a Tool for Research (IFCS 2009)* , University of Technology, Dresden, Dresden, Germany, March 14th 2009 (Plenary Invited Speaker).

Esposito Vinzi, V., and Lauro, C. (2003). PLS regression and classification. In *Proceedings of the PLS'03 International Symposium*, DECISIA, France, pp. 45–56.

Esposito Vinzi, V., and Russolillo, G. (2010). Partial least squares path modeling and regression. In E. Wegman, Y. Said, and D. Scott (Eds.), *Wiley interdisciplinary reviews: computational statistics*. New York: Wiley.

Esposito Vinzi, V., Trinchera, L., Squillacciotti, S., and Tenenhaus, M. (2008). Rebus-pls: a response-based procedure for detecting unit segments in pls path modeling. *Applied Stochastic Models in Business and Industry (ASMBI), 24*, 439–458.

Fornell, C. (1992). A national customer satisfaction barometer: the swedish experience. *Journal of Marketing, 56*, 6–21.

Fornell, C., and Bookstein, F. L. (1982). Two structural equation models: LISREL and PLS appliead to consumer exit-voice theory. *Journal of Marketing Research, 19*, 440–452.

Hahn, C., Johnson, M., Herrmann, A., and Huber, F. (2002). Capturing customer heterogeneity using a finite mixture PLS approach. *Schmalenbach Business Review, 54*, 243–269.

Hanafi, M. (2007). PLS path modeling: computation of latent variables with the estimation mode B. *Computational Statistics, 22*, 275–292.

Hensler, J., and Fassott, G. (2010). Testing moderating effects in PLS path models: An illustration of available procedures. In V. Esposito Vinzi, W. Chin, J. Henseler, and H. Wang (Eds.), *Handbook Partial Least Squares*. Heidelberg: Springer.

Jöreskog, K. (1971). Simultaneous factor analysis in several populations. *Psychometrika, 57*, 409–426.

Kaplan, D. (2000). *Structural equation modeling: foundations and extensions*. Thousands Oaks, California: Sage.

Lohmöller, J. (1987). LVPLS program manual, version 1.8, Technical report. Zentralarchiv für Empirische Sozialforschung, Universität Zu Köln, Köln.

Lohmöller, J. (1989). *Latent variable path modeling with partial least squares*. Heildelberg: Physica-Verlag.

Lyttkens, E., Areskoug, B., and Wold, H. (1975). The convergence of NIPALS estimation procedures for six path models with one or two latent variables. Technical report, University of Göteborg.

Ringle, C., and Schlittgen, R. (2007). A genetic algorithm segmentation approach for uncovering and separating groups of data in PLS path modeling. In *'PLS'07: 5th International Symposium on PLS and Related Methods*, Oslo, Norway, pp. 75–78.

Ringle, C., Wende, S., and Will, A. (2005). Customer segmentation with FIMIX-PLS. In T. Aluja, J. Casanovas, V. Esposito Vinzi, A. Morineau, and M. Tenenhaus Eds., *Proceedings of PLS-05 International Symposium*, SPAD Test&go, Paris, pp. 507–514.

Ringle, C., Wende, S., and Will, A. (2010). Finite mixture partial least squares analysis: Methodology and numerical examples. In V. Esposito Vinzi, W. Chin, J. Henseler, and H. Wang (Eds.), *Handbook Partial Least Squares*. Heidelberg: Springer.

Sanchez, G., and Aluja, T. (2006). Pathmox: a PLS-PM segmentation algorithm, *in* V. Esposito Vinzi, C. Lauro, A. Braverma, H. Kiers, and M. G.Schmiek (Eds.), *Proceedings of KNEMO 2006*, number ISBN 88-89744-00-6, Tilapia, Anacapri, p. 69.

Sanchez, G., and Aluja, T. (2007). A simulation study of PATHMOX (PLS path modeling segmentation tree) sensitivity. In *5th International Symposium - Causality explored by indirect observation*, Oslo, Norway.

Sörbom, D. (1974). A general method for studying differences in factor means and factor structures between groups. *British Journal of Mathematical and Statistical Psychology, 27*, 229–239.

Squillacciotti, S. (2005). Prediction oriented classification in PLS path modelling. In T. Aluja, J. Casanovas, V. Esposito Vinzi, A. Morineau, and M. Tenenhaus, (Eds.), *Proceedings of PLS- 05 International Symposium*, SPAD Test&go, Paris, pp. 499–506.

Tenenhaus, M. (1998). *La Régression PLS: théorie et pratique*. Paris: Technip.

Tenenhaus, M. (2008a). Component-based structural equation modelling. *Total Quality Management & Business Excellence, 19*, 871–886.

Tenenhaus, M. (2008b). 'Structural equation modelling for small samples. *HEC Paris: Jouy-en-Josas, Working paper, no. 885.*

Tenenhaus, M., and Esposito Vinzi, V. (2005). PLS regression, PLS path modeling and generalized procrustean analysis: a combined approach for PLS regression, PLS path modeling and generalized multiblock analysis. *Journal of Chemometrics, 19*, 145–153.

Tenenhaus, M., and Tenenhaus, A. (2009). A criterion-based PLS approach to SEM. In *3rd Workshop on PLS Developments*, ESSEC Business School of Paris, France, May 14th 2009.

Tenenhaus, M., Amato, S., and Esposito Vinzi, V. (2004). A global goodness-of-fit index for PLS structural equation modelling. *Proceedings of the XLII SIS Scientific Meeting*, Vol. Contributed Papers, CLEUP, Padova, pp. 739–742.

Tenenhaus, M., Esposito Vinzi, V., Chatelin, Y., and Lauro, C. (2005). PLS path modeling. *Computational Statistics and Data Analysis, 48*, 159–205.

Tenenhaus, M., Mauger, E., and Guinot, C. (2010). Use of ULS-SEM and PLS-SEM to measure a group effect in a regression model relating two blocks of binary variables. In V. Esposito Vinzi, W. Chin, J. Henseler, and H. Wang (Eds.), *Handbook Partial Least Squares*. Heidelberg: Springer.

Thurstone, L. L. (1931). *The theory of multiple factors*. Ann Arbor, MI: Edwards Brothers.

Trinchera, L. (2007). Unobserved Heterogeneity in structural equation models: a new approach in latent class detection in PLS path modeling. PhD thesis, DMS, University of Naples.

Trinchera, L., and Esposito Vinzi, V. (2006). Capturing unobserved heterogeneity in PLS path modeling. In *Proceedings of IFCS 2006 Conference*, Ljubljana, Sloveny.

Trinchera, L., Squillacciotti, S., and Esposito Vinzi, V. (2006). PLS typological path modeling : a model-based approach to classification. In V. Esposito Vinzi, C. Lauro, A. Braverma, H. Kiers, and M. G.Schmiek (Eds.), *Proceedings of KNEMO 2006*, ISBN 88-89744-00-6, Tilapia, Anacapri, p. 87.

Tukey, J. W. (1964). Causation, regression and path analysis. In *Statistics and mathematics in biology*. New York: Hafner.

Wedel, M., and Kamakura, W. A. (2000). , *Market segmentation – conceptual and methodological foundations*, 2 edn. Boston: Kluwer.

Wertz, C., Linn, R., and Jöreskog, K. (1974). Intraclass reliability estimates: Testing structural assumptions. *Educational and Psychological Measurement, 34*(1), 25–33.

Wold, H. (1966). Estimation of principal component and related models by iterative least squares. In P. R. Krishnaiah (Ed.), *Multivariate analysis*, (pp. 391–420). New York: Academic Press.

Wold, H. (1975a). PLS path models with latent variables: the nipals approach. In H. M. Blalock, A. Aganbegian, F. M. Borodkin, R. Boudon, and V. Cappecchi (Eds.), *Quantitative sociology: international perspectives on mathematical and statistical modeling*. New York: Academic Press.

Wold, H. (1975b). Modelling in complex situations with soft infromation. *Third World Congress of Econometric Society*, Toronto, Canada.

Wold, H. (1975c). Soft modeling by latent variables: the nonlinear iterative partial least squares approach. In J. Gani (Ed.), *Perspectives in probability and statistics, papers in honor of M. S. Bartlett* (pp. 117–142). London: Academic Press.

Wold, H. (1980). Model construction and evaluation when theoretical knowledge is scarce. In J. Kmenta, & J. B. Ramsey (Eds.), *Evaluation of econometric models*, pp. 47–74.

Wold, H. (1982). Soft modeling: the basic design and some extensions. In K. G. Jöreskog, and H. Wold, (Eds.), *Systems under indirect observation*, Part II (pp. 1–54). Amsterdam: North-Holland.

Wold, H. (1985). Partial least squares. In S. Kotz, and N. L. Johnson, (Eds.), *Encyclopedia of Statistical Sciences*, Vol. 6 (pp. 581–591). New York: Wiley.

Wold, S., Martens, H., and Wold, H. (1983). The multivariate calibration problem in chemistry solved by the PLS method. In: A. Ruhe, & B. Kagstrom (Eds.), *Proceedings of the Conference on Matrix Pencils, Lectures Notes in Mathematics*. Heidelberg: Springer.

Chapter 3
Bootstrap Cross-Validation Indices for PLS Path Model Assessment

Wynne W. Chin

Abstract The goal of PLS path modeling is primarily to estimate the variance of endogenous constructs and in turn their respective manifest variables (if reflective). Models with significant jackknife or bootstrap parameter estimates may still be considered invalid in a predictive sense. In this chapter, the objective is to shift from that of assessing the significance of parameter estimates (e.g., loadings and structural paths) to that of predictive validity. Specifically, this chapter examines how predictive indicator weights estimated for a particular PLS structural model are when applied on new data from the same population. Bootstrap resampling is used to create new data sets where new R-square measures are obtained for each endogenous construct in a model. The weighted summed (WSD) R-square represents how well the original sample weights predict when given new data (i.e., a new bootstrap sample). In contrast, the simple summed (SSD) R-square examines the predictiveness using the simpler approach of unit weights. Such an approach is equivalent to performing a traditional path analysis using simple summed scale scores. A relative performance index (RPI) based on the WSD and SSD estimates is created to represent the degree to which the PLS weights yield better predictiveness for endogenous constructs than the simpler procedure of performing regression after simple summing of indicators. In addition, a Performance from Optimized Summed Index (PFO) is obtained by contrasting the WSD R-squares to the R-squares obtained when the PLS algorithm is used on each new bootstrap data set. Results from two studies are presented. In the first study, 14 data sets of sample size 1,000 were created to represent two different structural models (i.e., medium versus high R-square) consisting of one endogenous and three exogenous constructs across seven different measurement scenarios (e.g., parallel versus heterogenous loadings). Five-hundred bootstrap cross validation data sets were generated for each of 14 data sets. In study 2, simulated data based on the population model conforming to the same scenarios in study 1 were used instead of the bootstrap samples in part to examine the accuracy of the bootstrapping approach. Overall, in contrast to Q-square which examines

W.W. Chin
Department of Decision and Information Sciences, Bauer College of Business,
University of Houston, TX, USA
e-mail: wchin@uh.edu

V. Esposito Vinzi et al. (eds.), *Handbook of Partial Least Squares*, Springer Handbooks of Computational Statistics, DOI 10.1007/978-3-540-32827-8_4,

predictive relevance at the indicator level, the RPI and PFO indices are shown to provide additional information to assess predictive relevance of PLS estimates at the construct level. Moreover, it is argued that this approach can be applied to other same set data indices such as AVE (Fornell C, Larcker D, J Mark Res 18:39–50, 1981) and GoF (Tenenhaus M, Amato S, Esposito Vinzi V, Proceedings of the XLII SIS (Italian Statistical Society) Scientific Meeting, vol. Contributed Papers, 739–742, CLEUP, Padova, Italy, 2004) to yield RPI-AVE, PFO-AVE. RPI-GoF, and PFO-GoF indices.

3.1 Introduction

PLS path modeling is a components based methodology that provides determinate construct scores for predictive purposes. Its goal is primarily to estimate the variance of endogenous constructs and in turn their respective manifest variables (if reflective). To date, a large portion of the model validation process consists of parameter inference where significance of estimated parameters are tested (Chin 1998). Yet, models with significant jackknife or bootstrap parameter estimates may still be considered invalid in a predictive sense. In other words, to what extent will the estimated weights from the PLS analysis predict in future situations when we have new data from the same underlying population of interest? If we develop a consumer based satisfaction scale to predict brand loyalty, for example, will the weights derived to form the satisfaction scale be as predictive. In this chapter, the objective is to shift the focus from that of assessing the significance or accuracy of parameter estimates (e.g., weights, loadings and structural paths) to that of predictive validity. Specifically, this chapter presents a bootstrap re-sampling process intended to provide a sense as to how efficacious the indicator weights estimated for a particular PLS structural model are in predicting endogenous constructs when applied on new data.

Predictive sample reuse technique as developed by Geisser (1974) and Stone (1975) represent a synthesis of cross-validation and function fitting with the perspective "that prediction of observables or potential observables is of much greater relevance than the estimation of what are often artificial constructs-parameters" (Geisser 1975, p. 320). For social scientists interested in the predictive validity of their models, the Q-square statistic has been the primary option. This statistic is typically provided as a result of a blindfolding algorithm (Chin 1998, pp. 317–318) where portions of the data for a particular construct block (i.e., indicators by cases for a specific construct) are omitted and cross-validated using the estimates obtained from the remaining data points. This procedure is repeated with a different set of data points as dictated by the blindfold omission number until all sets have been processed. Two approaches have been used to predict the holdout data. A communality-based Q-square takes the construct scores estimated for the target endogenous construct (minus the holdout data) to predict the holdout data. Alternatively, a redundancy-based Q-square uses the scores for those antecedent constructs that are modeled as directly impacting the target construct. In both instances, a Q-square relevance measure is obtained for the endogenous construct in question. This relevance measure is generally considered more informative than the R-square and the Average Variance Extracted statistics since the latter two have the inherent bias of being assessed on the same data that were used to estimate its parameters and thereby raises the issue of data overfitting.

As Chin (1995, p. 316) noted over a decade ago "alternative sample reuse methods employing bootstrapping or jackknifing have yet to be implemented." This still seems to be the case. Moreover, the Q-square measure is meant to help assess predictive validity at the indicator level, while there is still need for indices that help provide information regarding the predictive validity of a PLS model at the construct level. With that in mind, this chapter presents a bootstrap reuse procedure for cross-validating the weights derived in a PLS analysis for predicting endogenous constructs. It is meant to answer questions concerning the value of the weights provided in a PLS analysis as it relates to maximizing the R-square of the key dependent constructs of a model.

Standard cross validation involves using a sample data set for training followed by test data set from the same population to evaluate predictiveness of the model estimates. As Picard and Cook (1984, p. 576) noted in the context of regression models is that "when a model is chosen because of qualities exhibited by a particular set of data, predictions of future observations that arise in a similar fashion will almost certainly not be as good as might naively be expected. Obtaining an adequate estimator of MSE requires future data and, in the extreme, model evaluation is a long-term, iterative endeavor. To expedite this process, the future can be constructed by reserving part of the present, available data." Their approach is to split the existing data into two part (not necessarily of equal size) to see how the fitted model in part one performs on the reserved set for validation. Such an approach has been applied in chemometrics to determine the number of components in a PLS models (Du et al. 2006; Xu et al. 2004; Xu and Liang 2001). This approach is argued as a consistent method in determining the number of components when compared to the leave-one-out cross validation, but requires more than 50% of samples left out to be accurate (Xu and Liang 2001), although it can underestimate the prediction ability of the model selected if a large percentage of samples are left out for validation (Xu et al. 2004).

Here we differ by using the original sample set as the training set to estimate a given PLS model and then employ bootstrap re-sampling to create new data sets. The indicator weights derived from the original sample set are used on the new bootstrap samples and R-square measures are examined for each endogenous construct in the model. The weighted summed (WSD) R-square represents how well the original sample weights predict given new data (i.e., a new bootstrap sample). As comparison, we also calculate the Simple Summed (SSD) R-square which reflects the predictiveness using the simpler approach of unit weights. Such an approach is equivalent to what many social scientists normally do – that being to create unit weighted composite scores for each construct in order to run a traditional path analysis. The relative performance index (RPI) based on the WSD and SSD R-squares can then calculated to represent the degree to which the PLS weights from the original sample provide greater predictiveness for endogenous constructs than the simpler procedure of performing regression after simple summing of indicators.

For each bootstrap sample set, a standard PLS run can also be completed. The R-squares obtained from running the model in question on each bootstrap data set would represent optimized summed (OSD) R-squares as dictated by the PLS algorithm and thus should generally be greater than the WSD or SSD R-squares. A performance from optimized summed index (PFO) can then be obtained by contrasting the WSD to the OSD R-squares.

3.2 General Procedure

The specific steps for calculating the RPI and PFO indices are as follows[1]:

1. Take original sample set model run, record original sample weights and R-square for each endogenous construct in the model.
2. Create N bootstrap samples where each sample will be used to obtain three different R-squares for each endogenous construct (i.e., OSD, WSD, and SSD R-squares).
3. For each bootstrap sample, run PLS algorithm and record the R-square for each endogenous construct. This will be labeled the optimized summed (OSD) R-square.
4. Standardize each bootstrap sample data and apply the original sample weights to calculate the WSD set of construct scores. Unit weights are applied to calculate the SSD set of construct scores.
5. To obtain the WSD and SSD R-squares, replace each construct in the graph with the single indicator from your calculation in step 4. Estimate and record R-square twice. The R-square resulting from the use the weights from the original run will be labeled the Weighted Summed (WSD) R-square. The third R-square represents the baseline level of unit weights and is labeled the Simple Summed (SSD) R-square.
6. Calculate relative performance index (RPI) of using original samples weights (WSD R-square) over simple summed regression. (SSD R-square).

$$\text{RPI} = \frac{100 * (\text{WSD R-square} - \text{SSD R-square})}{\text{SSD R-square}}.$$

7. Calculate Performance from PLS optimized summed (PFO) by examining how the WSD R-square differs from the OSD R-square.

$$\text{PFO} = \frac{100 * (\text{OSD R-square} - \text{WSD R-square})}{\text{WSD R-square}}.$$

[1] This is based on the assumption that the default unit variance, no location algorithm is employed.

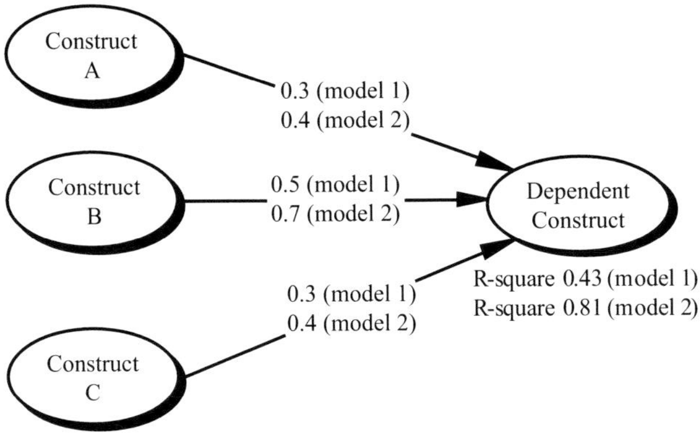

Fig. 3.1 Models used to generate data

3.3 Study 1

To test how the WSD, SSD, OSD, and two new indices RPI and PFO perform, 14 data sets of sample size 1,000 were generated to reflect 2 underlying models (see Fig. 3.1). Each model consists of one dependent construct and three independent constructs. Model 1 represents a medium predictive model with an R-square of 0.43 while model 2 has standardized paths that result in a higher R-square of 0.81. Six indicators were created for each construct. For each model, data sets for seven case scenarios were created. These case settings also used by Chin et al. (2003) in their simulation of interaction effects represents varying levels of homogeneity for each set of indicators as well as reliability (See column 1, Table 3.1). The first setting represents a baseline with homogeneous indicators all set at a standardized loading of 0.70. The expectation is that PLS estimated weights should not provide any substantive improvements over a simple summed approach. Setting 7, in comparison, is quite heterogeneous and lower in reliability with two indicator loadings set at 0.7, two at 0.5, and two at 0.3. Composite reliabilities (Werts et al. 1974) and average variance extracted (AVE) (Fornell and Larcker 1981) for each setting are presented in Tables 3.1 and 3.2. All data where generated from an underlying normal distribution.

Five-hundred bootstrap runs were performed for each of the 14 data sets and summary statistics are provided in Tables 3.1 and 3.2. Not surprisingly, Table 3.1 reflecting the medium R-square model 1 demonstrates that as the overall reliability of the indicators drop, the mean estimated R-square also lowers. Figure 3.2 provides a plot of these estimates. Interestingly enough, we see that the WSD R-squares are quite close to the OSD estimates. The SSD R-squares, as expected, only matches the other two estimates for the case of identical loadings (i.e., setting 1). Approximately the same pattern also appears for model 2 (see Fig. 3.3 and Table 3.2). But in this instance the relationship between the mean R-squares and the population

Table 3.1 Statistics for 500 bootstrap resamples for medium R-square Model 1 of 0.43

	Composite reliability	Average variance extracted		R-square OSD	R-square WSD	R-square SSD	RPI	PFO	Q-square OSD	Q-square SSD
Setting 1: All loadings set at 0.7	0.85	0.49	Mean	0.332	0.328	0.324	1.28	−1.20	0.08	0.07
			SE	0.025	0.025	0.025	0.34	0.59	0.03	0.03
			T-stat	13.511	13.199	13.107	3.77	−2.05	2.64	2.52
Setting 2: 3 loadings set 0.7, 3 at 0.3	0.68	0.29	Mean	0.250	0.242	0.184	31.97	−3.03	−0.02	−0.04
			SE	0.022	0.023	0.021	6.48	1.59	0.02	0.03
			T-stat	11.286	10.627	8.773	4.94	−1.91	−0.78	−1.39
Setting 3: 2 loadings set 0.8, 4 at 0.3	0.64	0.27	Mean	0.282	0.274	0.165	66.58	−2.86	0.01	−0.03
			SE	0.021	0.022	0.021	11.34	1.97	0.02	0.02
			T-stat	13.731	12.356	8.048	5.87	−1.46	0.28	−1.67
Setting 4: 1 loading set 0.8, 5 at 0.3	0.52	0.18	Mean	0.229	0.219	0.135	63.52	−4.52	−0.04	−0.07
			SE	0.021	0.023	0.020	11.21	2.67	0.07	0.04
			T-stat	10.986	9.578	6.643	5.67	−1.69	−0.57	−1.89
Setting 5: 2 loadings set 0.7, 4 at 0.3	0.59	0.22	Mean	0.190	0.180	0.123	48.11	−5.43	−0.06	−0.08
			SE	0.020	0.021	0.020	12.73	2.99	0.03	0.04
			T-stat	9.426	8.521	6.101	3.78	−1.82	−1.93	−2.19
Setting 6: 2 loadings set at 0.6, 4 at 0.3	0.54	0.18	Mean	0.164	0.153	0.124	23.90	−7.10	−0.08	−0.09
			SE	0.020	0.020	0.019	7.52	3.05	0.02	0.02
			T-stat	8.341	7.621	6.609	3.18	−2.33	−4.42	−4.52
Setting 7: 2 loadings set at 0.7, 2 at 0.5, and 2 at 0.3	0.67	0.28	Mean	0.259	0.252	0.219	15.08	−2.90	−0.01	−0.03
			SE	0.023	0.023	0.023	3.43	1.34	0.05	0.05
			T-stat	11.365	10.720	9.472	4.40	−2.16	−0.24	−0.53

Standardized paths set at 0.3, 0.5, and 0.3 for constructs A, B, and C respectively

Table 3.2 Statistics for 500 bootstrap resamples for high R-square Model 2 of 0.81

	Composite reliability	Average variance extracted		R-square OSD	R-square WSD	R-square SSD	RPI	PFO	Q-square OSD	Q-square SSD
Setting 1: All loadings set at 0.7	0.85	0.49	Mean	0.603	0.601	0.599	0.37	−0.37	0.31	0.30
			SE	0.020	0.020	0.020	0.14	0.20	0.03	0.03
			T-stat	30.179	29.767	29.457	2.67	−1.87	8.92	8.84
Setting 2: 3 loadings set 0.7, 3 at 0.3	0.68	0.29	Mean	0.519	0.515	0.406	26.80	−0.86	0.16	0.13
			SE	0.021	0.022	0.024	3.46	0.68	0.05	0.04
			T-stat	24.687	23.473	17.147	7.75	−1.26	3.51	3.14
Setting 3: 2 loadings set 0.8, 4 at 0.3	0.64	0.27	Mean	0.500	0.495	0.345	43.99	−0.98	0.14	0.09
			SE	0.022	0.023	0.024	5.52	0.99	0.03	0.03
			T-stat	23.080	21.551	14.177	7.97	−1.00	4.68	2.74
Setting 4: 1 loading set 0.8, 5 at 0.3	0.52	0.18	Mean	0.331	0.323	0.193	68.23	−2.56	0.02	−0.01
			SE	0.021	0.023	0.021	10.08	2.18	0.03	0.04
			T-stat	15.658	13.774	9.088	6.77	−1.17	0.62	−0.34
Setting 5: 2 loadings set 0.7, 4 at 0.3	0.59	0.22	Mean	0.399	0.393	0.298	32.21	−1.68	0.07	0.05
			SE	0.024	0.025	0.024	4.38	1.21	0.04	0.03
			T-stat	16.826	15.955	12.442	7.36	−1.39	1.94	1.46
Setting 6: 2 loadings set at 0.6, 4 at 0.3	0.54	0.18	Mean	0.322	0.313	0.266	17.78	−2.68	0.02	0.01
			SE	0.021	0.022	0.021	3.58	1.31	0.03	0.03
			T-stat	15.098	14.376	12.717	4.97	−2.05	0.72	0.35
Setting 7: 2 loadings set at 0.7, 2 at 0.5, and 2 at 0.3	0.67	0.28	Mean	0.409	0.403	0.337	19.85	−1.51	0.09	0.07
			SE	0.022	0.023	0.023	3.15	0.93	0.03	0.04
			T-stat	18.362	17.324	14.693	6.31	−1.62	2.58	1.79

Standardized paths set at 0.3, 0.5, and 0.3 for constructs A, B, and C respectively

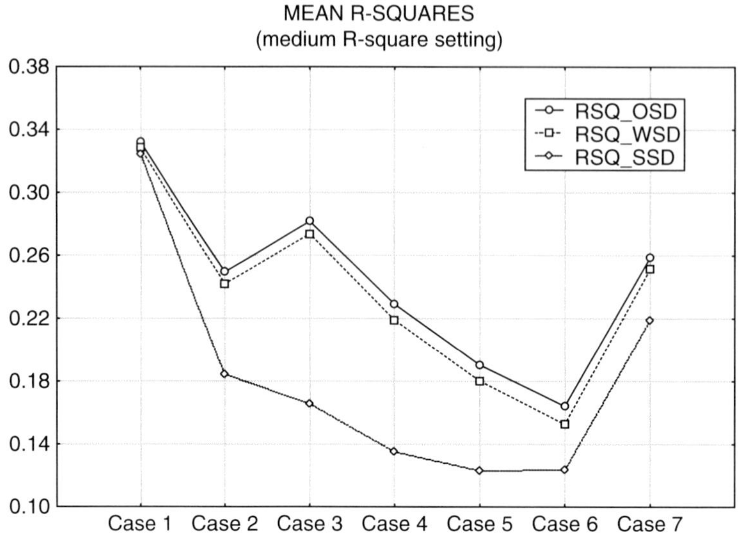

Fig. 3.2 Mean comparison of 500 bootstrap samples for Model 1

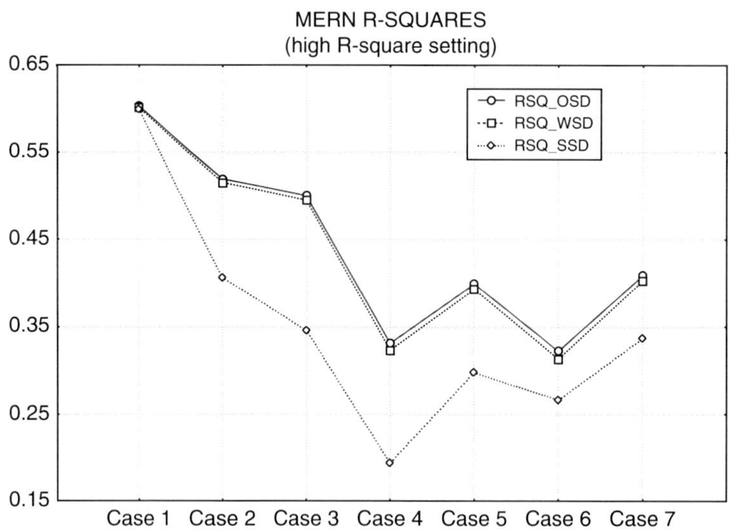

Fig. 3.3 Mean comparison of 500 bootstrap samples

based average variance extracted and composite reliability is more apparent. For example, case setting 5 has a higher average communality and scale reliability than case settings 4 and 6. These differences, in turn, are reflected in better estimates of the structural paths and higher mean R-squares.

Figure 3.4 provides a plot of the RPI across the two models and seven settings. Both model yielded somewhat similar results with model 2 being slightly

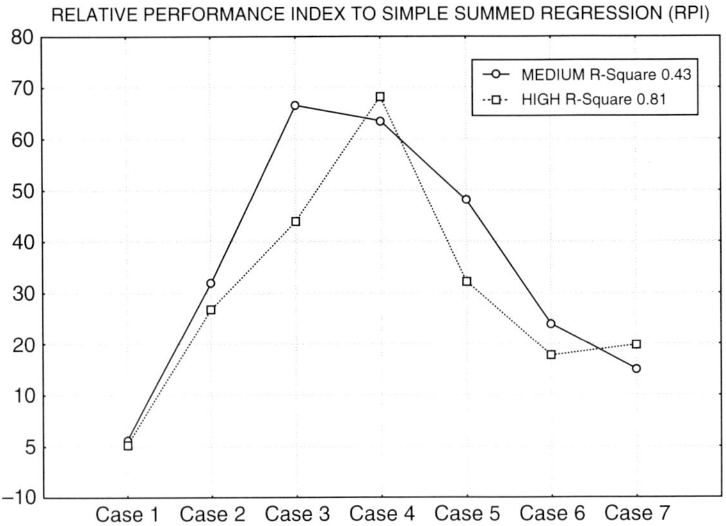

Fig. 3.4 Mean of RPI for 500 bootstrap samples

more consistent with the average communality of its indicators. Again, in agreement with the earlier R-square results, the average RPI for the baseline setting is not substantively different from zero. The relative percentage improvement over simple summed path analysis did reach 68% in the case of setting 4. As comparison, redundancy based Q-squares were also estimated for each data set. Figures 3.5 and 3.6 are plots of models 1 and 2 respectively using OSD and SSD weights. Since the objective of this measure is for evaluating predictiveness as the indicator level, we see the Q-square tends to drop as the indicator reliabilities go lower. On average, the OSD based Q-squares are slightly higher and follows the pattern of the mean R-squares. But the differences were not that dramatic. We also note that when the higher structural paths are higher as in Model 2, the Q square becomes more in line with the magnitude of the composite reliability and communality of the construct. For example, case 5 is now higher than either cases 4 or 6. This reflects the stronger linkage for the antecedent constructs in conjunction with the reliability of the indicators in predicting individual item responses. But, as expected, it provides little information on the strength of relationship at the construct level. The plot of PFO (see Fig. 3.7) in conjunction with the plot of the RPI provides a sense as to how well the PLS model performs. For case settings 3 through 5, for example, we see that the PLS supplied weights provide improvements over unit weighting regression in the range of 50%. In terms of the distance from the PLS optimized OSD R-square, the performance of the PLS estimated weights was never more than 5% from the optimized.

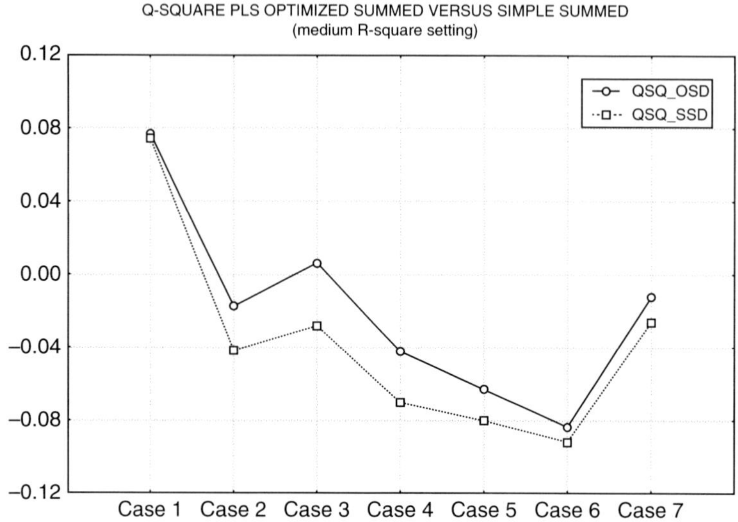

Fig. 3.5 Q-square comparison (omission distance of 7)

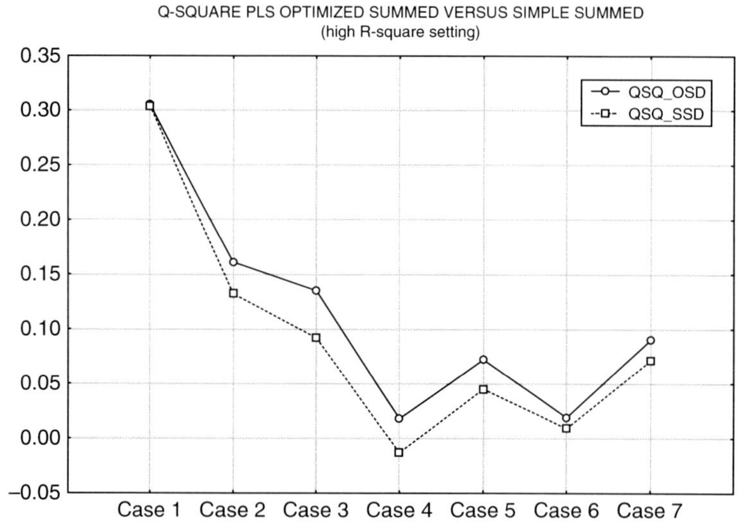

Fig. 3.6 Q-square comparison (omission distance of 7)

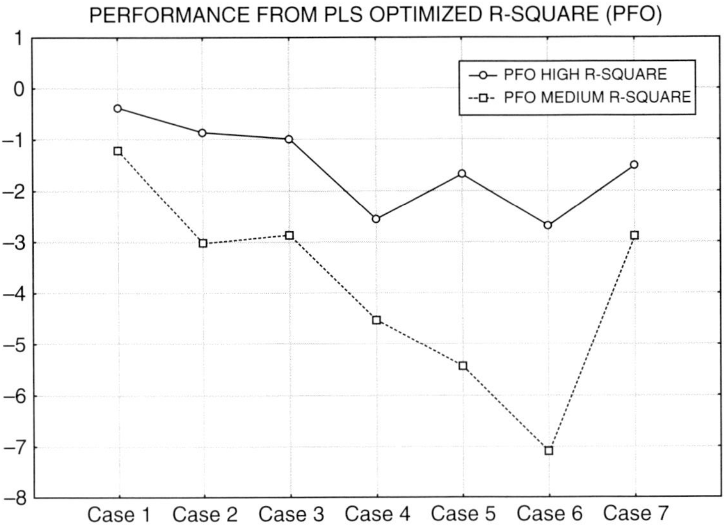

Fig. 3.7 Mean comparisons of PFO for 500 bootstrap samples

3.4 Study 2

For study 2, the same model and setting were applied. In fact the same 14 data sets with their associated weights were used. But instead of cross-validating based on bootstrap resamples, 500 simulated data sets reflecting the underlying population model were generated. In essence, instead of using bootstrapping to mirror the endeavor of obtaining 500 new data sets, we actually go and obtain new data. Thus, we can see how well the earlier bootstrapping approximates (i.e., mirrors) that of actual data. Tables 3.3 and 3.4 provide the summary results while Figs. 3.7 and 3.8 present the combined plots of the RPI and PFO estimates obtained from the earlier bootstrapped data along with the simulated data for this study. The results show that the RPI estimates using bootstrapping is quite similar to the simulated data. The medium R-square scenario tends to be more inflated than the higher R-square scenario for case setting 3 through 5. Overall, except for case setting 4, we see the strong convergence on the estimates for RPI. For the PFO statistic, we again see the simulation results follow a similar pattern to the bootstrap results. Two slight departures are found for case setting 4 and 6 for the medium R-square simulated data. Overall, it may be concluded that the bootstrap data did come close to reflecting the underlying population (Fig. 3.9).

Table 3.3 Statistics for 500 simulated samples for medium R-square model of 0.43

		R-square OSD	R-square WSD	R-square SSD	RPI	PFO
Setting 1: All loadings set at 0.7	Mean	0.318	0.314	0.314	−0.11	−1.31
	SE	0.024	0.025	0.025	0.33	0.57
	T-stat	13.060	12.707	12.720	−0.34	−2.28
Setting 2: 3 loadings set 0.7, 3 at 0.3	Mean	0.253	0.244	0.200	22.49	−3.66
	SE	0.023	0.023	0.023	5.53	1.69
	T-stat	11.044	10.416	8.873	4.07	−2.17
Setting 3: 2 loadings set 0.8, 4 at 0.3	Mean	0.267	0.261	0.180	46.33	−2.22
	SE	0.023	0.024	0.022	9.12	1.71
	T-stat	11.818	10.975	8.031	5.08	−1.30
Setting 4: 1 loading set 0.8, 5 at 0.3	Mean	0.194	0.173	0.118	48.01	−10.67
	SE	0.020	0.021	0.019	10.87	4.03
	T-stat	9.757	8.125	6.315	4.42	−2.65
Setting 5: 2 loadings set 0.7, 4 at 0.3	Mean	0.212	0.202	0.152	33.67	−4.71
	SE	0.022	0.023	0.021	8.44	2.42
	T-stat	9.748	8.989	7.353	3.99	−1.95
Setting 6: 2 loadings set at 0.6, 4 at 0.3	Mean	0.167	0.150	0.127	19.18	−9.79
	SE	0.020	0.020	0.019	6.78	3.43
	T-stat	8.265	7.477	6.692	2.83	−2.85
Setting 7: 2 loadings set at 0.7, 2 at 0.5, and 2 at 0.3	Mean	0.237	0.226	0.198	14.73	−4.45
	SE	0.023	0.023	0.022	3.98	1.61
	T-stat	10.449	9.931	9.057	3.70	−2.77

Standardized paths set at 0.3, 0.5, and 0.3 for constructs A, B, and C respectively

Table 3.4 Statistics for 500 simulated samples for high R-square model of 0.81

		R-square OSD	R-square WSD	R-square SSD	RPI	PFO
Setting 1: All loadings set at 0.7	Mean	0.592	0.589	0.590	−0.07	−0.43
	SE	0.018	0.019	0.019	0.15	0.21
	T-stat	32.117	31.734	31.787	−0.48	−2.09
Setting 2: 3 loadings set 0.7, 3 at 0.3	Mean	0.465	0.462	0.374	23.51	−0.77
	SE	0.023	0.024	0.025	3.67	0.73
	T-stat	19.962	19.133	15.112	6.41	−1.06
Setting 3: 2 loadings set 0.8, 4 at 0.3	Mean	0.490	0.482	0.335	44.26	−1.56
	SE	0.022	0.023	0.023	5.23	1.04
	T-stat	22.672	20.948	14.316	8.47	−1.50
Setting 4: 1 loading set 0.8, 5 at 0.3	Mean	0.352	0.345	0.219	58.41	−1.93
	SE	0.020	0.022	0.021	8.26	1.89
	T-stat	17.245	15.393	10.254	7.07	−1.02
Setting 5: 2 loadings set 0.7, 4 at 0.3	Mean	0.388	0.374	0.286	31.21	−3.60
	SE	0.024	0.025	0.025	4.77	1.42
	T-stat	16.371	15.105	11.488	6.54	−2.53
Setting 6: 2 loadings set at 0.6, 4 at 0.3	Mean	0.299	0.281	0.237	18.87	−5.90
	SE	0.024	0.025	0.024	4.12	2.02
	T-stat	12.387	11.370	9.873	4.58	−2.92
Setting 7: 2 loadings set at 0.7, 2 at 0.5, and 2 at 0.3	Mean	0.433	0.428	0.369	16.25	−1.20
	SE	0.023	0.024	0.024	3.06	0.85
	T-stat	18.909	17.960	15.347	5.31	−1.41

Standardized paths set at 0.3, 0.5, and 0.3 for constructs A, B, and C respectively

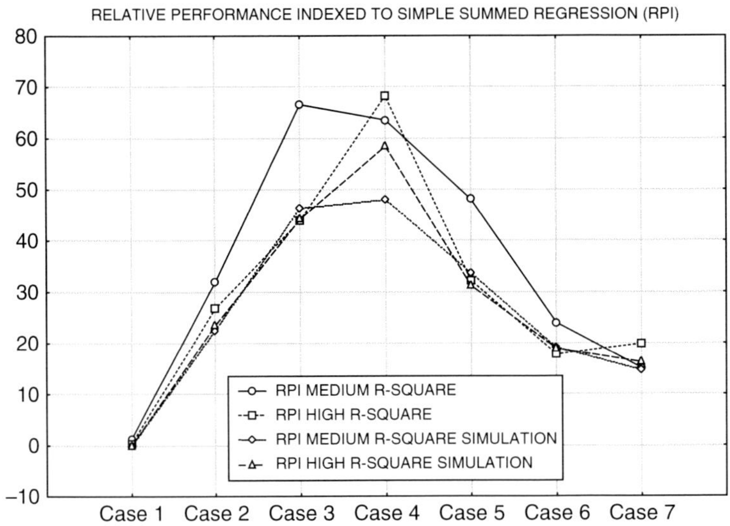

Fig. 3.8 Mean comparison of RPI for 500 bootstrap samples and simulated data

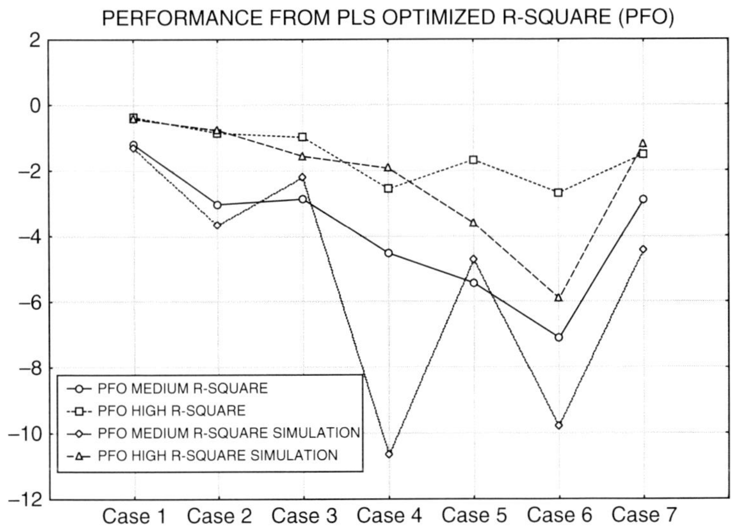

Fig. 3.9 Mean comparison of PFO for 500 bootstrap samples and simulated data

3.5 Discussion and Conclusion

This chapter has presented two new bootstrapped cross-validation indices designed to assess how well PLS estimated weights perform at predicting endogenous constructs. It uses bootstrap resampling to assess both the relative improvement over a simple summed path analytical strategy (i.e., RPI) and its proximity to the PLS optimized estimates (i.e., PFO). More importantly, it provides an alternative to model R-squares that PLS skeptics argue may be capitalizing on chance. The results presented here are encouraging, yet highlight some key points. In general, the cross validated R-squares are quite close to the PLS estimates as reflected in the PFO numbers reported here. Conversely, the RPI estimates show many instances where PLS makes a substantial improvement over unit weighted regression. But, if one expects the indicators used in measuring an underlying construct are relatively homogenous in their loadings, we should expect this belief will be corroborated by having a small RPI (i.e., close to zero). Low RPIs in general would suggest that a simple summed path analysis would generate similar results. But with greater measurement variability, the RPI can be useful in providing information on the relative improvement from using PLS estimates. As an example, case setting 3 for high R-square model 2 scenario (with 2 loadings of 0.8 and 4 at 0.3) show that the mean WSD R-square of approximately 0.5 provides a 43% improvement over unit weighted scales and is within 1% of the OSD estimates. This chapter also showed that while the Q-square measures provide similar patterns to the mean R-squares, it provides limited information on the value of PLS for maximizing the construct level relationships.

Overall, this chapter only scratches the surface of bootstrap cross validation and, as in the case of any study, a word of caution must be sounded before strong generalizations are made. First, both smaller and larger sample sizes should be examined along with varying the data distributions to match different levels of non-normality. In this study, all data were generated from an underlying normal distribution. If the data were assumed or estimated to be non-normal, significance testing of the indices may require a percentile or BCA tests with concomitant increase in bootstrap sample size (Efron and Tibshirani 1993). Moreover, the models examined in this chapter are relatively simplistic which is contrary to the level of complexity that PLS can ideally be applied. Furthermore, while the six indicator model was used to match those of previous studies, additional tests on the performance of indices for two, four and eight indicators would seem reasonable. Finally, the RPI and PFO indices should be considered part of the toolkit for researchers in appraising their models. Other measures based on the original sample set such as the communality of a block of measures (i.e., AVE), Q-square, and Goodness of Fit (GoF) (i.e., which is the geometric mean of a model's average estimated R-square with the average communality of measures used) do provide additional diagnostic value. One goal for the future would logically be to link these sample based measure or even other alternatives yet to be presented in a similar fashion done in this chapter with R-square. For example, bootstrap cross validation equivalents of the indices presented in this chapter using GoF (i.e., RPI-GoF and PFO-GoF), which shifts the focus away from only one single endogenous construct would be a logical next step for those interested in a global cross validation index since GoF is "meant as an index for validating a PLS model globally" (Tenenhaus et al. 2005).

References

Chin, W. W. (1995). Partial least squares is to LISREL as principal components analysis is to common factor analysis. *Technology Studies, 2*, 315–319.

Chin, W. W. (1998). The partial least squares approach for structural equation modeling. In G. A. Marcoulides (Ed.), *Modern methods for business research* (pp. 295–336). New Jersy: Lawrence Erlbaum.

Chin, W. W., Marcolin, B. L., & Newsted, P. R. (2003). A partial least squares latent variable modeling approach for measuring interaction effects: Results from a monte carlo simulation study and electronic mail emotion/adoption study. *Information Systems Research, 14*(2), 189–217.

Du, Y.-P, Kasemsumran, S., Maruo, K., Nakagawa, T., & Ozaki, Y. (2006). Ascertainment of the number of samples in the validation set in Monte Carlo cross validation and the selection of model dimension with Monte Carlo cross validation. *Chemometrics and Intelligent Laboratory Systems, 82*, 83–89.

Efron, B., & Tibshirani, R. J. (1993). *An introduction to the bootstrap* (monographs on statistics and applied probability #57). New York: Chapman & Hall.

Fornell, C., & Larcker, D. (1981). Evaluating structural equation models with unobservable variables and measurement error. *Journal of Marketing Research, 18*, 39–50.

Geisser, S. (1974). A predictive approach to the random effect model. *Biometrika, 61*(1), 101–107.

Geisser, S. (1975). The predictive sample reuse method with applications. *Journal of the American Statistical Association, 70*, 320–328.

Picard, R. R., & Cook, R. D. (1984). Cross-validation of regression models. *Journal of the American Statistical Association, 79*(387), 573–585.

Stone, M. (1975). Cross-validatory choice and assessment of statistical predictions. *Journal of the Royal Statistical Society, Series B, 36*(2), 111–133.

Tenenhaus, M., Amato, S., & Esposito Vinzi, V. (2004). A global Goodness-of-Fit index for PLS structural equation modelling. In *Proceedings of the XLII SIS (Italian Statistical Society) Scientific Meeting, vol. Contributed Papers* (pp. 739–742). Padova, Italy: CLEUP.

Tenenhaus, M., Esposito Vinzi, V., Chatelin, Y. M., & Lauro, C. (2005). *PLS path modelling, computational statistics and data analysis* (Vol. 48, No. 1, pp. 159–205). The Netherlands: North-Holland.

Xu, Q.-S., & Liang, Y.-Z. (2001). Monte Carlo cross valiation. *Chemometrics and Intelligent Laboratory Systems, 56*, 1–11.

Xu, Q.-S, Liang, Y.-Z, & Du, Y.-P. (2004). Monte Carlo cross-validation for selecting a model and estimating the prediction error in multivariate calibration. *Journal of Chemometrics, 18*, 112 120.

Werts, C. E., Linn, R. L., & Jöreskog, K. G. (1974). Intraclass reliability estimates: Testing structural assumptions. *Educational and Psychological Measurement, 34*(1), 25–33.

Chapter 4
A Bridge Between PLS Path Modeling and Multi-Block Data Analysis

Michel Tenenhaus and Mohamed Hanafi

Abstract A situation where J blocks of variables $X_1, \ldots, X_J$ are observed on the same set of individuals is considered in this paper. A factor analysis approach is applied to blocks instead of variables. The latent variables (LV's) of each block should well explain their own block and at the same time the latent variables of same order should be as highly correlated as possible (positively or in absolute value). Two path models can be used in order to obtain the first order latent variables. The first one is related to confirmatory factor analysis: each LV related to one block is connected to all the LV's related to the other blocks. Then, PLS path modeling is used with mode A and centroid scheme. Use of mode B with centroid and factorial schemes is also discussed. The second model is related to hierarchical factor analysis. A causal model is built by relating the LV's of each block X_j to the LV of the super-block X_{J+1} obtained by concatenation of $X_1, \ldots, X_J$. Using PLS estimation of this model with mode A and path-weighting scheme gives an adequate solution for finding the first order latent variables. The use of mode B with centroid and factorial schemes is also discussed. The higher order latent variables are found by using the same algorithms on the deflated blocks. The first approach is compared with the MAXDIFF/MAXBET Van de Geer's algorithm (1984) and the second one with the ACOM algorithm (Chessel and Hanafi, 1996). Sensory data describing Loire wines are used to illustrate these methods.

Introduction

In this paper, we consider a situation where J blocks of variables are observed on the same set of n individuals. The block X_j contains k_j variables . All these variables are supposed to be centered and are often standardized in practical applications. We

M. Tenenhaus
Department SIAD, HEC School of Management, 1 rue de la Libération, 78351 Jouy-en-Josas, France
e-mail: tenenhaus@hec.fr

M. Hanafi
Unité Mixte de Recherche (ENITIAA-INRA) en Sensométrie et Chimiométrie, ENITIAA, Rue de la Géraudière – BP 82225, Nantes 44322, Cedex 3, France
e-mail: hanafi@enitiaa-nantes.fr

V. Esposito Vinzi et al. (eds.), *Handbook of Partial Least Squares*, Springer Handbooks of Computational Statistics, DOI 10.1007/978-3-540-32827-8_5,

can follow a factor analysis approach on tables instead of variables. We suppose that each block X_j is summarized by m latent variables (LV's) plus a residual X_{jm}. Each data table is decomposed into two parts:

$$X_j = \left[F_{j1}p_{j1}^T + \cdots + F_{jm}p_{jm}^T\right] + \left[X_{jm}\right]$$

The first part of the decomposition is $F_{j1}p_{j1}^T + \cdots + F_{jm}p_{jm}^T$ where the F_{jh}'s are n-dimension column vectors and the p_{jh}'s are k_j-dimension column vectors. The latent variables (also called scores, factors or components) $F_{j1}, \ldots, F_{jm}$ should well explain the data table X_j and, at the same time, the correlations between the scores of same order h should be as high as possible in absolute value, or in positive value to improve interpretation. These scores play a similar role as the common factors in factor analysis (Morrison 1990). The second part of the decomposition is the residual X_{jm} which represents the part of X_j not related to the other blocks in a m dimensions model, i.e., the specific part of X_j. The residual X_{jm} is the deflated block X_j of order m.

To obtain first order latent variables that well explain their own blocks and are at the same time well correlated, covariance-based criteria have to be used. Several existing strategies can be used, among them the MAXDIFF/MAXBET (Van de Geer 1984) and ACOM (Chessel and Hanafi 1996) algorithms or other methods (Hanafi and Kiers 2006). In the present paper, it is shown how to use PLS path modeling for the analysis of multi-block data with these objectives. PLS path modeling offers two path models that can be used to obtain the first order latent variables. In the first strategy, the LV of each block is connected to all the LV's of the other blocks in such a way that the obtained path model is recursive (no cycles). This is a confirmatory factor analysis model with one factor per block (Long 1983). Then, PLS path modeling is used with mode A and centroid scheme. In the second strategy, a hierarchical model is built by connecting each LV related to block X_j to the LV related to the super-block X_{J+1}, obtained by concatenation of $X_1, \ldots, X_J$. PLS estimation of this model with mode A and path-weighting scheme gives an adequate solution for finding the first order latent variables. The use of mode B with centroid and factorial schemes is also discussed for both strategies. The higher order latent variables are found by using the same algorithms on the deflated blocks. These approaches will be compared to the MAXDIFF/MAXBET and ACOM algorithms.

Sensory data about Loire wines will be used to illustrate these methods. PLS-Graph (Chin 2005) has been used to analyze these data and the ouputs of this software will be discussed in details.

4.1 A PLS Path Modeling Approach to Confirmatory Factor Analysis

A causal model describing the confirmatory factor analysis (CFA) model with one factor per block is given in Fig. 4.1.

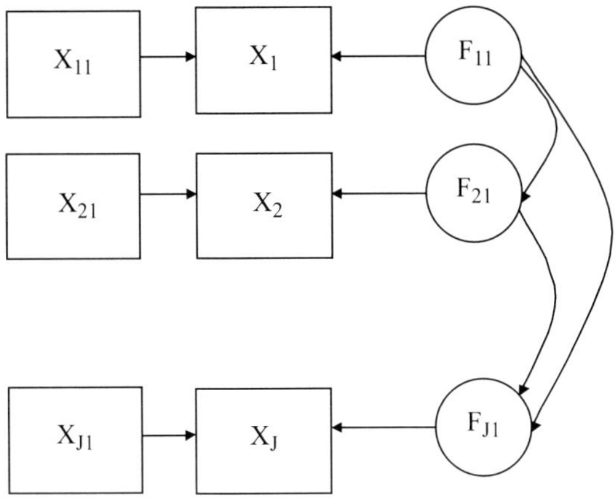

Fig. 4.1 Path model for confirmatory factor analysis

The general PLS algorithm (Wold 1985) can be used for the analysis of multi-block data (Lohmöller 1989; Tenenhaus et al. 2005). In usual CFA models, the arrows connecting the latent variables are double-headed. But in PLS, the link between two latent variables is causal: the arrow connecting two latent variables is unidirectional. So it is necessary to select, in the general PLS algorithm, options that don't take into account the directions of the arrows, but only their existence. This is the case for the centroid and factorial schemes of the PLS algorithm. The directions of the arrows have no importance, with the restriction that the complete arrow scheme must be recursive (no cycle).

The general PLS algorithm is defined as follows for this specific application. The indices 1 for first order weights and latent variables have been dropped out for improving the legibility of the paper.

4.1.1 External Estimation

Each block X_j is summarized by the standardized latent variable

$$F_j = X_j w_j$$

4.1.2 Internal Estimation

Each block X_j is also summarized by the latent variable

$$z_j = \sum_{k=1,k\neq j}^{J} e_{jk} F_k$$

where the coefficients e_{jk} are computed following two options:

- Factorial scheme: the coefficient e_{jk} is the correlation between F_j and F_k.
- Centroid scheme: the coefficient e_{jk} is the sign of this correlation.

A third option exists, the path-weighting scheme, but is not applicable for the specific path model described in Fig. 4.1 because it takes into account the direction of the arrows.

4.1.3 Computation of the Vector of Weights w_j Using Mode A or Mode B Options

For mode A

The vector of weights w_j is computed by PLS regression of z_j on X_j, using only the first PLS component:

$$w_j \propto X_j^T z_j \tag{4.1}$$

where $\propto$ means that the left term is equal to the right term up to a normalization. In PLS path modeling, the normalization is chosen so that the latent variable $F_j = X_j w_j$ is standardized.

For mode B

The vector of weights w_j is computed by OLS regression of z_j on X_j:

$$w_j \propto (X_j^T X_j)^{-1} X_j^T z_j \tag{4.2}$$

The PLS algorithm

The algorithm is iterative. We begin by an arbitrary choice of weights w_j. In the software PLS-Graph (Chin 2005), the default is to choose all the initial weights equal to 1. We get the external estimations, then the internal ones, choosing between the factor and centroid schemes. Using equation (4.1) if mode A is selected or (4.2) if mode B is preferred, we get new weights. The procedure is iterated until convergence which is always observed in practice.

4.1.4 Some Considerations on the Criteria

For mode A and centroid scheme

Using optimality properties of PLS regression, we can deduce that the weight vector w_j is obtained in two steps:

1. By maximizing the criterion

$$Cov(X_j \widetilde{w}_j, \sum_{k=1,k\neq j}^{J} e_{jk} F_k) \tag{4.3}$$

subject to the constraints $\|\widetilde{w}_j\| = 1$ for all j.

2. By normalizing $\widetilde{w}_j$ in order to obtain a standardized latent variable F_j:

$$w_j = \frac{\widetilde{w}_j}{s_j}$$

where s_j is the standard deviation of $X_j\widetilde{w}_j$.

Criterion (4.3) can also be written as

$$\sum_{k=1,k\neq j}^{J} e_{jk} Cov(X_j\widetilde{w}_j, F_k) = \sqrt{Var(X_j\widetilde{w}_j)} \sum_{k=1,k\neq j}^{J} \left|Cor(X_j\widetilde{w}_j, X_k\widetilde{w}_k)\right| \quad (4.4)$$

We may conclude that PLS path modeling of the causal model of Fig. 4.1, with mode A and centroid scheme, aims at maximizing the following global criterion

$$\sum_{j=1}^{J} \sqrt{Var(X_j\widetilde{w}_j)} \sum_{k=1,k\neq j}^{J} \left|Cor(X_j\widetilde{w}_j, X_k\widetilde{w}_k)\right| \quad (4.5)$$

subject to the constraints $\left\|\widetilde{w}_j\right\| = 1$ for all j. Therefore, we may conclude that the choice of mode A and centroid scheme leads to latent variables that are well explaining their own block and are well correlated (in absolute value) with the other blocks. The properties and the solution of this optimization problem are currently investigated and will be reported elsewhere.

The higher order latent variables are obtained by replacing the blocks X_j by the deflated blocks X_{jm} in the algorithm. Therefore, the latent variables related to one block are standardized and uncorrelated.

For mode B with centroid and factorial schemes

Using two different approaches and practical experience (i.e., computational practice), Mathes (1993) and Hanafi (2007) have shown that use of mode B with centroid scheme leads to a solution that maximizes the criterion

$$\sum_{j,k} \left|Cor(X_j w_j, X_k w_k)\right| \quad (4.6)$$

In the same way, they have concluded that use of mode B with factorial scheme leads to a solution that maximizes the criterion

$$\sum_{j,k} Cor^2(X_j w_j, X_k w_k) \quad (4.7)$$

This last criterion corresponds exactly to the "SsqCor" criterion of Kettenring (1971). Hanafi (2007) has proven the monotone convergence of criteria (4.6) and (4.7) when the Wold's algorithm is used instead of the Lohmöller's one.

The proof of the convergence of the PLS algorithm for mode A is still an open question for a path model with more than two blocks. Nevertheless, when mode B and centroid scheme are selected for each block, Hanafi and Qannari (2005) have proposed a slight modification of the algorithm in order to guarantee a monotone convergence. The modification consists in the replacement of the internal estimation

$$z_j = \sum_{k=1,k\neq j}^{J} sign(Cor(F_j, F_k)) \times F_k$$

by

$$z_j = \sum_{k=1}^{J} sign(Cor(F_j, F_k)) \times F_k$$

This modification does not influence the final result.

The MAXDIFF/MAXBET algorithm

Van de Geer introduced the MAXDIFF method in 1984. It comes to maximize the criterion

$$\begin{aligned} &\sum_{j,k=1,k\neq j}^{J} Cov(X_j\widetilde{w}_j, X_k\widetilde{w}_k) \\ &= \sum_{j,k=1,k\neq j}^{J} \sqrt{Var(X_j\widetilde{w}_j)}\sqrt{Var(X_k\widetilde{w}_k)}Cor(X_j\widetilde{w}_j, X_k\widetilde{w}_k) \end{aligned} \tag{4.8}$$

subject to the constraints $\|\widetilde{w}_j\| = 1$ for all j.

The MAXBET method is a slight modification of the MAXDIFF algorithm. In MAXBET, the following criterion

$$\begin{aligned} &\sum_{j,k=1}^{J} Cov(X_j\widetilde{w}_j, X_k\widetilde{w}_k) \\ &= \sum_{j=1}^{J} Var(X_j\widetilde{w}_j) + \sum_{j,k=1,k\neq j}^{J} \sqrt{Var(X_j\widetilde{w}_j)}\sqrt{Var(X_k\widetilde{w}_k)}Cor(X_j\widetilde{w}_j, X_k\widetilde{w}_k) \end{aligned} \tag{4.9}$$

is maximized instead of (8).

Let's describe the MAXBET algorithm. The algorithm is iterative:

1. Choose arbitrary weight vectors $\widetilde{w}_j$ with unit norm.
2. For each j, the maximum of (4.9) is reached by using PLS regression of $\sum_{k=1}^{J} X_k\widetilde{w}_k$ on X_j. Therefore new weight vectors are defined as

$$\widetilde{w}_j = \frac{X_j^T \sum_{k=1}^{J} X_k \widetilde{w}_k}{\left\| X_j^T \sum_{k=1}^{J} X_k \widetilde{w}_k \right\|}$$

3. The procedure is iterated until convergence.

Proof of the monotonic convergence of the MAXBET algorithm has been initially proposed by Ten Berge (1988). Chu and Watterson (1993) completed this previous property by showing that the MAXBET algorithm always converges. Hanafi and Ten Berge (2003) showed that the computation of the global optimal solution is guaranteed in some specific cases.

The MAXDIFF algorithm is similar to the SUMCOR algorithm (see table 1 below) with the covariance criterion replacing the correlation criterion. It would be rather useful to maximize criteria like

$$\sum_{j,k=1,k\neq j}^{J} \left| Cov(X_j \widetilde{w}_j, X_k \widetilde{w}_k) \right|$$

or

$$\sum_{j,k=1,k\neq j}^{J} Cov^2(X_j \widetilde{w}_j, X_k \widetilde{w}_k)$$

subject to the constraints $\left\| \widetilde{w}_j \right\| = 1$ for all j. The second criterion has recently been introduced by Hanafi and Kiers (2006) as MAXDIFF B criterion. The first criterion appears new. The computation of the solution for both criteria can be performed by using one monotonically convergent general algorithm proposed by Hanafi and Kiers (2006).

4.2 The Hierarchical PLS Path Model

It is rather usual to introduce a super-block X_{J+1} obtained by concatenation of the original blocks $X_1, \ldots, X_J : X_{J+1} = [X_1, \ldots, X_J]$. The hierarchical model proposed by Wold (1982) is described in Fig. 4.2. In this section too, the index 1 is removed for first order weights and latent variables.

Lohmöller (1989) has studied the use of mode A and of the path-weighting scheme for estimating the latent variables of the causal model described in Fig. 4.2. He has shown that a solution of the stationary equations related to this model is obtained for the first standardized principal component Y_{J+1} of the super-block X_{J+1} and for variables Y_j's defined as the standardized fragments of Y_{J+1} related to the various blocks X_j. In practice, he has noted that the PLS algorithm converges toward the first principal component.

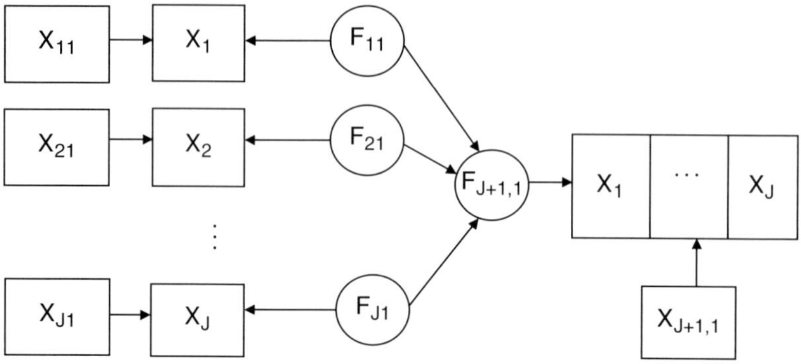

Fig. 4.2 Path model for hierarchical model

Lohmöller has called "Split Principal Component Analysis" this calculation of the first principal component and of its fragments. Let's describe the PLS algorithm for this application.

4.2.1 Use of Mode A with the Path-Weighting Scheme

1. The latent variable F_{J+1} is equal, in practical applications, to the first standardized principal component of the super-block X_{J+1}.
2. The latent variable $F_j = X_j w_j$ is obtained by PLS regression of F_{J+1} on block X_j, using only the first PLS component:

$$w_j \propto X_j^T F_{J+1}$$

So, it is obtained by maximizing the criterion

$$Cov(X_j \widetilde{w}_j, F_{J+1}) \tag{4.10}$$

subject to the constraint $\|\widetilde{w}_j\| = 1$, and standardization of $X_j \widetilde{w}_j$:

$$F_j = X_j w_j,$$

where $w_j = \widetilde{w}_j / s_j$ and s_j is the standard deviation of $X_j \widetilde{w}_j$.
3. We can check that the correlation between F_j and F_{J+1} is positive:

$$w_j \propto X_j^T F_{J+1} \Rightarrow F_j^T F_{J+1} = w_j^T X_j^T F_{J+1} \propto w_j^T w_j > 0$$

4. The ACOM algorithm of Chessel and Hanafi (1996) consists in maximizing the criterion

$$\sum_{j=1}^{J} Cov^2(X_j w_j, X_{J+1} w_{J+1}) \tag{4.11}$$
$$= \sum_{j=1}^{J} Var(X_j w_j) Var(X_{J+1} w_{J+1}) Cor^2(X_j w_j, X_{J+1} w_{J+1})$$

subject to the constraints $\|w_j\| = \|w_{J+1}\| = 1$.

It leads to the first principal component $X_{J+1}w_{J+1}$ of X_{J+1} and to the first PLS component in the PLS regression of $X_{J+1}w_{J+1}$ on X_j. This is exactly the solution that has been obtained above for the hierarchical path model with mode A and path-weighting scheme, up to a normalization. This leads to latent variables that are at the same time well explaining their own block and as positively correlated as possible to the first principal component of the whole data table. The higher order latent variables are obtained by replacing the blocks X_j by the deflated blocks X_{jm} in the algorithm.

4.2.2 Use of Mode B with Centroid and Factorial Schemes

Using the results by Mathes (1993) and Hanafi (2005) on the stationnary equations of the PLS algorithm, and practical experience, it is possible to conclude that use of mode B with centroid scheme leads to a solution that maximizes the criterion

$$\sum_{j=1}^{J} Cor(X_j w_j, X_{J+1} w_{J+1}) \tag{4.12}$$

Furthermore, the optimal solution has the following property:

$$X_{J+1} w_{J+1} \propto \sum_{j=1}^{J} X_j w_j \tag{4.13}$$

This is exactly the SUMCOR criterion proposed by Horst (1961). A known property of this method is that a solution that maximizes (4.12) also maximizes

$$\sum_{j,k=1}^{J} Cor(X_j w_j, X_k w_k) \tag{4.14}$$

In Tenenhaus et al. (2005), it was also shown that use of mode B with factorial scheme leads to a solution that maximizes the criterion

$$\sum_{j=1}^{J} Cor^2(X_j w_j, X_{J+1} w_{J+1})$$

This is exactly the criterion used by Carroll (1968) for generalized canonical correlation analysis.

4.3 Multi-block Analysis Methods and PLS Path Modeling

Several methods for analyzing multi-block data sets, related to PLS path modeling, have been proposed in this paper. It is useful to clarify the place of these methods among the most well known methods for multi-block analysis. In Table 4.1, we summarize methods which optimize a criterion and give, when the case applies, their PLS equivalences. Let's give some explanations on the criteria appearing in table 4.1:

(a) $\lambda_{first}\left[Cor(F_j, F_k)\right]$ is the first eigenvalue of block LV correlation matrix.
(b) $\lambda_{last}\left[Cor(F_j, F_k)\right]$ is the last eigenvalue of block LV correlation matrix.
(c) $\widehat{F}_j$ is the prediction of F in the regression of F on block X_j.
(d) The reduced block number j is obtained by dividing the block X_j by the square root of $\lambda_{first}\left[Cor(x_{jh}, x_{j\ell})\right]$.
(e) The transformed block number j is computed as $X_j[(1/n)X_j^T X_j]^{-1/2}$.

Methods 1–7 are all generalizations of canonical correlation analysis. Method 1 has to be preferred in cases where positively correlated latent variables are sought. The other methods 2–7 will probably give very close results in practical situations. Consequently, PLS path modeling, applied to a confirmatory or hierarchical model, leads to useful LV's summarizing the various blocks of variables.

Methods 8–11 are generalizations of PLS regression. Methods 8 and 9 are only interesting when positively correlated latent variables are sought.

Methods 12 and 14–16 have a common point: the auxiliary variable is the first principal component of a super block obtained by concatenation of the original blocks, or of transformed blocks to make them more comparable. Three of them have a PLS solution. As mode A is equivalent to a PLS regression with one component, it is worth noticing that these methods can be applied in a situation where the number of variables is larger than the number of individuals. Furthermore identical latent variables are obtained when block principal components are used instead of the original variables.

As a final conclusion for this theoretical part, we may consider that PLS path modeling appears to be a unified framework for Multi-block data analysis.

Table 4.1 Multi-block analysis methods with a criterion to be optimized and PLS approach

Method	Criterion	PLS path model	Mode	Scheme
(**1**) SUMCOR (Horst 1961)	$Max \sum_{j,k} Cor(F_j, F_k)$ or $Max \sum_j Cor\left(F_j, \sum_k F_k\right)$	Hierarchical	B	Centroid
(**2**) MAXVAR (Horst 1961) or GCCA (Carroll 1968)	$Max \{\lambda_{first}[Cor(F_j, F_k)]\}$ (**a**) or $Max \sum_j Cor^2(F_j, F_{j+1})$	Hierarchical	B	Factorial
(**3**) SsqCor (Kettenring 1971)	$Max \sum_{j,k} Cor^2(F_j, F_k)$	Confirmatory	B	Factorial
(**4**) GenVar (Kettenring 1971)	$Min \{\det[Cor(F_j, F_k)]\}$			
(**5**) MINVAR (Kettenring 1971)	$Min \{\lambda_{last}[Cor(F_j, F_k)]\}$ (**b**)			
(**6**) Lafosse (1989)	$Max \sum_j Cor^2\left(F_j, \sum_k F_k\right)$			
(**7**) Mathes (1993) or Hanafi (2005)	$Max \sum_{j,k} \lvert Cor(F_j, F_k)\rvert$	Confirmatory	B	Centroid
(**8**) MAXDIFF (Van de Geer, 1984 & Ten Berge, 1988)	$Max_{all\ \lVert w_j \rVert = 1} \sum_{j \neq k} Cov(X_j w_j, X_k w_k)$			
(**9**) MAXBET (Van de Geer, 1984 & Ten Berge, 1988)	$Max_{all\ \lVert w_j \rVert = 1} \sum_{j,k} Cov(X_j w_j, X_k w_k)$			
(**10**) MAXDIFF B (Hanafi and Kiers 2006)	$Max_{all\ \lVert w_j \rVert = 1} \sum_{j \neq k} Cov^2(X_j w_j, X_k w_k)$			
(**11**) (Hanafi and Kiers 2006)	$Max_{all\ \lVert w_j \rVert = 1} \sum_{j \neq k} \lvert Cov(X_j w_j, X_k w_k)\rvert$			
(**12**) ACOM (Chessel and Hanafi 1996) or Split PCA (Lohmöller 1989)	$Max_{all\ \lVert w_j \rVert = 1} \sum_j Cov^2(X_j w_j, X_{j+1} w_{j+1})$ or $Min_{F,p_j} \sum_j \left\lVert X_j - F p_j^T \right\rVert^2$	Hierarchical	A	Path-weighting
(**13**) CCSWA (Hanafi et al., 2006) or HPCA (Wold et al., 1996)	$Max_{all\ \lVert w_j \rVert = 1, Var(F)=1} \sum_j Cov^4(X_j w_j, F)$ or $Min_{\lVert F \rVert = 1} \sum_j \left\lVert X_j X_j^T - \lambda_j F F^T \right\rVert^2$			
(**14**) Generalized PCA (Casin 2001)	$Max \sum_j R^2(F, X_j) \sum_h Cor^2\left(x_{jh}, \hat{F}_j\right)$ (**c**)			
(**15**) MFA (Escofier and Pagès 1994)	$Min_{F,p_j} \sum_j \left\lVert \frac{1}{\sqrt{\lambda_{first}\left[Cor(x_{jh}, x_{jl})\right]}} X_j - F p_j^T \right\rVert^2$	Hierarchical (applied to the reduced X_j) (**d**)	A	Path-weighting
(**16**) Oblique maximum variance method (Horst 1965)	$Min_{F,p_j} \sum_j \left\lVert X_j \left(\frac{1}{n} X_j^T X_j\right)^{-1/2} - F p_j^T \right\rVert^2$	Hierarchical (applied to the transformed X_j) (**e**)	A	Path-weighting

4.4 Application to Sensory Data

In this section we are going to present in details the application on a practical example of one method described in the previous sections: PLS confirmatory factor analysis with mode A and centroid scheme. We will also mention more briefly the MAXDIFF/MAXBET algorithms and PLS hierarchical model with mode A and path-weighting scheme. On these data they practically yield the same latent variable estimates as the PLS confirmatory factor analysis. We have used sensory data about wine tasting that have been collected by C. Asselin and R. Morlat and are fully described in Escofier and Pagès (1988). This section can be considered as a tutorial on how to use PLS-Graph (Chin 2005) for the analysis of multi-block data.

4.4.1 Data Description

A set of 21 red wines with Bourgueil, Chinon and Saumur origins are described by 27 variables grouped into four blocks:

X_1 = *Smell at rest*

Rest1 = smell intensity at rest, Rest2 = aromatic quality at rest, Rest3 = fruity note at rest, Rest4 = floral note at rest, Rest5 = spicy note at rest

X_2 = *View*

View1 = visual intensity, View2 = shading (from orange to purple), View3 = surface impression

X_3 = *Smell after shaking*

Shaking1 = smell intensity, Shaking2 = smell quality, Shaking3 = fruity note, Shaking4 = floral note, Shaking5 = spicy note, Shaking6 = vegetable note, Shaking7 = phenolic note, Shaking8 = aromatic intensity in mouth, Shaking9 = aromatic persistence in mouth, Shaking10 = aromatic quality in mouth

X_4 = *Tasting*

Tasting1 = intensity of attack, Tasting2 = acidity, Tasting3 = astringency, Tasting4 = alcohol, Tasting5 = balance (acidity, astringency, alcohol), Tasting6 = mellowness, Tasting7 = bitterness, Tasting8 = ending intensity in mouth, Tasting9 = harmony

Two other variables are available and will be used as illustrative variables: (1) the global quality of the wine and (2) the soil with four categories, soil 3 being the reference one for this kind of wine. These data have already been analyzed by PLS and GPA in Tenenhaus and Esposito Vinzi (2005).

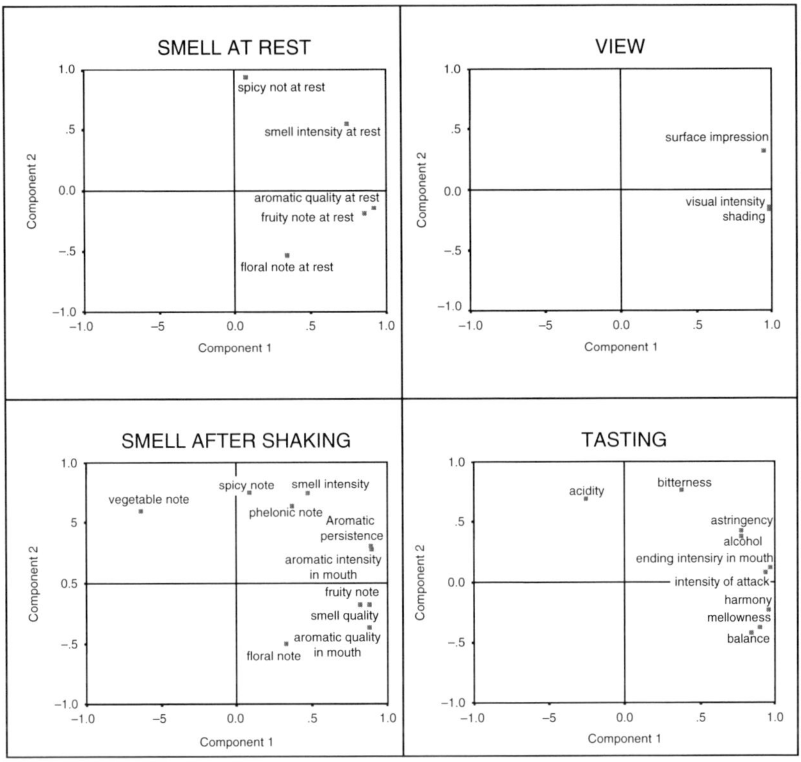

Fig. 4.3 Loading plots for PCA of each block

4.4.2 *Principal Component Analysis of Each Block*

PCA of each block is an essential first step for the analysis of multi-block data. The loading plots for each block are given in Fig. 4.3. The View block is one-dimensional, but the other blocks are two-dimensional.

4.4.3 *PLS Confirmatory Factor Analysis*

We have used the PLS-Graph software (Chin 2005), asking for mode A, centroid scheme and two dimensions.

4.4.3.1 Study of Dimension 1

The causal model is described in Fig. 4.4. The correlations between the first order latent variables are given in Table 4.2 and the other results in Tables 4.3, 4.4 and 4.5.

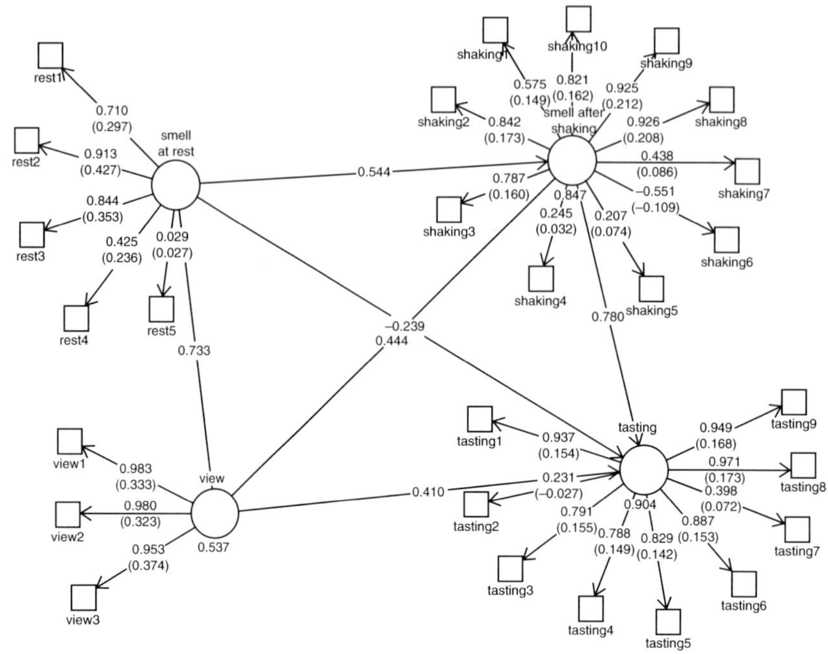

Fig. 4.4 PLS confirmatory factor analysis for wine data (dim. 1)

Table 4.2 Correlations between the first order latent variables

	Smell at rest	View	Smell after shaking	Tasting
Smell at rest	1.000			
View	0.733	1.000		
Smell after shaking	0.870	0.843	1.000	
Tasting	0.739	0.892	0.917	1.000

Table 4.3 Results for the first dimension (Inner model)

Inner Model		
Block	Mult. RSq[(a)]	AvCommun[(b)]
Smell at rest	0.7871	0.4463
View	0.8077	0.9449
Smell after shaking	0.9224	0.4646
Tasting	0.9039	0.6284
Average	0.8553	0.5692

(a) R^2 of each LV with all the other LVs, not a standard output of PLS-Graph

(b) Average gives the average of block communalities weighted by the number of MV by block

Table 4.4 Results for the first dimension (Outer model)

	Outer Model		
Variable	Weight[(a)]	Loading[(b)]	Communality[(c)]
Smell at rest			
rest1	0.2967	0.7102	0.5044
rest2	0.4274	0.9132	0.8340
rest3	0.3531	0.8437	0.7118
rest4	0.2362	0.4247	0.1804
rest5	0.0268	0.0289	0.0008
View			
view1	0.3333	0.9828	0.9660
view2	0.3229	0.9800	0.9604
view3	0.3735	0.9531	0.9085
Smell after shaking			
shaking1	0.1492	0.5745	0.3300
shaking2	0.1731	0.8422	0.7094
shaking3	0.1604	0.7870	0.6194
shaking4	0.0324	0.2448	0.0599
shaking5	0.0735	0.2069	0.0428
shaking6	−0.1089	−0.5515	0.3042
shaking7	0.0857	0.4377	0.1916
shaking8	0.2081	0.9263	0.8581
shaking9	0.2119	0.9250	0.8556
shakin10	0.1616	0.8214	0.6748
Tasting			
tasting1	0.1537	0.9373	0.8786
tasting2	−0.0270	−0.2309	0.0533
tasting3	0.1545	0.7907	0.6252
tasting4	0.1492	0.7883	0.6215
tasting5	0.1424	0.8292	0.6876
tasting6	0.1529	0.8872	0.7872
tasting7	0.0719	0.3980	0.1584
tasting8	0.1733	0.9709	0.9426
tasting9	0.1678	0.9494	0.9013

a. Weights of standardized original MV for LV 1 construction
b. Correlation between original MV and LV 1
c. Communality = R^2 between MV and first LV

The communalities are the square of the correlations between the manifest variables and the first dimension latent variable of their block. The four latent variables F_{j1} are well correlated with the variables related to the first principal components of each block.

The quality of the causal model described in figure 4 can be measured by a Goodness-of-Fit (GoF) index. It is defined by the formula

Table 4.5 First dimension latent variables

	Latent variables			
	Smell at rest	View	Smell after shaking	Tasting
2EL	0.224	0.522	0.146	−0.425
1CHA	−0.904	−1.428	−1.060	−0.730
1FON	−0.946	−0.721	−0.653	−0.176
1VAU	−2.051	−2.136	−2.303	−2.290
1DAM	2.290	0.742	1.460	0.963
2BOU	−0.391	0.966	0.325	0.801
1BOI	1.029	0.338	0.937	0.815
3EL	−0.533	0.105	0.255	0.433
DOM1	−0.796	0.292	0.185	0.121
1TUR	−0.980	−0.458	−0.521	−0.527
4EL	0.436	−0.007	0.522	0.536
PER1	0.639	1.151	0.400	0.506
2DAM	0.975	0.764	0.915	0.929
1POY	0.204	1.327	0.522	1.174
1ING	0.648	0.557	0.592	0.632
1BEN	0.248	−0.286	0.007	0.245
2BEA	1.055	0.067	1.428	0.297
1ROC	−0.355	−0.374	−0.098	−0.149
2ING	−1.660	−2.606	−2.559	−2.961
T1	0.791	0.604	−0.135	−0.375
T2	0.076	0.579	−0.365	0.180

$$\begin{aligned} GoF(1) &= \sqrt{\frac{1}{\sum_{j=1}^{p} k_j} \sum_{j=1}^{J} \sum_{k=1}^{k_j} Cor^2(x_{jk}, F_{j1})} \times \sqrt{\frac{1}{J} \sum_{j=1}^{J} R^2(F_{j1}; \{F_{k1}, k \neq j\})} \\ &= \sqrt{AvCommun(1)} \times \sqrt{Average\ Mult.RSq(1)} \\ &= \sqrt{0.5692 \times 0.8553} = 0.6977 \end{aligned} \tag{4.15}$$

where $AvCommun(1)$ and $Average\ Mult.RSq(1)$ are given in table 4.3.

The first term of the product measures the quality of the outer model and the second term the one of the inner model. The GoF index for the model described in figure 4 and for dimension 1 is equal to 0.6977.

Using the bootstrap procedure of PLS-Graph (results not shown), we have noticed that the weights related to rest5 (Spicy note at rest), shaking4 (Floral note), shaking5 (Spicy note), shaking7 (Phenolic note), tasting2 (Acidity) and tasting7 (Bitterness) are not significant ($|t| < 2$). It may be noted on figure 3 that these items are precisely those that are weakly contributing to component 1 and highly contributing to component 2 in the PCA of each block.

4.4.3.2 Study of Dimension 2

The second order latent variables are now computed on the deflated blocks X_{j1}. The results built on these blocks, but expressed in term of the original variables, are shown on Fig. 4.5. We obtain a new set of latent variables $F_{j2}, j = 1, \ldots, 4$. The correlations between the LV are given in table 4.6. We may notice that the second latent variable for the view block is weakly correlated to the other second order latent variables. The other results are given in Tables 4.7 and 4.8. The average communalities express the proportion of variance of each block explained by the two block latent variables.

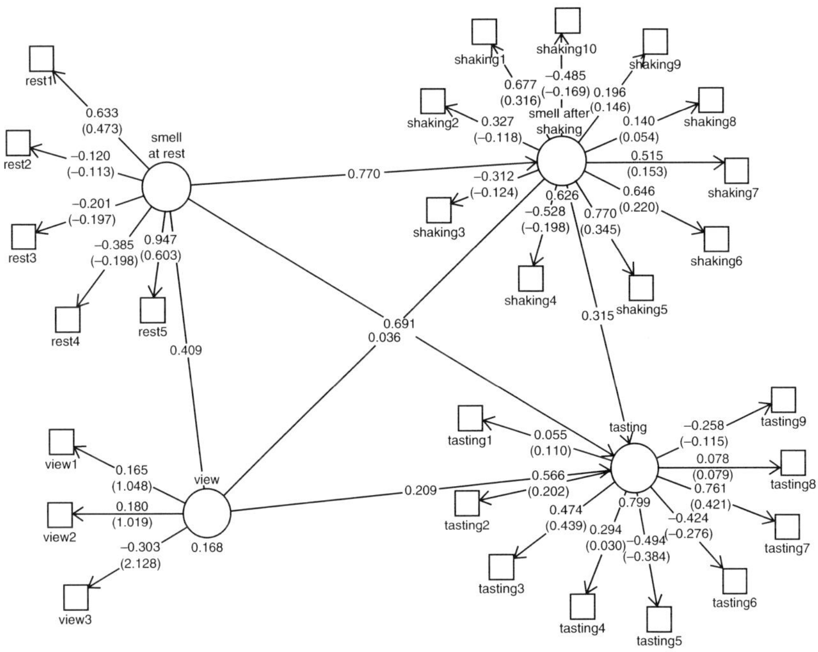

Fig. 4.5 PLS confirmatory factor analysis for wine data (dim. 2)

Table 4.6 Correlations between the second order latent variables

	Smell at rest	View	Smell after shaking	Tasting
Smell at rest	1.000			
View	0.409	1.000		
Smell after shaking	0.791	0.354	1.000	
Tasting	0.854	0.185	0.787	1.000

Table 4.7 Results for the second dimension (Inner model)

Inner Model		
Block	Mult.RSq	AvCommun
Smell at rest	0.8071	0.7465
View	0.2972	0.9953
Smell after shaking	0.6844	0.7157
Tasting	0.7987	0.8184
Average	0.6469	0.7867

Table 4.8 Results for the second dimension (Outer model)

Outer Model			
Variable	Weight(a)	Loading(b)	Communality(c)
Smell at rest			
rest1	0.4729	0.6335	0.9057
rest2	−0.1128	−0.1202	0.8484
rest3	−0.1971	−0.2014	0.7524
rest4	−0.1977	−0.3845	0.3283
rest5	0.6032	0.9469	0.8975
View			
view1	1.0479	0.1648	0.9932
view2	1.0192	0.1798	0.9927
view3	−2.1285	−0.3026	1.0000
Smell after shaking			
shaking1	0.3161	0.6772	0.7886
shaking2	−0.1179	−0.3269	0.8162
shaking3	−0.1235	−0.3120	0.7168
shaking4	−0.1977	−0.5283	0.3390
shaking5	0.3449	0.7701	0.6359
shaking6	0.2199	0.6459	0.7214
shaking7	0.1529	0.5153	0.4572
shaking8	0.0537	0.1401	0.8777
shaking9	0.1459	0.1961	0.8940
shaking10	−0.1686	−0.4853	0.9103
Tasting			
tasting1	0.1096	0.0554	0.8817
tasting2	0.2017	0.5658	0.3735
tasting3	0.4391	0.4739	0.8498
tasting4	0.0302	0.2935	0.7076
tasting5	−0.3838	−0.4943	0.9319
tasting6	−0.2756	−0.4239	0.9668
tasting7	0.4213	0.7611	0.7376
tasting8	0.0789	0.0781	0.9487
tasting9	−0.1145	−0.2578	0.9678

a. Weights of standardized original MV for LV2 construction
b. Correlation between original MV and LV2
c. Communality = R^2 between MV and two first LV's

Table 4.9 Results for the second dimension latent variables

	Latent variables			
	Smell at rest	View	Smell after shaking	Tasting
2EL	−0.436	0.556	−0.744	−0.943
1CHA	−0.909	−0.019	−1.077	−0.866
1FON	0.168	−0.091	−0.687	−1.012
1VAU	0.695	1.018	0.504	1.393
1DAM	0.171	−0.085	−0.313	−0.157
2BOU	−0.067	0.260	−0.953	0.443
1BOI	−0.145	0.219	−0.174	−0.171
3EL	0.625	1.540	1.631	−0.008
DOM1	0.008	−0.291	−0.470	−0.506
1TUR	−0.708	−0.595	−0.176	−0.294
4EL	0.199	−0.990	0.258	0.615
PER1	0.174	1.933	0.386	0.279
2DAM	−0.932	−0.981	−0.085	−0.939
1POY	−0.704	1.156	−0.011	−0.673
1ING	−0.448	−1.636	−0.489	0.217
1BEN	−0.309	−0.417	−1.150	−0.713
2BEA	−1.599	−1.967	0.029	−0.275
1ROC	−0.236	−1.298	−0.809	−0.096
2ING	−0.699	0.747	−0.543	−1.125
T1	2.112	0.088	2.396	1.950
T2	3.039	0.854	2.477	2.882

The GoF index for this second model is defined as:

$$
\begin{aligned}
GoF(2) &= \sqrt{\frac{1}{\sum_{j=1}^{p} k_j} \sum_{j=1}^{J} \sum_{k=1}^{k_j} Cor^2(x_{jk}, F_{j2}) \times} \sqrt{\frac{1}{J} \sum_{j=1}^{J} R^2(F_{j2}; \{F_{k2}, k \neq j\})} \\
&= \sqrt{AvCommun(2) - AvCommun(1)} \times \sqrt{Average\ Mult.RSq(2)} \\
&= \sqrt{(0.7867 - 0.5692) \times 0.6469} = 0.3751
\end{aligned}
$$

where $AvCommun(2)$ and $Average\ Mult.RSq(2)$ are given in table 4.7. This formula comes from the definition of $AvCommun(2)$ and from the fact that the latent variables F_{j1} and F_{j2} are uncorrelated:

$$
\begin{aligned}
&AvCommun(2) \\
&= \frac{1}{\sum_{j=1}^{p} k_j} \sum_{j=1}^{J} \sum_{k=1}^{k_j} R^2(x_{jk}; F_{j1}, F_{j2}) \\
&= \frac{1}{\sum_{j=1}^{p} k_j} \sum_{j=1}^{J} \sum_{k=1}^{k_j} Cor^2(x_{jk}, F_{j1}) + \frac{1}{\sum_{j=1}^{p} k_j} \sum_{j=1}^{J} \sum_{k=1}^{k_j} Cor^2(x_{jk}, F_{j2})
\end{aligned}
$$

The data can be visualized in a global component space by using the first principal component of the four first order components $F_{11}, F_{21}, F_{31}, F_{41}$ and the first principal component of the three second order components F_{12}, F_{32}, F_{42}. We have not used the second component of the view block because this component is not related with the other second components (Table 4.6). This graphical display is given in Fig. 4.6. The loading plot is given in Fig. 4.7. The various mapping (F_{j1}, F_{j2}) are given in Fig. 4.8.

Discussion

From global criterion (4.5), tables 2 and 6, PLS Confirmatory factor analysis comes here to carry out a kind of principal component analysis on each block such that the same order components are as positively correlated as possible. So, for each dimension h, the interpretations of the various block components $F_{jh}, j = 1, \ldots, J$ can be related.

In table 4.10 and in figure 4.7 the "Smell at rest", "View", "Smell after shaking" and "Tasting" loadings with the global components are displayed. It makes sense as the correlations of the variables with the block components and the global components are rather close. The global quality judgment on the wines has been displayed as an illustrative variable. This loading plot is quite similar to the one obtained by multiple factor analysis (Escofier and Pagès 1988, p. 117). So, we may keep their interpretation of the global components.

The first dimension is related with "Harmony" and "Intensity". For this kind of wine, it is known that these wine characteristics are closely related. The second dimension is positively correlated with "Bitterness", "Acidity", "Spicy" and "Vegetable" notes and negatively correlated with "Floral" note. Soil however is

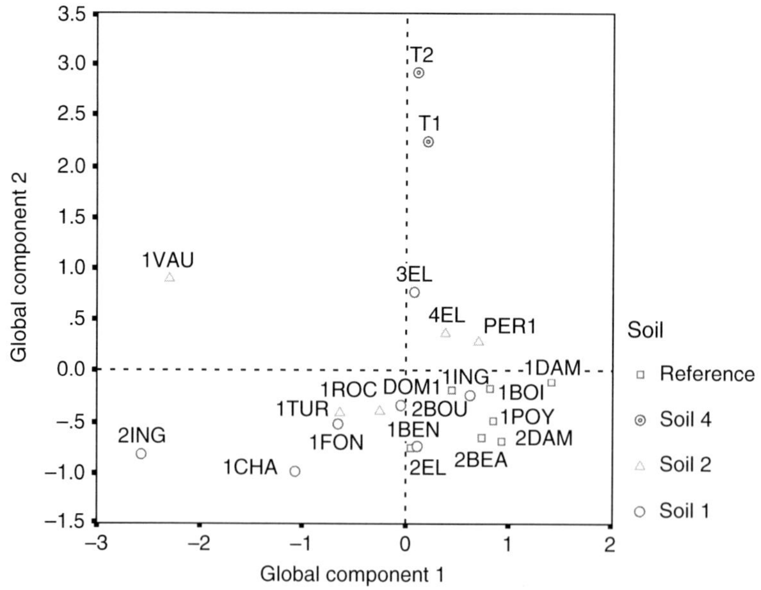

Fig. 4.6 Wine and soil visualization in the global component space

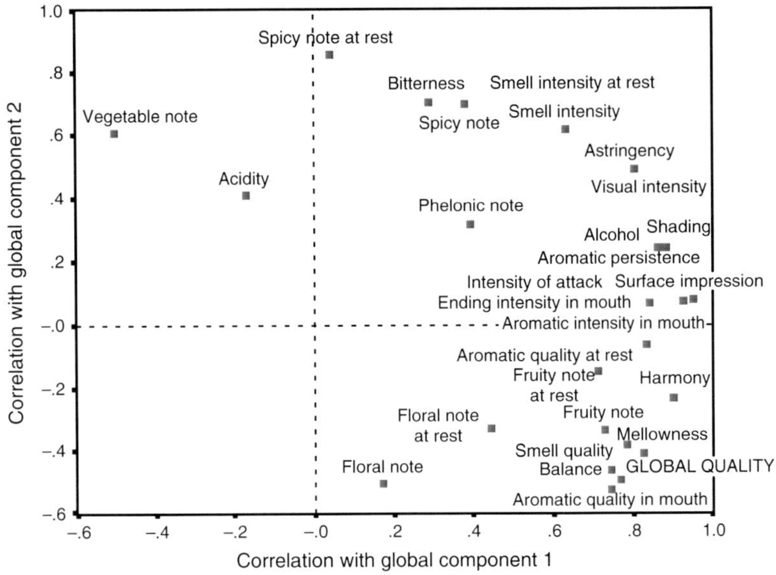

Fig. 4.7 Loading plot for the wine data

very predictive of the quality of wine: an analysis of variance of the global quality judgment on the soil factor leads to $F = 5.327$ with p-value $= .009$. This point is illustrated in figure 6. All the reference soils are located in the "good" quadrant. It can also be noted that the second dimension is essentially due to two wines from soil 4: T1 and T2. They are in fact the same wine presented twice to the tasters. In an open question on aroma recognition, aromas "mushrooms" and "underwood" were specifically mentioned for this wine.

4.4.4 *Use of the MAXDIFF/MAXBET Algorithms*

On this example, the PLS confirmatory factor analysis model and MAXDIFF/MAXBET give practically the same latent variables for the various blocks. The correlations between the latent variables on the same block for both approaches are all above .999. So it is not necessary to go further on this approach.

4.4.5 *Use of Hierarchical PLS Path Model*

The causal model estimated with mode A and path-weighting scheme is described in figure 4.9. The correlations between the latent variables are given in table 4.11.

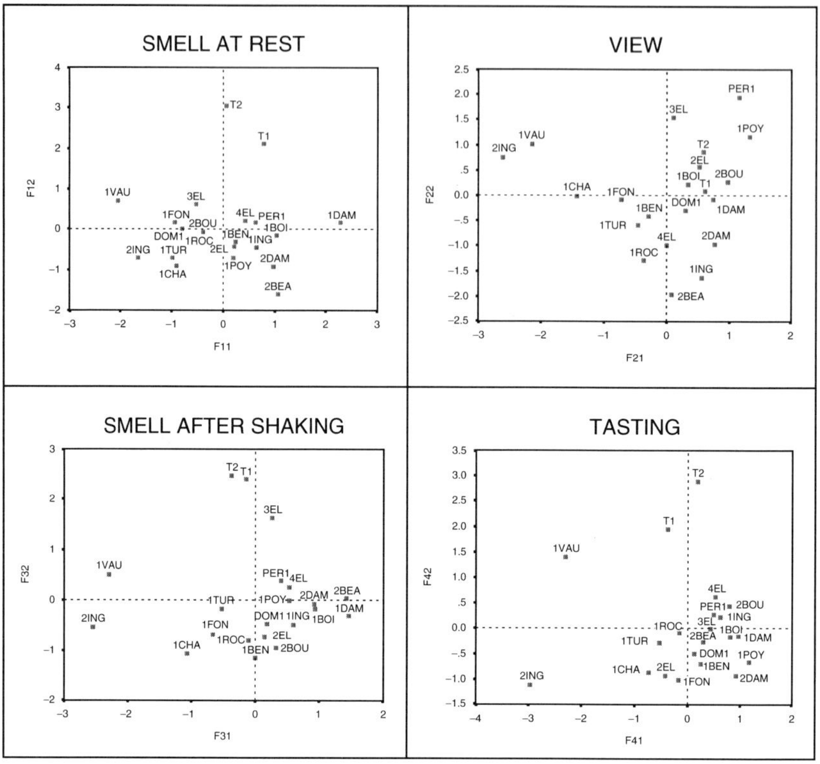

Fig. 4.8 Wine visualization with respect to four aspects

On this example, the PLS hierarchical model and the PLS confirmatory factor analysis model give the same latent variables for the various blocks. The correlations between the latent variables on the same block for both approaches are all above .999. The correlation between the first principal component of the four first order components of the PLS confirmatory factor analysis and the global score of the hierarchical PLS path model is equal to .995. So it is not necessary to go further on this approach.

4.5 Conclusion

There were two objectives in this paper. The first one was to show how PLS path modeling is a unified framework for the analysis of multi-block data. The second one was to give a tutorial on the use of PLS-Graph for multi-block data analysis. We can now give some guidelines for the selection of a method. There are three types of methods with respect to the unified general framework: (1) generalized canonical correlation analysis, (2) generalized PLS regression and (3) split-PCA.

Table 4.10 Correlations between the original variables and the global components 1 and 2 (Variables clearly related to second dimension are in italic)

	Correlation with global component 1	Correlation with global component 2
Smell intensity at rest (Rest1)	0.60	0.68
Aromatic quality at rest (Rest2)	0.83	−0.07
Fruity note at rest (Rest3)	0.71	−0.15
Floral note at rest (Rest4)	0.44	−0.33
Spicy note at rest (Rest5)	*0.04*	*0.86*
Visual intensity (View1)	0.88	0.24
Shading (View2)	0.86	0.24
Surface impression (View3)	0.95	0.08
Smell intensity (Shaking1)	0.63	0.62
Smell quality (Shaking2)	0.78	−0.38
Fruity note (Shaking3)	0.73	−0.34
Floral note (Shaking4)	*0.17*	*−0.50*
Spicy note (Shaking5)	*0.29*	*0.70*
Vegetable note (Shaking6)	*−0.50*	*0.61*
Phelonic note (Shaking7)	0.39	0.32
Aromatic intensity in mouth (Shaking8)	0.92	0.02
Aromatic persistence in mouth (Shaking9)	0.93	0.14
Aromatic quality in mouth (Shaking10)	0.74	−0.53
Intensity of attack (Tasting1)	0.84	0.07
Acidity (Tasting2)	*−0.17*	*0.41*
Astringency (Tasting3)	0.80	0.49
Alcohol (Tasting4)	0.78	0.22
Balance (Tasting5)	0.77	−0.50
Mellowness (Tasting6)	0.83	−0.41
Bitterness (Tasting7)	*0.38*	*0.70*
Ending intensity in mouth (Tasting8)	0.93	0.07
Harmony (Tasting9)	0.90	−0.23
GLOBAL QUALITY	0.74	−0.46

If the main objective is to obtain high correlations in absolute value between factors, mode B has to be preferred and methods number 2, 3, or 7 mentioned in table 1 will probably give very close results. If positive correlations are wished, then method number 1 is advised: PLS-graph appears to be a software where SUMCOR Horst's algorithm is available. For data with many variables and high multicolinearity inside the blocks, it is preferable (and mandatory when the number of variables is larger than the number of individuals) to use a generalized PLS regression method. The ACOM Chessel & Hanafi's algorithm seems to be the most attractive one and is easy to implement with PLS-graph (hierarchical PLS path model with mode A and path weighting scheme). Furthermore, ACOM will give the same results using the original MV's or the block principal components. That means that ACOM can still be used when the number of variables is extremely high. Multi-block analysis is very common in sensory analysis. We have given a detailed application in this field. We have commented the various outputs of PLS-Graph so that the reader should be

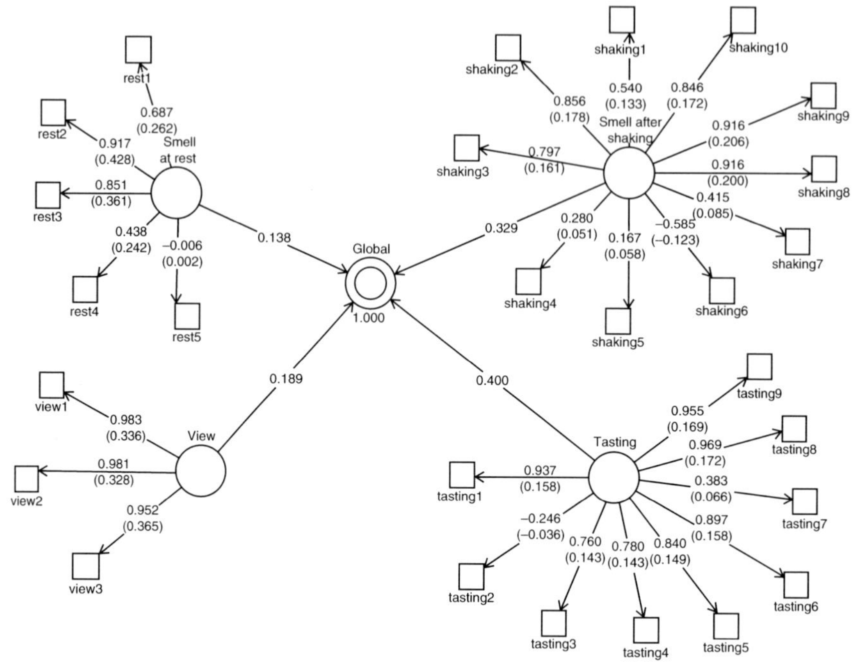

Fig. 4.9 Hierarchical PLS path modeling of the wine data

Table 4.11 Correlations between the first LV's for the hierarchical PLS model

	Smell at rest	View	Smell after shaking	Tasting
Smell at rest	1.000			
View	0.726	1.000		
Smell after shaking	0.866	0.828	1.000	
Tasting	0.736	0.887	0.917	1.00
Global	0.855	0.917	0.972	0.971

able to re-apply these methods for him(her)self. As a final conclusion to this paper, we mention our conviction that PLS path modeling will become a standard tool for multi-block analysis. We hope that this paper will contribute to reach this objective.

References

Carroll, J. D. (1968). A generalization of canonical correlation analysis to three or more sets of variables", *Proceedings of the 76th Convenction of American Psychological Association*, pp. 227–228.

Casin, P. A. (2001). A generalization of principal component analysis to K sets of variables. *Computational Statistics & Data Analysis*, *35*, 417–428.

Chin, W. W. (2005). *PLS-Graph User's Guide, C.T. Bauer College of Business*. USA: University of Houston.

Chessel, D. & Hanafi M. (1996). Analyses de la Co-inertie de K nuages de points. *Revue de Statistique Appliquée*, *44*(2), 35–60.

Chu, M. T. & Watterson, J. L (1993). On a multivariate eigenvalue problem, Part I : Algebraic theory and a power method. *SIAM Journal on Scientific Computing*, *14*,(5), 1089–1106.

Escofier, B. & Pagès J. (1988). *Analyses factorielles simples et multiples*. Paris: Dunod.

Escofier, B. & Pagès J. (1994). Multiple factor analysis, (AFMULT package), *Computational Statistics & Data Analysis*, *18*, 121–140.

Hanafi, M. (2007). PLS path modelling: Computation of latent variables with the estimation mode B. *Computational Statistics*, *22*(2), 275–292.

Hanafi, M. & Kiers H. A. L (2006). Analysis of K sets of Data with differential emphasis between and within sets. *Computational Statistics and Data Analysis*, *51*(3), 1491–1508.

Hanafi, M., Mazerolles G., Dufour E. & Qannari E. M. (2006). Common components and specific weight analysis and multiple co-inertia analysis applied to the coupling of several measurement techniques. *Journal of Chemometrics*, *20*, 172–183.

Hanafi, M. & Qannari E. M. (2005). An alternative algorithm to PLS B problem. *Computational Statistics and Data Analysis*, *48*, 63–67.

Hanafi, M. & Ten Berge, J. M. F (2003). Global optimality of the MAXBET Algorithm. *Psychometrika*, *68*(1), 97–103.

Horst, P. (1961). Relations among m sets of variables, *Psychometrika*, *26*, 126–149.

Horst, P. (1965). *Factor analysis of data matrices*. Holt, Rinehart and Winston: New York.

Kettenring, J. R. (1971). Canonical analysis of several sets of variables. *Biometrika*, *58*, 433–451.

Lafosse, R. (1989). Proposal for a generalized canonical analysis. In R. Coppi & S. Bolasco (Eds.), *Multiway data analysis* (pp. 269–276). Amsterdam: Elsevier Science.

Lohmöller, J. B. (1989). *Latent Variables Path Modeling with Partial Least Squares*, Physica-Verlag, Heildelberg.

Long, J. S. (1983). *Confirmatory Factor Analysis*, Series: Quantitative Applications in the Social Sciences, Thousand Oaks, CA: Sage.

Mathes, H. (1993). Global optimisation criteria of the PLS-algorithm in recursive path models with latent variables. In K. Haagen, D. J. Bartholomew, M. Deister (Eds), *Statistical modelling and latent variables*. Amsterdam: Elsevier.

Morrison, D. F. (1990). *Multivariate statistical methods*, New York: McGrawHill.

Ten Berge, J. M. F (1988). Generalized approaches to the MAXBET and the MAXDIFF problem with applications to canonical correlations. *Psychometrika*, *42*, 1, 593–600.

Tenenhaus, M. & Esposito Vinzi V. (2005). PLS regression, PLS path modeling and generalized Procustean analysis: a combined approach for multiblock analysis. *Journal of Chemometrics*, *19*, 145–153.

Tenenhaus, M., Esposito Vinzi V., Chatelin Y.-M., Lauro C. (2005). PLS path modeling. *Computational Statistics & Data Analysis*, *48*, 159–205.

Van de Geer, J.P. (1984). Linear relations among K sets of variables, *Psychometrika*, *49*, 1, 79–94.

Wold, H. (1982). Soft modeling: The basic design and some extensions, in *Systems under indirect observation, Part 2*, K.G. Jöreskog & H. Wold (Eds), North-Holland, Amsterdam, 1–54.

Wold, H. (1985). Partial Least Squares. In *Encyclopedia of Statistical Sciences*, vol. 6, S. Kotz and N.L. Johnson (Eds), John Wiley & Sons, New York, 581–591.

Wold, S., Kettaneh N. & Tjessem K. (1996). Hierarchical multiblock PLS and PC models for easier model interpretation and as an alternative to variable selection. *Journal of Chemometrics*, *10*, 463–482.

Chapter 5
Use of ULS-SEM and PLS-SEM to Measure a Group Effect in a Regression Model Relating Two Blocks of Binary Variables

Michel Tenenhaus, Emmanuelle Mauger, and Christiane Guinot

Abstract The objective of this paper is to describe the use of unweighted least squares (ULS) structural equation modeling (SEM) and partial least squares (PLS) path modeling in a regression model relating two blocks of binary variables, when a group effect can influence the relationship. Two sets of binary variables are available. The first set is defined by one block X of predictors and the second set by one block Y of responses. PLS regression could be used to relate the responses Y to the predictors X, taking into account the block structure. However, for multi-group data, this model cannot be used because the path coefficients can be different from one group to another. The relationship between Y and X is studied in the context of structural equation modeling. A group effect A can affect the measurement model (relating the manifest variables (MVs) to their latent variables (LVs)) and the structural equation model (relating the Y-LV to the X-LV). In this paper, we wish to study the impact of the group effect on the structural model only, supposing that there is no group effect on the measurement model. This approach has the main advantage of allowing a description of the group effect (main and interaction effects) at the LV level instead of the MV level. Then, an application of this methodology on the data of a questionnaire investigating sun exposure behavior is presented.

M. Tenenhaus
Department SIAD, HEC School of Management, 1 rue de la Libération, 78351 Jouy-en-Josas, France
e-mail: tenenhaus@hec.fr

E. Mauger
Biometrics and Epidemiology unit, CERIES, 20 rue Victor Noir, 92521 Neuilly sur Seine, France
e-mail: emmanuelle.mauger@ceries-lab.com

C. Guinot
Biometrics and Epidemiology unit, CERIES, 20 rue Victor Noir, 92521 Neuilly sur Seine, France
e-mail: christiane.guinot@ceries-lab.com
and
Computer Science Laboratory, Ecole Polytechnique, University of Tours, France

V. Esposito Vinzi et al. (eds.), *Handbook of Partial Least Squares*, Springer Handbooks of Computational Statistics, DOI 10.1007/978-3-540-32827-8_6,

5.1 Introduction

The objective of this paper is to describe the use of unweighted least squares structural equation modeling (ULS-SEM) and partial least squares path modeling (PLS-SEM) in a regression model relating a response block Y to a predictor block X, when a group effect A can affect the relationship. A structural equation relates the response latent variable (LV) η associated with block Y to the predictor latent variable ξ associated with block X, taking into account the group effect A. In usual applications, the group effect acts on the measurement model as well as on the structural model. In this paper, we wish to study the impact of the group effect on the structural model only, supposing that there is no group effect on the measurement model. This constraint is easy to implement in ULS-SEM, but not in PLS-SEM. This approach has the main advantage of allowing a description of the group effect (main and interaction effects) at the LV level instead of the manifest variable level. We propose a four-step methodology: (1) Use of ULS-SEM with constraints on the measurement model, (2) LV estimates are computed in the framework of PLS: the outer LV estimates $\widehat{\xi}$ and $\widehat{\eta}$ are computed using mode A and, as inner LV estimates, the ULS-SEM LVs, (3) Analysis of covariance relating the dependent LV $\widehat{\eta}$ to the independent terms $\widehat{\xi}$, A (main effect) and $A * \widehat{\xi}$ (interaction effect), and (4) Tests on the structural model, using bootstrapping.

These methods were applied on the data of a questionnaire investigating sun exposure behavior addressed to a cohort of French adults in the context of the SU.VI.MAX epidemiological study. Sun protection behavior was described according to gender and class of age (less than 50 at inclusion in the study versus more or equal to 50). This paper illustrates the various stages in the construction of latent variables, also called scores, based on qualitative data.

5.2 Theory

Chin, Marcolin and Newsted(2003) proposed to use the PLS approach to relate the response block Y to the predictor block X with a main effect A and an interaction term $A * X$ added to the model as described in Fig. 5.1. In this example, the group variable A has two values, and A_1 and A_2 are two dummy variables describing these values. Ping (1995) has studied the same model in the LISREL context. A path model equivalent to the one described in Fig. 5.1 is given in Fig. 5.2, where the redundant manifest variables have been removed. This model in Fig. 5.2 seems easier to estimate using ULS procedure than the model shown in Fig. 5.1, after removal of the redundant MVs: a negative variance estimate has been encountered in the presented application for the Fig. 5.1 model, and not for the Fig. 5.2 model. The study of the path coefficients related to the arrows connecting $X * A_1$, $X * A_2$ and A to Y in Fig. 5.2 gives some insight on the main group effect A and on the interaction effect $A * X$. However, this model can be misleading because the blocks $X * A_h$

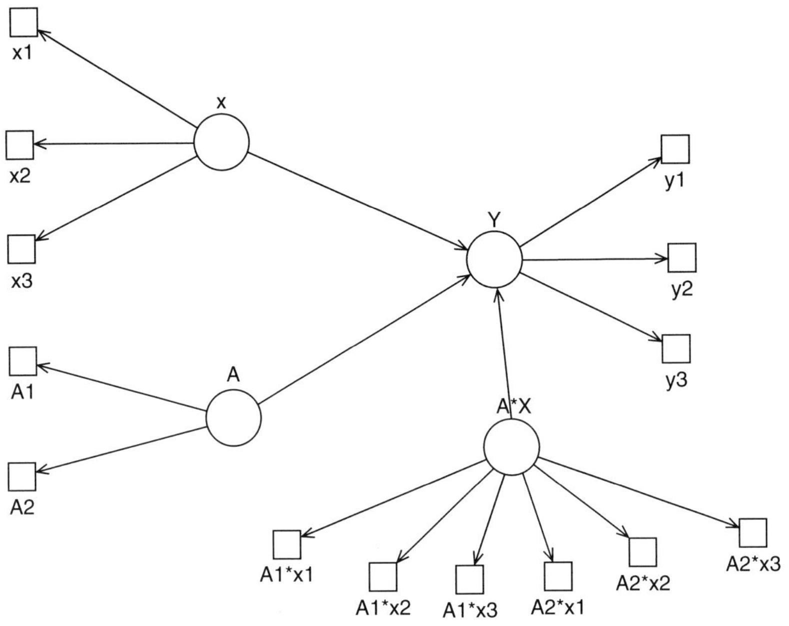

Fig. 5.1 Two-block regression model with a group effect (Ping 1995; Chin et al. 2003)

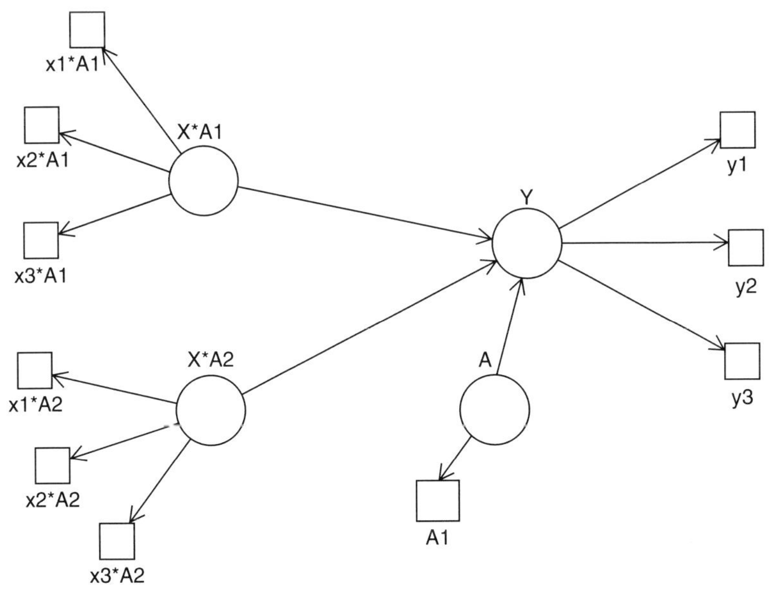

Fig. 5.2 Two-block regression model with main effect and interaction [*Group effect for measurement and structural models*]

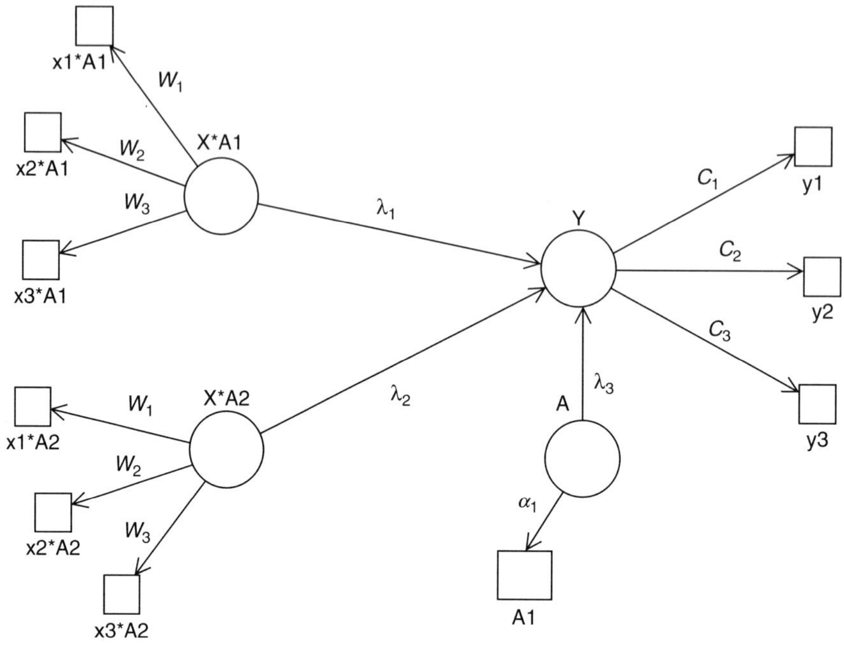

Fig. 5.3 Two-block regression model with main effect and interaction [*Group effect for the structural model only*]

do not represent the product of the group effect A with the latent variable related to the X block. In this model the influence of the group effect A on the measurement and the structural models are confounded. Henseler and Fassot (2006) propose a two-stage PLS approach: (1) Computing the LV scores $LV(X)$ and $LV(Y)$ using PLS on the model described in Fig. 5.1 without the interaction term and (2) Using the LV scores to carry out an analysis of covariance of $LV(Y)$ on $LV(X)$, A and $A * LV(X)$. In this paper, we propose a methodology to compute the LV scores taking into account the interaction term.

The main hypothesis that we need to do in this paper is that there is no group effect on the measurement model. The regression coefficients w_{jh} in the regression equations relating the MVs to their LVs are all equal among the $X * A_h$ blocks. This model is described in Fig. 5.3. These equality constraints cannot be obtained with PLS-Graph Chin (2005) nor with other PLS softwares. But, a SEM software like AMOS 6.0 Arbuckle (2005) could be used to estimate the path coefficients subject to these equality constraints with the ULS method.

5.2.1 Outer Estimate of the Latent Variables in the PLS Context

Using the model described in Fig. 5.3, it is possible to compute the LV estimates in a PLS way using the ULS-SEM weights w_j. For each block, the weight w_j is equal to the regression coefficient of ξ_h, LV for the block $X * A_h$, in the regression of the manifest variable $X_j A_h$ on the latent variable ξ_h:

$$\frac{Cov(X_j A_h, \xi_h)}{Var(\xi_h)} = \frac{Cov(w_j \xi_h + \varepsilon_{jh}, \xi_h)}{Var(\xi_h)} = w_j \tag{5.1}$$

Therefore, in each block, these weights are proportional to the covariances between the manifest variables and their LVs. With mode A, using the ULS-SEM latent variables as LV inner estimates, the LV outer estimate $\widehat{\xi}_h$ for block $X * A_h$ is given by the variable

$$\widehat{\xi}_h \propto \sum_j w_j (X_j A_h - \overline{X_j A_h}) \tag{5.2}$$

where $\propto$ means that the left term is equal to the right term up to a normalization to unit variance. This approach is described in Tenenhaus et al. (2005).

When all the X variables have the same units and all the weights w_j are positive, Fornell et al. (1996) suggest computing the LV estimate as a weighted average of the original MVs:

$$\widehat{\widehat{\xi}}_h = \sum_j \widehat{w}_j X_j A_h = \widehat{\xi} A_h \tag{5.3}$$

where $\widehat{w}_j = w_j / \sum_k w_k$ and $\widehat{\xi} = \sum_j \widehat{w}_j X_j = X\widehat{w}$. The LV estimate has values between 0 and 1 when the X variables are binary.

In the same way, the LV outer estimate for block Y is given by

$$\widehat{\eta}_h \propto \sum_k c_k (Y_k - \overline{Y_k}) \tag{5.4}$$

When all the weights c_k are positive, they are normalized so that they sum up to 1. We obtain, keeping the same notation for the "Fornell" η LV estimate,

$$\widehat{\eta}_h = \sum_k \widehat{c}_k Y_k = Y\widehat{c} \tag{5.5}$$

where $\widehat{c}_k = c_k / \sum_\ell c_\ell$. This LV has also values between 0 and 1 when the Y variables are binary.

5.2.2 Use of Multiple Regression on the Latent Variables

The structural equation of Fig. 5.3, relating η to ξ and taking into account the group effect A, is now estimated in the PLS framework by using the OLS multiple regression:

$$\begin{aligned}\widehat{\eta} &= \beta_0 + \beta_1 A_1 + \beta_2' \widehat{\xi} A_1 + \beta_3' \widehat{\xi} A_2 + \varepsilon \\ &= \beta_0 + \beta_1 A_1 + \beta_2' \widehat{\xi} A_1 + \beta_3' \widehat{\xi} (1 - A_1) + \varepsilon \\ &= \beta_0 + \beta_1 A_1 + \beta_3' \widehat{\xi} + (\beta_2' - \beta_3') \widehat{\xi} A_1 + \varepsilon\end{aligned} \tag{5.6}$$

The regression equation of $\widehat{\eta}$ on $\widehat{\xi}$, taking into account the group effect A, is finally written as follows:

$$\widehat{\eta} = \beta_0 + \beta_1 A_1 + \beta_2 \widehat{\xi} + \beta_3 \widehat{\xi} A_1 + \varepsilon \tag{5.7}$$

Consequently, there is a main group effect if the regression coefficient of A_1 is significantly different from zero and an interaction effect if the regression coefficient of $\widehat{\xi} A_1$ is significantly different from zero.

This approach can be generalized without difficulties if the group effect has more than two categories. In this approach ULS-SEM is only used to produce weights w and c that lead to the latent variables $\widehat{\xi}$ and $\widehat{\eta}$. The regression coefficients of model (5.7) are estimated by ordinary least squares (OLS), independently of the ULS-SEM parameters.

5.2.3 *Use of Bootstrap on the ULS-SEM Regression Coefficients*

Denoting the latent variables for the model in Fig. 5.3 as follows:

- η is the LV related to block Y
- ξ_1 is the LV related to block $X * A_1$
- ξ_2 is the LV related to block $X * A_2$
- ξ_3 is the LV related to block A

the theoretical model related to the model shown in Fig. 5.3 can be described by (5.8):

$$\eta = \lambda_1 \xi_1 + \lambda_2 \xi_2 + \lambda_3 \xi_3 + \delta \tag{5.8}$$

The test for a main effect A is equivalent to the test $H_0 : \lambda_3 = 0$. The test for an interaction effect $X * A$ is equivalent to the test $H_0 : \lambda_1 = \lambda_2$. Confidence intervals of the regression coefficients of model (5.8) can be constructed by bootstrapping using AMOS 6.0. These intervals can be used to test the main group effect and the interaction effect.

5.3 Application

5.3.1 *Introduction*

Ultraviolet radiations are known to play a major role in the development of skin cancers in humans. Nevertheless, in developed countries an increase in sun exposure has been observed over the last fifty years due to several sociological factors: longer holidays duration, traveling facilities and tanning being fashionable. To estimate the risk of skin cancer occurrence and of skin photoageing related to sun exposure behavior, a self-administered questionnaire was specifically developed in the context of the SU.VI.MAX cohort Guinot et al. (2001). The SU.VI.MAX study (SUppléments en VItamines et Minéraux Anti-oXydants) is a longitudinal cohort study conducted in France, which studies the relationship between nutrition and health through the main chronic disorders prevalent in industrialized countries. It involves a large sample of middle-age men and women right across the country recruited in a "free-living" adult population Hercberg et al. (1998). The study objectives, design and population characteristics have been described elsewhere Hercberg et al. (1998b). The information collected on this cohort offers the opportunity to conduct cross-sectional surveys using self-reported health behavior and habits questionnaires, such as those used to study the sun exposure behavior of French adults Guinot et al. (2001).

5.3.2 Material and Methods

Dermatologists and epidemiologists contributed to the definition of the questionnaire, which was in two parts, the first relating to sun exposure behavior over the past year and the second to sun exposure behavior evaluated globally over the subjects' lifetime. The questionnaire was addressed in 1997 to the 12,741 volunteers who were included in the cohort. Over 64% of the questionnaires were returned and analyzed (8,084 individuals: 4,825 women and 3,259 men).

In order to characterize the sun exposure of men and women, various synthetic variables characterizing sun exposure behavior were previously generated Guinot et al. (2001). Homogeneous groups of variables related to sun exposure behavior were obtained using a variable clustering method. Then, a principal component analysis was performed on these groups to obtain synthetic variables called "scores". A first group of binary variables was produced to characterize sun protection behavior over the past year (block Y with 6 variables). A second group of binary variables was produced to characterize lifetime sun exposure behavior (block X: 11 variables): intensity of lifetime sun exposure (4 variables), sun exposure during mountain sports (2 variables), sun exposure during nautical sports (2 variables), sun exposure during hobbies (2 variables), and practice of naturism (1 variable).

The objective of this research was to study the relationship between sun protection behavior over the past year of the individuals and their lifetime sun exposure behavior taking into account the group effects gender and class of age.

The methodology used was the following.

Firstly, the possible effect of gender has been studied. This analysis was carried out in four parts:

1st part. Because of the presence of dummy variables, the data are not multinormal. Therefore, ULS-SEM was carried out using AMOS 6.0 with the option *Method* = *ULS*. So, two weight vectors were obtained: a weight vector c for the sun protection behavior over the past year and a weight vector w for lifetime sun exposure behavior.

2nd part. Using these weights, two scores were calculated: one for the sun protection behavior and one for lifetime sun exposure behavior.

3rd part. Then, to study the possible gender effect on sun protection behavior over the past year, an analysis of covariance was conducted using PROC GLM (SAS software release 8.2 (SAS Institute Inc, 1999)) on lifetime sun exposure behavior score, gender and the interaction term between gender and lifetime sun exposure behavior score.

4th part. Finally, the results of the last testing procedure were compared with those obtained using the regression coefficient confidence intervals for model (5.8) calculated by bootstrapping (ULS-option) with AMOS 6.0.

Secondly, the possible effect of age was studied for each gender using the same methodology.

5.3.3 Results

The results are presented as follows. The relationship between sun protection behavior over the past year and lifetime sun exposure behavior has been studied, firstly with the gender effect (step 1), and secondly with the age effect for each gender (step 2a and step 2b). Finally, three different "*lifetime sun exposure*" scores were obtained, as well as three "*sun protection over the past year*" scores.

Step 1. Effect of Gender

ULS-SEM allowed to obtained weights c for the sun protection behavior over the past year and weights w for the lifetime sun exposure behavior. The AMOS results are shown in Fig. 5.4.

Then, the scores were calculated using the normalized weights on the original binary variables. The sun protection behavior over the past year was called "*Sun protection over the past year score* 1" (normalized weight vector $c1$ shown in Table 5.1). For example, the value $c11 = 0.24$ was obtained by dividing the original weight 1.00 (shown in Fig. 5.4) by the sum of all the $c1$ weights $(4.22 = 1.00 + 0.84 + \cdots + 0.46)$. The lifetime sun exposure behavior score was called "*Lifetime sun exposure score* 1" (normalized weight vector $w1$ shown in Table 5.2).

To study the possible effect of gender on sun protection behavior, an analysis of covariance was then conducted relating the "*Sun protection over the past year score* 1" to the "*Lifetime sun exposure score* 1", "*Gender*" and the interaction term "*Gender**Lifetime sun exposure score* 1". The results of this analysis are given in Table 5.3.

The LV "*Sun protection over the past year score* 1" is significantly related to the "*Lifetime sun exposure score* 1" (t-test = 9.61, p<0.0001), to "*Gender*"

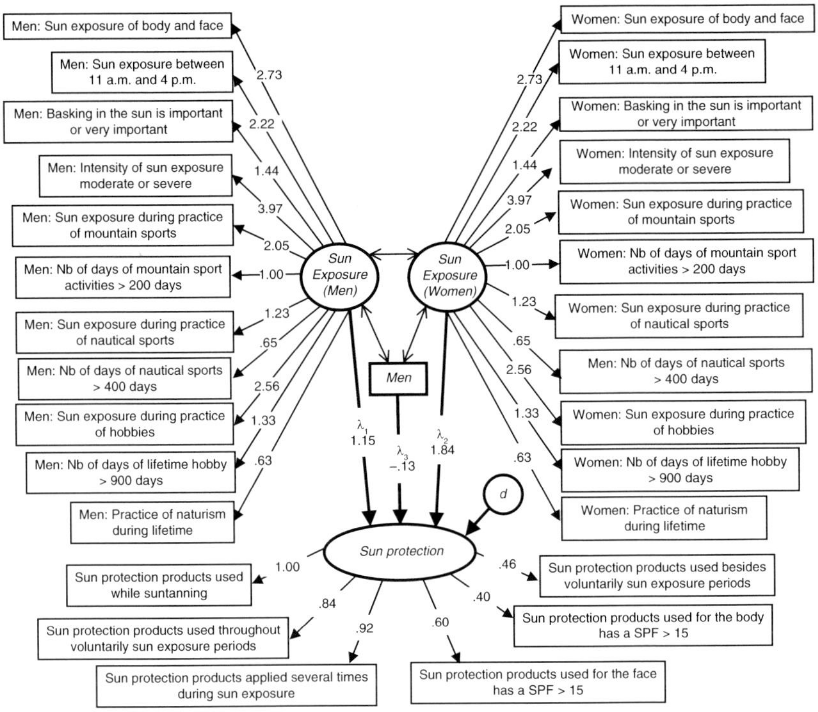

Fig. 5.4 Two block regression model for relating sun protection behavior over the past year to the lifetime sun exposure behaviors with a gender effect acting on the structural model and not on the measurement model

Table 5.1 Normalized weight vector $c1$ for the "*Sun protection over the past year score* 1" (Effect of gender)

+	0.24	If sun protection products used while sun tanning
+	0.20	If sun protection products used throughout voluntarily sun exposure periods
+	0.22	If sun protection products applied regularly several times during sun exposure periods
+	0.14	If the sun protection product used for the face has a SPF[a] over 15
+	0.09	If the sun protection product used for the body has a SPF[a] over 15
+	0.11	If sun protection products used besides voluntarily sun exposure periods

[a] SPF: Sun Protection Factor

(t-test = 8.15, p<0.0001) and to "*Gender*Lifetime sun exposure score* 1" (t-test = 4.87, p<0.0001). Generally, men tend to use less sun protection products than women; furthermore, this difference between men and women increases as lifetime sun exposure increases.

These results are confirmed by the bootstrap analysis of model (5.8) given in Table 5.4. The 95% Confidence Interval (CI) for the regression coefficient λ_3 is [−0.181, −0.084]. Therefore there is a significant "*Gender*" effect. The 95% CI for the regression coefficients λ_1 and λ_2 do not overlap. Therefore we may conclude

Table 5.2 Normalized weight vector $w1$ for the "*Lifetime sun exposure score* 1" (Effect of gender)

+	0.14	If sun exposure of the body and the face
+	0.11	If sun exposure between 11 a.m. and 4 p.m.
+	0.07	If basking in the sun is declared important or extremely important
+	0.20	If self-assessed intensity of sun exposure is declared moderate or severe
+	0.10	If sun exposure during practice of mountain sports
+	0.05	If the number of days of lifetime mountain sports activities > 200 days[a]
+	0.06	If sun exposure during practice of nautical sports
+	0.03	If the number of days of lifetime nautical sports activities > 400 days[a]
+	0.13	If sun exposure during practice of hobbies
+	0.07	If the number of days of lifetime hobby activities > 900 days[a]
+	0.03	If practice of naturism during lifetime

[a] Median value of the duration was used as a threshold for dichotomisation

Table 5.3 SAS output of analysis of covariance for "*Sun protection over the past year score* 1" on "*Lifetime sun exposure score* 1" (score_x1_protect), gender and interaction

Parameter		Estimate	Standard Error	t Value	Pr>|t|
Intercept		0.0729460737 B	0.01213456	6.01	<.0001
Score_x1_protect		0.2473795070 B	0.02574722	9.61	<.0001
GENDER	Women	0.1269948620 B	0.01557730	8.15	<.0001
GENDER	Men	0.0000000000 B	–	–	–
Score_x1_protec*GENDER	Women	0.1613712617 B	0.03316612	4.87	<.0001
Score_x1_protec*GENDER	Men	0.0000000000 B	–	–	–

Table 5.4 AMOS output for 95% CI of regression coefficients for Fig. 5.4 "*Sun protection over the past year score* 1" on "*Lifetime sun exposure score* 1"

Coefficients				Estimate	Lower	Upper
λ_1	***Sun exposure (men)***	→	***Sun protection***	1.155	0.950	1.360
λ_2	***Sun exposure (women)***	→	***Sun protection***	1.839	1.596	2.080
λ_3	***Men***	→	***Sun protection***	0.129	−0.181	−0.084

that $\lambda_1 \neq \lambda_2$. There is a significant interaction effect "*Gender*Sun Exposure*" on "*Sun Protection*". But this last approach does not produce any p-value.

Step 2. Effect of Age

As the relationship between the sun protection behavior over the past year and the lifetime sun exposure depends on gender, the effect of age was studied for each gender. Thus, the variable age was dichotomized. This variable called "*Age50*" is equal to 0 if less than 50 at inclusion in the SU.VI.MAX study and equal to 1 if more or equal to 50. However, in Figs. 5.5 and 5.6, this variable is called "*Age≥50*" to make the interpretation easier.

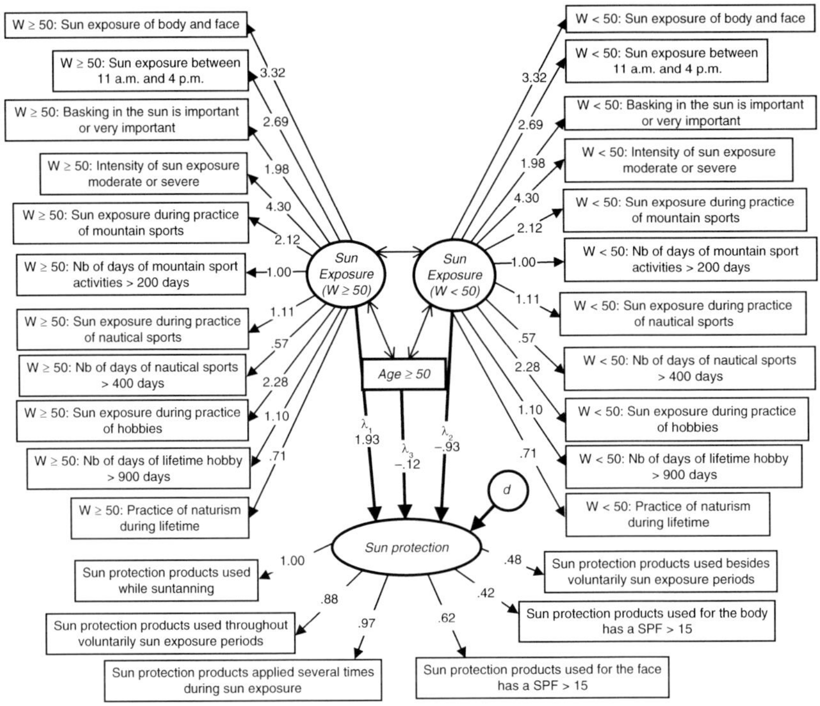

Fig. 5.5 Two block regression model for relating sun protection behavior over the past year to the lifetime sun exposure behaviors for women with an age effect acting on the structural model and not on the measurement model

Step 2a. Effect of age for women

Using the same methodology, a sun protection over the past year score ("*Sun protection over the past year score* 2") and a lifetime sun exposure score ("*Lifetime sun exposure score* 2") were obtained (normalized weights shown in Tables 5.5 and 5.6, in columns $c2$ and $w2$, respectively; normalized weights in $c1$ and $w1$ are the same as in Tables 5.1 and 5.2 and are given here again for comparison purpose). The AMOS results are shown in Fig. 5.5.

Then, to study the age effect on the sun protection over the past year for women, an analysis of covariance was conducted relating the "*Sun protection over the past year score* 2" to the "*Lifetime sun exposure score* 2", "*Age50*" and the interaction term "*Age50*Lifetime sun exposure score* 2". The results are given in Table 5.7. The LV "*Sun protection over the past year score* 2" is significantly related to "*Lifetime sun exposure score* 2" (t-test = 15.3, $p < 0.0001$) and to "*Age50*" (t-test = −4.95, $p < 0.0001$), but not to the interaction term (t-test = 0.43, p = 0.6687). Women less than 50 tend to use more sun protection products than women over or equal to 50.

These results are confirmed by the bootstrap analysis of model (5.8) given in Table 5.8. The 95% CI for the regression coefficient λ_3 is [−0.187, −0.049].

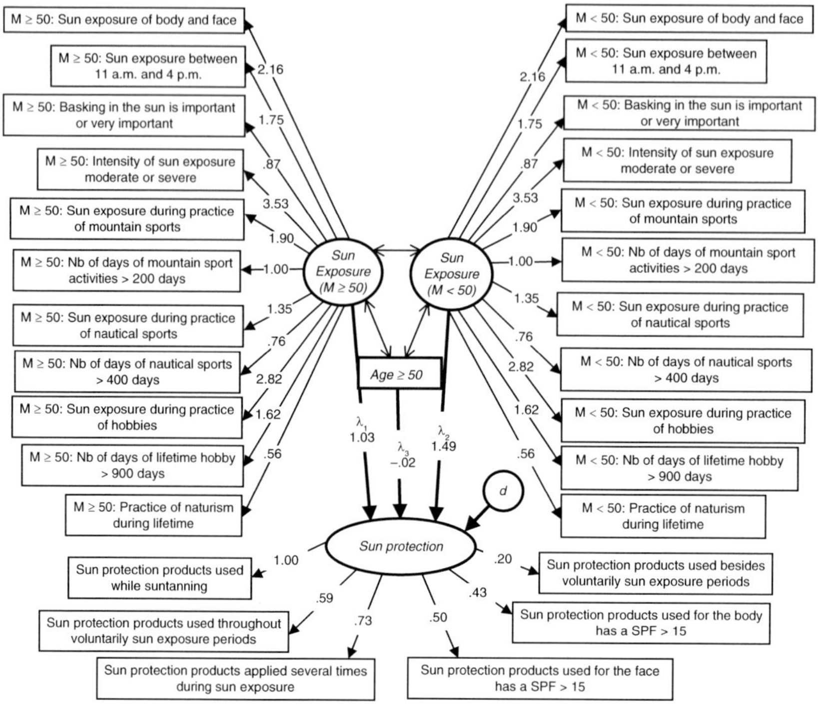

Fig. 5.6 Two block regression model for relating sun protection behavior over the past year to the lifetime sun exposure behaviors for men with an age effect acting on the structural model and not on the measurement model

Table 5.5 Normalized weights for the sun protection behavior over the past year scores

	$c1$	$c2$	$c3$
Sun protection products used while suntanning	0.24	0.23	0.29
Sun protection products used throughout voluntarily sun exposure periods	0.20	0.20	0.17
Sun protection products applied regularly several times during sun exposure	0.22	0.22	0.21
Sun protection product used for the face has a SPF over 15	0.14	0.14	0.14
Sun protection product used for the body has a SPF over 15	0.09	0.10	0.12
Sun protection products used besides voluntarily sun exposure periods	0.11	0.11	0.06

Therefore there is a significant "*Age50*" effect. The 95% CI for the regression coefficients λ_1 and λ_2 do overlap. Therefore we may conclude that $\lambda_1 = \lambda_2$. There is no significant interaction effect "*Age50*Sun Exposure*" on "*Sun Protection*" *f*or women.

Step 2b. Effect of age for men

The normalized weights $c3$ for computing the sun protection over the past year score ("*Sun protection over the past year score* 3") are given in Table 5.5, and

Table 5.6 Normalized weights for the lifetime sun exposure scores

	*w*1	*w*2	*w*3
Sun exposure of the body and the face	0.14	0.16	0.12
Sun exposure between 11 a.m. and 4 p.m.	0.11	0.13	0.10
Basking in the sun important or extremely important	0.07	0.09	0.05
Self-assessed intensity of sun exposure moderate or severe	0.20	0.20	0.19
Sun exposure during practice of mountain sports	0.10	0.10	0.10
Number of days of mountain sports activities > 200 days	0.05	0.05	0.05
Sun exposure during practice of nautical sports	0.06	0.05	0.07
Number of days of nautical sports activities > 400 days	0.03	0.03	0.04
Sun exposure during practice of hobbies	0.13	0.11	0.15
Number of days of lifetime hobby activities > 900 days	0.07	0.05	0.09
Practice of naturism during lifetime	0.03	0.03	0.03

Table 5.7 SAS output of analysis of covariance for "*Sun protection over the past year score* 2" on "*Lifetime sun exposure score* 2" (score_x2_protect_women), age and interaction

Parameter		Estimate	Standard Error	t Value	Pr>|t|
Intercept		0.2275434886 B	0.01269601	17.92	<.0001
score_x2_protect_women		0.4056820480 B	0.02651644	15.30	<.0001
age50	+50 years	−0.1090996349 B	0.02206033	−4.95	<.0001
age50	−50 years	0.0000000000 B	–	–	–
score_x2_protect_women*age50	+50 years	0.0197448010 B	0.04613877	0.43	0.6687
score_x2_protect_women*age50	−50 years	0.0000000000 B	–	–	–

Table 5.8 AMOS output for 95% CI of regression coefficients for figure 5 "*Sun protection over the past year score* 2" on "*Lifetime sun exposure score* 2" (for women)

Coefficients				Estimate	Lower	Upper
λ1	***Sun exposure (women ≥ 50)***	→	***Sun protection***	1.929	1.546	2.280
λ2	***Sun exposure (women < 50)***	→	***Sun protection***	1.929	1.628	2.223
λ3	***Age ≥ 50***	→	***Sun protection***	−0.124	−0.187	−0.049

the normalized weights $w3$ for lifetime sun exposure score ("*Lifetime sun exposure score* 3") in Table 5.6. The AMOS results are shown in Fig. 5.6.

The results of the analysis of covariance relating the "*Sun protection over the past year score* 3" to the "*Lifetime sun exposure score* 3", "*Age50*" and the interaction term "*Age50*Lifetime sun exposure score* 3" are shown in Table 5.9. The LV "*Sun protection over the past year score* 3" is significantly related to "*Lifetime sun exposure score* 3" (t-test = 8.28, p<0.0001) and tends to be related to the interaction term "*Age50*Sun protection over the past year score* 3" (t-test $= -1.95$, p $= 0.05$), but not to "*Age50*" (t-test $= -1.41$, p $= 0.1576$).

The sun protection behavior of men less than 50 tends to increase more rapidly with the level of lifetime sun exposure than the one of men over or equal to 50.

The 95% CI for the regression coefficient λ_3 is $[-0.104, 0.075]$. Therefore there is no significant "*Age50*" effect. The 95% CI for the regression coefficients λ_1 and λ_2 do overlap. We may conclude that $\lambda_1 = \lambda_2$ and, therefore, that there is no

Table 5.9 SAS output of analysis of covariance for "*Sun protection over the past year score* 3" on "*Lifetime sun exposure score* 3" (score_x3_protect_men), age and interaction

Parameter		Estimate	Standard Error	t Value	Pr>$\|t\|$
Intercept		0.1070220420 B	0.01677764	6.38	<.0001
score_x3_protect_men		0.2962098256 B	0.03577279	8.28	<.0001
age50	+50 years	−0.0315335922 B	0.02230895	−1.41	0.1576
age50	−50 years	0.0000000000 B	–	–	–
score_x3_protect_men*age50	+50 years	−0.0923178930 B	0.04726879	−1.95	0.0509
score_x3_protect_men*age50	−50 years	0.0000000000 B	–	–	–

These results are partially confirmed by the bootstrap analysis of model (5.8) given in Table 5.10

Table 5.10 AMOS output for 95% CI of regression coefficients for figure 6 "*Sun protection over the past year score* 3" on "*Lifetime sun exposure score* 3" (for men)

Coefficients				Estimate	Lower	Upper
λ1	***Sun exposure (men*** $\geq$ 50)	→	***Sun protection***	1.034	0.728	1.395
λ2	***Sun exposure (men*** < 50)	→	***Sun protection***	1.485	1.136	1.895
λ3	***Age*** $\geq$ 50	→	***Sun protection***	−0.017	−0.104	0.075

significant interaction effect "*Age50*Sun Exposure*" on "*Sun Protection*" for men. But this procedure does not control the risk α.

On the other hand, the previous procedure tends to give too small p-values because the LV estimates have been constructed to optimize the relationship. A bootstrap procedure on the first approach will probably give more reliable results.

5.4 Discussion

A major issue is the stability of the scores.

The lifetime sun exposure weights (Table 5.6), lead to scores highly correlated for women and men:

For women: *Cor*("*Lifetime sun exposure score* 1", "*Lifetime sun exposure score* 2") = 0.99.

For men: *Cor*("*Lifetime sun exposure score* 1","*Lifetime sun exposure score* 3") = 0.99.

The weights obtained for the sun protection over the past year scores are summarized in Table 5.5. The correlations between the scores are all above 0.99:

For women: *Cor*("*Sun protection over the past year score* 1","*Sun protection over the past year score* 2") = 0.99.

For men: *Cor*("*Sun protection over the past year score* 1","*Sun protection over the past year score* 3") = 0.99.

5.5 Conclusion

A software like AMOS is oriented toward the estimation of the path coefficients of a structural equation model, when a software like PLS-Graph is oriented toward the production of latent variables or scores. It is possible to use the results of AMOS to construct scores using the PLS methodology. In this paper, we propose a way to take into account interactions in the structural equations, independently from the measurement model. This procedure follows four parts:

1 Use of AMOS to compute the weights of the manifest variables subject to the constraint that the weights related to the same manifest variable are equal in the various groups.
2 Computation of the PLS LV estimates using the weights issued from AMOS.
3 Study of the interaction through an analysis of covariance relating the response block latent variable to the predictor block latent variable, the group effect and the interaction crossing the predictor block latent variable and the group effect.
4 Use of the bootstrapping possibilities of AMOS 6.0 to produce confidence intervals of the structural equation regression coefficients.

When the manifest variables are numerical, it is recommended to use the maximum likelihood (ML) option of AMOS. When the manifest variables are binary (it is the case in this paper) the unweighted least squares (ULS) should be preferred as ML estimation supposes multinormality.

In this paper, this methodology has been applied on a large dataset, and its simplicity and efficiency have been well demonstrated. Its generalization to more complex path models should be straightforward.

Acknowledgements The authors wish to thank Pr Denis Malvy and Dr Khaled Ezzedine for their comments and advices. We thank the staff who participated in or assisted practically in carrying out the SU.VI.MAX study, in particular Pr Serge Hercberg and Dr Pilar Galan.

References

Arbuckle, J. L. (2005). *AMOS 6.0*. Spring House, PA: AMOS Development Corporation.

Chin, W. W. (2005). *PLS-Graph user's guide*. Houston, USA: C.T. Bauer College of Business, University of Houston.

Chin, W. W., Marcolin, B. L., & Newsted, P. R. (2003). A Partial Least squares latent variable modeling approach for measuring interaction effects: results from a Monte Carlo simulation study and an electronic-mail emotion/adoption study. *Information Systems Research, 14*(2), 189–217.

Fornell, C., Johnson, M. D., Anderson, E. W., Cha, J., & Bryant B. E. (1996). The American Customer satisfaction index: nature, purpose and findings. *Journal of Marketing, 60*, 7–18.

Guinot, C., Malvy, D., Latreille, J., Preziosi, P., Galan, P., Vaillant, L., Tenenhaus, M., Hercberg, S., & Tschachler, E. (2001). Sun exposure behaviour of a general adult population in France. In: J. Ring, S. Weidinger, U. Darsow (Eds.), *Skin and Environment – Perception and Protection* (pp. 1099–1106), 10th EADV Congress, Munich. ISBN 88-323-1411-8, Monduzzi editore S.p.A., Bologna.

Henseler, J., & Fassot, G. (2006). Testing moderating effects in PLS path models. In: Esposito Vinzi, V., Chin, W. W., Henseler, J., & Wang, H. (Eds.), *Handbook PLS and Marketing*. Berlin, Heidelberg, New York: Springer.

Hercberg, S., Galan, P., Preziosi, P., Roussel, A.-M., Arnaud, J., Richard, M.-J., Malvy, D., Paul-Dauphin, A., Briançon, S., & Favier, A. (1998). Background and rationale behind the SU.VI.MAX. study, a prevention trial using nutritional doses of a combination of antioxidant vitamins and minerals to reduce cardiovascular diseases and cancers. *International Journal for Vitamin and Nutrition Research, 68*(1), 3–20.

Hercberg, S., Preziosi, P., Briançon, S., Galan, P., Triol, I., Malvy, D., Roussel, A.-M., & Favier, A. (1998). A primary prevention trial using nutritional doses of antioxidant vitamins and minerals in cardio-vascular diseases and cancers in a general population: "The SU.VI.MAX. study" - Design, methods and participants characteristic. *Controlled Clinical Trials, 19*(4), 336–351.

Hercberg, S., Galan, P., Preziosi, P., Bertrais, S., Mennen, L., Malvy, D., Roussel, A.-M., Favier, A., & Briançon, S. (2004). The SU.VI.MAX study: a randomized, placebo-controlled trial of the health effects of antioxidant vitamins and minerals. *Archives of Internal Medicine, 164*(21), 2325–2342.

Ping, R. A. (1995). A Parsimonious estimating technique for interaction and quadratic latent variables. *Journal of Marketing Research, 32*, 336–347.

SAS Institute Inc. (1999). *SAS/STAT® User's Guide*, Version 8. Cary, NC: SAS Institute Inc.

Tenenhaus, M., Esposito Vinzi V., Chatelin, Y. M., & Lauro, C. (2005). PLS path modeling. *Computational Statistics and Data Analysis, 48*(1), 159–205.

Chapter 6
A New Multiblock PLS Based Method to Estimate Causal Models: Application to the Post-Consumption Behavior in Tourism

Francisco Arteaga, Martina G. Gallarza, and Irene Gil

Abstract This study presents a new algorithm for estimating causal models based on multiblock PLS method. This new algorithm is tested in a particular post-consumption behavior with the aim of validating a complex system of relations between antecedents of value, perceived value, satisfaction and loyalty. The results are compared with the classical LVPLS method: both methods support the proposed structural relations, but the explained variance is slightly higher with the new algorithm.

6.1 Introduction

Partial least squares regression (PLS) and derived methods are successfully applied in a wide variety of scientific research areas, with some specific characteristics in specific applications. The PLS algorithm that is applied in causal modeling techniques and the PLS algorithm that is applied in chemometrics applications are very different in their implementation. The first usually apply the LVPLS method which computational aspects have been developed by Lohmöller (1987, 1989) and the second apply the non-linear iterative partial least squares (NIPALS) algorithm (Geladi and Kowalski 1986; Wold et al. 1987; Martens and Naes 1989).

F. Arteaga
Department of Statistics, Universidad Católica de Valencia San Vicente Martir, Guillem de Castro, 175, Valencia 46008, Spain
e-mail: francisco.arteaga@ucv.es

M. G. Gallarza
Faculty of Economics, Department of Marketing, Universitat de Valencia, Avenida de los Naranjos s/n, Valencia 46022, Spain
e-mail: martina.gallarza@uv.es

I. Gil
Department of Marketing, Universitat de València, Avenida de los Naranjos s/n, Valencia 46022, Spain
e-mail: irene.gil@uv.es

V. Esposito Vinzi et al. (eds.), *Handbook of Partial Least Squares*, Springer Handbooks of Computational Statistics, DOI 10.1007/978-3-540-32827-8_7,

The aim of this work is to adapt a multiblock PLS method used in chemometric applications (Gerlach et al. 1979; Frank et al. 1984; Frank and Kowalski 1985; Wold et al. 1987; Wangen and Kowalski 1988; Westerhuis and Coenegracht 1997; Westerhuis et al. 1998), to deal with the causal modeling estimation problem and to compare the estimations from the new method with the analogous from the classic LVPLS method. We are also interested in the performance of both methods.

The first part of this work starts with a primer on causal modeling, useful to introduce the problem nature and the notation; we then review the standard partial least squares path modeling algorithm (PLSPM) to estimate causal models and follow by reviewing the PLS regression method as it is used in other applications as chemometrics or pharmaceutical applications, from the two blocks PLS to the multiblock PLS method, and then we adapt the multiblock PLS method to deal with causal models, yielding the MultiBlock PLS path modeling method (MBPLSPM). In the second part, we compare both methods (PLSPM and MBPLSPM) over a post-consumption behavior application and, with the aim to confirm the conclusions obtained, we compare both methods over a simulated example. Finally we present the conclusions.

Notation is detailed in Appendix 1.

6.2 Causal Modeling

In our context a Causal Model consists of a set of concepts difficult to measure directly (value, loyalty, satisfaction ...) that present different linear relations between them. To study and confirm these relations we need to build a scale for each concept. A scale is a set of observable variables related to the concept that altogether gives us an indirect measure of it. Because of this, the concepts are called constructs and the variables from the scale are called indicators. The constructs are also called latent variables (LV) and the indicators are called manifest variables (MV).

The set of linear relations between the constructs is the so called *structural model* and the set of relations between each construct and its indicators is the so called *measurement model*.

In Fig. 6.1 we can see the structural model for a particular study of the post-consumption behavior with nine constructs and eleven linear relations.

In this work we assume that the manifest and the latent variables have zero mean and standard deviation one. Lohmöller (1987), in his software LVPLS 1.8, proposes new options for the PLS algorithm and, in particular, proposes the above mentioned standardization for the manifest variables when three conditions matches: the variable scales are comparable, the means are not interpretable and variance is not related whit variable importance. This standardization yields also in a simplest representation of the model equations.

We represent a causal model as a set of B standardized random latent variables, ξ_b, related between them in a way that some of them are caused by a subset of the other latent variables that are its precedents, i.e., we assume that each caused latent variable can be expressed as a linear combination of a subset of the other latent variables, plus a zero mean error term not correlated with them.

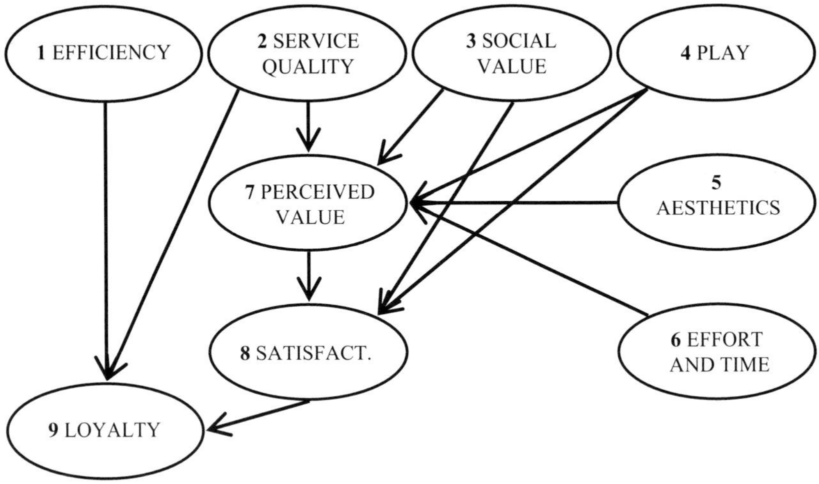

Fig. 6.1 Representation of the structural model for a particular study of the post-consumption behavior

The structural model leads to a set of linear equations relating the latent variables between them:

$$\xi_b = \sum_{i \neq b} \beta_{bi} \xi_i + \zeta_b. \tag{6.1}$$

The residual ζ_b is a zero mean random term not correlated with the latent variables ξ_i that cause ξ_b (*prediction specification* condition).

For the structural model in Fig. 6.1 we can write three structural equations:

$$\begin{array}{c} \xi_7 = \beta_{72}\xi_2 + \beta_{73}\xi_3 + \beta_{74}\xi_4 + \beta_{75}\xi_5 + \beta_{76}\xi_6 + \zeta_7 \\ \xi_8 = \beta_{83}\xi_3 + \beta_{84}\xi_4 + \beta_{87}\xi_7 + \zeta_8 \qquad \xi_9 = \beta_{91}\xi_1 + \beta_{92}\xi_2 + \beta_{98}\xi_8 + \zeta_9 \end{array}. \tag{6.2}$$

The latent variables are latent because they are not directly measurable and this is why we need a set of manifest variables (the above mentioned scale) that altogether give us an indirect measure of it. That is, knowing the value for the manifest variables associated to a latent variable, we can assign a value for the latent variable. Each latent variable has a set of manifest variables that are measured over n subjects; that is the reason why the data consist of B matrices $\mathbf{X}_1, \mathbf{X}_2, \ldots, \mathbf{X}_B$ with n rows and $k_1, k_2, \ldots, k_B$ columns respectively. For each matrix $\mathbf{X}_b$, we need to obtain a vector with the value of the bth latent variable for the n individuals. Each block of variables constitutes the measurable face of a latent variable and it is interesting the way in which the variables x_{bj} with $j = 1, 2, \ldots, k_b$ (block $\mathbf{X}_b$) are connected with the latent variable ξ_b. This is the so called measurement model.

There are three ways to relate each manifest variable with its latent variable, but for our proposal we only employ the so called reflective way where each manifest

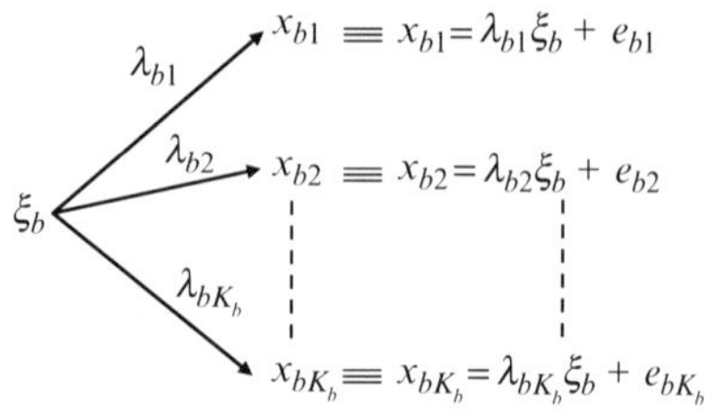

Fig. 6.2 In the reflective way each MV comes from a linear transformation of their LV, plus a zero mean random term not correlated with the LV

variable is related to its latent variable by a simple regression:

$$x_{bj} = \lambda_{bj}\xi_b + e_{bj}. \tag{6.3}$$

The residual e_{bj} is a zero mean random term not correlated with the latent variable ξ_b (again the *prediction specification* condition). This implies that in the reflective way each block of manifest variables is unidimensional (Hulland 1999) in the meaning of factor analysis, because all the manifest variables for a latent variable are linear transformations of the same latent variable plus a zero mean random term.

The name reflective is due to the fact that each manifest variable x_{bj} constitutes a reflect of its latent variable ξ_b.

In Fig. 6.2 we illustrate the reflective way for the bth latent variable ξ_b.

6.3 PLS Path Modeling

Lohmöller (1987, 1989) developed the computational aspects in the LVPLS software application. The algorithm employed in this work, that we call LVPLS method, is described by Guinot et al. (2001) and it is showed below.

The algorithm consists of alternating two types of estimation of the latent variables until they converge to the same results. The types of estimation are the external estimation (each latent variable is estimated from their manifest variables) and the internal estimation (each latent variable is estimated from the previous external estimation of the other latent variables).

In the algorithm the external estimation for the bth latent variable is denoted by $\mathbf{y}_b$ and the internal estimation is denoted by $\mathbf{z}_b$.

To begin, each external estimation $\mathbf{y}_b$ is made by assigning the first column of the $\mathbf{X}_b$ matrix ($\mathbf{X}_b$ is the n by K_b matrix that accommodates the value for the K_b variables measured over the n individuals).

The internal estimation $\mathbf{z}_b$ of ξ_b is defined as:

$$\mathbf{z}_b = \left(\sum_{j:\beta_{bj}\neq 0} d_{bj}\mathbf{y}_j + \sum_{j:\beta_{jb}\neq 0} d_{jb}\mathbf{y}_j \right)^* . \tag{6.4}$$

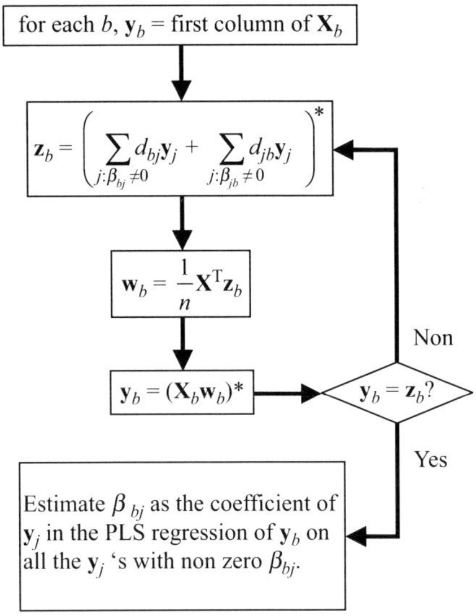

Fig. 6.3 Schedule for the LVPLS method, with the structural scheme for the internal estimation and with mode A for the weights estimation

In this formula and in the following text, the asterisk means that the variable in parentheses is standardized.

In (4) we divide the sum into two parts. In the first part d_{bj} is the regression coefficient of $\mathbf{y}_j$ in the multiple regression of $\mathbf{y}_b$ on all the $\mathbf{y}_j$'s related to the predecessors of ξ_b, and in the second part d_{jb} is the correlation between $\mathbf{y}_j$ and $\mathbf{y}_b$. This is the so called structural or path weighting scheme, proposed by Lohmöller (1987) in his software LVPLS 1.8.

The external estimation $\mathbf{y}_b$ of ξ_b is defined by:

$$\mathbf{y}_b = (\mathbf{X}_b \mathbf{w}_b)^* . \tag{6.5}$$

In (5), $\mathbf{w}_b$ are the weights that determinate the influence of each manifest variable in the construction of the latent variable $\mathbf{y}_b$. The coordinates of the vector $\mathbf{w}_b$ are the correlations between variables $\mathbf{x}_{bj}$ and the previous internal estimation $\mathbf{z}_b$. This way of calculating the weights is the so called "mode A," that is appropriated when the manifest variables are collinear as it is the case in the reflective way.

We alternate the internal and external estimations until they match (within desired precision).

When the algorithm ends we can calculate the path coefficients, that is, for each b, estimate the β_{bj} values as the coefficient of $\mathbf{y}_j$ in the single component PLS regression of $\mathbf{y}_b$ on all the $\mathbf{y}_j$'s related to the precedents of ξ_b.

In Fig. 6.3 we can see a scheme useful for a better understanding of the LVPLS algorithm.

6.4 PLS Regression Methods

In the previous section we describe the PLS path modeling method as it is used in marketing research: the LVPLS method. In the following sections we describe the multiblock PLS method and its adaptation to deal with causal models. We begin by describing the basic two-blocks PLS method (Sect. 6.4.1) for comparing it with the multiblock extension of the method (Sect. 6.4.2). Finally, we adapt the multiblock PLS method to deal with causal models, yielding the multiblock PLS path modeling method (MBPLSPM) (Sect. 6.4.3) as an alternative to the LVPLS above descript (Sect. 6.3).

6.4.1 Two-Blocks PLS Method

Partial least squares (PLS) is a regression method mainly developed by Herman Wold and co-workers (Wold 1982, 1985). Stone and Brooks (1990) show how PLS can be considered as a two stage process in which the set of k predictor variables are first linearly transformed into a new set of $A(A < K)$ factors which have maximal covariance with the response variable subject to them being orthogonal to each other. To known the history of PLS the reader can see Geladi (1988). For a tutorial on PLS refer to Geladi and Kowalski (1986).

In two blocks PLS method we start from two data matrices $\mathbf{X}$ and $\mathbf{Y}$. $\mathbf{X}$ is an $N \times K$ matrix and $\mathbf{Y}$ is an $N \times M$ matrix, without assumptions about N, K or M. In general, we can suppose that $\mathbf{X}$ and $\mathbf{Y}$ are centered (each column has zero mean) and scaled (the variance for each column is one) matrices. Figure 6.4 shows the algorithm and the arrow scheme for the PLS method. This is the known non-linear iterative partial least squares (NIPALS) algorithm (Geladi and Kowalski 1986; Wold et al. 1987; Martens and Naes 1989).

The data matrices $\mathbf{X}$ (descriptors) and $\mathbf{Y}$ (responses) are represented by their latent variables $\mathbf{t}$ and $\mathbf{u}$ respectively. The corresponding weights $\mathbf{w}$ and $\mathbf{c}$ are obtained by multiplying the latent variables through the specific matrix, being $\mathbf{w}$ normalized to length one. New values for the latent variables are obtained from the weights. This is repeated until convergence of $\mathbf{u}$. From this algorithm we obtain the latent variables $\mathbf{t}$ and $\mathbf{u}$ that summarize $\mathbf{X}$ and $\mathbf{Y}$ respectively maximizing their covariance. Loadings $\mathbf{p}$ are calculated to obtain the residuals on $\mathbf{X}$, $\mathbf{X}_{\mathrm{RES}} = \mathbf{X} - \mathbf{tp}^{\mathrm{T}}$, and we employ $\mathbf{c}$ to obtain the residuals on $\mathbf{Y}$, $\mathbf{Y}_{\mathrm{RES}} = \mathbf{Y} - \mathbf{tc}^{\mathrm{T}}$. The residual matrices $\mathbf{X}_{\mathrm{RES}}$ and $\mathbf{Y}_{\mathrm{RES}}$ can be used as the original $\mathbf{X}$ and $\mathbf{Y}$ matrices to obtain new $\mathbf{t}$ and $\mathbf{u}$ latent variables, but in our case we are only interested in the first set of latent variables (remember that the reflective way for the measurement model implies the unidimensionality of the blocks, in the meaning of factor analysis).

6.4.2 Multiblock PLS Method

When the number of variables is large, and additional information is available we can block the descriptor variables into several conceptually meaningful blocks, to improve the interpretability of the model.

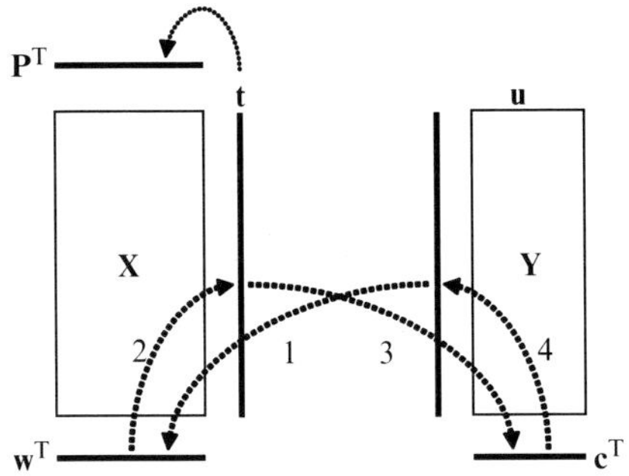

0. $\mathbf{u}$ = first column of $\mathbf{Y}$
1. $\mathbf{w} = \mathbf{X}^T\mathbf{u}$ $\mathbf{w} = \mathbf{w}/\|\mathbf{w}\|$
2. $\mathbf{t} = \mathbf{X}\mathbf{w}$
3. $\mathbf{c} = \mathbf{Y}^T\mathbf{t}$
4. $\mathbf{u} = \mathbf{Y}\mathbf{c} / \mathbf{c}^T\mathbf{c}$

...go to step 1 until $\mathbf{u}$ converges.

$\mathbf{p} = \mathbf{X}^T\mathbf{t}/\mathbf{t}^T\mathbf{t}$

Fig. 6.4 Algorithm and arrow scheme of the two blocks PLS method

Many multiblock PLS algorithms have been presented in the literature, Gerlach et al. (1979), Frank et al. (1984), Frank and Kowalski (1985), Wold et al. (1987), Wangen and Kowalski (1988),Westerhuis and Coenegracht (1997), Westerhuis et al. (1998) study and compare different variations from the original multiblock PLS algorithm.

In the multiblock PLS method, as in the previously descript two-blocks PLS method, we distinguish between the I descriptor blocks, $\mathbf{X}_i$, with $i = 1, \ldots, I$, and the response block, $\mathbf{Y}$. The latent variables for the ith descriptor block, with $i = 1, \ldots, I$, is denoted by $\mathbf{t}_i$ and for the response block by $\mathbf{u}$.

In Fig. 6.5 we can see the algorithm and the arrow scheme for the basic multiblock PLS algorithm.

In this algorithm a start latent variable $\mathbf{u}$ is regressed on each block $\mathbf{X}_i i = 1, \ldots, I$ to give the block variable weights $\mathbf{w}_i$, that are normalized to length one and multiplied through the block to give the block latent variable $\mathbf{t}_i$. The I latent variables are combined into the super block $\mathbf{T}$ and a two-blocks PLS cycle between $\mathbf{T}$ and $\mathbf{Y}$ is performed to give the combined weight $\mathbf{w}_\mathbf{T}$, which is also normalized to length one, and the combined latent variable $\mathbf{t}_\mathbf{T}$. We repeat this until convergence on $\mathbf{u}$.

6.4.3 Multiblock PLS Path Modeling Method

The presented multiblock PLS method can be adapted to deal with more general linear models. In particular it is interesting to considerer the possibility of blocking also the response variables and the existence of blocks of variables that are simultaneously descriptor variables and response variables as in the causal models above mentioned.

Wangen and Kowalski (1988) introduced a multiblock PLS algorithm that was based on an algorithm originally presented by Wold et al. (1984). We adapt the Wangen and Kowalski multiblock PLS method as an alternative to the LVPLS method, yielding the multiblock PLS path modeling (MBPLSPM) method.

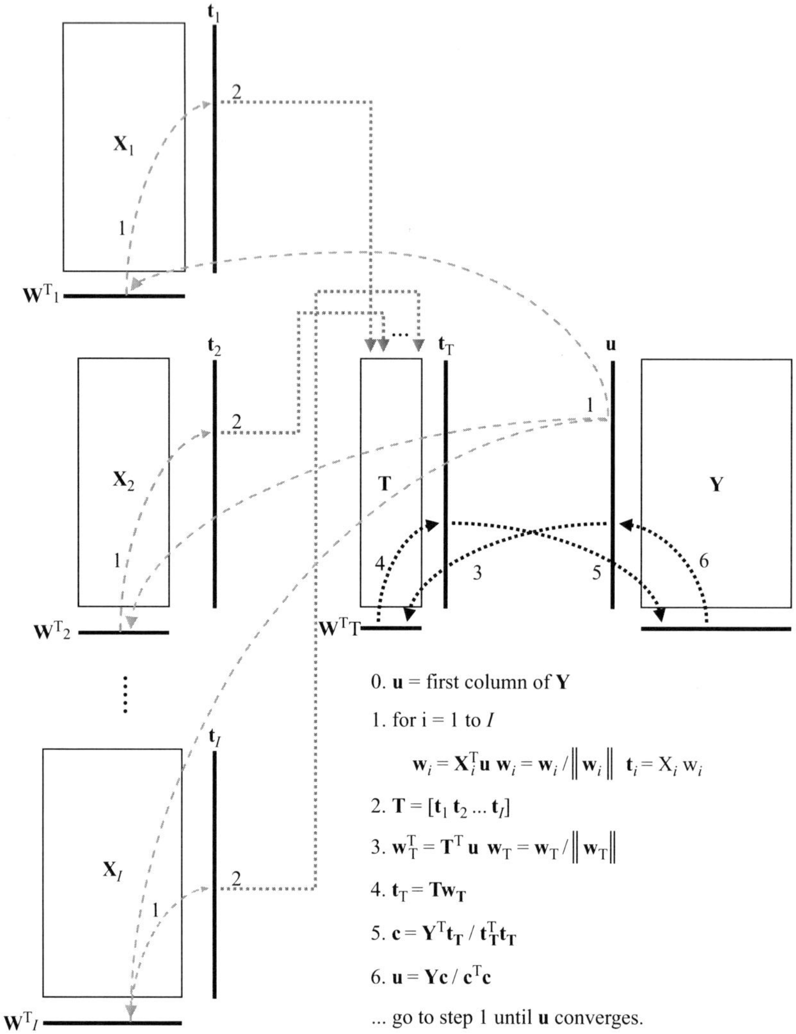

Fig. 6.5 Algorithm and arrow scheme of the multiblock PLS method

MBPLSPM is an extension of the PLS regression method useful to deal with causal models, where there can be more than one set of predictors and more complex relations: various predictor blocks, various predicted blocks and various blocks which are simultaneously predictor and predicted.

In two-blocks PLS and Multiblock PLS methods we denote by $\mathbf{X}_b$ the descriptor blocks and by $\mathbf{Y}$ the response blocks, also we call $\mathbf{t}_b$ the latent variable for a descriptor block $\mathbf{X}_b$ and by $\mathbf{u}$ the latent variable for a response block $\mathbf{Y}$. Now, in the MBPLSPM method, there can be several predictor blocks, several response blocks and several blocks that are simultaneously predictor and response blocks, because of this, in the MBPLSPM method we denote by B the overall number of blocks and by $\mathbf{X}_b, b = 1, \ldots, B$, the blocks. We assume that the blocks numeration in the model is logically specified from left $(1, 2, \ldots)$ to right $(B-2, B-1, B)$, that is, if a block i is the precedent of another block j, necessarily $i < j$. Left end blocks are defined as blocks that only predict, right end blocks are blocks that are predicted but do not predict and interior blocks both predict and are predicted. The left end blocks are also called exogenous blocks and the interior blocks altogether with the right end blocks are also called endogenous blocks. If we considerer the causal model in Fig. 6.1, we can see six left end or exogenous blocks (from block 1 to block 6), two interior blocks (blocks 7 and 8) and an unique right end block (block 9). The three last blocks are endogenous blocks.

In MBPLSPM each block $\mathbf{X}_b$ has associated two latent variables, $\mathbf{t}_b$ and $\mathbf{u}_b$, instead of one, as in the previous methods. The first latent variable, $\mathbf{t}_b$, summarizes the information contained in block $\mathbf{X}_b$ considering that it must predict the $\mathbf{X}_j$ blocks with $\beta_{jb} \neq 0$. The second latent variable, $\mathbf{u}_b$, summarizes the information contained in block $\mathbf{X}_b$ considering that it must be predicted by the $\mathbf{X}_j$ blocks with $\beta_{bj} \neq 0$. If $\mathbf{X}_b$ is a predictor block, we are only interested in $\mathbf{t}_b$, if $\mathbf{X}_b$ is a response block, we are only interested in $\mathbf{u}_b$, but if $\mathbf{X}_b$ is simultaneously a predictor and a response block, we are interested in both latent variables.

MBPLSPM Algorithm

Step 0. Initialization

For b increasing from 1 to B: $\mathbf{t}_b$ and $\mathbf{u}_b$ = the first column of $\mathbf{X}_b$

Step 1. Backward phase

For b decreasing from B to 1
if $\mathbf{X}_b$ predicts no blocks then: set $\mathbf{t}_b = \mathbf{u}_b$
if $\mathbf{X}_b$ predicts only the block $\mathbf{X}_{kb}$ then: $\mathbf{w}_b = \mathbf{X}_b^{\mathrm{T}}\mathbf{u}_{kb} \Rightarrow \mathbf{t}_b = (\mathbf{X}_b\mathbf{w}_b)^*$ (remember that the asterisk means that the variable in parentheses is standardized)
if $\mathbf{X}_b$ predicts $Nb > 1$ blocks then: $\mathbf{U}_b = [\mathbf{u}_{b1}, \mathbf{u}_{b2}, \ldots, \mathbf{u}_{bNb}]$

$$\mathbf{c}_{\mathrm{U}b} = \mathbf{U}_b^{\mathrm{T}}\mathbf{t}_b \Rightarrow \mathbf{u}_{\mathrm{U}b} = \mathbf{U}_b\mathbf{c}_{\mathrm{U}b} \Rightarrow \mathbf{w}_b = \mathbf{X}_b^{\mathrm{T}}\mathbf{u}_{\mathrm{U}b} \Rightarrow \mathbf{t}_b = (\mathbf{X}_b\mathbf{w}_b)^*$$

Step 2. Forward phase

For b increasing from 1 to B
if $\mathbf{X}_b$ is predicted by no other blocks, then: set $\mathbf{u}_b = \mathbf{t}_b$
if $\mathbf{X}_b$ is predicted by one block $\mathbf{X}_{kb}$, then: $\mathbf{c}_b = \mathbf{X}_b^{\mathrm{T}}\mathbf{t}_{kb} \Rightarrow \mathbf{u}_b = (\mathbf{X}_b\mathbf{c}_b)^*$
if $\mathbf{X}_b$ is predicted by $Nb > 1$ blocks, then: $\mathbf{T}_b = [\mathbf{t}_{b1}, \mathbf{t}_{b2}, \ldots, \mathbf{t}_{bNb}]$

$$\mathbf{w}_{\mathrm{T}b} = \mathbf{T}_b^{\mathrm{T}}\mathbf{u}_b \Rightarrow \mathbf{t}_{\mathrm{T}b} = \mathbf{T}_b\mathbf{w}_{\mathrm{T}b} \Rightarrow \mathbf{c}_b = \mathbf{X}_b^{\mathrm{T}}\mathbf{t}_{\mathrm{T}b} \Rightarrow \mathbf{u}_b = (\mathbf{X}_b\mathbf{c}_b)^*$$

After completing one backward plus forward cycle (Steps 1 and 2 respectively), all the right end blocks $\mathbf{u}_b$ vectors are tested for convergence. If, within desired precision, these $\mathbf{u}_b$ are the same as they were during the previous iteration, go to step 3, otherwise return to step 1.

Step 3. Path coefficients calculation

For each b, estimate the β_{bj} values as the coefficient of $\mathbf{t}_j$ in the PLS regression of $\mathbf{t}_b$ on all the $\mathbf{t}_j$'s related to the precedents of ξ_b.

In the multiblock PLS algorithm from Wangen and Kowalski (1988) the loading vectors **w** and **c** are standardized and the scores **t** and **u** are not. Nevertheless, following current causal modeling practice, we standardize only the latent variables **t** and **u**. This is also useful to make easier the comparison with the LVPLS algorithm. Westerhuis et al. (1998) study and compare different variations from the original multiblock PLS algorithm, but only with blocks that are exclusively predictor or predicted blocks, not including blocks that perform simultaneously both functions.

In a complete cycle of the algorithm (backward and forward phases) each block is taken into account in all their roles played. For instance, if we think in block 9 in our model from Fig. 6.1, the relevant part of the model is reduced to the submodel in Fig. 6.6.

Blocks 1, 2 and 8 are clearly related with block 9 because they predict it. Block 7 is also related with block 9 because when we estimate the latent variable for block 2, precedent of block 9, we need to take into account that block 2 is also precedent of block 7.

In backward phase we see that block 1 and block 8 only predict block 9 and then we estimate $\mathbf{t}_1$ and $\mathbf{t}_8$ from $\mathbf{u}_9$:

$$\mathbf{w}_1 = \mathbf{X}_1^{\mathrm{T}}\mathbf{u}_9 \Rightarrow \mathbf{t}_1 = (\mathbf{X}_1\mathbf{w}_1)^* \,,$$
$$\mathbf{w}_8 = \mathbf{X}_8^{\mathrm{T}}\mathbf{u}_9 \Rightarrow \mathbf{t}_8 = (\mathbf{X}_8\mathbf{w}_8)^* \,.$$

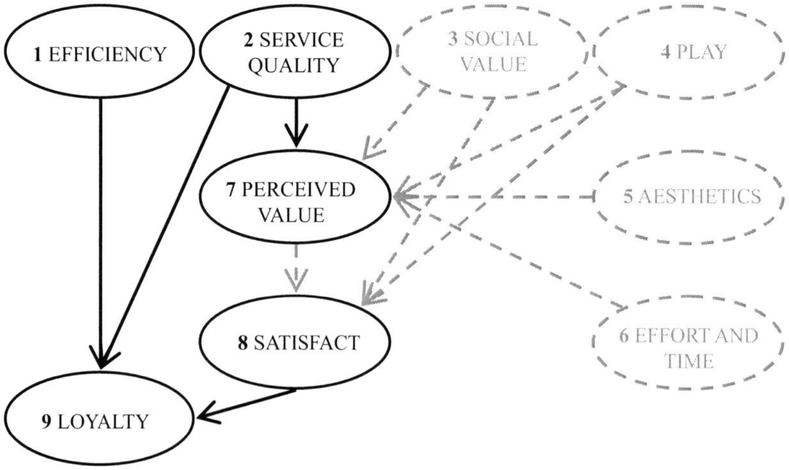

Fig. 6.6 Reduced model including only the blocks and paths that are explicitly related with block 9 in the MBPLSPM algorithm. Grey part of the figure corresponds to blocks and links not explicitly related with block 9 in the algorithm

Block 2 predicts block 9, but block 2 also predicts block 7, then we estimate $\mathbf{t}_2$ from $\mathbf{U}_2 = [\mathbf{u}_7\ \mathbf{u}_9]$:

$$\mathbf{c}_{U2} = \mathbf{U}_2^T\mathbf{t}_2 \Rightarrow \mathbf{u}_{U2} = \mathbf{U}_2\mathbf{c}_{U2} \Rightarrow \mathbf{w}_2 = \mathbf{X}_2^T\mathbf{u}_{U2} \Rightarrow \mathbf{t}_2 = (\mathbf{X}_2\mathbf{w}_2)^* .$$

In forward phase we see that block 9 is predicted by blocks 1, 2 and 8 and we estimate $\mathbf{u}_9$ from $\mathbf{T}_9 = [\mathbf{t}_1\ \mathbf{t}_2\ \mathbf{t}_8]$:

$$\mathbf{w}_{T9} = \mathbf{T}_9^T\mathbf{u}_9 \Rightarrow \mathbf{t}_{T9} = \mathbf{T}_9\mathbf{w}_{T9} \Rightarrow \mathbf{c}_9 = \mathbf{X}_9^T\mathbf{t}_{T9} \Rightarrow \mathbf{u}_9 = (\mathbf{X}_9\mathbf{c}_9)^* .$$

The previous description shows that we need both phases (backward and forward) to consider the different roles played by block 9 in relation to the other blocks in the structural model, and this is also true for the other blocks.

For better understanding the MBPLSPM algorithm, Figs. 6.7–6.9 show how the MBPLSPM algorithm deal with three different situations in a hypothetic causal model. The first is a single link (Fig. 6.7), the second consists of various precedents for a block (Fig. 6.8) and the last consists of various consequents for a block (Fig. 6.9).

In Fig. 6.8 the expression $\hat{b}_{T3} = \frac{\mathbf{u}_3^T\mathbf{t}_{T3}}{\mathbf{t}_{T3}^T\mathbf{t}_{T3}}$ comes from an ordinary least squares regression model but in our implementation we have changed this for a PLS regression model.

The expression $\hat{b}_{U1} = \frac{\mathbf{u}_{U1}^T\mathbf{t}_1}{\mathbf{t}_1^T\mathbf{t}_1}$ comes from an ordinary least squares regression model but in our implementation we have changed this for a PLS regression model.

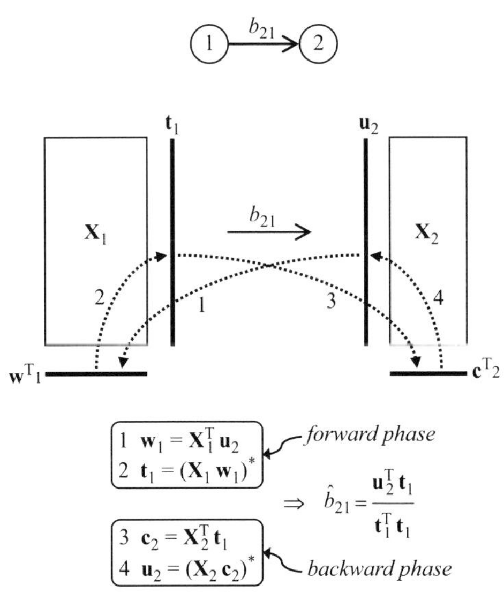

Fig. 6.7 MBPLSPM algorithm dealing with a single link. b_{21} is the estimation for the coefficient β_{21}

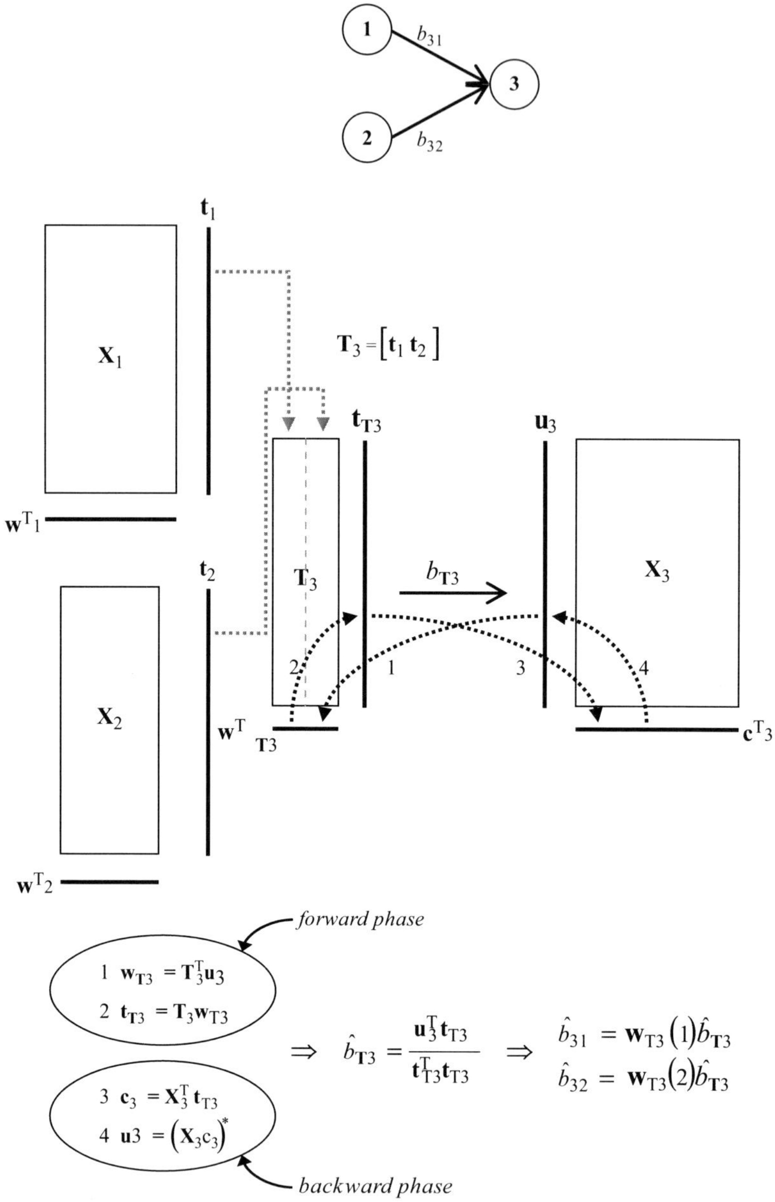

Fig. 6.8 MBPLSPM dealing with various precedents for a block. $\mathbf{w}_{T3}(j)$, with $j = 1, 2$, is the jth coordinate of vector $\mathbf{w}_{T3}$. b_{31} and b_{32} are the estimations for the coefficients β_{31} and β_{32} respectively

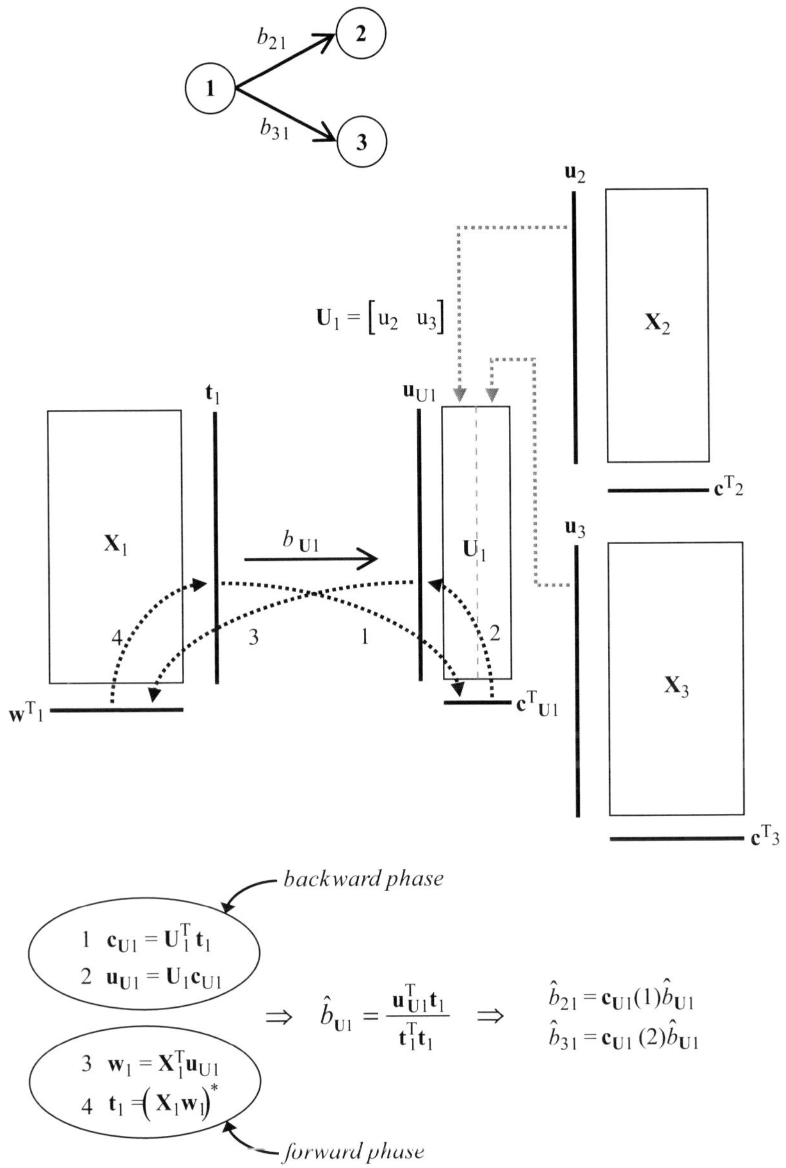

Fig. 6.9 MBPLSPM dealing with various consequents for a block. $\mathbf{c}_{U1}(j)$, with $j = 1, 2$, is the jth coordinate of vector $\mathbf{c}_{U1}$. b_{21} and b_{31} are the estimations for the coefficients β_{21} and β_{31} respectively

6.5 The Data Set

For the purposes of the paper, the empirical setting is grounded on services marketing research, where the efforts and discussions made for the last decades have enabled researchers to obtain a better understanding of the conceptual and methodological relationships between service quality, perceived value and customer satisfaction. Links and associations between service quality and customer satisfaction have been investigated deeply (Ngobo 1997; Giese and Cote 2000; Oliver 1997; Caruana et al. 2000; Brady et al. 2002; Grace and O'Cass 2005). Value has been a rather neglected aspect in customer's assessment of services during the nineties (Petrick 2002; Lin et al. 2005). Although since 2000, both academically and managerially the interest on value research has been deeply tackled. Consequently, in recent years, most modern theoretical proposals take discussion to a higher order of perceived value, where "value becomes a superordinate concept subsuming quality" (Oliver 1999, p. 58), or where "customer satisfaction management needs to be backed-up with in-depth learning about customer value" (Woodruff 1997, p. 139).

The range of empirical studies on methodological links among service quality, perceived value and customer satisfaction is very wide: Fig. 6.10 proposes a review that shows links and constructs. As in recent years a special interest on the loyalty behavior has emerged, we have also considered the loyalty construct, along with the other three classical constructs (quality, satisfaction and value). Since 1999, tourism services are one of the most preferred fields for exploring and assessing post consumption behavior (e.g. Oh 1999, 2000, 2003; Petrick et al. 2001; Gallarza and Gil 2006).

As Fig. 6.10 shows, generally, the link between quality and value provides the widest consensus among authors, quality being an input of value. Methodologically, the quality-satisfaction proposal is more common as Fig. 6.10 shows, with some remarkable exceptions such as Bolton and Drew (1991). Regarding the discussion on the relative superiority of value or satisfaction, although some authors propose a satisfaction-value link (e.g. Petrick et al. 2001; Chiou 2004), most of them consider value as the best and most complete antecedent of satisfaction (Fornell et al. 1996, McDougall and Levesque 2000; Babin and Kim 2001). Furthermore, the cumulative insights of services literature support the general notion that both value and satisfaction contribute to loyalty as positive behavioral intention (Parasuraman and Grewal 2000; Cronin et al. 2000). Consequently, even with some exceptions as Fig. 6.10 shows, we could say that a certain consensus exists in the literature regarding a natural chain between quality, value and satisfaction, which has led, in recent years, to customer loyalty as a final outcome. This theoretical chain of constructs has led to an important research on the assessment of links between quality, value and satisfaction, with causal modeling as a natural methodological approach.

In the present study, we explore relationships between classical constructs (service quality, satisfaction, perceived value and loyalty) as a post-consumption sequence. Additionally, given the multidimensional nature of consumption value (Sheth et al. 1991; Babin et al. 1994; Parasuraman and Grewal 2000) we postulate that positive and negative value dimensions can have positive and negative effects

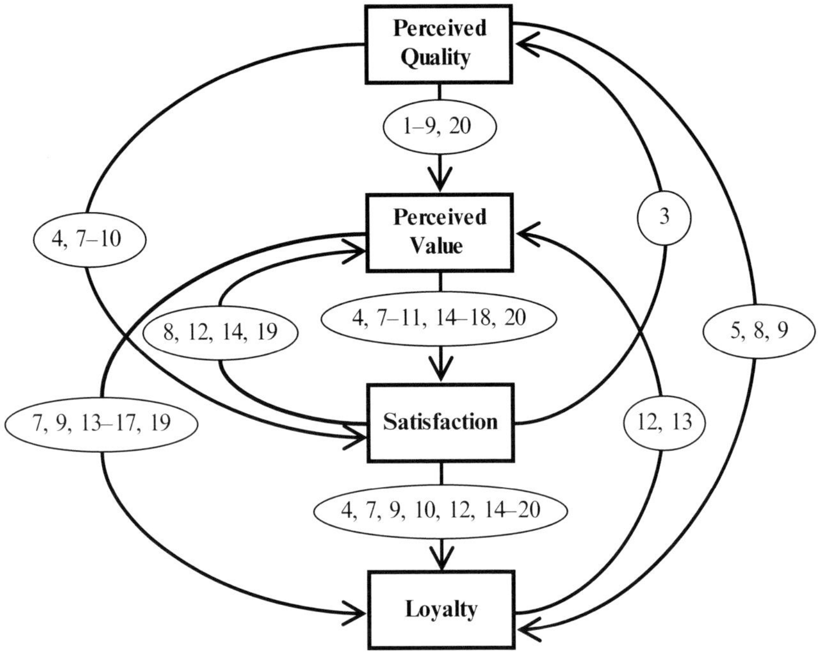

Authors:

1 Zeithaml (1988)	11 Babin and Kim (2001)
2 Dodds, Monroe and Grewal (1991)	12 Petrick et al. (2001)
3 Bolton and Drew (1991)	13 Oh (2003)
4 Fornell et al. (1996)	14 Chiou (2004)
5 Sirohi, McLaughlin and Wittink (1998)	15 Yang and Peterson (2004)
6 Oliver (1999)	16 Lin and Wang (2006)
7 Oh (1999)	17 Grace and O'Cass (2005)
8 Oh (2000)	18 Lin et al. (2005)
9 Cronin, Brady and Hult (2000)	19 Duman & Mattila (2005)
10 McDougall and Levesque (2000)	20 Gallarza & Gil (2006)

Fig. 6.10 Constructs and links in post consumption behavior: a review

on some of these constructs. Following Holbrook's theory, the eight cells of his typology could be considered as positive value inputs. Hence, among positive value dimensions, we consider *efficiency*, *quality*, *play*, and *aesthetics* as more representative of an individualistic approach of consumer behavior study and *social value* as a representation of all social interactions when consuming (Holbrook 1999; Oliver 1999).

Concerning the negative inputs, according to Gallarza and Gil (2006) we introduce *time and effort invested* as the main cost of consuming, directly related to *perceived value*.

According to Holbrook's typology, the affective vs cognitive nature of value dimensions allows us to consider potential links among cognitive antecedents (efficiency and service quality) and both loyalty and value and affective antecedents (social value and play) and both satisfaction and loyalty. More precisely, we considered the following links supported by the following hypothesis:

H1: Perceptions of efficiency are related positively to loyalty.
H2(a): Perceptions of service quality are related positively to perceived value.
H2(b): Perceptions of service quality are related positively to loyalty.
H3(a): Perceptions of play are related positively to perceived value.
H3(b): Perceptions of play are related positively to satisfaction.
H4: Perceptions of aesthetics are related positively to perceived value.
H5(a): Perceptions of social value are related positively to perceived value.
H5(b): Perceptions of social value are related positively to satisfaction.
H6: Perceptions of time and effort spent are related negatively to perceived value.

Additionally, according to the aforementioned discussion on the primacy of major consumer behavioral constructs, we propose a sequential relationship between perceived value, satisfaction and loyalty: so, we also postulated that:

H7: *Perceived value is related positively to satisfaction* and
H8: *Satisfaction is related positively to loyalty.*

A combination of sources was used in the construction of scales (see Appendix 2): Holbrook (1999) as a conceptual proposal on value dimensionality, literature review on tourism behavior and previous qualitative techniques (three in deep interviews and five group discussions).

Concerning endogenous variables, satisfaction was measured using a previously applied and reliable scale (Cronin et al. 2000). The perceived value scale came form the same source but an additional indicator was included, according to Zeithaml's definition of value as a trade-off between get and give elements. For the loyalty scale, according to tourism services literature we measured several behavioral intentions: we have considered both the visit to the same destination and to other destinations in the same area (Murphy et al. 2000), but also a positive word of mouth (Kozak and Rimmington 2000), both to the destination and to the agency (Petrick et al. 2001). We also conducted a pilot study among students and thus we made a few corrections and adjustment in the wording and structure of the questionnaire. A five-point Liker-type scale was used for the nine latent constructs (44 indicators). The data were taken from an academic investigation on the post-consumption behavior of a convenience sample of university students: 273 undergraduate students who travel in the spring break of their third and/or fifth year at university. See Fig. 6.1 for the representation of the structure described by the hypothesis.

6.6 The Causal Model Estimated

In this section we estimate the model with the known Lohmöller's LVPLS method and with the multiblock PLS Path Modeling method, that for short is called the MBPLS based method, or simply MBPLS. The aim is to compare the estimations and the performance of both methods.

In Table 6.1 we give the statistics for checking the unidimensionality of each block.

Except for the second eigenvalue of blocks Social Value and Loyalty, that are bigger than one, all these statistics lead to an acceptation of the unidimensionality of all but two blocks. Nevertheless, because the Cronbach's α is bigger than 0.70 for both blocks, we consider all the blocks as unidimensional ones.

The estimated coefficients for the structural relations, the β's, built from the hypothesis are show in Table 6.2.

We can see that there are not great differences between both sets of estimations as it happens with the outer weights (see Table 6.3), that are the factors used to build each latent variable from their manifest variables, and with the correlations between the manifest variables and their latent variables (see Table 6.4).

Table 6.1 Check for block unidimensionality

		1st principal component		2nd principal component	
	Cronbach's α	Eigenvalue	Expl. var. (%)	Eigenvalue	Expl. var. (%)
1 Efficiency	0.7348	2.259	56.47	0.796	19.90
2 Service quality	0.9418	6.151	68.34	0.868	9.64
3 Social value	0.7251	2.442	48.84	1.304	26.08
4 Play	0.8471	2.749	68.72	0.555	13.89
5 Aesthetics	0.8156	2.193	73.10	0.474	15.80
6 Effort and time spent	0.7962	3.193	45.61	0.899	12.84
7 Perceived value	0.8872	2.448	81.60	0.306	10.20
8 Satisfaction	0.8709	2.384	79.48	0.333	11.10
9 Loyalty	0.8043	3.050	50.84	1.337	22.28

Table 6.2 Estimated coefficients with both methods for the structural relations

From	To	Estimated coefficients	
		LVPLS	MBPLS
1 Efficiency	9 Loyalty	0.2362	0.2628
2 Service quality	7 Perceived value	0.1737	0.1730
2 Service quality	9 Loyalty	0.1133	0.1375
3 Social value	7 Perceived value	0.2998	0.2998
3 Social value	8 Satisfaction	0.2490	0.2483
4 Play	7 Perceived value	0.3330	0.3305
4 Play	8 Satisfaction	0.3361	0.3358
5 Aesthetics	7 Perceived value	0.1248	0.1249
6 Effort and time spent	7 Perceived value	0.1655	0.1765
7 Perceived value	8 Satisfaction	0.3417	0.3395
8 Satisfaction	9 Loyalty	0.5506	0.5133

Table 6.3 Outer weights for both methods

	Outer weight	
	LVPLS	MBPLS
Efficiency		
Efficiency 1	0.0152	0.0214
Efficiency 2	0.0200	0.0163
Efficiency 3	0.0221	0.0206
Efficiency 4	0.0225	0.0226
Service quality		
Service quality 1	0.0078	0.0077
Service quality 2	0.0076	0.0092
Service quality 3	0.0084	0.0086
Service quality 4	0.0082	0.0079
Service quality 5	0.0084	0.0076
Service quality 6	0.0083	0.0069
Service quality 7	0.0083	0.0081
Service quality 8	0.0087	0.0096
Service quality 9	0.0077	0.0078
Social value		
Social value 1	0.0208	0.0216
Social value 2	0.0208	0.0198
Social value 3	0.0184	0.0203
Social value 4	0.0130	0.0092
Social value 5	0.0116	0.0130
Play		
Play 1	0.0172	0.0157
Play 2	0.0177	0.0167
Play 3	0.0198	0.0192
Play 4	0.0184	0.0213
Aesthetics		
Aesthetics 1	0.0227	0.0177
Aesthetics 2	0.0240	0.0268
Aesthetics 3	0.0242	0.0259
Effort and time		
Effort and time 1	−0.0127	−0.0056
Effort and time 2	−0.0115	−0.0053
Effort and time 3	−0.0118	−0.0144
Effort and time 4	−0.0146	−0.0154
Effort and time 5	−0.0150	−0.0177
Effort and time 6	−0.0137	−0.0170
Effort and time 7	−0.0098	−0.0115
Perceived value		
Perceived value 1	0.0227	0.0234
Perceived value 2	0.0216	0.0204
Perceived value 3	0.0228	0.0234

(continued)

Table 6.3 (continued)

	Outer weight	
	LVPLS	MBPLS
Satisfaction		
Satisfaction 1	0.0220	0.0209
Satisfaction 2	0.0234	0.0224
Satisfaction 3	0.0226	0.0246
Loyalty		
Loyalty 1	0.0124	0.0114
Loyalty 2	0.0125	0.0113
Loyalty 3	0.0172	0.0154
Loyalty 4	0.0134	0.0158
Loyalty 5	0.0144	0.0179
Loyalty 6	0.0151	0.0132

Table 6.4 Correlations with both methods between manifest and latent variables

	Corr. with the LV	
	LVPLS	MBPLS
Efficiency		
Efficiency 1	0.566	0.647
Efficiency 2	0.745	0.700
Efficiency 3	0.825	0.801
Efficiency 4	0.839	0.832
Service quality		
Service quality 1	0.792	0.797
Service quality 2	0.767	0.782
Service quality 3	0.847	0.854
Service quality 4	0.827	0.818
Service quality 5	0.852	0.844
Service quality 6	0.843	0.832
Service quality 7	0.847	0.846
Service quality 8	0.878	0.879
Service quality 9	0.780	0.781
Social value		
Social value 1	0.837	0.850
Social valuc 2	0.836	0.832
Social value 3	0.741	0.763
Social value 4	0.525	0.478
Social value 5	0.466	0.459
Play		
Play 1	0.782	0.765
Play 2	0.800	0.791
Play 3	0.897	0.898
Play 4	0.833	0.854

(continued)

Table 6.4 (continued)

	Corr. with the LV	
	LVPLS	MBPLS
Aesthetics		
Aesthetics 1	0.820	0.779
Aesthetics 2	0.868	0.888
Aesthetics 3	0.876	0.889
Effort and time		
Effort and time 1	−0.668	−0.575
Effort and time 2	−0.607	−0.515
Effort and time 3	−0.619	−0.663
Effort and time 4	−0.767	−0.761
Effort and time 5	−0.791	−0.807
Effort and time 6	−0.719	−0.758
Effort and time 7	−0.514	−0.540
Perceived value		
Perceived value 1	0.897	0.900
Perceived value 2	0.911	0.906
Perceived value 3	0.902	0.904
Satisfaction		
Satisfaction 1	0.883	0.877
Satisfaction 2	0.904	0.900
Satisfaction 3	0.887	0.897
Loyalty		
Loyalty 1	0.745	0.715
Loyalty 2	0.656	0.618
Loyalty 3	0.809	0.771
Loyalty 4	0.681	0.738
Loyalty 5	0.658	0.722
Loyalty 6	0.712	0.681

The path coefficients (β_{ij}) have been estimated with both algorithms (see Table 6.2). For assessing significativity of the estimations, as we have no previous hypothesis about the data distribution, we used the bootstrap method (Efrom and Tibshirani 1993), taking 10,000 samples with replacement of 273 individuals from the original sample. It is shown in Table 6.5 the bootstrap confidence intervals for the estimated coefficients with both methods (from the 2.5 to the 97.5 percentile of the 10,000 values obtained for each coefficient with both methods).

We see again that the two methods lead to similar results, but in all the eleven cases the confidence interval width is smaller for the MBPLS based method, that is, there is less uncertainty for the estimated coefficients with the MBPLS based method than the calculated with the LVPLS method.

In order to compare the performance of both algorithms, it is interesting to calculate the explained variance percentage for the endogenous latent variables

Table 6.5 Bootstrap confidence intervals for the path coefficients with both methods

From	To	Bootstrap percentiles			
		LVPLS		MBPLS	
		$P_{2.5}$	$P_{97.5}$	$P_{2.5}$	$P_{97.5}$
1 Efficiency	9 Loyalty	0.1585	0.3157	0.1944	0.3406
2 Service quality	7 Perceived value	0.1129	0.2262	0.1144	0.2236
2 Service quality	9 Loyalty	0.0189	0.1982	0.0518	0.2146
3 Social value	7 Perceived value	0.2466	0.3484	0.2505	0.3457
3 Social value	8 Satisfaction	0.2053	0.2850	0.2072	0.2840
4 Play	7 Perceived value	0.2777	0.3816	0.2785	0.3742
4 Play	8 Satisfaction	0.3000	0.3726	0.3000	0.3708
5 Aesthetics	7 Perceived value	0.0569	0.1990	0.0650	0.1970
6 Effort and time spent	7 Perceived value	0.1035	0.2222	0.1307	0.2320
7 Perceived value	8 Satisfaction	0.2995	0.3859	0.2972	0.3815
8 Satisfaction	9 Loyalty	0.4577	0.6384	0.4256	0.5938

Table 6.6 Estimated R^2 for the endogenous constructs and its bootstrap confidence intervals

Latent variable	LVPLS			MBPLS		
	Estimated	Bootstrap percentiles		Estimated	Bootstrap percentiles	
	R^2 (%)	$P_{2.5}$(%)	$P_{97.5}$(%)	R^2(%)	$P_{2.5}$(%)	$P_{97.5}$(%)
7 Perceived value	48.07	38.29	58.10	49.42	40.93	59.57
8 Satisfaction	59.32	49.48	68.43	59.41	49.71	68.55
9 Loyalty	52.82	44.28	61.84	53.50	45.65	62.55

(R^2), which is slightly higher when applying the MBPLS based method than when applying the LVPLS algorithm (see Table 6.6).

Again the confidence interval width is smaller for the MBPLS based method than the correspondent to the LVPLS method.

In order to highlight the significativity of the slight difference between the R^2 value from the MBPLS and the LVPLS methods $\left(R^2_{MBPLS} - R^2_{LVPLS}\right)$ we can see the bootstrap distribution function for this difference (see Fig. 6.11).

As Fig. 6.11 shows, the explained variance for the 7th latent variable (perceived value) is higher with the MBPLS algorithm than with LVPLS algorithm for the 99.70% of the samples; for the 8th latent variable (satisfaction) this percentage reaches 70.34% and 93.58% for the 9th latent variable (loyalty).

As a conclusion, we can affirm that our hypothesis were all supported. The R^2 values are quite high, and every path is significant. Among predictors, the impact of play dimension on perceived value is quite high (0.33); but play dimension also provides a strong link with satisfaction (0.34), showing a clear prominence of the affective dimension of the tourism experience investigated. Social value is also relevant for the perceived value (0.30) and for satisfaction (0.25). The more cognitive antecedents (efficiency and service quality) are linked to loyalty behavior (0.26 and 0.14 respectively), as a willingness to recommend and/or repurchase. Besides,

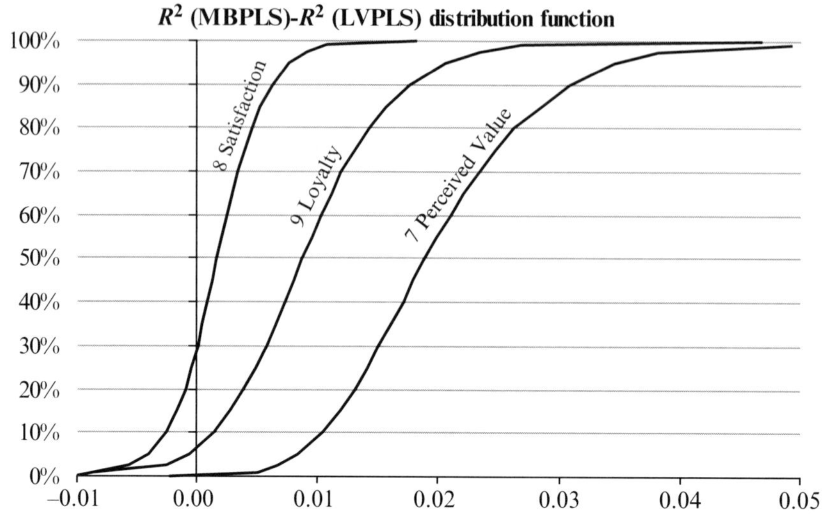

Fig. 6.11 Estimated distribution function for the R^2 difference between the two methods for the 10,000 bootstrap samples

affective antecedents (play and social value) are better predictors of both customer satisfaction and perceived value.

Concerning the proposal of investigating the quality–value–satisfaction–loyalty chain, the study indicates a clear pattern: perceived value is then a mediator between quality and satisfaction, satisfaction being the behavioral consequence of perceived value, and leading also to loyalty behavior. Thus, the consumer evaluation investigated is well modeled in a complex system where positive and negative value dimensions have effects on three behavioral constructs: satisfaction, perceived value and loyalty.

6.7 Simulated Data

In order to confirm the slight superiority of the MBPLS based method from the R^2 criterion we use simulated data from a hypothetic causal model (see Fig. 6.12).

The model has four latent variables and there are five structural relations summarized in three equations:

$$\begin{aligned} \xi_2 &= \beta_{21}\xi_1 + \zeta_2 \\ \xi_3 &= \beta_{31}\xi_1 + \beta_{32}\xi_2 + \zeta_3. \\ \xi_4 &= \beta_{41}\xi_1 + \beta_{43}\xi_3 + \zeta_4 \end{aligned}$$

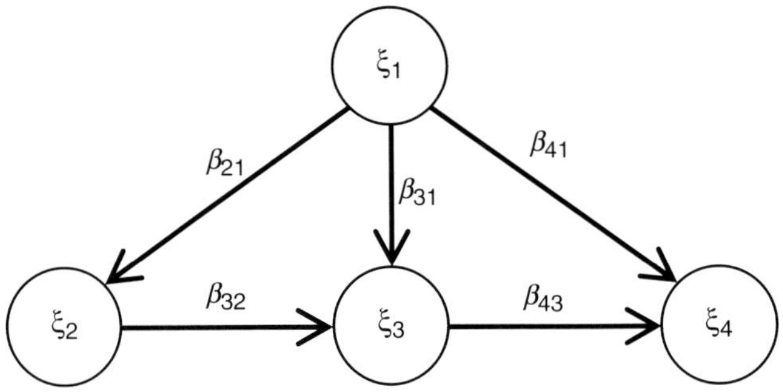

Fig. 6.12 Hypothetic model

To generate the four latent variables we follow the next process:

1. Select a value for each one of the five coefficients and calculate the covariance matrix **S**, being $Var(\xi_b) = 1$ for all b.
   By example, if $\beta_{21} = 0.7$; $\beta_{31} = 0.2$; $\beta_{32} = 0.6$; $\beta_{41} = 0.3$; $\beta_{43} = 0.7$, results:

$$\mathbf{S} = \begin{bmatrix} 1.000 & 0.700 & 0.620 & 0.734 \\ 0.700 & 1.000 & 0.740 & 0.728 \\ 0.620 & 0.740 & 1.000 & 0.886 \\ 0.734 & 0.728 & 0.886 & 1.000 \end{bmatrix}$$

2. Generate $n = 150$ samples from a multivariate normal distribution with zero mean and covariance matrix **S**, obtaining $\mathbf{L} = \{l_{ib}\}$, a 150 by 4 matrix with l_{ib} being the value of the b latent variable for the ith subject.
3. To build $\mathbf{X} = \begin{bmatrix} \mathbf{X}_1\ \mathbf{X}_2\ \mathbf{X}_3\ \mathbf{X}_4 \end{bmatrix}$, where $\mathbf{X}_b = \mathbf{l}_b \cdot \begin{bmatrix} 0.6\ 0.7\ 0.8\ 0.9 \end{bmatrix} + \mathbf{E}_b$, being $\mathbf{E}_b$ a 150 by 4 matrix which elements are normal with zero mean and standard deviation 0.9 and $\mathbf{l}_b$ the bth column of **L**.
4. To centre and autoscale the **X** matrix to obtain 16 columns (4 manifest variables for each one of the 4 latent variables) with zero mean and standard deviation one.
5. To apply the LVPLS and MBPLS methods on **X** and calculate the correspondent R^2 values for the three endogenous latent variables (ξ_2, ξ_3 and ξ_4).

Following the previous process we generate a set of four blocks of manifest variables for 150 individuals that verify the hypothesis for a casual model with the reflective way. We can now estimate the model with both methods and, in particular, we can measure the performance of the estimation from the R^2 statistic for the endogenous blocks.

We are interested in the distribution for the difference $R^2_{MBPLS} - R^2_{LVPLS}$, and for this we apply the Monte Carlo method, that is, we run the previous process, from step one to step four, 1,000 independent times, obtaining 1,000 independent values for the difference $R^2_{MBPLS} - R^2_{LVPLS}$, this is, a sample for the difference, from which we can estimate its distribution function from their percentiles (see Fig. 6.13).

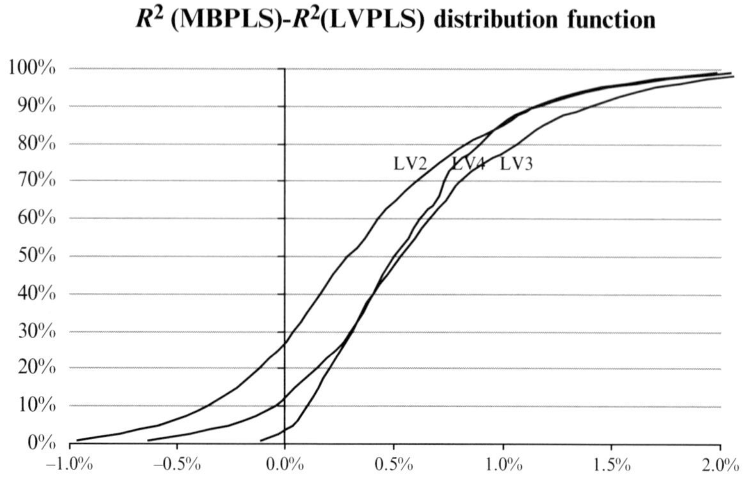

Fig. 6.13 Monte Carlo distribution function for the difference between the explained variance with both methods

As Fig. 6.13 shows, the explained variance for LV2 is higher with the MBPLS algorithm than with LVPLS algorithm for the 72.35% of the samples; for LV3 this percentage reaches 87.44% and 95.86% for LV4. This confirms that the MBPLS based method is slightly superior to the classic LVPLS under the R^2 criterion.

The hypothetic structural model from Fig. 6.10, the path coefficients selected in step 0 of the algorithm ($\beta_{21} = 0.7$, $\beta_{31} = 0.2$, $\beta_{32} = 0.6$, $\beta_{41} = 0.3$ and $\beta_{43} = 0.7$), the size of the blocks (150 by 4), the coefficients for the measure model (0.6, 0.7, 0.8 and 0.9) and the standard deviation for the residuals (0.9) selected in step 2 of the algorithm are a particular election. As the exposed results can be affected by this election, we have proved a variety of different situations (not shown), yielding that the slight superiority of the MBPLS based method over the classic LVPLS method is repeated in all cases.

6.8 Summary and Conclusions

In this work we have presented a new method to estimate causal models based on the multiblock PLS method from Wangen and Kowalski (1988). The new method has been compared with the classical LVPLS algorithm from Lohömoller (1985), using an academic investigation on the post-consumption tourism behavior of a particular profile of university students.

The results for both methods are quite similar (see Tables 6.2–6.4), but the explained percentage of variance (the R^2 coefficient) for the endogenous latent variables is slightly higher for the MBPLS based method (see Table 6.6).

From a bootstrap analysis we have built the confidence intervals for the estimated coefficients of the structural model (see Table 6.5) and for the R^2 coefficient correspondent to the endogenous latent variables (see Table 6.6), showing that in both cases the uncertainty is slightly smaller for the MBPLS based method.

We have also estimated the distribution function for the difference R^2 (*MBPLS*)$-$ R^2 (*LVPLS*) of the endogenous latent variables (see Fig. 6.10), showing that this difference is positive for a majority of the bootstrap samples: 99.7% for the 7th latent variable (perceived value), 70.34% for the 8th latent variable (satisfaction) and 93.5% for the 9th latent variable (loyalty).

To confirm these results we have built a hypothetic causal model fixing the coefficients for the structural model and simulating 1,000 independent sets of data adding normal distributed noise. We have applied both methods on each data set and have built the Monte Carlo distribution function for the difference R^2 (*MBPLS*)$-R^2$ (*LVPLS*) for the three endogenous latent variables (see Fig. 6.12), obtaining that the explained variance for LV2 is higher with the MBPLS algorithm than with LVPLS algorithm for the 72.35% of the samples; for LV3 this percentage reaches 87.44% and 95.86% for LV4, confirming the slightly superiority of the MBPLS based method to the classic LVPLS under the R^2 criterion.

Appendix 1 Notation

ξ_b	Unknown latent variable for the bth block
x_{bj}	jth measured variable for ξ_b
β_{ij}	path coefficient that indicates the influence of construct j over construct i
λ_{bj}	Unknown coefficient of ξ_b in the explained part of x_{bj} in reflective way
e_{bj}	White noise part of x_{bj} in reflective way, not correlated with ξ_b
ζ_b	White noise part of ξ_b not correlated with the precedents of ξ_b
$\mathbf{y}_b$	External estimation of the latent variable for block b in LVPLS method
$\mathbf{z}_b$	Internal estimation of the latent variable for block b in LVPLS method
d_{bj}	With non null β_{bj}: regression coefficient of $\mathbf{y}_j$ in the multiple regression of $\mathbf{y}_b$ on all the $\mathbf{y}_j$'s related to the predecessors of ξ_b
d_{jb}	With non null β_{jb}: correlation between $\mathbf{y}_j$ and $\mathbf{y}_b$
$\mathbf{X}$	Descriptor data in two-blocks PLS method
$\mathbf{Y}$	Response block in two-blocks and multiblock PLS methods
$\mathbf{X}_i$	Descriptor block in multiblock PLS method
$\mathbf{X}_b$	Descriptor and response blocks in the new MBPLSPM method
$\mathbf{t}$ $\mathbf{u}$ $\mathbf{t}_i$	Latent variable for $\mathbf{X}$, $\mathbf{Y}$ and $\mathbf{X}_i$, respectively
$\mathbf{t}_b\mathbf{u}_b$	Pair of latent variables for $\mathbf{X}_b$
$\mathbf{w}$ $\mathbf{w}_i$	Weight of variables in block $\mathbf{X}$ (two-blocks PLS) and $\mathbf{X}_i$ (multiblock PLS)
$\mathbf{w}_b$	Weight of variables in predictor block $\mathbf{X}_b$
$\mathbf{c}$	Weight of variables in block $\mathbf{Y}$
$\mathbf{c}_b$	Weight of variables in predicted block $\mathbf{X}_b$
$\mathbf{T}$	Super block containing all the $\mathbf{t}_i$'s in multiblock PLS method

$\mathbf{T}_b$	Super block containing all $\mathbf{t}_b$'s for blocks that predict $\mathbf{X}_b$ in MBPLSPM
$\mathbf{U}_b$	Super block containing all $\mathbf{u}_b$'s for blocks predicted by $\mathbf{X}_b$ in MBPLSPM
$\mathbf{w}_\mathbf{T}$	Weight of latent variables in super block $\mathbf{T}$
$\mathbf{t}_\mathbf{T}$	Super latent variable summarizing super block $\mathbf{T}$
$\mathbf{t}_{\mathbf{T}b}$	Super latent variable summarizing super block $\mathbf{T}_b$
$\mathbf{u}_{\mathbf{U}b}$	Super latent variable summarizing super block $\mathbf{U}_b$
n	Number of individuals in all blocks
K_b	Number of variables in block $\mathbf{X}_b$

Appendix 2 Scales and Sources Used

See Table 6.7.

Table 6.7 Scales and sources used in the questionnaire

Efficiency (5 items) Holbrook (1999), Heung and Qu (2000) + focus groups	Information received during the trip (maps, timetables, ...) was
	Infrastructures destination were
	Gastronomy at destination was
	Lodging facilities at destination where
Service quality (9 items) Cronin et al. (2000)	Provide service reliably, consistently and dependently
	Provide service in a timely manner
	Competent employees (knowledgeable and skillful)
	Approachable employees and easy to contact
	Courteous, polite and respectful employees
	Employees listen to me and we understood each other
	Employees were trustworthy, believable and honest
	Employees make the effort to understand my needs
	Employees were neat and clean
Social value (5 items) Adaptated from Sweeney and Soutar (2001) + focus groups	Reinforce my feeling of belonging to the group
	A better knowledge of my classmates
	Being socially accepted in the group
	Relationship with other tourists outside the group
	Relationship with residents
Play (4 items) Holbrook (1999), Babin and Kim (2001) + focus groups	I enjoyed the leisure (pubs, bars, ...)
	I enjoyed my free time
	The leisure was pleasurable
	I had fun in the destination

(continued)

Table 6.7 (continued)

Aesthetics (4 items) Adapted from Gallarza et al. (2002), Holbrook (1999) + focus groups	The city, its streets, buildings were ...
	Exhibitions, museums concerts were ...
	The beauty of the art (monuments) was ...
Time and effort spent (7 items) General tourism literature + focus groups	Cost of time planning and preparing
	Time spent in return trip
	Cost of time losses
	Cost associated with the time invested in the trip
	Opportunity cost associated with the trip
	Effort made for leaving tasks and works to do
	Mental effort made for leaving family and friends
Perceived value (3 items) Zeithaml (1988), Cronin et al. (2000)	Overall, the value of this experience is
	Comparing what I gave up and what I received ...
	The experience has satisfied my needs and wants
Satisfaction (3 items) Cronin et al. (2000)	My choice to purchase this trip was a wise one
	I did the right thing when I purchased this trip
	This experience is exactly what I needed
Loyalty (6 items) Adaptated from Murphy et al. (2000), Kozak and Rimmington (2000), Petrick et al. (2001)	Likelihood to return to same destination in next 5 years
	Likelihood to return to same area in next 5 years
	Likelihood to recommend the destination to friends and relatives
	Likelihood to recommend the agency to friends and relatives
	Same situation, same choice of agency
	Same situation, same choice of destination

Acknowledgements The authors are grateful to the Spanish Ministry of Science and Technology for financial support for this work (SEJ2007-66054).

References

Babin, B. J., Darden, W. R., & Griffin, M. (1994). Work and/or fun: Measuring hedonic and utilitarian shopping value. *Journal of Consumer Research*, *20*(Marzo), 644–656.

Babin, B. J., & Kim, K. (2001). International students travel behaviour: A model of the travel-related consumer/dissatisfaction process. *Journal of Travel and Tourism Marketing*, *10*(1), 93–106.

Bolton, R., & Drew, J. (1991). A multistage model of customers' assessments of service quality and value. *Journal of Consumer Research, 17*, 375–384.

Brady, M. K., Cronin, J. J., & Brand, R. R. (2002). Performance-only measurement of service quality: A replication and extension. *Journal of Business Research, 55*, 17–31.

Caruana, A., Money, A. H., & Berthon, P. R. (2000). Service quality and satisfaction – the moderating role of value. *European Journal of Marketing, 34*(11/12), 1338–1353.

Chiou, J. S. (2004). The antecedents of consumers' loyalty toward internet service providers. *Information and Management, 41*, 685–695.

Cronin, J. J., Brady, M. K., & Hule, G. T. M. (2000). Assessing the effects of quality, value and customer satisfaction on consumer behavioural intentions in service environments. *Journal of Retailing, 76*(2), 193–218.

Efrom, B., & Tibshirani, R. (1993). *Introduction to the bootstrap*. New York: Chapman-Hall.

Fornell, C., Johnson, M. D., Anderson, E. W., Cha, J., & Everitt Bryant, B. (1996). The American customer satisfaction index: Nature, purpose, and findings. *Journal of Marketing, 60*, 7–18.

Frank, I. E., Feikema, J., Constantine, N., & Kowalski, B. R. (1984). Prediction of product quality from spectral data using the partial least-squares method. *Journal of Chemical Information and Computer Science, 24*, 20–24.

Frank, I. E., & Kowalski, B. R. (1985). A multivariate method for relating groups of measurements connected by a causal pathway. *Analytica Chimica Acta, 167*, 51–63.

Gallarza, M. G., Gil, I., & Calderón, H. (2002). Destination image: Towards a conceptual framework. *Annals of Tourism Research, 29*(1), 56–78.

Gallarza, M. G., & Gil, I. (2006). Value dimensions, perceived value, satisfaction and loyalty: An investigation of university students' travel behavior. *Tourism Management, 27*(3), 437–452.

Geladi, P. (1988). Notes on the history and nature of partial least squares (PLS) modeling. *Journal of Chemometrics, 2*, 231–246.

Geladi, P., & Kowalski, B. R. (1986). Partial least-squares regression: A tutorial. *Analytica Chimica Acta, 185*, 1–17.

Gerlach, R. W., Kowalski, B. R., & Wold, H. (1979). Partial least squares path modeling with latent variables. *Analytica Chimica Acta, 112*, 417–421.

Giese, J. L., & Cote, J. A. (2000). Defining consumer satisfaction. *Academy of Marketing Science Review (on line), 00*(01), 1–34.

Grace, D. & O'Cass, A. (2005). An examination of the antecedents of re-patronage intentions across different retail store formats, *Journal of Retailing and Consumer Services, 12*, 227–243.

Guinot, C., Latreille, J., & Tenenhaus, M. (2001). PLS path modelling and multiple table analysis. Application to the cosmetics habits of women in Ile-de-France. *Chemometrics and Intelligent Laboratory Systems, 36*, 247–259.

Heung, V. C. S., & Qu, H. (2000). Hong Kong as a travel destination: An analysis of Japanese tourist' satisfaction levels, and the likelihood of them recommending Hong Kong to others. *Journal of Travel and Tourism Marketing, 9*(1/2), 57–80.

Holbrook, M. B. (1999). *Consumer value. A framework for analysis and research*. London: Routledge.

Hulland, J. (1999). Use of partial least squares (PLS) in strategic management research: A review of four recent studies. *Strategic Management Journal, 20*, 195–204.

Kozak, M., & Rimmington, M. (2000). Tourist satisfaction with Mallorca, Spain, as an off-season holiday destination. *Journal of Travel Research, 38*(3), 260–269.

Lin, H. H., & Wang, Y. S. (2006). An examination of the determinants of customer loyalty in mobile commerce contexts. *Information and Management, 43*, 271–282.

Lin, C. H., Sher, P. J., & Shih, H. Y. (2005). Past progress and future directions in conceptualizing customer perceived value. *International Journal of Service Industry Management, 16*(4), 318–336.

Lohmöller, J. B. (1987). *LVPLS program manual, version 1.8*, Zentralarchiv für Empirische Sozialforschung, Universität zu Köln, Köln.

Lohmöller, J. B. (1989). *Latent variables path modeling with partial least squares*. Heildelberg: Physica-Verlag.

McDougall, G. H. G., & Levesque, T. (2000). Customer satisfaction with services: Putting perceived value into the equation. *The Journal of Services Marketing*, *14*(5), 392–410.

Martens, H., & Naes, T. (1989). *Multivariate calibration*. New York: Wiley.

Murphy, P. E., Pritchard, M. P., & Smith, B. (2000). The destination product and its impact on traveller perceptions. *Tourism Management*, *21*, 43–52.

Ngobo, P.V. (1997). Qualité perçue et satisfaction des consommateurs: un état des recherches. *Revue Française de Marketing*, *163*, 67–69

Oh, H. (1999). Service quality, customer satisfaction, and customer value: A holistic perspective. *International Journal of Hospitality Management*, *18*(1), 67–82.

Oh, H. (2000). The effect of brand class, brand awareness, and price on customer value and behavioral intentions. *Journal of Hospitality and Tourism Research*, *24*(2), 136–162.

Oh, H. (2003). Price fairness and its asymmetric effects on overall price, quality, and value judgements: The case of an upscale hotel. *Tourism Management*, *24*, 397–399.

Oliver, R. L. (1997). *Satisfaction. A behavioural perspective on the consumer*. New York: McGraw Hill.

Oliver, R. L. (1999). Value as excellence in the consumption experience. In M. B. Holbrook (Ed.), *Consumer value. A framework for analysis and research*. London: Routledge.

Parasuraman, A., & Grewal, D. (2000). The impact of technology on the quality-value-loyalty chain: A research agenda. *Journal of the Academy of Marketing Science*, *28*(1), 168–174.

Petrick, J. F., Morais, D. D., & Norman, W. C. (2001). An examination of the determinants of entertainment vacationers' intentions to revisit. *Journal of Travel Research*, *40*, 41–48.

Petrick, J. F. (2002). Development of a multi-dimensional scale for measuring the perceived value of a service. *Journal of Leisure Research*, *34*(2), 119–134.

Sheth, J. N., Newman, B. I., & Gross, B. L. (1991). Why we buy what we buy: A theory of consumption values. *Journal of Business Research*, *22*, 159–170.

Stone, M., & Brooks, R. J. (1990). Continuum regression: Cross-validated sequentially constructed prediction embracing ordinary least squares, partial least squares and principal components regression (with discussion). *Journal of Royal Statistical Society Series B*, *52*, 237–269. Corrigendum (1992) 54, 906–907.

Sweeney, J., & Soutar, G. (2001). Consumer perceived value: The development of a multiple item scale. *Journal of Retailing*, *77*, 203–207.

Wangen, L. E., & Kowalski, B. R. (1988). A multiblock partial least squares algorithm for investigating complex chemical systems. *Journal of Chemometrics*, *3*, 3–20.

Westerhuis, J. A., & Coenegracht, P. M. J. (1997). Multivariate modeling of the pharmaceutical two-step process of wet granulation and tableting with multiblock partial least squares. *Journal of Chemometrics*, *11*, 379–392.

Westerhuis, J. A., Kourti, T., & MacGregor, J. F. (1998). Analysis of multiblock and hierarchical PCA and PLS models. *Journal of Chemometrics*, *12*, 301–321.

Woodruff, B. R. (1997). Customer value: The next source for competitive advantage. *Journal of the Academy of Marketing Science*, *25*(2), 139–153.

Wold, H. (1982). Soft modeling: The basic design and some extensions. In K. G. Jöreskog & H. Wold (Eds.), *Systems under indirect observation, Part 2* (pp. 1–54). Amsterdam: North-Holland.

Wold, S., Martens, H., & Wold, H. (1984). In S. Wold (Ed.), *MULDAST Proceedings*. Technical Report on Research Group for Chemometrics, Umeå University, Sweden.

Wold, H. (1985). Partial least squares. In S. Kotz & N. L. Johnson (Eds.), *Encyclopedia of statistical sciences* (Vol. 6, pp. 581–591). New York: Wiley.

Wold, S., Hellberg, S., Lundstedt, T., Sjostrom, M., & Wold, H. (1987) Proceedings of Symposium on PLS: Theory and Application. Frankfurt am Main, Germany.

Zeithaml, V. A. (1988). Consumer perceptions of price, quality, and value: A means-end model and synthesis of evidence. *Journal of Marketing*, *52*, 2–22.

Chapter 7
An Introduction to a Permutation Based Procedure for Multi-Group PLS Analysis: Results of Tests of Differences on Simulated Data and a Cross Cultural Analysis of the Sourcing of Information System Services Between Germany and the USA

Wynne W. Chin and Jens Dibbern

Abstract To date, multi-group comparison of Partial Least Square (PLS) models where differences in path estimates for different sampled populations have been relatively naive. Often, researchers simply examine and discuss the difference in magnitude of specific model path estimates from two or more data sets. When evaluating the significance of path differences, a t-test based on the pooled standard errors obtained via a resampling procedure such as bootstrapping from each data set is made. Yet problems can occur if the assumption of normal population or similar sample size is made. This paper provides an introduction to an alternative distribution free approach based on an approximate randomization test – where a subset of all possible data permutations between sample groups is made. The performance of this permutation procedure is tested on both simulated data and a study exploring the differences of factors that impact outsourcing between the countries of US and Germany. Furthermore, as an initial examination of the consistency of this new procedure, the outsourcing results are compared with those obtained from using covariance based SEM (AMOS 7).

7.1 Introduction

Partial Least Squares (PLS) modeling has been gaining attention among social scientists in recent years (e.g., Chin 1995; Chin and Higgins 1991; Fornell 1982; Mathieson 1991; Sambamurthy and Chin 1994). One of the reasons is that the

W. W. Chin
Department of Decision and Information Sciences, Bauer College of Business, University of Houston, TX, USA
e-mail: wchin@uh.edu

J. Dibbern
Department of Information Engineering, Institute of Information Systems, University of Bern, Engehaldenstr. 8, Room 204, 3012 Bern, Switzerland
e-mail: jens.dibbern@iwi.unibe.ch

V. Esposito Vinzi et al. (eds.), *Handbook of Partial Least Squares*, Springer Handbooks of Computational Statistics, DOI 10.1007/978-3-540-32827-8_8,

PLS approach, consistent with standard structural equation modeling precepts, provides the researcher with greater ability to predict and understand the role and formation of individual constructs and their relationships among each other (Chin 1998b; Hulland 1999). Moreover, PLS is often considered more appropriate than covariance-based modeling techniques like LISREL when the emphasis is prediction since it attempts to maximize the explained variance in the dependent construct. Furthermore, sample size requirements are considerably smaller than the minimum recommended for covariance-based techniques especially for complex models (Chin and Newsted 1999). In the case of multi-group structural equation modeling (MGSEM), advanced procedures for group comparison have been implemented in covariance-based SEM (e.g., as provided in AMOS 7.0). This approach, however, can pose high demands on data properties and sample size. Another less restrictive way of testing structural equation models across groups is the use of the component-based procedure, partial least squares (PLS).

To date, multi-group comparison of PLS models where differences in path estimates for different sampled populations have been relatively naive. Often, researchers simply examine and discuss the difference in magnitude of particular model path estimates for two or more data sets (e.g., Thompson et al. 1994). When assessing the significance of the differences, a t-test based on the pooled standard errors obtained via a resampling procedure such as bootstrapping from each sample is made (e.g., Keil et al. 2000). Yet problems can occur if the assumption of normal population distribution or similar sample size is not tenable. As an alternative distribution free approach, this paper will present the results of applying an approximate randomization test – where a subset of all possible data permutations between sample groups is made. In assessing the significance for a two-sided permutation test, we could examine whether the originally observed difference falls outside of the middle $n\%$ (e.g., 95 or 99 percentile) of the distribution of differences for the subset runs performed. But typically, a one-sided test is performed to examine the percentage of subset runs that are greater than the original observed difference. The performance of this permutation procedure is tested on both simulated data and a study exploring the differences of factors that impact outsourcing between the countries of US and Germany. Furthermore, for reasons of curiosity and in order to examine the consistency of this new procedure, the outsourcing results will be compared with those obtained from using covariance based SEM (AMOS 7).

7.2 The Permutation Procedure

Randomization, or permutation procedures are now the preferred tests of significance for non-normal data. These techniques are considered distribution-free tests in that they require no parametric assumptions. Randomization tests should not be viewed as alternatives to parametric statistical tests, rather they should be considered as those tests for that particular empirical form being examined. The availability of fast computers has made permutation tests increasingly feasible, even for large data sets. Since such methods require no particular assumptions concerning statistical distributions (with the exception of the important assumption of independent observations), permutation tests are increasingly applied even in the context of traditional statistical tests (e.g. correlation, t-tests, ANOVAs, etc.).

The procedure for a permutation test based on random assignment, as described by Edgington (1987) and Good (2000), is carried out in the following manner.

1. A test statistic is computed for the data (e.g., contrasting experimental treatment/control or nonexperimental groupings).
2. The data are permuted (divided or rearranged) repeatedly in a manner consistent with the random assignment procedure. With two or more samples, all observations are combined into a single large sample before being rearranged. The test statistic is computed for each of the resulting data permutations.
3. These data permutations, including the one representing the obtained results, constitute the reference set for determining significance.
4. The proportion of data permutations in the reference set that have test statistic test statistic values greater than or equal to (or, for certain test statistics, less than or equal to) the value for the experimentally obtained results is the P-value (significance or probability value). For example, if your original test statistic is greater than 95% of the random values, then you can reject the null hypothesis at $p < 0.05$.

Determining significance on the basis of a distribution of test statistics generated by permuting the data is characteristic of all permutation tests. When the basis for permuting the data is random assignment, that permutation test is often called a randomization test. This preceding definition is broad enough to include procedures called randomization tests that depend on random sampling as well as randomization. The modern conception of a randomization test, however, is a permutation test that is based on randomization alone, where it does not matter how the sample is selected.

A permutation test based on randomization, as Edgington (1987) notes "is valid for any kind of sample, regardless of how the sample is selected." This is an extremely important property because the use of nonrandom samples is common in surveys and experimentation and would otherwise invalidate the use of parametric statistical tables (e.g., t or F tables). Essentially, the random sampling assumption underlying these significance tables states that all possible samples of n cases within a specified population has the same probability of being drawn.

Statisticians going back to Sir Ronald Fisher (1936, p. 59, c.f., Edgington 1987) have indicated that the randomization test is the correct test of significance and that the corresponding parametric test is valid only to the extent the results yield the same statistical decision. Fisher, in particular, referred to the application of permuting the data to determine significance. But Efron and Tibshirani (1993, p. 202) noted that Fisher introduced the idea of permutation testing "more as a theoretical argument supporting Student's t-test than as a useful statistical method in its own right." With modern computational power available for permutation tests to be used on a routine basis, the reliance on parametric tests as an approximation is no longer necessary.

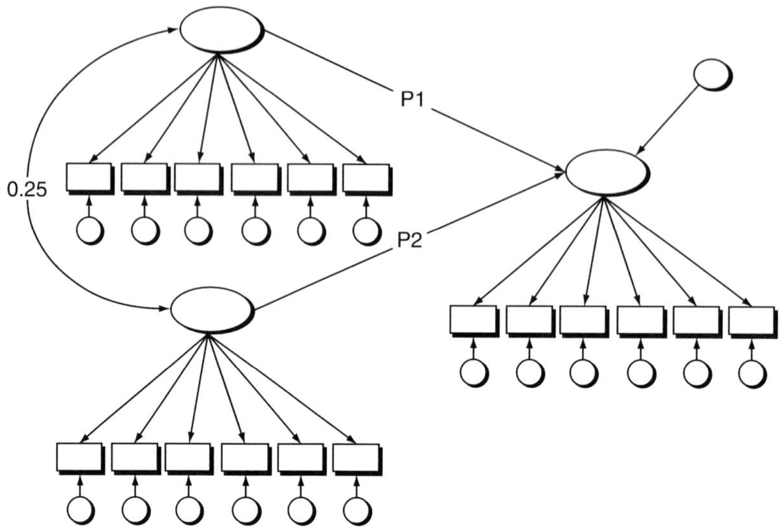

Fig. 7.1 Base model tested with structural paths $P1$ and $P2$ varied

Good (2000) clearly articulates that when samples are very large, decisions based on parametric tests like the t and F tests usually agree with decisions based on the corresponding permutation test. But with small samples, "the parametric test will be preferable IF the assumptions of the parametric test are satisfied completely" (Good 2000, p. 9). Otherwise, even for large samples, the permutation test is usually as powerful as the most powerful parametric test and may be more powerful when the test statistic does not follow the assumed distribution (Noreen 1989, pp. 32–41).

In this paper, we examine the two sample situation where two independent random samples $G1 = (m_1, m_2, \ldots, m_i)$ and $G2 = (n_1, n_2, \ldots, n_k)$ are drawn from potentially two different probability distributions D_{G1} and D_{G2}. The test statistic is the difference in the PLS parameter estimates such as $P1$ and $P2$ as seen in Fig. 7.1 (i.e., $e = P1 - P2$). Having observed sample sets G_1 and G_2, we test the null hypothesis H_0 of no difference between D_{G1} and D_{G2} (i.e., $H_0 : D_{G1} = D_{G2}$).

7.3 Monte Carlo Design

Figure 7.1 provides the basis for the Monte Carlo generated data. Two exogenous constructs, labeled X and Z, are created with a correlation of 0.25. Both are modeled to impact the endogenous construct Y. Six indicators were created as measures reflecting each construct. The standardized loadings were set at 0.6 for three indicators and 0.8 for the other three indicators. While not a full factorial design, the cells studied provides initial information to contrast varying structural path effect sizes with data normality (normal versus high kurtosis). In addition, asymmetry in

Table 7.1 Power for $p < 0.05$ significance level for path differences (percentages out of 1,000 runs)

		Path setting 1 Group 1 ($p1 = 0.5$, $p2 = .03$) Group 2 ($p1 = 0.3$, $p2 = .05$)	Path setting 2 Group 1 ($p1 = 0.7$, $p2 = .05$) Group 2 ($p1 = 0.5$, $p2 = .07$)	Path setting 3 Group 1 ($p1 = 0.6$, $p2 = .03$) Group 2 ($p1 = 0.3$, $p2 = .06$)
Data Setting 1	$N = 150$ (group 1), $N = 150$ (group 2)		82.0 ($p1$) 83.0 (p2)	90.3 ($p1$) 88.2 (p2)
Data Setting 2	$N = 150$ (group 1), $N = 75$ (group 2)		64.9 ($p1$) 68.3 ($p2$)	76.9 ($p1$) 76.5 ($p2$)
Data Setting 3	$N = 150$ (group 1) $N = 150$ (group 2) non-normal conditions	na ($p1$) 66.9 ($p2$) setting A	66.6 ($p1$) 67.0 ($p2$) setting B	78.8 ($p1$) 79.0 ($p2$) setting C

sample sizes for the two groups was also tested (150 cases for both versus 150 and 75 for groups 1 and 2 respectively). Data were generated using PreLis 2 (Jöreskog and Sörbom 1996). For non-normal data, the generalized Lambda distribution suggested by Ramberg et al. (1979) was used following the procedure described by Reinartz et al. (2002).

The structural paths were varied symmetrically with the effects for the two causal paths in group 1 the same, but reversed of group 2. Thus, for example, in the first effect treatment the standardized paths were set for $P1$ at 0.5 and $P2$ at 0.3 for the group 1 and reversed with $P1$ at 0.3 and $P2$ at 0.5 for group 2. This provided the opportunity to see the performance for two paths with the same effect size differences.

Table 7.1 presents the results for those cells analyzed. Each cell represents the results of running one million PLS analysis. This is due to the fact that 1,000 Monte Carlo sample sets were created for each cell to reflect that particular condition. Then 1,000 permutations were conducted for each sample to determine the p-value for the test statistic. The first two rows represent results using normal data, whereas the last row presents results using non-normal data. For the non-normal conditions, the item skewness ranged from 0.952 to 1.759 and kurtosis (see Table 7.2) ranged from 2.764 to 18.425.

The results in Table 7.1 provide us with an initial sense of the power for detecting structural path differences for different sample populations. As typical of power analysis, the sample and effect size was found to have an impact. For the first row, we see that the power for normal data where the population path difference is 0.2 was detected at the $p < 0.05$ level approximately 82% of the time. When the difference in path was increased to 0.3 (i.e., path setting 3), the power went up to 88 for $p2$ and 90.3 for $p1$. Conversely, the power dropped when the number of cases for the second group was lowered from 150 to 75 (i.e., data setting 2). Interestingly enough,

Table 7.2 Level of kurtosis for indicators used for the non-normal runs

	setting A		setting B		setting C	
	$g1$	$g2$	$g1$	$g2$	$g1$	$g2$
$X1$	8.286	6.176	5.705	6.412	5.356	6.176
$X2$	7.748	5.503	5.498	5.392	6.410	5.503
$X3$	7.176	8.151	4.908	5.964	6.970	8.151
$X4$	9.206	4.407	4.218	6.435	4.544	4.407
$X5$	8.144	4.295	3.842	6.830	4.205	4.295
$X6$	8.068	3.880	4.555	6.220	3.784	3.880
$Z1$	6.927	5.405	4.863	4.775	6.741	5.405
$Z2$	5.345	7.502	7.297	5.754	5.392	7.502
$Z3$	5.178	5.545	5.580	5.552	7.350	5.545
$Z4$	7.566	4.483	3.841	4.211	3.628	4.483
$Z5$	6.160	5.126	4.232	6.195	3.738	5.126
$Z6$	6.517	5.667	3.726	4.308	3.978	5.667
$Y1$	5.713	5.028	5.292	5.823	7.525	5.028
$Y2$	5.999	4.672	4.489	6.165	4.896	4.672
$Y3$	5.249	5.248	9.645	6.161	4.990	5.248
$Y4$	4.847	2.874	2.765	3.610	4.092	2.874
$Y5$	4.786	3.850	2.818	3.962	3.721	3.850
$Y6$	4.690	3.056	2.974	3.909	3.899	3.056

Table 7.3 Power at $p < 0.05$ significance level for loading differences of 0.2 (percentage out of 1,000 runs for six loadings)

0.8 vs. 0.6 (normal)	0.8 vs. 0.6 (normal)	0.8 vs. 0.6 (non normal)	0.9 vs. 0.6 (non normal)
Group 1 = 150, Group 2 = 150	Group 1 = 150, Group 2 = 75	Group 1 = 150, Group 2 = 75	Group 1 = 150, Group 2 = 75
85.0 – 90.5	76.1 – 77.4	51.2 – 52.1	89.4 – 92.3

this same drop in power can also be achieved if the data was highly non-normal (i.e., data setting 3). Finally, it seems it is not simply the effect size, but also the overall magnitude of predictiveness that may make a difference. In a separate run (not presented in the table), we kept both path differences equal at 0.3, but changed the model to represent more substantive paths (i.e., 0.7 and 0.4 versus 0.6 and 0.3). The power increased a corresponding 20%.

The power to detect standardized path loading differences of 0.2 were also examined (see Table 7.3). Overall, the power ranged from 76 to 90 in the normal data settings. Under high non-normality, the power dropped to the 50 percentile range. But when the effect size was increased to 0.3 population difference, the power dramatically improved moving into the 89.4–92.3 range.

Taken together, these results are suggestive of the countervailing impact that asymmetry in group sample sizes, degree of non-normality, difference in magnitude of path effects, and overall predictiveness of the model have upon each other. In other words, while asymmetry in group sample sizes is expected to lower the power to detect structural path differences, a more predictive model, on average, may moderate this effect. Ideally, we would like high predictive models with normal data and sample sizes of 150 or higher for each group.

7.4 Cross-Cultural Analysis of an Information Systems Outsourcing Model

We now provide a didactic example of the use of the PLS based permutation procedure in a cross cultural context. The example includes the testing of a model that explains why companies outsource the development and maintenance of software applications to external vendors. Over the past 15 years, the practice of information systems (IS) outsourcing has grown significantly. Many industry watchers have attributed this growth to the first IS outsourcing mega deal in 1989, when Kodak decided to outsource major parts of their IS infrastructure to IBM, DEC and Businessland in a 10-year, $250 million deal (Dibbern et al. 2004). However, in spite of the fact that the outsourcing market has grown globally, there are a number of obvious differences between countries. First of all, when looking at the overall amount of money that is spent for IS services, it soon becomes apparent that the U.S. is still the leading country in terms of IS outsourcing expenditures with three times more money spent on IS outsourcing than Germany (Murphy et al. 1999; OECD 2000) as an example. Second, there are significant differences between countries in terms of what IS functions are being outsourced (Apte et al. 1997; Barthelemy and Geyer 2001). This phenomenon is essentially attributed to the increasing practice of selective outsourcing. That is, rather than outsourcing their entire IS department, firms prefer to outsource part or all of particular IS functions, such as data center operations, help desk services or applications development.

Thus, the question is raised as to why such national differences do exist. Is the sourcing decision fundamentally different between countries (i.e., is it motivated or restricted by different factors?) and, if yes, why so? Most research on IS outsourcing has been conducted in a single country. Indeed the majority of research is U.S.-based and it is hard to say to what extent these findings are generalizable across countries. The few studies with a cross-national perceptive are purely descriptive (Apte et al. 1997; Barthelemy and Geyer 2001).

7.4.1 Theoretical Framework

Figure 7.2 presents a graphical representation of the theoretical model to be tested. This model suggests that the decision to outsource application services is influenced by three distinct sets of variables: efficiency variables, effectiveness variables as well as social influences and other constraints. In addition, firm size similar to other studies is included as a control variable. The discussion below elaborates upon each

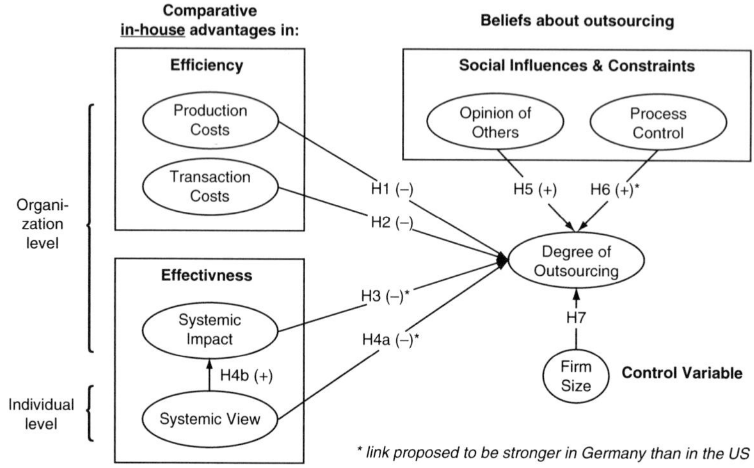

Fig. 7.2 Theoretical framework on IS sourcing

set of factors and explains why the strength of certain linkages is expected to differ between German and U.S. organizations.

7.4.2 *Efficiency Factors*

Production Costs. Previous empirical research on IS outsourcing has shown that cost reduction is one of the major objectives for IS outsourcing (c.f. Dibbern et al. 2004) where an external vendor can realize higher economies of scale because of its ability to provide the same type of service for multiple customers. At the same time however, it is one of the major reasons why some companies decide to keep there IS function in-house or to bring it back in-house (Dibbern et al. 2003; Hirschheim and Lacity 2000). Thus, overall, the decision of whether it is more production cost efficient to insource an IS function or to outsource it to an external vendor should be made on a case to case basis (c.f. Ang and Straub 1998).

Transaction Costs. In addition to production costs, however, transaction costs should not be neglected (Ang and Straub 1998). Transaction costs are all costs in terms of time, effort, and money spent that arise when delegating tasks of an IS function to one or more agents. The magnitude of these transaction costs may also vary between insourcing and outsourcing, and hence it is important to be clear which sourcing arrangement is more transaction cost efficient.

The argument that the make-or-buy decision should be guided by both transaction and production cost considerations can be traced back to transaction cost theory, which considered the sum of production and transaction cost differences between the firm and the market (Williamson 1981). Thus, as reflected in paths H1 and H2, the higher the comparative costs of outsourcing is relative to the firm, the less a particular application service is outsourced.

7.4.3 Effectiveness Factors

Focusing solely on *efficiency*, however, neglects the fact that the *output* of the IS work could be significantly influenced by the sourcing choice as well. Empirical findings have shown that some organizations change their current sourcing arrangement for strategic intents (DiRomualdo and Gurbaxani 1998; McLellan et al. 1995). The precondition for strategic impacts are variations in the *effectiveness* of the IS function.

Systemic Impact. For reaching a high level of IS effectiveness, it is often argued that beyond producing application software whose features and capabilities meet the needs of the users, it is even more important to ensure that an organization's application software fits synergistically with other IS functions such as data center operations, network design and maintenance, user support and telecommunications services. It is often hard to separate the effectiveness of the application software from that of the overall IS (c.f. Hamilton and Chervany 1981; Pitt et al. 1995). Accordingly, as tested via path H3, it is important for an organization to examine whether the systemic impact of application services is higher in-house or with an external vendor.

Systemic View. In line with the arguments made above and with the resource-based view (Wade and Hulland 2004), IS workers that feel responsible not only for their own work, but also for how their work relates to the work of others, may be viewed as valuable resources. IS executives, when evaluating and comparing alternative sourcing options, may well consider whether their choice leads to IS workers with more of an integrative view of the firm. This is reflected in path H4a, which suggests that the more systemic the view of in-house employees as opposed to outsourced workers in performing application services, the less these services are outsourced. Path H4b, in a similar vein, suggests that the impact of the application development and maintenance work on overall systems performance is better achieved in-house, if an organization's own employees have more of a systemic view than the personnel of an external service provider.

7.4.4 Social Influences and Constraints

Opinion of Influential Others. The preceding factors are based on the assumption that the sourcing decision represents a rational decision based on efficiency and effectiveness criteria. This view has been partially contradicted by other studies that show an organizations sourcing decision can be influenced by various social influences and constraints (Lacity and Hirschheim 1993; Lacity and Hirschheim 1995). Overall, these studies support the view that the opinion of others could have a profound impact on the sourcing decision of organizations and this is tested via path H5.

Outsourcing Process Control. A final main factor that may explain variations in the degree of outsourcing application services is extent to which organizations have control (i.e. unlimited power of direction) over all necessary activities associated with outsourcing an IS function to an external service provider. These influences may limit the ability of the main decision makers to act strictly relationally. Accordingly, one would expect that the less the implementation of an outsourcing decision is constrained by various forces, the easier it is for an organization to outsource application services. Path H6 tests for this impact. Finally, in accordance with previous studies on IS sourcing, firm size is added as a control variable and tested via path H7 (Ang and Straub 1998; Sobol and Apte 1995).

7.4.5 Proposed Cultural Differences

The preceding net of hypotheses (see Fig. 7.2) may be viewed as a mid-range theory that seeks to explain variations in the extent to which organizations outsource application services. The question for this study is whether the relationships between constructs are the same in Germany and the U.S., or whether country specific factors affect the generalizability of the proposed linkages. One way of approaching this question is (1) to identify those cultural dimensions that were found to differ between Germany and the U.S. in previous cross-cultural research, (2) to select those dimensions that have an impact on the mid-range theory, and (3) to develop propositions about how selected linkages will differ between Germany and the U.S. (based on Lytle et al. 1995).

In following this procedure, three candidates have been identified that may account for cross-cultural variation in the theoretical framework. Two of them are cross-cultural dimensions that refer to relationship characteristics between societal members, while the third refers to more general patterns of institutions and social systems (Lytle et al. 1995).

The first dimension is *individualism-collectivism* based on a large scale survey of approximately 116,000 respondents from 50 different cultural regions worldwide (Hofstede 1980). The U.S. sample showed the highest individualism ranking of all the countries, while Germany ranking above the average but significantly lower on the index scale (rank 15 from 50; index 67 as opposed to 91 from the U.S.) (Hofstede 1983, 1991). Two of seven categories identified by Triandis (1996) are (1) the people's concern about how their decisions could affect others in their collectivity; and (2) the belief in the correspondence of ones own outcomes, both positive and negative, with the outcome of others. These two aspects of collectivism can be seen to be closely related to two constructs in our theoretical model, namely systemic impact and systemic view.

Another cultural dimension that is closely related to the aspect of systemic view is the *analytical versus integrative view*. This dimension was extracted by another cross cultural study that included about 1000 intercultural trainee programs, plus a survey of about 30,000 managers of 30 organizations with locations in 50 different countries (Hampden-Turner and Trompenaars 1993; Trompenaars and Hampden-Turner 1994). The analytical view reflects the extent to which a firm is perceived as a collection of tasks, functions, people, and machines rather than as a group of related persons working together (an integrative viewpoint). Overall, Germany showed a higher tendency towards an integrative view of an organization than the U.S.

Taking these preceding cultural dimensions together, it can be argued that in nations such as Germany, where members of organizations show a tendency towards collectivism and have more of an integrative view of the organization, it matters greatly for managers to consider how the overall IS function will be affected by the sourcing choice. By contrast, managers in countries, such as the U.S., where individual performance is valued higher than collective action, and where managers have more of an analytical view of the organization, the systemic impact of the sourcing choice may reside to the background. This leads to the following proposition:

P1: The negative relationship between comparative in-house advantages in systemic impact and the degree of outsourcing (H3-) is stronger in Germany than in the U.S.

Moreover, German IS managers may be more inclined to consider whether in-house personnel or the staff of external vendors shows more of a systemic view in doing their work:

P2: The negative relationship between comparative systemic view advantages of in-house workers and the degree of outsourcing (H4a-) is stronger in Germany than in the U.S.

Third, in Germany there are a number of unique legal and legitimized institutional constraints that do not exist in the same form in the U.S. For example, in Germany, the protection of employee interests is codified in law. Employee interests are legally supported by the works constitution act ("Betriebsverfassungsgesetz BetrVG") that guarantees the right of employee participation and codetermination ("Mitbestimmung") in social, economic, and personnel matters (Richardi 1990).

Overall, these restrictions suggest that in Germany, major organizational decisions, such IS outsourcing, where personnel and social affairs are affected, are more participative than in the U.S. Accordingly, German managers may be more sensitive to consider the extent to which they have control over the outsourcing process when deciding on IS sourcing than their U.S. colleagues:

P3: The impact between the extent to which IS managers believe that they have control over the outsourcing process and the degree of applications outsourcing is stronger in Germany than in the U.S.

7.5 Method

7.5.1 Data

Data for this study was gathered via a mailed questionnaire survey. Only companies with more than 500 employees were considered. The questionnaires were administered to the highest ranking IS executives of organizations in the USA and Germany. Overall, 180 usable questionnaires were returned. Since the survey included both questions about the development and maintenance of software applications, the sample for this study includes 278 decisions on the sourcing of software applications in Germany and 82 cases in the U.S.

7.5.2 Measures

Each of the constructs from our model was measured with a block of indicators (questionnaire items). Whenever possible, existing measures from prior empirical studies were adopted. An overview of the constructs and exemplified measurement items is provided in Table 7.4. Most of the items were measured on a (positive to negative) five point Likert scale ranging from "strongly agree" to "strongly disagree", with "neither agree nor disagree" as a mid-point. For measures of the *degree of outsourcing*, respondents were asked to provide percentages ranging form 0% to 100%. For the construct *opinion of others*, the semantic differential approach to measurement was adopted (Osgood et al. 1957), where each response is located on an evaluative bipolar (negative to positive) dimension, using a seven point Likert scale. All blocks of indicators were formulated in the reflective mode (Chin 1998a; Chin and Newsted 1999; Fornell 1989). The unit of analysis was the respective application service. The respondents had to answer each question for both the development and the maintenance of application software.

7.6 Analysis and Results

In the following, the results of the model testing for both the U.S. and Germany will be presented. This includes the test of (1) the measurement model and (2) the structural model in both countries, as well as (3) the test of differences in the structural paths between both countries.

7.6.1 Results of Partial Least Squares Estimation

Measurement Model. In order to check whether the indicators of each construct measure what they are supposed to measure, tests for convergent and discriminant validity were performed in both the U.S. and German sample. Before doing any multigroup comparisons, it is always important to first establish the measures perform adequately in both data samples.

In terms of convergent validity (Bagozzi and Phillips 1982), both indicator reliability and construct reliability were assessed (Peter 1981). *Indicator reliability* was examined by looking at the construct loadings. All loadings are significant at the 0.01 level and above the recommended 0.7 parameter value (Significance tests were conducted using the bootstrap routine with 500 resamples (Chin 1998b).

Table 7.4 Questionnaire measures

Construct	Source	Sample Item
Degree of Outsourcing	Based on Dibbern and Heinzl (2004); Teng et al. (1995)	For each of the two IS functions, please estimate the average percentage currently allocated to external service providers in terms of 1. ... the functions total budget (from 0 to 100%) 2. ... total person working days. 3. ... total number of people that participate in doing the work.
Comparative production cost advantage	Based on Ang and Straub (1998)	In doing the actual work required for each of the IS functions 1. ... our internal staff works more cost efficient than an external service provider. 2. ... we can realize higher economies of scale internally than an external service provider.
Comparative transaction cost advantage	Based on Ang and Straub (1998)	When delegating i.e. transferring tasks of the particular IS function 1. ... the costs incurred in negotiating, managing and coordinating are lower within the firm than in case of contracting with an external service provider. 2. ... less transaction costs are incurred for internal employees than when using an external service provider.
Comparative systemic impact advantage	Informed by the notion of task interdependence (Pfeffer and Salancik 1978; Thompson 1967)	If this IS function is not performed in-house but externally, 1. ... the integration of this IS function into the overall IS function of our organization is weakened. 2. ... the synergetic effects to other IS functions will be threatened. 3. ... the overall performance of our entire IS function will be greatly affected.
Comparative systemic view advantage	See above plus the individualism-collectivism categorization by Hui and Triandis (1986)	In doing the actual work required for each of the IS functions, our own employees tend much more than personnel of external service providers to 1. ... have a systems view of the organization. 2. ... have an organization wide perspective of how work in different areas effect one another. 3. ... consider the task interdependencies in our organization. 4. ... have an integrated view of the organization.
Outsourcing Process Control	Based on Ajzen (1991); Ajzen and Fishbein (1980)	When it comes to outsourcing this IS function to an external service provider 1. ... our organization can act unrestrictedly. 2. ... there are no impediments to our organization.

(continued)

Table 7.4 (continued)

Construct	Source	Sample Item
External Influences	Based on Ajzen (1991); Ajzen and Fishbein (1980)	Persons or groups whose opinion is important to our organization think that outsourcing this particular IS function is 1. ... bad - good (-3 to +3). 2. ... negative - positive. 3. ... harmful - beneficial. 4. ... foolish - wise. 5. ... illogical - logical. 6. ... worthless - valuable.
Firm size	Based on Ang and Straub (1998)	Please estimate your organization's overall number of employees.

Table 7.5 Indicator and construct reliability

Construct	Item	Germany			USA		
		Loading	CR	AVE	Loading	CR	AVE
Degree of Outsourcing	Out1	0.96	0.97	0.93	0.95	0.97	0.91
	Out2	0.96			0.98		
	Out3	0.96			0.94		
Production Cost Advantage	Pc1	0.85	0.86	0.75	0.92	0.90	0.82
	Pc3	0.88			0.89		
Transaction Cost Advantage	Tc1	0.90	0.85	0.74	0.70	0.83	0.71
	Tc4	0.82			0.97		
System Impact Advantage	Impact1	0.89	0.91	0.78	0.92	0.94	0.85
	Impact2	0.89			0.90		
	Impact3	0.86			0.94		
System View Advantage	EmplOrit1	0.77	0.91	0.71	0.77	0.91	0.73
	EmplOri2	0.87			0.77		
	EmplOri3	0.83			0.91		
	EmplOri4	0.89			0.89		
Opinion of Others	Other1	0.92	0.97	0.82	0.93	0.98	0.87
	Other2	0.93			0.92		
	Other3	0.92			0.93		
	Other4	0.89			0.97		
	Other5	0.88			0.96		
	Other6	0.89			0.90		
Process Control	CoPro1	0.94	0.93	0.87	1.00	0.93	0.87
	CoPro2	0.94			0.86		

Construct reliability and validity was tested using two indices: (1) the *composite reliability* (CR) and (2) the *average variance extracted* (AVE). All the estimated indices were above the threshold (Bagozzi and Yi 1988) of 0.6 for CR and 0.5 for AVE (see Table 7.5). Finally, the *discriminant validity* of the construct items

Table 7.6 PLS crossloadings for U.S. sample

	PC	TC	firm size	Out	SysImp	Control	SysView	ExtInfl
Pc1	**0.92**	0.39	0.02	0.36	0.53	0.01	0.17	0.30
Pc3	**0.89**	0.47	0.02	0.31	0.59	0.02	0.36	0.33
Tc1	0.31	**0.70**	0.02	0.11	0.31	0.15	0.34	0.25
Tc4	0.46	**0.97**	0.02	0.30	0.36	0.07	0.35	0.20
NoAll	0.02	0.02	**1.00**	0.16	0.10	0.04	0.06	0.17
Out1	0.28	0.19	0.25	**0.95**	0.08	0.00	0.06	0.29
Out2	0.36	0.33	0.11	**0.98**	0.19	0.02	0.01	0.32
Out3	0.41	0.27	0.11	**0.94**	0.25	0.04	0.01	0.37
Impact1	0.62	0.40	0.16	0.22	**0.92**	0.17	0.37	0.34
Impact2	0.50	0.31	0.00	0.16	**0.90**	0.11	0.30	0.44
Impact3	0.56	0.35	0.09	0.14	**0.94**	0.07	0.44	0.40
CoPro1	0.01	0.10	0.04	0.02	0.13	**1.00**	0.10	0.01
CoPro2	0.11	0.10	0.03	0.00	0.09	**0.86**	0.04	0.03
EmplOri1	0.19	0.28	0.12	0.09	0.34	0.19	**0.77**	0.28
EmplOri2	0.34	0.44	0.05	0.03	0.31	0.01	**0.84**	0.28
EmplOri3	0.25	0.38	0.12	0.11	0.40	0.04	**0.91**	0.35
EmplOri4	0.19	0.23	0.08	0.08	0.35	0.12	**0.89**	0.28
Other1	0.32	0.25	0.17	0.28	0.39	0.05	0.28	**0.93**
Other2	0.35	0.21	0.23	0.28	0.37	0.07	0.24	**0.92**
Other3	0.31	0.12	0.14	0.24	0.42	0.05	0.31	**0.93**
Other4	0.36	0.27	0.15	0.36	0.42	0.05	0.34	**0.97**
Other5	0.34	0.26	0.17	0.34	0.39	0.02	0.41	**0.96**
Other6	0.26	0.21	0.08	0.37	0.37	0.10	0.36	**0.90**

was assured by looking at the cross-loadings. They are obtained by correlating the component scores of each latent variable with both their respective block of indicators and all other items that are included in the model (Chin 1998b). In Tables 7.6 and 7.7, in the Appendix, the cross loadings for both the USA and Germany are presented. The loadings on their respective constructs are shadowed. Moving across the rows reveals that each item loads higher on its respective construct than on any other construct. Going down a column also shows that a particular constructs loads highest with its own item. Taken together, this implies discriminant validity for both samples.

Structural Model. Having gained confidence that the measures work appropriate for both the U.S. and German sample, the next step is to test the explanatory power of the entire model on IS sourcing as well as the predictive power of the independent variables in both countries. The explanatory power is examined by looking at the *squared multiple correlations* (R^2) of the main dependent variable, the degree of IS outsourcing. As can be inferred from Fig. 7.3, in Germany 33% ($R^2 = 0.33$) of the variation in the degree of outsourcing are explained by the independent variables, while in the U.S. 27% ($R^2 = 0.27$) are accounted for. The hypotheses are tested by examing the magnitude of the standardized parameter estimates between constructs together with the corresponding t-values that indicate the level of significance.

Table 7.7 PLS crossloadings for German sample

	PC	TC	firm size	Out	SysImp	Control	SysView	ExtInfl
Pc1	**0.85**	0.57	0.05	0.34	0.40	0.16	0.44	0.25
Pc3	**0.88**	0.44	0.10	0.38	0.49	0.12	0.42	0.33
Tc1	0.53	**0.90**	0.12	0.36	0.33	0.14	0.33	0.30
Tc4	0.45	**0.82**	0.03	0.27	0.40	0.06	0.39	0.29
NoAll	0.09	0.07	**1.00**	0.01	0.03	0.01	0.13	0.00
Out1	0.40	0.36	0.03	**0.96**	0.41	0.05	0.35	0.36
Out2	0.41	0.37	0.01	**0.96**	0.43	0.04	0.38	0.32
Out3	0.38	0.36	0.02	**0.96**	0.41	0.04	0.37	0.38
Impact1	0.51	0.41	0.03	0.38	**0.89**	0.24	0.46	0.21
Impact2	0.46	0.36	0.03	0.41	**0.89**	0.14	0.44	0.28
Impact3	0.40	0.34	0.02	0.35	**0.86**	0.16	0.41	0.22
CoPro1	0.16	0.11	0.03	0.05	0.18	**0.97**	0.17	0.05
CoPro2	0.14	0.12	0.07	0.03	0.21	**0.90**	0.17	0.05
EmplOri1	0.34	0.36	0.08	0.23	0.40	0.15	**0.77**	0.18
EmplOri2	0.47	0.39	0.17	0.38	0.41	0.08	**0.87**	0.31
EmplOri3	0.41	0.31	0.03	0.33	0.39	0.22	**0.83**	0.17
EmplOri4	0.44	0.34	0.14	0.33	0.46	0.15	**0.89**	0.19
Other1	0.37	0.37	0.03	0.34	0.28	0.07	0.25	**0.92**
Other2	0.35	0.35	0.03	0.33	0.27	0.07	0.27	**0.93**
Other3	0.33	0.31	0.01	0.34	0.27	0.03	0.22	**0.92**
Other4	0.26	0.28	0.03	0.33	0.22	0.09	0.22	**0.89**
Other5	0.23	0.25	0.03	0.31	0.22	0.06	0.18	**0.88**
Other6	0.27	0.32	0.04	0.34	0.21	0.02	0.22	**0.89**

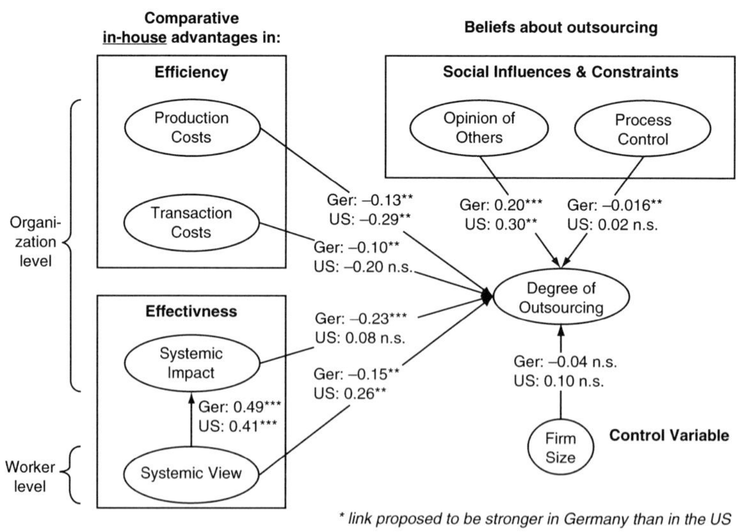

Fig. 7.3 Structural model findings for Germany and the U.S.

Table 7.8 PLS results for structural model and group comparisons

Independent Variable	Dependent Variable	Hypo-thesis	Germany $n = 278$		USA $n = 82$		Country Difference	
			Path	P-values	Path	P-values	Path	P-value
Production cost advantage	Degree of Outsourcing	$H1(-)$	−0.13**	3.1	−0.29**	2.0	0.17 n.s.	13.0
Transaction cost advantage	Degree of Outsourcing	$H2(-)$	−0.10**	4.8	−0.20 n.s.	10.9	0.10 n.s.	25.2
Systemic impact advantage	Degree of Outsourcing	$H3(-)$	−0.23***	<0.1	0.08 n.s.	29.5	**−0.31****	**2.5**
Systemic view advantage	Degree of Outsourcing	$H4a(-)$	−0.15**	2.4	0.26**	1.7	**−0.40*****	**0.3**
Systemic view advantage	Systemic impact advantage	$H4b(+)$	0.49***	<0.1	0.41***	<0.1	0.08 n.s.	17.1
External influence	Degree of Outsourcing	$H5(+)$	0.20***	<0.1	0.30**	1.0	−0.10 n.s.	20.9
Outsourcing Process Control	Degree of Outsourcing	$H6(+)$	−0.16**	1.3	0.02 n.s.	41.7	**−0.18***	**7.9**
Firm size	Degree of Outsourcing	$H7$	−0.04 n.s.	14.0	0.10 n.s.	25.8	−0.14 n.s.	12.0

t-values were obtained through the bootstrap routine (Chin 1998b). An overview of the results can be inferred from Table 7.8. Moreover, Fig. 7.3 shows a graphical representation of the findings for Germany and the U.S.

The findings show solid support for the efficiency and effectiveness hypotheses in Germany. All of the path coefficients show the expected negative sign and are significant at the 0.05 (**) or 0.01 (***) level. Notably, perceived comparative in-house advantages in the systemic impact have the strongest impact ($H3 : -0.23$, $t = 3.67$). The impact of Social Influences & Constraints is less consistent. While solid support can be found for the impact of influential others on the degree of outsourcing ($H5 : 0.20$, $t = 3.93$), the link between decision control and outsourcing is negative instead of positive as predicted in the model. Moreover, firm size has no impact. In the U.S., the opposite was found, that comparative advantages of in-house workers in the systemic view are positively related to the degree of outsourcing and not negatively, as predicted. Moreover, in contrast to Germany, no evidence can be found for the significant impact of comparative transaction cost advantages and systemic impact advantages, as well as for decision control and firm size.

Significance of Group Differences. The question is, however, whether the observed differences between Germany and the U.S. are significant and whether those differences are in line with the proposed cultural differences ($P1 - P3$). This can be inferred from the right column of Table 7.8. It shows the level of probability with which the hypotheses that the parameter estimates equal zero (i.e., that the Null-hypothesis) is true. This probability (scaled from 0 to 100) is also called critical distance and should be limited to 1% ($P < 1$), 5% ($P < 5$), or 10% ($P < 10$) (Mohr 1991).

The results show that the path coefficient from systemic impact advantage to degree of outsourcing (H3) in the structural model for Germany is significantly stronger ($P = 2.5$) than the corresponding path in the structural model for the U.S., supporting P1 at the 0.05 level of significance. Moreover, the link between outsourcing process control and degree of outsourcing is significantly stronger ($P = 7.9$) in Germany than in the U.S., supporting P3 at the 0.1 level of significance. Finally, P2 is supported partially. It was proposed that the negative link between systemic view advantage and degree of outsourcing were stronger in Germany than in the U.S. However, the results show that not the strength, but the direction of that link is significantly different between Germany and the U.S. It is negative in Germany, while positive in the U.S.

Given the results of our earlier simulation, we might conjecture that the asymmetry in sample size between Germany and U.S. may impact the p-value estimate for P3. While it was found to be significant at the 0.1, it would not be at the 0.05 level of significance. The Germany size at $n = 278$ is larger than our simulated size of 150 as was the U.S. sample of 82 being slightly larger than the 75 setting we tested. At an exact 150 versus 75 group sample difference, recall that we found the power to range from 65 to 68. Thus, we might conjecture that had the U.S. sample been closer to 150, we would have obtained a multi-group p-value at 0.05.

7.6.2 Results of AMOS Estimation

The AMOS results of the structural model for Germany and the U.S., as well as the test results for country differences in the structural model are depicted in Table 7.9. The focus is on comparing the level of significance for the differences in structural paths as provided by AMOS with those from PLS. The comparison reveals strong agreement between the PLS and AMOS results. Just like in PLS, only the relationships from H3, H4, and H6 show significant differences between both countries. There are only differences in the level of significance, e.g. the country difference for the path coefficient from systemic view advantage to degree of outsourcing is significant at the 0.01 level in Germany ($P = 0.3$) and at the 0.1 level in the U.S. ($P = 8.9$).

Table 7.9 Amos results for structural model and group comparisons

Independent Variable	Dependent Variable	Hypothesis	Germany $n = 278$		USA $n = 82$		Country Difference	
			Path	P-value	Path	P-value	Path	P-value
Production cost advantage	Degree of Outsourcing	$H1(-)$	−0.16 n.s.	60.2	−0.20 n.s.	25.8	0.04 n.s.	100.0
Transaction cost advantage	Degree of Outsourcing	$H2(-)$	−0.10 n.s.	71.3	−0.38*	4.9	0.29 n.s.	37.1
Systemic impact advantage	Degree of Outsourcing	$H3(-)$	−0.27***	<0.1	0.05 n.s.	65.2	**−0.32****	**1.1**
Systemic view advantage	Degree of Outsourcing	$H4a(-)$	−0.10 n.s.	37.2	0.39**	0.7	**−0.49***	**8.9**
Systemic view advantage	Systemic impact advantage	$H4b(+)$	0.60***	<0.1	0.48***	<0.1	0.12 n.s.	52.7
External influence	Degree of Outsourcing	$H5(+)$	0.18**	0.4	0.29*	1.1	−0.12 n.s.	100.0
Outsourcing Process Control	Degree of Outsourcing	$H6$	−0.19**	0.2	−0.004 n.s.	92.7	**−0.18****	**4.0**
Firm size size	Degree of Outsourcing	$H7$	−0.04 n.s.	39.6	0.03 n.s.	72.8	−0.08 n.s.	40.3

7.7 Discussion and Summary

This paper has presented results from two PLS based MGSEM studies. First, it provides initial insights into how this new procedure for multi-group comparison using PLS performs with simulated data. This was intended to provide an initial sense of the sample sizes required to achieve adequate power. Second, it empirically provides a didactic example of a confirmatory test on cross-cultural differences related to IS outsourcing. Specifically, we provide an example of how social scientists might introduce three propositions on differences between two countries.

In terms of the cross cultural results, we showed that some of the factors that explain variations in the degree of application software outsourcing are the same in both countries, while other influences differ significantly between both countries.

Commonalities. In both the U.S. and German sample, differences in production costs between in-sourcing and outsourcing as well as the opinion of influential others have a significant impact on the sourcing of application services. Both findings are in line with the empirical literature on IS outsourcing. The results also show that it is not a strictly rational decision process that occurs within the boundaries of the IS department, but rather a participative process that recognizes the opinion of external others.

Country Differences. While efficiency matters both in the U.S. and Germany, effectiveness criteria were found to be treated differently. First of all, while perceived in-house advantages in the *systemic impact* of an IS function were found to impede the extent to which application services are outsourced in Germany, the relationship was found to be irrelevant in the U.S. This obvious country difference is consistent with our perspective that German managers have more of an integrative view of the organization, where the firm is viewed as a group of related persons working together. By contrast, U.S. managers may see the firm as a collection of tasks, functions, people, and machines that can be changed and exchanged more flexibly, without leading to severe consequences for overall firm performance (Hampden-Turner and Trompenaars 1993, p. 18).

Second, the results show that in both countries, *systemic view* is an important predictor of the extent to which application services are outsourced, however, with different directional impacts. Germany, with a more integrative view and collectivist culture is less likely (more negative path) to outsource an IS function if they perceive a systemic view advantage exists for their company employees relative to outsourced workers. In contrast, the collectivist nature is likely viewed potentially as a hindrance in the U.S. The analytical nature of the U.S. workforce emphasizes compartmentalized effort and rotation/shifting of workers when required. Thus, the more systemic or collectivistic a CIO may perceive his or her company to be, the greater the desire to minimize this culture through the use of an external workforce.

Another relationship that was found to be culturally sensitive is the link between *outsourcing decision control* and degree of outsourcing. It was proposed, that a higher level of perceived control over the outsourcing process would be positively related with the degree of outsourcing and that this link would be stronger in Germany than in the U.S. Interestingly, there was a significant difference in the impact of that link between Germany and the U.S. But unexpectedly, that link was positive, instead of negative in Germany, while insignificant in the U.S. In other words, German organizations show a higher level of outsourcing if IS managers do not believe that they have full control over all necessary activities associated with outsourcing. A similar reversed link, albeit in a different organizational context, was also found in the study from Cordano and Frieze Hanson (2000, p. 637). From their point of view, this finding may be explained by the limited power of managers, which hinders them to act in accordance with their beliefs.

Overall, the PLS MGSEM analysis is shown to provide useful information for researchers interested in applied areas such as cross cultural studies. Using this technique, we were able to determine that cultural differences play a substantial role in IS sourcing decisions and that it is necessary to recognize that behavioral and institutional differences between countries can significantly limit the generalizability of mid-range theories of IS sourcing. In terms of our Monte Carlo simulation, the results, while not surprising, provides a sense of how the effect size, sample size, normality, and magnitude of prediction impacts the ability to detect an effect. A future study might involve a more complete assessment of the effect of asymmetry in the sample size between the two groups with the combined cases fixed at the same number. Furthermore, we'd recommend a comparison of how the PLS algorithm compares with a simple summed regression. Our initial test with an asymmetric sample set of 150 and 75, non-normal condition, and 0.7 and 0.4 path differences resulted in the PLS algorithm providing a 10 percent higher level in statistical power.

In summary, this paper attempted to illustrate the appropriateness of using a new non-parametric procedure for conducting MGSEM analysis using PLS. As noted earlier, such an approach employing randomization tests should not be viewed as alternatives to parametric statistical tests, rather they should be considered as those tests for that particular empirical form being examined. Thus, normal theory MGSEM may be viewed as approximations. This is an extremely important property in the case of both data distributions and nonrandom samples common in surveys, which would otherwise invalidate the use of parametric statistical tables (e.g., t or F tables). Nevertheless, in the case of our outsourcing data set, we did find remarkably similar results with the AMOS analysis, which provides greater confidence in a methodological convergent validity sense. Unfortunately, due to page and analytical constraints, comparison of our Monte Carlo results with those obtained using AMOS or similar covariance based MGSEM analysis was not performed. What would be useful in the future is to generate such data conforming to a model with varying levels of non-normality (both leptokurtic and platykurtic and left and right skewed) to see how both methods perform.

References

Ajzen, I. (1991). The theory of planned behavior. *Organizational behavior and human decision processes*, *50*, 179–211.

Ajzen, I., & Fishbein, M. (1980). *Understanding attitudes and predicting social behavior*. Englewood Cliffs, NY: Prentice Hall.

Ang, S., & Straub, D. W. (1998). Production and transaction economies and IS outsourcing: a study of the U.S. banking industry. *MIS Quarterly*, *22*(4), 535–552.

Apte, U. M., Sobol, M. G., Hanaoka, S., Shimada, T., Saarinen, T., Salmela, T., & Vepsalainen, A. P. J. (1997). IS outsourcing practices in the USA, Japan and Finland: a comparative study. *Journal of Information Technology*, *12*, 289–304.

Bagozzi, R. P., & Phillips, L. (1982). Representing and testing organizational theories: a holistic construal. *Administrative Science Quarterly*, *27*, 459–489.

Bagozzi, R. P., & Yi, Y. (1988). On the evaluation of structural equation models. *Journal of the Academy of Marketing Science*, *16*, 74–94.

Barthelemy, J., & Geyer, D. (2001). IT outsourcing: evidence from France and Germany. *European Management Journal*, *19*(2), 195–202.

Chin, W. W. (1998a). Issues and opinion on structural equation modeling. *MIS Quarterly*, *22*(1), VII–XVI.

Chin, W. W. (1998b). The partial least squares approach for structural equation modeling. In G. A. Marcoulides (Ed.), *Modern methods for business research* (pp. 295–336). Hillsdale, NJ: Lawrence Erlbaum Associates.

Chin, W. W., & Gopal, A. (1995). Adoption intention in GSS: relative importance of beliefs. *The data base for advances in information systems*, *26*(2), 42–63.

Chin, W. W., & Newsted, P. R. (1999). Structural equation modeling: analysis with small samples using partial least squares. In R. Hoyle (Ed.), *Statistical strategies for small sample research* (pp. 307–341). Thousand Oaks, CA: Sage.

Compeau, D. R., & Higgins, C. A. (1991). *A social cognitive theory perspective on individual reactions to computing technology*. International conference on information systems (ICIS), New York, 187–198.

Cordano, M., & Frieze Hanson, I. (2000). Pollution reduction preferences of U.S. environmental managers: applying Ajzen's theory of planned behavior. *Academy of Management Review*, *43*(4), 627–641.

Dibbern, J., Goles, T., Hirschheim, R. A., & Jayatilaka, B. (2004). Information systems outsourcing: a survey and analysis of the literature. *The DATA BASE for Advances in Information Systems*, *35*(4), 6–102.

Dibbern, J., & Heinzl, A. (2001). Outsourcing der informationsverarbeitung im mittelstand: test eines multitheoretischen kausalmodells. *Wirtschaftsinformatik*, *43*(4), 339–350.

Dibbern, J., Heinzl, A., & Leibbrandt, S. (2003). Interpretation des sourcing der informationsverarbeitung: hintergründe und grenzen ökonomischer einflussgrößen. *Wirtschaftsinformatik*, *45*(5), 533–540.

DiRomualdo, A., & Gurbaxani, V. (1998). Strategic intent for IT outsourcing. *Sloan Management Review*, Summer, 67–80.

Edgington, E. S. (1987). *Randomization tests*. New York: Marcel Dekker Inc.

Efron, B., & Tibshirani, R. J. (1993). *An introduction to the bootstrap*. In monographs on statistics and applied probability, No. 57. New York: Chapman & Hill.

Fornell, C. (1989). The blending of theoretical empirical knowledge in structural equations with un-observables. In H. Wold (Ed.), *Theoretical empiricism: a general rationale for scientific model-building* (pp. 153–174). New York: Paragon House.

Fornell, C., & Bookstein, F. (1982). Two structural equation models: LISREL and PLS applied to consumer exit-voice theory. *Journal of Marketing Research*, *19*, 440–452.

Good, P. (2000). *Permutation tests, a practical guide to resampling methods for testing hypotheses*. New York: Springer.

Hamilton, S., & Chervany, N. L. (1981). Evaluating information system effectiveness part I: comparing evaluation approaches. *MIS Quarterly*, *5*(3), 55–69.

Hampden-Turner, C., & Trompenaars, A. (1993). *The seven cultures of capitalism: value systems for creating wealth in the United States, Japan, Germany, France, Britain, Sweden, and the Netherlands*. New York: Doubleday.

Hirschheim, R. A., & Lacity, M. C. (2000). The myths and realities of information technology in-sourcing. *Communications of the ACM*, *43*(2), 99–107.

Hofstede, G. (1980). *Culture's consequences, international differences in work-related values*. Beverly Hills: Sage.

Hofstede, G. (1983). National culture in four dimensions. *International Studies of Management and Organization*, *13*(2), 46–74.

Hofstede, G. (1991). *Cultures and organizations: software of the mind*. London: McGraw-Hill.

Hui, C. H., & Triandis, H. C. (1986). Individualism-collectivism: a study of cross-cultural research. *Journal of Cross-Cultural Psychology*, *17*, 222–248.

Hulland, J. (1999). Use of partial least squares (PLS) in strategic management research: a review of four recent studies. *Strategic Management Journal*, *20*(2), 195–204.

Jöreskog, K., & Sörbom, D. (1996). *PRELIS 2: User's reference guide*. Chicago: Scientific Software International.

Murphy, C., Ker, S., & Chen, S. (1999). IDC: U.S. and worldwide outsourcing markets and trends, 1998–2003. International data corporation, No. 19322

Keil, M., Tan, B. C. Y., Wei, K. -K., Saarinen, T., Tuunainen, V., & Wassenaar, A. (2000). A cross-cultural study on escalation of commitment behavior in software projects. *MIS Quarterly*, *24*(2), 299–325.

Lacity, M. C., & Hirschheim, R. A. (1993). The information systems outsourcing bandwagon. *Sloan Management Review*, *35*(1), 73–86.

Lacity, M. C., & Hirschheim, R. A. (1995). *Beyond the information systems outsourcing bandwagon : the insourcing response*. Chichester, New York: Wiley.

Lytle, A. L., Brett, J. M., Barsness, Z. I., Tinsley, C. H., & Maddy, J. (1995). A paradigm for confirmatory cross-cultural research in organizational behavior. In B. M. Staw and L. L. Cummings (Eds.), *Research in organizational behavior* (Vol. 17, pp. 167–214). Greenwich, CT: JAI.

Mathieson, K. (1991). Predicting user intentions: comparing the technology acceptance model with the theory of planned behavior. *Information Systems Research*, *2*(3), 173–191.

McLellan, K. L., Marcolin, B. L., & Beamish, P. W. (1995). Financial and strategic motivations behind IS outsourcing. *Journal of Information Technology*, *10*, 299–321.

Mohr, L. B. (1991). *Understanding significance testing, series: quantitative applications in the Social Sciences*. Newbury Park, CA: Sage.

Noreen, E. W. (1989). *Computer intensive methods for testing hypotheses, an introduction*. New York: Wiley.

OECD (2000). *Information technology outlook: 2000*. Paris: OECD.

Osgood, C. E., Suci, G. J., & Tannenmann, P. H. (1957). *The measurement of meaning*. Urbana, IL: University of Illinois Press.

Peter, J. (1981). Reliability: a review of psychometric basics and recent marketing practices. *Journal of Marketing Research*, *16*, 6–17.

Pfeffer, J., & Salancik, G. R. (1978). *The external control of organizations: a resource dependence perspective*. New York: Harper & Row.

Pitt, L. F., Watson, R. T., & Kavan, C. B. (1995). Service quality: a measure of information systems effectiveness. *MIS Quarterly*, *19*(2), 173–187.

Ramberg, J. S., Dudewicz, E. J., Tadikamalla, P. R., Pandu, R., & Mykytka, E.F. (1979). A probability distribution and its uses in fitting data. *Technometrics*, *21*, 201–214.

Reinartz, W. J., Echambadi, R., & Chin, W. W. (2002). Generating non-normal data for simulation of structural equation models using Mattson's method. *Multivariate Behavioral Research*, *37*(2), 227–244.

Richardi, R. (1990). Labour Right. In C. E. Poeschel (Ed.), *Handbook of German business management* (Vol. 2, pp. 1278–1290). Stuttgart: Verlag.

Sambamurthy, V., & Chin, W. W. (1994). The effects of group attitudes toward alternative GDSS designs on the decision-making performance of computer-supported groups. *Decision Sciences*, *25*(2), 215–241.

Sobol, M. G., & Apte, U. M. (1995). Domestic and global outsourcing practices of America's most effective IS users. *Journal of Information Technology*, *10*, 269–280.

Teng, J. T. C., Cheon, M. J., & Grover, V. (1995). Decisions to outsource information systems functions: testing a strategy-theoretic discrepancy model. *Decision Sciences*, *26*(1), 75–103.

Thompson, R. L. (1967). *Organizations in action*. New York: McGraw-Hill.

Thompson, R. L., Higgins, C. A., & Howell, J. M. (1994). Influence Of experience on personal computer utlization: testing a conceptual model. *Journal of Management Information Systems*, *11*(1), 167–187.

Triandis, H. C. (1996). The psychological measurement of cultural syndromes. *American Psychologist*, *51*(4), 407–415.

Trompenaars, A., & Hampden-Turner, C. (1994). *Riding the waves of culture : understanding diversity in global business*. Burr Ridge, Ill: Irwin Professional Pub.

Wade, M., & Hulland, J. (2004). Review: The resource-based view and information systems research: review, extension, and suggestions for future research. *MIS Quarterly*, *18*(1), 107–142.

Williamson, O. E. (1981). The Economics of organization: the transaction cost approach. *American Journal of Sociology*, *87*(3), 548–577.

Chapter 8
Finite Mixture Partial Least Squares Analysis: Methodology and Numerical Examples

Christian M. Ringle, Sven Wende, and Alexander Will

Abstract In wide range of applications for empirical data analysis, the assumption that data is collected from a single homogeneous population is often unrealistic. In particular, the identification of different groups of consumers and their appropriate consideration in partial least squares (PLS) path modeling constitutes a critical issue in marketing. In this work, we introduce a finite mixture PLS software implementation which separates data on the basis of the estimates' heterogeneity in the inner path model. Numerical examples using experimental as well as empirical data allow the verification of the methodology's effectiveness and usefulness. The approach permits a reliable identification of distinctive customer segments along with characteristic estimates for relationships between latent variables. Researchers and practitioners can employ this method as a model evaluation technique and thereby assure that results on the aggregate data level are not affected by unobserved heterogeneity in the inner path model estimates. Otherwise, the analysis provides further indications on how to treat that problem by forming groups of data in order to perform a multi-group path analysis.

8.1 Introduction

Structural equation modeling (SEM) and path modeling with latent variables (LVP) are applied in marketing research to measure complex cause-effect relationships (Fornell and Larcker 1981; Steenkamp and Baumgartner 2000). Covariance structure

C.M. Ringle
University of Hamburg Institute for Industrial Management and Organizations
Von-Melle-Park 5, 20146 Hamburg, Germany
e-mail: c.ringle@smartpls.de
and
University of Technology Sydney Centre for Management and Organisation Studies
PO Box 123, Broadway NSW 2007, Australia

S. Wende and A. Will
University of Hamburg Institute for Industrial Management and Organizations
Von-Melle-Park 5, 20146 Hamburg, Germany
e-mail: s.wende@smartpls.de, mail@alexwill.de

V. Esposito Vinzi et al. (eds.), *Handbook of Partial Least Squares*, Springer Handbooks of Computational Statistics, DOI 10.1007/978-3-540-32827-8_9,

analysis (CSA) (Jöreskog 1978) and partial least squares analysis (PLS) (Lohmöller 1989) constitute the two corresponding, yet different (Schneeweiß 1991), statistical techniques for estimating such models. An important research issue in SEM and LVP is the measurement of customer satisfaction (Fornell et al. 1996; Hackl and Westlund 2000), which is closely related to the requirement of identifying distinctive customer segments (ter Hofstede et al. 1999; Wu and Desarbo 2005).

In SEM, segmentation can be achieved based on the heterogeneity of scores for latent variables in the structural model (DeSarbo et al. 2006). Jedidi et al. (1997) pioneer this field of research and propose a procedure that blends finite mixture models and the expectation-maximization (EM) algorithm (McLachlan and Krishnan 2004; Wedel and Kamakura 2000). However, this technique extends CSA but is inappropriate for PLS path modeling. For this reason, Hahn et al. (2002) propose the finite mixture partial least squares (FIMIX-PLS) approach that joins a finite mixture procedure with an EM algorithm specifically regarding the ordinary least squares (OLS)-based predictions of PLS. Sarstedt (2008) reviews existing segmentation techniques for PLS path modeling and concludes that FIMIX-PLS can currently be viewed as the most comprehensive and commonly used approach to capture heterogeneity in PLS path modeling.

Building on the guiding articles by Jedidi et al. (1997) and Hahn et al. (2002), this paper presents FIMIX-PLS as it is implemented for the first time in a statistical software application (SmartPLS; Ringle et al. 2005). Thereby, this methodology for segmenting data based on PLS path modeling results is made broadly applicable for research in marketing, management and other social sciences disciplines. This kind of analysis is typically performed in two stages. In the first step, FIMIX-PLS (see Chap. 8.2) is applied for different numbers of classes using standard PLS path modelling estimates. If distinctive groups of observations in the overall set of data cause heterogeneity in the inner PLS path model estimates, FIMIX-PLS results permit detection of this heterogeneity and provide implications how to treat it by segmentation. In the second step (see ex post analysis in Chap. 8.3), an explanatory variable must be uncovered that entails both, similar clustering of data, as indicated by evaluated FIMIX-PLS outcomes, and interpretability of the formed groups of observations. Then, correspondingly separated sets of data are used as new input for segment-specific LVP computations with PLS facilitating multigroup analysis (Chin and Dibbern 2010, Chap. 8.7). Both analytical steps frame a comprehensive application of the FIMIX-PLS approach and are carried out by numerical examples with experimental (see Chap. 8.4) and empirical data (see Chap. 8.5) in this paper. The numerical examples reveal some important methodological implications that have not been addressed, yet.

As segmentation is a key element for marketers to form and improve their targeted marketing strategies, these analyses allow us to demonstrate the potentials of FIMIX-PLS for identifying homogeneous clusters of consumers with regard to the benefits they seek or in their response to marketing programs. This research is important to expand the methodological toolbox for analyzing LVP with PLS. Like the confirmatory tetrad analysis to empirically test whether a measurement model is reflective or formative (Gudergan et al. 2008), researchers and practitioners should employ FIMIX-PLS as a standard procedure to evaluate their PLS path modeling results. They thereby assure that outcomes on the aggregate data level are not affected by unobserved heterogeneity in the inner path model estimates. Otherwise, the analysis provides further indications on how to treat that problem by forming groups of data in order to perform a multi-group path analysis. Significantly distinctive group-specific path model estimations impart further differentiated interpretations of PLS modeling results and may foster the origination of more effective (marketing) strategies (Rigdon et al. 2010; Ringle et al. 2010a; Sarstedt et al. 2009).

8.2 Methodology

The first methodological step is to estimate path models by applying the basic PLS algorithm for LVP (Lohmöller 1989). Then, FIMIX-PLS is employed as formally described and discussed by its developers (Hahn et al. 2002) using the estimated scores of latent variables and their modified presentation of relationships in the inner model (see Table 8.7 in the appendix for a description of all of the symbols used in the equations presented in this paper):

$$B\eta_i + \Gamma\xi_i = \zeta_i \tag{8.1}$$

Segment-specific heterogeneity of path models is concentrated in the estimated relationships between latent variables. FIMIX-PLS captures this heterogeneity. The distributional function for each segment is defined as follows, assuming that η_i is distributed as a finite mixture of conditional multivariate normal densities $f_{i|k}(\cdot)$:

$$\eta_i \sim \sum_{k=1}^{K} \rho_k f_{i|k}(\eta_i|\xi_i, B_k, \Gamma_k, \Psi_k) \tag{8.2}$$

Substituting $f_{i|k}(\eta_i|\xi_i, B_k, \Gamma_k, \Psi_k)$ results in the following equation:

$$\eta_i \sim \sum_{k=1}^{K} \rho_k \left[\frac{|B_k|}{\sqrt[M]{2\pi}\sqrt{|\Psi_k|}} e^{-\frac{1}{2}(B_k\eta_i + \Gamma_k\xi_i)'\Psi_k^{-1}(B_k\eta_i + \Gamma_k\xi_i))} \right] \tag{8.3}$$

It is sufficient to assume multivariate normal distribution of η_i. Equations (8.4) and (8.5) represent an EM-formulation of the likelihood function and the log-likelihood (lnL) as the corresponding objective function for maximization:

$$L = \prod_i \prod_k [\rho_k f(\eta_i|\xi_i, B_k, \Gamma_k, \Psi_k)]^{z_{ik}} \tag{8.4}$$

$$lnL = \sum_i \sum_k z_{ik} ln(f(\eta_i|\xi_i, B_k, \Gamma_k, \Psi_k)) + \sum_i \sum_k z_{ik} ln(\rho_k) \tag{8.5}$$

The EM algorithm is used to maximize the likelihood in this model in order to ensure convergence. The "expectation" of (8.5) is calculated in the E-step, where z_{ik} is 1 if subject i belongs to class k (or 0 otherwise). The relative segment size ρ_k, the parameters ξ_i, B_k, Γ_k and Ψ_k of the conditional probability function are given, and provisional estimates (expected values) for z_{ik} are computed as follows according to Bayes' theorem:

$$E(z_{ik}) = P_{ik} = \frac{\rho_k f_{i|k}(\eta_i|\xi_i, B_k, \Gamma_k, \Psi_k)}{\sum_{k=1}^{K} \rho_k f_{i|k}(\eta_i|\xi_i, B_k, \Gamma_k, \Psi_k)} \tag{8.6}$$

Equation (8.5) is maximized in the M-step. Initially, new mixing proportions ρ_k are calculated by the average of adjusted expected values P_{ik} that result from the previous E-step:

$$\rho_k = \frac{\sum_{i=1}^{I} P_{ik}}{I} \tag{8.7}$$

Thereafter, optimal parameters for B_k, Γ_k, and Ψ_k are determined by independent OLS regression (one for each relationship between latent variables in the inner model). ML estimators of coefficients and variances are assumed to be identical to OLS predictions. The following equations are applied to obtain the regression parameters for endogenous latent variables:

$$Y_{mi} = \eta_{mi} \tag{8.8}$$

$$X_{mi} = (E_{mi}, N_{mi})' \tag{8.9}$$

$$E_{mi} = \begin{cases} \{\xi_1, \ldots, \xi_{A_m}\}, A_m \geq 1, a_m = 1, \ldots, A_m \wedge \xi_{a_m} \text{ is regressor of } m \\ \emptyset \text{ else} \end{cases} \tag{8.10}$$

$$N_{mi} = \begin{cases} \{\eta_1, \ldots, \eta_{B_m}\}, B_m \geq 1, b_m = 1, \ldots, B_m \wedge \eta_{b_m} \text{ is regressor of } m \\ \emptyset \text{ else} \end{cases} \tag{8.11}$$

The closed form OLS analytic formulation for τ_{mk} and ω_{mk} is given as follows:

$$\tau_{mk} = \left((\gamma_{a_m mk}), (\beta_{b_m mk})\right)' = \left[\sum_i P_{ik}(X'_{mi} X_{mi})\right]^{-1} \left[\sum_i P_{ik}(X'_{mi} Y_{mi})\right] \tag{8.12}$$

$$\omega_{mk} = \text{ cell } (m \times m) \text{ of } \Psi_k = \frac{\sum_i P_{ik}(Y_{mi} - X_{mi}\tau_{mk})(Y_{mi} - X_{mi}\tau_{mk})'}{I\rho_k} \tag{8.13}$$

The M-step computes new mixing proportions. Independent OLS regressions are used in the next E-step iteration to improve the outcomes for P_{ik}. Based on an a priori specified convergence criterion, the EM-algorithm stops whenever the lnL hardly improves (see Fig. 8.1). This is more a measure of lack of progress than a measure of convergence, and there is evidence that the algorithm is often stopped too early (Wedel and Kamakura 2000).

When applying FIMIX-PLS, the EM-algorithm monotonically increases lnL and converges towards an optimum. Experience shows that FIMIX-PLS frequently

```
// initial E-step
set random starting values for P_ik; set last_lnL = V; set 0 < S < 1

repeat do
begin
    // the M-step starts here
    ρ_k = (Σ_{i=1}^{I} P_ik) / I  ∀k
    determine B_k, Γ_k, Ψ_k, ∀k
    calculate current_lnL
    Δ = current_lnL − last_lnL

    // the E-step starts here
    if Δ ≥ S then
    begin
        P_ik = ρ_k f_{i|k}(η_i|ξ_i, B_k, Γ_k, Ψ_k) / Σ_{k=1}^{K} ρ_k f_{i|k}(η_i|ξ_i, B_k, Γ_k, Ψ_k)  ∀i, k
        last_lnL = current_lnL
    end
end
until Δ < S
```

Fig. 8.1 The FIMIX-PLS algorithm

stops in local optimum solutions, caused by multimodality of the likelihood, so that the algorithm becomes sensitive to starting values. Moreover, the problem of convergence in local optima seems to increase in relevance whenever component densities are not well separated or the number of parameters estimated is large and the information embedded in each observation is limited (Wedel and Kamakura 2000). This results in relatively weak updates of membership probabilities in the E-step. Some examples of simple strategies for escaping local optima include initializing the EM-algorithm from a wide range of (random) values or using sequential clustering procedures, such as K-means, to obtain an appropriate initial partition of data. If alternative starting values of the algorithm result in different local optima, then the solution with the maximum value of likelihood is recommended as best solution. An issue for future research is to address concerns whether this kind of an unsystematically selected solution reaches the global optimum.

Another crucial aspect is that FIMIX-PLS only applies mixtures to the regressions in the inner model while this is not possible for the outer model. The algorithm's static use of latent variable scores does not entail dynamically forming new groups of data and computing group-specific outer and inner PLS path model estimates in every iteration compared to a prediction oriented segmentation algorithm presented by Squillacciotti (2010), Chap. 9. Eventhough, computational experiment for various data constellations show that FIMIX-PLS performs better or equally well compared with those alternative PLS segmentation approaches such as PLS-GAS (Ringle et al. 2010b), PLS-TPM and REBUS-PLS (Esposito Vinzi et al. 2007). In FIMIX-PLS, one regression equation for each segment captures the predictor-outcome relationships at the same time that the uncovered segments are captured in the inner model and, thus, reliably accounts for heterogeneity in the relationships of latent variables as demonstrated by two numerical in the Chaps. 8.4 and 8.5. Although, FIMIX-PLS results ought not instantaneously be analyzed and interpreted. In a second analytical step, the ex-post analysis (see the following chapter), an explanatory variable must be identified that allows forming groups of data as indicated by FIMIX-PLS. Then, these a-priori segmented data is used as new inputs for PLS estimations providing group-specific latent variables scores as well as results for the outer and inner measurement models. By this means, concerns on the subject of static utilization of latent variable scores are relaxed and turned into a key advantage of this segmentation approach (Sarstedt and Ringle 2010). FIMIX-PLS is generally applicable for all kinds of PLS path models regardless of whether measurement models for latent variables are operationalized as formative or reflective (see the numerical example in Chap. 8.4).

8.3 Segmentation and Ex Post Analysis

When applying FIMIX-PLS, the number of segments is unknown and the identification of an appropriate number of K classes is not straightforward. A statistically satisfactory solution does not exist for several reasons (Wedel and Kamakura 2000), i.e., mixture models are not asymptotically distributed as chi-square and do not allow for the likelihood ratio statistic. For this reason, Hahn et al. (2002) propose the repeated operation of FIMIX-PLS with consecutive numbers of latent classes K (e.g., 1–10) and to compare the class-specific outcomes for criteria such as the lnL, the Akaike information criterion ($AIC_K = -2lnL + 2N_K$), the consistent AIC ($CAIC_K = -2lnL + (ln(I) + 1)N_K$) or the Bayesian Information Criterion ($BIC_K = -2lnL + ln(I)N_K$). The results of these heuristic measures and their comparison for different numbers of classes provide evidence about an appropriate number of segments. Moreover, an entropy statistic (EN), limited between 0 and 1, indicates the degree of separation in the individually estimated class probabilities (Ramaswamy et al. 1993):

$$EN_K = 1 - \frac{\left[\sum_i \sum_k -P_{ik} ln(P_{ik})\right]}{I ln(K)} \tag{8.14}$$

The quality of separation of the derived classes will improve the higher EN is. Values of EN above 0.5 imply estimates for P_{ik} that permit unambiguous segmentation. Thus, this criterion is especially relevant for identifying and clustering different types of customers in the field of marketing.

Given these assumptions, FIMIX-PLS is only applicable for additional analytic purposes, if an explanatory variable can be identified. An explanatory variable must facilitate both a-priori clustering of data, as indicated by the evaluated FIMIX-PLS results, and interpretability of the distinctive groups. This kind of analysis is essential for exploiting FIMIX-PLS findings for PLS path modeling, and it is the most challenging analytical step to accomplish. Hahn et al. (2002) suggest an ex post analysis of the estimated probabilities of membership using an approach proposed by Ramaswamy et al. (1993). The additional findings can be used to a-priori group data (e.g., into "younger customers" and "older customers") as well as to compute and analyze the LVP for each segment. The following numerical examples, which use experimental and empirical data, document this approach.

8.4 Example Using Experimental Data

Suppose that a market researcher has formulated a LVP on theoretically well developed cause-effect relationships. The researcher suspects, however, that an unobserved moderating factor accounts for Heterogeneity or that the data belongs to a finite number of segments. In such situations, theoretical assumptions can be used to identify a-priori moderating factors that account for consumer heterogeneity in PLS path model. This kind of strategy is not feasible in many marketing applications (Jedidi et al. 1997), and it gives rise to analytical techniques like FIMIX-PLS.

SmartPLS 2.0 (Ringle et al. 2005) is the first statistical software application for (graphical) path modeling with latent variables employing both the basic PLS algorithm (Lohmöller 1989) as well as FIMIX-PLS capabilities for the kind of segmentation proposed by Hahn et al. (2002). Applying this statistical software module to experimental data for a marketing-related path model demonstrates the potentials of the methodology for PLS-based research. In terms of heterogeneity in the inner model, it might be desirable to identify and describe price sensitive consumers (Kim et al. 1999) and consumers who have the strongest preference for another particular product attribute (Allenby et al. 1998), e.g., quality. Thus, the path model for our numerical example with experimental data has one endogenous latent variable, *Satisfaction*, and two exogenous latent variables, *Price* and *Quality*, in the inner model (DeSarbo et al. 2001; Dillon et al. 1997). The used experimental set of data consist of the following equally sized segments:

- Price-oriented customers (segment 1) – this segment is characterized by a strong relationship between *Price* and *Satisfaction* and a weak relationship between *Quality* and *Satisfaction*.
- Quality-oriented customers (segment 2) – this segment is characterized by a strong relationship between *Quality* and *Satisfaction* and a weak relationship between *Price* and *Satisfaction*.

Instead of using single item constructs, each exogenous latent variable (*Price* and *Quality*) has five indicators, and the endogenous latent variable (*Satisfaction*) is measured by three manifest variables (Sarstedt and Wilczynski 2009). We use the correlation matrix in Table 8.1 to generate experimental data. This matrix is partially adopted with changed variable names from Albers and Hildebrandt (2006) who compare, among other aspects, results of formative and reflective operationalized PLS path models with experimental data. A Monte Carlo simulation is performed employing the SEPATH module of the software application STATISTICA 7.1 to

Table 8.1 Correlation of manifest variables

	Price1	Price2	Price3	Price4	Price5	Quality1	Quality2	Quality3	Quality4	Quality5	Satisfaction1	Satisfaction2	Satisfaction3
Price1	1.00												
Price2	0.03	1.00											
Price3	−0.01	0.15	1.00										
Price4	0.06	−0.05	0.10	1.00									
Price5	−0.01	0.08	0.06	0.56	1.00								
Quality1	−0.02	0.07	0.10	−0.05	−0.06	1.00							
Quality2	−0.03	−0.05	−0.02	0.06	−0.01	0.12	1.00						
Quality3	0.07	0.02	0.01	0.01	−0.04	0.24	0.57	1.00					
Quality4	−0.02	−0.04	0.02	0.00	−0.05	0.29	0.49	0.53	1.00				
Quality5	0.05	−0.02	0.00	−0.02	−0.01	0.13	0.20	0.29	0.27	1.00			
Satisfaction1	0.15	0.14	0.19	−0.02	0.01	0.08	0.08	0.03	0.06	−0.02	1.00		
Satisfaction2	0.19	0.11	0.16	0.04	0.01	0.10	0.04	0.02	0.04	−0.01	0.85	1.00	
Satisfaction3	0.09	0.14	0.15	0.01	0.04	0.11	0.01	0.00	0.03	0.00	0.89	0.83	1.00

generate manifest variable scores. The first one hundred case values are computed for a strong relationship of 0.9 between *Price* and *Satisfaction* and a week relationship of 0.1 between *Quality* and *Satisfaction* in the inner path model (segment 1). Correspondingly, another one hundred cases reflect the characteristics of the quality-oriented segment 2 so that the full set of experimental data includes 200 cases.

PLS path modelling permits both, formative as well as reflective operationalization of latent variables' measurement model with manifest variables (Lohmöller 1989; Ringle et al. 2009). The choice depends on the theoretical foundation and interpretation of cause-effect relationships (Diamantopoulos and Winklhofer 2001; Jarvis et al. 2003; Gudergan et al. 2008; Rossiter 2002). Consequently, FIMIX-PLS must properly perform for this experimental set of data using three different examples of outer measurement models:

- Reflective case – all latent variables have reflective indicators.
- Formative case – all latent variables have formative indicators.
- Mixed case – the exogenous latent variables have a formative while the latent endogenous variable has a reflective measurement model.

To begin with, we use reflective measurement model for all three latent variables. FIMIX-PLS employs the estimates of the standard PLS procedure for this numerical example with experimental data in order to process the latent variable scores for $K = 2$ classes. The standard PLS inner model weights in Table 8.2 show that both constructs, *Price* and *Quality*, have a relatively high effect on *Satisfaction* resulting in a substantial R^2 of 0.465. An overview of results is provided by Table 8.8 in the appendix. However, it is quite misleading to instantaneously examine and further interpret these good estimates for a PLS path model.

The application of FIMIX-PLS permits additional analysis that lead to different conclusions. This procedure identifies two equally sized groups of data that exhibit segment-specific path coefficients with the same characteristics as expected for the experimental set of data (see Table 8.2). Attributable to the experimental design, segment-specific regression variances are very low for the latent endogenous variable *Satisfaction* (0.170 for segment 1 and 0.149 for segment 2) resulting in corresponding outcomes for R^2 at a high level for each segment. Among other results, SmartPLS 2.0 provides the final probability of membership P_{ik} of each case to fit into one of the two classes. More than 80% of the cases are assigned to the class they have been intended to belong to in accordance with the design of data generation in this numerical example. An EN above 0.5 indicates a good separation of data.

Table 8.2 Inner model weights

	Price → Satisfaction	Quality → Satisfaction
Standard PLS	0.538	0.450
FIMIX-PLS segment 1	0.899	0.009
FIMIX-PLS segment 2	0.113	0.902

In the second analytical step, we test the FIMIX-PLS results for segment-specific PLS analysis. The FIMIX-PLS probabilities of membership allow splitting the experimental set of data into two groups. These two sets of data are then separately used as input matrices for manifest variables to estimate the path model for each group with PLS. The FIMIX-PLS results for segment-specific relationships in the inner model are essentially re-establish by this supplementary analysis. While the lower relationship in the inner path model for each group of price- or quality-oriented consumers remains at a value around 0.1, the higher relationship is at a value close to 0.9 and R^2 is around 0.8 in both cases. An overview of these result is given by Table 8.8 in the appendix.

FIMIX-PLS reliably identifies two a-priori formed segments in this numerical example with experimental data and reflective operationalization of latent variables in the PLS path model. However, the question remains, if the methodology also properly performs for path models with formative measurement model. For this reason, all three latent variables are measured with formative indicators and, in the mixed case, *Price* and *Quality* have a formative measurement model while *Satisfaction* has reflective indicators. The standard inner PLS path model estimates as well as the FIMIX-PLS results for two segments in these additional analysis (for the formative and the mixed case) are at the same level as indicated for the reflective case. Then, in the second analytical step, we split the experimental set of data according to the FIMIX-PLS probabilities of membership P_{ik} into two sets of data that are then used as new input matrices for groups specific PLS path model estimates. The computations also provide almost the same estimates for the inner path model relationships and the R^2 of *Satisfaction* as described before for the reflective case (see Tables 8.9 and 8.10 in the appendix).

As a result from these numerical examples with experimental data, we further substantiate the earlier stated rationale that FIMIX-PLS is capable to identify and treat heterogeneity of inner path model estimates by segmentation no matter if latent variables have formative or reflective measurement models. The corresponding group-specific PLS analysis are important for marketers to further differentiate interpretations of the path model resulting in more specific recommendations for the use of the marketing-mix instruments to effectively target each group of consumers.

8.5 Marketing Example Using Empirical Data

When researchers work with empirical data and do not have a-priori segmentation assumptions to capture unobserved heterogeneity in the inner PLS path model relationships, FIMIX-PLS is often not as clear-cut as demonstrated in the foregoing example that is based on experimental data. Until now, research efforts to apply FIMIX-PLS and to assess its usefulness for expanding methodological instruments in marketing was restricted by the unavailability of a statistical software application for this kind of analysis. Since such functionalities are provided as presented in Chap. 8.2, extensive use of FIMIX-PLS with empirical data in future research ought to furnish additional findings about the methodology and its applicability. For this reason, we make use of that technique for a marketing-related path model and empirical data from Gruner+Jahr's "Brigitte Communication Analysis 2002".

Gruner+Jahr is one of the leading publishers of printed magazines in Germany. They have been conducting their communication analysis survey every other year since 1984. In the survey, over 5,000 women answer numerous questions on brands in different product categories and questions regarding their personality. The women represent a cross section of the German female population. We choose answers to questions on the Benetton fashion brand name (on a four-point scale from "low" to "high") in order to use the survey as a marketing-related example of FIMIX-PLS-based customer segmentation. We assume that Benetton's aggressive and provocative advertising in the 1990s resulted in a lingering customer heterogeneity that is more distinctive and easier to identify compared with other fashion brands in the Communication Analysis Survey (e.g., Esprit or S.Oliver).

The scope of this paper does not include a presentation of theoretically hypothesized LVP and its PLS-based estimation with empirical data (Bagozzi 1994; Hansmann and Ringle 2005). Consequently, we do not provide a discussion if one ought use CSA or PLS to estimate the cause-effect relationship model with latent variables (Bagozzi and Yi 1994), a line of reasoning if the measurement models of latent variables should be operationalized as formative or reflective (Diamantopoulos and Winklhofer 2001; Rossiter 2002) or an extensive presentation of the survey data. Our goal is to demonstrate the applicability of FIMIX-PLS to empirical data for a reduced cause-effect relationship model on branding (Yoo et al. 2000) that principally guides all kinds of LVP analysis in marketing employing this segmentation technique.

The PLS path model for Benetton's brand preference consists of one latent endogenous *Brand preference* variable, and two exogenous latent variables, *Image* and *Person*, in the inner model. All latent variables are operationalized via a reflective measurement model. Figure 8.2 illustrates the path model with the latent variables and the particular manifest variables from Gruner+Jahr's "Brigitte Communication Analysis 2002" employed. The basic PLS algorithm (Lohmöller 1989) is applied for estimating that LVP using the SmartPLS 2.0 (Ringle et al. 2005) software application.

We follow the suggestions given by Chin (1998a) and Henseler et al. (2009) for arriving at a brief evaluation of results. All relationships in the reflective measurement model have high factor loadings (the smallest loading has a value of 0.795). Moreover, results for the average variance extracted (AVE) and ρ_c are at good levels (see Table 8.11 in the appendix). The exogenous latent *Image* variable (weight of 0.423) exhibits a strong relationship to the endogenous latent *Brand preference* variable. The influence of the exogenous latent *Person* variable is considerably weaker (weight of 0.177). Both relationships are statistically significant [tested with the bootstrapping procedure using individual sign change (Tenenhaus et al. 2005)]. The endogenous latent variable *Brand preference* has a R^2 of 0.239 and, thus, is at a moderate level for PLS path models.

The FIMIX-PLS module of SmartPLS 2.0 is applied for customer segmentation based on the estimated scores for latent variables. Table 8.3 shows heuristic

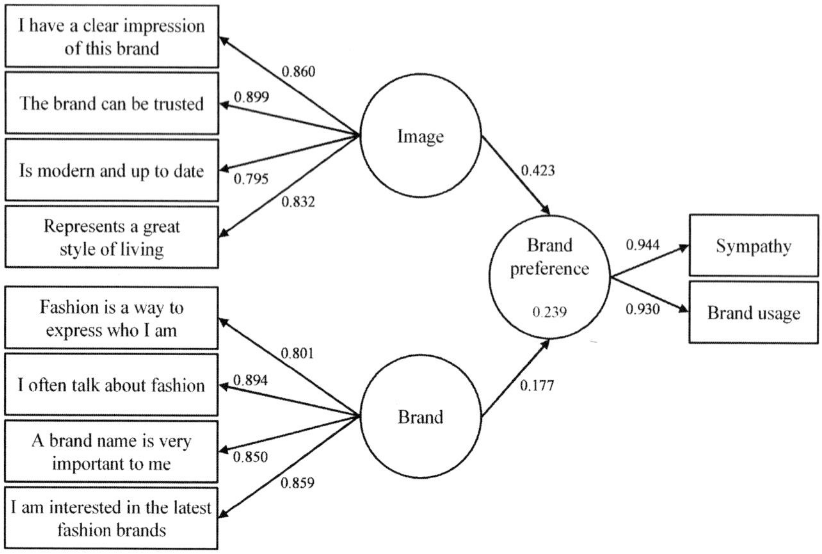

Fig. 8.2 The brand preference model

Table 8.3 Evaluation of FIMIX-PLS results

Number of latent classes	*lnL*	*AIC*	*BIC*	*CAIC*	*EN*
$K = 2$	−713.233	1448.466	1493.520	1493.545	0.501
$K = 3$	−942.215	1954.431	2097.784	2097.863	0.216
$K = 4$	−1053.389	2192.793	2450.830	2450.972	0.230
$K = 5$	−1117.976	2441.388	2846.874	2847.097	0.214

FIMIX-PLS evaluation criteria for alternative numbers of classes K. According to these results, the choice of two latent classes seems to be appropriate for customer segmentation purposes, especially in terms of EN. Compared to EN of 0.43 arrived at in the only other proficient FIMIX-PLS segmentation presented thus far in literature by Hahn et al. (2002), our EN result of 0.501 also reaches a proper level indicating well separable groups of data.

Table 8.4 presents the FIMIX-PLS results for two latent classes. In a large segment (relative size of 0.809), the explained variance of the endogenous latent *Brand preference* variable is at a relatively weak level for PLS models ($R^2 = 0.108$). The variance is explained by the exogenous latent *Image* variable, with its weight of 0.343, and the exogenous latent *Person* variable, with its weight of 0.177. A smaller segment (relative size of 0.191) has a relatively high R^2 for *Brand preference* (value of 0.930). The influence of the *Person* variable does not change much for this segment. However, the weight of the *Image* variable is more than twice as high and has a value of 0.759. This result reveals that the preference for Benetton is explained to a high degree whenever the image of this brand is far more important than the individuals' personality.

Table 8.4 FIMIX-PLS disaggregate results for two latent classes

	$K = 1$	$K = 2$
Relative segment size	0.809	0.191
R^2 (for *Brand preference*)	0.108	0.930
Path *Image* to *Brand preference*	0.343	0.759
Path *Person* to *Brand preference*	0.177	0.170

Table 8.5 A-priori segmentation based on *I like to buy fashion designers' perfumes*

	Segment 1	Segment 2
R^2 (for *Brand preference*)	0.204	0.323
Image → *Brand preference*	0.394	0.562
Person → *Brand preference*	0.164	0.104

The next step of FIMIX-PLS involves the identification of a certain variable to form and characterize the two uncovered customer segments. For this reason, we conducted an ex post analysis for finite mixture models according to the approach proposed by Ramaswamy et al. (1993). Among several possible indicators examined, the most significant explanatory variable are: *I am very interested in the latest fashion trends*, *I get information about current fashion from magazines for women*, *Brand names are very important for sports wear* and *I like to buy fashion designers' perfumes* (t-statistics ranging from 1.462 to 2.177). These variables may be appropriate for explaining the segmentation of customers into two classes.

Table 8.5 shows PLS results using the *I like to buy fashion designers' perfumes* variable for an a-priori customer segmentation into two classes. Both corresponding outcomes for segment-specific LVP estimations (see Table 8.12 in the appendix) satisfy the relevant criteria for model evaluation (Chin 1998a; Henseler et al. 2009). Segment 1 represents customers that are not interested in fashion designers' perfumes (relative size of 0.777). By contrast, segment 2 (relative size of 0.223) is characterized by female consumers that are attracted to Benetton and who would enjoy using Benetton products in other product categories, such as perfumes. From a marketing viewpoint, these customers are very important to fashion designers who want to plan for brand extensions.

Except for the *I like to buy fashion designers' perfumes* variable, the other four variables identified in the ex post analysis to explain the two classes (with reasonable t-statistics) do not offer much potential for a meaningful a-priori separation of data into two groups and segment-specific PLS path modeling. The corresponding results are at similar levels as the estimates for the full set of data. We therefore consider reasonable alternatives and test the *Customers' age* variable for an a-priori segmentation of Benetton's brand preference LVP. The ex post analysis of FIMIX-PLS results does not furnish evidence for the relevance of this variable (t-statistic of 0.690). Yet, when creating a customer segment for females over age 28 (segment 1; relative segment size: 0.793) and for younger women (segment 2; relative segment size: 0.207), we do achieve a result (see Table 8.6) that is nearly identical to the a-priori segmentation using *I like to buy fashion designers' perfumes*. The evaluation of results (Chin 1998a; Henseler et al. 2009) substantiates that the PLS path model estimates are acceptable for each segment (see Table 8.13 in the appendix).

Table 8.6 A-priori segmentation based on *Customers' age*

	Segment 1	Segment 2
R^2 (for *Brand preference*)	0.172	0.356
Image → *Brand preference*	0.364	0.559
Person → *Brand preference*	0.158	0.110

The findings that we present for the technique to uncover explanatory variables proposed by Ramaswamy et al. (1993) depict indistinct outcomes for PLS path modeling. Consequently, reliable procedures for the identification of fitting explanatory variables in the ex post analysis are required and future research must advance on this essential issue for the applicability of FIMIX-PLS.

Another implication addresses the FIMIX-PLS segment-specific estimates for relationships in the inner model and R^2 of endogenous latent variables. The procedure must be executed for successively increased numbers of classes and the outcomes for evaluation criteria must be compared in order to determine an appropriate number of segments. However, segment-specific FIMIX-PLS results are often improper for interpretation when a certain number of classes is exceeded. In most cases, the standardized weights in the inner model are at values higher than one and/or the unexplained variance of endogenous latent variables exceeds the value of one (or becomes negative). These kinds of outcomes indicate that the heterogeneity in the inner path model cannot appropriately be segmented by FIMIX-PLS for the chosen number of classes and that the analysis of additional classes may be stopped. Thus, these findings allow to further improve this methodology. Hahn et al. (2002) suggest limiting segment-specific FIMIX-PLS estimates between reasonable bounds. Future research must determine if such bounds for FIMIX-PLS computation impart useful improvements of the methodology regarding the identification of an adequate number of segments.

Our numerical example that uses empirical data demonstrates that FIMIX-PLS reliably identifies distinctive groups of customers. The larger segment tendencyally exhibits comparable results to the overall PLS path model estimates. Thus, this group of individuals is not subject for obtaining additional conclusions. In contrast, the smaller segment with a substantial relationship between *Image* and *Brand preference* is of high relevance from a marketing perspective. For these women, *Brand preference* of Benetton is foremost explained by aspects that are potentially under control of marketing activities that aim at creating an exclusive *Image* for the brand. Characteristics of the individual *Person* that are more difficult to influence by marketers are not an important issue for Benetton's brand preference in this segment of consumers. Furthermore, two kinds of explanatory variables are uncovered to form and characterize these two groups of data. Females who would like to buy Benetton's perfume or, alternatively, younger female consumers account for the smaller group of data. Hence, the specific PLS path model outcomes for the a priori formed smaller group of customers are particularly important for originating marketing strategies with regard to potential brand extensions or Benetton's target group of customers.

8.6 Summary

FIMIX-PLS allows us to capture unobserved heterogeneity in the estimated scores for latent variables in path models by grouping data. This is advantageous to a-priori segmentation because homogeneous segments are explicitly generated for the inner path model relationships. The procedure is broadly applicable in business research. For example, marketing-related path modeling can exploit this approach for distinguishing certain groups of customers.

In the first numerical example involving experimental data, FIMIX-PLS reliably identifies and separates the two a-priori created segments of price- and quality-oriented customers no matter what kind of outer measurement model, reflective or formative, is employed. The second numerical example of a marketing-related path model for Benetton's brand preference is based on empirical data, and it also demonstrates that FIMIX-PLS reliably identifies an appropriate number of customer segments if distinctive groups of customers exist that cause heterogeneity within the inner model. In this case, FIMIX-PLS enables us to identify and characterize: (1) a large segment of customers that shows similar results when compared to the original model estimation as well as (2) a smaller segment of customers that is highly important for marketing programs revealing a strong relationship between *Image* and *Brand preference*.

We accordingly conclude that the methodology offers valuable capabilities to extend and further differentiate PLS-based analysis of LVP in order to develop targeted marketing strategies (Rigdon et al. 2010; Ringle et al. 2010a). Under extreme circumstances, poor standard PLS results for the overall set of data, caused by the heterogeneity of estimates in the inner model, may result in significant estimates of the inner relationships and substantial values for R^2 of endogenous latent variables for at least one group after segmentation. (Sarstedt and Ringle 2010; Sarstedt et al. 2009). Researchers and practitioners should employ FIMIX-PLS as a standard procedure to evaluate their PLS path modeling results. They thereby assure that outcomes on the aggregate data level are not affected by unobserved heterogeneity in the inner path model estimates. Otherwise, the analysis provides further indications on how to treat that problem by forming groups of data. Significantly distinctive group-specific path model estimations impart further differentiated interpretations of PLS modeling results and may foster the origination of more effective (marketing) strategies.

The initial application and critical review of this new segmentation technique for partial least squares path modeling finally allows us to unveil and discuss some of the problematic aspects (Ringle 2006) and to address significant areas of future research. As pointed out in the foregoing chapters, advances on the problem of local optimum solutions, not interpretable FIMIX-PLS estimates as well as a reliable procedure to identify explanatory variables in the ex post analysis are crucial for the applicability of this approach. In addition, extensive simulations with experimental data and broad use of empirical data are required to further exemplify how FIMIX-PLS provides additional findings for PLS path modeling.

8.7 Appendix

8.7.1 Description of Symbols

Table 8.7 Table of symbols

Symbol	Description
A_m	number of exogenous variables as regressors in regression m
a_m	exogenous variable a_m with $a_m = 1, \ldots, A_m$
B_m	number of endogenous variables as regressors in regression m
b_m	endogenous variable b_m with $b_m = 1, \ldots, B_m$
$\gamma_{a_m mk}$	regression coefficient of a_m in regression m for class k
$\beta_{b_m mk}$	regression coefficient of b_m in regression m for class k
τ_{mk}	$((\gamma_{a_m mk}), (\beta_{b_m mk}))'$ vector of the regression coefficients
ω_{mk}	cell $(m \times m)$ of Ψ_k
c	constant factor
$f_{i\|k}(\cdot)$	probability for case i given a class k and parameters $(\cdot)$
I	number of cases or observations
i	case or observation i with $i = 1, \ldots, I$
J	number of exogenous variables
j	exogenous variable j with $j = 1, \ldots, J$
K	number of classes
k	class or segment k with $k = 1, \ldots, K$
M	number of endogenous variables
m	endogenous variable m with $m = 1, \ldots, M$
N_k	number of free parameters defined as $(K-1) + KR + KM$
P_{ik}	probability of membership of case i to class k
R	number of predictor variables of all regressions in the inner model
S	stop or convergence criterion
V	large negative number
X_{mi}	case values of the regressors for regression m of individual i
Y_{mi}	case values of the regressant for regression m of individual i
z_{ik}	$z_{ik} = 1$, if the case i belongs to class k; $z_{ik} = 0$ otherwise
ζ_i	random vector of residuals in the inner model for case i
η_i	vector of endogenous variables in the inner model for case i
ξ_i	vector of exogenous variables in the inner model for case i
B	$M \times M$ path coefficient matrix of the inner model
Γ	$M \times J$ path coefficient matrix of the inner model
Δ	difference of $current_{lnL}$ and $last_{lnL}$
B_k	$M \times M$ path coefficient matrix of the inner model for latent class k
Γ_k	$M \times J$ path coefficient matrix of the inner model for latent class k
Ψ_k	$M \times M$ matrix for latent class k containing the regression variances
ρ	$(\rho_1, \ldots, \rho_K)$, vector of the K mixing proportions of the finite mixture
ρ_k	mixing proportion of latent class k

8.7.2 PLS Path Modeling Results for Experimental Data

Table 8.8 Overview of experimental PLS modeling results (reflective case)

Analysis with reflective measurement models	PLS results for the full set of experimental data			Group 1: PLS results for a-priori segmented experimental data			Group 2: PLS results for a-priori segmented experimental data		
	Price (loadings)	Quality (loadings)	Satisfaction (loadings)	Price (loadings)	Quality (loadings)	Satisfaction (loadings)	Price (loadings)	Quality (loadings)	Satisfaction (loadings)
Price 1	0.854			0.848			0.730		
Price 2	0.872			0.888			0.740		
Price 3	0.843			0.829			0.712		
Price 4	0.860			0.882			0.871		
Price 5	0.890			0.898			0.941		
Quality 1		0.862			0.738			0.847	
Quality 2		0.864			0.639			0.843	
Quality 3		0.892			0.599			0.893	
Quality 4		0.880			0.977			0.862	
Quality 5		0.903			0.684			0.893	
Satisfaction 1			0.877			0.892			0.855
Satisfaction 2			0.861			0.867			0.872
Satisfaction 3			0.907			0.919			0.882
Price -> Satisfaction			0.538			0.873			0.169
Quality -> Satisfaction			0.450			0.077			0.898
AVE	0.746	0.775	0.777	0.756	0.547	0.797	0.646	0.754	0.756
ρ_c	0.936	0.945	0.913	0.939	0.854	0.777	0.900	0.939	0.903
R^2			0.465			0.777			0.843

Table 8.9 Overview of experimental PLS modeling results (formative case)

Analysis with formative measurement models	PLS results for the full set of experimental data			Group 1: PLS results for a-priori segmented experimental data			Group 2: PLS results for a-priori segmented experimental data		
	Price (weights)	Quality (weights)	Satisfaction (weights)	Price (weights)	Quality (weights)	Satisfaction (weights)	Price (weights)	Quality (weights)	Satisfaction (weights)
Price 1	−0.116			0.127			−0.654		
Price 2	0.079			0.090			−0.115		
Price 3	0.017			0.257			−0.778		
Price 4	0.458			0.326			0.531		
Price 5	0.641			0.342			0.928		
Quality 1		−0.005			−0.074			−0.001	
Quality 2		−0.137			−0.584			0.160	
Quality 3		0.059			−0.907			0.320	
Quality 4		0.779			0.948			0.362	
Quality 5		0.354			0.422			0.289	
Satisfaction 1			0.502			0.348			0.382
Satisfaction 2			0.167			0.279			0.411
Satisfaction 3			0.449			0.485			0.357
Price − > Satisfaction			0.527			0.875			0.065
Quality − > Satisfaction			0.406			0.052			0.892
R^2			0.470			0.784			0.835

Table 8.10 Overview of experimental PLS modeling results (mixed case)

Analysis with formative and reflective measurement models	PLS results for the full set of experimental data			Group 1: PLS results for a-priori segmented experimental data			Group 2: PLS results for a-priori segmented experimental data		
	Price (weights)	Quality (weights)	Satisfaction (loadings)	Price (weights)	Quality (weights)	Satisfaction (loadings)	Price (weights)	Quality (weights)	Satisfaction (loadings)
Price 1	−0.066			0.168			0.647		
Price 2	0.037			0.071			−0.102		
Price 3	0.016			0.248			−0.787		
Price 4	0.436			0.303			0.539		
Price 5	0.659			0.352			0.927		
Quality 1		−0.098			−0.102			0.005	
Quality 2		−0.109			−0.602			0.158	
Quality 3		0.078			−0.901			0.322	
Quality 4		0.785			0.952			0.359	
Quality 5		0.384			0.330			0.287	
Satisfaction 1			0.881			0.892			0.852
Satisfaction 2			0.862			0.867			0.876
Satisfaction 3			0.902			0.919			0.880
Price -> Satisfaction			0.513			0.873			0.063
Quality -> Satisfaction			0.418			0.053			0.893
AVE			0.777			0.797			0.756
ρ_c			0.913			0.922			0.903
R^2			0.464			0.781			0.835

8.7.3 PLS Path Modeling Results for the Example with Empirical Data

Table 8.11 Overview of empirical PLS path modeling results

	PLS results for the full set of empirical data		
	Image	Person	Brand Preference
I have a clear impression of this brand	0.860		
This brand can be trusted	0.899		
Is modern and up to date	0.795		
Represents a great style of living	0.832		
Fashion is a way to express who I am		0.801	
I often talk about fashion		0.894	
A brand name is very important to me		0.850	
I am interested in the latest trends		0.859	
Sympathy			0.944
Brand usage			0.930
AVE	0.718	0.725	0.881
ρ_c	0.910	0.913	0.937
R^2			0.239
Image − > Brand preference			0.423
Person − > Brand preference			0.177
Relative segment size			1.000

Table 8.12 Segment-specific PLS results for *I like to buy...*

	Group 1: PLS results for a-priori segmented empirical data based on the explanatory variable *I like to buy fashion designers' perfumes*			Group 2: PLS results for a-priori segmented empirical data based on the explanatory variable *I like to buy fashion designers' perfumes*		
	Image	Person	Brand Preference	Image	Person	Brand Preference
I have a clear impression of this brand	0.851			0.873		
This brand can be trusted	0.894			0.906		
Is modern and up to date	0.783			0.832		
Represents a great style of living	0.821			0.870		
Fashion is a way to express who I am		0.761			0.788	
I often talk about fashion		0.878			0.645	
A brand name is very important to me		0.851			0.747	
I am interested in the latest trends		0.823			0.718	
Sympathy			0.949			0.906
Brand usage			0.930			0.934
AVE	0.703	0.687	0.882	0.758	0.528	0.847
ρ_c	0.904	0.898	0.938	0.926	0.816	0.917
R^2			0.204			0.323
Image – > Brand preference			0.394			0.562
Person – > Brand preference			0.164			0.104
Relative segment size			0.777			0.223

Table 8.13 Segment-specific PLS results for *Customers' age*

	Group 1: PLS results for a-priori segmented empirical data based on the explanatory variable *Customers' age*			Group 1: PLS results for a-priori segmented empirical data based on the explanatory variable *Customers' age*		
	Image	Person	Brand Preference	Image	Person	Brand Preference
I have a clear impression of this brand	0.860			0.850		
This brand can be trusted	0.879			0.928		
Is modern and up to date	0.777			0.888		
Represents a great style of living	0.807			0.897		
Fashion is a way to express who I am		0.769			0.849	
I often talk about fashion		0.894			0.834	
A brand name is very important to me		0.867			0.752	
I am interested in the latest trends		0.831			0.818	
Sympathy			0.936			0.958
Brand usage			0.925			0.940
AVE	0.692	0.708	0.866	0.794	0.663	0.900
ρc	0.900	0.906	0.928	0.939	0.887	0.948
R^2			0.172			0.356
Image -> Brand preference			0.364			0.559
Person -> Brand preference			0.158			0.110
Relative segment size			0.793			0.207

References

Albers, S., & Hildebrandt, L. (2006). Methodische Probleme bei der Erfolgsfaktorenforschung – Messfehler, versus reflektive Indikatoren und die Wahl des Strukturgleichungs-Modells. *Schmalenbachs Zeitschrift fr betriebswirtschaftliche Forschung*, *58*(1), 2–33.

Allenby, G. M., Arora, N., & Ginter, J. L. (1998). On the heterogeneity of demand. *Journal of Marketing Research*, *35*(3), 384–389.

Bagozzi, R. P. (1994). Structural equation models in marketing research: basic principles. In R. P. Bagozzi (Ed.), *Principles of marketing research* (pp. 317–385). Oxford: Blackwell.

Bagozzi, R. P., & Yi, Y. (1994). Advanced topics in structural equation models. In R. P. Bagozzi, (Ed.), *Principles of marketing research* (pp. 1–52). Oxford: Blackwell.

Chin, W. W. (1998). The partial least squares approach to structural equation modeling. In G. A. Marcoulides, (Ed.), *Modern methods for business research* (pp. 295–358). Mahwah: Lawrence Erlbaum Associates.

Chin, W. W., & Dibbern, J. (2010). A permutation based procedure for multi-group PLS analysis: results of tests of differences on simulated data. In replace V. Esposito Vinzi, W. W. Chin, J. Henseler, & H. Wang, (Eds.), *Handbook of computational statistics (Vol. II): Partial Least Squares: Concepts, Methods and Applications*, 171–193. Berlin: Springer.

DeSarbo, W. S., Anthony Benedetto, C., Jedidi, K., & Song, M. (2006). Identifying sources of heterogeneity for empirical deriving strategic types: a constrained finite-mixture structural-equation methodology. *Management Science*, *62*(6), 909–924.

DeSarbo, W. S., Jedidi, K., & Sinha, I. (2001). Customer value analysis in a heterogeneous market. *Strategic Management Journal*, *22*(9), 845–857.

Diamantopoulos, A., & Winkelhofer, H. (2001). Index construction with formative indicators: an alternative to scale development. *Journal of Marketing Research*, *38*(2), 269–277.

Dillon, W. R., White, J. B., Rao, V. R., & Filak, D. (1997). Good science: use structural equation models to decipher complex customer relationships. *Marketing Research*, *9*(4), 22–31.

Esposito Vinzi, V., Ringle, C.M., Squillacciotti, S., Trinchera, L. (2007): Capturing and Treating Unobserved Heterogeneity by Response Based Segmentation in PLS Path Modeling: A Comparison of Alternative Methods by Computational Experiments, ESSEC Research Center, Working Paper No. 07019, ESSEC Business School Paris-Singapore, Cergy Pontoise Cedex 2007.

Fornell, C., Johnson, M. D., Anderson, E. W., Jaesung, C., & Bryant, B. E. (1996). The american customer satisfaction index: nature, purpose, and findings. *Journal of Marketing*, *60*(4), 7–18.

Fornell, C., & Larcker, D. F. (1981). Structural equation models with unobservable variables and measurement error: algebra and statistics. *Journal of Marketing Research*, *18*(3), 328–388.

Gudergan, S., Ringle, C.M., Wende, S., Will, A. (2008): Confirmatory tetrad analysis in PLS path modeling, in: Journal of Business Research (JBR), Volume 61 (2008), Issue 12, pp. 1238–1249.

Hackl, P., & Westlund, A. H. (2000). On structural equation modeling for customer satisfaction measurement. *Total Quality Management*, *11*(4/5/6), 820–825.

Hahn, C., Johnson, M. D., Herrmann, A., & Huber, F. (2002). Capturing customer heterogeneity using a finite mixture PLS approach. *Schmalenbach Business Review*, *54*(3), 243–269.

Henseler, J., Ringle, C. M., Sinkovics, R. R.: The use of partial least squares path modeling in international marketing, in: Sinkovics, R. R., Ghauri, P. N. (eds.), Advances in International Marketing, Vol. 20, Bingley 2009, pp. 277–320.

Hansmann, K., & Ringle, C. M. (2005). Strategies for cooperation. In T. Theurl & E. C. Meyer, (Eds.), *Competitive advantage of the cooperatives' networks* (pp. 131–152). Aachen: Shaker.

Jarvis, C. B., MacKenzie, S. B., & Podsakoff, P. M. (2003). A critical review of construct indicators and measurement model misspecification in marketing and consumer research. *Journal of Consumer Research*, *30*(2), 199–218.

Jedidi, K., Jagpal, H. S., & DeSarbo, W. S. (1997). Finite-mixture structural equation models for response-based segmentation and unobserved heterogeneity. *Marketing Science*, *16*(1), 39–59.

Jöreskog, K. G. (1978). Structural analysis of covariance and correlation matrices. *Psychometrika, 43*(4), 443–477.
Kim, B., Srinivasan, K., & Wilcox, R. T. (1999). Identifying price sensitve consumers: the relative merits of domogrphic vs. purchase pattern information. *Journal of Retailing, 75*(2), 173–193.
Lohmöller, J. (1989). *Latent variable path modeling with partial least squares*. Heidelberg: Verlag.
McLachlan, G. J., & Krishnan, T. (2004). *The EM algorithm and extensions*. Chichester: Wiley.
Ramaswamy, V., DeSarbo, W. S., Reibstein, D. J., & Robinson, W. T. (1993). An empirical pooling approach for estimating marketing mix elasticities with PIMS data. *Marketing Science, 12*(1), 103–124.
Rigdon, E.E., Ringle, C.M., Sarstedt, M.: Structural Modeling of Heterogeneous Data with Partial Least Squares, in: Malhotra, N.K. (ed.), Review of Marketing Research, Volume 7, Armonk, 2010, forthcoming.
Ringle, C. M., Wende, S., & Will, A. (2005). SmartPLS 2.0 (beta). http://www.smartpls.de.
Ringle, C. M., Sarstedt, M., Mooi, E. A. (2009a): Response-based segmentation using FIMIX-PLS: Theoretical foundations and an application to American customer satisfaction index data, in: Stahlbock, R., Crone, S. F., Lessmann, S. (eds.): Annals of Information Systems, Special Issue on Data Mining, Vol. 8, Berlin-Heidelberg 2010, 19–49.
Rossiter, J. R. (2002). The C-OAR-SE procedure for scale development in marketing. *International Journal of Research in Marketing, 19*(4), 305–335.
Sarstedt, M. (2008): A review of recent approaches for capturing heterogeneity in partial least squares path modelling. *Journal of Modelling in Management, 3*(2): 140–161.
Sarstedt, M., & Wilczynski, P. (2009). More for Less? A Comparison of Single-Item and Multi-Item Measures. Business Administration Review, *69*(2), 211–227.
Sarstedt, M., Schwaiger, M., & Ringle, C. M. (2009). Do we fully understand the critical success factors of customer satisfaction with industrial goods? - Extending Festge and Schwaigers model to account for unobserved heterogeneity, in: Journal of Business Market Management (JBM), Volume 3 (2009), Issue 3, pp. 185–206.
Sarstedt, M., Ringle, C. M. (2010). Treating unobserved heterogeneity in PLS path modelling: A comparison of FIMIX-PLS with different data analysis strategies, in: Journal of Applied Statistics, Vol. 37, 2010, in press.
Schneeweiß, H. (1991). Models with latent variables: LISREL versus PLS. *Statistica Neerlandica, 45*(2), 145–157.
Squillacciotti, S. (2010). Prediction oriented classification in PLS path modeling. In V. Esposito Vinzi, W. W. Chin, J. Henseler, & H. Wang, (Eds.), *Handbook of Computational Statistics (Vol. II): Partial Least Squares: Concepts, Methods and Applications*, 219–234. Berlin: Springer.
Steenkamp, J., & Baumgartner, H. (2000). On the use of structural equation models for marketing modeling. *International Journal of Research in Marketing, 17*(2/3), 195–202.
Tenenhaus, M., Vinzi, V. E., Chatelin, Y., & Lauro, C. (2005). PLS path modeling. *Computational Statistics & Data Analysis, 48*(1), 159–205.
Ter Hofstede, F., Steenkamp, J. E. M., & Wedel, M. (1999). International market segmentation based on consumer-product relations. *Journal of Marketing Research, 36*(1), 1–17.
Wedel, M., & Kamakura, W. (2000). *Market segmentation: conceptual and methodological foundations* (2nd ed.). Dordrecht: Kluwer.
Wu, J., & Desarbo, W. S. (2005). Market segmentation for customer satisfaction studies via a new latent structure multidimensional scaling model. *Applied Stochastic Models in Business and Industry, 21*(4/5), 303–309.
Yoo, B., Donthu, N., & Lee, S. (2000). An examination of selected marketing mix elements and brand equity. *Academy of Marketing Science Journal, 28*(2), 195–211.

Chapter 9
Prediction Oriented Classification in PLS Path Modeling

Silvia Squillacciotti

Abstract Structural Equation Modelling methods traditionally assume the homogeneity of all the units on which a model is estimated. In many cases, however, this assumption may turn to be false; the presence of latent classes not accounted for by the global model may lead to biased or erroneous results in terms of model parameters and model quality. The traditional multi-group approach to classification is often unsatisfying for several reasons; above all because it leads to classes homogeneous only with respect to external criteria and not to the theoretical model itself.

In this paper, a prediction-oriented classification method in PLS Path Modelling is proposed. Following PLS Typological Regression, the proposed methodology aims at identifying classes of units showing the lowest distance from the models in the space of the dependent variables, according to PLS predictive oriented logic. Hence, the obtained groups are homogeneous with respect to the defined path model. An application to real data in the study of customers' satisfaction and loyalty will be shown.

9.1 Introduction

PLS Path Modeling has become one of the reference statistical methodologies in the analysis of customer satisfaction. It allows to build latent variables (such as customer satisfaction, or perceived value) from a number of manifest variables measuring the unobserved complex constructs. The scores for these variables can be computed, thus allowing to build and compare indexes of satisfaction and loyalty among individuals and in time. The model can be estimated through a "soft modeling" technique, which avoids some of the main drawbacks found in the Maximum-Likelihood approach (SEM-ML) (Jöreskog 1970), namely the restrictive distributional hypotheses on the observed variables (Tenenhaus et al. 2005). This is certainly an advantage when working on data from marketing surveys. Hence, PLS Approach to Structural Equation Modeling is an alternative statistical methodology to the Maximum

S. Squillacciotti
EDF R&D, Département ICAME, 1 avenue du Général de Gaulle, 92141 Clamart, France
e-mail: silvia.fagnoni@edf.fr

V. Esposito Vinzi et al. (eds.), *Handbook of Partial Least Squares*, Springer Handbooks of Computational Statistics, DOI 10.1007/978-3-540-32827-8_10,

Likelihood Approach. The two techniques are more complementary than competing, and the choice of one rather than the other should depend on the purpose of the analysis and the research context (Jöreskog and Wold 1982; Lohmöller 1989). In three specific cases, however, PLS may be preferable to ML: when the sample size is very small, when the data to be analyzed is not multinormal, and when the complexity of the model to be estimated may lead to improper or non-convergent results (Bagozzi and Yi 1994). In marketing applications, such as in the analysis of the drivers of satisfaction and of its links to loyalty, data are very rarely multinormal. Moreover, PLS shows the interesting feature of allowing the computation of "scores" for the latent variables, which can serve as indexes for the underlying latent concept (for example as a satisfaction index).

EDF (*Electricité de France*) is greatly concerned with the analysis of customer satisfaction and with modeling customers' drivers to satisfaction and loyalty. The European energy market is undergoing a great number of major changes. Many European countries have already witnessed the transition from a monopolistic market to a competitive one. In France, the energy market is open for all customer segments except residential customers, who will be free to choose their energy supplier from July 2007. The study of customer satisfaction through adapted models allows to find out which elements may lead to customer satisfaction or customer non-satisfaction and, hence, to the decision of switching to a new energy provider.

The definition of a unique model, however, although allowing the global identification of the main drivers of customer satisfaction, may "hide" differences in customer behavior. As underlined in Ozcan (1998), marketing managers are interested in finding ways to exploit opportunities resulting from heterogeneity in customers' behavior when defining their strategical and tactical business decisions. When customers do have different behaviors, models accounting for this heterogeneity allow the definition of targeted and more efficient strategies.

The traditional approach to segmentation in Structural Equation Modeling consists in estimating separate models for customer segments which have been obtained either by assigning customers to a priori segments on the basis of demographic or consumption variables, or through a cluster analysis on the original variables. None of these approaches, however, is to be considered satisfactory. A limitation to this "a priori" approach is that very rarely heterogeneity may be captured by well-known observable variables (Hahn et al. 2002). Clustering procedures, on the other hand, may be problematic since traditional cluster analysis assumes independence among variables; preliminary data reduction techniques may also lead to statistical problems (Jedidi et al. 1997). Apart from the statistical considerations, a priori segmentation is not conceptually acceptable since no causal structure among the variables is postulated. Units should be clustered according to all the available information, hence in relation with the defined model. In other words, a model-based clustering method should be used, where the obtained clusters are homogeneous with respect to the structural causal relationships.

A new technique for the identification of groups homogeneous with respect to the defined model in the framework of PLS Approach to Structural Equation Modeling (SEM-PLS) is proposed in this paper. The proposed technique, PLS Typological

Path Modeling (PLS-TPM) allows to take into account the predictive purpose of PLS techniques when defining the classes. Differently from existing model-based classification methods, PLS-TPM requires no distributional assumptions on observed and/or latent variables.

Structural Equation Modeling (SEM) is often applied in marketing research, especially in the analysis of customer satisfaction, in order to measure complex cause-effect relationships. Two statistical methodologies exist for the estimation of such models: SEM-ML (*Maximum Likelihood Approach to Structural Equation Modeling*), also known as LISREL (*LInear Structural RELations* (Jöreskog 1970)) approach, and PLS (*Partial Least Squares* (Wold 1975)). The following paragraph gives an overview of model-based classification techniques in the framework of ML-SEM, while paragraph 3 deals with model-based classification in PLS methods. FIMIX-PLS, which generalizes finite mixture models to a PLS framework, is described in paragraph 3.1. PLS-TPM, the methodology proposed by the author, is introduced in paragraph 3.2.

An empirical application of PLS-TPM to data from a satisfaction survey for an English energy provider is shown in Chap. 3. Finally, Chap. 4 describes the major research issues in model-based classification in a PLS framework.

9.2 SEM-ML and Classification

When group membership is known a priori, traditional standard multi-group methods (Jöreskog 1971, 1973; Sörböm 1974) can be used in order to account for heterogeneity. Basically this technique consists in computing separate models for each segment, where segments have been defined according to the available *a priori* information. Segments can be either defined according to prior knowledge on their homogeneity according to external variables (such as socio-demographic or consumption variables) or on the basis of a cluster analysis. Unfortunately, background variables such as demographic or psychographic descriptors are rarely sufficient to allow to form groups *a priori*. On the other hand, cluster analysis, besides showing a number of statistical drawbacks (Hahn et al. 2002), is conceptually unfit to the available data structure since it ignores the available information on the relationships among the variables in the model.

A more sophisticated approach is given by finite mixture models in SEM (DeSarbo and Cron 1988; Jones and Maclachlan 1992; Jedidi et al. 1997). Segmentation is performed by taking into account the defined model and the implied relationships among variables. In this approach to classification, the data is supposed to be the result of the mixture of two or more populations mixed in different proportions. In other words, each subject is supposed to belong to a segment, each segment being characterized by a different covariance structure. Hence, data arise from a mixture of distributions, and the aim is to estimate the probability that each subject belongs to each of these sub-populations. Distributions that are more frequently used are the multivariate normal or multinomial distribution. The technique is based on the EM algorithm, which allows the estimation of the posterior probabilities. The posterior probabilities represent a fuzzy classification of the observations in the K segments based on the postulated measurement and structural models.

The main drawbacks of the methodology concern the distributional assumptions, the risk of encountering a local optimum in the iterations and the identification of the number of classes. The distributional assumptions are required in order to ensure the model identification. They may however be a problematic constraint, especially in marketing applications where data are rarely normal and more frequently highly skewed. The risk of falling into a local optimum may be resolved by choosing different starting points for the iterations. Finally, the number of classes K is supposed to be known. However, in a totally exploratory approach, the number of classes is very rarely known a priori and is, instead, to be worked out by the analysis. The solution proposed in Jedidi et al. (1997) consists in performing the Finite Mixture Structural Equation Model with different possible values of K and comparing several global measures of fit such as Akaike's Information Criterion (AIC) or Bayesian Information Criterion (BIC). However, in terms of quality of results, the finite-mixture approach outperforms traditional sequential procedures combining cluster analysis and multi-group SEM, as it has been demonstrated in a simulation study in Görz et al. (2000).

9.3 PLS and Classification

Traditionally, classification in PLS Path Modeling has been performed through multi-group analysis: groups are defined according to prior knowledge, background variables, or external analyses. Separate PLS path models are then estimated for each group, and the results are compared in order to identify, if possible, the differences among the groups. The existence of groups showing internally homogeneous structural models may eventually be validated by means of a partial analysis criterion as shown in Amato and Balzano (2003).

9.3.1 FIMIX PLS: The PLS Finite Mixture Models

Recently, a different approach to classification in PLS Path Modeling has been proposed: the Finite Mixture Partial Least Squares Approach (FIMIX-PLS) (Hahn et al. 2002; Ringle et al. 2005), which generalizes the finite mixture approach to PLS Path Modeling. For further details and a complete description of the algorithm, cf. Chap. 8. The methodology begins with the estimation of the path model through the traditional PLS-PM algorithm. The information concerning the heterogeneity of individual behaviors is supposed to be contained in the structural relationships between the latent variables. The model requires the assumption of multivariate normal distribution only for the endogenous latent variables η_i. This assumption is sufficient since the endogenous variables are expressed as function of the exogenous variables ξ_i. The likelihood of the model is maximized through the EM algorithm.

As shown in (Hahn et al. 2002), although the EM algorithm monotonically increases the lnL and converges towards an optimum, there is a risk of it converging towards local optima. This risk increases when group densities are not well separated, when there is a high number of parameters to be estimated and when the information contained in each observation is limited. Since the convergence of the EM algorithm depends on the starting values (Wedel and Kamakura 2000) a possible solution is to initialize the algorithm with different values or to obtain an initial partition through clustering procedures (for example k-means) (Ringle et al. 2005).

As in the original finite mixture models described in paragraph 2, in FIMIX-PLS the number of groups is a priori unknown, and the identification of an optimal number of classes K is not straightforward. The proposed solution to this problem follows what has already been said concerning "classical" Finite Mixture Models, and consists in running FIMIX-PLS several times with different possible choices of values for K. The choice of the best partition will be based on criteria such as the lnL, the AIC or the BIC indicator. Moreover, an indicator of the degree of separation for the estimated individual class probabilities, as defined in Ramaswamy et al. (1993) is available (EN_k). This statistic varies between 0 and 1 (1 indicates a perfect separation among classes, whereas values very close to 0 indicate that segments are "fuzzy" and hardly interpretable). Segmentations can be considered unambiguous with values of EN higher than 0,5 (Ringle et al. 2005).

The main problematic issue in FIMIX-PLS is basically the one described for the original Finite Mixture Models, related to the EM algorithm: namely the risk of convergence in local optima. To that we may add the difficulty of accepting, under a strictly theoretical point of view, the imposition of a distributional assumption on the endogenous latent variable in the framework of PLS Modeling. Finally, FIMIX-PLS is characterized by static outer models for the groups: in order to ensure the convergence of the procedure, the outer models (i.e. the loadings) are kept constant over all the classes. This problem is solved by adding a further step in the overall classification procedure: the "external" analysis. This step consists in searching for available external descriptive variables leading to the same partition as the one identified through FIMIX-PLS. Once the variable(s) identified, traditional multi-group analysis is performed over the groups, i.e. new local models are estimated, each having its own outer model. We may however remark that experience has shown how rarely few external variables allow to univocally recover the same groups as those identified by a model-based procedure.

9.3.2 *PLS Typological Path Modeling (PLS-TPM)*

PLS-TPM is a generalization of PLS Typological Regression to a PLS Path modeling framework, i.e. where variables may be grouped in more than two blocks and blocks are supposed to be linked by means of causal paths. The relationships existing between PLS Path Modeling and PLS Regression have been widely discussed in literature (Tenenhaus 1998; Sampson et al. 1989). A brief overview of classification methods in the framework of PLS Regression is however required in order to better understand the context of this chapter and of the proposed methodology.

9.3.2.1 Classification in PLS Regression

In PLS Regression, classification has traditionally been performed through the SIMCA (*Soft Independent Modeling of Class Analogy*) approach (Wold et al. 1984). The technique consists in performing a first PLS Regression over all the units in the data set. Units are then assigned to different classes according to their positions on the extracted PLS components, and one local model is estimated for each class. The class membership for a new unit can be determined according to its distance from the PLS Regression model in the independent variable space (*DModX, N*).

In a discrimination approach, PLS Discriminant Analysis (PLS-DA) (Sjöström et al. 1986) searches for the PLS components allowing the best separation of the classes. This methodology basically consists of a PLS Regression where the dependent variables are the indicators of the class membership.

The two above methods are however affected by two major drawbacks. In SIMCA, the predictive purpose of PLS Regression seems to be of minor importance in the definition of the classes: classes are defined once and for all and are not optimized with respect to the model's predictive performance since the prediction of a new unit's class membership depends on the unit's distance from the local models in the *independent* variables space. In PLS Discriminant Analysis, instead, the only allowed dependent variable is the one containing the class membership information.

More recently, two truly model-based approaches to classification in PLS regression have been proposed: PLS clusterwise regression (Preda and Saporta 2005) and PLS Typological Regression (PLS-TR) (Esposito Vinzi and Lauro 2003).

As in clusterwise linear regression (Charles 1977; Spaeth 1979), in PLS clusterwise regression the points in each clusters are supposed to be generated according to a linear regression relation. The aim of the method is to simultaneously find both the optimal partition of the data and the regression coefficient for each local regression model. Hence, in order to maximize the overall fit, the parameters to be estimated are the number of classes, the regression coefficients for each cluster and the variance of the residuals within each class. The goal is obtained by minimizing a function of both the partition and the regression parameters.

Since the estimation of the local models may become a difficult task if the number of units is low with respect to the number of variables, the use of PLS regression is particularly useful in this context.

PLS Typological Regression (PLS-TR) (Esposito Vinzi and Lauro 2003). allows a classification of the statistical units in the framework of a traditional PLS Regression taking into account the predictive purpose of PLS. As in the SIMCA approach, a PLS Regression is performed over all the units, which are then assigned to different classes according to the results of an ascending hierarchical classification on the retained PLS components and a local model is estimated for each class. While SIMCA stops here, in PLS-TR the distance of each unit from each local model is computed in the dependent variables space, following the *DModY,N* index given in (9.1) (Tenenhaus 1998), where e^2_{kij} is the square of the i-th residual on the j-th dependent variable for the PLS model relative to group k, J is the number of dependent variables, m_k is the number of retained components in the k-th group, and $Rd(\mathbf{T_k}, \mathbf{y_j})$ is the portion of the j-th dependent variable variance explained by the components of group k ($\mathbf{T_k}$ being the matrix containing the component scores for the k-th local model).

$$DModY, N_k = \sqrt{\frac{\frac{\sum_{j=1}^{J}\left[e^2_{kij}/Rd\left(\mathbf{T}_k, \mathbf{y}_j\right)\right]}{(J-m_k)}}{\frac{\sum_{i=1}^{I_k}\sum_{j=1}^{J}\left[e^2_{kij}/Rd\left(\mathbf{T}_k, \mathbf{y}_j\right)\right]}{(I_k-m_k-1)(J-m_k)}}} \tag{9.1}$$

The index *DmodY,N* in (9.1) represents the distance of the i-th unit from the estimated PLS Regression model in the dependent variable space. Performing a classification according to such distance measure leads to defining classes optimizing the predictive capacity of the local models.

An iterative process begins: each unit is assigned to the closest local model according to the distance computed in (9.1). If there are any changes in the class composition, the local models are re-estimated, the distances computed and the units eventually re-assigned to a new class. The algorithm stops at convergence, i.e. when there is no change in the class compositions from one step to the following. A compromise model is then computed, which allows the description of the whole set of units taking into account the existence of different local models. The final local models are hence optimized with respect to their predictivity. Moreover, this technique may also be used when classes are known a priori in order to validate their existence and their composition.

9.3.2.2 PLS Typological Path Modeling (PLS-TPM)

The model-based approach to classification in PLS Path Modeling proposed in this paper consists of an extension of PLS Typological Regression to the situation where more than two variable blocks are available. When dealing with more than two variable blocks, and wishing to take into account the links among all variable blocks, PLS Path Modeling is more appropriate than PLS Regression. This is the case for example in the analysis of customer satisfaction or customer loyalty, where many latent variables are supposed to interact and impact on satisfaction and/or intentional loyalty.

The iterative algorithm starts with the estimation of the global PLS Path Model (over the entire sample). According to the results of the global model, classes are defined. Local models are then estimated (one for each class), and a measure of the distance of each unit from each local model is computed. Units are then re-assigned to the class corresponding to the closest local model: if this causes any changes in the composition of the classes, the local models are re-estimated and the distances are computed once again. When there is no change in the composition of the classes from one step to the following, the obtained local models are compared in terms of predictivity (R^2) and of intensity of the structural links on the final endogenous latent variables.

The distance indicator used for the assignment of the units to the classes is an adaptation of the *DmodY, N* index used in PLS Typological Regression. The distance is given in the following (9.2):

$$D_k = \sqrt{\frac{\frac{\sum_{j=1}^{J}\left[e_{kij}^2 / Rd\left(\xi, y_j\right)\right]}{(J - A_k)}}{\frac{\sum_{i=1}^{I_k}\sum_{j=1}^{J}\left[e_{kij}^2 / Rd\left(\xi, y_j\right)\right]}{(I_k - A_k - 1)(J - A_k)}}} \tag{9.2}$$

where:

ξ is the generic latent variable;
e_{kij}^2 is the residual of the "redundancy" model, i.e. the regression of the final endogenous manifest variables over the exogenous latent variables;
$Rd\left(\xi, y_j\right)$ is the redundancy index for the final endogenous manifest variables for group k;
I_k is the number of units in group k;
A_k is the number of exogenous latent variables in the local model for group k;
J is the number of final endogenous manifest variables.

As in Finite Mixture Models and in FIMIX-PLS, in a purely exploratory approach groups are not known a priori. The initial assignment of the units to classes requires therefore the definition of the number K of classes as well as the criterion for assigning each unit to a class. Obviously, in a model-based classification, this criterion should not derive from external analyses but from the global model itself. In PLS Typological Regression the membership of each unit is decided according to the results of a clustering procedure on the extracted global model components. Such a procedure cannot however be extended to PLS Path Modeling given the deep conceptual differences between the Regression components and the Path Modeling latent variables. PLS Regression components, obtained by means of a deflative procedure (Tenenhaus 1998), represent statistically independent syntheses of the same variable blocks, whereas latent variables are usually uni-dimensional combinations of different variable blocks, often strongly correlated one to the other. Different options are possible: the initial sample may be progressively partitioned in K classes according to the distances of the units from the global model (units showing high distances being considered as badly represented by the global model), or units may be randomly assigned to a pre-defined number of classes K. The procedure is the following:

Step 1: Estimation of the PLS Path Model over the entire sample;
Step 2: Imputation of the units to the K classes according to the global PLS model results;
Step 3: Estimation of the K local models;
Step 4: Computation of the distance D_k as defined in (7);
Step 5: Attribution of each unit to the closest local model: if there is any change in the composition of the classes, repeat steps 3, 4, and 5, otherwise move to step 6.
Step 6: Comparison of the final local models.

In a validation approach, where classes are known to exist *a priori* both in number and in composition, the procedure may simply be used to verify if the external segmentation criterion also leads to homogeneity in the estimated local models.

Differently from FIMIX-PLS, PLS-TPM requires no distributional assumptions either on manifest or latent variables, according to PLS theory. Moreover, both the inner model and the outer model are dynamically updated each time that the computation of the distances lead to a redefinition of the groups memberships. Hence, at each step, each class has different component scores, structural coefficients, outer weights and loadings.

9.4 Empirical Application

9.4.1 Presentation of the Empirical Data and of the Path Model

The described methodology, PLS Typological Path Modeling, will be illustrated using data from a satisfaction survey on customers of an English energy provider. The sample includes 791 customers, chosen so as to represent the global churn proportion in the population. Among all the respondents to the survey, in fact, 133 customers have eventually switched to a different energy provider. Knowing that in the population of the energy provider customers the churn rate is around 17%, the remaining 83% of the sample (658 customers) have been chosen randomly so as to be representative of certain criteria in the population (namely socio-demographic criteria and the length of the relationship with the provider). There were 26 manifest variables for the path model, divided into 6 blocks:

- Price (2 manifest variables)
- Service (4 manifest variables)
- Communication (4 manifest variables)
- Billing (9 manifest variables)
- Image (9 manifest variables)
- Satisfaction (one manifest variable).

The model chosen to represent the drivers of satisfaction is shown in Fig. 9.1.

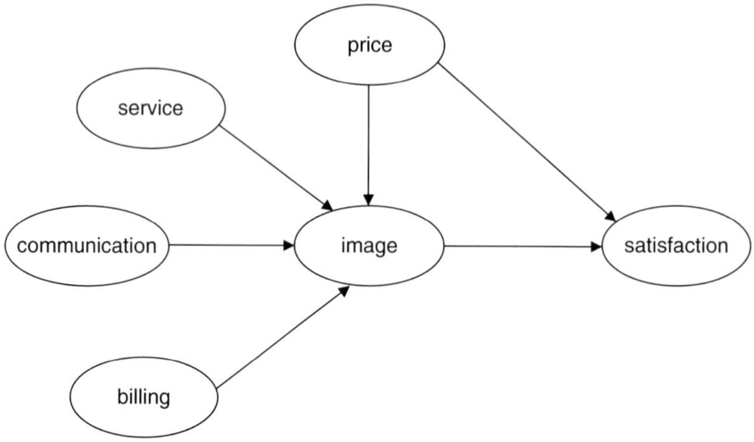

Fig. 9.1 Representation of the satisfaction model

There is one final endogenous variable (satisfaction), one intermediate endogenous variable (image) and four exogenous variables (price, service, communication, and billing). All blocks are supposed to be reflective. The represented model is a simplified version of the model actually in use, not shown for industrial secrecy reasons. In choosing this model, it has been supposed that the main driver for churn (effective and not intentional loyalty) is satisfaction: a satisfied customer will choose to stay with his present provider, whereas an unsatisfied customer will more probably "switch" to competitors. Satisfaction, on the other hand, is influenced by other latent constructs which indirectly, through satisfaction, affect churn (price, perceived image, perceived quality of the service, communication, billing).

In a traditional approach to classification in PLS Path Models, one would have computed two separate models, one for "loyal" customers, and the other for "switching" customers. This procedure, however, does not guarantee that the chosen external variable leads to homogeneous models in a predictive sense. Through PLS Typological Path Modeling we wish to find out classes which are homogenous with respect to the proposed satisfaction model (i.e. in each class customers should have the same main drivers for satisfaction). Classes can then be characterized by the dichotomous variable representing churn (customer has switched/customer has not switched).

9.4.2 Results for the Overall PLS Model and for PLS-TPM Segmented Data

First of all, the parameters for the global model shown in Fig. 9.1 have been estimated. Results (structural coefficients and R^2) are shown in Table 9.1. The PLS Path Models have all been estimated through SPAD® (version 6.0.1) PLS Path Modeling module.

Table 9.1 PLS Path Modeling results for the global model. All coefficients are significant with *Prob* > |T| = 0.000

Block	Factor	Regression coefficient	Student's T
Image	$R^2 = 0.76$		
	Communication	0.28	8.78
	Billing	0.24	9.53
	Service	0.19	6.54
	Price	0.29	9.37
Satisfaction	$R^2 = 0.39$		
	Price	−0.26	−5.76
	Image	0.39	8.64

Table 9.2 PLS path modeling results for the initial local model for class 1. All coefficients are significant with *Prob* > |T| = 0.000

Block	Factor	Regression coefficient	Student's T
Image	$R^2 = 0.78$		
	Communication	0.31	7.52
	Billing	0.26	7.44
	Service	0.21	5.634
	Price	0.22	5.37
Satisfaction	$R^2 = 0.42$		
	Price	−0.23	−3.73
	Image	0.45	7.33

Table 9.3 PLS path modeling results for the initial local model for class 2. All coefficients are significant with *Prob* > |T| = 0.000

Block	Factor	Regression coefficient	Student's T
Image	$R^2 = 0.73$		
	Communication	0.24	4.87
	Billing	0.21	5.93
	Service	0.18	3.91
	Price	0.35	7.68
Satisfaction	$R^2 = 0.34$		
	Price	−0.30	−4.44
	Image	0.32	4.74

In a totally exploratory approach to the research of classes, we have chosen to randomly assign customers to $K = 2$ classes of approximately the same size. In step 1, class 1 includes 390 customers and class 2, 401 customers. Tables 9.2 and 9.2 show the results for the initial local models respectively for class 1 and 2.

In a prediction-oriented classification, we expect the final local models to be at least as predictive (in terms of R^2) as the starting local models (Tables 9.2 and 9.3) and more predictive than the global one (Table 9.1).

Table 9.4 PLS path modeling results for the final local model for class 1. All coefficients are significant with *Prob* > |T| = 0, 000

Block	Factor	Regression coefficient	Student's T
Image	$R^2 = 0.81$		
	Communication	0.25	4.62
	Billing	**0.29**	6.82
	Service	0.19	4.52
	Price	**0.29**	5.96
Satisfaction	$R^2 = 0.86$		
	Price	−0.65	−16.53
	Image	0.31	7.83

Table 9.5 PLS path modeling results for the final local model for class 2. All coefficients are significant with *Prob* > |T| = 0.000

Block	Factor	Regression coefficient	Student's T
Image	$R^2 = 0.78$		
	Communication	**0.34**	7.12
	Billing	0.19	597
	Service	**0.29**	4.57
	Price	0.18	3.72
Satisfaction	$R^2 = 0.57$		
	Price	−0.38	−5.65
	Image	**0.59**	6.37

In the subsequent steps, distances between the units and the local models according to (7) have been computed in SAS ® (version 8) [SAS99]. After 29 iterations the final local model results for classes 1 (221 customers) and 2 (570 customers) are shown in Tables 9.4 and 9.5.

Latent variables showing the higher impacts on satisfaction are in bold. Customer satisfaction in class 1 seem to be more strongly influenced by items connected to money (price and billing). In class 2, instead, satisfaction seems to depend more on less "material" drivers such as communication, perceived quality of service and image. Both local models show higher predictivity (R^2) for satisfaction than both the initial local models and the global one. The local model for class 2, however, is less predictive for satisfaction ($R^2 = 0, 57$). This may be due to a higher heterogeneity in this class (570 units vs. 221 in class 1).

The characterization of the classes with respect to the external variable (switched/ not switched) leads to the following results: 107 lost customers (80% of all lost customers are in class 1), and 26 lost customers (20% of all lost customers) are in class 2. Although the low number of lost customers (representing 17% of the total sample) leads to be cautious in the characterization of the classes, we may conclude that customers more likely to churn are more sensible to monetary issues such as price and billing.

9.5 Conclusions and Future Research

PLS-TPM is a really model-based, prediction-oriented classification technique, allowing both the discovery of previously unknown classes and the validation of the existence of supposed ones. According to PLS theory, it does not require any distributional assumption for observed or latent variables.

However, a number of open problems are currently under investigation and represent interesting future research directions. The first subject to deal with concerns the choice of the number of classes: the empirical application shows a lower predictivity for the final model for class 2, which may be due to the existence of a third class, hidden by the initial choice of two classes.

The choice of the number of classes in model-based classification methods represents a frequently dealt with issue in literature, also when limiting our overview to a PLS framework. In PLS model-based regression techniques described in paragraph 3.2 the solution to the problem is rather straightforward. In PLS-TR the ascendant hierarchical classification on the PLS components allows easily to identify the number of classes according to the classical statistical criteria. In PLS clusterwise regression, the number of classes is identified simultaneously with the local model coefficients by minimization of a criterion which is function of both the partition and the local model parameters. In a SEM-PLS framework, instead, the solution is less immediate. In FIMIX-PLS the problem is solved by running the algorithm several times with different choices of k and by selecting the partition leading to the best separated classes. In practical applications however the choice is often not so simple since some values of the number of classes may lead to impossible solutions (for example variances higher than 1) or to excessively unbalanced class effectives. However, in PLS-TPM the choice is even harder. Although it is possible to run the method with different values for k, no indication may be given on how to choose the best partition, except the consistency of the obtained solutions (interpretable results for the local models, balanced classes, acceptable R^2, etc.).

Although not affected by the same convergence problems as FIMIX methods based on the EM algorithm, the convergence of the procedure is under study: it should be verified if the final partition is robust to the initial assignment of the units to the classes (be it random or based on a priori knowledge).

Finally, a more theoretical research issue concerns the definition itself of model-based heterogeneity: in FIMIX-PLS, "segment specific heterogeneity of path models is concentrated in the estimated relationships between latent variables" (Ringle et al. 2005). In other words, the information concerning the differences between the groups is supposed to be carried only in the different intensities of the structural coefficients. In PLS-TPM, instead, heterogeneity is not only concentrated in the structural model, but takes into account the overall theoretical model (structural and outer model). This feature allows to identify classes which, although having similar path coefficients, may differ with respect to the importance of the manifest variables in the blocks.

Both methods, however are based on the strong assumption that the structural model remains the same for all groups. This may not always be true: differences among groups may also depend on the existence, in each group, of a different structural model (for example some latent variables may not be linked in one group). Comparison among groups with different structural models may be problematic, especially in a PLS framework, given the absence of a global quality indicator (such as chi-square in ML-SEM). Moreover, this issue requires further investigations in PLS-TPM with respect to the redefinition of the distance proposed in (9.2).

References

Amato, S. Balzano, S. (2003). Exploratory approaches to group comparison. *Proceedings of the PLS'03 International Symposium*, DECISIA, France.

Bagozzi, R. P., Yi, Y. (1994). Advanced topics in structural equation models, In R. P. Bagozzi (Ed.) *Principles of Marketing Research*, pp. 317–385.

Charles, C. (1977). Régression Typologique et reconnaissance des formes. Thèse de Doctorat, Université Paris IX.

DeSarbo, W. S., Cron, W. L. (1988). A maximum likelihood methodology for clusterwise linear regression. *Journal of Classification*, **5**, 249–282.

Esposito Vinzi, V., Lauro, C. (2003) PLS regression and classification, *Proceedings of the PLS'03 International Symposium*, DECISIA, France.

Görz, N., Hildebrandt, L., Annacker, D. (2000). Analyzing multigroup data with structural equation models. Discussion Papers of Interdisciplinary Research Project 373, Paper provided by Humboldt Universitat zu Berlin in its series Quantifikation und Simulation Okonomischer Prozesse; **11**. ISSN 1436–1086.

Hahn, C., Johnson, M., Herrmann, A., Huber, F. (2002). Capturing customer heterogeneity using a finite mixture PLS approach. *Schmalenbach Business Review*, **54**, 243–269.

Jedidi, K., Harshanjeet, S. J., De Sarbo W. S. (1997). STEMM: A general finite mixture structural equation model. *Journal of Classification*, **14**, 23–50.

Jones, P., Maclachlan, G. (1992). Fitting finite mixture models in a regression context. *Australian Journal of Statistics*, **32**, 233–240.

Jöreskog, K. G. (1970). A general method for analysis of covariance structure. *Biometrika*, **57**, 239–251.

Jöreskog, K. G. (1971). Simultaneous factor analysis in several populations. *Psychometrika*, **36**, 409–426.

Jöreskog, K. G. (1973). A general method for estimating a linear structural equation system. In A.S. Goldberger & O. D. Duncan (Eds.), *Structural Equation Models in the Social Sciences*, New York: Academic Press, 85–112.

Jöreskog, K. G., Wold, H. (1982). The ML and PLS techniques for modelling with latent variables: Historical and comparative aspects. Amsterdam, North Holland

Lohmöller, J.-B. (1989). Latent variable path modelling with partial least squares. Heidelberg: Physica-Verlag.

Ozcan, K. (1998). Modelling consumer heterogeneity in marketing science. Working Paper, Department of Marketing – University of Michigan Business School.

Preda, C., Saporta, G. (2005). Clusterwise PLS regression on a stochastic process. *Computational Statistics and Data Analysis*, **49**, 99–108.

Ramaswamy, V., DeSarbo, W. S., Reibstein, D. J., Robinson, W. T. (1993). An empirical pooling approach for estimating marketing mix elasticities with PIMS data. *Marketing Science*, **12**, 130–124.

Ringle, C. M., Wende, S., Will, A. (2005). Customer segmentation with FIMIX-PLS. *Proceedings of the PLS'05 International Symposium*, France: SPAD Test&go.

SAS Institute Inc. (1999). SAS/STAT ® User's Guide, Version 8, Cary, NC: SAS Institute Inc.

Sörböm, D. (1974). A general method for studying differences in factor means and factor structure between groups. *British Journal of Mathematical and Statistical Psychology*, **28**, 138–151.

Sampson, P. D., Streissguth, A. P., Barr, H. M., Bookstein, F. L. (1989). Neurobehavioural effects of prenatal alcohol: Part II. *Partial Least Squares, Neurotoxicology and teratology*, **11**, 477–491.

Sjöström, M., Wold, S., Södeström, B. (1986). PLS discriminant plots. *Proceedings of PARC in Practice*, North Holland: Elsevier.

Spaeth, H. (1979). Clusterwise linear regression. *Computing*, **22**, 367–373.

Tenenhaus, M., Esposito Vinzi, V., Chatelin, Y.-M., Lauro, C. (2005). PLS path modeling. *Computational Statistics and Data Analysis*, **48**, 159–205.

Tenenhaus, M. (1998). La Régression PLS: Théorie et Pratique. (Ed.) Paris: Technip.

Wold, S., Albano, C., Dunn, W. J., Edlund, U., Esbensen, K., Geladi, P. et al. (1984). Multivariate data analysis in chemistry. *SIAM Journal of Scientific and Statistical Computing*, **5**, 735–744.

Wedel, M., Kamakura, W. A. (2000). Market segmentation: Conceptual and methodological foundations. Boston: Kluwer.

Wold, H. (1975). Soft Modelling by latent variables: The Non-Linear Iterative Partial Least Squares (NIPALS) approach. In J. Gani (Ed.), Perspectives in probability and statistics: Papers in honour of M.S. Bartlett on the occasion of his sixty-fifth birthday, pp. 117–142. London: Applied Probability Trust, Academic.

Chapter 10
Conjoint Use of Variables Clustering and PLS Structural Equations Modeling

Valentina Stan and Gilbert Saporta

Abstract In PLS approach, it is frequently assumed that the blocks of variables satisfy the assumption of unidimensionality. In order to fulfill at best this hypothesis, we use clustering methods of variables. We illustrate the conjoint use of variables clustering and PLS structural equations modeling on data provided by PSA Company (Peugeot Citroën) on customers' satisfaction. The data are satisfaction scores on 32 manifest variables given by 2,922 customers.

10.1 Clustering of Variables

There are two main methodologies: hierarchical methods and direct partitioning methods. Hierarchical methods are either agglomerative or divisive. Partitioning methods usually require that the number of groups should be defined beforehand and will not be used here.

A good partition is such that the variables of the same class are correlated as much as possible.

We will use here algorithms which provide clusters which are as unidimensional as possible, and where correlations between variables of the same clusters are larger than correlations between variables of different clusters. This means that blocks of variables should be as homogeneous as possible, but are not independent.

One may distinguish two cases, depending on whether the sign of the correlation coefficient is important or not (i.e. if negative values show a disagreement between variables).

V. Stan
Groupe ESSCA Angers, 1 Rue Lakanal – BP 40348, 49003 Angers, Cedex 01, France
e-mail: valentina_titu@yahoo.fr

G. Saporta
Conservatoire National des Arts et Métiers, Chaire de Statistique Appliquée, case 441, 292 rue Saint Martin, 75141 Paris, Cedex 03, France
e-mail: gilbert.saporta@cnam.fr

V. Esposito Vinzi et al. (eds.), *Handbook of Partial Least Squares*, Springer Handbooks of Computational Statistics, DOI 10.1007/978-3-540-32827-8_11,

10.1.1 Agglomerative Hierarchical Clustering Methods

10.1.1.1 Methods Derived from Clustering of Statistical Units (Nakache and Confais 2005)

Various dissimilarity measures can be used, based on the usual correlation coefficient like:

$1 - r_{jj'}$ or $1 - |r_{jj'}|$ if the sign of the correlation is not important; $s_{jj'} = \cos^{-1}(r_{jj'})$.

Then we use the following strategies of aggregation: single linkage, average linkage, complete linkage, Ward's criteria etc.

10.1.1.2 The VARHCA Method (Vigneau and Qannari 2003)

Let $C_1, C_2 \ldots . C_k$ be k blocks (or clusters) of manifest variables and $Y_1, Y_2 \ldots . Y_k$ the standardized latent variables (first principal component) associated respectively with each cluster. Manifest variables are centred, but not necessarily standardized. The following hierarchical procedure aims at locally optimizing the criterion T defined by:

$$T = n \sum_{r=1}^{k} \sum_{j=1}^{p} \delta_{rj} \operatorname{cov}^2 \left(x_j, Y_r\right) \quad where \quad \delta_{rj} = \begin{cases} 1 \text{ if } x_j \in C_r \\ 0 \text{ otherwise} \end{cases}$$

– At the first level of the hierarchy, each variable forms a cluster by itself; then, $T_0 = \sum_{j=1}^{p} \operatorname{var}\left(x_j\right)$;

– At level i, one merges the two clusters giving the minimal variation of T:

$\Delta T = T_{i-1} - T_i = \lambda_1^{(A)} + \lambda_1^{(B)} - \lambda_1^{(A \cup B)}$ where $\lambda_1^{(A)}, \lambda_1^{(B)}, \lambda_1^{(A \cup B)}$ are the largest eigenvalues of the covariance matrices of the variables in clusters A, B and $A \cup B$.

10.1.2 Cutting Trees

The resulting tree should be cut at a suitable level to get a partition. We use here a criterion of unidimensionality of the groups to obtain this cut. Starting from the root of the tree, we first realize a cut in 2 classes and verify the hypothesis of unidimensionality by using the Cronbach's α or the Dillon–Goldstein's ρ. If these values are close to 1, then the hypothesis of unidimensionality is accepted. Otherwise, we proceed to a cut at the following level of the tree, and so on. We repeat the procedure until we obtain classes satisfying the unidimensionality criteria.

10.1.3 Divisive Methods

SAS VARCLUS procedure is one of the best known. At first step one performs a PCA with all manifest variables. If there is only one principal component with an eigenvalue greater than 1, there is only one cluster.

Otherwise one considers the first two principal components: each manifest variable is associated with the principal component to which it is the closest, in regard to the squared linear correlation coefficient, thus forming two groups of variables. If the second eigenvalue of a group is greater than 1, this group is divided in turn, according to the same method, and so on, until each group has only one principal component.

10.2 Application to Structural Equation Modeling

Let p variables be observed upon n units. The p variables are partitioned in J subsets or blocks of k_j variables which are presumed to be pertinent for describing the phenomenon. Each of these blocks is designed to describe a theme of the general phenomenon. We shall designate these blocks by X_j and we shall consider them as matrices with dimension $(n \times k_j)$ (Tenenhaus et al. 2005).

In the following, we shall always suppose that each block is associated with only one latent variable (unidimensionality). In order to obtain unidimensional blocks, we propose to use some of the clustering methods, previously presented in Sect. 10.1. Therefore we can identify the blocks by the same name as their latent variable. The latent variable corresponding to the X_j block will be designated by ξ_j.

In the following, we study the specific case where there are no pre-defined causal relationships between the latent variables. We use the blocks obtained by each method to build the causality scheme.

With the help of experts we propose relationships between latent variables with the aim of explaining the general satisfaction of the customers, and we therefore establish the inner model. To choose the best model from many, we use the global quality criterion developed by Amato et al. (2004):

$$GoF = \sqrt{\overline{communality} \times \overline{R^2}}$$

where $\overline{communality}$ is the average of the communality of each block and measures the quality of the external model. $\overline{R^2}$ is the average of R^2 for each endogenous latent variable.

The R^2 measures the quality of the inner model and is calculated for each endogenous variable according to latent variables which explain it.

The software used is PLSX module of SPAD.

10.3 Practical Application

10.3.1 The Questionnaire

The data obtained are satisfaction scores scaled between 1 and 10 on 32 services for a car. 2,922 customers participated. Manifest variables are the followings (Table 10.1).

Table 10.1 Manifest variables

Variable	
General satisfaction	Sat01h
General quality	Sat02h
Quality–price ratio	Sat03h
Absence of small, irritating defects	Sat04h
Absence of noise and vibrations	Sat05h
General state of the paintwork	Sat06h
Robustness of commands, buttons	Sat33h
Solidity and robustness	Sat08h
Lock, door and window mechanisms	Sat09h
Inside space and seat modularity	Sat34h
Inside habitability	Sat11h
Dashboard: quality of materials and finishing	Sat12h
Insider: quality of mat. and finishing	Sat13h
Front seat comfort	Sat14h
Driving position	Sat15h
Visibility from driver's seat	Sat16h
Radio–CD-ROM	Sat17h
Heating–ventilation	Sat18h
Boot capacity	Sat19h
Security	Sat20h
Braking	Sat21h
Acceleration	Sat22h
Handling	Sat23h
Suspension comfort	Sat24h
Silence in rolling	Sat25h
Maniability	Sat26h
Direction	Sat27h
Gears	Sat28h
Mechanic reliability	Sat29h
Oil consumption	Sat30h
Mechanic's efficiency in solving problems	Sat31h
Maintenance cost and repairs	Sat32h

10.3.2 Clustering Variables

We have used $1 - r_{jj'}$ as distance. We have applied 6 clustering methods of variables: single linkage, average linkage, complete linkage, Ward's criterion, VARCLUS and VARHCA. Single linkage and average linkage did not provide well separated clusters, so they are eliminated.

For Ward's criterion, the tree shows that a partition in 8 classes is reasonable and for complete linkage in 6 classes. The partition obtained by cutting VARHCA tree into 7 clusters is here exactly the same as the partition given by VARCLUS. The Cronbach's α coefficients show that the obtained blocks are unidimensional.

In the following, we present the blocks for complete linkage, Ward's criterion VARCLUS and VARHCA:

In Table 10.2 we can observed that the blocks "solidity" and "driving quality" are identical for all methods. "General satisfaction" has the same composition for complete linkage, VARCLUS and VARHCA, but partition issued from Ward's criterion is more logical, according to experts. By comparison with the other methods, complete linkage groups in a single block the variables which form the blocks "interior design," "driving comfort," "interior comfort" in Ward's criterion, VARCLUS and VARHCA. For VARCLUS and VARHCA, the variables which are associated to the block "maintenance" in Ward's criterion and complete linkage, are in the same block with "quality–price ratio". Complete linkage is the only method which realizes a distinct block for the variable "quality-price ratio."

The tree for Ward's criterion (partition in 8 classes):

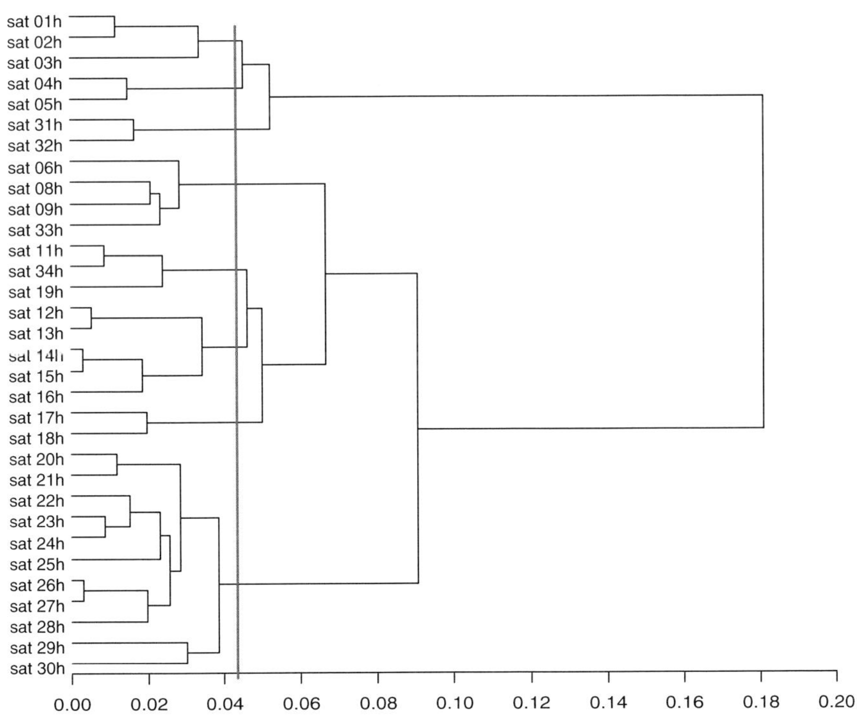

Table 10.2 Blocks of manifest variables after Ward's criterion, complete linkage, VARCLUS and VARHCA

Ward's criterion						Complete linkage						VARCLUS and VARHCA					
Block	MV	Block	MV	Block	MV	Block	MV	Block	MV	Block	MV	Block	MV	Block	MV	Block	MV
General satisfaction (Gs)	Sat01h	Solidity (Sd)	Sat06h	Driving quality (Dq)	Sat20h	General satisfaction (Gs)	Sat01h	Comfort (Cf)	Sat11h	Driving quality (Dq)	Sat20h	General satisfaction (Gs)	Sat01h	Solidity (Sd)	Sat06h	Driving quality (Dq)	Sat20h
	Sat02h		Sat08h		Sat21h		Sat02h		Sat34h		Sat21h		Sat02h		Sat08h		Sat21h
	Sat03h		Sat09h		Sat22h		Sat04h		Sat12h		Sat22h		Sat04h		Sat09h		Sat22h
Construct quality (Cq)	Sat04h		Sat33h		Sat23h		Sat05h		Sat13h		Sat23h		Sat05h		Sat33h		Sat23h
	Sat05h	Driving comfort (Dc)	Sat12h		Sat24h	Quality – price ratio (Qpr)	Sat03h		Sat14h		Sat24h	Quality – price (Qp)	Sat03h	Driving comfort (Dc)	Sat12h		Sat24h
Maintenance (Mn)	Sat31h		Sat13h		Sat25h	Maintenance (Mn)	Sat31h		Sat15h		Sat26h		Sat31h		Sat13h		Sat25h
	Sat32h		Sat14h		Sat26h		Sat32h		Sat16h		Sat27h		Sat32h		Sat14h		Sat26h
Interior design (Id)	Sat11h		Sat15h		Sat27h	Solidity (Sd)	Sat06h		Sat19h		Sat25h	Interior design (Id)	Sat11h		Sat15h		Sat27h
	Sat34h		Sat16h		Sat28h		Sat08h		Sat17h		Sat28h		Sat19h		Sat16h		Sat28h
	Sat19h	Interior comfort (Ic)	Sat17h		Sat29h		Sat09h		Sat18h		Sat29h		Sat34h	Interior comfort (Ic)	Sat17h		Sat29h
			Sat18h		Sat30h		Sat33h				Sat30h				Sat18h		Sat30h

The tree for complete linkage (partition in 6 classes):

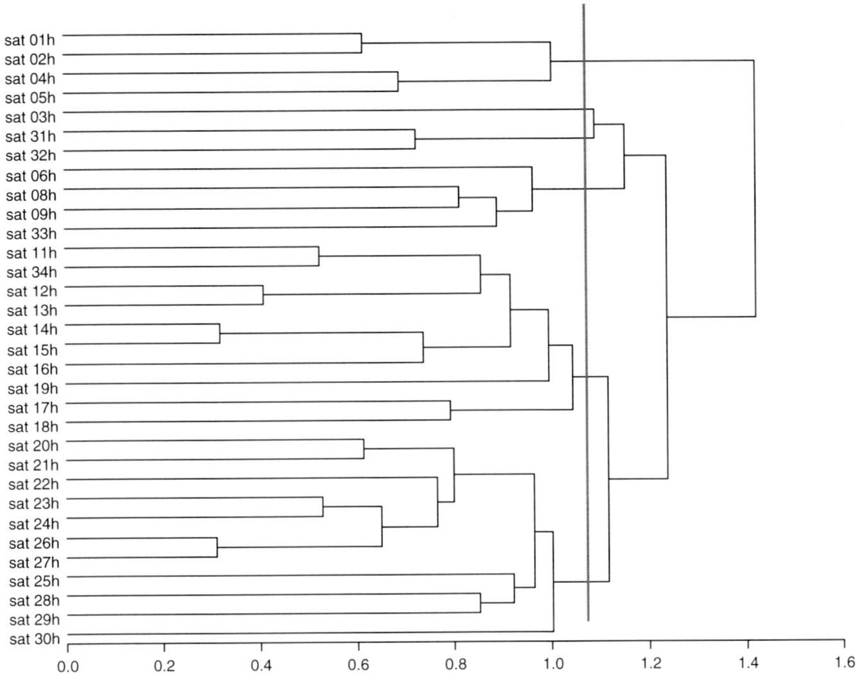

The tree for VARHCA et VARCLUS (partition in 7 classes)

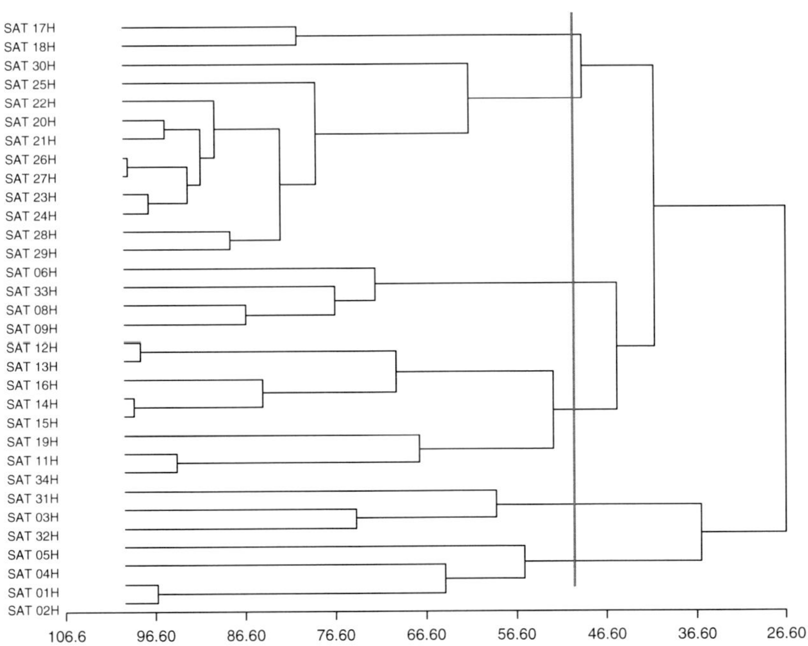

10.3.3 PLS Structural Models

The clustering techniques provide blocks but not the relationships between them.

With the help of experts we then propose relations between blocks, so as to explain the latent variable "general satisfaction". The following figures give the 3 causality schemes (Figs. 10.1–10.3):

The values of Amato's criterion GoF are:

- For Ward's criterion: $GoF = 0.48$
- For complete linkage: $GoF = 0.42$
- For VARCLUS: $GoF = 0.47$

Ward's clustering gives the best result and will be selected.

10.3.4 Results and Interpretations

10.3.4.1 The Measurement Model

After convergence of the PLS algorithm, one obtains the final weights which allow us to link the manifest variables with the latent variables. An example for "general satisfaction":

$$Gs \;=\; 0,22\,Sat01h + 0,57\,Sat02h + 0,48\,Sat03h.$$

This table presents only correlations larger than the mean of the absolute values (0.3333) (Table 10.3).

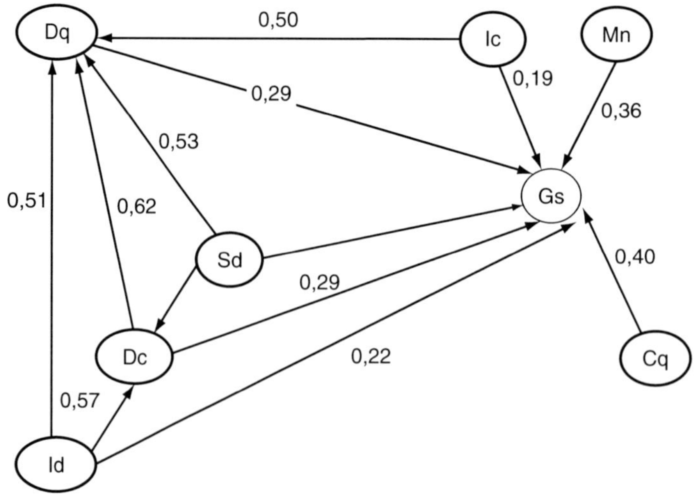

Fig. 10.1 Causality scheme after Ward's clustering

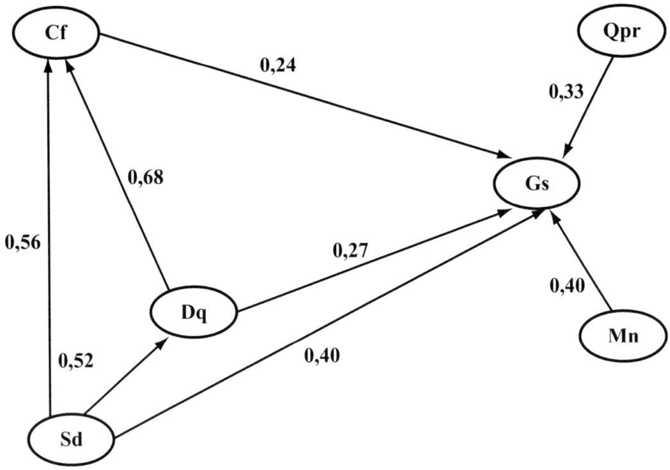

Fig. 10.2 Causality scheme after complete linkage clustering

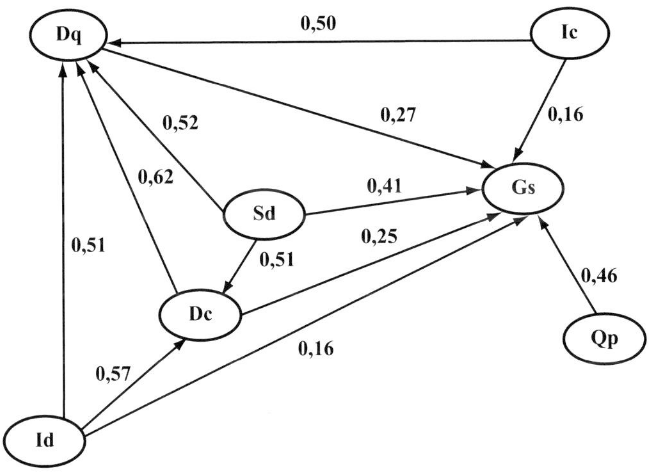

Fig. 10.3 Causality scheme after VARCLUS or VARCHA clustering

Analyzing the correlations, we observe that all latent variables are well correlated with their own manifest. So, the manifest variables "describe" their latent appropriately and the blocks are therefore validated.

10.3.4.2 The Structural Model

The R^2 coefficients between connected latent variables are:

$$R^2\,(Driving\ comfort; Sd,\ Id) = 0.42$$

$$R^2\,(Driving\ quality; Sd,\ Id,\ Dc,\ Ic) = 0.5$$

$$R^2\,(General\ satisfaction; Cq,\ Mn,\ Sd,\ Id,\ Dc,\ Ic,\ Dq) = 0.27$$

Table 10.3 Correlations between manifest and latent variables

Variables	General satisfaction	Construct quality	Maintenance	Solidity	Interior design	Driving comfort	Interior comfort	Driving quality
Sat01h	0.6442	0.3588						
Sat02h	0.8706	0.4011		0.3731				
Sat03h	0.7397							
Sat04h	0.3667	0.8780						
Sat05h		0.8449						
Sat31h		0.3828	0.8739					
Sat32h			0.8332					
Sat06h				0.6534		0.3428		
Sat08h				0.7867	0.3558	0.4223		0.4605
Sat09h				0.7057		0.3493		0.3707
Sat33h				0.7061		0.3420		
Sat11h				0.3597	0.8801	0.5249		0.4442
Sat34h				0.4039	0.8286	0.4816		0.4088
Sat19h					0.7015	0.3651		0.3774
Sat12h				0.4308	0.4684	0.7711	0.3480	0.4782
Sat13h				0.4305	0.4502	0.7903	0.3396	0.4522
Sat14h				0.3756	0.4351	0.8122	0.3461	0.4786
Sat15h				0.3914	0.4611	0.8283	0.3851	0.5367
Sat16h				0.3444	0.3971	0.6595	0.3403	0.4455
Sat17h						0.3508	0.8110	0.3895
Sat18h				0.3434	0.3506	0.4086	0.8665	0.4514
Sat20h				0.4589	0.4924	0.5299	0.4713	0.7315
Sat21h				0.3909	0.3453	0.3952	0.3760	0.6739
Sat22h					0.3349	0.3944	0.3349	0.6757
Sat23h				0.3737	0.3685	0.4458	0.3690	0.7716
Sat24h				0.3685	0.3647	0.4789	0.3379	0.7362
Sat25h						0.3840		0.6218
Sat26h				0.3908	0.3724	0.4791	0.3593	0.7837
Sat27h				0.3902	0.3594	0.4880	0.3751	0.7841
Sat28h				0.3573		0.4048		0.6396
Sat29h								0.5690
Sat30h								0.4743

For "general satisfaction," the R^2 coefficient generated by the other latent variables is 27%, and we consider that as satisfactory because there are 2,922 individuals (Table 10.4).

The correlations between the latent variables are given below:

Analyzing the correlations between the latent variables, we can see that to improve "driving quality", the producer should concentrate on "driving comfort" (correlation coefficient = 0.62), on the "solidity" (0.53) and on the "interior design" (0.51).

In order to obtain a good "driving comfort", the producer could concentrate on "interior design" (0.57) and on "solidity" (0.51).

Given the causality scheme, the determination of "general satisfaction" is a complex procedure in which almost all the latent variables are directly involved.

Table 10.4 The correlations between latent variables

	General satisfaction	Construct quality	Maintenance	Solidity	Interior design	Driving comfort	Interior comfort	Driving quality
General satisfaction	1.0000							
Construct quality	0.4041	1.0000						
Maintenance	0.3576	0.3503	1.0000					
Solidity	0.3722	0.3407	0.2914	1.0000				
Interior design	0.2237	0.0988	0.1979	0.4217	1.0000			
Driving comfort	0.2928	0.1539	0.2266	0.5119	0.5729	1.0000		
Interior comfort	0.1854	0.1233	0.2301	0.3951	0.3812	0.4542	1.0000	
Driving quality	0.2943	0.2023	0.3071	0.5257	0.5085	0.6180	0.5029	1.0000

"Construct quality" is the most important variable for the "general satisfaction" (correlation coefficient $= 0.40$) and the less important is the "interior comfort" (0.19).

Consequently, in order to increase the general satisfaction, the producer should concentrate first on the "construct quality" and then on the "solidity", "maintenance", "driving quality", "driving comfort", "interior design" and "interior comfort".

The equation is as follows:

$$Gs = 0.26\,Cq + 0.19\,Mn + 0.15\,Sd + 0.03\,Id + 0.10\,Dc - 0.03\,Ic + 0.04\,Dq.$$

10.4 Conclusions

Variables clustering provide a simple way of obtaining unidimensional blocks in structural equation modeling, when prior knowledge of blocks is not available.

It must be underlined that this study did not follow the logical sequence of steps of the PLS approach: the construction of a model by experts, the construction of a questionnaire using this model, and the collection of customer data using this questionnaire.

In our case, the process is inverted: we have tried to build a model using data that had already been collected. This fact has obviously effects on the final results which cannot be measured.

By means of clustering methods of variables, we established the external model. According to Amato's criterion, Ward's clustering was choosen as the best technique for our data set. But we observe that the values of this criterion for the 3 models are very close.

For the chosen model, a hierarchy of the influence of the latent variables on general satisfaction can be established using the structural model:

I. Construct quality; II. Solidity; III. Maintenance; IV. Driving quality, V. Driving comfort, VI. Interior design, VII. Interior comfort.

The results obtained are satisfactory: $R^2 = 27\%$ for a large sample of almost 3,000 respondents.

References

Amato, S., Esposito Vinzi, V., & Tenenhaus, M. (2004). A global goodness-of-fit index for PLS structural equation modeling. *Oral Communication to PLS Club, HEC School of Management, France, March 24.*

Nakache, J. P., & Confais J. (2005) Approche pragmatique de la classification. *Ed. Technip, Paris.*

Tenenhaus, M., Esposito Vinzi, V., Chatelin, Y.-M., & Lauro, C. (2005). PLS path modeling. *Computational Statistics and Data Analysis, 48*(1), 159–205.

Vigneau, E., & Qannari, E. M. (2003). Clustering of variables around latent component – application to sensory analysis. *Communications in Statistics, Simulation and Computation, 32*(4), 1131–1150.

Chapter 11
Design of PLS-Based Satisfaction Studies

Kai Kristensen and Jacob Eskildsen

Abstract In this chapter we focus on design of PLS structural equation modeling with respect to satisfaction studies in general. Previous studies have found the PLS technique to be affected by things as the skewness of manifest variables, multicollinearity between latent variables, misspecification, question order, sample size as well as the size of the path coefficients (Cassel et al. 1999; Auh et al. 2003; Eskildsen and Kristensen 2005; Kristensen and Eskildsen 2005a,b). In this chapter we expand on these contributions in order to provide the reader with recommendations on all aspects included in designing PLS-based satisfaction studies.
The recommendations are based on an empirical PLS project conducted at the Aarhus School of Business, Center for Corporate Performance. Within this project five different studies have been conducted that cover a variety of aspects of designing PLS-based satisfaction studies.
The data used in subsequent sections comes from a variety of sources. In relation to the empirical PLS project at the Aarhus School off Business the following five different studies have been conducted:

- Scale study
- Empirical experiment
- Simulation study – data collection
- Simulation study – missing values
- Empirical study of model specification for a customer satisfaction model

11.1 Data Collection Considerations

When planning a satisfaction study one of the first things to consider will be the practical aspects of data collection. Among these we find things like the design of the questionnaire and the sampling method.

K. Kristensen and J. Eskildsen
School of Business, University of Aarhus, Haslegaardsvej 10, 8210 Aarhus V, Denmark
e-mail: kak@asb.dk, Eskildsen@asb.dk

V. Esposito Vinzi et al. (eds.), *Handbook of Partial Least Squares*, Springer Handbooks of Computational Statistics, DOI 10.1007/978-3-540-32827-8_12,

It has for some time been assumed that the practical design of a satisfaction study may influence the results. Auh, Salisbury and Johnson have for instance suggested that there may be an order effect of the variables (Auh et al. 2003). Other studies have suggested that factors like the intro text and presentation of the interviewer as well as sampling technique may also have an effect.

To analyze this, an experimental design was set up to test the influence that the practical design of a satisfaction study may have on the results. Recently it has been suggested that there may be an order effect of the variables (Auh et al. 2003) and local studies in the Nordic countries have suggested that factors like the intro text and presentation of the interviewer may also have an effect. Furthermore it has been suggested that the data collection technique may also influence the results.

The theoretical framework used as the basis for this study is the so-called Reputation Excellence (REEX) Index shown in Fig. 11.1 (Eskildsen et al. 2004b; Kristensen and Eskildsen 2005a).

The Reputation Excellence Index is estimated on the basis of 18 generic statements that cover the areas shown in Table 11.1.

In order to carry out the test an experimental design was set up with the following eight factors:

Sampling method

- Telephone
- Postal

Presentation of researcher

- Company itself
- The University

Intro text

- Corporate level
- Local level

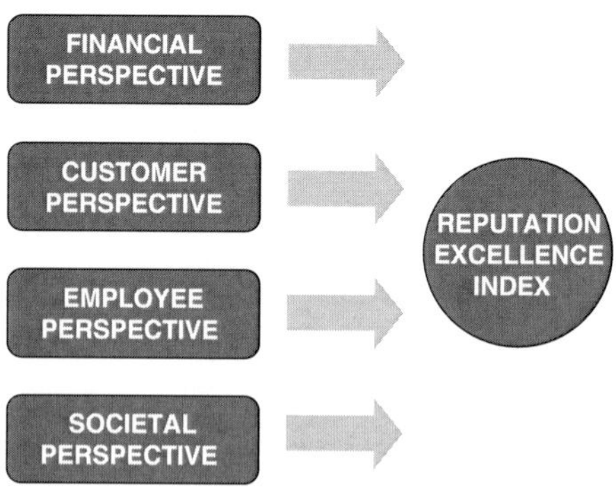

Fig. 11.1 The REEX model

Table 11.1 Generic REEX statements

Perspective	Statements
Financial perspective	Financially sound
	Financially successful
Customer perspective	Value for money
	Quality of products
	Quality of service
	Innovative
Employee perspective	Competent employees
	Competent management
	Good working conditions
Society perspective	Trustworthiness
	Cares for the environment
	Social commitment
REEX	A well-run company
	Customer-oriented company
	Good place to work
	A responsible company
	Likes the company
	Trust the company

Order of questions on questionnaire

- Endogenous first
- Exogenous first

A half fraction of a 2^4 design was used; i.e. eight combinations were used. The combinations were chosen according to an orthogonal main effects plan. For each combination approximately 125 respondents were interviewed and for each combination the latent model was estimated using PLS estimation. Data were analyzed on both aggregate level and on individual subject level using profile analysis.

11.1.1 Intro Text

Any questionnaire used for satisfaction studies will contain an intro text, which explains the background and context of the study. In relation to this we have in a number of Nordic customer satisfaction and similar studies experienced that there is a tendency that the results may be affected by the way that you introduce the respondent to your study. Especially we have seen that for larger companies, results seem to depend on whether you ask the respondent to focus on the head office or on the local outlet when answering the questions.

In order to put this to a more rigorous test we have in the study included two versions of the intro text: One focusing on the corporate level and one focusing on the local level.

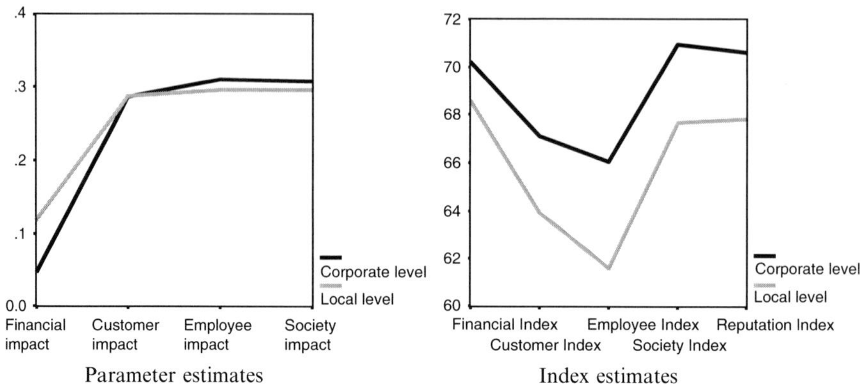

Fig. 11.2 Path coefficients and index estimates for the intro text factor

As described previously we used a simple society satisfaction model to do the test. This model has four exogenous variables corresponding to the right hand side of the European Foundation for Quality Management (EFQM) Excellence model (financial results, customer results, employee results and society results) and one endogenous variable describing company reputation. Each latent variable has approximately five manifest indicators.

In Fig. 11.2 above parameter estimates for the model and average index estimates are given for the eight runs of the experiment. To the left you will find the parameter estimates and to the right the index estimates.

The conclusion concerning the parameters is quite clear. When performing a profile analysis, the p-value for parallel profiles is 0.388, and the p-value for coinciding profiles is 0.519. Both are far from being significant. Hence the conclusion is that, for the parameter estimates, the profiles are identical. The intro text does not influence the parameter estimates.

When it comes to the index estimates the picture is a little unclear. The profiles are indeed parallel ($p = 0.825$) which clearly can be seen from the diagram, but there seems to be a tendency that the answers for corporate level have been shifted upwards. To be strict the shift is not significant ($p = 0.122$), but since the sample size is rather small (eight runs at the aggregate level) and since we have observed this tendency in other studies, we have decided not to reject the hypothesis totally.

Furthermore, theoretically, it makes good sense that the corporate levels should be higher than the local levels. Reputation at the corporate level is clearly based on advertising campaigns, TV commercials, newspaper articles and the like, while reputation at local level will be much more influenced by your own actual experience.

Thus our recommendation is to be careful when formulating the intro text and when comparing results from studies with different texts. We are not in doubt that

the model structure will be the same but we are still in doubt whether the index levels will be affected.

11.1.2 Interviewer Presentation

Another aspect of the sampling, which might influence the results, is the way the interviewer introduces himself to the respondent. In our case we have experimented with two ways of doing this presentation. Either the interviewer introduces himself as coming from the university or he introduces himself as coming from the company in question.

Our hypothesis is that respondents are more critical when the interviewer comes from a university than if he comes from the company in question.

Our results for the aggregate sample can be found in Fig. 11.3. In this case we observe a structure, which is quite similar to the one we found for the intro text factor. There is no effect for the parameter estimates (p = 0.578 for parallel profiles and p = 0.707 for coinciding profiles). Hence, we conclude that interviewer presentation does not affect the parameter estimates.

When it comes to the index level estimates we find parallel profiles (p = 0.287), but once again we come close to significance (p = 0.136) when it comes to the test of a parallel shift.

To investigate this a little further we analyzed the index estimates for the entire sample (1,000 interviews). The result of this was that we indeed find a significant shift (p = 0.047) indicating that the index level is in general lower when the interviewer is introduced as coming from a university.

Once again we conclude that parameter estimates are not affected but when it comes to index levels we must be careful when comparing estimates coming from data that have been collected with different presentation of the interviewer.

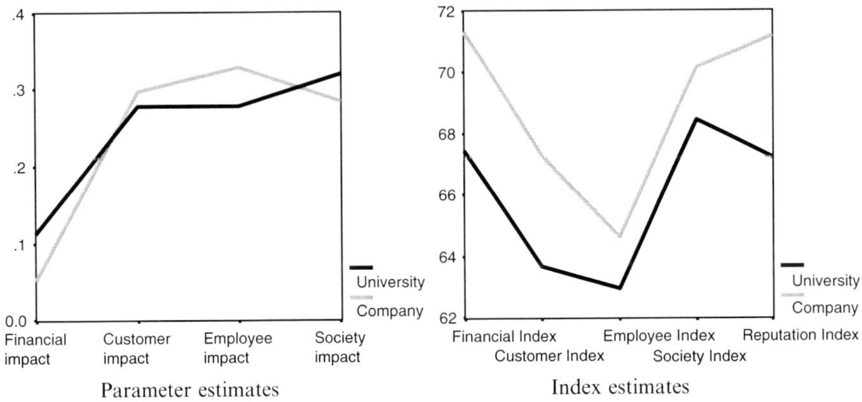

Fig. 11.3 Path coefficients and index estimates for the presentation factor

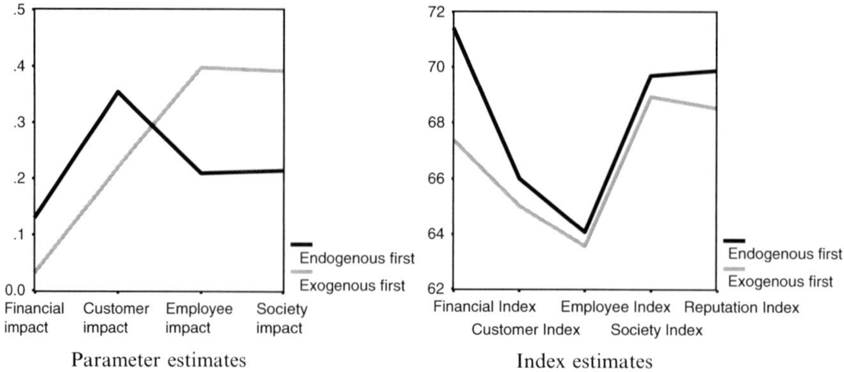

Fig. 11.4 Path coefficients and index estimates for the order factor

11.1.3 Question Order

As previously mentioned Auh et al. (2003) have indicated that there might be an order effect of variables when analyzing customer satisfaction. In order to test this in more detail we included an order effect factor in our experiment. In half of the cases the endogenous variables were listed first in the questionnaire and in the other half the exogenous variables were first.

Results are given in Fig. 11.4.

In this case the index profiles are not different ($p = 0.477$), but as it clearly appears from the figure the parameter estimates are far from being equal ($p = 0.001$). The difference is not a simple one since we are not just talking about a parallel shift. The entire structure is changing. In the case of the exogenous first the employee and social factors are the most important drivers, but in the case of the endogenous first the customer factor is dominating the rest of the drivers.

This is a very unfortunate situation. We believe that a parallel shift in the index levels is possible to live with from a decision point of view in a company, but a situation where the drivers are changing place is a very different matter. We cannot from a statistical point of view tell which solution is the correct one (if any), but the result definitely tells the researcher that it is extremely important to have a subject matter discussion of the order of variables before a study is launched. It also tells the researcher that he must be very careful when comparing results from different studies with different order of the variables.

11.1.4 Data Collection Method

The last factor included in the experiment was the data collection method. It is often assumed that postal and telephone interviews will produce different results and we decided to test this assumption in our customer satisfaction environment.

The results are given in Fig. 11.5.

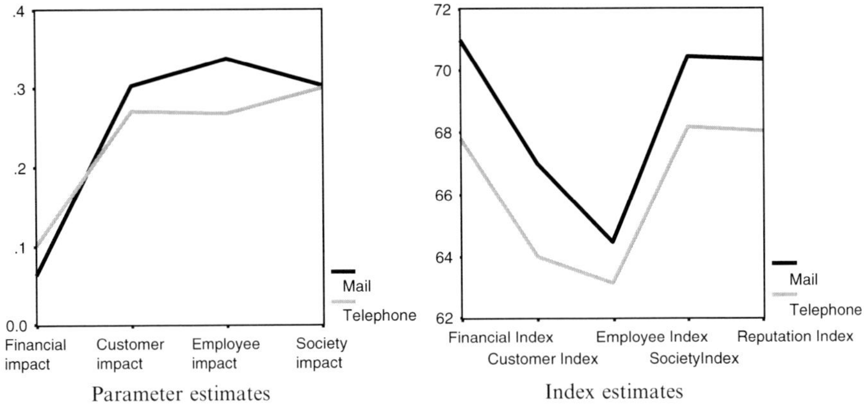

Fig. 11.5 Path coefficients and index estimates for the data collection factor

In neither case we can find any significant difference between the postal and telephone surveys. They seem to produce the same driver structure and the same index levels. This is certainly a very satisfying conclusion since we in practice very often have to mix samples from different sources.

11.2 Data Considerations

Apart from sampling considerations we, when planning a satisfaction study, also have to pay attention to the size of the collected dataset and the appearance of the data. Hence we have studied in more detail the effect of the sample size and the effect of the choice of different scales when asking questions. The results of these studies are summarized in the following two sections.

11.2.1 Sample Size

In order to study the effect of the sample size among other things a study was set up (Kristensen and Eskildsen 2005b). This study is a simulation study and the basic structural equation model applied in the simulation study is shown in Fig. 11.6. This model is thought to be a simplified replicate of the European Performance Satisfaction Index (EPSI) Rating framework.

The simulation stage was conducted in two stages. The first stage was a screening stage where the irrelevant factors were identified. In the second stage we focused on the factors having an impact on the outcome of a structural equation modeling analysis (Kristensen and Eskildsen 2005b).

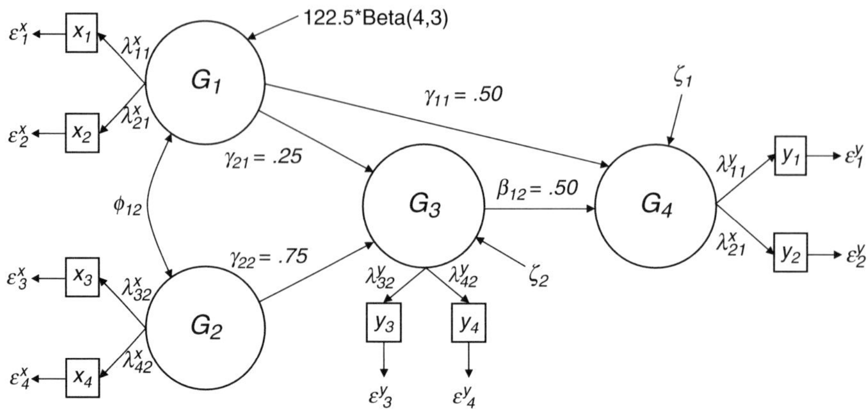

Fig. 11.6 The simulation model

In the screening stage we applied an orthogonal main effect plan with 7 factors in 27 runs with 25 replications for each run. Each replication has a number of observations varying between 50, 250 and 1,000. In this stage the following factors was included:

- Exogenous distribution (Beta vs. Normal)
- Multicollinearity between latent exogenous variables
- Indicator validity (bias – G1)
- Indicator reliability (standard deviation within a block – G1)
- Structural model specification error
- Sample size
- Number of manifest indicators in each block

In the second stage we focused on the resulting four most interesting factors being multicollinearity, reliability, sample size and the number of manifest indicators. This resulted in a full factorial design with 4 factors in 54 runs with 25 replications for each run – a total of 585.000 observations. The four factors were included in the second stage study with the following levels:

- Multicollinearity: $\rho = [0.2;0.8]$.
- Reliability (G1): $\sigma = [1; 10; 20]$.
- Sample size: n = [50; 250; 1,000].
- Number of manifest indicators: p = [2; 4; 6].

In order to assess the effect of the four factors the following response variables have been retained for all 25 replications for each of the 54 runs:

- Absolute bias of indices
- Standard deviation of indices
- Bias of path coefficients
- Standard deviation of path coefficients
- R^2, AVE and RMSE.

The measure RMSE is given as:

$$RMSE = \sqrt{\frac{\sum_{i=1}^{n}\sum_{j=1}^{n}\left(\Psi_{ij}-\Psi(\theta)_{ij}\right)^2}{n(n-1)}},\ i \neq j \ \wedge \ \Psi(\theta) = \Omega R_{XY}\Omega'$$

Ψ: correlations matrix between the manifest variables
$\Psi(\theta)$: parameterized correlations matrix between the manifest variables
Ω: matrix of loadings between latent and manifest variables according to model structure
R_{XY}: Latent correlations matrix

The final response variable for each of the 54 runs is an average of the response variables from the 25 replications (Kristensen and Eskildsen 2005b).

In Fig. 11.7 the effect that the sample size has on the mean absolute bias of the latent constructs.

The absolute bias of indices is here found to be decreasing with an increasing sample size. The indices are however not the only part of the model affected by the sample size.

In Fig. 11.8 below we show the effect of the sample size on the relative bias of γ_{21} the path coefficient between the latent variables G1 and G3 in Fig. 11.6.

From Fig. 11.8 it appears as if the benefit of increasing the sample size more or less fades out when the sample size reaches 250.

The general recommendation for practitioners is therefore that a sample size of 250 is sufficient to ensure a reasonable level of bias of the path coefficient in a PLS structural equation model (Kristensen and Eskildsen 2005b).

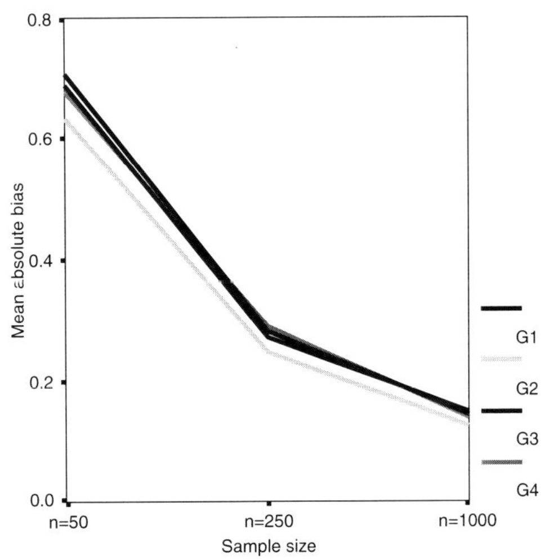

Fig. 11.7 Absolute bias of indices: The effect of sample size

Fig. 11.8 Mean relative bias of γ_{21}

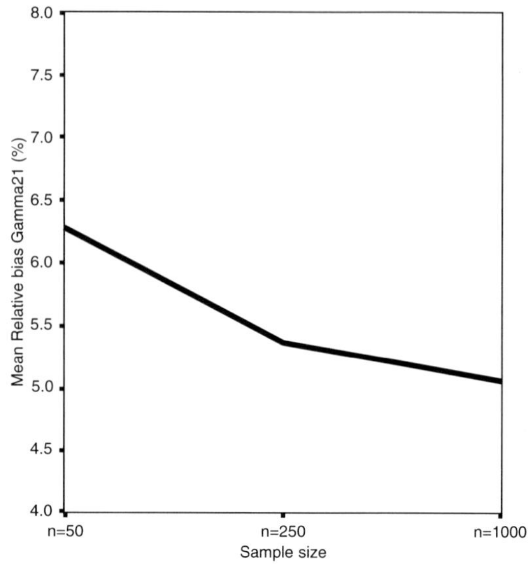

11.2.2 Choice of Scale

The debate about the optimal rating scale has been going on since the emergence of these instruments. Research has among other things focused on the number of categories to apply (Green and Rao 1970; Matell and Jacoby 1972; Ramsay 1973; Cox 1980; Givon and Shapira 1984; Preston and Colman 2000), reliability and validity (Bending 1954; Ramsay 1973; Flamer 1983), type of scale (Menezes and Elbert 1979; Ofir et al. 1987), usage of a middle and a "don't know" category (Guy and Norvell 1977; Cox 1980; O'Muircheartaigh et al. 2001) and demographic differences with respect to response behavior (Baumgartner and Steenkamp 2001; Ewing et al. 2002; Ueltschy et al. 2004). Among the few consistent results across these studies are that a "don't know" category is preferable (O'Muircheartaigh et al. 2001), that there are demographic differences with respect to response behavior (Baumgartner and Steenkamp 2001; Ewing et al. 2002; Ueltschy et al. 2004) and that the semantic differential scale is preferable (Ofir et al. 1987). It is therefore difficult to assert the effect that the scale can have in customer satisfaction studies.

In order to test the possible effect of scale choice on the results of customer satisfaction studies a controlled experiment was set up. Under totally identical conditions two samples were drawn from the same population. The only difference between the samples was that in the first sample a 5-point semantic differential scale was used and in the second the standard EPSI rating 10-point semantic differential scale was used. The questionnaires were the standard EPSI Rating questionnaires with differing scale length. The size of the samples was 545 for the 10-point scale and 563 for the 5-point scale. Mean values and standard deviations for the seven latent variables

Table 11.2 Results from the rating scale experiment

Latent Variable	Mean		Significance	Standard deviation		Significance
	Ten points	Five points	Two tailed	Ten points	Five points	Two tailed
Image	63.6	64.0	0.740	18.1	19.5	0.069
Expectations	73.3	75.1	0.128	19.2	20.1	0.476
Products	64.2	64.3	0.879	19.1	20.5	0.274
Service	66.9	66.4	0.703	21.2	23.4	0.014
Value	54.4	54.4	0.958	19.7	22.4	0.005
Satisfaction	65.2	65.2	0.970	19.3	21.5	0.013
Loyalty	57.5	58.7	0.355	21.7	23.6	0.054

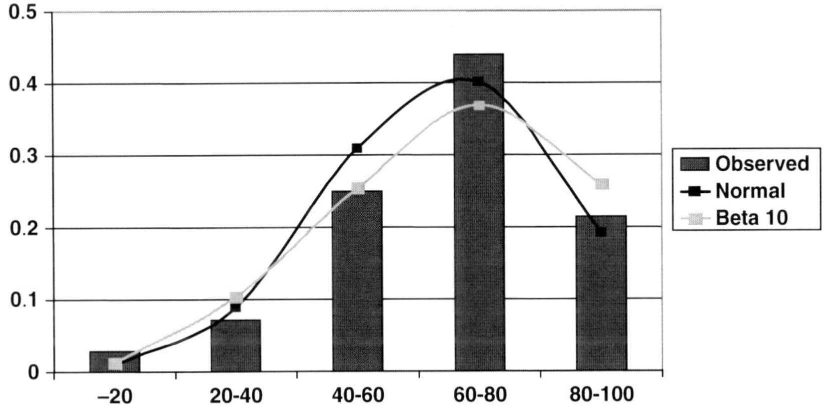

Fig. 11.9 Comparison of observed and theoretical distributions

in the EPSI rating framework on the two scales standardized to 0–100 are shown in Table 11.2.

From Table 11.2 it is evident that there is no significant difference between the mean values of the aggregate variables. This means that the choice of scale has no influence on the level of the customer satisfaction index or the loyalty index. As expected the standard deviation of the 10-point scale is smaller than the standard deviation of the 5-point scale with Image, Expectations and Products as possible exceptions. The difference is on the average approx. 10% and the reason for this difference is that the underlying distributions are discrete.

A comparison of the observed distribution of the 10-point scale with theoretical distributions of satisfaction is shown in Fig. 11.9. The beta distribution or the doubly truncated normal distribution seems to give the closest approximation to the distribution but even here there is a significant difference in both cases.

Regardless of the choice of scale it is well known that there are demographic differences with respect to age, gender, education and the degree of urbanization when it comes to customer satisfaction. Customer satisfaction is increasing with age, decreasing with education as well as decreasing with the degree of urbanization. Furthermore women are in general more satisfied than men.

However, we did not find any interaction effects between the choice of scale and age, gender, education and the degree of urbanization. It is therefore our general conclusion that the demographic interpretation of customer satisfaction studies will not be seriously affected by the choice of scale.

Our general conclusion is that a 10-point scale is preferable to a 5-point scale. This is due to a smaller standard deviation and to the fact that an increasing number of points will bring the scale closer to a continuous scale and thus closer to the assumption of most of the statistical techniques used by the practitioner. On the other hand we did not find any differences between the mean values of the standardized 5- and 10-point scales. In practice this means that it will be possible to compare results from satisfaction studies using these different scales.

11.3 Data Analysis Considerations

11.3.1 Treatment of Missing Values

Methods for Handling Missing Values

Previous research has investigated the consequences of methods for handling nonresponse in structural equation modeling based on traditional likelihood estimation (Brown 1994; Olinsky et al. 2003).

These studies generally find that pair-wise or list-wise deletion as well as mean substitution is outperformed by the more sophisticated techniques such as EM substitution (Brown 1994; Olinsky et al. 2003). Similar findings have been reported when it comes to traditional regression analysis as well as other multivariate statistical techniques (Afifi and Elashoff 1966; Beale and Little 1975; Bello 1993, 1995; Schafer 1997; Allison 2002). Little has however been done in relation to PLS.

Based on these previous findings we have therefore included the following techniques in a simulation study (Kristensen and Eskildsen 2005a):

- Pairwise deletion
- Mean substitution
- Regression based substitution
- EM substitution

These techniques are all available in the SPSS Missing Value module. Furthermore previous research has shown that the quality of the imputations can be negatively affected if the assumption of missing at random is breached (Greenless et al. 1982). We will therefore incorporate this into the simulation study as well. Finally it has been shown that the Missing Value Package in SPSS may deliver biased results (von Hippel 2004) and since our PLS algorithm is based on SPSS we will also be able to assess whether or not these biases has an effect on the outcome of a PLS analysis.

The basic structural equation model applied is the same as the one previously applied (see Fig. 11.6). All the manifest variables are generated as a function of the latent variables plus an error term that is (0.10) normally distributed.

A full experimental design was conducted, i.e. 24 runs with 25 replicates in each and each replicate consisting of 250 observations yielding a total of 150.000 observations. The following factors were thus studied (Kristensen and Eskildsen 2005a):

- The probability of missing values for one group of manifest variables (x1). (5%; 10%; 20%)
- The conditional probability of missing values for another group of manifest variables within the same latent variable (x2) given the first group. (25%; 50%)
- The method of handling missing values (Pairwise deletion; Mean substitution; Regression; EM algorithm)

In order to assess the effect of the three factors the following response variables have been retained for all 25 replications for each of the 54 runs:

- Absolute bias of indices
- Standard deviation of indices
- Bias of path coefficients
- Standard deviation of path coefficients
- R^2, AVE and RMSE.

The final response variable for each of the 24 runs is an average of the response variables from the 25 replications.

Results

In the missing value simulation study our analysis is based on four basic research assumptions or hypothesis coming from the existing missing value literature (Kristensen and Eskildsen 2005a):

1. We assume that the RMSE of both latent variable estimates and parameter estimates will depend heavily on both the fraction of missing values and the missing value strategy.
2. We assume that the effect is not limited to the latent variable with missing values but is transmitted to the endogenous variables, but to a smaller extent.
3. We assume that the RMSE is affected by the interaction between the fraction of missing values and the missing value strategy.
4. We assume that the EM and regression techniques outperform mean value substitution and pair-wise deletion.

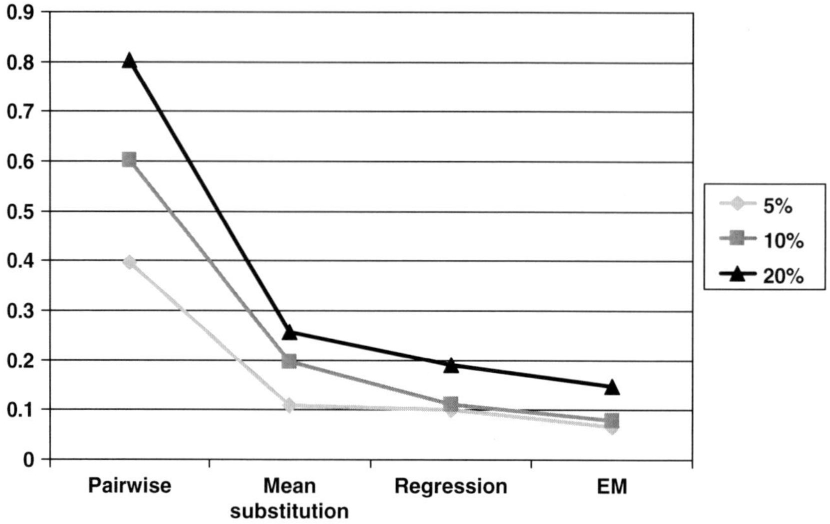

Fig. 11.10 RMSE for a latent exogenous variable with missing values as a function of the fraction of missing values and the latent variable strategy

Some of the results of our analysis of the latent variable estimates are shown in Fig. 11.10 above. First of all it will be seen that the size of the fraction of missing values has a tremendous effect on the precision. As a rule of thumb we may say that when the fraction is doubled the RMSE is increased by 50%. It will also be seen that when it comes to the latent variable estimate (mean value) all techniques including mean value substitution perform better than doing nothing (pairwise deletion), and in this particular case there is no statistical difference between the techniques. The reduction in RMSE is around 75% when applying a technique instead of just using pairwise deletion.

When it comes to the RMSE of the standard deviation we see another picture. This appears from Fig. 11.11. Again we find a significant effect of fraction and strategy as well as an interaction. This time, however, mean value substitution is the big problem. It performs significantly worse than all other techniques including pairwise deletion. Regression and EM are by far the preferred techniques.

We find similar effects for endogenous variables. The effects are, however, very small and they are not of any practical importance in our case.

When we look at the parameter estimates we find results very similar to the ones described above. In all cases analyzed we have significant effects for the fraction of missing values, the missing value technique, and the interaction between the two. The size of the conditional probability is of small importance.

As an illustration we give the values forβ_{12}, the impact from the first to the second endogenous variable.

In this case we again find that mean substitution is the worst performer. The difference between the rest is very small.

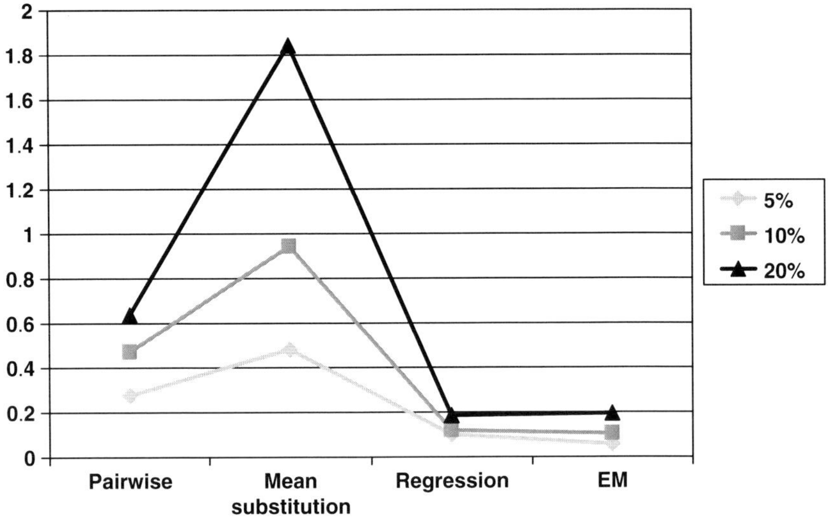

Fig. 11.11 RMSE for the standard deviation of latent exogenous variable with missing values as a function of the fraction of missing values and the latent variable strategy

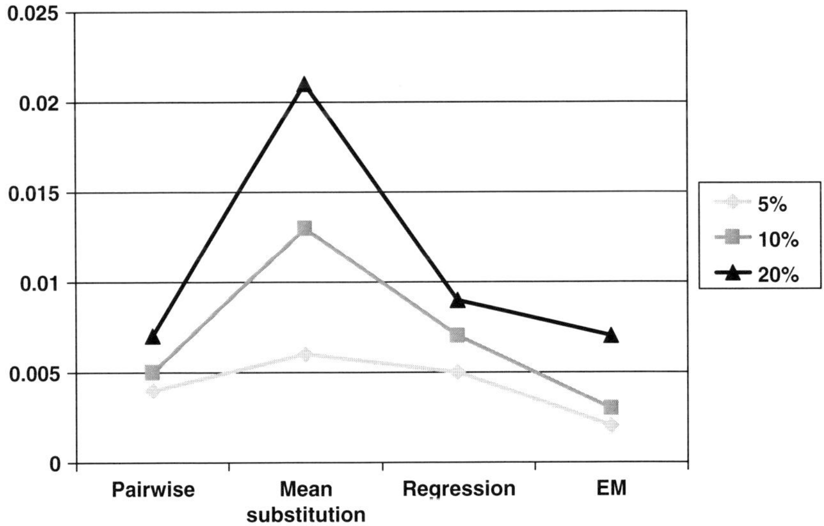

Fig. 11.12 RMSE for the impact estimate from one endogenous variable to the next

Results for the standard deviation are found in Fig. 11.12.

The well-known pattern is found once again, but in this case the EM algorithm is outperforming all other techniques. In addition we find a slightly higher importance of the conditional probability.

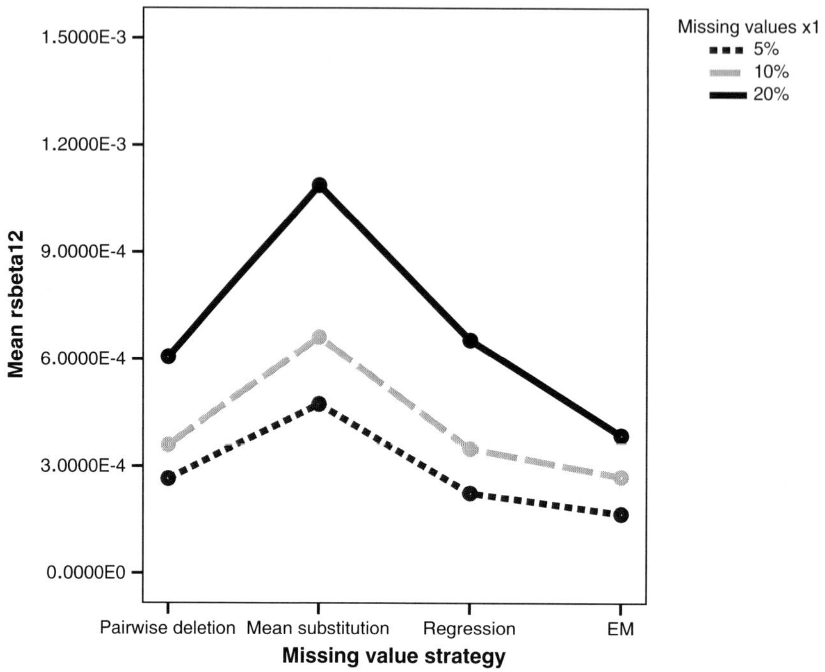

Fig. 11.13 RMSE for the standard deviation of parameter beta12

So far our results confirm our hypothesis. In general the regression and EM techniques are outperforming other techniques, and it seems as if the EM algorithm due to the interaction is the most preferable technique when the fraction of missing values is increasing (within the limits that we have analyzed).

As a final example of the results let us look at the model fit by analyzing the first coefficient of determination in the model, that is the regression from the two exogenous variables to the first endogenous variable. The results are shown in Fig. 11.14.

There is a tendency that mean substitution is exaggerating the explanatory power. This exaggeration is an increasing function of the fraction of missing values. Again the regression technique and the EM algorithm are performing well and are producing results very close to the true result.

Concluding remarks concerning missing values

In order to give some guidelines for the choice of technique this study has compared four different methods of handling missing values in a customer satisfaction like PLS model.

The results show that the regression technique and the EM algorithm in general are outperforming the other techniques. There is, however, a tendency that the

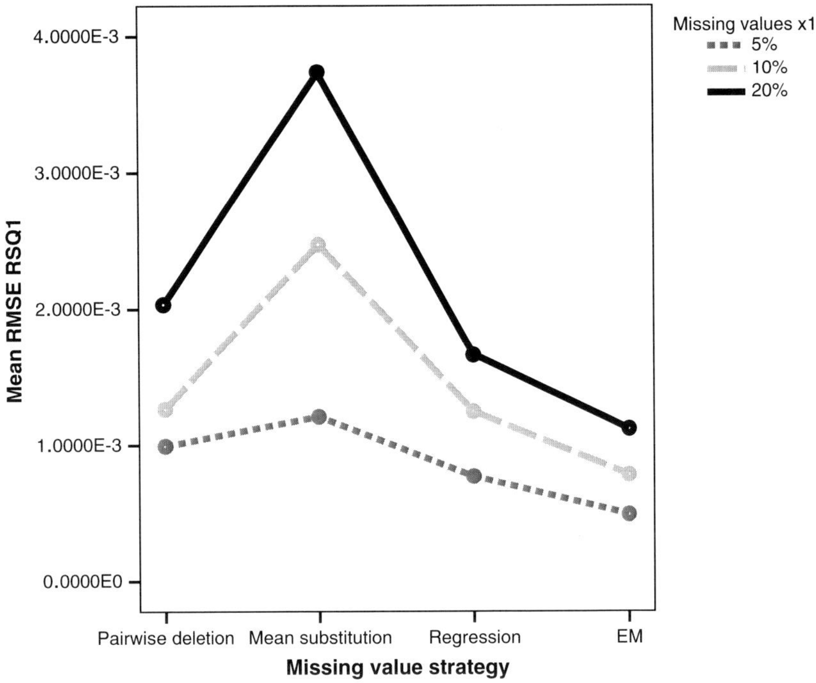

Fig. 11.14 Precision of model fit as a function of fraction of missing values and the missing value technique

EM algorithm is to be preferred over the regression technique when the fraction of missing values is increasing. For small fractions of missing values the two techniques are not significantly different.

The mean value substitution is in general a bad technique. It produces way too small estimates of the standard deviation of both the latent variables and the parameter estimates. Furthermore use of the mean substitution technique will produce bad estimates of the explanatory power of the model. Based on this our recommendation is in general to use the EM algorithm for correcting missing values.

In this study we did not analyze the relatively recent multiple imputation strategies, due to the fact that they have not yet been implemented in the most common commercial computer programs like SPSS and SAS. However, due to the close relationship between these techniques and the EM strategy we are relatively comfortable in concluding that these techniques will be highly efficient and probably even more efficient than the EM technique.

In general the level of missing values experienced with the EPSI Rating customer satisfaction framework is not problematic although the level is high for some manifests. The level of missing values in different branches of industries is shown in Table 11.3.

Questions containing comparisons are in general creating a problem. This problem is of the same importance for both men and women. Furthermore the ability to

Table 11.3 The level of missing values

Supermarkets	Banks	Automobiles	Petrol Stations
Below 10% for all items	Relative comparisons are problematic. 40–50% missing values	19 out of 22 items have missing values below 5%	13 out of 22 have missing values below 10% 8 have missing values between 10% and 20%

do comparisons is decreasing with age and increasing with education. In general the number of missing values is low except for

- Image (Questions about corporate creativity)
- Products (Questions about technical aspects, e.g. the internet and similar technical matters. Likewise comparisons to other products give problems)
- Service (Questions about comparisons to other companies)

In general the population may be segmented into three groups when we are talking about missing values:

- A small group (approx. 2%), which in general has a large number of missing values on all questions.
- A relatively large group (approx. 38%), which has missing, values when it comes to comparisons. This group does not have any problems regarding the rest of the questions.
- A very large group (approx. 60%) where the number of missing values is low and where the causes seem to be random.

In general the number of missing values is lower for women than for men. The only area where we observe the opposite is questions about value (e.g. value for money).

Thus, our study of the Danish EPSI Rating material has shown that in general the fraction of missing values for most questions and most industries is below 20%. In a few cases the fraction may be a high as 50%, but in these cases the missing values are to be considered as structural and not random. For these cases the problem has to be remedied by reformulating the questions. In the random cases the problem can be dealt with by using one of the existing missing value correction techniques. On the other hand the 2% group mentioned in the segmentation poses a problem. This group has a very large number of missing values. The validity and reliability of the answers from this group should in general be questioned, and we propose that this group is simply excluded from the analysis in order to prevent contamination of the rest of the dataset.

11.3.2 Number of Indicators

In order to evaluate the consequences of the concept of consistency at large the number of indicators was included as a factor in the simulation study described in Sect. 11.2.1. In Fig. 11.15 the effect of the number of indicators on the absolute bias of the latent indices is shown.

The absolute bias of indices is here found to be decreasing with an increasing number of manifest indicators. This supplement the results described in Sect. 11.2.1, where the absolute bias of indices was found to be decreasing with an increasing sample size. All in all the results with respect to the absolute bias of the latent variables can be summated to the rough rules of thumb shown in Table 11.4 (Kristensen and Eskildsen 2005b).

In our opinion these rough rules of thumb can serve as guidelines for any practitioner that wishes to increase the precision of the latent variables in a PLS structural equation model (Kristensen and Eskildsen 2005b).

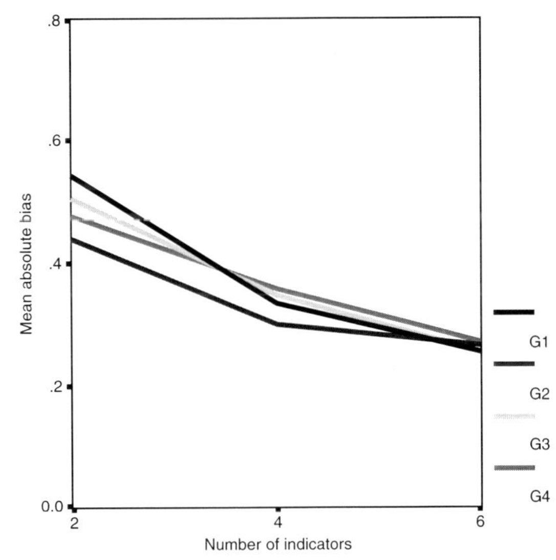

Fig. 11.15 Absolute bias of indices: The effect of the number of indicators

Table 11.4 Rules of thumb

Let σ be the standard deviation of the manifest variables, n the sample size, and p the number of indicators, then:
$\text{BIAS}(k\sigma, n, p) = k\text{BIAS}(\sigma, n, p)$
$\text{BIAS}(\sigma, kn, p) = \frac{1}{\sqrt{k}}\text{BIAS}(\sigma, n, p)$
$\text{BIAS}(\sigma, n, kp) = \frac{1}{k}\text{BIAS}(\sigma, n, p)$

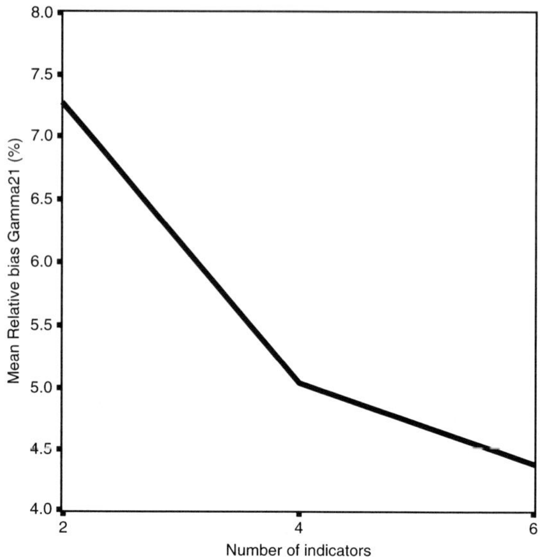

Fig. 11.16 Mean relative bias of γ_{21}: The effect of the number of indicators

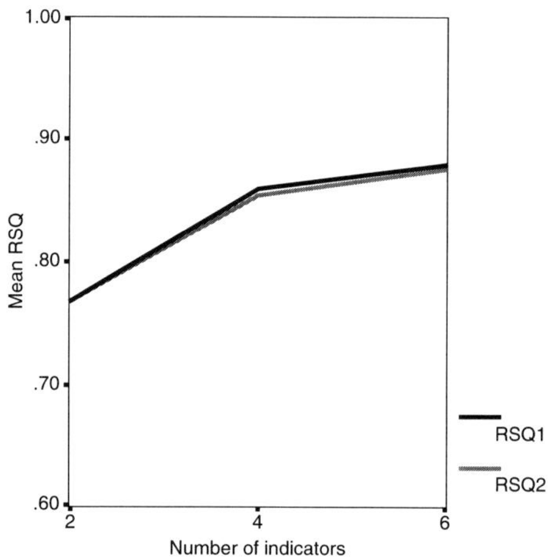

Fig. 11.17 Degree of explanation: The effect of the number of indicators

The number of indicators also has an impact on other aspects of a PLS model. In Fig. 11.16 the impact that the number of indicators has on the mean relative bias of γ_{21} is shown.

From Fig. 11.16 it appears as if the benefit of increasing the number of indicators decreases when the number of indicators reaches 4. The same goes for the degree of explanation (R^2) which is shown in Fig. 11.17.

The general recommendation for practitioners is therefore that a sample size of 250 and 4 indicators for each latent variable is sufficient to ensure a reasonable level of bias of the path coefficient in a PLS structural equation model (Kristensen and Eskildsen 2005b).

11.3.3 Question Reliability

Although many satisfaction studies are based on generic measurement system such as the EPSI Rating framework the reliability of the individual constructs differs across industries and furthermore some of the manifest variables are may need rethought in relation to specific industries. Table 11.5 highlights some of our experiences from different industries.

As Table 11.5 indicates that "comparison to ideal" from the satisfaction construct tend to have a much lower level than the other manifests in some industries. In general the level of "comparison to ideal" is lower than the other manifests from the satisfaction construct but it is not equally problematic in all branches of industry.

In order to evaluate the consequences of different levels of indicator reliability we can again turn towards the results from the simulation study described in Sect. 11.2.1. Figure 11.6 shows the effect of indicator reliability in the mean absolute bias of the constructs.

As Fig. 11.18 shows the mean absolute bias seems to increase linearly with decreasing indicator reliability. Decreasing indicator reliability also has an effect on the on the degree of explanation (R^2) which is evident from Fig. 11.19.

Not only does the R^2 decrease when the indicator reliability decreases it seems to do so disproportionally.

All in all indicator reliability has an enormous influence on all measured responses, i.e. bias, standard deviation and fit measures. Furthermore several cases of two-factor interaction with multicollinearity, sample size, and the number of indicators were found (Kristensen and Eskildsen 2005b).

Table 11.5 Reliability and choice of manifests

Industry	Experiences
Supermarkets	– In the satisfaction construct the "comparison to ideal" may cause a problem. Much lower level than the two other questions – The value for money indicator and the assortment indicator may cause a problem since they reflect the type of supermarket – The question about opening hours which is classified as belonging to the service block should possibly be re-classified
Banks	– In the satisfaction construct the "comparison to ideal" may cause a problem. Much lower level than the two other questions
Automobiles	– Reasonable reliability – No need for changes
Petrol Stations	– High reliability – No need for changes

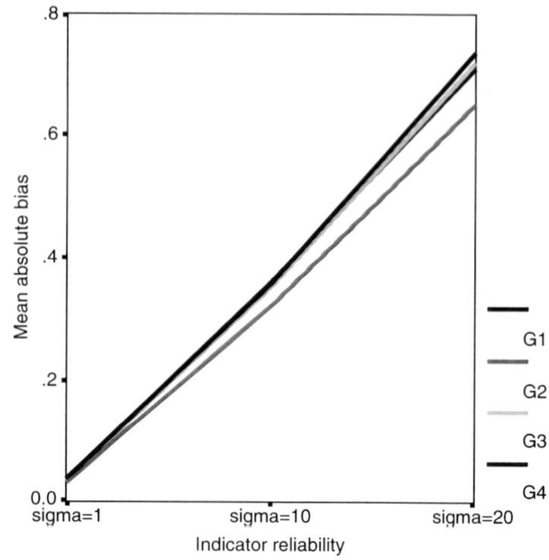

Fig. 11.18 Absolute bias of indices: The effect of indicator reliability

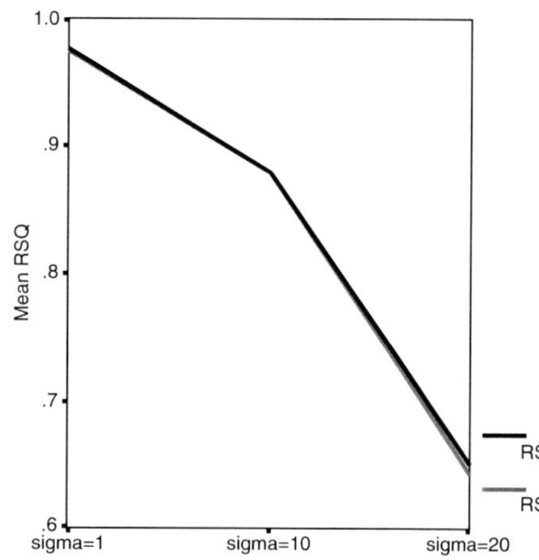

Fig. 11.19 Degree of explanation: The effect of indicator reliability

11.3.4 Multicollinearity

From our experience multicollinearity between the exogenous latent variables is rather high in most satisfaction studies. The degree of multicollinearity is however not the same for all industries and the sets of variables primarily affected is not the same across industries either. Here some of the general observations regarding the degree of multicollinearity in the EPSI Rating framework:

- Banks: Correlations between 0.54 (expectation and service) and 0.82 (product and service).
- Petrol stations: Correlations between 0.42 (expectations and service) and 0.69 (product and service).
- Automobiles: Correlations between 0.48 (expectations and service) and 0.85 (product and service).

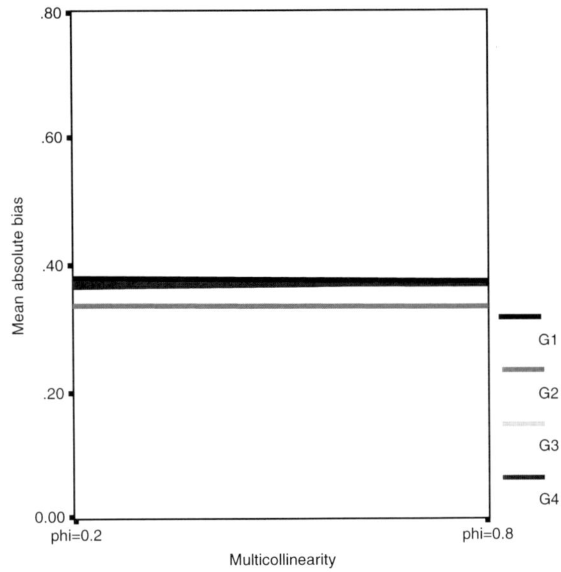

Fig. 11.20 Absolute bias of indices: The effect of multicollinearity

- Mobile telephones: Correlations between 0.44 (expectations and service) and 0.76 (product and service).
- Supermarkets: Correlations between 0.52 (expectations and image) and 0.71 (image and product).

To get an insight into the effect of multicollinearity we can draw on the results from the simulation study described in Sect. 11.2.1. In Fig. 11.20 the effect of multicollinearity on the absolute bias of indices is shown.

From this figure it is evident that multicollinearity between the latent variables is without importance for the estimated indices. It has however a significant but small impact on the bias of the path coefficients and a significant effect on all standard deviations (Kristensen and Eskildsen 2005b).

11.4 Considerations Concerning Specification of a Customer Satisfaction Model

Since its introduction in 1999 the EPSI Rating framework has become one of the most popular frameworks for modeling customer satisfaction in Europe. The primary result of interest for businesses is the level of the seven indices in the EPSI Rating framework and this has been the focus of quite a number of studies (Fornell 1992; Fornell et al. 1996; Eskildsen et al. 2000, 2003; Kristensen et al. 2001; Juhl et al. 2002; Selivanova et al. 2002; Kristensen and Westlund 2003).

There are however two different specifications of the EPSI Rating framework. One of them is used to estimate the EPSI Rating framework in Denmark (DKI) and the other is used to estimate the EPSI Rating framework in Sweden (SKI).

In 1989, Sweden became the first country in the world to establish a uniform, cross-company and cross-industry methodology for measuring customer satisfaction and customer loyalty. This national measurement instrument for customer satisfaction and customer loyalty is called the Swedish Customer Satisfaction Barometer (SCSB).

SCSB was adopted and adapted for use in the American Customer Satisfaction Index (ACSI) in 1994 and the successful experiences of the Swedish and American customer satisfaction indices inspired moves towards establishing a uniform methodology for measuring customer satisfaction and customer loyalty in Europe.

Based on the recommendations from a feasibility study (Grigoroudis and Siskos 2004) and by the work provided by the ECSI (European Customer Satisfaction Index) Technical Committee (ECSI Technical Committee 1998) the EPSI Rating framework for measuring customer satisfaction and customer loyalty was designed. A pilot study was conducted in 1999 and measurements have so far been imple mented in a small set of industries in a sample of the European countries.

As previously stated there are now two different specifications of the EPSI Rating framework. One of them is used to estimate the EPSI Rating framework in Denmark (DKI) and the other is used to estimate the EPSI Rating framework in Sweden (SKI). The two different specifications are shown in the figure below.

The experiences from the EPSI Rating initiative have shown that both model specifications are associated with both advantages as well as disadvantages.

The SKI specification only has one exogenous variable. This means that the only area in which a company is with respect to its image since the model captures the structural part of the variation in the remaining variables according to predictor specification (Wold 1980, 1985; Fornell and Cha 1994). The DKI specification on the other hand has the advantage of having four exogenous variables.

The DKI specification is however rather unstable meaning that quite a lot of the inner relationships is insignificant in the individual analyses. This is typically not a problem to the same degree with the SKI specification (Eskildsen and Kristensen 2005).

The ideal specification would thus be one that could combine the advantages of the DKI and the SKI specification.

In a recent study it has been argued that the image variable in the EPSI framework should be modeled as an outcome of customer satisfaction (Johnson et al. 2001). This argument is based on the notion that corporate image will have been affected by the customers' most recent experiences but this can be said for all the other variables in the EPSI framework as well. If we are to keep a non-recursive structure the only viable option is to model image as a mediating variable between the actual consumption experience and the cumulative post-consumption evaluation of value and satisfaction (Fig. 11.21).

Furthermore the study was argued that the image variable could be used instead of measuring expectations. The argument is that since the pre purchase expectations are collected post-purchase what is really being collected is the customers perception of corporate image (Johnson et al. 2001). The problem with this argument is that the image construct in the EPSI Rating framework is much broader than the expectation construct which only encompasses expectations with products and services and

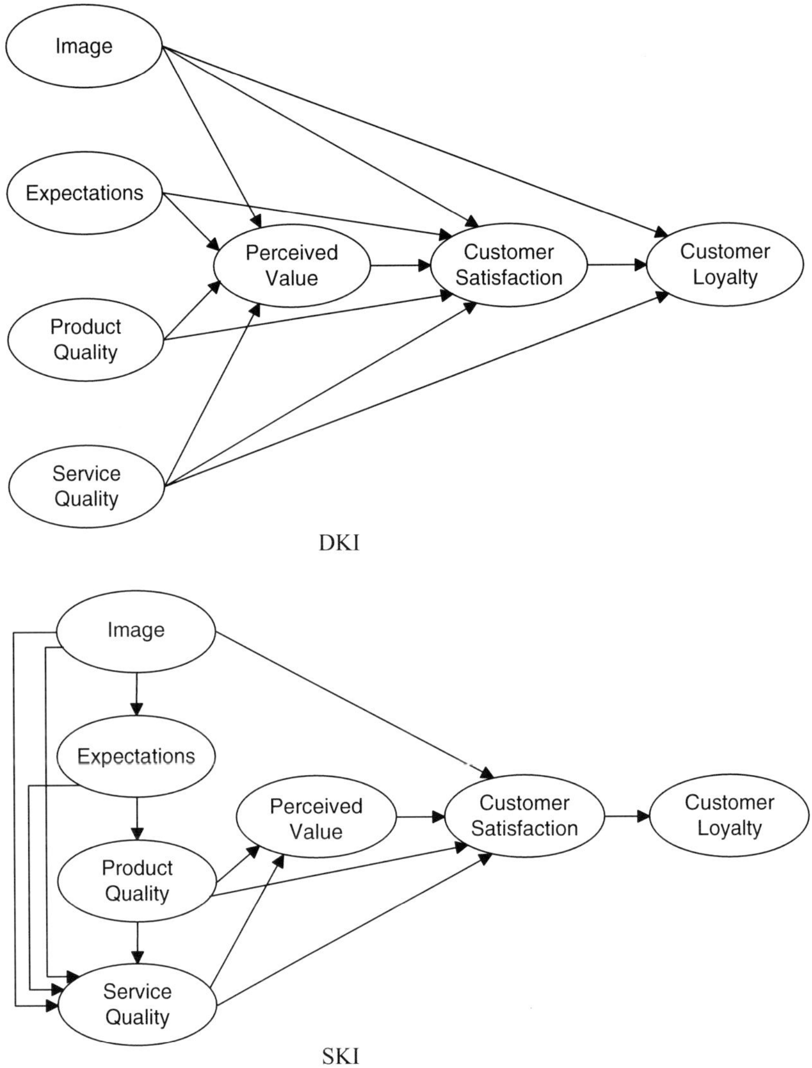

Fig. 11.21 The two different model specifications

nothing more. It would therefore probably make more sense to keep the expectation variable as an antecedent to value and corporate image.

Previous studies has shown that there is a strong relationship between product /service quality and value and satisfaction (Eskildsen et al.2004a). If we are to model image as a mediating variable between the actual consumption experience and the cumulative post-consumption evaluation of value and satisfaction then product /service quality must affect image as well. The authors' proposal for a common EPSI Rating specification is shown in Fig. 11.22 (Eskildsen and Kristensen 2005).

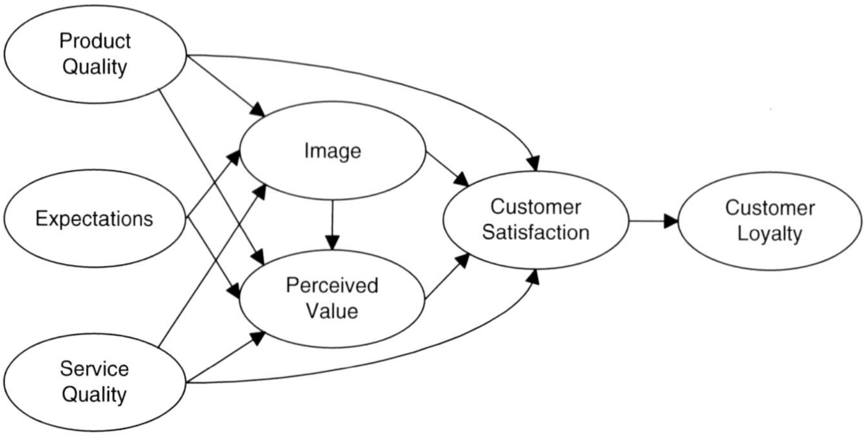

Fig. 11.22 New model specification

In order to test the consequences of different model specifications an empirical study was set up and the data used in the analysis comes from the 2004 EPSI Rating study in Denmark in the following industries:

- mobile phone operators
- supermarkets
- banks
- mortgage providers

All in all there are 15 companies included in the study with approx. 250 responses for each company. For each company three different model specifications have been estimated using PLS. The latent variables are identical in each of the three specifications and the relationship between latent and manifest variables are reflective by nature.

In order to assess the standard deviation of the outer weights bootstrapping was applied and 1,000 replications were conducted for each company in the analysis.

The overall results from this empirical study are given in Table 11.6 below. As the table indicates there are no major differences between the three model specifications when it comes to the customer satisfaction index as well as the models' ability to explain the customer satisfaction construct.

Table 11.6 Results for the 3 model specifications

	DKI	SKI	New Model
Index – customer satisfaction	72.02	71.78	71.66
R^2 – customer satisfaction	0.708	0.705	0.705
R^2 – average	0.661	0.585	0.639
GoF	0.688	0.647	0.676

Table 11.7 Results for the outer weights (bootstrapped)

Industry Statement	Average outer weight	Average maximum difference	Average standard deviation	Maximum ratio d/std.
Supermarkets				
Overall satisfaction	0.4018	0.0043	0.0258	0.3553
Fulfillment of expectations	0.4040	0.0041	0.0254	0.3818
Comparison to ideal	0.4299	0.0034	0.0265	0.2560
Banking				
Overall satisfaction	0.3813	0.0027	0.0155	0.4491
Fulfillment of expectations	0.3684	0.0022	0.0143	0.3255
Comparison to ideal	0.3902	0.0045	0.0159	0.6855
Mobile telecom				
Overall satisfaction	0.3708	0.0046	0.0167	0.5659
Fulfillment of expectations	0.3940	0.0036	0.0171	0.3078
Comparison to ideal	0.3966	0.0031	0.0179	0.3607
Morgage providers				
Overall satisfaction	0.3869	0.0038	0.0184	0.2452
Fulfillment of expectations	0.3783	0.0026	0.0195	0.4730
Comparison to ideal	0.4107	0.0034	0.0198	0.2370

Furthermore the new model specification seems to capture the ability of the DKI specification in relation to the Tennenhaus/Vinzi overall goodness of fit index (GoF) as well as the average degree of explanation in the model (Eskildsen and Kristensen 2005).

Although there are no apparent differences with respect to the reported indices the outer weights for the three manifest variables may very well be different in the three specifications. Whether or not such differences were found in the study case can be seen from Table 11.7.

As Table 11.7 indicates the average maximum differences among the outer weights within each industry are very small. The maximum ratio between the differences and the standard deviation were found to be only 0.6855 and there are thus no significant differences between the outer weights from the three different model specifications.

Another important aspect in this context is how the results of the companies will change if the model specification is altered. From Table 11.6 it is evident that the customer satisfaction index is virtually the same no matter what specification is applied.

This is however not the case when it comes to the relative overall effect of the image construct on customer satisfaction. This is shown in Fig. 11.23 below.

From this figure it is clear that the relative overall effect of the image construct is reduced when it becomes an endogenous latent variable. This is something that needs to be considered before changing the way the EPSI Rating framework is specified (Eskildsen and Kristensen 2005).

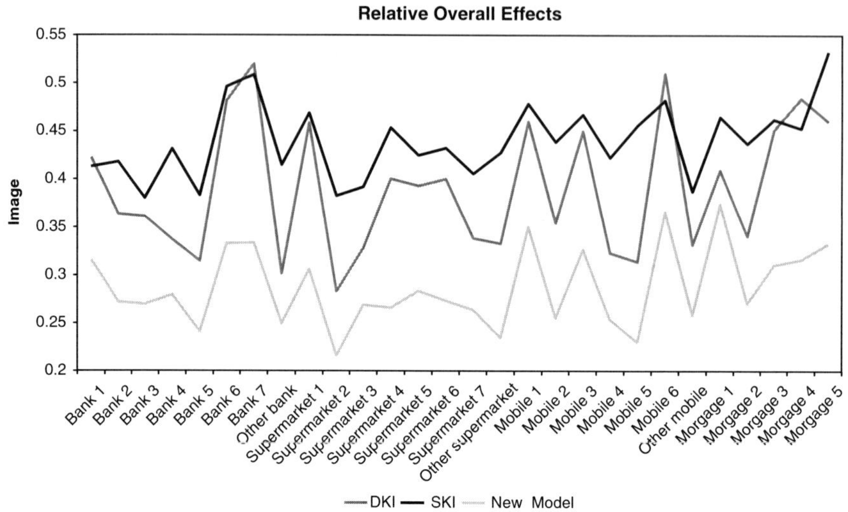

Fig. 11.23 Relative overall effects for Image

In the proposed model the relative overall effect of service and product quality is higher than in the DKI and SKI specifications. This means that the focus of companies' will shift from image to the product and services offered to the customers.

11.5 Concluding Remarks

The five studies at the Centre for Corporate Performance have ended up in a number of conclusions and recommendations that we sum up here. We hope that they may be of help to researchers doing satisfaction studies in the future:

- Sampling Considerations
  - Be careful when formulating the intro text. Choice of intro text may influence the index levels.
  - Be careful when introducing the interviewer to the respondent. If the interviewer is introduced as coming from the company itself it may increase index estimates.
  - The order of questions on the questionnaire has a major influence on estimation results. Both indices and path coefficients. We recommend that endogenous variables are always put first in the questionnaire. Both in order to establish a standard, but also in order to get "top of mind" evaluations of the endogenous variables.

- Data Considerations
  - A sample size of around 250 is in general sufficient for overall (national) customer satisfaction studies.
  - A 10-point scale is preferable to a 5-point scale. A 10-point scale produces more accurate results, but the two scales are in general comparable.
  - The EM algorithm is preferable when correcting for missing values. We cannot recommend mean substitution.
  - The number of indicators for each latent variable should be around 4.
  - Indicator reliability is very important. Hence the researcher should look for questions with as low a standard deviation as possible.
  - In general multicollinearity does not affect the indices. It has a limited effect on the bias of the path coefficients and effect on all standard deviations. We have experienced that when multicollinearity is severe PLS Regression is a good correction technique. We have in a number of empirical studies seen that it outperforms both PC Regression and Ridge Regression. In the near future we are going to analyze this further in a simulation study.
- Model Specification
  - A new EPSI specification may be required. We suggest that Image becomes endogenous. This does not affect the index estimates but it will of course change the overall importance of the remaining exogenous variables.

In general PLS is very suitable for doing satisfaction studies. It is very robust and is thus well suited for the type of data that we experience in satisfaction studies. We believe that paying attention to the practical aspects mentioned in this article will further contribute to the applicability of the technique.

References

Afifi, A. A., & Elashoff, R. M. (1966). Missing observations in multivariate statistics I. Review of the literature. *Journal of the American Statistical Association, 61*, 595–604.

Allison, P. D. (2002). *Missing data.* Thousand Oaks, CA: Sage.

Auh, S., Salisbury, L. C., & Johnson, M. D. (2003). Order effects in customer satisfaction modelling. *Journal of Marketing Management, 19*, 379–400.

Baumgartner, H., & Steenkamp, J.-B. E. M. (2001). Response styles in marketing research: a cross-national investigation. *Journal of Marketing Research, 38*, 143–156.

Beale, E. M. L., & Little, R. J. A. (1975). Missing values in multivariate statistical analysis. *Journal of the Royal Statistical Society, 61*, 595–604.

Bello, A. L. (1993). Choosing among imputation techniques for incomplete multivariate analysis. *Communications in Statistics, 22*, 853–877.

Bello, A. L. (1995). Imputation techniques in regression analysis: Looking closely at their implementation. *Computational Statistics & Data Analysis, 20*, 45–57.

Bending, A. W. (1954). Reliability and the number of rating scale categories. *Journal of Applied Psychology, 61*, 38–40.

Brown, R. L. (1994). Efficacy of the indirect approach for estimating sructural equation models with missing data: A comparison of five methods. *Structural Equation Modeling, 1*, 287–316.

Cassel, C., Hackl, P., & Westlund, A. H. (1999). Robustness of partial least-squares method for estimating latent variable quality structures. *Journal of Applied Statistics, 26*(4), 435–446.

Cox, E. P. (1980). The optimal number of response alternatives for a scale. *Journal of Marketing Research, 17*(4), 407–422.

ECSI Technical Committee. (1998). European customer satisfaction index: foundation and structure for harmonized national pilot projects. *Report, October.*

Eskildsen, J. K., & Kristensen, K. (2005). The consequences of different model specifications when estimating national customer satisfaction indices using PLS. In *PLS '05*, Barcelona.

Eskildsen, J. K., Kristensen, K., Juhl, H. J., & Østergaard, P. (2004a). The drivers of customer satisfaction & loyalty. *Total Quality Management & Business Excellence, 15*(5–6), 859–868.

Eskildsen, J. K., Kristensen, K., Vestergaard, T., & Schmidt-Laugesen (2004b). *Corporate reputation as a strategic performance measure.* Edinburgh, Scotland: Performance Measurement and Management: Public and Private.

Eskildsen, J. K., Kristensen, K., & Westlund, A. H. (2003). The predictive power of intangibles. *Measuring Business Excellence, 7*(2), 46–54.

Eskildsen, J. K., Martensen, A., Grønholdt, L., & Kristensen, K. (2000). Benchmarking student satisfaction in higher education based on the ECSI methodology. *Sinergie - Rapporti di ricerca, 9*(18), 385–402.

Ewing, M. T., Caruana, A., & Zinkhan, G. M. (2002). On the cross-national generalisability and equivalence of advertising response scales developed in the USA. *International Journal of Advertising, 21*, 323–343.

Flamer, S. (1983). Assessment of the multitrait-multimethod matrix validity of Likert scales via confirmatory factor analysis. *Multivariate Bevavioral Research, 18*, 275–308.

Fornell, C. (1992). A national customer satifaction barometer: the Swedish experience. *Journal of Marketing, 56*(1), 6–21.

Fornell, C., & Cha, J. (1994). Partial least squares. In: R. P. Bagozzi (Ed.), *Advanced methods of marketing research.* Cambridge: Blackwell.

Fornell, C., Johnson, M. D., Anderson, E. W., Cha, J., & Bryant, B. E. (1996). The American customer satisfaction index: nature, purpose and findings. *Journal of Marketing, 60*(4), 7–18.

Givon, M. M., & Shapira, Z. (1984). Response to rating scales: a theoretical model and its application to the number of categories problem. *Journal of Marketing Research, 21*, 410–419.

Green, P. E., & Rao, V. (1970). Rating scales and information recovery – how many scales and response categories to use? *Journal of Marketing, 34*(3), 33–39.

Greenless, J. S., Reece, W. S., & Zieschang, K. D. (1982). Imputation of missing values when the probability of response depends on the variable being imputed. *Journal of the American Statistical Association, 77*(378), 251–261.

Grigoroudis, E., & Siskos, Y. (2004). A survey of customer satisfaction barometers: some results from the transportation-communications sector. *European Journal of Operational Research, 152*, 334–353.

Guy, R. F., & Norvell, M. (1977). The neutral point on a Likert Scale. *The Journal of Psychology, 95*, 199–204.

Johnson, M. D., Gustafsson, A., Andreassen, T. W., Lervik, L., & Cha, J. (2001). The evolution and future of national customer satisfaction index models. *Journal of Economic Psychology, 22*, 217–245.

Juhl, H. J., Kristensen, K., & Østergaard, P. (2002). Customer satisfaction in European food retailing. *Journal of Retailing and Consumer Services, 9*, 327–334.

Kristensen, K., & Eskildsen, J. K. (2005a). Missing values, partial least squares and the estimation of customer satisfaction. *PLS '05*, Barcelona.

Kristensen, K., & Eskildsen, J. K. (2005b). PLS structural equation modelling for customer satisfaction measurement: some empirical and theoretical results. In: F. W. Bliemel, A. Eggert, G. Fassott, & J. Henseler (Eds.), *Handbuch PLS-Pfadmodellierung - Methode, Anwendung, Praxisbeispiele* (pp. 5–24), Schäffer-Poeschel.

Kristensen, K., Eskildsen, J. K., & Juhl, H. J. (2001). Benchmarking as a strategic tool. In: M. Christensen (Ed.), *Benchmarking across boundaries (in danish)* (pp. 1–19). Copenhagen: Jurist og Økonomforbundets forlag.

Kristensen, K., & Westlund, A. H. (2003). Valid and reliable measurements for sustainable non-financial reporting. *Total Quality Management, 14*(2), 161–170.

Matell, M. G., & Jacoby, J. (1972). Is there an optimal number of alternatives for Likert-scale items? *Journal of Applied Psychology, 60*, 10–13.

Menezes, D., & Elbert, N. F. (1979). Alternative semantic scaling formats for measuring store image: An evaluation. *Journal of Marketing Research, 16*, 80–87.

O'Muircheartaigh, C., Krosnick, J. A., & Helic, A. (2001). Middle alternatives, acquiescence, and the quality of questionnaire data. The Harris School University of Chicago: Paper Series 10.3. USA.

Ofir, C., Reddy, S. K., & Bechtel, G. G. (1987). Are semantic response scales equivalent? *Multivariate Bevavioral Research, 22*, 21–38.

Olinsky, A., Chen, S., & Harlow, L. (2003). The comparative efficacy of imputation methods for missing data in structural equation modelling. *European Journal of Operational Research, 151*, 53–79.

Preston, C. C., & Colman, A. M. (2000). Optimal number of response categories in rating scales: reliability, validity, discriminating power, and respondent preferences. *Acta Psychologica, 104*, 1–15.

Ramsay, J. O. (1973). The effect of number of categories in rating scales on precision of estimation of scale values. *Psychometrika, 38*, 513–532.

Schafer, J. L. (1997). *Analysis of incomplete multivariate data.* Boca Raton, FL: Chapman & Hall.

Selivanova, I., Hallissey, A., Letsios, A., & Eklöf, J. (2002). The EPSI rating initiative. *European Quality, 9*(2), 10–25.

Ueltschy, L. C., Laroche, M., Tamilia, R. D., & Yannopoulos, P. (2004). Cross-cultural invariance of measures of satisfaction and service quality. *Journal of Business Research, 57*, 901–912.

von Hippel, P. T. (2004). Biases in SPSS 12.0 missing value analysis. *The American Statistician, 58*(2), 160–164.

Wold, H. (1980). Model construction and evaluation when theoretial knowledge is scarce: theory and application of partial least squares. In: J. Kmenta, & J. B. Ramsey (Eds.), *Evaluation of econometric models* (pp. 47–74). New York: Academic.

Wold, H. (1985). Partial least squares. In: S. Kotz, & N. Johnson (Eds.), *Encyclopedia of statistical sciences* (pp. 581–591). New York: Wiley.

Chapter 12
A Case Study of a Customer Satisfaction Problem: Bootstrap and Imputation Techniques

Clara Cordeiro, Alexandra Machás, and Maria Manuela Neves

Abstract Bootstrap is a resampling technique proposed by Efron (The Annals of Statistics 7:1–26, 1979). It has been used in many fields, but in case of missing data studies one can find only a few references.

Most studies in marketing research are based on questionnaires, that, for several reasons present missing responses. The missing data problem is a common issue in market research. Here, a customer satisfaction model following the ACSI barometer from Fornell (Journal of Marketing 60(4):7–18, 1996; The American customer satisfaction index: methodology report. Michigan: University of Michigan Business School, 1998) will be considered. Sometimes not all customers experience all services or products. Therefore, we may have to deal with missing data, taking the risk of reaching non-significant impacts of these drivers on Customer Satisfaction and resulting in inaccurate inferences. To estimate the main drivers of Customer Satisfaction, Structural Equation Models methodology is applied (Peters and Enders, Journal of Targeting Measurement and Analysis for Marketing 11(1):81–95, 2002).

For a case study in mobile telecommunications several missing data imputation techniques were reviewed and used to complete the data set. Bootstrap methodology was also considered jointly with imputation techniques to complete the data set. Finally, using Partial Least Squares (PLS) algorithm we could compare the above procedures. It suggests that bootstrapping before imputation can be a promising idea.

C. Cordeiro
Department of Mathematics, FCT, University of Algarve, Campus de Gambelas, 8005-139 Faro, Portugal
e-mail: ccordei@ualg.pt

A. Machás
Polytechnic Institute of Lisbon, Escola Superior de Comunicação Social Campus de Benfica do IPL, 1549-014 Lisboa, Portugal
e-mail: amachas@escs.ipl.pt

M. M. Neves
Department of Mathematics, Instituto Superior de Agronomia, Technical University of Lisbon (TULisbon), Tapada da Ajuda, 1349-017 Lisboa, Portugal
e-mail: manela@isa.utl.pt

V. Esposito Vinzi et al. (eds.), *Handbook of Partial Least Squares*, Springer Handbooks of Computational Statistics, DOI 10.1007/978-3-540-32827-8_13,

12.1 Introduction

From the last years, organizations have given a special attention to monitor satisfaction scores. A wide spread of Customer Satisfaction Models (CSM) (Smith and Smith-Langfield 2004) has been developed to produce strategic insights using several methodologies to collect data and to estimate adequate models. Organizations gain conscious of the correlation between customer satisfaction and loyalty to their financial results and market share. In our days, the best practices in business organizations, taking globalization and high competition environment, goes beyond the excellence of quality of products and services to customers. Strategic marketing planning includes targets for satisfaction indexes and customer retention (Chan et al. 2003).

Statisticians have joined efforts with marketers to develop advanced models of satisfaction that include the evaluation of all touch points that each customer experiences when establishes an interaction with the organization.

For all these reasons, internal and personalized models have been implemented and gained a high magnitude inside the organization. For the referred importance, there is a huge need of obtaining a major validity and reliability, in order to reach higher credibility and confidence in CSM.

Bring up better methods to estimate Customer Satisfaction (CS) is one of the latest issues. Structural Equation Models (SEM) provide a combination of confirmatory factorial analysis and regression models, to be applied to data obtained from a CS survey when a theoretical model can not be considered.

Many articles have already compared the robustness of Maximum Likelihood methods (like LISREL or AMOS) with Partial Least Squares (PLS) (Vinzi et al. 2003), i.e., should we estimate SEM based on a covariance structure or on a variance structure of data (Dijkstra 2000; Tenenhaus 2003; Loughlin and Coenders 2004)?

The data quality and the SEM design are now more complex and even more crucial. Only ensuring the good quality of data, SEM and CSM can be considered a true strategic instrument for the organizations and the base for a marketing planning.

The main objective of this article is to bring into discussion a situation that occurs in survey studies and can affect the quality of estimators and the validation of the model- *the missing data problem*. Missing values emerge from the non-response on our customer satisfaction questionnaire. Sometimes there are some difficulties to answer certain top of mind questions or perceptions. Some questions have a higher probability for non-response due to a certain social-demographic profile of the individual or other cultural aspects besides the non-experience in certain issues asked in the questionnaire.

Most data analysis procedure were not designed for dealing with missing data. Various imputation methods are used in practice. In this study we discuss and compare several methods presented in the literature to treat missing data in market research, see Sect. 12.2.

The bootstrap methodology has been considered by Efron (1994) and Shao and Sitter (1996) for missing data. Here a different approach is considered: bootstrap followed by an imputation technique is applied to a mobile telecommunication case study.

12.2 Missing Data and Imputation Techniques

The problem of missing data or incomplete data is frequently found in many data bases. The missing values arise due to a lack of response or missing response. Such as *no value*, *not admissible* and so on. Many studies in marketing research are based on questionnaires. Some practitioners classify the item *non-response* in the questionnaire as a missing value. The non-response pattern of these missing values is a haphazard one and is said to be *missing completely at random* (MCAR) if the probabilities of observing some components and unobserving the others do not depend neither on the observed data nor on the unobserved data. That is, if the probability that the observations are missing is not related to either Y_0 or Y_m, where Y_0 represents the observed values in Y(the input data matrix) and Y_m represents the missing values in Y. So,

$$P(Y|Y_0) = P(Y|Y_m).$$

Missing data creates difficulties in statistical analysis because the techniques used are not designed for them, therefore, missing data reduces statistical power. This implies that, estimates calculated from incomplete data set can be biased. However, many statistical studies contain data structures with partially observed data. The sources of missingness of some of the variables may vary from the totally randomness to the strong dependence on the true values of the variables.

In order to solve the problem of missing data have been proposed some missing data techniques or data imputation algorithms for transforming the incomplete data to a complete data set. These algorithms fill out the missing data values, by examining the range of probable values for each variable and calculating many possible values randomly. So, using these methods we end up with a credible data set and the results often produce more accurate estimates and/or can be dealt with by the usual computer programs.

Listwise and Pairwise are methods that can be used with any statistical analysis. They are quick and easy to implement. These methods delete the cases or variables with missing data. They are not considered in this work because a large amount of data is lost, the sample size decreases, and thus, reduces statistical power. Our preference is for methods that do not discard information and allow to complete the data set.

Rubin (1976) was the first to describe the mechanism that result in missing observations and to handle with incomplete data. Rubin proposed drawing multiple random imputations of the missing data rather than a single best-fit imputation. Variability of results between the randomly imputed data sets can then be used to assess the true accuracy of an estimate.

Several imputation methods are presented in literature Roth (1994) and Schafer and Graham (2002), some of them are summarized above:

Estimation Methods: These methods use the information based on non-missing data to replace the missing responses. The most used is the Mean Imputation (Mean). This method uses the available observations and fill the missing values with the calculated mean of those observations. It is simple to implement but underestimates the variance and inflates the adjusted R^2 value. It distorts the underlying distribution of the sample and it only works for quantitative features (for qualitative features, the mode would be a good alternative). Besides this, it is a good option when the data are both missing completely at random and normally distributed.

Nearest Neighbor Method: The missing values are replaced by the values of a Nearest Neighbor. The method tries to find the complete cases most similar to the incomplete case. The similarity can be measured in many ways, such as k-Nearest Neighbor method (KNN), SRPI (simple response pattern imputation), correlation (KNNc), etc. The k-Nearest Neighbor works by finding the k complete cases most similar/nearest to the incomplete case where the similarity is measured by some distance parameter.

Hot-Deck Imputation (HDI): This strategy has become popular in survey research. The principle is that researchers should replace a missing value with a value selected with replacement from the sample observed. Hence, the data set that will yield the imputed score is termed *hot* because is currently in use by the computer. Proponents argue that hot-deck imputation tends to increase accuracy over simple techniques strategies because missing data values are replaced by realistic values. The values sampled preserve the distributional data characteristics opposite to the estimative method. This approach is particularly helpful when data are missing in certain patterns.

The disadvantages of Hot-Deck procedure is: there is little theoretical or empirical work to determine their accuracy; and the number of classification variables may become unmanageable in large surveys.

Multiple Imputation (MI): The statistical method most used to deal with missing data is the *multiple imputation* (Rubin 1976). This technique replaces each missing value with a pointer of m values. The m values ($3 \leq m \leq 10$) come from m possible scenarios or imputation procedures based either on the observed information or on historical or posterior follow-up registers.

Multiple Imputation has the advantage of using complete-data methodologies for the analysis and allows to reproduce the uncertainty due to the sampling variability assuming that the reasons for non-response are known as well as the variability due to the uncertainty about the reasons of non-response.

A big disadvantage is that this method is difficult to implement and is computationally demanding.

Maximum Likelihood (ML): It is a method that estimates the parameters of a model by the value that maximizes the likelihood function. This method begins with the likelihood function for the data, witch is the probability of obtaining that particular sample of the data given the chosen probability model.

Expectation-Maximization (EM): When the full-data model and the ignorability assumption are correct, all information about the parameters is contained in the observed-data likelihood, but this expression can be complicate and special tools are required. Dempster et al. (1977) proposed the Expectation-Maximization (EM) algorithm. This is a general method for obtaining the maximum likelihood estimates of the parameters in missing data. It can be summarized in four steps:

1. Replace missing values by estimated values.
2. Estimate parameters of the variables distribution.
3. Re-estimate missing values assuming that the new parameters estimates are true.
4. Re-estimate parameters in an iterative procedure until convergence.

This converge could be very slow if there is a huge amount of missing data.

The last three methods assume a model for the missing data and that inferences are based on the likelihood function. The parameters are estimated using parametric approaches like the maximum likelihood and they are computationally demanding.

The importance of choosing the best imputation method to handle missing data refers basically to the possible solutions that a researcher has at his disposal. So, the consequences for not choosing the best imputation method are reflected in the quality of the estimators and the quality of the model.

12.3 Bootstrap Methodology

Resampling methods are an important tool in statistical inference and they became more popular since the advance of computers. Bootstrap is one of these resampling methods. The bootstrap is a computer-intensive method that provides answers to a large class of statistical inference problems. This methodology was introduced by Efron (1979) with the objective of providing solutions in situations where the traditional methods failed. The bootstrap involves the drawing of samples with replacement from the original data or from an appropriate model, usually for purposes of making inferences about how samples behave when draw from a population (see Efron and Tibshirani 1993; Mooney and Duval 1993; Davison and Hinkley 1997 for more details).

The principal reference for resampling in missing data problems is given by Efron (1994). Efron discusses the relationship between bootstrap and the theory of multiple imputation (Rubin 1976) and presents computationally efficient ways of executing them. In this article, Efron shows some differences and similarities between bootstrap and the imputation approaches.

The missing data techniques described in the previous section, were used and compared via empirical measures when a bootstrap procedure was previously realized. The whole computation procedures were implemented using *R software* and their packages.

In bootstrap application we considered each sampled unit as a vector of responses and the *procedure* considered is the following:

Step 1: Matrix rows are resampled with replacement
Step 2: A bootstrap matrix is obtained
Step 3: An imputation technique is used (using each method described in Sect. 12.2)

Step 4: For the complete matrix and for each variable the mean is computed
Step 5: Go to step 1

The process is repeated r times and finally a new mean using the values of step 4 is obtained and the initial matrix is completed with the values.

Performing a missing data technique, for example the *Hot-Deck Imputation*, the bootstrap can be carried out without auxiliary information. So the HDI imputes Y_m by an i.i.d. sample select from Y_o with replacement. The key in applying the bootstrap to complex survey data with imputed non-respondents is that the bootstrap data set should be imputed in the same way as the original data set.

12.4 Case Study

The data come from a market survey conducted for the mobile telecommunication sector in Portugal. About 250 mobile telecom customers were interviewed in 2003, for each of the 3 mobile companies. Criterions for respondents selection, sample size and queries that integrated the CSM followed the ACSI model. Image was also included as in the ECSI/EPSI model (Fornell et al. 1996; 1998). The ACSI model assesses customer satisfaction and loyalty. Perceived value, perceived quality, expectations and image represent the antecedents of customer satisfaction.

Structural Equation Models (SEM) methodology is applied to estimate the main drivers of Customer Satisfaction (CS) in a mobile telecommunication sector. CS is considered to be explained by a set of latent variables reflecting dimensions of brand perception and real experience with products and services provided by the mobile operator. Although we are dealing with latent variables that reflect measurement indicators, this does not mean that all customer will experience all services and all products. Therefore, we may have to deal with missing data, taking the risk of reaching non significant impacts of these drivers on CS or resulting in inaccurate inferences.

Partial Least Squares (PLS) algorithm (Chin and Todd 1995; Chin 1998) used to estimated SEM model, lead to the best estimators when we are dealing with small samples and free distribution, compared with other methodologies based on Maximum Likelihood. PLS was used to estimate the Customer Satisfaction Model (CSM) for mobile telecommunication sector, which includes an inner structural model and an outer measurement model.

But, sometimes there is a problem with PLS application. This algorithm does not work in the presence of missing data, so in this case the data must be fulfill whenever a missing value is found. The Mean Imputation is the *ad hoc* procedure adopted for the ECSI/EPSI model, but there are many others missing data imputation techniques (see Sect. 12.2).

The bootstrap resampling method is considered to give an answer to the problem of small sample size in SEM, in case of missing observations. Bootstrap and the imputation methods are going to work together in an extensive computer performance.

In the case study, the missing data imputation is done separately for each of three mobile brands considered and for each of the scenarios described above.

Scenario 1: Let $X = (x_{ij})$ denote a $(n \times k)$ rectangular data matrix, the input data matrix, were n is the number of individuals and k the number of variables in the questionnaire. The X matrix have a 10% non-response rate.

Scenario 2: Now, consider another matrix $M = (m_{ij})$, $(n \times k)$, having a MCAR pattern defined as $m_{ij} = 0$ if it is missing, and $m_{ij} = 1$ otherwise. A new matrix $Y = (y_{ij})$, $n \times k$, is constructed with X data, that is, if $m_{ij} = 0$ then $y_{ij} = NA$, otherwise, if $m_{ij} = 1$ then $y_{ij} = x_{ij}$. In this case, the non-response rate is must higher (50%) and Y is the input matrix.

The missing data techniques described in the Sect. 12.2, were used for the two scenarios and also used after a bootstrap procedure. Using the results, we intend to obtain estimates for the missing data in order to perform the PLS methodology, that is know as an efficient method, reducing the biases of estimators (Cassel et al. 2000).

The bootstrap procedure described in Sect. 12.3 was repeated r = 5,000 times and at the end, the missing values in matrix X or Y, are replaced with new estimates. Then, the PLS estimation is performed.

The PLS output gives many statistical tools for evaluating the model adjustment such as: communalities, redundancies, residual variance, percentage of variance explained (R^2), etc. The results for these last two statistical measures denoted by R_S^2 and Var_S are the following:

At a Customer Satisfaction level and for *scenario 1*, values of R_S^2 and Var_S are similar for all missing data methods (Tables 12.1 and 12.2). So for this case, it seems that the use of the bootstrap does not bring up any improvement in this study.

Table 12.1 R_S^2 in scenario 1

Methods	Without bootstrap	With bootstrap
Mean	59.8	59.8
ML	59.8	60.2
MI	62.8	59.7
KNN	63.6	59.8
KNNc	62.6	59.8
HDI	56.7	59.8

Table 12.2 Av. Residual Var_S in scenario 1

Methods	Without bootstrap	With bootstrap
Mean	0.76	0.76
ML	0.75	0.75
MI	0.73	0.76
KNN	0.72	0.76
KNNc	0.74	0.75
HDI	0.84	0.76

Table 12.3 R_S^2 in scenario 2

Methods	Without bootstrap	With bootstrap
Mean	35.6	35.7
ML	36.2	34.7
MI	60.1	36.2
KNN	26.3	35.6
KNNc	77.5	35.9
HDI	21.3	35.6

Table 12.4 Av. Residual Var_S in scenario 2

Methods	Without bootstrap	With bootstrap
Mean	0.66	0.68
ML	0.67	0.68
MI	0.58	0.68
KNN	0.81	0.66
KNNc	0.63	0.66
HDI	0.55	0.66

For the worst case in analysis (50% non-response), Tables 12.3 and 12.4, there are some differences among the missing data techniques. The bootstrap used with Multiple Imputation and the k-Nearest Neighbor, based in the correlation (KNNc), does not give good results. On the other hand, the k-Nearest Neighbor and Hot-Deck Imputation present better results with bootstrap application. Indeed, the R_S^2 is higher and the residual variance is lower compared with the other imputation methods.

No one of these methods are a perfect solution for missing data problems. But, to use the bootstrap, can be an idea to improve the performance of these methods. Some research is in progress, namely some simulations covering several percentages of missing values are now being realized.

12.5 Conclusion

For estimating a standard CSM with missing data using PLS, the Mean Imputation is the *ad hoc* procedure adopted for ECSI/EPSI model. Here is proposed to use the bootstrap methodology followed by several imputation techniques and results are compared on base of two empirical measures.

The simplest nonparametric bootstrap approach is used. The rows in the data are resampled with replacement from the original data. A nonparametric bootstrap matrix is obtained and an estimate is calculated for each variable in the study.

For a real data matrix with a 10% missing data rate several missing data imputation methods has been performed combined with resampling techniques. Another missing data rate was simulated and the previous methods were used.

It seems that combining data imputation methods and resampling methods may conduct to improve the results. An extensive simulation study is in progress.

References

Cassel, C., Hackl, P., & Westhund, A. (2000). On measurement of tangible Assets, a study of robusteness of Partial Least Squares. *Total Quality Management, 11*, 897–907.

Chan, L., Hui, Y., Lo, H., & Tse, S., et al. (2003). Consumer satisfaction index: New practice and findings. *European Journal of Marketing, 37*, 872–898.

Chin, W. (1998). In G. Marcoulides (Ed.), *The partial least squares approach for structural equation modeling*. Mahwah: Lawrence Erlbaum.

Chin, W., & Todd, P. (1995). On the use, usefulness, and ease of use of structural equation modeling in MIS research: a note of caution. *MIS Quarterly, 19*(2), 237–246.

Davison, A., & Hinkley, D. (1997). *Bootstrap methods and their application*. Cambridge Series in Statistical and Probabilistic Mathematics. Cambridge: Cambridge University Press.

Dempster, A. P., Laird, N. M., & Rubin, D. B. (1977). Maximum likelihood estimation from incomplete data via the EM algorithm. *Journal of the Royal Statistical Society, B, 39*, 1–38.

Dijkstra, T. (1983). Some comments on Maximum likelihood and partial least squares methods. *Journal of Econometrics, 22*, 67–90.

Efron, B. (1979). Bootstrap methods: another look at the Jackknife. *The Annals of Statistics, 7*, 1–26.

Efron, B. (1994). Missing Data, imputation, and the bootstrap. *Journal of American Statistical Association, 89*(426), 463–475.

Efron, B., & Tibshirani, R. J. (1993). *An introdution to the bootstrap*. London: Chapman and Hall.

Fornell, C., Anderson, E., Johnson, M., Bryant, B., & Cha, J. (1996). The American Customer Satisfaction Index: Nature, purpose and findings. *Journal of Marketing, 60*(4), 7–18.

Fornell, C., Johnson, M., Anderson, E., Cha, J., & Bryant, B. (1998). *The American customer satisfaction index: methodology report*. Michigan: University of Michigan Business School.

Loughlin, C., & Coenders, G. (2004). Estimation of the European customer satisfaction index: maximum likelihood versus partial least squares. Application to postal services. *Total Quality Management and Business Excellence, 17*, 1231–1255 .

Mooney, C., & Duval, R. (1993). *Bootstrapping: a nonparametric approach to statistical inference*. New York: Sage.

Peters, C., & Enders, C. (2002). A primer for the estimation of structural equation models in the presence of missing data: Maximum likelihood algorithms. *Journal of Targeting Measurement and Analysis for Marketing, 11*(1), 81–95.

Roth, P. (1994). Missing data: a conceptual review for applied psychologists. *Personnel Psychology, 47*(3), 537–560.

Rubin, D. (1976). Inference and missing data. *Biometrika, 63*(3), 581–592.

Schafer, J., & Graham, J. (2002). Missing data: our view of the state of the art. *Psychological Methods, 7*(2), 147–177.

Shao, J., & Sitter, R. (1996). Bootstrap for Imputed Survey Data. *Journal of the American Statistical Association, 91*(435), 1278–1288.

Smith, D., & Smith-Langfield, K.: Structural Equation Modeling in Management accounting research: citical analysis and opportunities. *Journal of Accounting Literature, 23*, 49–86.

Tenenhaus, M. (2003). Comparison between PLS and LISREL approaches for structural equation modeling: application to the measure of customer satisfaction. In: DECISIA (Ed.) PLS and Related Methods. Proceedings of the PLS03 International Symposium.

Vinzi, V.E., Lauro, C., & Tenenhaus, M. (2003). PLS Path Modeling. In: DECISIA (Ed.), *PLS and related methods*. Proceedings of the PLS03 International Symposium.

Chapter 13
Comparison of Likelihood and PLS Estimators for Structural Equation Modeling: A Simulation with Customer Satisfaction Data

Manuel J. Vilares, Maria H. Almeida, and Pedro S. Coelho

Abstract Although PLS is a well established tool to estimate structural equation models, more work is still needed in order to better understand its relative merits when compared to likelihood methods. This paper aims to contribute to a better understanding of PLS and likelihood estimators' properties, through the comparison and evaluation of these estimation methods for structural equation models based on customer satisfaction data. A Monte Carlo simulation is used to compare the two estimation methods. The model used in the simulation is the ECSI (European Customer Satisfaction Index) model, constituted by 6 latent variables (image, expectations, perceived quality, perceived value, customer satisfaction and customer loyalty). The simulation is conducted in the context of symmetric and skewed response data and formative blocks, which constitute the typical framework of customer satisfaction measurement. In the simulation we analyze the ability of each method to adequately estimate the inner model coefficients and the indicator loadings. The estimators are analyzed both in terms of bias and precision. Results have shown that globally PLS estimates are generally better than covariance-based estimates both in terms of bias and precision. This is particularly true when estimating the model with skewed response data or a formative block, since for the model based on symmetric data the two methods have shown a similar performance.

13.1 Introduction

Covariance based methods are undoubtedly the most well-known methods to estimate Structural Equation Models (SEM), with the result that many social researchers use the terms (SEM and covariance based methods) synonymously.

M.J. Vilares and P.S. Coelho
ISEGI – New University of Lisbon, Campus de Campolide, 1070-032 Lisbon, Portugal
e-mail: mjv@isegi.unl.pt, psc@isegi.unl.pt

M.H. Almeida
Faculty of Economics, New University of Lisbon, Campus de Campolide, 1099-032 Lisbon, Portugal
e-mail: mhalmeida@fe.unl.pt

V. Esposito Vinzi et al. (eds.), *Handbook of Partial Least Squares*, Springer Handbooks of Computational Statistics, DOI 10.1007/978-3-540-32827-8_14,

Partial Least Squares (PLS) methods constitute one alternative to estimating SEM. However in spite of the growing usage of PLS methods in several fields (for instance in customer satisfaction measurement), these methods are still often seen as ad hoc algorithms that have generally not been formally analysed (McDonald 1996).

Several authors (Chin 1998; Fornell and Bookstein 1982) argue that PLS presents several advantages when compared to covariance based methods. In fact it is argued that some conditions should be met in order for these later methods to produce consistent parameter estimates, namely the data should follow a specific multivariate distribution and have independence of observations. Moreover, indicators are typically required to be reflective and unique case values for latent variables cannot be obtained.

In contrast, beyond being based on simpler algorithms, PLS methods do not require any assumptions regarding the joint distribution of indicators or the independence of observations. On the other hand, unique case values for the latent variables can be estimated. Also indicators can be modeled in either direction (i.e. formative or reflective).

However there is neither a formal proof nor a simulation study in the framework of a realistic model that show these advantages of PLS techniques over covariance based methods. The main goal of this paper is to fill such gap. In Fornell and Bookstein (1982) the two kinds of estimation methods are compared using survey data, but authors used an extremely simple model (with only three latent variables: two endogenous and one exogenous) and only focused on the conditions that may cause improper maximum likelihood (ML) estimates. In Cassel et al. (2000) a simulation study is conducted to access the performance of PLS estimates. The authors have used a simplified version of ECSI (European Customer Satisfaction Index) model and accessed the robustness of PLS estimators in the presence of multicolinearity between manifest or latent variables, in presence of skewness of manifest variables, and in the presence of misspecification (erroneous omission of manifest and latent variables). Nevertheless, covariance-based estimates are not obtained in this study and therefore the relative merits of PLS can not be determined.

In our paper, we will study the effects of two assumptions: the symmetry of the distribution and the reflective modeling of the indicators. Thus we compare how the two kinds of methods (PLS and covariance based methods) perform both when these assumptions hold and when they are violated, i.e. when the distribution of the observations is skewed and some indicators follow a formative scheme. We shall perform this analysis in the framework of the ECSI (European Customer Satisfaction Index) model. In fact, the interest in the performance of the two methods in the context of skewness of response and the formative nature of some blocks in the model is particularly justified when we deal with customer satisfaction data. This formative nature of blocks is common in marketing applications and according to Hulland (1998) tends to result in slightly better overall model quality. Fornell and Bookstein (1982) and Tenenhaus (2003, pp. 253–4), give criteria for choosing between a reflective and a formative scheme.

The structure of the remaining part of the paper is as follows. Section 13.2 presents the ECSI model. Two different estimation procedures for structural equation models are presented in Sect. 13.3: Subsects. 13.3.1 and 13.3.2 present the covariance based methods and PLS, respectively. The organization of the simulation study, including the data generating process is shown in Sect. 13.4. Section 13.5 presents and analysis the main results obtained in this simulation. The paper concludes with the discussion presented in Sect. 13.6.

13.2 The ECSI Model

The ECSI model is a framework that aims to harmonize the national customer satisfaction indices in Europe. It was an adaptation of the Swedish Customer Satisfaction Barometer (Fornel 1992) and of the ACSI-American Customer Satisfaction Index (Fornell et al. 1998). The ECSI model is presented in detail in ECSI (1998) and some of the more relevant issues discussed there are briefly presented in this section. A comparison between ECSI and ACSI models can be found in Vilares and Coelho (2005).

The ECSI model is composed of two sub-models: the structural model and the measurement model. The structural model includes the relations between the latent or non-observable variables and is represented in Fig. 13.1. Customer satisfaction is the central variable of this model, having as antecedents or drivers the corporate image of the company, customer expectations, perceived quality of products and services and perceived value. The main consequent of customer satisfaction as specified by the model is customer loyalty. The coefficients of the structural model (also named path coefficients) give the direct impact on a latent variable when there

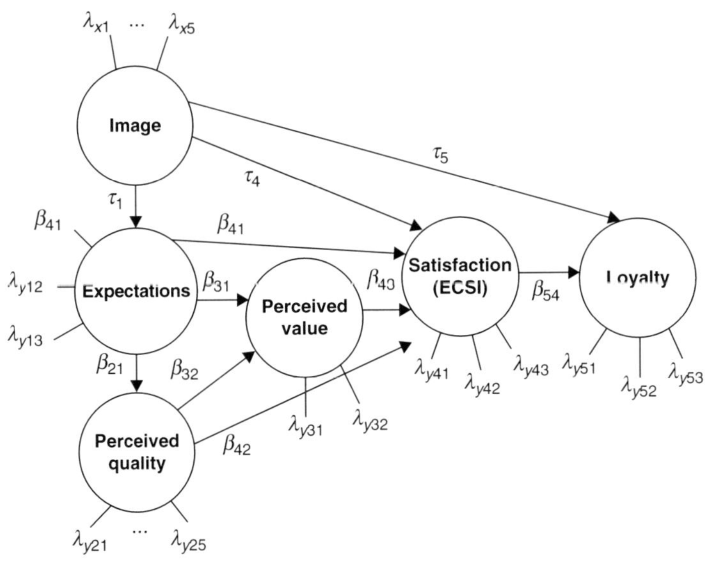

Fig. 13.1 The ECSI model

is a change in an antecedent latent variable. The model therefore consists of one exogenous latent variable (image) and five endogenous variables.

The measurement model defines the relations between the latent variables and the observed indicators or manifest variables. As it is well known, one may have three kinds of measurement models:

- A reflective model when the observed indicators are assumed to be the reflex of the latent variables (the arrow is directed to the observed indicator from its latent variable)
- A formative model when the observed indicators are assumed to cause or form the latent variables (the arrows are directed to the latent variables from their indicators)
- A mixed model when some of the latent variables (in general, the exogenous ones) use a formative model, while others adopt a reflective model

All the indicators are obtained through the administration of a questionnaire to customers, using a scale from 1 to 10 where the value 1 expresses a very negative customer perception and the value 10 a very positive one.

The ECSI structural and measurement models may be formally described by the equations shown in appendix.

13.3 Estimation Procedures

The model to estimate is composed of (13.2) to (13.4) if one adopts a reflective scheme for the measurement model, or by (13.2), (13.5) and (13.6) if the formative scheme is used.

The estimation of ECSI model as well as of other Structural Equation Models faces several difficulties, of which we emphasize:

- The latent variables are not observed
- The measurement indicators that correspond to responses to a customer satisfaction questionnaire may not follow a normal distribution. The frequency distribution of these indicators is in general not symmetric and typically presents skewness to the right
- The measurement variables often present some level of multicollinearity
- Some blocks can hardly be seen as reflective. This is the case of the exogenous latent variable (image), where theory behind the measurement model suggests that the latent variable may be of a formative nature, i.e. the indicators may be viewed as the cause of the latent variable

Two families of methods have been used to estimate this type of models: the PLS methods and the covariance-based methods. We will present in this section a brief introduction of these two groups of methods.

13.3.1 Covariance Based Methods

This group of methods is the most widely adopted. According to Ridgon (1998), the different covariance based methods are variations on the minimization of a common general discrepancy function:

$$F = (S - \Sigma)^T W^{-1} (S - \Sigma) \tag{13.1}$$

Where S is a vector of the unique (non-redundant) elements of the sample covariance or correlation matrix, Σ is a parallel vector of elements from the model-implied matrix, and W is a matrix of weights. Different estimation methods correspond to different matrices W. The two most widely used estimation methods are the Generalized Least Squares (GLS) (with W as the variance and covariance matrix of the residuals) and the Maximum Likelihood (ML) Method (that uses the fitting function $\ln|\Sigma| + trace(S/\Sigma) - \ln(|S|) - p$, with p being the number of indicators).

There are several programs available that are able to minimize different discrepancy functions. Among these programs the most well known is LISREL (Linear Structural Relations) that is a ML implementation. LISREL is so much associated with the estimation of structural equation models that it is often confused with SEM itself.

ML methods produce asymptotically unbiased, consistent and efficient estimators under the empirical conditions that the indicators follow a multivariate normal distribution, the sample is large and independence of observations exists (Bollen 1989).

When these assumptions are violated, these methods may produce, according to several authors (Fornell and Bookstein 1982; Chin 1998) improper solutions such as negative variance estimates. Moreover, these methods do not provide unique values for individual case values of latent variables, since there is an infinite set of possible scores that are consistent with the parameter estimates. Finally, all the indicators must be treated in a reflective manner because the model otherwise would create a situation where we are unable to explain the covariances of all indicators, which is the rationale for this approach (Chin 1998).

13.3.2 PLS Methods

PLS for structural equation modeling may be seen as a distribution free method, since no assumptions are made about the distribution of measurement variables or about the independence of observations. The PLS approach has two stages: the first estimates the observations of the latent variables (case-values) with an iterative scheme. The second estimates the parameters of the structural equations and measurement model. In opposition to covariance-based methods, PLS aims to minimize the residual variance of the dependent variables (both latent and measurement variables).

PLS is supported by an iterative process that iterates between two approximations to the latent variables: the inner approximation and the outer approximation. In each iteration the outer approximation produces an estimate for each latent variable as a weighted mean of their manifests. The inner approximation produces another estimate for the latent variables. Here each variable is obtained as a combination of the external approximation of the other latent variables directly connected to it. Various weighting schemes have been used in this context, the best well-known being: the centroid, the factor and the path weighting schemes. The two estimations are iterated until convergence is reached (using a stopping rule based on relative change from previous iteration). From this process each latent variable is determined both by the inner and outer structure, in such a way that both the inner residual variances from (13.2) and the outer residual variances from (13.3)–(13.4) or (13.5)–(13.6) are minimized.

Finally, case values for the latent variables are obtained, allowing using Ordinary Least Squares (OLS) to estimate, in a non-iterative way, the structural model coefficients, the measurement model loadings, as well as mean scores and location parameters for all variables. The adoption of OLS is possible since model (13.2) is recursive and as a consequence the matrix of the parameters of endogenous variables is triangular.

The case values of the latent variables are inconsistent due to the fact that they are estimated as aggregates of the observed or manifest variables (cf. outside approximation), which include a measurement error. The bias of estimates will tend to zero as the number of indicators per block and sample size both increase. This limiting case is termed "consistency at large" and this property has been argued as a justification for using PLS as an estimation method to estimate LISREL parameters in case where the number of manifest variables is large (Schneeweiss 1990).

PLS is the estimation method adopted for estimating ECSI model. There are several presentations of the PLS methodology in this framework (Cassel et al. 2000; ECSI 1998). More general descriptions of the PLS methodology may be found in Fornell and Cha (1994), Lomöller (1989), Tenenhaus (2003) and Vilares and Coelho (2005).

13.4 The Simulation Study

13.4.1 Introduction

Due to the complexity of SEM models and PLS and covariance-based methods, the analysis of their relative merits and their robustness when some of their assumptions are violated can hardly be assessed in an analytical form. This is a fertile ground for the use of simulation studies.

Thus, we use a Monte Carlo simulation with three models: one where all blocks are reflective and the measurement variables show a symmetric distribution (referred as base model); a model in which the measurement variables continue to show a symmetric distribution, but the block corresponding to the exogenous latent variable (image) is formative; and finally a model of reflective nature, but where measurement variables show an asymmetric right skewed distribution. As already remarked, the two chosen violations were intended to be typical of customer satisfaction data that frames the study.

The simulation aims to analyze the quality of PLS and ML estimates of structural model coefficients (matrices $\boldsymbol{\beta}$ and $\boldsymbol{\tau}$) and of measurement model coefficients (Λ_y, Λ_x and λ_ξ) in the context of the three variants: base model, formative model and skewed data. The PLS and ML estimators of model coefficients are analyzed in terms of bias and precision (as measured by the mean square errors). The bias of an estimator of a generic coefficient β_{ij} is obtained as $B_{\beta_{ij}} = K^{-1}\sum_{k=1}^{K}(\hat{\beta}_{ij,k}-\beta_{ij})$ and the mean square error by $MSE_{\beta_{ij}} = K^{-1}\sum_{k=1}^{K}(\hat{\beta}_{ij,k} - \beta_{ij})^2$, where K represents the number of replicates in the simulation and $\hat{\beta}_{ij,k}$ the estimate of β_{ij} obtained with replicate k by one estimation method (PLS or ML). The simulation was run using the SAS system. The PLS approach was implemented using a SAS macro and the ML estimation using CALIS procedure.

The simulation error regarding the estimators' biases is small and varies between 0.00 and 0.02 when estimating the indicators loadings and between 0.00 and 0.10 when approaching the structural coefficients.

13.4.2 The Data Generating Process

The starting point of our simulation is the ECSI model (cf. Fig. 13.1). Data are generated according to the ECSI model, where we have assumed that the coefficients of both models (structural and measurement models) are known. Thus the postulated structural model is:

$$
\begin{aligned}
\eta_1 &= 0.9\xi_1 + \nu_1\\
\eta_2 &= 0.8\eta_1 + \nu_2\\
\eta_3 &= 0.3\eta_1 + 0.7\eta_2 + \nu_3\\
\eta_4 &= 0.3\xi_1 + 0.1\eta_1 + 0.4\eta_2 + 0.3\eta_3 + \nu_4\\
\eta_5 &= 0.3\xi_1 + 0.7\eta_4 + \nu_5
\end{aligned}
$$

where ξ_1 is the exogenous variable image, and $\eta_1 - \eta_5$ are endogenous variables that represent customer expectations, perceived quality, perceived value, customer satisfaction and customer loyalty.

The measurement models for the endogenous variables are of reflective form, assuming the following values for the parameters: $\lambda_{1j} = 1.2, 0.8, 1.0$ for $j = 1, 2, 3$, $\lambda_{2j} = 0.8, 1.1, 1.0, 0.7, 0.9$ for $j = 1, \ldots, 5$, $\lambda_{3j} = 1.2, 0.75$ for $j = 1, 2$, $\lambda_{4j} = 1.1, 0.8, 0.6$ for $j = 1, 2, 3$ and $\lambda_{5j} = 0.9, 0.7, 0.6$ for $j = 1, 2, 3$. For the base model and the skewed data model the measurement scheme for the exogenous variable is reflective assuming the following values for parameters:

$$\lambda_{xj} = 1.00, 0.75, 1.15, 0.90, 0.80 \text{ for } j = 1, \ldots, 5.$$

For the formative model, we have adopted a formative scheme for the exogenous latent variable:

$$\xi_1 = 0.4x_1 + 0.25x_2 + 0.15x_3 + 0.1x_4 + 0.1x_5 + \delta.$$

The values for inner and outer model coefficients were chosen in order to be as similar as possible to the ones that would be obtained with real world data. For that we have observed typical estimates of model coefficients obtained thought the estimation of ECSI model applied to different companies and sectors, and postulated a model structure consistent with those estimates.

For the base model the cases of the exogenous latent variable (ξ_1) were generated using a $\beta(4, 4)$ symmetric distribution in the interval [1, 10] and all the errors both in the inner and outer models (ν_i, ε_{ij} and δ) were generated using a $\beta(3, 3)$ symmetric distribution in the interval [−1.5, 1.5]. For the formative model we used the same symmetric distributions $\beta(4, 4)$ and $\beta(3, 3)$, but we started by generating the cases of the 5 measurement variables associated with the exogenous block. Finally, in the skewed data model we used a right skewed distribution $\beta(10, 4)$ both for the cases of the exogenous latent variable (ξ_1) and the errors (ν_i, ε_{ij} and δ). In the three models the values of the measurement variables were converted into scores in a ten point scale 1–10, which is the scale used in ECSI and ACSI questionnaires. Figure 13.2 shows the frequency distribution of generated data for the exogenous variable both in the symmetric (a) and skewed data context (b).

For the simulation we have generated 1,000 data sets of 250 observations for each one of the three models, resulting in a total of 750,000 observations for the 21 measurement variables. The choice of a sample size of 250 observations was motivated by the fact that this is the sample size used in ECSI and ACSI to estimate the satisfaction models at company level. For each one of the 1,000 replicates the

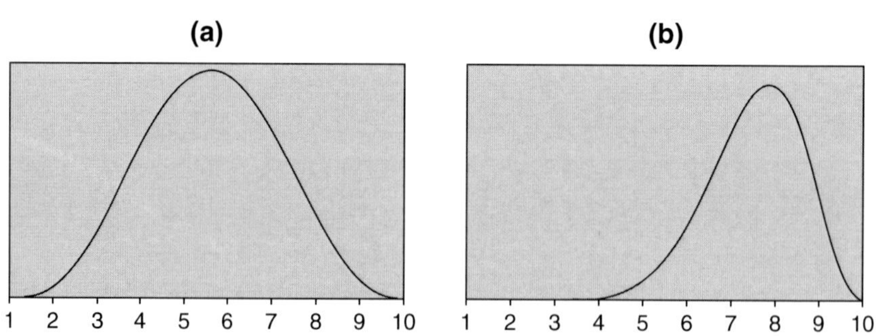

Fig. 13.2 Frequency distributions for the exogenous variable (*Image*): (**a**) beta distribution $\beta(4, 4)$; (**b**) beta distribution $\beta(10, 4)$

three subsets of 250 observations (generated by three different models) are used to estimate inner and outer models coefficients ($\boldsymbol{\beta}$, $\boldsymbol{\tau}$, Λ_y, Λ_x and λ_ξ) using both ML and PLS methods. For each model and estimation procedure the 1,000 sets of estimates are then compared with true values and used to obtain ML and PLS estimators biases and mean square errors.

13.5 Simulation Results

The goal of our simulation study is to analyze and compare Maximum Likelihood and PLS estimators' properties for structural equation models (SEM) based on customer satisfaction data, both in terms of bias and precision.

Simulation results are shown in Tables 13.1 and 13.2. Table 13.1 shows the bias of model parameters (loadings and inner model coefficients) both for the PLS and ML techniques and for the three model formulations (base model, formative exogenous latent variable and skewed response data). The precision of the estimates is assessed through the mean square errors of these parameters that are presented in Table 13.2. In both tables the simple averages of the absolute bias and the averages of mean square errors for the inner and outer structures are also shown.

13.5.1 Base Model

Looking at Table 13.1 it is possible to observe that when the base model is adopted, the PLS and ML estimators presented on average similar biases when estimating the outer structure. The average of absolute bias is 0.10 for PLS and 0.11 for ML. PLS shows smaller biases for most of the indicator loadings. The only exceptions are those that belong to the Image block (exogenous variable) and to the Loyalty block (the last latent variable in the model) where PLS shows a poor performance in terms of bias.

In the inner structure context, the superiority of PLS performance in terms of bias is increased significantly as the absolute bias reaches on average 0.07 compared to 0.12 reached by ML. This method is better than PLS only in the estimation of the coefficients that relate to the following latent variables: Expectations with Satisfaction (β_{41}), Value with Satisfaction (β_{43}), Image with Loyalty (τ_5) and Satisfaction with Loyalty (β_{54}).

Of special interest is the fact that while there is a tendency of PLS to overestimate loadings, a negative bias tends to be observed in the same structure when the likelihood estimator is used. In contrast, we can find some evidence of the opposite situation in the inner structure context, with a tendency of underestimation for PLS and overestimation for ML.

Mean square errors for the base model can be found in the first two columns of Table 13.2. The PLS estimator shows better precision than the ML estimator in

Table 13.1 Bias

Parameter	Base model PLS	Base model ML	Image Formative PLS	Image Formative ML	Skewed data PLS	Skewed data ML
			Indicator loadings			
λ_{x1}	0.08	0.04	–	–	0.08	0.14
λ_{x2}	0.06	0.03	–	–	0.08	0.11
λ_{x3}	0.09	0.05	–	–	0.07	0.17
λ_{x4}	0.07	0.04	–	–	0.08	0.13
λ_{x5}	0.07	0.03	–	–	0.08	0.12
λ_{11}	−0.01	−0.16	−0.01	−0.63	−0.02	0.17
λ_{12}	0.00	−0.11	−0.01	−0.42	0.01	0.11
λ_{13}	−0.01	−0.13	−0.01	−0.52	−0.01	0.14
λ_{21}	0.09	0.09	0.09	−0.19	0.10	0.15
λ_{22}	0.12	0.12	0.12	−0.26	0.10	0.21
λ_{23}	0.11	0.11	0.11	−0.23	0.10	0.20
λ_{24}	0.08	0.08	0.08	−0.16	0.10	0.14
λ_{25}	0.10	−0.19	0.10	0.03	0.10	−0.13
λ_{31}	0.02	−0.23	0.01	−0.44	0.00	−0.34
λ_{32}	0.01	−0.15	0.01	−0.27	0.03	−0.21
λ_{41}	0.20	−0.26	0.19	−1.01	0.18	−0.39
λ_{42}	0.15	−0.19	0.14	−0.74	0.15	−0.28
λ_{43}	0.11	−0.14	0.11	−0.55	0.13	−0.21
λ_{51}	0.32	−0.08	0.31	−0.83	0.30	−0.01
λ_{52}	0.25	−0.06	0.25	−0.64	0.25	−0.01
λ_{53}	0.21	−0.05	0.21	−0.55	0.23	−0.01
Average (abs)	0.10	0.11	0.11	0.47	0.10	0.16
			Inner model coefficients			
τ_1	−0.03	0.21	−0.26	1.55	−0.01	0.06
β_{21}	0.05	−0.17	−0.06	−0.22	0.06	0.00
β_{31}	0.04	0.04	−0.01	0.00	0.06	0.25
β_{32}	−0.15	0.29	−0.17	0.12	−0.17	0.52
τ_4	−0.04	0.14	−0.17	0.92	−0.02	0.33
β_{41}	0.06	0.03	0.11	2.07	0.07	0.11
β_{42}	−0.09	0.16	−0.17	1.93	−0.09	0.32
β_{43}	−0.08	0.02	−0.06	1.71	−0.09	0.07
τ_5	0.07	0.06	−0.09	2.35	0.10	0.11
β_{54}	−0.15	−0.12	−0.07	0.03	−0.18	−0.22
Average (abs)	0.07	0.12	0.12	1.09	0.09	0.20

the estimation of all the inner coefficients and the majority of the outer coefficients. On average, PLS mean square errors reach 0.02 and 0.01 in the outer and inner structures against the ML mean square errors that are on average 0.03 and 0.05, respectively. As with bias, it is in the inner model estimation that PLS shows a better relative performance.

Table 13.2 Mean Square Error

Parameter	Base model PLS	Base model ML	Image Formative PLS	Image Formative ML	Skewed data PLS	Skewed data ML
			Indicator loadings			
λ_{x1}	0.01	0.00	–	–	0.01	0.04
λ_{x2}	0.00	0.00	–	–	0.01	0.02
λ_{x3}	0.01	0.00	–	–	0.01	0.05
λ_{x4}	0.01	0.00	–	–	0.01	0.03
λ_{x5}	0.01	0.00	–	–	0.01	0.03
λ_{11}	0.00	0.03	0.00	0.43	0.00	0.17
λ_{12}	0.00	0.02	0.00	0.19	0.00	0.08
λ_{13}	0.00	0.02	0.00	0.29	0.00	0.12
λ_{21}	0.01	0.02	0.01	0.06	0.01	0.06
λ_{22}	0.02	0.03	0.01	0.12	0.01	0.12
λ_{23}	0.01	0.03	0.01	0.10	0.01	0.10
λ_{24}	0.01	0.01	0.01	0.05	0.01	0.05
λ_{25}	0.01	0.04	0.01	0.07	0.01	0.03
λ_{31}	0.00	0.13	0.00	0.26	0.00	0.18
λ_{32}	0.00	0.05	0.00	0.10	0.00	0.07
λ_{41}	0.04	0.08	0.04	1.02	0.03	0.20
λ_{42}	0.02	0.04	0.02	0.54	0.02	0.11
λ_{43}	0.01	0.02	0.01	0.30	0.02	0.06
λ_{51}	0.10	0.01	0.10	0.68	0.09	0.08
λ_{52}	0.06	0.01	0.06	0.41	0.06	0.05
λ_{53}	0.05	0.01	0.04	0.30	0.05	0.04
Average	0.02	0.03	0.02	0.31	0.02	0.08
			Inner model coefficients			
τ_1	0.00	0.05	0.07	2.80	0.00	0.04
β_{21}	0.00	0.03	0.00	0.10	0.00	0.09
β_{31}	0.01	0.02	0.00	0.03	0.01	0.19
β_{32}	0.03	0.13	0.03	0.16	0.03	0.44
τ_4	0.01	0.05	0.03	1.00	0.00	0.19
β_{41}	0.01	0.02	0.02	4.89	0.01	0.09
β_{42}	0.01	0.13	0.03	5.24	0.01	0.25
β_{43}	0.01	0.03	0.01	3.51	0.01	0.07
τ_5	0.01	0.02	0.01	8.26	0.01	0.06
β_{54}	0.03	0.03	0.01	0.02	0.04	0.09
Average	0.01	0.05	0.02	2.60	0.01	0.15

Moreover, we conclude that the PLS estimators are generally more efficient not only due to smaller absolute biases but also due to smaller estimator variances. The exceptions are again the loadings belonging to the first and last blocks (Image and Loyalty) where the mean square errors of PLS exceed the ones obtained thought ML. This is exclusively explained by the higher bias already reported.

13.5.2 Formative Measurement Model

Biases for PLS and ML estimators when a formative measurement model is adopted for the exogenous latent variable (Image) are shown in 3rd and 4th columns of Table 13.1. Note that ML estimates were obtained considering all blocks as reflective and, therefore, ignoring the formative nature of the exogenous variable, while PLS estimators explicitly consider this formative block. This may constitute an "unfair" context for the benchmark of ML and PLS estimators, but allows to understand the effect of ignoring the formative nature of a block on estimators properties. As the exogenous block is modelled differently within each of the estimation methods (ML and PLS), in Tables 13.1 and 13.2 results regarding the estimation of indicators loadings for this block are omitted. For PLS it can be seen that biases for outer coefficients are similar to the ones observed in the reflective context. For the inner coefficients some tendency for a bias increase is observed, having an average absolute bias of 0.12 against 0.07 in the base model. Also, the higher bias increases observed in the inner structure estimation are usually associated with the formative block (Image). PLS estimates continue to show a tendency for the overestimation of the outer structure and the underestimation of the inner structure.

On the other hand, the bias of the ML estimation is drastically increased both in the outer and inner structure. In the outer structure the average absolute bias is now four times higher and in the inner structure it has increased from 0.12 to 1.09. Also ML has originated some extremely biased estimates reaching 2.35 for the Image-Loyalty impact. These extreme estimates are generally associated with impacts involving the Image block, which is not surprising due to the impossibility of ML methods to consider the formative nature of this block (cf. Sect. 13.3.1). Finally, we can observe that there is a downward bias tendency for ML in the outer structure and for overestimation of the structural path estimates. Remember that this pattern is the opposite of that observed in PLS estimates. As a result of these behaviors PLS significantly outperforms the ML method in terms of bias when estimating both structures.

Regarding the mean square errors in the formative model (columns 3 and 4 of Table 13.2) it can be observed that of the PLS estimators remain with a precision similar to the one observed in the base model when estimating both the inner and outer structure. However the mean square errors of the ML estimators increase severely, especially in the inner structure, where the mean square error reaches in average the value 2.6. Moreover, ML estimates produce more extreme values, with the mean square error reaching 8.26 for the image-loyalty impact. Consequently, the precision of ML estimates when approximating both inner and outer structures is without exception drastically worse than PLS estimates. In accordance with what was remarked in Sect. 13.3.1, we have also obtained some improper solutions (negative variance estimates) with ML methods applied to data generated with a formative block. In those cases we have imposed restrictions to the estimation, setting a lower bound of zero for all variance estimates.

13.5.3 Skewed Response Data

This simulation study also analyses the impact of skewed response data on PLS and ML estimation when a reflective measurement model for the first block is adopted.

Last two columns of Table 13.1 show the bias of PLS and ML estimators when we use skewed response data. Looking at the outer coefficients results, we can observe that the average of the absolute bias of PLS estimators remains near 0.10 as in the base model. In fact, PLS estimates for the indicator loadings show similar biases when using symmetric or skewed response data. On the other hand, ML shows an increase in average absolute bias (from 0.11 to 0.16). Now, ML estimates only outperform PLS when approaching the loadings of loyalty indicators (the last block in the model). A similar pattern arises when analysing inner model estimates. Bias for PLS estimates now show a modest increase when compared to the base model (0.07 to 0.09). The bias for ML in the estimation of inner model coefficients (0.20 on average) is much higher than the one observed with PLS and increases significantly when compared to the base model (from 0.12 to 0.20). Also, ML estimates have shown a higher bias increase than PLS when we move from base model to the skewed data model. We can observe that the general overestimation bias tendency of PLS in the outer structure and the general underestimation tendency of PLS in the structural path estimates are still noticeable with asymmetric response data. ML also shows an overestimation tendency when approaching inner model coefficients.

The MSE results for the skewed data model are shown in last columns of table 13.2. We conclude that while the PLS estimators assure on average the same precision in the base model and in the skewed data model (0.02 and 0.01 respectively in the outer and inner coefficients), the ML estimators are not able to maintain the same precision level when we move to skewed data. In the outer structure, the mean square error changes from 0.03 to 0.08 and in the inner structure from 0.05 to 0.15. This degradation in precision is significant, although not as high as the one originated by the formative model. This evolution takes place not only due to the bias increase, but also because the variance of the ML estimators increases significantly more than the PLS estimators in the presence of skewed data, making the ML estimation less efficient. Consequently, when we move from symmetric to skewed data the advantage of PLS over ML in terms of precision is significantly increased.

Note that the models used in the simulation included different number of indicators. In fact we have included a block with two indicators (perceived value), three blocks with three indicators (expectations, satisfaction and loyalty) and two blocks with five indicators (image and perceived quality). Globally it can be seen that PLS always produces good estimates for perceived value loadings. This is an interesting result, since PLS is presented as being "consistent at large" requiring both sample size and the number of indicators per block to increase in order to approach the true values. So, globally, with the sample sizes typically used when estimating customer satisfaction models (about 250 observations), PLS also seems to show robustness to estimating the loadings in blocks with a small number of measurement variables.

13.6 Discussion

Although the covariance-based procedures are by far the most well known techniques among structural equation modeling, the PLS approach can also be a very useful tool that can be applied by researchers.

Our paper gives some insights into the quality of PLS estimation when applied to a structural equation model representing customer satisfaction data. We have postulated a model similar to the ECSI model composed by 6 latent variables (Image, Expectations, Quality, Value, Satisfaction and Loyalty). Within a simulation study we have evaluated both PLS and ML estimates in terms of bias and precision when estimating the inner and outer model coefficients. We have used a reflective with symmetric data as base model, but also two variants: one where the exogenous variable (Image) is formative and other where data is obtained with a right-skewed distribution. These are both situations that are typical of customer satisfaction data.

Results have shown that globally PLS estimates are generally better than ML estimates both in terms of bias and precision. Nevertheless, in the base model (reflective model with symmetric data) the quality of the two estimation methods is very similar, especially in what regards the estimation of indicator loadings. On the other hand, it is when a formative latent variable is introduced that PLS method shows the most significant gains when compared to the covariance-based method.

We have also concluded that within the generated simulation PLS approach was very robust both to the inclusion of formative blocks and skewed data. In the formative model, PLS have shown modest increase in bias and mean square errors specially when estimating outer structure. Nevertheless, the robustness of the PLS estimation procedure was even more conspicuous with respect to skewed response distributions. In our simulation both precision and bias remained almost unchanged, or have shown very modest increases when compared to the results generated by symmetric data.

On the other hand, the ML estimators were much more sensitive to the various potential deficiencies in data and in the model specification. When asymmetric data is used and especially if a formative block is used, the quality of the estimates decreases drastically. This result underlines the fact that when a formative block is wrongly taken as reflective ML may render poor and sometimes extreme estimates for some model coefficients.

This is not a surprise since the covariance-based approach for SEM, typically makes the underlying assumption that the observed variables follow a normal multivariate distribution. In addition, it only allows for reflective measurement schemes.

It should also be emphasized that globally, and for the sample size used in the simulation (250 observations), PLS also seems to show a high robustness for estimating indicator loadings in blocks with a small number of measurement variables. Obviously, this conclusion can't be generalized without further research.

Finally, it is also noticeable that generally PLS has shown a tendency for overestimation of the outer model and for underestimation of the inner model structure. The ML method showed exactly the opposite tendency, with overestimation of the path coefficients and a general underestimation of the indicator loadings. The only exception was in the context of skewed data where the ML method no longer showed a systematic underestimation of the outer model.

Note that PLS is usually presented as a useful tool when the primary interest is to obtain indicator weights and produce the predictions of the latent variables. On the other hand covariance-based methods are usually presented as useful when the interest is to obtain model coefficients. Since, in this simulation, we compare the ability of the two methods to reproduce the inner model coefficients and the outer model loadings, it can be argued that it runs in a framework that favors the covariance-based methods. Therefore, we may conclude that PLS can be used as an alternative to ML estimators even when the primary interest of the research relates to obtaining and interpreting model coefficients and sample sizes are relatively large. Moreover, it appears that practitioners should use PLS as a preferable choice over ML methods when using skewed data and one or more blocks in the model can be considered as having a formative nature, situations that are typical in marketing research framework. In this later contexts PLS significantly outperforms ML estimators, and practitioners using PLS methods may more likely obtain accurate coefficient estimates and achieve a better understanding of the structures underlying the data they have collected.

A major limitation of this simulation is the fact that we have not considered different levels of skewness in the data set. Further research should be done in order to understand how different levels of skewness in the measurements variables affect the properties of the two estimators (PLS and ML). Also, some work should be done in understanding the estimators' properties with small sample sizes. In fact, although satisfaction models in the context of national customer satisfaction indexes are usually estimated with a sample size consistent with the one used in our simulation, in the industry context it is usual to try to estimate models for market segments where the available sample sizes are significantly smaller. Finally, future work should also access the performance of both methods, ML and PLS, in the presence of multicolinearity and or model misspecification. In fact, with real world applications, erroneous omissions of model coefficients or manifest and latent variables are common. Also erroneous inclusions of non-existent relationships between variables may arise. This is fertile ground to a more in dept study of ML and PLS performance.

Appendix

Structural Model

The model consists of five equations (i.e. the same number of endogenous variables) that can be written in a compact form as:

$$\eta = \beta\eta + \tau\xi + \nu \tag{13.2}$$

where η is a vector (5×1) of endogenous latent variables (all except Image), ξ is the exogenous latent variable (Image), β and τ are impact matrices and ν is a vector

(5 × 1) of specification errors. We shall assume the usual properties about these errors (zero mean, homoscedasticity and zero covariance between the errors).

More specifically the matrices of structural coefficients $\boldsymbol{\beta}$ and $\boldsymbol{\tau}$ are the following:

$$\boldsymbol{\beta} = \begin{bmatrix} 0 & 0 & 0 & 0 & 0 \\ \beta_{21} & 0 & 0 & 0 & 0 \\ \beta_{31} & \beta_{32} & 0 & 0 & 0 \\ \beta_{41} & \beta_{42} & \beta_{43} & 0 & 0 \\ 0 & 0 & 0 & \beta_{54} & 0 \end{bmatrix}, \boldsymbol{\tau} = \begin{bmatrix} \tau_1 \\ 0 \\ 0 \\ \tau_4 \\ \tau_5 \end{bmatrix}.$$

with ξ: image; η_1: customer expectations; η_2: perceived quality of products and services; η_3: perceived value; η_4 customer satisfaction; η_5: customer loyalty.

Measurement Model

When the model is considered reflective, the equations are:

$$\boldsymbol{y} = \Lambda_y \boldsymbol{\eta} + \boldsymbol{\varepsilon} \tag{13.3}$$

$$\boldsymbol{x} = \Lambda_x \boldsymbol{\xi} + \boldsymbol{\delta} \tag{13.4}$$

$$E(\boldsymbol{\varepsilon}) = E(\boldsymbol{\delta}) = E(\boldsymbol{\varepsilon}|\boldsymbol{\eta}) = E(\boldsymbol{\delta}|\boldsymbol{\xi}) = \mathbf{0}.$$

where $\boldsymbol{y}' = (y_1, y_2, \ldots, y_p)$ and $\boldsymbol{x}' = (x_1, x_2, \ldots, x_q)$ are the manifest endogenous and exogenous variables, respectively. Λ_y and Λ_y are the corresponding parameters matrices (loadings) and ε and δ are specification errors.

Representing by $\boldsymbol{y}_1' = (y_{i1}, \ldots, y_{iH_i})$ the vector of manifest variables related to the latent endogenous variable η_i and by $\boldsymbol{x}' = (x_1, \ldots, x_G)$ the vector of manifest variables related to the latent exogenous variable ξ, we can also write the model in the form:

$$y_{ij} = \lambda_{y_{ij}} \eta_i + \varepsilon_{ij}, i = 1, \ldots, 5; \; j = 1, \ldots, H_i$$

$$x_j = \lambda_{x_j} \xi + \delta_j, \; j = 1, \ldots, G$$

where H_i is the number of manifest variables associated with variable η_i and G the number of manifest variables associated with variable ξ.

If the model is considered formative, we have for the endogenous variable η_i and for the exogenous variable ξ the following equations:

$$\eta_i = \sum_{l=1}^{H} \lambda_{\eta_{il}} y_l + \delta_{\eta_i}, i = 1, \ldots, 5 \tag{13.5}$$

$$\xi = \sum_{l=1}^{G} \lambda_{\xi l} x_l + \delta_\xi \tag{13.6}$$

where $\lambda_{\eta_{il}}$ and $\lambda_{\xi l}$ are coefficients of the formative model and δ_{η_i} and δ_{ξ} are specification errors.

References

Bollen, K. A. (1989). Structural equations with latent variables. New York: Wiley.

Cassel, C., Hackl, P., & Westlund, A. (2000). On measurement of intangibles assets: a study of robustness of partial least squares. *Total Quality Management, 7*, 897–907.

Chin, W. W. (1998). The partial least squares approach to structural equation modeling. In G. A. Marcoulides (Ed.), *Modern methods for business research*, Chap.10. New Jersey: Lawrence Erlbaum Associates.

Dijkstra, T. (1983). Some comments on maximum likelihood and partial least squares methods. *Journal of Econometrics, 22*, 67–90.

ECSI: European Customer Satisfaction Index. (1998). Report prepared for the ECSI Steering Committee.

Fornell, C. (1992). A national customer satisfaction barometer: the Swedish experience. *Journal of Marketing, 56*, 6–21.

Fornell, C., & Bookstein, F. L. (1982). The two structural equation models: LISREL and PLS applied to customer exit-voice theory. *Journal of Marketing Research, 19*, 440–452.

Fornell, C., & Cha, J. (1994). Partial least squares. In R. P. Bagozzi (Ed.), *Advanced methods of marketing research*. MA, Cambridge: Blackwell.

Fornell, C., Johnson, M. D., Anderson, E. W., Cha, J., & Everitt Bryant, B. (1998). The American customer satisfaction index: nature, purpose and findings. *Journal of Marketing, 60*, 7–18.

Hulland, J. (1998). Use of partial least squares in strategic management research. *Comportamento Organizacional e Gesto, 4*(1), 181–202.

Lomöller, J. B. (1989). Latent variable path modelling with partial least squares. Heidelberg: Physica.

Mcdonald, R. P. (1996). Path analysis with composite variables. *Multivariate Behavioral Research, 31*(2), 239–270.

Ridgon, R. E. (1998). Structural equation modelling. In G. A. Marcoulides (Ed.), *Modern methods for business research*. MahWah, NJ: Lawrence Erlbaum Associates.

Schneeweiss, H. (1990). Models with latent variables: LISREL versus PLS. *Contempororary Mathematics, 112*, 33–40.

Tenenhaus, M. (1999). Lapproche PLS. *Revue de Statistique Applique, 47*(2), 5–40.

Tenenhaus, M. (2003). Comparison between PLS and LISREL approaches for structural equation models. In Vilares, Tenenhaus, Coelho Vinzi, Morinau (Eds.), *PLS and related methods*. Proceedings of the PLS03 International Conference.

Vilares, M. J., & Coelho, P. S. (2005). A Satisfação e a Lealdade do Cliente - Metodologias de Avaliação, Gestão e Análise. Escolar Editora.

Chapter 14
Modeling Customer Satisfaction: A Comparative Performance Evaluation of Covariance Structure Analysis Versus Partial Least Squares

John Hulland, Michael J. Ryan, and Robert K. Rayner

Abstract Partial least squares (PLS) estimates of structural equation model path coefficients are believed to produce more accurate estimates than those obtained with covariance structure analysis (CVA) using maximum likelihood estimation (MLE) when one or more of the MLE assumptions are not met. However, there exists no empirical support for this belief or for the specific conditions under which it will occur. MLE-based CVA will also break down or produce improper solutions whereas PLS will not. This study uses simulated data to estimate parameters for a model with five independent latent variables and one dependent latent variable under various assumption conditions. Data from customer satisfaction studies were used to identify the form of typical field-based survey distributions. Our results show that PLS produces more accurate path coefficients estimates when sample sizes are less than 500, independent latent variables are correlated, and measures per latent variable are less than 4. Method accuracy does not vary when the MLE multinormal distribution assumption is violated or when the data do not fit the theoretical structure very well. Both procedures are more accurate when the independent variables are uncorrelated, but MLE estimations break down more frequently under this condition, especially when combined with sample sizes of less than 100 and only two measures per latent variable.

J. Hulland
Katz Business School, University of Pittsburgh, Pittsburgh, PA 15260, USA
e-mail: jhulland@katz.pitt.edu

M.J. Ryan
Ross School of Business, University of Michigan,
P.O. Box 105, Bass Harbor, ME 04653, USA
e-mail: mikeryan@umich.edu

R.K. Rayner
Market Strategies International, 20255 Victor Parkway, Suite 400, Livonia, MI 48152, USA
e-mail: bob_rayner@marketstrategies.com

V. Esposito Vinzi et al. (eds.), *Handbook of Partial Least Squares*, Springer Handbooks of Computational Statistics, DOI 10.1007/978-3-540-32827-8_15,

14.1 Introduction

Causal modeling techniques allow researchers to simultaneously examine theory and measures. They permit the explicit inclusion of measurement error and an ability to incorporate abstract and unobservable constructs (Fornell 1982). Bagozzi (1980) suggests that causal models provide researchers with four key benefits: (1) they make the assumptions, constructs, and hypothesized relationships in a theory explicit; (2) they add a degree of precision to a theory, since they require clear definitions of constructs, operationalizations, and functional relationships; (3) they permit a more complete representation of complex theories; and (4) they provide a formal framework for constructing and testing both theories and measures.

The most commonly employed causal modeling techniques involve covariance structure analysis (CVA), using software such as LISREL (Jöreskog 1973; Jöreskog and Sörbom 1993), AMOS (Arbuckle 1994), and EQS (Bentler 1995). However, CVA approaches are poorly suited to dealing with small data samples (Fornell 1982) and can yield non-unique or otherwise improper solutions or simply break down, particularly when sample sizes are small or the number of indicators per construct is low (Fornell and Bookstein 1982). An alternative causal modeling approach known as partial least squares (PLS) avoids some of these limitations (Barclay et al. 1995; Hulland 1999; Lohmöller 1989; Lohmöller and Wold 1984; Wold 1966, 1980, 1982, 1985; Chin 1998; Dijkstra 1993).

Given the assumptions of multivariate normality, large sample sizes, and well-specified models (Bagozzi 1980), covariance-based coefficients (typically estimated using a maximum likelihood estimation – or MLE – technique) should more closely match true model parameters than estimates derived with PLS. The opposite should be true when the MLE assumptions are not met (Wold 1982). Since the MLE assumptions can often be difficult to achieve in both theoretical and applied marketing research (e.g., customer satisfaction) contexts, Wold's assertion has implications for the accuracy of parameter estimates obtained using either of the two approaches. To our knowledge, no one has produced empirical evidence showing the degree to which assumption violations favor either PLS or MLE-based CVA. The primary focus of this paper is to specifically address this issue, using simulated data, but based on distributions typically encountered in the field.

The remainder of the paper is organized as follows. First, we review the two techniques, followed by a detailed discussion of the conditions under which either PLS or CVA should produce more accurate path coefficient estimates. We then describe our simulation study, designed to systematically assess the effects of various factors (e.g., sample size, estimation technique and number of indicators per construct) on causal model parameter estimates, followed by a presentation of the results. Finally, we discuss our findings and their implications for future research.

Table 14.1 Comparison of partial least squares and covariance-based analysis

Criterion	*PLS*	*CVA*
Objective	Prediction oriented	Parameter oriented
Approach	Variance based	Covariance based
Assumptions	Nonparametric	Parametric
Parameter estimates	Consistent at large	Consistent
Latent variable scores	Explicitly estimated	Indeterminate
Model complexity allowed	High complexity	Small to moderate
Minimum sample size	20–100	200–800

Note: Adapted from Chin and Newsted (1998)

14.2 Background

Path analysis and causal modeling, introduced by Wright (1934), has been adopted and refined in many disciplines. For example, Goldberger (1973) in economics, Asher (1976) and McArdle (1980) in political science, James et al. (1982) in psychology, Duncan (1966) in sociology, and Jöreskog (1970) in statistics all contributed to early causal modeling development[1]. Collectively, this work has become known as latent variable structural equation modeling using maximum-likelihood estimation procedures or as covariance structure analysis (CVA). Structural equation modeling is based on the notions that valid construct operationalization requires multiple measures, conceptual models must precede empirical testing, and theoretical variables should be ordered and their effects decomposed. Applications are ubiquitous in the social sciences with Lenk's (2000) citation search returning 2,463 entries.

Marketing and customer satisfaction researchers have been traditional users and developers of CVA models (e.g., Bagozzi 1977, 1980, 1982, 1984; Bagozzi and Yi 1989, 1994; Bagozzi et al. 1991; Beardon et al. 1982; Fornell 1982, 1983, 1992; Fornell and Larcker 1981; Hulland et al. 1996; Lenk 2000; Sharma et al. 1989). PLS has also seen a number of marketing applications (e.g., Barclay 1991; Fornell 1992; Green et al. 1995; Johnson and Fornell 1987; Plouffe et al. 2001a, b; Qualls 1987; Smith and Barclay 1997; Zinkhan et al. 1987) and some extensions (Bagozzi and Yi 1989; Bagozzi et al. 1991; Fornell and Bookstein 1982). The basic PLS and MLE submodels are well known (Bagozzi 1980; Chin and Newsted 1998; Fornell and Cha 1993) and will not be repeated here.

The variance-based approach of PLS shifts emphasis from theory testing to predictive modeling, since the objective of PLS is to maximize prediction in the endogenous constructs rather than explain the covariances of all of the indicators used in a model (Fornell 1989; Falk and Miller 1992). As summarized in Table 14.1, researchers believe that PLS is best suited to situations where the primary modeling objective is prediction, when the focus is on explaining variance, when parametric assumptions do not hold, when explicit latent variable scores are desired, when model complexity is high, and when sample sizes are small.

[1] For recent applications in Psychology see MacCallum and Austin (2000).

14.3 Expected Effects

PLS is acknowledged to be more robust than MLE, but is believed to produce biased estimates, overestimating measurement model coefficients and underestimating path coefficients. On the other hand, it is widely held that MLE should produce more accurate results, especially if its assumptions are largely met.[2] PLS estimates are known to be biased. In some cases, the differences resulting from use of the two different approaches will be small. For example, Fornell and Bookstein (1982) reported high correlations between PLS and MLE-based CVA estimates. Nonetheless, it remains an empirical matter to determine whether MLE-based CVA or PLS will produce more accurate coefficient estimates under a variety of different assumption violations.

14.3.1 Model Assumptions Thought to Affect Accuracy

MLE maximizes the probability of observing the data given a hypothesized model assuming multivariate normality of variables. It is important to note that normality violations impact parameter estimates, not merely inferential statistics. PLS, on the other hand, uses a series of interdependent OLS regressions to minimize residual variances without making any distributional assumptions. Consequently, PLS should be more accurate than MLE-based CVA if variables distributions are skewed and the increased accuracy is not offset by the bias inherent in PLS. Since skewed distributions are often found in field-based surveys of customer satisfaction, this suggests that more frequent use of PLS by satisfaction researchers may be warranted.

The multivariate normality assumption can be relaxed with elliptical estimation or asymptotic distribution-free estimation (Browne 1984). However, these procedures typically require sample sizes in excess of 200 (depending on the number of parameter estimates). Large samples are typically not used by academic researchers, especially in experiments, and may be unavailable, prohibitively expensive, or impractical to obtain due to multiple segments in commercial research. Yet a substantial body of evidence shows that causal models estimated using covariance-based approaches and based on small samples often lead to poor model fit and inadmissible solutions (e.g., Anderson and Gerbing 1984; Boomsma 1983; Chou and Bentler 1995; Dillon et al. 1987; Hu and Bentler 1995). MacCallum et al. (1996) have suggested that under exploratory conditions minimum samples sizes of 200 to 400 cases are needed.

---

[2] The accuracy depends on data characteristics. As noted by Schneeweiss (1990, p. 38), PLS will generate consistent parameter estimates when "blocks of manifest variables are related to each other." Conversely, PLS estimates based on data drawn from a covariance-based model may be inconsistent.

The PLS procedure is believed to be preferable when sample sizes are small. Wold (1985) reports estimation results based on a sample of 10 cases. Fornell and Bookstein (1982) used a sample of 24 cases with 28 manifest variables included in their model. In contrast, a large number of parameter estimates relative to sample size can lead to non-convergent or improper solutions in MLE-based CVA (Gerbing and Anderson 1987). For example, Ryan and Rayner (1998) found frequent MLE failures using simulated data for a model with 15 latent variables when the sample size was less than 1,000. Since PLS does not suffer from non-convergent or improper solutions, it does not break down when estimating large models with small samples. However, it remains to be seen if PLS produces more accurate estimates than MLE-based CVA given convergence and proper solutions and, if so, what sample size is needed before MLE-based CVA outperforms PLS.

PLS provides the best prediction of a specified set of variable relationships without requiring the strong measurement models demanded by MLE-based CVA. For example, Ryan and Rayner (1998) found consistent MLE failures when models contained only two measures per latent variable. They also found that when measurement model loadings approached .70, PLS estimates were more accurate than MLE estimates, whereas the results converged when loadings were .90. It appears, therefore, that one would favor PLS estimates when model measurement is not strong. However, it is unknown how this effect would behave under different sample sizes or normality violations or whether model fit and colinearity conditions produce different empirical results. To address these unanswered questions, we conducted an extensive simulation study, as described in the next section.

14.4 Simulation Study

A series of Monte Carlo simulations were used to study the effects of various design factors on alternate measures of path estimation accuracy. The design factors that were varied across cells are the following: estimation approach (PLS versus AMOS), sample size (50, 100, 200, 500, and 1,000), data distribution (MVN versus extreme), number of measurement items per construct (2, 4, and 6), correlations among the independent constructs (low versus high), and R-squared for the dependent construct (low versus high). The study design is full-factorial, with 240 separate design cells and 50 replications conducted per cell.

To aid interpretability, other factors are held constant across all conditions. Loadings between measures and constructs are set at 0.70 in all cases, and only reflective epistemic relationships are used. All models incorporate the same structure: five independent constructs are causally linked to a single dependent construct (Fig. 14.1 shows the structure of a typical model). The true path values between the independent constructs and the dependent construct are set to be 0.35, 0.2, 0.2, 0.2, and 0.05. In our experience, these values are typical of the relative magnitudes of effects that are observed in many applied research contexts (e.g., customer satisfaction studies).

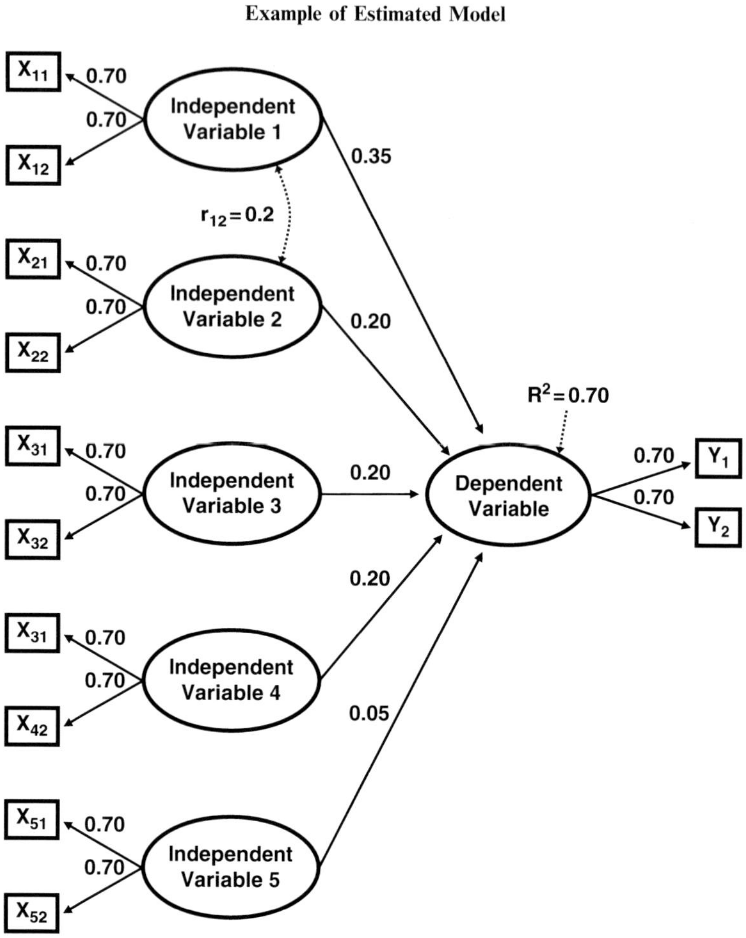

Fig. 14.1 Example of estimated model

14.4.1 Model Performance

Three commonly reported measures are used to assess how well the models estimate the path coefficients between the independent and dependent constructs: root mean square error (RMSE), mean absolute deviation (MAD), and bias. All three of these measures are based on comparisons between the estimated and true path values. In addition, the number of estimation failures for each design cell is noted.

14.4.2 Data Generation

To aid interpretability across design cells, we assume that the constructs in Fig. 14.1 (the five IVs and the dependent variable) are "standardized" variables (that is, they have zero means and standard deviations of one), with the specified correlation structure among the IVs (0.2 or 0.8).[3] Consequently, the path coefficients are standardized regression coefficients (or beta weights).

Initially, fifty separate random samples were generated using MATLAB for each of 60 multivariate normal (MVN) design cells (5 sample sizes by 2 levels of correlation among the IVs by 2 levels of R-squared values by 3 levels number of measurement items per construct). For a given design cell, data sets for the simulations were built up by a process that started with the generation of five IVs with the desired sample size and correlation structure. Of course the actual sample values were not exactly mean zero and standard deviation one, and so, in the next step these variables were standardized, in the usual way.

Then, the dependent variable was computed as a weighted average of the five standardized variables, using path = (0.35, 0.20, 0.20, 0.20, 0.05) as weights, but with the addition of an error component that was drawn from a univariate normal distribution. Specifically, the dependent variable was determined according to:

$$\begin{aligned} \mathrm{DV} &= 0.35^*\mathrm{IV}_1 + 0.20^*\mathrm{IV}_2 + 0.20^*\mathrm{IV}_3 + 0.20^*\mathrm{IV}_4 + 0.05^*\mathrm{IV}_5 + \mathrm{e} \\ &= \mathrm{model} + \mathrm{e}, \end{aligned}$$

where e is an error term with standard deviation chosen to satisfy the equation

$$\begin{aligned} \mathrm{R-squared} &= \mathrm{var(model)/var(DV)} \\ &= \mathrm{var(model)}/\left[\mathrm{var(model)} + \mathrm{var(e)}\right]. \end{aligned}$$

Once the DV was computed, it too was standardized. However, this had the effect of reducing the expected size of the path coefficients by a constant value that depends on C (the 5×5 correlation matrix for the IVs) and the R-squared value for the model. To see this, note that standardization of DV involves multiplication of the DV equation by the factor 1/sqrt(var(DV)). Also, we have:

$$\mathrm{var(model)} = \mathrm{path'} \times \mathrm{C} \times \mathrm{path}.$$

Therefore, in order to have all of the path estimates comparable directly to the true path coefficients, estimates were weighted by the constant (path' $\times$ C $\times$ path/R-squared)$^{0.5}$ before computing deviations. Finally, the values for the observed items (the X's and Y's in Fig. 14.1) were computed from the standardized constructs, with coefficients set equal to 0.7 in all cases. Before estimation, the observed items were also standardized.

[3] For example, if the intra-correlations among the IVs is 0.8, the 5×5 correlation matrix, C, has diagonal values of 1.0 and off-diagonal values of 0.8.

14.4.3 Design Factors

Estimation Method: The generated data were analyzed using AMOS 4.0 and PLS Graph version 3.0 (Chin 2001).

Sample Size: Random samples of 50, 100, 200, 500, or 1,000 cases were generated for individual design cells. These values were chosen since many published rules of thumb for deciding on a minimum sample size recommend at least 150 to 200 observations. Sample sizes of 50 and 100 fall below such thresholds, while sample sizes of 500 and 1,000 fall above them. Use of a 200 case sample size allows us to also assess how well the estimation methods perform for the most frequently recommended minimum sample size.

Between – (IV) Construct Correlations: For each design cell, the underlying independent constructs were generated to have expected intra-correlations of either 0.2 or 0.8. These values span the range of intra-correlation values among the IVs typically found in practical applications.

Distribution of variables: The data sets originally generated for the MVN cell conditions were transformed directly into extreme distribution data sets, so that the results obtained across the two types of distributions would be directly comparable. This transformation was done in two stages. First, we examined consumers' responses to satisfaction measures from commercial market research drawn from three different industries: information technology, electrical utilities, and healthcare. (In all three cases, multiple measures of satisfaction were used. In the first two industries, 0–10 scales anchored at both ends were used, while in healthcare, 1–5 scales anchored at both ends were used.) Specifically, we looked at the skewness and kurtosis values associated with the distributions of these responses across the three industries. Table 14.2 summarizes our findings, both by industry and overall.

Using skewness and kurtosis values of 1.2 and 4.8 (respectively), the power transformation approach recommended by Fleishman (1978) was then employed to create the extreme distribution data sets.[4] By using this approach, we ensured both that the extreme distribution results can be compared directly to the MVN results and that these results are typical of what a customer satisfaction researcher would expect in the field.

Table 14.2 Average Kurtosis and Skewness values for commercial research measures of customer satisfaction in the IT, electrical utility, and healthcare industries

Industry	*Skewness*	*Kurtosis*
Information technology (3)	−0.810	3.471
Electrical utility (2)	−1.088	3.932
Healthcare (3)	−1.635	6.986
Average	−1.178	4.796

[4] Fleishman's approach requires positive values for both skewness and kurtosis. Since skewness is symmetric about a mean of zero, we used the absolute value obtained from the empirical data reported in Table 2.

Number of Items per Construct: The number of indicators per construct was set at 2, 4 or 6 for an individual design cell. All constructs (both dependent and independent) in a particular cell were constrained to have the same number of indicators.

R-Squared: A cursory review of studies published in the past decade, taken together with our own experience, suggests that customer models typically report R-square values in the range 0.3 to 0.7. We used the two extreme values from this range in our simulation design.

14.5 Results

The RMSE, MAD, and bias performance measures calculated for the 50 replication runs conducted for each of the 240 design cells were used as inputs to three separate MANOVA assessments (one each for the sets of RMSE, MAD, and bias measures). For each MANOVA, five separate dependent measures were used, representing each of the five path coefficients. Initially, the MANOVAs included all main effects as well as all higher order interaction terms that included the design variable representing the estimation approach employed (i.e., AMOS or PLS). It quickly became apparent that the four-way and five-way interaction terms in all three MANOVAs were not significant, and these were dropped from subsequent analyses. The models were then further trimmed to eliminate any three-way interaction terms not significant in any of the three MANOVAs. F-values for the various components retained in the resulting trimmed models are reported in Table 14.3. (All main effect and two-way interaction terms were retained in these models, whether significant or not.)

14.5.1 Main Effects

As shown in Table 14.3, most of the main effects of the design factors on the RMSE, MAD and bias performance measures are significant. The mean values for these performance measures across the different levels of the design factors are reported in Table 14.4. In addition, Table 14.4 reports the number of times we were unable to obtain a model solution for each main effect level.

Somewhat surprisingly, whether the data follow an MVN or extreme distribution has no effect on model estimation performance. In contrast, the estimation approach used, the correlations between the independent constructs, the number of indicators, and the sample size are all observed to have a strong main effect on model performance. Finally, the R-squared valued has a significant effect on MAD, but not on either RMSE or bias. Each of these findings is described in more detail below.

Estimation technique: RMSE, MAD, and bias are all significantly higher when AMOS is used to estimate the models than when PLS is employed. Furthermore,

Table 14.3 Summary of trimmed model F-values (based on Wilk's Λ) for RMSE, MAD, and Bias measures

Component	RMSE	MAD	Bias
NORMAL	0.70	0.75	0.82
AMOS	27.32***	26.30***	36.02***
HIIVCORR	31.72***	35.04***	38.29***
HIRSQUARE	1.74	2.73*	2.00
NUMIND	3.15***	4.75***	5.97***
SAMSIZE	14.73***	17.33***	14.55***
AMOS×NORMAL	0.54	0.40	0.47
AMOS×HIIVCORR	25.08***	24.12***	25.15***
AMOS×HIRSQUARE	0.98	1.43	2.58*
AMOS×NUMIND	3.00***	3.49***	3.74***
AMOS×SAMSIZE	9.36***	10.30***	12.86***
AMOS×HIIVCORR×HIRSQUARE	1.62	2.34**	2.43**
AMOS×HIIVCORR×NUMIND	2.38***	2.78***	2.59***
AMOS×HIIVCORR×SAMSIZE	8.48***	8.67***	9.19***

Notes:
* $p < 0.05$, ** $p < 0.01$, *** $p < 0.001$
NORMAL = 1 if data are MVN, 0 if extreme distribution
AMOS = 1 if MLE was used to estimate the model, 0 if PLS was used
HIIVCORR = 1 if the IV correlation was 0.8, 0 if it was 0.2
HIRSQUARE = 1 if the DV R-square value was 0.7, 0 if it was 0.3
NUMIND = 2, 4, or 6; and
SAMSIZE = 50, 100, 200, 500, or 1,000

all model estimation failures occur with AMOS.[5] In contrast, model solutions are always obtained with PLS.

Independent Construct Correlations: As would be expected, when the correlations between the IVs are low, RMSE, MAD, and bias are all significantly lower than when the inter-IV correlations are high. However, most of the estimation failures using AMOS occur when the inter-IV correlations are low.

Model R-Square: MAD is observed to be significantly lower when a higher R-square is employed. RMSE and bias are also better under a higher R-square, but the differences in these values versus the lower R-square condition are not significant.

Number of Indicators: A significant and systematic decrease in RMSE, MAD, and bias is observed as the number of indicators used to measure each construct increases from 2 to 4, and then to 6. Furthermore, the vast majority of model estimation failures using AMOS occur when only two indicators are used; conversely, AMOS always works when 6 indicators are employed.

Sample Size: As the sample size is increased, model performance systematically and significantly improves for all three performance measures. Moreover,

[5] Out of the 6000 models we attempted to run using AMOS, 235 could not be successfully estimated, representing a failure rate of 3.9%. As will be described later, these failures occur under specific design conditions rather than uniformly across all conditions.

Table 14.4 Average RMSE, MAD, Bias Values, and total number of estimation failures, by main effect

Main effect	*n*	*RMSE*	*MAD*	*Bias*	*Percentage of failures*
NORMAL = 0	120	0.341	0.207	0.120	2.03
NORMAL = 1 (MVN)	120	0.266	0.169	0.091	1.88
AMOS = 0 (PLS)	120	0.104	0.085	−0.019	0
AMOS = 1	120	0.502	0.291	0.231	3.92
HIIVCORR = 0	120	0.083	0.065	−0.007	3.40
HIIVCORR = 1	120	0.524	0.312	0.218	0.52
HIRSQUARE = 0	120	0.359	0.225	0.137	2.08
HIRSQUARE = 1	120	0.248	0.151	0.075	1.83
NUMIND = 2	80	0.437	0.260	0.166	5.68
NUMIND = 4	80	0.273	0.175	0.096	0.20
NUMIND = 6	80	0.201	0.130	0.055	0
SAMSIZE = 50	48	0.682	0.406	0.299	5.79
SAMSIZE = 100	48	0.406	0.242	0.151	3.29
SAMSIZE = 200	48	0.242	0.155	0.075	0.58
SAMSIZE = 500	48	0.117	0.080	0.009	0.12
SAMSIZE = 1000	48	0.070	0.057	−0.005	0

Notes:
1. n in this Table refers to the number of deign cells, with 50 replications run for each cell
2. The percentage of failures reported is the proportion of model estimations that were unsuccessful out of the 50n attempted. For example, out of the 6,000 (50 × 120) normal distribution models that we attempted to run, a total of 122 were not successful

the incidence of AMOS model failure systematically declines as the sample size is increased. However, some model estimation failures result even with a sample size of 500.

14.5.2 Two-way Interactions

Three of the five interactions between estimation technique and the other design factors are significant across all three model performance measures, as reported in Table 14.3. Somewhat surprisingly, there is not a significant interaction between chosen estimation technique and data distribution (i.e., the AMOS × NORMAL interaction term is not significant). In addition to the three interactions that are significant across all three performance measures, the estimation technique – R-square interaction term (i.e., AMOS × HIRSQUARE) has a significant effect on bias (only). Figure 14.2 visually summarizes the nature of the three significant two-way interactions affecting RMSE.[6]

Estimation technique by IV correlation: As panel (a) in Fig. 14.2 shows, when the inter-IV correlation is low, both AMOS and PLS result in similar RMSE values. However, when the inter-IV correlation is high, the AMOS model estimates have a

[6] For all three performance measures, the effects of the two-way and three-way interactions were very similar, so in our discussion we focus only on the RMSE results.

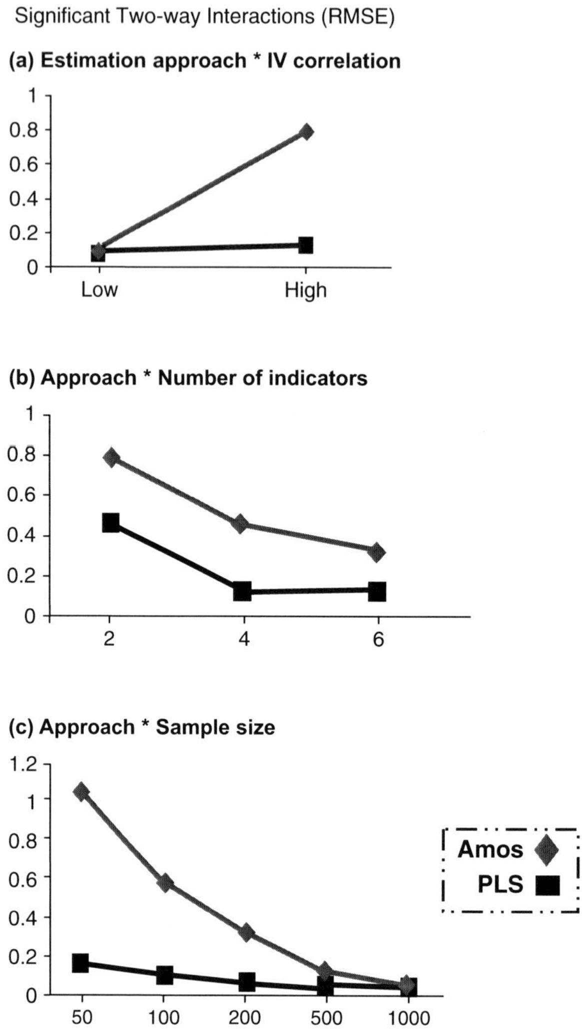

Fig. 14.2 Significant two-way interactions (RMSE). (**a**) Estimation approach × IV correlation. (**b**) Approach × Number of indicators. (**c**) Approach × Sample size

much higher level of RMSE. In contrast, the RMSE values obtained using PLS are quite similar regardless of the level of inter-IV correlation.

Estimation technique by number of indicators: As noted previously, RMSE systematically declines as the number of indicators per construct is increased. However, as shown in Fig. 14.2b, combined with this main effect there is also a systematic effect of estimation technique on RMSE. Specifically, the AMOS model parameter estimates lead to consistently higher RMSE values than PLS, regardless of the number of indicators per construct.

Estimation technique by sample size: As shown in Fig. 14.2c, RMSE is relatively insensitive to sample size when PLS is used. In contrast, when AMOS is employed RMSE depends critically on sample size, with very large errors (i.e., greater than 1.0 on average) observed for a sample size of 50.

14.5.3 Three-way Interactions

Two three-way interactions have a significant effect on all three sets of estimation performance measures: (1) estimation technique by number of indicators by IV correlation (AMOS × HIIVCORR × NUMIND), and (2) estimation technique by sample size by IV correlation (AMOS × HIIVCORR × SAMSIZE). These interactions are visually summarized (for RMSE) in Fig. 14.3. In addition, the three-way interaction between estimation technique, IV correlation, and model R-square (AMOS × HIIVCORR × HIRSQUARE) has a significant effect on MAD and bias, but not RMSE.

Estimation technique by IV correlation by number of indicators: Fig. 14.3a shows that RMSE is relatively stable (and low) when PLS is used (regardless of the level of inter-IV correlation involved) or when AMOS is used under the low inter-IV correlation condition. However, use of AMOS when the IV correlations are high leads to higher RMSE values, particularly when fewer indicators per construct are used. Model estimation failure using AMOS is also much higher when the inter-IV correlation is high and the number of indicators is small.

Estimation technique by IV correlation by sample size: Similarly, RMSE is relatively invariant when PLS is used, or when AMOS is used under the low inter-IV correlation condition. However, when the inter-IV correlation is high and AMOS is employed, RMSE is strongly dependent on sample size. Extreme errors are observed when the sample size is small (n = 50, 100), but even when moderate sample sizes (n = 200, 500) are used, AMOS leads to notably larger RMSE values than does PLS when the inter-IV correlation is high.

Estimation technique by IV correlation by sample size: Both the MAD and bias (but not RMSE) measures are strongly affected when AMOS is used under the high inter-IV correlation condition, regardless of whether R-square is low or high, but both are more severely affected when the model R-square is low. In contrast, the estimation performance measures are relatively stable regardless of model R-square whenever PLS is used or when AMOS is employed in the low between-IV correlation condition.

14.6 Discussion

14.6.1 Multivariate Normal Versus Skewed Distributions

The results suggest that the type of distribution is not a factor in deciding whether to use PLS or an MLE-based approach to causal modeling. Neither failure rate nor

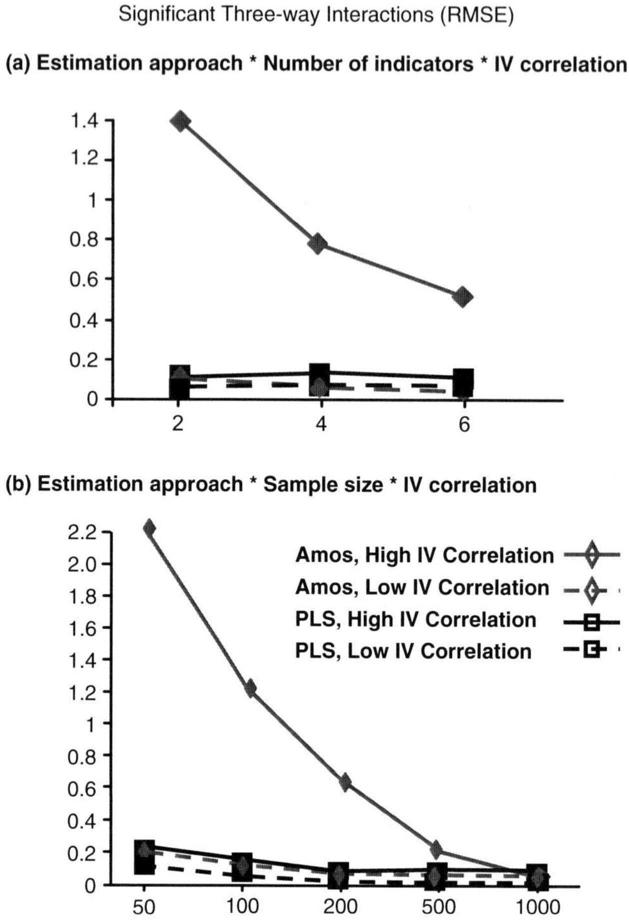

Fig. 14.3 Significant three-way interactions (RMSE). (**a**) Estimation approach × Number of indicators × IV correlation. (**b**) Estimation approach × Sample size × IV correlation

parameter estimation accuracy vary significantly across normal or extremely skewed distributions, whether distribution is considered separately or in combination with other factors. This result is surprising, since the distribution free assumption of PLS in estimating parameters is frequently touted as an advantage of using it rather than MLE. Our results suggest that normality should not be a consideration in choosing which procedure to use.

14.6.2 Model Fit and Parameter Estimate Accuracy

The degree of model fit (i.e., R-square) is a minor factor, affecting MAD to some extent whether considered separately or in combination with other factors. Our results, which suggest that either method will yield comparably accurate estimates whether or not the data strongly fit the theoretical structure, run counter to conventional wisdom.

14.6.3 Sample Size, Independent Variable Correlation, and Number of Measures

Our finding that small sample size and a limited number of indicators per construct result in high model inadmissibility (when the independent constructs are only weakly correlated with one another) or extremely biased parameter estimates (when the independent constructs are highly inter-correlated) is consistent with a study by Marsh and Hau (1998). They recommend the use of at least four or five indicators per construct, particularly when sample sizes drop below 100. Our results suggest that even with six indicators per construct, MLE-based CVA produces relatively high estimate errors when the independent constructs are highly correlated with one another (a condition not studied by Marsh and Hau).

PLS errors are considerably lower than those generated by MLE-based CVA when 2 or 4 indicators per construct are used, when sample sizes are below 500, and when high between-IV correlations exist. In contrast, PLS errors remain relatively constant in the face of high correlations down to sample sizes of 100 and across 2, 4, or 6 indicators. Thus, PLS can be thought of as an estimation technique that is highly robust across different model characteristics.

14.7 Conclusion

For the situations modeled here, PLS appears to generally produce more accurate structural equation path estimates than does MLE. The accuracy gap widens considerably in favor of PLS with 4 or 2 indicators and sample sizes below 500 in the face of high correlations. It appears that sample sizes of at least 1,000, low correlations among the independent variables, and 6 indicators per construct are needed to produce accurate MLE path estimates.

One main reason for this may be that even if the model specification is exactly correct, and the errors are normally distributed, the coefficient estimates from MLE – for relatively small samples – can have relatively larger standard errors than those from PLS because of the relation:

$$\text{MSE} = \text{Variance} + (\text{Bias})^2$$

Thus, PLS can perform better for small sample situations; while there is a slight bias to PLS estimates, the variance of PLS estimates can be much smaller than the variance of the MLE estimates. In such cases, the total MSE will be smaller for PLS than for MLE estimation.

It must be acknowledged that the findings upon which these conclusions are drawn are based on a rather simple causal model. Use of more complex models (e.g., more independent variables, the addition of moderator variables, multiple dependent variables, direct and indirect paths) would seem likely to exacerbate the differences reported here as they entail more parameters relative to sample size and create the need for long questionnaires that may introduce other elements of error and bias (e.g., measurement error, sampling bias). On the other hand, related studies both in this book (i.e., Vilares et al. 2010; Barraso et al. 2010) and elsewhere (e.g., Chin et al. 2003) also show empirical support for use of PLS over MLE models in specific situations (particularly in situations involving small samples and a smaller number of indicators per construct).

In summary, PLS appears to be the preferred estimation method when the researcher's purpose is either to guide management in the allocation of scarce resources or to determine the relative relationships among latent variables in a theoretical structure. In addition, PLS models do not break down as do MLE models, especially when sample sizes are 100 or less, there are 2 indicators per construct, and independent variable correlations are low.

Acknowledgements John Hulland is Associate Professor of Marketing at the Katz Graduate School of Business, University of Pittsburgh. Michael J. Ryan is Professor Emeritus at the Ross School of Business, University of Michigan. Robert K. Rayner is Vice President, Market Strategies, Inc. The authors appreciate financial support for this work from the Richard Ivey School of Business, the Katz Graduate School of Business, and the Social Sciences and Humanities Council of Canada. We also appreciate the very helpful comments made on earlier versions of the paper by Michel Wedel and Fred Feinberg. Finally, we are extremely grateful for the study design and estimation assistance provided by Sungjin Hong.

References

Anderson, J. C., & Gerbing, D. W. (1984). The effect of sampling error on convergence, improper solutions, and goodness-of-fit indices for maximum likelihood confirmatory factor analysis. *Psychometrika*, *49*, 155–173.

Arbuckle, J. (1994). AMOS – Analysis of moment structures. *Psychometrika*, *59*(1), 135–137.

Asher, H. B. (1976). *Causal modeling*. Beverly Hills: Sage

Bagozzi, R. P. (1977). Structural equation lModels in experimental research. *Journal of Marketing Research*, *14*, 209–226.

Bagozzi, R. P. (1980). *Causal models in marketing*. New York, NY: Wiley.

Bagozzi, R. P. (1982). Introduction to the special issue on causal modeling. *Journal of Marketing Research*, *19*, 403.

Bagozzi, R. P. (1984). A prospectus for theory construction in marketing. *Journal of Marketing*, *48*, 11–29.

Bagozzi, R. P., & Yi, Y. (1989). On the use of structural equation models in experimental designs. *Journal of Marketing Research*, *26*, 271–284.

Bagozzi, R. P, & Yi, Y. (1994). Advanced topics in structural equation models. In R. P. Bagozzi (Ed.), *Advanced Methods of Marketing Research* (pp. 1–78). Cambridge: Blackwell Business.

Bagozzi, R. P., Yi, Y., & Singh, S. (1991). On the use of structural equation models in experimental designs: two extensions. *International Journal of Research in Marketing*, *8*, 125–140.

Barclay, D. W. (1991). Interdepartmental conflict in organizational buying: the impact of the organizational context. *Journal of Marketing Research*, *28*, 145–159.

Barclay, D. W., Higgins, C. A., & Thompson, R. (1995). The partial least squares (pls) approach to causal modeling: personal computer adoption and use as an illustration. *Technology Studies*, *2*, 285–323.

Barroso, C., Carrión, G. C., & Roldán, J. L. (2010). Applying maximum likelihood and PLS on different sample sizes: studies on SERVQUAL model and employee behavior model. In V. Esposito Vinzi, W. Chin, J. Henseler, & H. Wang (Eds.), *Handbook Partial Least Squares*. Berlin: Springer.

Beardon, W. O., Sharma, S., & Teel, J. (1982). Sample size effects on chi square and other statistics used in evaluating causal models. *Journal of Marketing Research*, *19*, 425–30.

Bentler, P. M. (1995). *EQS structural equations program manual*. Encino, California: Multivariate Software.

Boomsma, A. (1983). On the robustness of LISREL (maximum likelihood estimation) against small sample size and non-normality. Amsterdam: Sociometric Research Foundation.

Chin, W. W. (1998). The partial least squares approach for structural equation modeling. In G. A. Marcoulides (Ed.), *Modern Methods for Business Research* (pp. 295–336). Mahwah, NJ: Erlbaum.

Chin, W. W. (2001). *PLS-Graph Manual Version 3.0*, http://disc-nt.cba.uh.edu/chin.

Chin, W. W., & Newsted, P. R. (1998). Structural equation modeling analysis with small samples using partial least squares. In R. H. Hoyle (Ed.), *Statistical strategies for small sample research* (pp. 307–341). Newberry Park, California: Sage.

Chin, W. W., Marcolin, B. L., & Newsted, P. R. (2003). A partial least squares latent variable modeling approach for measuring interaction effects: results from a monte carlo simulation study and an electronic-mail emotion/adoption study. *Information System Research*, *14*(2), 189–217.

Chou, C. P., & Bentler, P. M. (1995). Estimates and tests in structural equation modeling. In R. H. Hoyle (Ed.), *Structural equation modeling: concepts, issues, and applications* (pp. 37–55). Newberry Park, California: Sage.

Dillon, W. R., Kumar, A., & Mulani, N. (1987). Offending estimates in covariance structure analysis: comments on the causes of and solutions to heywood cases. *Psychological Bulletin*, *101*, 126–135.

Dijkstra, T. (1983). Some comments on maximum likelihood and partial least squares methods. *Journal of Econometrics*, *22*, 67–90.

Duncan, O. D. (1966). Path analysis sociological examples. *American Journal of Sociology*, *72*, 1–16.

Falk, R. F., & Miller, N. B. (1992). *A Primer for Soft Modeling*. Akron, Ohio: University of Akron Press.

Fleishman, A. I. (1978). A method for simulating non-normal distributions. *Psychometrika*, *43*(4), 521–532.

Fornell, C. (1982). *A second generation of multivariate analysis*, Vol. 1, New York: Praeger.

Fornell, C. (1983). Issues in the application of covariance structure analysis. *Journal of Consumer Research*, *9*, 443–448.

Fornell, C. (1989). The blending of theoretical and empirical knowledge in structural equations with unobservables. In H. Wold (Ed.), *Theoretical Empiricism* (pp. 153–173). New York: Paragon.

Fornell, C. (1992). A national customer satisfaction barometer: the swedish experience. *Journal of Marketing*, *56*, 6–21.

Fornell, C., & Cha, J. (1993). Partial least squares, Unpublished working paper. Ann Arbor: University of Michigan Business School.

Fornell, C., & Larcker, D. (1981). Evaluating structural equation models with unobservable variables and measurement error. *Journal of Marketing Research*, *18*, 39–50.

Fornell, C., & Bookstein, F. (1982). Two structural equation models: LISREL and PLS applied to consumer exit-voice theory. *Journal of Marketing Research, 19*, 440–452.

Gerbing, D. W., & Anderson, J. C. (1987). Improper solutions in the analysis of covariance structures: their interpretability and a comparison of alternative respecifications. *Psychometrika, 52*, 99–111.

Goldberger, A. S. (1973). Structural equation models: an overview. In A. S. Goldberger, & O. D. Duncan (Eds.), *Structural equation models in the social sciences* (pp. 1–18). New York: Seminar Press.

Green, D. H., Barclay, D. W., & Ryans, A. B. (1995). Entry strategy and long-term performance: conceptualization and empirical examination. *Journal of Marketing, 59*(October), 1–16.

Hu, L. T., & Bentler, P. M. (1995). Evaluating model fit. In R. H. Hoyle (Ed.), *Structural Equation Modeling Concepts, Issues, and Applications* (pp. 76–99). Newberry Park, California: Sage.

Hulland, J. S. (1999). Use of partial least squares (PLS) in strategic management research: a review of four recent studies. *Strategic Management Journal, 20*, (2), 195–204.

Hulland, J. S., Chow, Y. H., & Lam, S. Y. (1996). Use of causal models in marketing research: a review. *International Journal of Research in Marketing, 13*(2), 181–197.

James, L. R., Mulaik, S. A., & Brett, J. M. (1982). *Causal analysis: assumptions, models, and data*. Beverly Hills: Sage.

Johnson, M., & Fornell, C. (1987). The nature and methodological implication of the cognitive representation of products. *Journal of Consumer Research, 14*, 214–228.

Jöreskog, K. G. (1970). A general method for analysis of covariance structures. *Biometrika, 57*, 239–251.

Jöreskog, K. G. (1973). A general method for estimating a linear structural equation system. In A. S. Goldberger, & O. D. Duncan (Eds.), *Structural equation models in the social sciences* (pp. 85–112). New York: Seminar Press.

Jöreskog, K. G., & Sörbom, D. (1993). *LISREL 8: Structural Equation Modeling with the SIMPLIS Command Language*. Hillsdale, N.J.: Lawrence Erlbaum Associates.

Lenk, P. (2000). *Bayesian path analysis*, Unpublished working paper. Ann Arbor: University of Michigan Business School.

Lohmöller, J.-B. (1989). *Latent variable path modeling with partial least squares*. Heidelberg: Verlag.

Lohmöller, J.-B., & Wold, H. (1984). Introduction to PLS estimation of path models with latent variables, including some recent developments on mixed scales variables. In G. Melischek, K. Rosengren, & J. Stappers (Eds.), *Cultural indicators: an international symposium*. Wein, Germany: Sonderbruck.

MacCallum, R. C., & Austin, J. T. (2000). Applications of structural equation modeling in psychological research. *Annual Review of Psychology*, 202–220.

MacCallum, R. C., Browne, M. W., & Sugawara H. M. (1996). Power analysis and determination of sample size for covariance structure modeling. *Psychological Methods, 1*, 130–149.

Marsh, H. W., & Hau, K.-T. (1998). Confirmatory factor analysis: strategies for small sample sizes. In R. H. Hoyle (Ed.), *Statistical strategies for small-sample research* (pp. 251–284). Newbury Park, California: Sage.

McArdle, J. J. (1980). Causal modeling applied to psychonomic systems simulation. *Behavioral Research Methods and Instrumentation, 12*, 193–209.

Plouffe, C. R., Vandenbosch, M., & Hulland, J. (2001). Intermediating technologies and multi-group adoption: a comparison of consumer and merchant adoption intentions towards a new electronic payment system. *Journal of Product Innovation Management, 18*, 65–81.

Plouffe, C., Hulland, J. S., & Vandenbosch, M. (2001). Research Report: richness versus parsimony in modeling technology adoption decisions – understanding merchant adoption of a smart card-based payment system. *Information Systems Research, 12*(2), 208–222.

Qualls, W. J. (1987). Household decision behavior: the impact of husbands' and wives' sex role orientation. *Journal of Consumer Research, 14*, 264–279.

Ryan, M. J., & Rayner, B. (1998). Estimating structural equation models with PLS. Paper presented at the American Marketing Association Advanced Research Techniques Forum, Keystone, CO.

Schneeweiss, H. (1990). Models with latent variables: LISREL versus PLS. *Contemporary Mathematics*, *112*, 33–40.

Sharma, S., Durvasula, S., & Dillon W. R. (1989). Some results on the behavior of alternative covariance structure estimation procedures in the presence of non-normal data. *Journal of Marketing Research*, *26*, 214–221.

Smith, J. B., & Barclay, D. W. (1997). The effects of organizational differences and trust on the effectiveness of selling partner relationships. *Journal of Marketing*, *61*, 3–21.

Vilares, M. J., Almeida, M. H., & Coelho, P. S. (2010). Comparison of likelihood and PLS estimators for structural equation modeling: a simulation with customer satisfaction data. In V. Esposito Vinzi, W. Chin, J. Henseler, & H. Wang (Eds.), *Handbook Partial Least Squares*. Berlin: Springer.

Wold, H. (1966). Nonlinear estimation by iterative least squares procedures. In F. N. David (Ed.), *Festschrift for J. Neyman: Research Papers in Statistics* (pp. 411–444). London: Wiley.

Wold, H. (1980). Model construction and evaluation when theoretical knowledge is scarce: theory and application of partial least squares. In J. Kmenta, & J. B. Ramsey (Eds.), *Evaluation of Econometric Models*. New York: Academic Press.

Wold, H. (1982). Soft modeling: the basic design and some extensions. In K. G. Joreskog, & H. Wold (Eds.), *Systems Under Indirect Observations: Causality, Structure, Prediction*, Part 2, (pp. 1–54). North-Holland: Amsterdam.

Wold, H. (1985). Systems analysis by partial least squares. In P. Nijkamp, L. Leitner, & N. Wrigley (Eds.), *Measuring the Unmeasurable* (pp. 221–251). Dordrecht, Germany: Marinus Nijhoff.

Wright, S. (1934). The method of path coefficients. *Annals of Mathematical Statistics*, *5*, 161–215.

Zinkhan, G. M., Joachimsthaler, E., & Kinnear, T. C. (1987). Individual differences and marketing decision support system usage and satisfaction. *Journal of Marketing Research*, *24*, 208–214.

Chapter 15
PLS in Data Mining and Data Integration

Svante Wold, Lennart Eriksson, and Nouna Kettaneh

Abstract Data mining by means of projection methods such as PLS (projection to latent structures), and their extensions is discussed. The most common data analytical questions in data mining are covered, and illustrated with examples.

(a) Clustering, i.e., finding and interpreting "natural" groups in the data
(b) Classification and identification, e.g., biologically active compounds vs inactive
(c) Quantitative relationships between different sets of variables, e.g., finding variables related to quality of a product, or related to time, seasonal or/and geographical change

Sub-problems occurring in both (a) to (c) are discussed.

(1) Identification of outliers and their aberrant data profiles
(2) Finding the dominating variables and their joint relationships
(3) Making predictions for new samples

The use of graphics for the contextual interpretation of results is emphasized.

With many variables and few observations (samples) – a common situation in data mining – the risk to obtain spurious models is substantial. Spurious models look great for the training set data, but give miserable predictions for new samples. Hence, the validation of the data analytical results is essential, and approaches for that are discussed.

S. Wold, N. Kettaneh
NNS Consulting, 42 Pine Hill Rd, Hollis, NH 03049, USA
e-mail: sbwold@gmail.com, nouna@nnsconsulting.com

L. Eriksson
Umetrics Inc, 17 Kiel Ave, Kinnelon, NJ 07405, USA
e-mail: Lennart.eriksson@umetrics.com

V. Esposito Vinzi et al. (eds.), *Handbook of Partial Least Squares*, Springer Handbooks of Computational Statistics, DOI 10.1007/978-3-540-32827-8_16,

15.1 Introduction

15.1.1 General Considerations and Scope of Chapter

Data mining and integration refer to the use of historical data from large, multiple and diverse databases for some specific purpose, usually classification or predictive modeling (regression). *Data mining* involves a wide variety of data analytical approaches, ranging from traditional exploratory data analysis to the application of a collection of fancy data analytical tools to massive amounts of messy and incomplete data sets. *Data integration* entails the amalgamation of data of different shapes from multiple parts of a system, usually sitting in separate data bases, in an attempt to provide the basis for a holistic understanding of the whole system.

Our objective is to discuss data mining and data integration by means of projection methods, notably PLS, getting an efficient, simple and interpretable approach to this difficult problem area. We shall advocate the use of principal component analysis (PCA) for the analysis of one data table and projections to latent structures (PLS) for the analysis of the two-block (X/Y) problem. A thorough discussion of the two-block PLS method as used in chemometrics is found in Wold et al. (2001).

Extensions of the basic PCA and PLS methodologies play an important role in data mining and data integration. These methods involve hierarchical PCA & PLS (Eriksson et al. 2002; Wold et al. 1996), batch PLS (Wold et al. 1998) and orthogonal PLS (OPLS) (Trygg and Wold 2002).

The transparent properties of projection methods and their extensions to the handling of missing data, noise in X & Y, and multicollinearities while still providing results that are meaningful and have a faithful graphical representation – make these methods ideal for data mining and integration. For example, the scores of a PCA or PLS model provide summaries of the original variables that are optimal in a certain sense. Hence, they are ideal for information transfer from one block of data to another and hence provide a straight forward mechanism for data integration.

15.1.2 Data Analytical Challenges

Historical data often reside in large, multiple and diverse databases, and it is usually far from trivial to prepare the data for analysis. This involves synchronizing and linking different records, to achieve data providing a reliable, representative and interpretable picture of the whole system or process. Aligning and shifting data consumes considerable effort, while the actual data analysis, is relatively straightforward.

As an example, one of the greatest challenges pharmaceutical companies face when rolling out process analytical technology (PAT) relates to the ability to organize and concatenate measured data. The approach in pharmaceutical manufacturing has very much been to store data securely in vast data bases but rarely, if ever, retrieve and use them. Indeed, previous regulatory environments did not provide incentives for analysis of manufacturing processes because implementing improvements would have required costly and time consuming re-validation. The current state of pharmaceutical data infrastructure reflects this situation. However, integrating, synchronizing and aligning data from all relevant sources is a pre-requisite before any data analysis can begin.

We emphasize that this chapter deals with the problems encountered after the data collection phase, which may be the most difficult and critical in itself. We here describe data mining and data integration based on PCA and PLS. However, as pointed out below, data mining and data integration are terms carrying different connotations and meanings to different people. Depending on the context, the research field, the data analytical culture, etc., the emphasis of the data analysis may vary, and different data analytical tools including hard to interpret "black boxes" are used (Hand 1998; Buydens et al. 1999).

15.1.3 A Few Words on Data Mining

Data mining is a multi-faceted task and in many cases the goals are unclear. Large, scattered data sets are common (see Fig. 15.1 for an example). Following Hand (1998), we shall here discuss the typical data mining problem as one with the following characteristics:

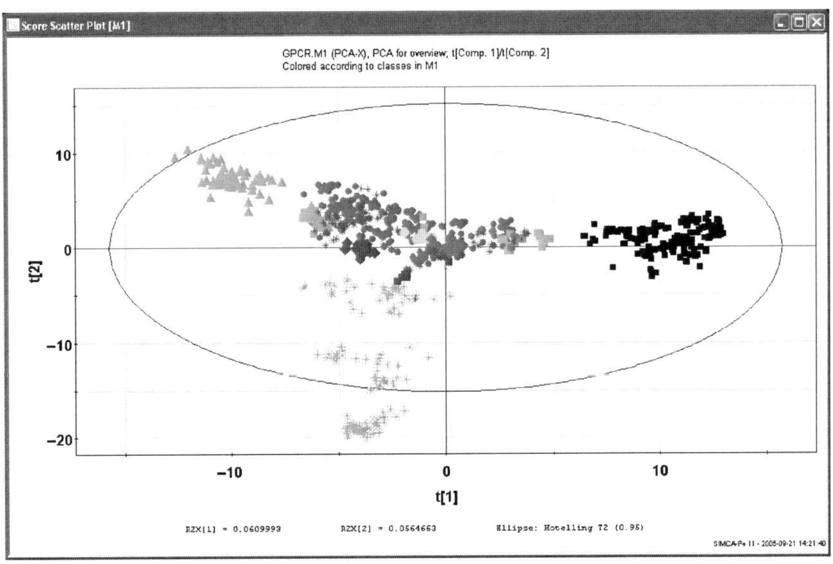

Fig. 15.1 First score plot of a PCA model of a GPCR (G-protein coupled receptors) data set. Each point is one GPCR sequence. In total, there are 897 sequences, distributed across 12 classes. Each GPCR is characterized by 675 sequence descriptor variables. The smallest class has 4 members and the largest 302. The dataset shows strong groupings. Any data mining effort must consider the clustering and the varying cluster size

- *Large datasets (databases)* – many records, cases, observations, and possibly also many variables, indicators, features, etc.;
- *Secondary analysis* – the data were collected for another objective than the one underlying the present analysis;
- *Heterogeneous data* – the data are clustered in unknown ways, and relationships between variables may change between clusters;
- *Non-independence* – both observations (rows) and variables (columns) are often dependent and/or correlated, but in different ways and degrees in different parts of the data;
- *Selection bias* – different categories of cases have different amounts of data recorded;
- *Drift in the data* – data measured at different times have different means and variances and relationships;
- *Non-numeric data* – qualitative variables, or just pieces of unorganized text are often mixed with quantitative variables.

Data mining also has a certain common set of objectives that occur in various combinations. These objectives can be translated to finding models or patterns, where the former are seen as global (stable in time) and the latter being temporary and of short duration (Hand 1998). We see no principal difference between models that are local in time or local in, for instance, geographical space, and hence we will discuss local models and patterns as synonymous.

Hence, we can define a number of "data mining unit operations" (DUOs), which are similar to those of ordinary data analysis, but often modified to take into account the peculiarities of the data mining situation, as listed above.

These DUOs may be categorized as:

a) data cleaning and pre-processing;
b) overview and cluster analysis;
c) classification and discrimination;
d) relationships and predictive modeling.

These DOUs are amplified upon below (see Sect. 15.2).

15.1.4 A Few Words on Data Integration

Data integration seeks to integrate data from multiple parts of a system or process, so that the essence of the system or process can be readily understood (see Fig. 15.2 for an example). The main data analytical objectives are usually to reveal:

- how the different blocks of data are related (correlated),
- which parts of the system or process provide overlapping information,
- which data blocks provide unique information, and
- which sources of data are most useful from a predictive and interpretative viewpoint.

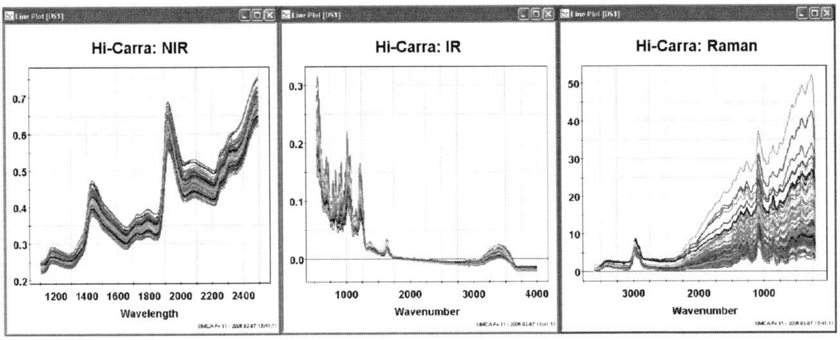

Fig. 15.2 Raw spectra from the Carrageenan dataset. Measurements from three distinct data blocks, i.e. spectra from three different instruments, are to be integrated

One relevant example is systems biology, which tries to gather information from a multitude of compartments in a biological system. Another example is PAT in pharmaceutical manufacturing. Here the aim is to collate relevant process information ranging from raw material, intermediate measurements to final conditions, in order to enhance product quality and lower production costs.

The different data structures to integrate and analyze may vary considerably, from simple two-block (X/Y) problems to the analysis of multi-block (X/Y/Z/...) and multi-way (batches) problems, and extensions thereof. Due to its intrinsic multi-block nature, the problems of data integration are often well addressed by hierarchical modeling techniques. With this approach to data integration, the objective is primarily to achieve a condensed overview of the system or process, an overview which may or may not have a predictive focus on final endpoint data (Y-data).

Hierarchical extensions of the PCA and PLS methods are especially apt at handling data from many sources. Score vectors of models of each source are used as the variables in "upper" models. These hierarchical (blocked) PCA and PLS models, are much simpler than the multi-block PLS models used in the social and economic sciences. Our experience is that this use of scores of blocks of data as variables in "upper" models is an indispensable tool in data mining and data integration. Both for providing summaries of the blocks, and for sequential information transfer from one block to another when the blocks are nested chronologically.

Part of the data integration may also be to uncover which information is common in several data blocks, and which is unique to a specific block. The recent extensions of two-block PLS, namely OPLS and O2-PLS, answer this type of questions by separating the variation in each block into parts that are related to the other block, and parts that are orthogonal to the other block (Trygg and Wold 2002).

In many systems, information (correlation) can flow in more than one direction; for example, it is known that genes influence the metabolic output (genes $\rightarrow$ metabolites), but metabolites also influence the expression of the genes (metabolites $\rightarrow$ genes). Thus, with the O-PLS methods, it is possible not only to focus on the correlation between the different data tables, but also to capture the non-correlating information among these data tables.

Moreover, in very complex data integration problems, with vast arrays of data, possibly also containing noisy variables and missing data, we have found the combination of hierarchical modeling and OPLS to be very powerful. The hierarchical models provide a compact summary of the different data blocks which can then be interrogated and contrasted using OPLS and O2-PLS.

15.1.5 Design and Sampling

As in any data analytical activity, good experimental design and sampling are crucial to ensure representative and meaningful data. With the large numbers of observations usually prevailing in data mining and data integration applications, sampling according to simple multivariate designs is particularly useful (Eriksson et al. 2004).

In many applications, including the analysis of large molecular data sets, a cluster analysis followed by multivariate designs in the separate clusters works well (Eriksson et al. 2000). An example of multivariate design in spectroscopy was presented by Svensson (Svensson et al. 1997). These multivariate designs, usually based on the scores of a PCA or PLS model, aid in selecting a representative subset of the observations. An important new design type in this area is the D-optimal onion design family (Olsson et al. 2004a, b; Eriksson et al. 2004).

Reducing a large number of (sometimes unevenly scattered) observations to only those close to the points of a suitable experimental design, results in a more balanced, more informative and easier data analysis. This approach is in contrast to the standard data mining philosophy of analyzing all data with the hope of mining a few valuable nuggets of information. A by-product of applying design to reduce data is that outliers and other anomalies are quickly filtered out, solving another common problem encountered in data mining.

15.1.6 Organization of the Remaining Parts of the Chapter

Separate sections are devoted to work-flow descriptions and accounts of pertinent examples. Thus, in Sect. 15.2 the work-flow of data mining is rolled out, which is followed by a data mining application (Sect. 15.3). Similarly, Sect. 15.4 (work-flow) and Sect. 15.5 (application) discuss and exemplify the data integration concept. Finally, Sect. 15.6 provides an interesting example where the two concepts are combined. The last section, Sect. 15.7, provides concluding remarks.

15.2 Work-Flow of Data Mining

This section describes an appropriate work-flow for data mining. We re-iterate that we start from the point where the data have been collected, but the data mining has yet to begin. In general one has the choice between looking at all the data, or a small sub-sample. In addition, if the models are to be used for predictive purposes, it makes sense to select a representative test set and set this aside for subsequent validation.

15.2.1 Outliers, Trimming and Winsorising

The larger the dataset, the greater the likelihood that the data are dominated by outliers. These are data points that lie a long way away from the bulk of the data and often severely distorting the results of the analysis. Hence, outliers may also compromise the model interpretability. Since, the vast majority of large datasets do contain outliers, these need to be identified and eliminated from the training set before the modeling begins.

With both PCA and PLS, severe outliers are easily identified in score plots (or Hotelling's T^2 plots if the number of components exceeds 2 or 3). Moderate outliers are identified by examining the row residual standard deviations of X, often called DModX or distance to the model. The severe outliers are extreme cases and must be addressed. Moderate outliers, on the other hand, have a smaller impact (lower leverage) on the model, and need not be excluded from the analysis.

With large and complex data sets, however, the repetitive use of PCA and PLS for data cleaning becomes tedious, and then faster approaches are needed. The simple approaches of trimming and winsorising remove most, if not, all of the serious outliers and hence are particularly useful for pre-processing large data sets (Kettaneh et al. 2005).

Trimming and winsorising involve the separate sorting of each variable and removing or modifying a small percentage of the extreme values (typically between 1 and 5%). Note that only the extreme elements of a single variable are modified at each step – the whole observation (row) is not removed. With trimming, the extreme elements are simply set to "missing" introducing between 2% and 10% of missing values in the data. With winsorising, the extreme elements are replaced with values closer to the mean, e.g. 3 standard deviations (computed in a robust way) or the "last good value" with process data.

15.2.2 Representative and Diverse Data (the Training Set)

It is important to recognize that any model must be based on a representative set of observations to be interpretable. A model based on a dataset exhibiting particular properties may not necessarily be predictively viable with respect to data with different properties. The training set, used to build the model, and the test set, used to validate the model, must contain observations that are similar chemically, biologically, technically, etc. This must also be so for the prediction set for which routine large-scale predictions will be made.

With the large numbers of observations usually available in a data mining application, sampling according to simple multivariate designs is very useful, see e.g., (Kettaneh et al. 2005; Wold et al. 2004). In many cases, including large molecular data sets, a cluster analysis followed by separate multivariate designs in the separate clusters works well (Eriksson et al. 2000). These multivariate designs are usually based on the scores of a PCA or PLS of the trimmed/winsorized data set, selecting a subset of the observations according to a D-optimal or onion design (similar to space filling), see e.g., (Olsson et al. 2004a, b).

15.2.3 Selection of Test Data (Validation Set, Prediction Set)

Also, to create a diverse and representative test set, sampling according to a suitable experimental design is a practical approach. As discussed above, such a design is usually based on the scores of an initial PCA or PLS model of the dataset. With manufacturing data, however, the test set occurs naturally with each additional unit or batch produced, thus providing a strong validation of the goodness of the model.

15.2.4 Centering and Scaling

The standard procedure in most data analyses is to standardize variables by subtracting their averages and dividing by their standard deviations. This shifts the data to the origin in multidimensional space and gives each variable equal importance. This also corresponds to working with the correlation matrix as opposed to the covariance matrix.

However, in some datasets there are other more natural reference points. For example, in processes the set points of controlled variables provide obvious and interpretable reference values. For uncontrolled process variables, using averages may still be the best approach. In QSAR, the unsubstituted compound (H) is often a natural reference point while in biological trials the average of the "controls" may be preferred.

Analogously, range scaling (dividing by the permissible range) is sometimes warranted with process and biomedical data. This references all variation to the maximum allowed, facilitating interpretation of the results.

In data mining it may make sense to apply centering and scaling as described above first, followed by a separate centering and scaling of any clusters that emerge. This again makes the cluster models and their differences easier to interpret.

15.2.5 Overview and Cluster Analysis

An initial PCA or PLS analysis followed by a 2D or 3D plot of the scores often serves as a good overview of the dataset and shows how the observations are grouped. Such groups may indicate a discrete change in some process condition, which eventually might be linked to variable product quality. For large datasets the score plots are often cluttered, and a more formal cluster analysis may be warranted (Maitra 2001; Naes and Mevik 1999). In the present examples, no cluster analysis seems to be needed.

15.2.6 Classification and Discriminant Analysis

Many data mining applications concern classification and discriminant analysis, i.e. to recognize pre-defined groups (classes) in the data, finding out if these are well separated and by which variables, and how well additional observations not contained in the training set can be classified. The most popular data analysis methods seem to be classification trees (CT), linear and quadratic discriminant analysis (LDA and QDA), neural networks (NN), and support vector machines (SVM). With the exception of SVM, all of these methods are based on a regression-like step with inversion of the variance covariance matrix, hence requiring independent variables, which is never the case in data mining. There seems to be a fixation to methods which need more observations than variables and with such methods a preliminary variable reduction step is therefore required. This runs the risk of discarding critical information, as well as ending up with spurious models.

This is somewhat surprising given the multivariate and correlated nature of data mining datasets for which methods such as PLS-DA and SIMCA (based on one PCA or PLS model per class) were specifically developed (Albano et al. 1978; Sjöström et al. 1986). These methods work perfectly well even when there are more variables than observations and so no prior variable selection is required.

A major advantage of PLS compared with LDA, QDA, NN, and SVM is the *model* that PLS makes of the X-space. The loading maps and similarities between variables in the loading space indicate which variables have similar information content and hence may be mechanistically related. Analogously, the score plot of the X-model shows similarities and dissimilarities between the observations (cases, samples) and sub-groups of the classes related to secondary patterns due to, e.g., sex, age, nutritional status, etc. Thus, PCA and PLS models yield much richer results than the other methods, leading the scientist in new directions and stimulating creative thinking (Munck 2005).

15.2.7 Relationships and Predictive Modeling

The search for and estimation of quantitative relationships in data mining is very similar to classification. Again, regression-like methods dominate the scene – regression trees (RT), step-wise multiple regression (SWMLR), neural networks (NN), and support vector machines (SVM). Again, given the properties of the datasets in question, the popularity of these methods makes little sense. The assumption of independence of the X-variables, is rarely realistic.

The major advantage of getting an interpretable model of X, with scores, loadings, and residuals is the same in this type of analysis as in classification discussed above. Hence, PLS is the preferred choice. Finally, the ease with which PLS models can be extended to hierarchical models and non-linear models while retaining interpretability and predictive power makes PLS a suitable tool for data mining.

15.3 A Chemical Data Mining Example: ChemGPS

15.3.1 Background and Objectives

The objective with this investigation is the characterization of molecular structures for compound subset selection and prioritization. We are indebted to Prof. Johan Gottfries at AstraZeneca R&D, Mölndal, Sweden, for granting us permission to use this dataset, which was originally published in Oprea and Gottfries (2001).

Some 10 years ago, combinatorial chemistry evolved as a potentially expeditious route to large sets of molecules with promising pharmacological activity. Using combinatorial approaches, it became possible to synthesize chemical libraries containing in the order of 10^2–10^9 compounds, i.e., to maximize the number of compounds within a given time-frame (Oprea and Gottfries 2001). Today, however, it is realized that there is no strict correlation between the number of compounds and their joint information content. Rather, it is important *which* compounds are made – their ensemble should provide optimal information. This necessitates some kind of multivariate design as the guiding principle for their selection.

A set of informative compounds needs to be based on their molecular similarity and diversity. The compounds must be dissimilar enough to provoke a change in the pharmacological activity profile, yet sufficiently similar to show the same pharmacological mechanism, and hence be possible to model in a single QSAR model. Multivariate characterization and design is an approach whereby a set of representative and diverse compounds can be constructed or selected. A key step in this approach is the application of PCA to the X-matrix containing a multitude of measured and computed molecular descriptors.

Unfortunately, many corporate databases of compounds represent scattered data rendering the selection of diverse subsets more difficult than it should be (Fig. 15.3). This is because the compounds are often heterogeneous and contain strong groups, plus outliers, discontinuities, and other undesirable properties. An initial PCA is useful in the sense that it will often point to such problematic structure in the data. A pre-requisite for subset selection using, e.g. D-optimal onion design, is a homogenous dataset devoid of outliers and subgroups.

The clustered and unbalanced nature of such databases does not only arise because of the varying molecular architecture of different pharmaceutical projects,

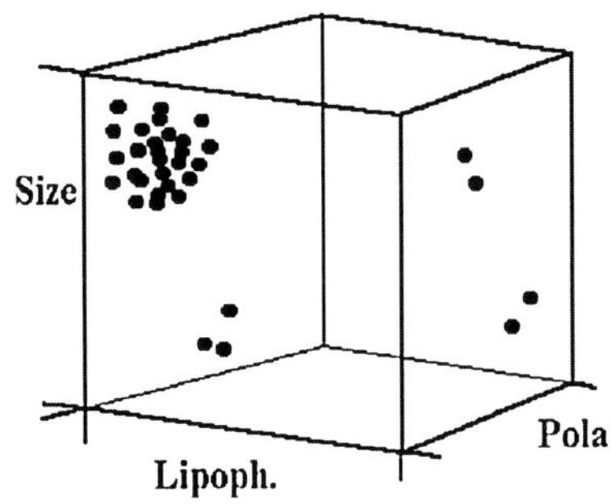

Fig. 15.3 Compound databases in pharmaceutical industry often contain grouped and unbalanced data

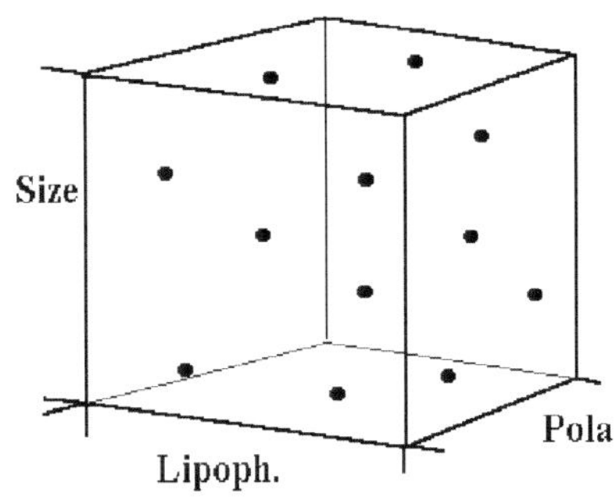

Fig. 15.4 The use of statistical molecular design (SMD) aims at creating a representative and diverse set of compounds for testing

but is primarily due to the lack of properly designed test compounds. Statistical molecular design (SMD) is an efficient tool to select a fairly balanced data set from unbalanced databases of chemical and biological molecular data (Fig. 15.4). The basic principles of SMD are simple – after an initial PCA or PLS analysis of the larger data set, D-optimal or similar designs in the scores are used to select a subset with the best possible balance and representativity (Linusson et al. 2000; Eriksson et al. 2003).

15.3.2 The Need for a Stable Drugspace

As noticed by Gottfries and Oprea, there is often a need to recurrently update a local projection model. There are several potential reasons for this, e.g. (1) the inclusion of new compounds, (2) the incorporation of new chemical descriptors, or (3) a desire to merge data from different sources. In addition, local models pertaining to relatively similar groups of compounds are often difficult to contrast, in particular because the descriptors used in the different models are not necessarily the same.

In order to reduce the burden of model updating, Gottfries and Oprea set out to create a drug-space which would remain stable over a longer period of time and so require less updating. According to their vision, such a drug-space would display more global, as opposed to local, character. And, most importantly, it would

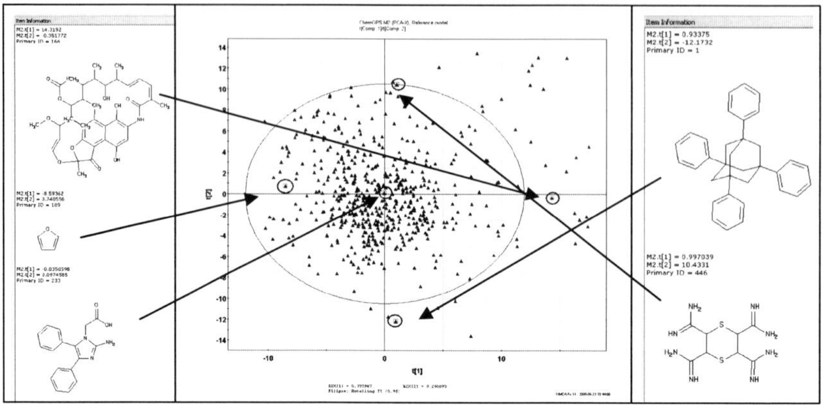

Fig. 15.5 Score plot of the two first principal components of the ChemGPS data set. Each triangle represents one compound. This plot accounts for 60% (34% + 26%) of the total variance in the chemical descriptors. By using the SMILES notation to assist the interpretation of the PCA scores, it is easily realized that the first score vector (*along the horizontal direction*) discriminates between small (*left*) and large (*right*) compounds, and that the second score vector (*along the vertical direction*) separates between hydrophilic (*top*) and hydrophobic (*bottom*) compounds

support prediction and classification of new compounds via interpolation rather than extrapolation.

This new approach to a "fixed drugspace" was called ChemGPS. It is based on defining a set of "core" molecules, i.e. representative drugs, which are used to compute a principal components model, in this context called a principal property model. This model is based on variables being the most important properties (lipophilicity, polarisability, charge, flexibility, etc.) of the investigated molecules. Additionally, one then defines a set of "satellite" molecules, which are intentionally positioned beyond the drug-space defined by the "core" molecules. Such satellite molecules exhibit some extreme properties but still contain druglike structural fragments.

Originally, the ChemGPS model was developed from a set of 423 core and satellite structures using 72 chemical descriptors. The most recent edition, has grown to include 552 compounds. During the prediction phase, ChemGPS positions novel chemical structures in the drug-space. By evaluating the closest neighbors of the predicted sample, the analyst gains a rapid insight into whether the new molecule is likely to function as a drug or not.

One advantage of the ChemGPS model is that the score values are comparable across a large number of chemicals, and do not change much as new structures are addressed (unless, of course, radically new molecules are used to update the model). Hence, this tool may serve as a reference system for comparing multiple compound libraries, and for keeping track of previously inspected regions of the chemical space. The first score plot of the current version of the ChemGPS model is shown in Fig. 15.5. The SMILES (Simplified Molecular Input Line Entry Specification) notation associated with each compound gives a rapid interpretation of the PCA model.

15.3.3 ChemGPS and Its Use in Data Mining

The PCA model of the ChemGPS dataset can be used for data mining purposes in the sense that it provides a coherent basis for the selection of a representative and diverse subset of compounds. As can be seen from Fig. 15.5, the compounds are not divided into major clusters but are rather spread throughout the chemical space. This can be attributed to the fact that the compounds were deliberately selected to span the chemical space of orally active drugs (Oprea and Gottfries 2001).

The ChemGPS model can be used in different ways depending on the objective of the investigation; two possibilities are:

- to select a small subset of compounds spanned by the training set;
- to identify new compounds with promising pharmaceutical properties, based on a selection from the training set and prediction of the locations of a test set of new compounds.

Figure 15.6 suggests the basic principles for the selection of compounds according to the two possibilities above.

15.3.4 Discussion of Example

The objective of the ChemGPS model is to provide a consistent mapping device that avoids extrapolations when estimating and positioning the properties of a new set of compounds or druglike organic molecules. Unlike "conventional" applications of data mining in chemical space, where all data for all compounds in a database are analyzed, this approach relies on the modeling of a subset of compounds deliberately selected to cover as many aspects as possible of the chemical space spanned by orally active drugs. This reference set of compounds is homogenous, i.e. does not contain strong sub-clustering, and provides a fixed universe of orally active drugs.

When new compounds become available in a research project they can be projected into this fixed universe and their properties be readily understood by looking at the property profiles of compounds situated in the vicinity of the new compounds. Sometimes outliers are encountered in such a prediction process. If there are several persistent outliers "of the same type", this suggests that the new compounds contain new information which the ChemGPS system is not trained for. Outliers that hold interesting or extreme properties in some chemical descriptors can therefore be used to expand the ChemGPS model.

If the number of predicted compounds is large, and it is of relevance to select a smaller number for continued exploration and testing, then statistical molecular design can be applied.

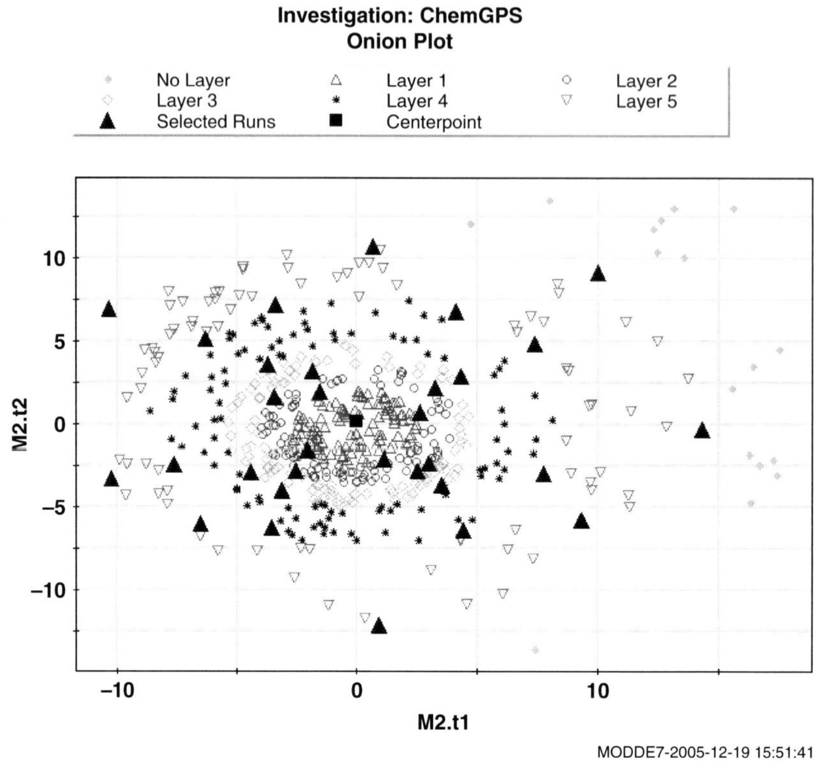

Fig. 15.6 Selection of a representative subset of compounds from the ChemGPS training set. The selection was made using an onion design (Olsson et al. 2004a, b). Equivalently, a set of new compounds with diverse properties could have been identified by laying out an onion design, or similar design, in a set of predicted t-scores including information also for a test set of compounds. By reducing the large number of (the sometimes unevenly scattered) observations to only those close to the points of a suitable experimental design, one obtains a more balanced, informative, and easier to analyze dataset

15.4 Workflow of Data Integration

As discussed in the introduction to this chapter, perhaps the most challenging phase of data integration is data formatting, i.e. the selection, alignment and organization of acquired data, such that the various measurements and calculated indices, possibly drawn from different databases, are matched up and address the same observation (time point, sample, object, ...) in a meaningful and transparent way. We do not discuss the data formatting phase itself, but start from the point when the data are appropriately arranged and ready for multi- and megavariate data analysis.

15.4.1 Pre-processing of Data

The main steps of data mining were listed in Sect. 15.2. These principles apply in data integration as well. Thus, pre-processing of data in terms of trimming, winsorizing, centering, scaling, filtering, transformation, etc., is the mandatory first step. A key objective of data pre-processing is to prepare and shape the data into a form that is suitable for data analysis, a form which should enhance the efficiency of the ensuing modeling.

15.4.2 Obtaining an Overview of the Data

The first modeling step is often a PCA overview model of the entire data set. This will provide a general feeling for what is going on in the dataset, its structure, whether there are outliers, time trends, groupings, etc.

15.4.3 Ascertaining Homogeneity and Representativity

As with any type of modeling, in data mining and integration, homogeneity and representativity are essential for the training set as well as the test and prediction sets. A homogenous dataset with little clustering and no high leverage points (influential outliers) increases the chances of obtaining a viable model.

Representativity implies that the training set contains all the sources of variation that are expected to influence the system or process over time, and that it covers the range of the data expected in the future. This is achieved with sound sampling strategies and/or DOE. In batch modeling (example 3, see Sect. 15.6) it is crucial to ensure that sufficiently many good batches are included so that the model can be trained on the mechanistically and practically relevant variation. There should also be additional good and bad batches set aside as a test set to verify the predictive ability of the batch model.

15.4.4 Block-Modeling Using the Hierarchical Approach

After the initial overview, removal of outliers, division of observations into pertinent training and test sets, etc., it may be advisable to compute a local PCA model, and/or local PLS model if Y-variables are available, for each block of data. The score vectors arising from each of these local models can then be combined to form a new block of data which is analyzed to explore the relationships between the blocks. This is the essence of the hierarchical modeling approach to megavariate data (Eriksson et al. 2006b, Chap. 24).

15.4.5 Linking and Contrasting Blocks of Data, and Information Transfer

Depending on the objective of the data integration, the score variables drawn from the various block models can be used differently. Some of the options are as follows:

- *Linking blocks of data*: Here, the score vectors from the lower levels of the hierarchical model are simply joined together in one X-matrix and then co-analyzed using PCA or PLS. This analysis will highlight how the different score variables co-vary, i.e. the degree of information overlap between the different blocks. When external Y-data are available, such as in QSAR or at the batch level of batch modeling, PLS or OPLS are used. In that case, the focus of the data analysis is predictive modeling.
- *Contrasting blocks of data*: Within the framework of this data analytical objective, the set of score vectors derived from each of the local models are not put together as one combined matrix, but are compared and contrasted using either OPLS or O2-PLS. In this case, the aim is to reveal the unique information residing within each block, and also which parts of the system or process provide similar information.
- *Information transfer between blocks of data*: Since the score variables are optimal summaries of the original variables, they may also be used for transferring the information from one block of data to another. This may be of interest in process modeling and monitoring where one part of a process is followed sequentially in time by another part, which is followed by a third, and so on, and the output information at one stage represents the input to the next stage. Here, the variation in the process parameters and raw material properties at the first part/stage of the process can be summarized by a PCA model. The first few scores of this model are then "sent" further downstream as memory parameters and are glued together with the next set of process measurements acquired at the second part/stage. A new PCA or PLS model is then computed and the first few score vectors of this new model are in turn joined with the process and raw material data obtained for the next process step. In this way, the important process information gathered upstream in a process is conveniently exported to the later stages of the same process.

15.4.6 Predictive Modeling

With the relationships among the different blocks of data investigated and understood, the final goal of the data analysis is typically to build quantitative predictions of Y-variables. This step can be carried out using PLS or OPLS. For instance, it may be of relevance to know which block of data gives the best predictions of the dependent data. The general comments voiced in Sect. 15.2.7 apply here too.

15.5 A Spectroscopic Data Integration Example: Carrageenan

15.5.1 Background and Objective

Carrageenans are polysaccharides extracted from seaweed. They are used as gelling and thickening agents in a wide range of products, including food, pharmaceuticals and cosmetics. Many different types of carrageenans exist, each having different gelling and thickening properties. In commercial production, the raw material (seaweed) contains a mixture of carrageenan types and hence the final product is also a mixture of types. It is therefore important for the carrageenan manufacturers to know the composition of their products in order to direct specific products to appropriate application areas, and, if necessary, perform chemical modification prior to release.

We here develop a calibration model of carrageenan composition based on spectral data of three types, namely near infrared (NIR), IR and Raman. A secondary objective was to understand which type of spectral data provided the best basis for predicting carrageenan content in real samples. We are indebted to Jan Larsen, CPKelco A/S, Denmark, and co-workers for granting us permission to use this dataset. This original reference is Dyrby et al. (2004).

15.5.2 Mixture Design to Get Representative Y-Data in the Training and Test Sets

A five-constituent mixture design in six levels was laid out resulting in 128 powder samples. The five Y-variables represent the relative amounts of each of the five carrageenan types (Lambda, Kappa, Iota, Mu and Nu) in each mixture sample. As the five production samples used for mixing the 128 samples were not totally pure, the Y-variables are not exactly at the design levels of 0, 20, 40, 60, 80 and 100%. Furthermore, the Carrageenan types Mu and Nu are not found in natural seaweed at higher purity than about 20%. One fifth of the samples (26 samples) was set aside as a test set (Dyrby et al. 2004) leaving 102 samples in the training set (calibration set).

15.5.3 The Three Blocks of Spectral Data

For each of the 128 samples, three blocks of spectral data were acquired (Fig. 15.2):

- NIR spectra, 1,100–2,500 nm, 699 variables
- IR spectra, 550–4,000 cm −1, 662 variables
- Raman spectra, 3,600–200 cm −1, 3,401 variables

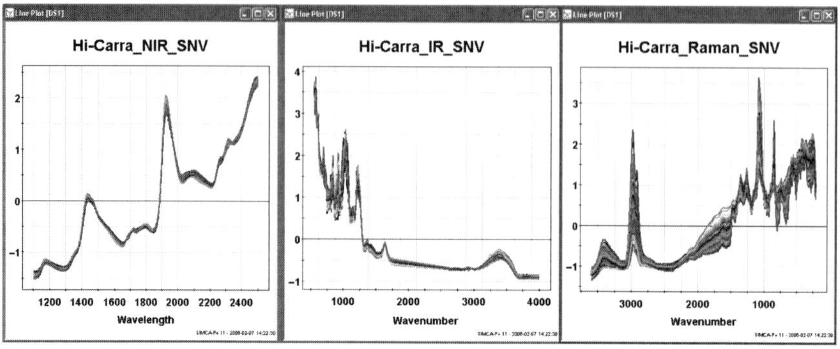

Fig. 15.7 Standard normal variate (SNV) pre-processed spectral data of the Carrageenan application. Compare with Fig. 15.2

As seen, the Raman block is more than five times wider than the IR block. If these blocks were modeled together, the Raman block might dominate the smaller blocks for purely numerical reasons. The raw spectra were plotted above in Fig. 15.2.

15.5.4 Pre-Processing of Spectral Data

Prior to the data analysis, the spectra were SNV transformed to remove baseline variation (Barnes et al. 1989) and mean-centered. The SNV filtered data are plotted in Fig. 15.7. In comparison with the raw spectra (cf. Fig. 15.2), the filtered spectra are much better matched.

15.5.5 Hierarchical Model Structure

The hierarchical model is outlined in Fig. 15.8. Three OPLS models will be developed at the base level, one for each spectral block versus the same Y-block. This enables a good categorization of the base level score variables, i.e., to uncover which are predictive for Y and which are not. The top level modeling will then be pursued in two directions, one that further examines the Y-predictive score variables, and another which concentrates on the Y-orthogonal variation only (Fig. 15.8).

15.5.6 Scaling of Data

Different scalings were used for the base and top level models. In all base level models, Pareto scaling was used (Wold et al. 1993). The reason for this was that

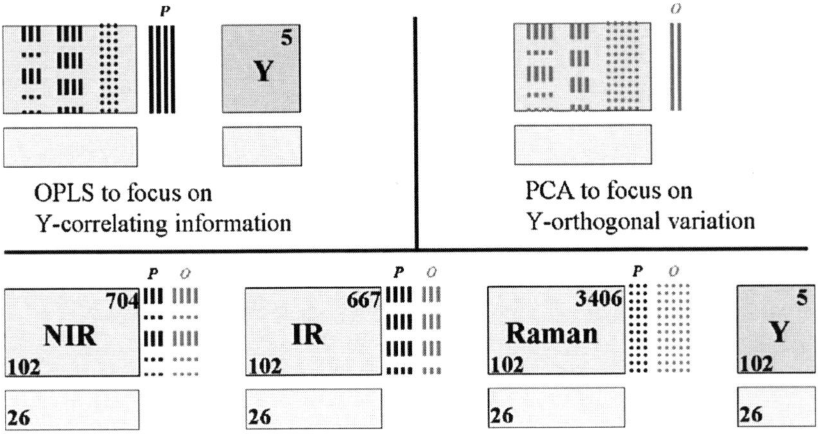

Fig. 15.8 Hierarchical model structure used when separating Y-predictive and Y-orthogonal spectral variation. The lower level consists of three OPLS models. On the upper level there are two models, the first is an OPLS model targeting the Y-predictive spectral variation and the second is a PCA model addressing the Y-orthogonal spectral variation

inside each X-block variables of similar origin were used. In all top level models, Unit Variance scaling was used. This to make scores of different origin (arising from different spectral measurement techniques) comparable.

15.5.7 Base Level OPLS Models

The three base level OPLS models had 3 predictive and 4 orthogonal components (NIR block), 4 predictive and 3 orthogonal (IR block), and 3 predictive and 5 orthogonal (Raman block). The predictive components use 76.8 (NIR), 72.6 (IR) and 78.6% (Raman) of the X-variance. Between 95% (NIR and IR) and 84% (Raman) of the variation in the Carrageenan proportions (Y) are explained by the different models. Hence, the Raman spectra contain less information about Y than the NIR and IR spectra.

15.5.8 Top Level OPLS Model Summarizing the 10 Y-Predictive Score Variables

Next, the 10 (3 + 4 + 3) predictive score variables – originating from the three base level OPLS models – were concatenated to form the top level predictive X-matrix (Fig. 15.8). The top level OPLS model comprised 4 predictive components displaying an explained and predicted Y-variance of 98%, i.e., slightly higher values than the corresponding ones of the three base level models. Figure 15.9 illustrates

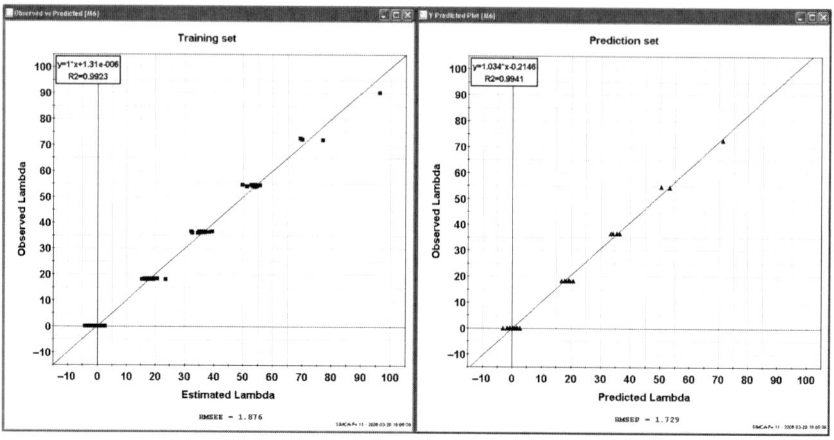

Fig. 15.9 Prediction results for the top level model for one representative response (Lambda). (*left*) Relationship between observed and predicted proportions of Lambda constituent for the training set samples. (*right*) Relationship between observed and predicted proportions of Lambda constituent for the prediction set samples

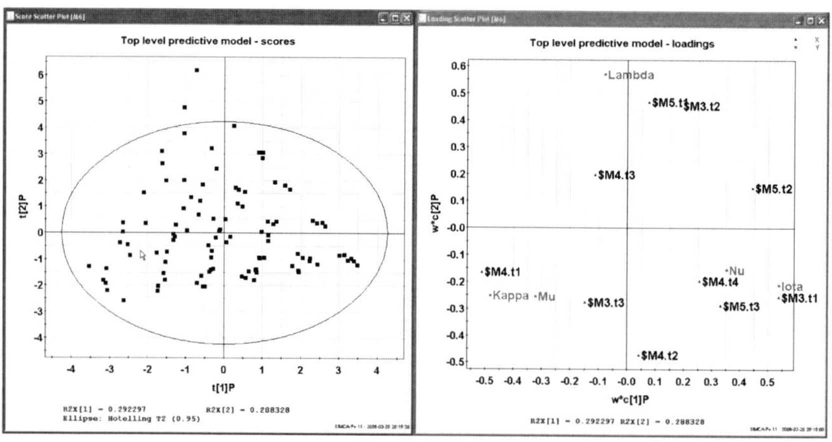

Fig. 15.10 Scores (*left*) and loadings (*right*) of the first two components of the top level Carrageenan OPLS model. The mixture design structure of the underlying design is easy to see. In the loading plot $M3 refers to the NIR base level model, $M4 to the IR model, and $M5 to the Raman model

the predictive power of the model, and Fig. 15.10 provides plots of its scores and loadings.

When evaluating the estimation (training set) and prediction (prediction set) errors of Y (no detailed results provided), two interesting observations can be made. First and foremost, the prediction results of the prediction set well match the result for the training set, i.e., the RMSEP and RMSEE values are similar. This suggests the top level model is not overtrained and has sound predictive power (see also Fig. 15.9). Secondly, the fifth and last response, relating to the proportions of Nu, is more challenging to model and predict than the other four Y-variables.

The structure of the underlying mixture design is clearly evident in Fig. 15.10 (scores t_1 vs t_2). The extreme vertices of the simplex design are defined by the amounts of the Lambda, Kappa and Iota carrageenan types. Recall that these species were varied in the largest ranges, i.e., from 0 to 100% in the nominal metric (see Sect. 15.5.2).

Interestingly, the loadings reveal that each of the three predictive NIR score variables accounts for one of these three main constituents. The Mu and Nu constituents, which were varied in smaller proportions (between 0 and 20% on the nominal scale), dominate the model in the third and fourth components. In particular, the Nu carrageenan species is almost exclusively described by the fourth score of the IR model.

15.5.9 Top Level PCA Model Summarizing the 12 Y-Orthogonal Score Variables

The top level PCA model – investigating the Y-orthogonal variation in X only – comprised two components explaining 33% of the variance. The scores of these two components are plotted in Fig. 15.11 together with the first two orthogonal scores of the three base level OPLS models. A systematic shift among the observations is seen in the top level model, which is clearly due to structure in the NIR and IR data, but not the Raman data (Fig. 15.11).

The line plot of the first component of this "orthogonal" top level PCA model shows a strong time trend (Fig. 15.12). Hence, there is a trend in the spectral data that is not associated with changes in Y. This trend is weak during the two first days of the measurement campaign, while the scores of day 3 are markedly higher than for the other three days. The group contribution plot shown in Fig. 15.13 confirms that this shift is due to variation in the NIR and IR spectra, but not the Raman spectra.

Figure 15.14 displays how features seen in the upper level group contribution plot (Fig. 15.13) can be related back to the spectral domain; the line plot of the loading p_{4O} of the base level NIR model suggests that the main area of Y-orthogonal spectral variation is located around 1970 nm (i.e., in the water region).

15.5.10 Discussion of Example

As seen in the carrageenan analysis, integrating and contrasting blocks of data from different sources is greatly facilitated by the hierarchical modeling approach. The obvious advantages being simplicity and transparency. Other advantages are:

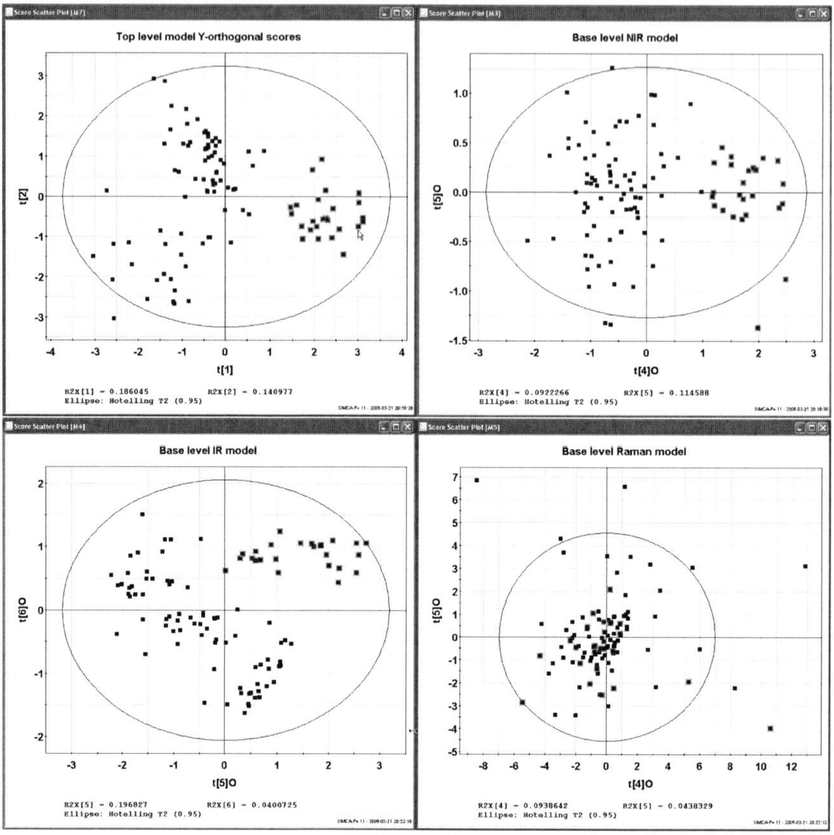

Fig. 15.11 (*upper left*) Score plot of the two first scores of the top level PCA model of the Y-orthogonal base model scores. (*upper right*) Score plot of the two first orthogonal scores of the base level NIR model. (*lower left*) Score plot of the two first orthogonal scores of the base level IR model. (*lower right*) Score plot of the two first orthogonal scores of the base level Raman model. The same observations (samples) are marked throughout all four plots

- blocks of very different widths (i.e., containing radically different number of variables) are given equal opportunity to influence the top level model;
- different observations can be outliers in different base level models;
- the loading plot of the top level model is less cluttered and easier to interpret but the information represented by the scores is still easily accessible.

The hierarchical models – based on the three OPLS lower level models – enables a partitioning of the spectral data into predictive and orthogonal components. A detailed analysis of the orthogonal variation revealed systematic shifts in the NIR and IR spectra between the different days of the sampling campaign. A close-up interpretation suggests this day-to-day shift to be linked to varying moisture content. NIR and IR measurements are more affected by such variations than Raman spectroscopy.

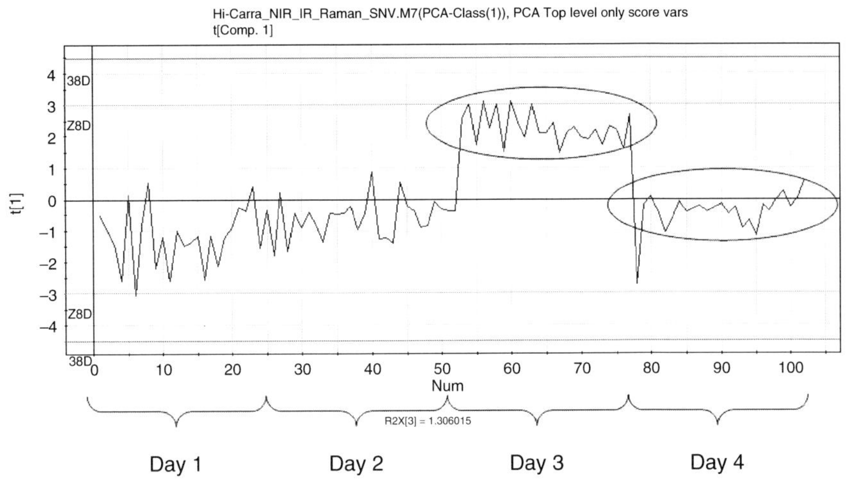

Fig. 15.12 Score line plot of t_1 of the hierarchical PCA model of the Y-orthogonal base model scores. Systematic shifts between the different sampling days are obvious

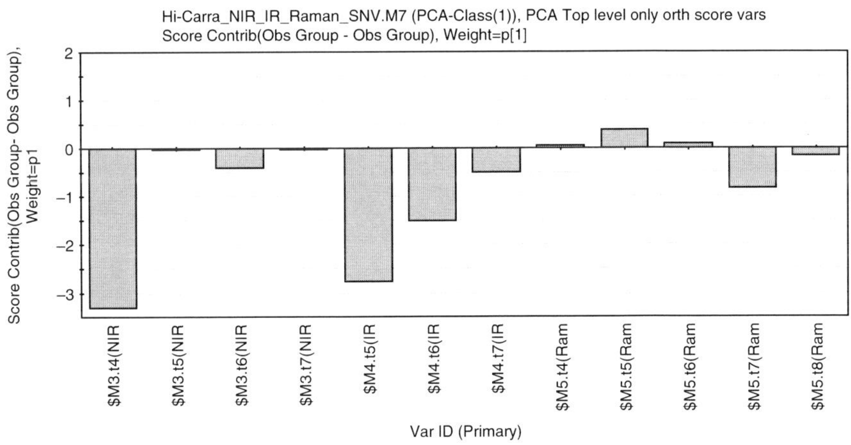

Fig. 15.13 Group contribution plot between the day 3 and day 4 samples. This plot confirms that the "jump" seen in t_1 arises because of anomalies in the NIR and IR spectra

15.6 Combining the Concepts: The Novartis PAT Example

15.6.1 Background

The last dataset employed to illustrate some of the concepts (read: data mining and data integration) described in this chapter comes from a feasibility study of production data of Novartis in Suffern, NY, USA. We are grateful to James Cheney, John Sheehan, and Fritz Erni for granting us permission to show this example.

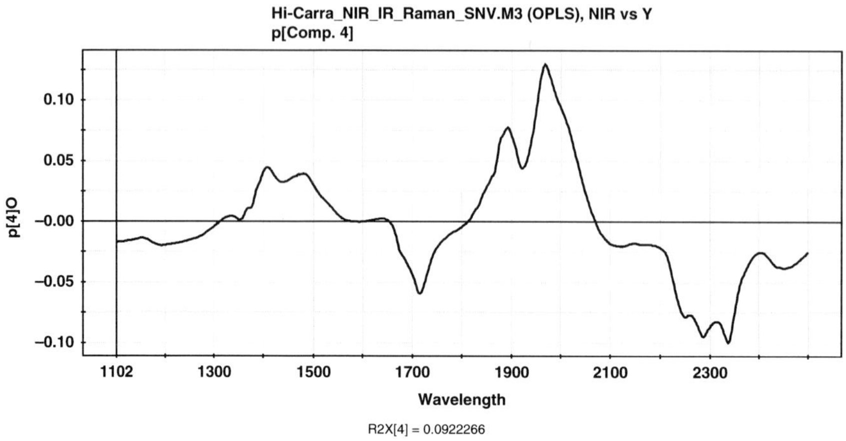

Fig. 15.14 Line plot of p_{4o} – the loading of the first **Y**-orthogonal component – of the base level NIR model. This plot suggests the region around 1970 nm to contribute a large fraction of the non-correlating variation

In this pharmaceutical production process, the manufactured tablets recently showed a tendency to dissolve too slowly (dissolution rate less than 90). Process variables and analytical data of the raw materials were available for a period of just over 2 years. In the first phase of the investigation, only raw material data plus dynamic process data of the dryer were included because the latter was suspected of causing the dissolution problems.

The process data comprise $N_{tot} = 5743$ observations with 6 process variables (e.g. temperature, air flow, etc.) + process time for each. These observations are divided into $N_{bat} = 313$ batches, which in turn are divided into 2 phases (process steps). In addition, data for seven raw materials (excipients) are available for each batch, as well as summary data for other process steps (granulation, solvent addition, film coating), giving 298 variables in total. Finally, 21 quality measurements are available for each batch, where the dissolution rate is the most important.

Note that these data reside in different databases and are unaligned and unsynchronized so a key part of the data analysis is to ensure that the batches are comparable and that the right raw material data are aligned with the right batch.

15.6.2 *Workflow to Synchronize Data*

In this case, the data have different dimensions and sizes and it is therefore necessary to align them to put them on a comparable footing. Getting the different types of data into a single data structure was made according to the principles of hierarchical and batch-wise analysis of process data outlined in Wold et al. (1996, 1998). Figure 15.15 shows an overview of the models' structure.

- *Step 1*. First, the data of the two phases were arranged in accordance with Fig. 15.16. Two separate PLS models were made, one for each phase (step) of the process using local process time as y (Fig. 15.16).
- *Step 2*. Subsequently, the original process variables were chopped up in one piece per batch, transposed, aligned by linear time warping, and used as descriptors of the process dynamics. The resulting matrix, with one line per batch, was then subjected to a PLS analysis with y = dissolution rate. This gave 2 lower level models (Fig. 15.15).
- *Step 3*. The data for each raw material (one row per batch) were fitted in separate PLS models with y = dissolution rate. This gave an additional 7 lower level models (Fig. 15.15).

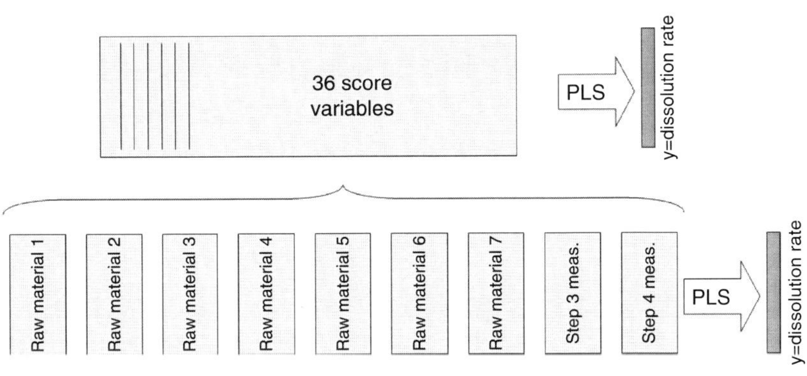

Fig. 15.15 Overview of the modeling workflow to get a synchronized data structure. The structure seen is a combination of batch analysis and hierarchical modeling. Each row is one batch

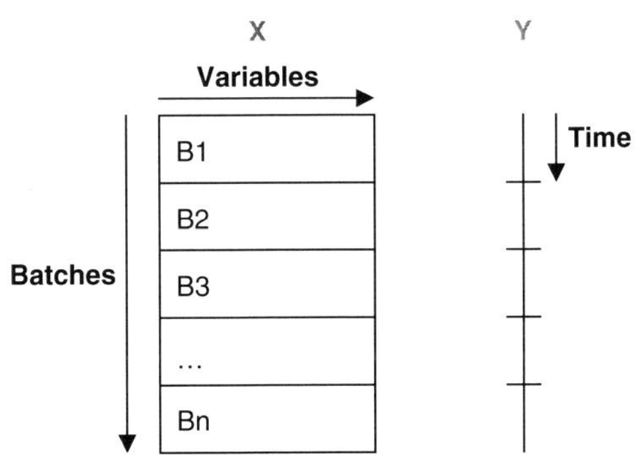

Fig. 15.16 The initial three-way data table of Batch data is unfolded by preserving the direction of the variables. This gives a two-way matrix with N × J rows and K columns. Each row contains data points x_{ijk} from a single batch observation (batch i, time j, variable k). If regression is made against local batch time, the resulting PLS scores reflect linear (t_1), quadratic (t_2), and cubic (t_3) relationships to local batch time

- *Step 4*. The scores resulting from the 9 lower level PLS models developed in steps 2 and 3, in total 36 score variables, were used as variables in the "top hierarchical" batch model, again with y = dissolution rate (Fig. 15.15).

The net result of the first three steps was that all the data were combined into a single, consistent and synchronized structure. This allows the PLS estimation of the relationship between y = dissolution rate and X = all data from the process, including raw materials, summarized steps, and the full dynamics of the dryer.

15.6.3 Nine Base Level Batch Models

The PLS analysis was, as described in Sect. 15.6.2, divided into two hierarchical levels. At the base level, 9 separate PLS models were built, each with the same y-variable (dissolution rate), while the X-variables were from the two process steps and the 7 raw materials. These 9 base models resulted in between 3 and 6 components each and a total of 36 score vectors. The models had R^2X values of between 0.62 and 0.93 with an average of 0.72. The average R^2Y was 0.15.

15.6.4 Top Level Batch Model

The 36 base level score vectors were then used as X-variables (UV-scaled) in the top level hierarchical PLS model. The resulting two-component PLS model had $R^2X = 0.25$, $R^2Y = 0.44$ and $Q^2Y = 0.39$. The X-scores are plotted in Fig. 15.17 and the relationship between observed and predicted dissolution rate is shown in Fig. 15.18. The score plot clearly contains clusters and, by coloring by time, it becomes evident that later batches with poor dissolution properties are clustered to the left (negative end of t_1).

The model was validated by a permutation test (Fig. 15.19), and, more dramatically, by a final model including more complete process data and its real on-line predictions (Fig. 15.20). The top model PLS weights of the two components are plotted in Fig. 15.21. Several raw materials as well as one phase of the dryer are indicated as important for the changes in y = dissolution rate.

15.6.5 Discussion and Epilogue

This example of hierarchical PLS modeling/process data mining demonstrated that much of the process quality variation could be explained and understood. This led to the start of a much larger project looking at all relevant data for both off-line modeling and on-line monitoring.

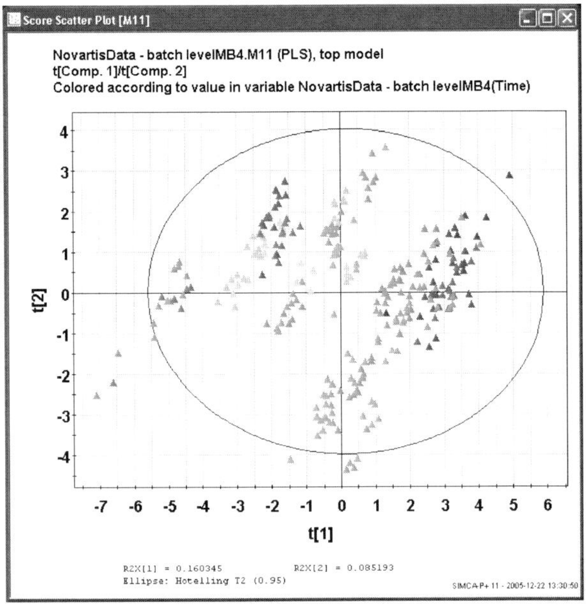

Fig. 15.17 Score scatter plot of t_1/t_2. A high proportion of late batches with poor dissolution properties are clustered in the left-hand region of the plot

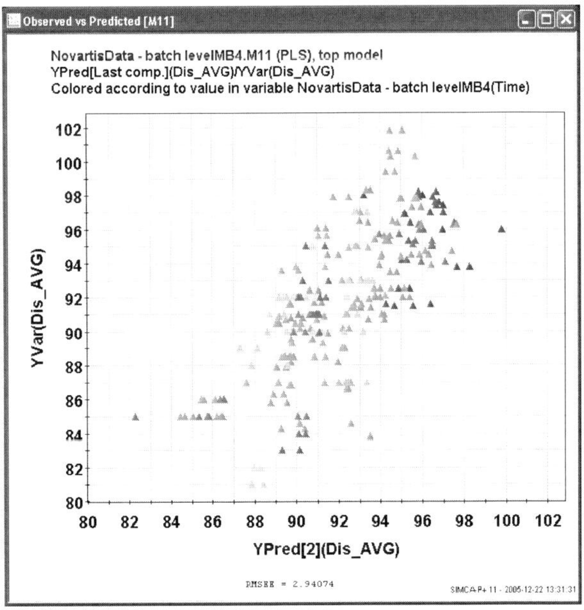

Fig. 15.18 Agreement between measured and predicted dissolution rate for the training set

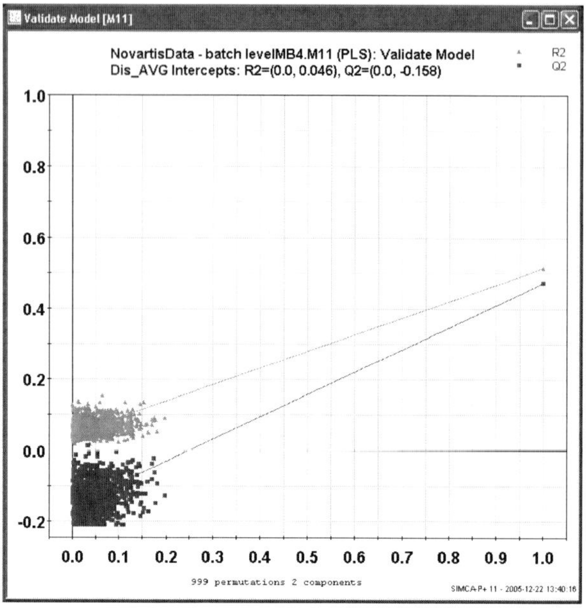

Fig. 15.19 Validation plot based on 999 permutations of the y-vector followed by the fitting of a PLS model between X (unperturbed) and the permuted y. The horizontal axis shows the correlation between the original and the permuted y, and the vertical axis the values of R2 (*upper line, light*) and Q2 (*lower line, dark*). The plot where all Q2 values of the permuted y models are below zero is a clear indication that the original model does not happen by coincidence

About a year after the first feasibility study, a system collecting, integrating, aligning, and synchronizing all process and raw material measurements was in place at Novartis, Suffern. A final PLS model was developed from all these data giving a remarkably good (and validated) relationship with dissolution rate (Fig. 15.20). This model was put on-line and has faithfully predicted dissolution rate for several months now (large points in Fig. 15.20).

In summary, this shows that good models can be obtained, even when dealing with very complex processes, provided that relevant data are available from all pertinent parts of the process and that these data are properly aligned and analyzed.

15.7 Summary and Discussion

Process data provide a very satisfactory testing ground for data mining and data integration because, although the data are numerous, they are often of high precision and the rewards are substantial. The PLS analysis of process data, both dynamic and point data, provides models that are interpretable and easily validated. The hierarchical approach to PLS modeling provides additional useful tools for data formatting and alignment, making it possible to integrate data of different shapes and types in the same final model.

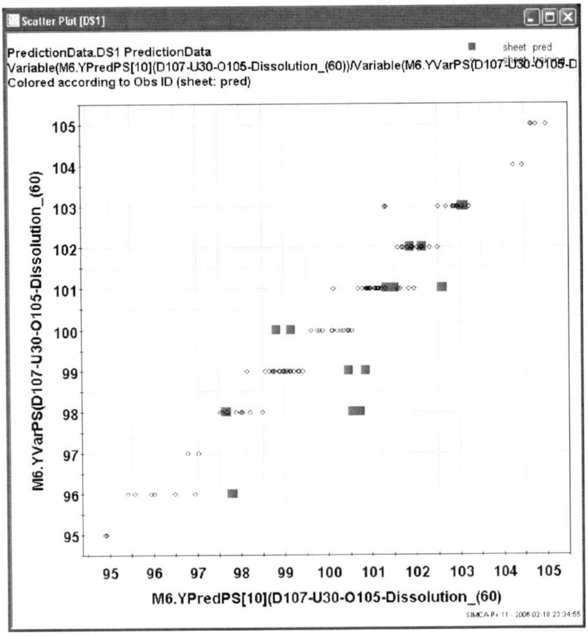

Fig. 15.20 Observed y = dissolution rate on the vertical axis plotted versus the predicted y from the top hierarchical PLS model later developed from more complete process data. Large points are predictions for new batches. R2 = 0.82

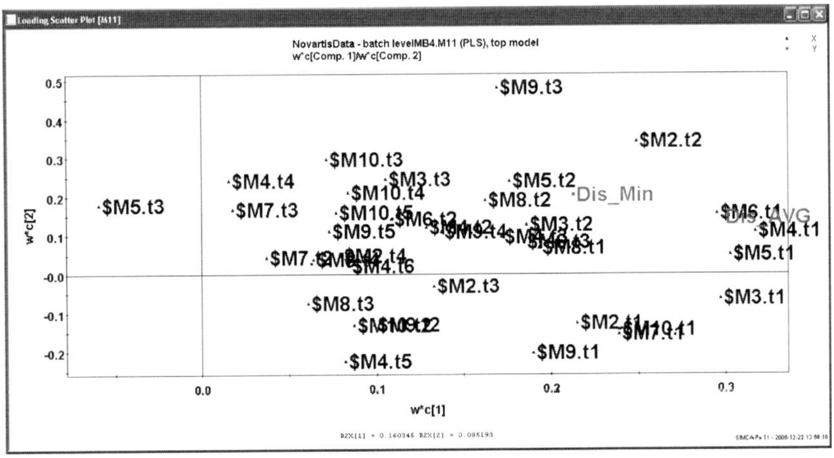

Fig. 15.21 PLS-weights (w) for the first (*horizontal axis*) and second components (*vertical axis*) of the hierarchical top PLS model

Process data are usually fairly linear since manufacturing processes are well controlled and have a limited range of variation. In other areas such as structure-activity modeling, however, non-linear PLS models have been shown to compare well with other non-linear approaches. Supported by appropriate multivariate cluster analyses (not shown here), PLS, OPLS and hierarchical and non-linear extensions form a comprehensive tool-box for successful data mining and integration.

Faced with large unbalanced databases containing clusters and gross outliers, a combination of appropriate pre-processing and DOE can often facilitate a difficult modeling situation and hence contribute to a successful data mining and integration.

In summary, PLS (and PCA) are suitable for data mining because they can deal with complex, collinear, incomplete, noisy, and numerous data as they are. This is in contrast to regression-like methods such as linear discriminant analysis, regression and classification trees, and neural networks, which break down when the variables are collinear and too numerous.

Also, the modeling of the X-block(s) by PLS provides and additional source for diagnostics (outliers, process upsets, drift, trends) and understanding. The X-matrix contains the data describing the process, the spectra, the molecular properties, and the interpretation of the patterns in loading plots and score plots is invaluable for the understanding of the investigated system or process.

With computer intensive data analytical tools such as cross-validation and jack-knifing, and permutation tests, rigorous inference and predictions can be made as easily for large and complex data sets as by the use of classical statistics for small and well behaved data sets. This opens the way for the efficient and profitable use of the information inherent in complex data to better understand our complex world, and perhaps sometimes even improve it.

References

Albano, C., Dunn, W. G., III, Edlund, U., Johansson, E., Nordén, B., & Sjöström, M., et al. (1978). Four levels of pattern recognition. *Analytica Chemica Acta, 103*, 429–443.

Barnes, R. J., Dhanoa, M. S., & Lister, S. J. (1989). Standard normal variate transformation and de-trending of near-infrared diffuse reflectance spectra. *Applied Spectroscopy, 43*, 772–777.

Buydens, L. M. C., Reijmers, T. H., Beckers, M. L. M., & Wehrens, R. (1999). Molecular data mining: A challenge for chemometrics. *Chemometrics and Intelligent Laboratory Systems, 49*, 121–133.

Dyrby, M., Petersen, R. V., Larsen, J., Rudolf, B., Nørgaard, L., & Engelsen, S. B. (2004). Towards on-line monitoring of the composition of commercial carrageenan powders. *Carbohydrate Polymers, 57*, 337–348.

Eriksson, L., Johansson, E., Müller, M., & Wold, S. (2000). On the selection of training set in environmental QSAR when compounds are clustered. *Journal of Chemometrics, 14*, 599–616.

Eriksson, L., Johansson, E., Lindgren, F., Sjöström, M., & Wold, S. (2002). Megavariate analysis of hierarchical QSAR data. *Journal of Computer-Aided Molecular Design, 16*, 711–726.

Eriksson, L., Jaworska, J., Worth, A., Cronin, M., McDowell, R. M., & Gramatica, P. (2003). Methods for reliability, uncertainty assessment, and applicability evaluations of classification and regression based QSARs. *Environmental Health Perspectives, 111*, 1361–1375.

Eriksson, L., Antti, H., Gottfries, J., Holmes, E., Johansson, E., Lindgren, F., et al. (2004). Using chemometrics for navigating in the large data sets of genomics. *Proteomics and Metabonomics, Analytical and Bioanalytical Chemistry, 380*, 419–429.

Eriksson, L., Johansson, E., Kettaneh-Wold, S., Trygg, J., Wikström, C., & Wold, S. (2006a). Multi- and megavariate data analysis. Part I. Basic principles and applications. Second revised and enlarged edition, Umetrics Academy, 2006. ISBN-10: 91–973730–2–8. ISBN-13: 978–91–973730–2–9.

Eriksson, L., Johansson, E., Kettaneh-Wold, S., Trygg, J., Wikström, C., & Wold, S. (2006b). Multi- and megavariate data analysis. Part II. Advanced applications and method extension. Second revised and enlarged edition, Umetrics Academy, 2006. ISBN-10: 91–973730–3–6. ISBN-13: 978–91–973730–3–6.

Hand, D. J. (1998). Data mining: Statistics and more? *American Statistician, 52*, 112–188.

Kettaneh, N., Berglund, S., & Wold, S. (2005). PCA and PLS with very large data sets. *Computational Statistics & Data Analysis, 48*, 69–85.

Linusson, A., Gottfries, J., Lindgren, F., & Wold, S. (2000). Statistical molecular design of building blocks for combinatorial chemistry. *Journal of Medicinal Chemistry, 43*, 1320–1328.

Maitra, R. (2001). Clustering massive data sets with applications in software metrics and tomography. *Technometrics, 43*, 336–346.

Munck, L. (2005). The revolutionary aspect of exploratory chemometric technology. The universe and the biological cell as computers. A plea for cognitive flexibility in mathematical modelling. Gylling: Narayana Press.

Naes, T., & Mevik, B. H. (1999). The flexibility of fuzzy clustering illustrated by examples. *Journal of Chemometrics, 13*, 435–444.

Olsson, I. M., Gottfries, J., & Wold, S. (2004a). D-optimal onion design in statistical molecular design. *Chemometrics and Intelligent Laboratory Systems, 73*, 37–46.

Olsson, I. M., Gottfries, J., & Wold, S. (2004b). Controlling coverage of D-optimal onion designs and selections. *Journal of Chemometrics, 18*, 548–557.

Oprea, T. I., & Gottfries, J. (2001). Chemography: The art of navigating in chemical space. *Journal of Combinatorial Chemistry, 3*, 157–166.

Sjöström, M., Wold, S., & Söderström, M. (1986). PLS discriminant plots, In E. S. Gelsema & L. N. Kanal (Eds.), Pattern recognition in practice II (pp. 461–470). North-Holland: Elsevier Science.

Svensson, O., Josefsson, M., & Langkilde, F. W. (1997). Classification of chemically modified celluloses using a near-infrared spectrometer and soft independent modeling of class analogy. *Applied Spectroscopy, 51*, 1826–1835.

Trygg, J., & Wold, S. (2002). Orthogonal projections to latent structures (OPLS). *Journal of Chemometrics, 16*, 119–128.

Wold, S., Johansson, E., & Cocchi, M. (1993). PLS. In H. Kubinyi (Ed.), 3D-QSAR in drug design, theory, methods, and applications (pp. 523–550). Ledien: ESCOM Science.

Wold, S., Kettaneh, N., & Tjessem, K. (1996). Hierarchical multiblock PLS and PC models for easier model interpretation and as an alternative to variable selection. *Journal of Chemometrics, 10*, 463–482.

Wold, S., Kettaneh, N., Fridén, H., & Holmberg, A. (1998). Modeling and diagnostics of batch processes and analogous kinetic experiments. *Chemometrics and Intelligent Laboratory Systems, 44*, 331–340.

Wold, S., Sjöström, M., & Eriksson, L. (2001). PLS-regression: A basic tool of chemometrics. *Chemometrics and Intelligent Laboratory Systems, 58*, 109–130.

Wold, S., Josefson, M., Gottfries, J., & Linusson, A. (2004). The utility of multivariate design in PLS modeling. *Journal of Chemometrics, 18*, 156–165.

Chapter 16
Three-Block Data Modeling by Endo- and Exo-LPLS Regression

Solve Sæbø, Magni Martens, and Harald Martens

Abstract In consumer science it is common to study how various products are liked or ranked by various consumers. In this context, it is important to check if there are different consumer groups with different product preference patterns. If systematic consumer grouping is detected, it is important to determine the person characteristics which differentiate between these consumer segments, so that they can be reached selectively. Likewise it is important to determine the product characteristics that consumer segments seem to respond differently to.

Consumer preference data are usually rather noisy. The products × persons data table ($\mathbf{X}_1$) usually produced in consumer preference studies may therefore be supplemented with two types of background information: a products × product-property data table ($\mathbf{X}_2$) and a person × person-property data table ($\mathbf{X}_3$). These additional data may be used for stabilizing the data modeling of the preference data $\mathbf{X}_1$ statistically. Moreover, they can reveal the product-properties that are responded to differently by the different consumer segments, and the person-properties that characterize these different segments. The present chapter outlines a recent approach to analyzing the three types of data tables in an integrated fashion and presents new modeling methods in this context.

S. Sæbø
Department of Chemistry, Biotechnology and Food Science (IKBM), Norwegian University of Life Sciences, P.O. Box 5003, 1432 Ås, Norway
e-mail: solve.sabo@umb.no

M. Martens
Norwegian Food Research Institute, Matforsk, 1430 Ås, Norway
e-mail: magni.martens@matforsk.no
and
Faculty of Life Sciences, Department of Food Science, University of Copenhagen, Rolighedsvej 30, 1958 Frederiksberg C, Denmark

H. Martens
Norwegian Food Research Institute, Matforsk, 1430 Ås, Norway
e-mail: harald.martens@matforsk.no
and
Faculty of Life Sciences, Department of Food Science, University of Copenhagen, Rolighedsvej 30, 1958 Frederiksberg C, Denmark
and
Norwegian University of Life Sciences, IKBM/CIGENE, P.O. Box 5003, 1432 Ås, Norway

V. Esposito Vinzi et al. (eds.), *Handbook of Partial Least Squares*, Springer Handbooks of Computational Statistics, DOI 10.1007/978-3-540-32827-8_17,

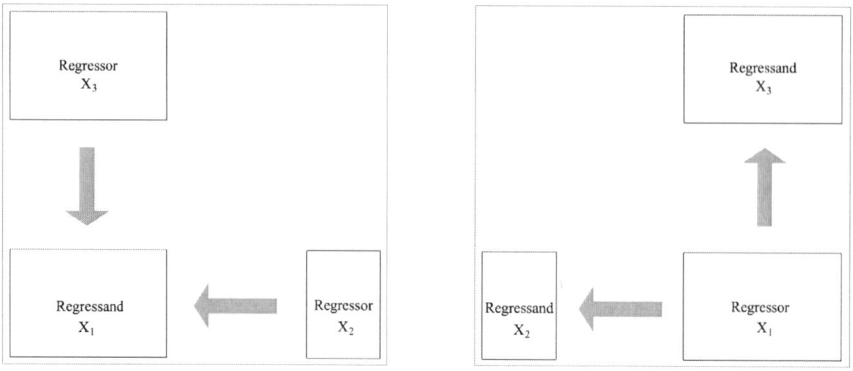

Fig. 16.1 The L-shaped matrix systems of endo-LPLSR (*left*) and exo-LPLSR (*right*). In both cases the columns of $\mathbf{X}_1$ and the columns of $\mathbf{X}_3$ are related, as well as the rows of $\mathbf{X}_1$ and the rows of $\mathbf{X}_2$. In endo-LPLSR the corner matrix $\mathbf{X}_1$ is the regressand, whereas it serves as a regressor in exo-LPLSR

16.1 Introduction

The LPLSR approach was presented by Martens et al. (2005) as a method for exploring consistent patterns of co-variation between three data matrices arranged in an L-shaped ("corner-shaped") system, where $\mathbf{X}_2$ and $\mathbf{X}_3$ give additional descriptions of the rows and of the columns in $\mathbf{X}_1$, respectively. The LPLSR is a horizontal-and-vertical extension of the ordinary PLS regression (PLSR) (Wold et al. 1983), and is a formalized version of an early development by Wold et al. (1987), which was not pursued by those authors. The version of LPLSR presented at PLS'01 at Capri, Italy in 2001, is here named "endo-LPLSR", where "endo" reflects the inward-pointed regression of a single response matrix from two outer regressors as illustrated in Fig. 16.1 (left). Martens et al. (2005) used endo-LPLS regression to model consumers' liking data of apples, and later Mejlholm and Martens (2006) adopted the method in a study of Danish beer liking data.

A subset of this consumer science data set on beer liking will be used here for the purpose of comparing the endo-LPLSR with a new, but related method, here called "exo-LPLSR". The exo-LPLSR gives more emphasis on the consumer liking data $\mathbf{X}_1$, at the expense of the product- and consumer descriptors $\mathbf{X}_2$ and $\mathbf{X}_3$. The exo-LPLSR approach shares the L-shaped or corner-shaped structure with the endo-version, but is characterized by a simultaneous outward ("exo") regression of two regressands from a single regressor matrix, as shown in Fig. 16.1 (right). We have arranged the matrices differently for endo- and exo-LPLSR since it is conventional to place the regressand to the left of the regressor.

The direction of regression (endo or exo) may be based on causal assumptions, or merely a choice of convenience if the purpose is data exploration. Since the data matrices involved will serve different purposes for the two methods (regressand or regressor), they will simply be denoted by $\mathbf{X}_1$ (corner matrix), $\mathbf{X}_2$ and $\mathbf{X}_3$ to avoid any confusion (Fig. 16.2). The function of each matrix should be clear from

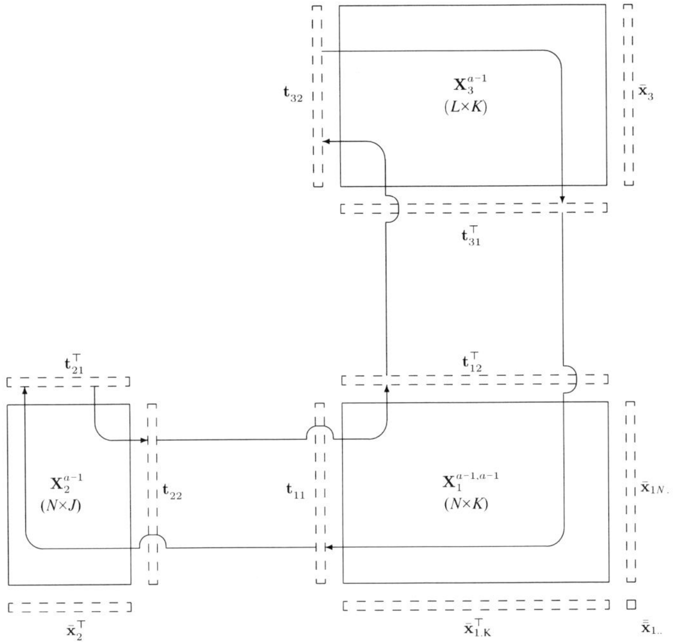

Fig. 16.2 The t-vectors are identified by the iterative NIPALS algorithm. The mean vectors are used in the initial centering of the matrices as given in (16.1)–(16.3)

the context. The reversed L-shape in Fig. 16.1 (right) will be used throughout the remainder of this chapter for both LPLSR-methods, since the usual endo-LPLSR setup can be reached by swopping $\mathbf{X}_2$ and $\mathbf{X}_3$ and transposing the matrices.

The endo-LPLSR algorithm, as it is presented in Martens et al. (2005), may be based on a singular value decomposition (SVD) of a matrix product of $\mathbf{X}_1$ $(N \times K)$, $\mathbf{X}_2$ $(N \times J)$ and $\mathbf{X}_3$ $(L \times K)$. In fact, also the exo-LPLSR algorithm may be worked out from a SVD of the same three matrices. As an alternative to SVD the iterative NIPALS algorithm can be used to extract singular vectors from some data matrix $\mathbf{X}$, as outlined in Martens and Martens (2001) for a Principal Component Analysis. SVD- and NIPALS- extractions of singular vectors have also been used in parallel in PLSR for the analysis of the relationship between two data matrices $\mathbf{Y}$ and $\mathbf{X}$. The singular vectors used in PLSR are those of the matrix product $\mathbf{X}^\top\mathbf{Y}$. Also the endo- and exo- LPLSR algorithms may alternatively be based on NIPALS extractions of latent vectors.

In the following NIPALS-based LPLSR algorithms are presented since this best illustrates the differences and similarities between the endo- and the exo- versions. However, the relations to the SVD solutions will be remarked. For instance, if both $\mathbf{X}_2$ and $\mathbf{X}_3$ are non-informative (e.g. equal to identity $N \times N$ and $K \times K$ matrices), the exo-LPLSR reduces to a PCA of $\mathbf{X}_1$ (after double-centering). If only $\mathbf{X}_2$ or $\mathbf{X}_3$ is non-informative, the exo-LPLSR reduces to a two-block PLSR of $\mathbf{X}_2$ ("Y") vs $\mathbf{X}_1$ ("X"), or $\mathbf{X}_3^\top$ ("Y") vs $\mathbf{X}_1^\top$ ("X").

Some alternatives to LPLSR do exist for analysis of such three-block data. The elements in $\mathbf{X}_1$ may be strung out as an $(NK \times 1)$ vector and regressed on an $(NK \times (J+L))$ matrix representing $\mathbf{X}_2$ and $\mathbf{X}_3$ as main effects; this model can also be extended to include multiplicative interactions. The weakness of this approach is that it may preclude consumer/product segmentation, as described by Martens et al. (2005). A two-step approach for using the $\mathbf{X}_3$-information together with the $\mathbf{X}_1$ and $\mathbf{X}_2$ data is to fold the information from $\mathbf{X}_3$ with $\mathbf{X}_1^\top$ in order to give the $\mathbf{X}_3$-information a dimension N, common with $\mathbf{X}_1$ and $\mathbf{X}_2$. Kubberød et al. (2002) first estimated the reduced-rank regression coefficient matrix $\mathbf{B}_{1,3}$ between $\mathbf{X}_3^\top$ and $\mathbf{X}_1^\top$ by PLSR, and secondly regressed both $\mathbf{X}_1$ and $\mathbf{B}_{1,3}^\top$ on $\mathbf{X}_2$ by a second PLSR. Thybo et al. (2002) used a similar two-step approach, but in order to simplify the analysis, they replaced the regression coefficient matrix $\mathbf{B}_{1,3}$ by the matrix of correlation coefficients $\mathbf{R}_{1,3}$ between the N rows in $\mathbf{X}_1$ and the L columns in $\mathbf{X}_3$, correlated over K elements. In either case, the three-block interpretations were meaningful. But these two-step procedures are cumbersome and somewhat nontransparent with regard to optimization criterion and variable weighting. The optimization criterion used for model estimation using a two-step method must necessarily also be a two-step criterion. The LPLSR is more transparent in that respect since it can be based on latent vectors extracted from a single matrix product of all three data matrices. This is in analogy to how the two-step criterion of Principal Component Regression relates to the one-step criterion of PLSR in two-block modeling.

This chapter is organized as follows: Section 16.2 describes the mathematics of the two LPLSR methodologies, in terms of data preprocessing (16.2.1), the endo-LPLSR (16.2.2), the exo-LPLSR (16.2.3), and outlines associated topics in model validation (16.2.4) and visualization (16.2.5). In Sect. 16.3 these methodologies are applied to the consumer science data set on beer liking, in terms of data description (16.3.1) and a comparison of endo- and exo-LPLSR (16.3.2). Section 16.4 puts the LPLSR developments into the more general framework of Domino-PLS regression (Martens 2005) and outlines methods to improve the methodologies further.

16.2 LPLS Regression

16.2.1 Data Pre-processing and Extraction of Latent Variables

Preceding the extraction of latent vectors, the columns of X_2 and the rows of X_3 are typically centered by:

$$\mathbf{X}_2^0 = \mathbf{X}_2 - \mathbf{1}_N \bar{\mathbf{x}}_2^\top \tag{16.1}$$

$$\mathbf{X}_3^0 = \mathbf{X}_3 - \bar{\mathbf{x}}_3 \mathbf{1}_K^\top , \tag{16.2}$$

where $\bar{\mathbf{x}}_2$ is the J-vector of column-means of $\mathbf{X}_2$, and $\bar{\mathbf{x}}_3$ is the L-vector of row-means of $\mathbf{X}_3$. The corner matrix $\mathbf{X}_1$ is subject to a double centering across both rows and columns:

$$\mathbf{X}_1^{00} = \mathbf{X}_1 - \mathbf{1}_N \bar{\mathbf{x}}_{1.K}^{\mathrm{T}} - \bar{\mathbf{x}}_{1N.} \mathbf{1}_K^{\mathrm{T}} + \mathbf{1}_N \bar{\bar{x}}_{1..} \mathbf{1}_K^{\mathrm{T}}, \tag{16.3}$$

where $\bar{\mathbf{x}}_{1.K}$ is the K-vector of column means, $\bar{\mathbf{x}}_{1N.}$ is the N-vector of row means and $\bar{\bar{x}}_{1..}$ is the overall mean of $\mathbf{X}_1$, respectively. If the column variables of $\mathbf{X}_2$ and the row variables of $\mathbf{X}_3$ are internally very different in scale, it may be natural to perform a standardization to yield a common variance equal to 1 before proceeding.

LPLSR is here presented as algorithms with sequential extractions of latent structures. At each extraction step the iterative NIPALS algorithm is used to identify latent structures. In analogy with ordinary PLSR, it is assumed that a relatively small set A of latent structures is sufficient for capturing the majority of the variability in the response variables.

Set a ($a = 1, \ldots, A$) of latent variables is identified by iteratively projecting the data matrices onto a set of vectors $\mathbf{t}_{11}$, $\mathbf{t}_{12}$, $\mathbf{t}_{21}$, $\mathbf{t}_{22}$, $\mathbf{t}_{31}$ and $\mathbf{t}_{32}$, as shown in Fig. 16.2. The iterative algorithm starts out by choosing an arbitrary J-vector $\mathbf{t}_{21}$ onto which $\mathbf{X}_2^{a-1}$ is projected, giving the N-vector $\mathbf{t}_{22}$:

$$\mathbf{t}_{22} = \mathbf{X}_2^{a-1} \mathbf{t}_{21} (\mathbf{t}_{21}^{\mathrm{T}} \mathbf{t}_{21})^{-1}$$

As indicated by arrows in Fig. 16.2, the NIPALS-iterations continue by projecting $\mathbf{X}_1^{a-1,a-1}$ onto $\mathbf{t}_{22}$ to construct $\mathbf{t}_{12}$. Further, the vectors $\mathbf{t}_{32}$, $\mathbf{t}_{31}$ and $\mathbf{t}_{11}$ are found by the appropriate projections, and finally the first round is completed by projecting $\mathbf{X}_2^{a-1}$ onto $\mathbf{t}_{11}$ to give an update of $\mathbf{t}_{21}$. These steps are repeated until minimal change is traced in the latent t-vectors. The set of t-vectors found upon convergence is defined as set a of latent variables, $\mathbf{t}_{11}^a, \ldots, \mathbf{t}_{32}^a$. These six t-vectors (two for each of the three matrices) will, depending on the chosen LPLSR method, serve as either weights or bi-linear modeling parameters.

The A sets of such latent variables form the basis for the bi-linear models defined in both LPLSR-approaches, as described next.

16.2.2 Endo-LPLSR

The basic steps of the endo-LPLSR algorithm described below are illustrated in Fig. 16.3. This version of endo-LPLSR is based on a sequential NIPALS extraction of latent variables. Auxiliary loadings are then estimated to ensure proper deflation at each step.

The endo-LPLSR algorithm

For latent vectors extraction $a = 1, \ldots, A$

1. Find t-vectors $\mathbf{t}_{22}^a$ and $\mathbf{t}_{31}^a$ by the iterative algorithm described above and as shown in Fig. 16.2, cycling through $\mathbf{X}_1^{a-1,a-1}$, $\mathbf{X}_2^{a-1}$ and $\mathbf{X}_3^{a-1}$. Let $\mathbf{T}_{22} = (\mathbf{t}_{22}^1, \ldots, \mathbf{t}_{22}^a)$ and $\mathbf{T}_{31} = (\mathbf{t}_{31}^1, \ldots, \mathbf{t}_{31}^a)$.

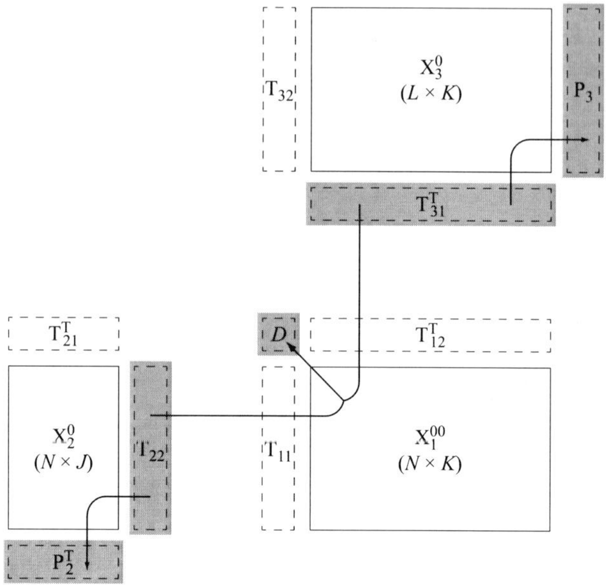

Fig. 16.3 The estimation steps of the regression parameters of the endo-LPLSR algorithm. At step a the T-variables and the P-loadings contain a columns, and $\mathbf{D}$ is a $(a \times a)$ matrix

2. Compute $\mathbf{X}_2$- and $\mathbf{X}_3$-loadings by projection onto orthogonal column matrices $\mathbf{T}_{22}$ and $\mathbf{T}_{31}$:

$$\mathbf{P}_2 = (\mathbf{X}_2^0)^\top \mathbf{T}_{22}(\mathbf{T}_{22}{}^\top \mathbf{T}_{22})^{-1} \tag{16.4}$$

$$\mathbf{P}_3 = \mathbf{X}_3^0 \mathbf{T}_{31}(\mathbf{T}_{31}{}^\top \mathbf{T}_{31})^{-1} , \tag{16.5}$$

and a kernel loadings matrix for $\mathbf{X}_1$, $\mathbf{D}$ $(a \times a)$, defined by

$$\mathbf{D} = (\mathbf{T}_{22}{}^\top \mathbf{T}_{22})^{-1} \mathbf{T}_{22}{}^\top \mathbf{X}_1^{00} \mathbf{T}_{31}(\mathbf{T}_{31}{}^\top \mathbf{T}_{31})^{-1}$$

3. Deflate the data matrices by the contribution of the scores identified to form residual matrices

$$\mathbf{X}_1^{aa} = \mathbf{X}_1^{00} - \mathbf{T}_{22}\mathbf{D}\mathbf{T}_{31}^\top$$
$$\mathbf{X}_2^{a} = \mathbf{X}_2^{0} - \mathbf{T}_{22}\mathbf{P}_2^\top$$
$$\mathbf{X}_3^{a} = \mathbf{X}_3^{0} - \mathbf{P}_3\mathbf{T}_{31}^\top$$

end

The double-centered response matrix $\mathbf{X}_1^{00}$ may upon completion of A extractions be expressed in terms of the latent components and the kernel loadings matrix $\mathbf{D}$:

$$\mathbf{X}_1^{00} = \mathbf{T}_{22}\mathbf{D}\mathbf{T}_{31}^\top + \mathbf{E}_1^A ,$$

where the E-matrix contains the residual variation in the observed variables which is not accounted for by the orthogonal latent variables in $\mathbf{T}_{22}$ and $\mathbf{T}_{31}$.

The model for $\mathbf{X}_1^{00}$ may alternatively be expressed in terms of the original variables:

$$\mathbf{X}_1^{00} = \mathbf{X}_2^0\mathbf{C}\mathbf{X}_3^0 + \mathbf{E}_1^A , \tag{16.6}$$

where $\mathbf{C}$ is a $(J \times L)$ matrix of regression coefficients estimated by

$$\begin{aligned}\hat{\mathbf{C}} &= \mathbf{V}_1\mathbf{D}\mathbf{V}_3^\top, \quad \text{where} \\ \mathbf{V}_1 &= \mathbf{T}_{21}(\mathbf{P}_1^\top\mathbf{T}_{21})^{-1} \\ \mathbf{V}_3 &= \mathbf{T}_{32}(\mathbf{P}_3^\top\mathbf{T}_{32})^{-1}\end{aligned}$$

In Martens et al. (2005) the reduced-rank linear model for $\mathbf{X}_2^{00}$ was expressed in terms of a set of latent variables defined from singular value decompositions of matrix products of type

$$\mathbf{G}_{endo} = \mathbf{X}_2^\top\mathbf{X}_1\mathbf{X}_3^\top$$

It can be shown that the first left-hand and the first right-hand singular vectors found by SVD($\mathbf{G}_{endo}^{a-1}$) (input for iteration a) are proportional to $\mathbf{t}_{21}^a$ and $\mathbf{t}_{32}^a$, and further that the $\mathbf{X}_2$- and $\mathbf{X}_3$-relevant latent variables are proportional to $\mathbf{t}_{22}^a$ and $\mathbf{t}_{31}^a$, respectively. Hence, the NIPALS and the sequential SVD versions of endo-LPLSR are equivalent given proper convergence in the NIPALS steps.

An alternative endo-LPLSR based on extracting all A sets of latent vectors simultaneously by SVD of $\mathbf{G}_{endo}$ is described in Martens et al. (2005) as an analogy to the PLSR of Bookstein et al. (1996). However, the maximum number of latent components which can be extracted is limited to the rank of $\mathbf{G}_{endo}$.

16.2.3 Exo-LPLSR

The basic idea of exo-LPLSR was proposed by Martens (2005) as a method for a bi-directional regression of two regressands from a single regressor. A sequential algorithm based on NIPALS extractions is presented below, and the basic steps are shown in Fig. 16.4.

The exo-LPLSR algorithm

For latent vectors extraction $a = 1, \ldots, A$

1. Find t-vectors $\mathbf{t}_{11}^a$ and $\mathbf{t}_{12}^a$ by the NIPALS algorithm as shown in Fig. 16.2, cycling through $\mathbf{X}_1^{a-1,a-1}$, $\mathbf{X}_2^{a-1}$ and $\mathbf{X}_3^{a-1}$. Let $\mathbf{T}_{11} = (\mathbf{t}_{11}^1, \ldots, \mathbf{t}_{11}^a)$ and $\mathbf{T}_{12} = (\mathbf{t}_{12}^1, \ldots, \mathbf{t}_{12}^a)$

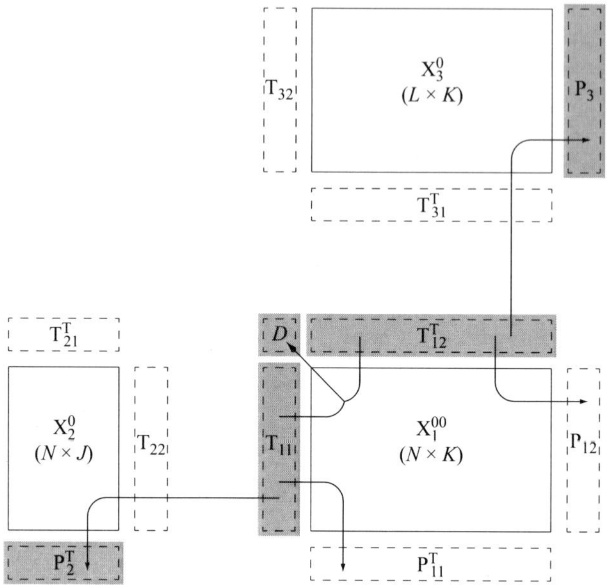

Fig. 16.4 The estimation steps of the regression parameters of the exo-LPLSR algorithm. At step a the T-variables and the P-loadings contain a columns, and **D** is a $(a \times a)$ matrix

2. Compute $\mathbf{X}_2$- and $\mathbf{X}_3$-loadings

$$\begin{aligned}\mathbf{P}_2 &= (\mathbf{X}_2^0)^\top \mathbf{T}_{11}(\mathbf{T}_{11}{}^\top \mathbf{T}_{11})^{-1}\\ \mathbf{P}_3 &= \mathbf{X}_3^0 \mathbf{T}_{12}(\mathbf{T}_{12}{}^\top \mathbf{T}_{12})^{-1},\end{aligned}$$

and a kernel loadings matrix $\mathbf{D}$ $(a \times a)$, defined by

$$\mathbf{D} = (\mathbf{T}_{11}{}^\top \mathbf{T}_{11})^{-1} \mathbf{T}_{11}{}^\top \mathbf{X}_1^{00} \mathbf{T}_{12}(\mathbf{T}_{12}{}^\top \mathbf{T}_{12})^{-1}.$$

Also construct $\mathbf{X}_1$-loadings in both $\mathbf{X}_2$- and $\mathbf{X}_3$-directions (used for construction of regression coefficients in (16.11)).

$$\begin{aligned}\mathbf{P}_{11} &= (\mathbf{X}_1^{00})^\top \mathbf{T}_{11}(\mathbf{T}_{11}{}^\top \mathbf{T}_{11})^{-1}\\ \mathbf{P}_{12} &= \mathbf{X}_1^{00} \mathbf{T}_{12}(\mathbf{T}_{12}{}^\top \mathbf{T}_{12})^{-1}.\end{aligned}$$

3. Deflate the data matrices by the contribution of the scores identified to form residual matrices

$$\begin{aligned}\mathbf{X}_2^a &= \mathbf{X}_2^0 - \mathbf{T}_{11}\mathbf{P}_2^\top\\ \mathbf{X}_3^a &= \mathbf{X}_3^0 - \mathbf{P}_3\mathbf{T}_{12}^\top\\ \mathbf{X}_1^{aa} &= \mathbf{X}_1^{00} - \mathbf{T}_{11}\mathbf{D}\mathbf{T}_{12}^\top\end{aligned}$$

end

The exo-LPLSR algorithm described above does not return orthogonal scores, that is, $\mathbf{T}_{11}^{\top}\mathbf{T}_{11}$ and $\mathbf{T}_{12}^{\top}\mathbf{T}_{12}$ are non-diagonal. To account for this $\mathbf{P}_2$ and $\mathbf{P}_3$ must be recomputed (step 3) and the total contribution from all a scores subtracted from the original centered matrices (step 5) at each iteration of the algorithm. (For endo-LPLSR the scores are orthogonal and the loadings may alternatively be computed sequentially.)

The bi-linear models for $\mathbf{X}_2^0$ and $\mathbf{X}_3^0$ after A extractions are given by

$$\mathbf{X}_2^0 = \mathbf{T}_{11}\mathbf{P}_2^{\top} + \mathbf{E}_2^A \tag{16.7}$$

$$\mathbf{X}_3^0 = \mathbf{P}_3\mathbf{T}_{12}^{\top} + \mathbf{E}_3^A. \tag{16.8}$$

For the purpose of prediction and interpretation it may be more convenient to express the models in terms of the original variables in $\mathbf{X}_1^{00}$:

$$\mathbf{X}_2^0 = \mathbf{X}_1^{00}\mathbf{B}_2 + \mathbf{E}_2^A \tag{16.9}$$

$$\mathbf{X}_3^0 = \mathbf{B}_3^{\top}\mathbf{X}_1^{00} + \mathbf{E}_3^A\,, \tag{16.10}$$

where the regression coefficients are estimated by

$$\hat{\mathbf{B}}_2 = \mathbf{T}_{31}(\mathbf{P}_{11}^{\top}\mathbf{T}_{31})^{-1}\mathbf{P}_2^{\top} \tag{16.11}$$

$$\hat{\mathbf{B}}_3 = \mathbf{T}_{22}(\mathbf{P}_{12}^{\top}\mathbf{T}_{22})^{-1}\mathbf{P}_3^{\top}. \tag{16.12}$$

The exo-way of modeling is reasonable if some degree of connectivity may be assumed between the two regressands. The latent score vector $\mathbf{t}_{11}$, used to model $\mathbf{X}_2$, is a linear combination of the columns of $\mathbf{X}_1$ where the weights are influenced by $\mathbf{X}_3$ through $\mathbf{t}_{31}$. Likewise, the weights used to construct the scores $\mathbf{t}_{12}$ for modeling $\mathbf{X}_3$ are influenced by $\mathbf{X}_2$ through $\mathbf{t}_{22}$.

The difference between endo- and exo-LPLSR for model formulation and parameter estimation is apparent when Figs. 16.3 and 16.4 are compared. The latent t-vectors used for modeling the regressand(s) are consequently those vectors defined as linear combinations of rows or columns of the regressor(s). This is in analogy with ordinary PLSR and, of course, necessary for prediction purposes.

Alternative exo-LPLSR approaches

Exo-LPLSR with sequential extraction of orthogonal scores may be achieved by alternatively deflating $\mathbf{X}_1$ by

$$\mathbf{X}_1^{aa} = \mathbf{X}_1^{a-1,a-1} - \mathbf{t}_{11}^a(\mathbf{p}_{11}^a)^{\top} - \mathbf{p}_{12}^a(\mathbf{t}_{12}^a)^{\top} + \mathbf{t}_{11}^a\mathbf{d}^a(\mathbf{t}_{12}^a)^{\top}\,, \tag{16.13}$$

where the loadings $\mathbf{p}_{11}^a$, $\mathbf{p}_{12}^a$ and d^a are computed as in step 4 of the exo-LPLSR algorithm, using only step a vectors $\mathbf{t}_{11}^a$ and $\mathbf{t}_{12}^a$. Here the total variation in $\mathbf{X}_1^{a-1,a-1}$

captured in both directions (rows and columns) by the scores $\mathbf{t}_{11}^a$ and $\mathbf{t}_{12}^a$, is subtracted. However, the two score-vectors may capture overlapping variability. This overlap is equal to the cross term $\mathbf{t}_{11}^a d^a (\mathbf{t}_{12}^a)^\top$ being added at the deflation step. Note the similarity between (16.13) and the double centering in the pre-processing step for $\mathbf{X}_1$ (16.3). Exo-LPLSR with orthogonal scores may be a better choice for visualization using correlation loadings plots, as discussed below in Sect. 16.2.5.

In its original form (Martens 2005) exo-LPLSR was defined through an SVD of the matrix product of type

$$\mathbf{G}_{exo} = \mathbf{X}_1 \mathbf{X}_3^\top \mathbf{X}_3 \mathbf{X}_1^\top \mathbf{X}_2 \mathbf{X}_2^\top \mathbf{X}_1.$$

The motivation for this approach was to combine PLS regressions in both $\mathbf{X}_2$- and $\mathbf{X}_3$-directions. The last part of this matrix product, $\mathbf{X}_1^\top \mathbf{X}_2 \mathbf{X}_2^\top \mathbf{X}_1$, is recognized as the basis for eigenvector extraction for a regular PLS regression between $\mathbf{X}_1$ and $\mathbf{X}_2$. Likewise the first part, $\mathbf{X}_1 \mathbf{X}_3^\top \mathbf{X}_3 \mathbf{X}_1^\top$, partially overlapping with the former, is a corresponding basis for PLS regression between $\mathbf{X}_1^\top$ and $\mathbf{X}_3^\top$.

Also for exo-LPLSR there is a direct correspondence between the SVD of $\mathbf{G}_{exo}$ and the latent t-vectors identified by NIPALS. The first left-hand singular vector of SVD($\mathbf{G}_{exo}^{a-1}$) is proportional to $\mathbf{t}_{11}^a$ and the first right-hand singular vector is proportional to $\mathbf{t}_{12}^a$. Hence, exo-LPLSR based on sequential NIPALS extractions or SVD extractions are equivalent.

The SVD approach may, on the other hand, facilitate simultaneous extraction of all A score-vectors. Define $\mathbf{G}_{exo}^0$ by

$$\mathbf{G}_{exo}^0 = \mathbf{X}_1^{00} (\mathbf{X}_3^{0\top} \mathbf{X}_3^0 \mathbf{X}_1^{00\top} \mathbf{X}_2^0 \mathbf{X}_2^{0\top}) \mathbf{X}_1^{00} = \mathbf{X}_1^{00} \mathbf{G}'_{exo} \mathbf{X}_1^{00}$$

Perform SVD on $\mathbf{G}_{exo}^0$ to obtain

$$\mathbf{G}_{exo}^0 = \mathbf{U}\mathbf{L}\mathbf{V}^\top$$

and define $\mathbf{T}_{11}$ as the first A columns of $\mathbf{U}$ and $\mathbf{T}_{12}$ as the first A columns of $\mathbf{V}$. The maximum number of latent vectors which can be extracted, is limited to the rank of $\mathbf{G}_{exo}^0$. The linear models for the centered regressands are subsequently expressed in terms of the orthogonal score-vectors as given by (16.7) with loadings computed as in step 3 of the exo-LPLSR algorithm. Finally, models expressed in terms of $\mathbf{X}_1^{00}$ are as given by (16.9), but with regression coefficients estimated by

$$\hat{\mathbf{B}}_2 = \mathbf{G}'_{exo} \mathbf{X}_1^{00} \mathbf{T}_{12} \mathbf{L}_A^{-1} \mathbf{T}_{11}^\top \mathbf{X}_2^0 \tag{16.14}$$

$$\hat{\mathbf{B}}_3 = \mathbf{G}'^{\top}_{exo} \mathbf{X}_1^{00\top} \mathbf{T}_{11} \mathbf{L}_A^{-1} \mathbf{T}_{12}^\top \mathbf{X}_3^{0\top}, \tag{16.15}$$

where $\mathbf{L}_A$ is the square sub-matrix of $\mathbf{L}$ with the A first (and largest) singular values on the diagonal. The coefficient estimates given by (16.14) will for $A > 1$ typically differ from those given by (16.11) since the scores of the two algorithms span different subspaces of the full variable space.

16.2.4 Cross-validation and Jackknifing

Predictors for the regressands in endo- and exo-LPLSR may be constructed from (16.6) and (16.9). This gives the opportunity to perform model validation on prediction performance, for instance, by test set prediction or by cross-validation (Stone 1974) . Further, just as in ordinary PLS-regression, cross-validation may be used to find the optimal model complexity through the number A of latent components. In endo-LPLSR cross-validation can be performed in three ways; 1) by holding out rows of $\mathbf{X}_2$ and $\mathbf{X}_1$, 2) By holding out columns of $\mathbf{X}_1$ and $\mathbf{X}_3$, or 3) by a combination of both holding out rows and columns. Usually the nature of the data will give guidance to how to perform the cross-validation. Typically either rows or columns of $\mathbf{X}_1$ represent random selections from some population (i.e. persons, objects), and often the purpose of cross-validation is to assess some general predictive property of the model for new objects from this population. In exo-LPLSR prediction may be performed in two directions, but it is reasonable to perform cross-validation separately for the two cases. An alternative to regular cross-validation is bootstrap validation (see e.g. Chap. 3).

In many implementations of PLS-regression the significance of regression coefficients is determined by jackknifing (Martens and Martens 2001), which conveniently can be performed without much extra computational cost during cross-validation. Jackknife-testing is straightforwardly implemented also in LPLSR as soon as a proper cross-validation setup has been identified.

16.2.5 Visualization of Model Fit

The fitted models from the various LPLSR approaches described above, should be evaluated to verify that the results are reasonable in light of prior knowledge of the phenomenon studied. The so-called correlation loadings plot is frequently used for graphical model evaluation (Martens and Martens 2001). Correlation loadings are unit-free loading-vectors corresponding to the loadings found in the LPLSR algorithms. For instance, the $\mathbf{X}_2$ loadings $\mathbf{P}_2$, as found in (16.4) for endo-LPLSR, are transformed to correlations $\mathbf{R}_2$ between the columns of $\mathbf{X}_2^0$ and the score vectors $\mathbf{T}_{22}$. For variable j in $\mathbf{X}_2^0$ the correlation loading along latent component a (element j,a in $\mathbf{R}_2$) is defined by

$$\mathbf{r}_{2,j}^a = \frac{\mathbf{x}_{2,j}^{0\ \top}\mathbf{t}_{22}^a}{(\mathbf{x}_{2,j}^{0\ \top}\mathbf{x}_{2,j}^0)^{1/2}(\mathbf{t}_{22}^{a\ \top}\mathbf{t}_{22}^a)^{1/2}} \tag{16.16}$$

Correspondingly, $\mathbf{R}_3$ are the correlation loadings defined as the correlations between the rows of $\mathbf{X}_3^0$ and the columns of $\mathbf{T}_{31}$. Further, correlation loadings $\mathbf{R}_{11}$ and $\mathbf{R}_{12}$ may be explored, although their analogues $\mathbf{P}_{11}$ and $\mathbf{P}_{12}$ are not used in the model fit. $\mathbf{R}_{11}$ is here defined as the correlations between the columns of $\mathbf{X}_1^{00}$ and the

$\mathbf{X}_2$-relevant score vectors $\mathbf{T}_{22}$, whereas $\mathbf{R}_{12}$ are the correlation between the rows of $\mathbf{X}_1^0$ and the $\mathbf{X}_3$- relevant scores $\mathbf{T}_{31}$. Correlation loadings for evaluating the exo-LPLSR fit are computed similarly, the only difference being that the scores $\mathbf{T}_{22}$ and $\mathbf{T}_{31}$ are replaced by $\mathbf{T}_{11}$ and $\mathbf{T}_{12}$, respectively. If the scores are orthogonal, the distance of the correlation loadings from the origin can be interpreted as the proportion of explained variance. In that respect the orthogonal exo-LPLSR should be used for visualizations, since the correlation loadings from the non-orthogonal version do not have this interpretation and may end up outside the unit circle.

Typically the correlation loadings along the two first latent components are plotted for all or some of $\mathbf{R}_2$, $\mathbf{R}_3$, $\mathbf{R}_{11}$ and $\mathbf{R}_{12}$ in the same plot. The correlation plot summarizes in an apprehensible manner the main systematic patterns of covariation between the three data matrices.

For data exploration it can be useful to construct correlation plots from both endo- and exo-LPLSR. If the endo- and the exo- analyses give rise to similar correlation loadings plots, the $\mathbf{X}_2$- and $\mathbf{X}_3$-relevant patterns in $\mathbf{X}_1$ should not be very different from the $\mathbf{X}_1$-relevant patterns in $\mathbf{X}_2$ and the $\mathbf{X}_1$-relevant patterns in $\mathbf{X}_3$, respectively. This two-way data exploration is analogous to how Martens and Martens (2001) used PLSR to regress both $\mathbf{X}$ on $\mathbf{Y}$ and $\mathbf{Y}$ on $\mathbf{X}$ in two block modeling. In cases where the corner-matrix in LPLSR represents empirical data (e.g. consumer response to products) and the off-corner matrices hold "design information" about consumers and products, the endo-LPLSR may be considered as a three-block generalization of ANOVA-modeling via PLSR. The exo-LPLSR, on the other hand, may represent a three-block generalization of PLS discriminant analysis.

16.3 Real Data Example

16.3.1 Beer Liking Data

After examining the theoretical properties of endo- and exo-LPLS regression in the previous sections, it may be enlightening to study a real data example. We shall use data from a Danish beer study (Mejlholm and Martens 2006) which constitute suitable data sets for comparing the two approaches from a data exploration point of view.

Nine commercially available Danish beers, selected to span a relevant space with respect to new and established products on the market, were used as samples. The chosen samples represented three types of beer namely; Lager (named L1, L2, L3 with %alcohol range 4.6-5.8), Strong lager (named S1, S2, S3 with %alcohol range 7.1-7.7) and Ale (named A1, A2, A3 with %alcohol range 5.7-7.3). From this, four beer sample characteristics were made for each of the nine samples: Design variables L, S, A as well as %alcohol.

The same beer samples were used for sensory profiling and a consumer liking test (Meilgaard et al. 1999). A trained panel consisting of nine members carried out the sensory profiling in an accredited sensory laboratory, evaluating the following nine sensory attributes concerning appearance, taste and flavor: color (darkness), body, carbonation, bitter, alcohol-, fruity-, floral-, spicy- and grainy/roasted- flavor. The intensity of the attributes was scored on a 15 cm unstructured line scale in three replicates. 38 consumers (18–59 years) dispersed on 29 males and 9 females participated in the consumer test. The nine samples were evaluated for overall liking on a 7-point hedonic scale (1 = dislike extremely; 7 = like extremely). Following the final sample evaluation each consumer completed a questionnaire concerning their background (15 variables): demographic information (Male, Female, Age 18–29, Age 30–59, Student, type of Working) and habits/attitudes towards beer ((frequency of usage (2/month, 4/month, 10/month, 16/month), favorite type of beer (LagerFavo, StrongFavo, AleFavo) and preferred temperature (FridgeTemp, CellarTemp)).

In summary, three data sets/blocks from the original beer study (Mejlholm and Martens 2006), with minor modifications, constitute the basis for the present data modeling by endo- and exo-LPLS regression. (With respect to the original study, one beer sample is taken out and minor consumer background variables are exchanged giving slightly different results in the original and present study.) More experimental details and results can be found in Mejlholm and Martens (2006). Referring to Fig. 16.1 (right) the $\mathbf{X}_1$ (9 × 38) block consists of consumer liking score for each of the nine samples; the $\mathbf{X}_2$ (9 × 13) block consists, for each of the nine samples, of nine sensory profiling variables (average data across panelists and replicates for each attribute) plus four sample characteristics (see above); and the $\mathbf{X}_3$ (15 × 38) block consists of consumer background data (group average data across 15 variables).

16.3.2 Comparing Results from Endo- and Exo-LPLSR

The endo-LPLSR (Fig. 16.3) and the exo-LPLSR (Fig. 16.4) were applied to the same set of input data. (For exo-LPLSR the orthogonal version was used.) In order to illustrate the difference in the optimization criteria between endo- and exo-LPLSR, the statistical validation (cross-validation/jackknifing) is skipped in this context. Instead the simple fit of the three data tables to the endo- and exo-LPLSR models by the first two LPLS components is reported (Table 16.1). In the table the decrease in the sums-of-squares of three tables is expressed in percent of their initial sums-of-squares (after mean-centering and scaling).

The table shows that exo-LPLSR explains more of $\mathbf{X}_1$ than the endo-LPLSR. This is as expected, since the exo-LPLSR defines the latent structures in terms of its

Table 16.1 Percent sum-of-squares in the three blocks explained by the first two components

	Comp. 1	Comp. 2
Endo-LPLSR		
$\mathbf{X}_1$	8	5
$\mathbf{X}_2$	37	12
$\mathbf{X}_3$	20	21
Exo-LPLSR		
$\mathbf{X}_1$	27	15
$\mathbf{X}_2$	31	9
$\mathbf{X}_3$	7	8

bi-linear components from $\mathbf{X}_1$, ($\mathbf{T}_{11}$, $\mathbf{T}_{12}$, Fig. 16.4), while the endo-LPLSR defines them from the $\mathbf{X}_2$- and $\mathbf{X}_3$-components $\mathbf{T}_{22}$ and $\mathbf{T}_{31}$ (Fig. 16.3). Moreover, it shows that the product descriptor data in $\mathbf{X}_2$ are in general better modeled than the person descriptor data in $\mathbf{X}_3$, in particular in the exo-LPLSR case.

Bi-linear models are generally well suited for graphical inspection of the main patterns of co-variation in data tables; this is also the case for the LPLSR methods. Figures 16.5 and 16.7 show the results for the endo- and exo-LPLSR, respectively.

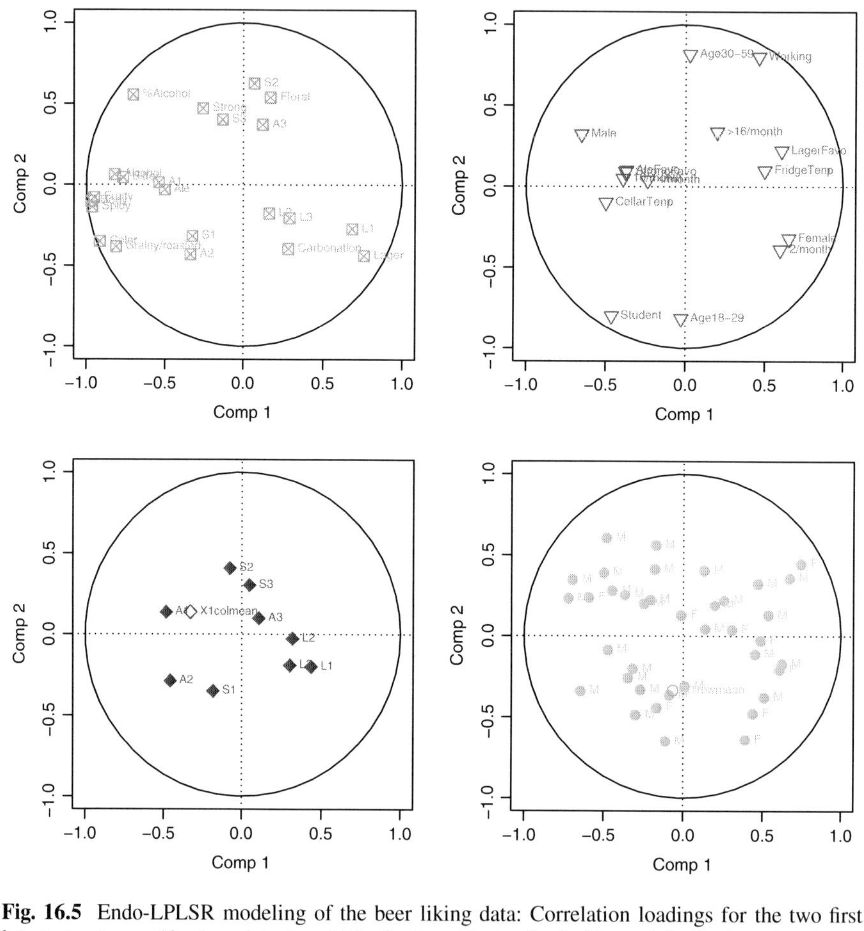

Fig. 16.5 Endo-LPLSR modeling of the beer liking data: Correlation loadings for the two first latent structures. *Abscissa*: latent variable #a=1, ordinate: #a=2. *Upper left*: Product descriptors (columns in $\mathbf{X}_2$) correlations to column 1 and 2 in $\mathbf{T}_{22}$. The 9 products are positioned by the correlations between the 9 × 9 identity matrix $\mathbf{I}_9$ and $\mathbf{T}_{22}$. *Upper right*: Person descriptors (rows in $\mathbf{X}_3$) correlations to $\mathbf{T}_{31}^{\top}$. *Lower left*: Product likings (rows in $\mathbf{X}_1$) correlations to $\mathbf{T}_{31}^{\top}$. The row of column means in the input $\mathbf{X}_1$, i.e. the average liking level for the different persons, is also correlated to $\mathbf{T}_{31}^{\top}$, and named "X1colmean". *Lower right*: Person likings (columns in $\mathbf{X}_1$) correlations to $\mathbf{T}_{22}$. The column of row means in the input $\mathbf{X}_1$, i.e. the average liking level for the different beers, is also correlated to $\mathbf{T}_{22}$, and named "X1rowmean"

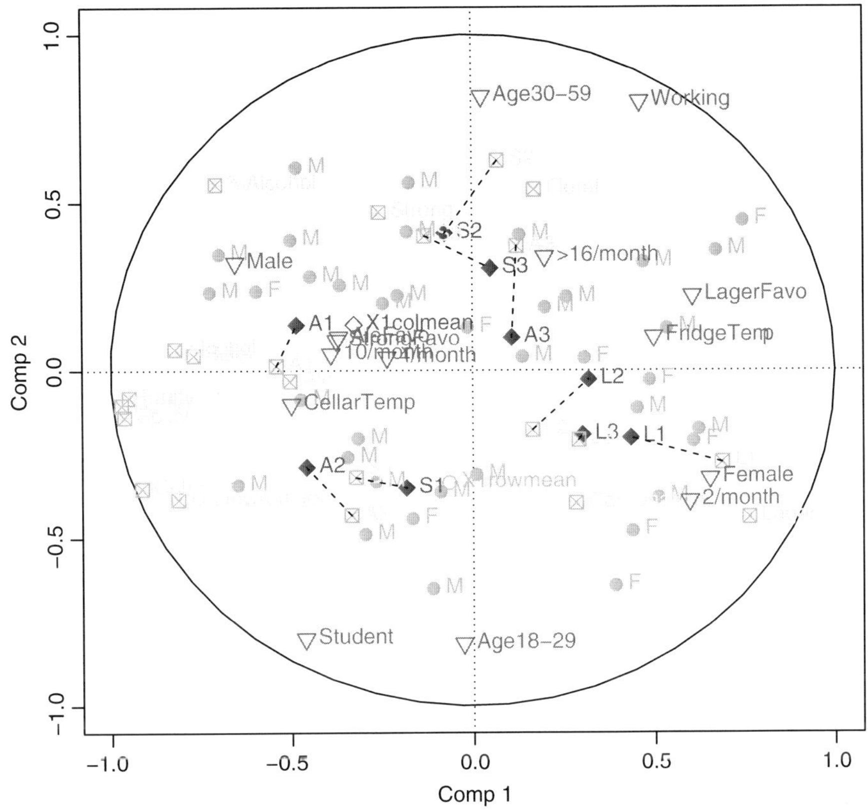

Fig. 16.6 Combination of all endo-LPLSR subplots of Fig. 16.5. The product indicators from the leftmost subfigures of Fig. 16.5 are connected by *dashed lines*

It is important to attain both a detailed interpretation of the patterns within each of the three data tables and an integrative overview of how the three tables' patterns are related. Therefore, the correlation loadings method of Martens and Martens (2001) was extended to all the component parameters in LPLSR, as described in Martens et al. (2005). Hence, the four subplots in Figs. 16.5 and 16.7 show results for the individual data tables, while Figs. 16.6 and 16.8 superimpose the subplots for overview.

Each of the axes in correlation loadings plots represent the simple correlation coefficient between a key LPLSR component vector and the corresponding vectors in the input data (rows or columns in the $\mathbf{X}_1^{00}$, $\mathbf{X}_2^0$ or $\mathbf{X}_3^0$ submitted to LPLSR, conf. (16.16)). In endo-LPLSR the key component vectors are defined as the first two vectors in $\mathbf{T}_{22}$ and $\mathbf{T}_{31}$ (Fig. 16.3), while in exo-LPLSR they are defined as the first two vectors in $\mathbf{T}_{11}$ and $\mathbf{T}_{12}$ (Fig. 16.4). The unit circle in the plots represent 100% explained variance using these two components, while the origin represent 0% explained variance.

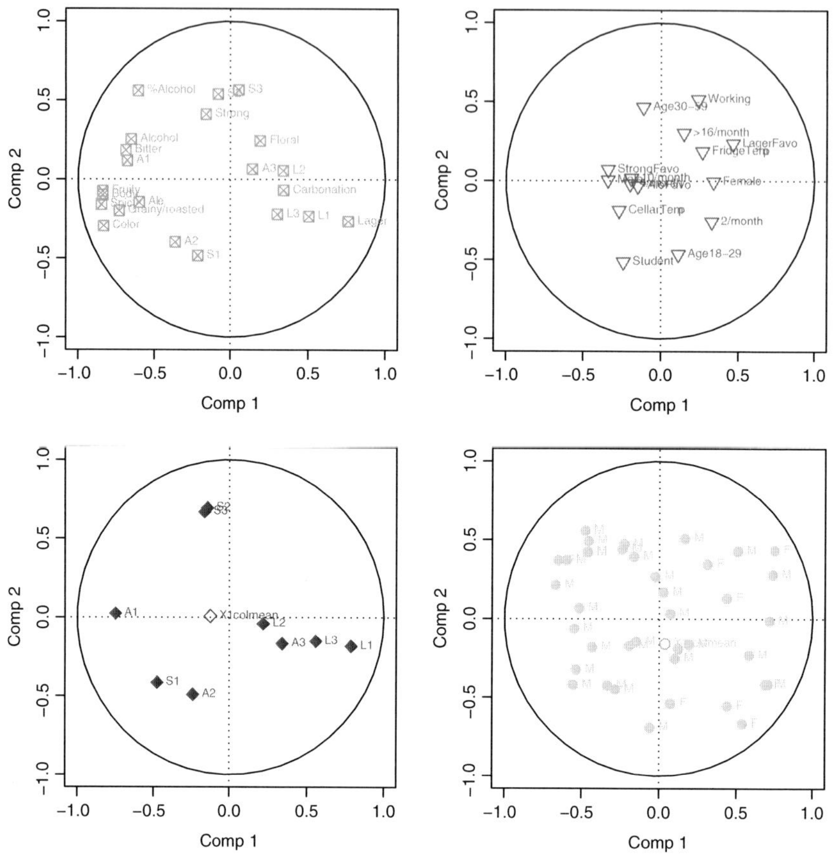

Fig. 16.7 Exo-LPLSR modeling of the beer liking data: Correlation loadings for the two first latent structures. *Abscissa*: latent variable #a=1, ordinate: #a=2. *Upper left*: Product descriptors (columns in $\mathbf{X}_2$) correlations to column 1 and 2 in $\mathbf{T}_{11}$. The 9 products are positioned by the correlations between the 9 × 9 identity matrix $\mathbf{I}_9$ and $\mathbf{T}_{11}$. *Upper right*: Person descriptors (rows in $\mathbf{X}_3$) correlations to $\mathbf{T}_{12}^{\top}$. *Lower left*: Product likings (rows in $\mathbf{X}_1$) correlations to $\mathbf{T}_{12}^{\top}$. The row of column means in the input $\mathbf{X}_1$, i.e. the average liking level for the different persons, is also correlated to $\mathbf{T}_{12}^{\top}$, and named "X1colmean". *Lower right*: Person likings (columns in $\mathbf{X}_1$) correlations to $\mathbf{T}_{11}$. The column of row means in the input $\mathbf{X}_1$, i.e. the average liking level for the different beers, is also correlated to $\mathbf{T}_{11}$, and named "X1rowmean"

This quantitative interpretation is possible because the LPLSR algorithms are here defined to ensure that the key LPLSR component vectors are orthogonal (i.e. $\mathbf{T}_{11}^{\top}\mathbf{T}_{11}$=diag, $\mathbf{T}_{12}^{\top}\mathbf{T}_{12}$=diag, $\mathbf{T}_{22}^{\top}\mathbf{T}_{22}$=diag, $\mathbf{T}_{31}^{\top}\mathbf{T}_{31}$=diag; otherwise some points might have ended outside the unit circle). Moreover, in order to maintain that the first two component patterns found in the different data tables really reflect the same two latent structures, the coupling between the structures should be weak. This means that the kernel loadings matrix $\mathbf{D}$ in Figs. 16.3 and 16.4 should be close to diagonal. Table 16.2 shows that this is the case for the present data; not only for the exo-LPLSR (where orthogonality was ensured by (16.13)), but also for the endo-LPLSR.

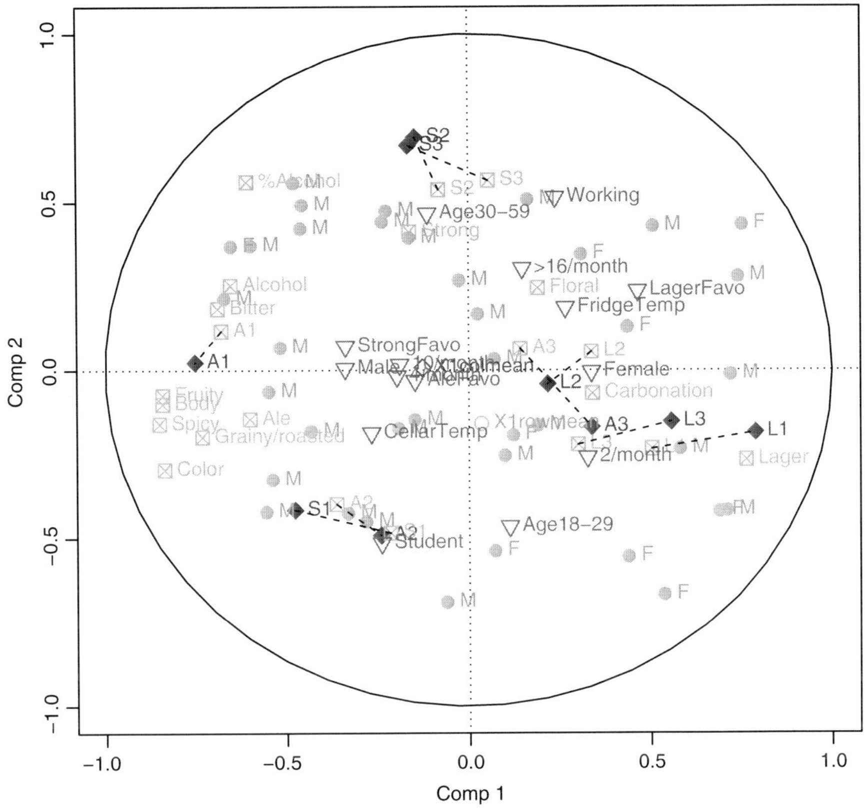

Fig. 16.8 Combination of all exo-LPLSR subplots of Fig. 16.7. The product indicators from the leftmost subfigures of Fig. 16.5 are connected by *dashed lines*

Table 16.2 The kernel loadings matrix **D** from endo-and exo-LPLSR of the Danish beer data with two latent components

	Comp. 1	Comp. 2
Endo-LPLSR		
Comp. 1	0.094	−0.010
Comp. 2	−0.003	0.128
Exo-LPLSR		
Comp. 1	0.119	0
Comp. 2	0	0.170

Hence, a conventional inspection of the individual correlation loadings plots as well as their superimposed overview plot is warranted.

Figure 16.5 (upper left) shows that the first component associated high levels of several sensory descriptors (alcoholic taste, bitter, fruity, spicy, body, color, grainy/roasted) with Ales and not Lager, while the second component associated high levels of the sensory descriptor floral and the chemical descriptor %Alcohol and lower levels of carbonation with Strong lager. Hence, the three Lagers L1, L2 and L3 appear to have high carbonation, low %Alcohol and alcoholic and floral taste; Ales A1 and A2 and Strong lager S1 are e.g. fruity, spicy and of dark color, while the Strong lagers S2 and S3 and Ale A3 have somewhat floral flavor, but little color.

The first component in Fig. 16.5 (upper right) seems to pit people, primarily female, who verbally claim to have Lager as favorite beer, like to drink beer at Fridge temperature and drink less than 2 beers/month, against primarily males who apparently have a tendency to have ale as favorite beer and prefer drinking beer at room temperature. The second component seems to distinguish older, working people from younger students.

The likings data in $\mathbf{X}_1$ is visualized in Fig. 16.5 (lower left). The first component primarily differentiates the three Lagers against Ales A1 and A2, while the second component primarily differentiates Strong lagers S2 and S3 against Strong lager S1. The "average liking of beer" (X1colmean) is to the left of the origin in the plot.

Figure 16.5 (lower right) shows how the 38 individual consumers span the two first components. There is a tendency for more males (M) to the left and more females (F) to the right. The "average person" with respect to beer liking lies below the origin in the plot.

The fit of the endo-LPLSR model is summarized in Fig. 16.6. It shows, for instance, (in the North-West/South-East direction) that males tend to prefer beer with higher alcohol and bitterness levels than women do. Further, women tend to drink less than 2 beers per month, claim that they have Lager as favorite, and in practice show that they prefer Lager (with high carbonation). However, the gender distinction is far from complete. Older people and working people tend to have higher liking for strong lager than students.

It is important not to lose track of the means used in the mean centering prior to the model fit. The figure shows that on the average, the liking of beer (X1colmean) seems to be higher for men than for women, and stronger for Ale than for Lager, i.e. those who really like beer tend to Male and to prefer Ale over Strong lager or Lager. There is conversely a tendency that the average beer-liking person (X1rowmean) tends to be a young student, and floral beers are not liked on the average. However, this just shows how the mean liking vectors project on the LPLSR solutions, these mean liking vectors should also be studied by themselves.

Figures 16.7 and 16.8 likewise summarize the exo-LPLSR solution. The results are rather similar to those of the endo-LPLSR in Figs. 16.5 and 16.6, although some differences may be noticed: First of all, the person descriptors $\mathbf{X}_3$ are generally closer to the origin and further from the unit circle in the exo-LPLSR than in the endo-LPLSR result, as expected from the theory and as summarized by Table 16.1. Secondly, the gender difference is far less pronounced in the exo-LPLSR solution. But the general pattern remains, e.g. that people who say that they prefer lager also in practice like lagers L1, L2 and L3 well (and sometimes even strong lagers S2 and S3), that students don't like Strong Lagers S2 and S3 etc.

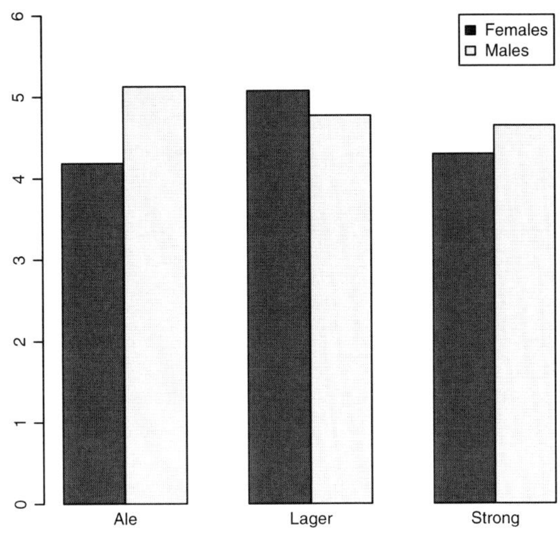

Fig. 16.9 The average liking score in $\mathbf{X}_1$ for the three beer types Ale, Lager, and Strong Lager for females and males

In summary we find that the general picture is rather similar for both LPLSR approaches with regard to the interpretation of the correlation loadings plots, which is an indication that there are consistent patterns of co-variation in the three data matrices (as discussed in Sect. 16.2.5). The consumer response data in this case represent a useful consensus between product- and consumer-descriptors. Still, the two approaches emphasize slightly different aspects of the data, and investigating both fits may give increased insight.

Finally, a word of caution regarding data centering: The analyst should bear in mind at all times that the patterns revealed in the correlation loadings plots are relative to the subtracted row and column means, and the patterns should be verified by studying plots of the raw data. For instance, the plots indicate that males score Ale and Strong Lager beers higher than Lager beer, whereas the opposite should apply to the females. This is more or less consistent with the mean values shown in Fig. 16.9. We may also be led to believe that males claim that they have Ale and Strong Lager as their favorite beers and strongly claim their disliking to Lager beer. However, in the consumer background data, almost 45% of the males hold Lager as their favorite over all other beers in the study. The picture of males preferring Ale and Strong Lager over Lager is generated by the fact that females vote stronger is disfavor of these beers than males do. No females claim they are in favor of Ale or Strong Lager, whereas the numbers for males are 7% and 21% respectively. Hence, it is important to verify the conclusions made from the LPLSR analysis on the centered data by plotting the raw data.

16.4 Outlook

Extracting and visualizing more than one latent variable makes it possible to overcome the traditional limitation of rank-1 blocks in reflexive path modeling. In market research it may be seen as a tool for developing path models in more than one direction, as demonstrated in this application. In biology it may e.g. be used for multivariate modeling of how ($\mathbf{X}_1$) expressions from K genes in N objects relate to ($\mathbf{X}_2$) J phenotypic descriptors in the same N objects in light of ($\mathbf{X}_3$) known functional relationships of the K genes wrt L biological pathways ("gene ontology") or L clusters of literature reference ("co-citation correlations"). In particular, the LPLSR modeling may be seen as a PLS path modeling in two directions - over objects (e.g. "products") and over variables (e.g. "persons"). Other extensions of path modeling are described in Chaps. 4–8.

The endo- and exo-LPLSR approaches may be combined in order to perform one-directional regressions e.g. from $\mathbf{X}_3$ via $\mathbf{X}_1$ to $\mathbf{X}_2$. This means that $\mathbf{X}_3$ serves as a regressor for $\mathbf{X}_1$, which in turn serves as a regressor for $\mathbf{X}_2$. For instance, the biological data on gene expression mentioned above should fit nicely into this framework, where the intention is to model phenotypic data in $\mathbf{X}_2$ using the gene expressions in $\mathbf{X}_1$ under the influence of the background information on gene dependencies in $\mathbf{X}_3$. Technically this may be achieved by using $\mathbf{t}_{31}$ as the score-vector for modeling $\mathbf{X}_3$, and at the same time it serves as weights for defining the scores $\mathbf{t}_{11}$ used for modeling $\mathbf{X}_1$ and $\mathbf{X}_2$. A refinement of this approach would be to use a weighted average of $\mathbf{t}_{31}$ and $\mathbf{t}_{12}$ as the weights defining the scores used to model $\mathbf{X}_1$ and $\mathbf{X}_2$. In this way the background information on gene dependencies may be taken into account to a varying extent.

The one-block PCA, the two-block PLSR and the three-block LPLSR modeling approaches may all be seen as special cases of the more general concept of "Domino-PLS" Martens (2005) , which provides a flexible framework for multivariate data modeling of very complex systems in various conceptual spaces. The eigen-analysis of various covariance matrices, e.g. $\mathbf{X}_2^\top \mathbf{X}_1 \mathbf{X}_3^\top$ (endo-LPLSR) or $\mathbf{X}_1 \mathbf{X}_3 \mathbf{X}_3^\top \mathbf{X}_1^\top \mathbf{X}_2^\top \mathbf{X}_2 \mathbf{X}_1$ (exo-LPLSR) links together various rows and columns in a way that find and visualize the major co-variation patterns in such complex data structures. This methodology can be combined with local re-weighting and clustering in order to balance global vs local modeling (e.g. global patterns common to several consumer groups vs local patterns of different consumer groups).

As the two-block PLSR has been extended in a number of ways, Domino-PLS may be modified in many ways. One interesting aspect is the removal of "irrelevant dimensions". If one or more of the data matrices in Domino-PLS contain types of variation that are orthogonal to its regressand or regressor "neighbor" in the model structure, their "fingerprints" may be identified and effects removed. As shown in Martens and Næs (1989) for the two-block case, "irrelevant" variation patterns in a matrix $\mathbf{X}_1$, varying independently of the valuable information in another matrix $\mathbf{X}_2$, may be identified by first regressing $\mathbf{X}_1$ on $\mathbf{X}_2$ and then analyzing the $\mathbf{X}_1$-residuals by PCA: The first, major PC K-dimensions loading(s) $\mathbf{P}_{1,Irrelevant}$ should then give good estimates of unknown interferences' patterns; their detrimental effect can be more or less eliminated from $\mathbf{X}_1$ by e.g. projection on $\mathbf{P}_{1,Irrelevant}$. Methods like Orthogonal Scatter Correction , Direct Orthogonalization and O-PLS are variations of this theme, similarly developed for the two-block case. Matrix $\mathbf{X}_2$ can similarly be cleaned of J-dimensional $\mathbf{X}_1$-irrelevant patterns, $\mathbf{P}_{2,Irrelevant}$. In the case of Domino-PLS, irrelevant variation patterns can be defined in several more ways as well, e.g.

in the four-block case between $\mathbf{X}_1$ and $\mathbf{X}_3$ (L- or N-dimensional) and between $\mathbf{X}_2$ and $\mathbf{X}_4$ (M- or N-dimensional). This opens up for possibilities for simplifying the data analysis of complex systems. If any of the input matrices are N-way, instead of just two-way, the Domino-PLS can be modified to handle N-way model extensions.

Acknowledgements Thanks to Ole Mejlholm for permission to use the beer data.

References

Bookstein, F. L., Sampson, P. D., Streissguth, A. P., & Barr, H. M. (1996). Exploiting redundant measurement of dose and developmental outcome: new methods from the behavioral teratology of alcohol. *Developmental Physchology, 32*, 404–415.

Kubberød, E., Ueland, Ø., Rødbotten, M., Westad, F., & Risvik, E. (2002). Gender specific preferences and attitudes towards meat. *Food Quality and Preference, 13*, 285–294.

Martens, H. (2005). Domino PLS: a framework for multi-directional Path Modelling. In *Proceedings of PLS'05 International Symposium*, Aluja, T., Casanovas, J., Esposito Vinzi, V., Morineau, A. and Tenenhaus, M. (eds), SPAD Test & Go Group, Paris, pp. 125–132.

Martens, H., Anderssen, E., Flatberg, A., Gidskehaug, L. H., Høy, M., Westad, F., Thybo, A., & Martens, M. (2005). Regression of a data matrix on descriptors of both its rows and of its columns via latent variables: L-PLSR. *Computational Statistics & Data Analysis, 48*, 103–123.

Martens, H., & Martens, M. (2001). Multivariate analysis of quality. An introduction (Monograph). New York: Wiley.

Martens, H., & Næs, T. (1989). Multivariate calibration. Chichester, UK: Wiley.

Meilgaard, M., Civille, G. V., & Carr, B. T. (1999). Sensory evaluation techniques, 3rd Edn. New York: CRC.

Mejlholm, O., & Martens, M. (2006). Beer identity in Denmark. *Food Quality and Preference, 17*, 108–115.

Stone, M. (1974). Cross-validatory choice and assessment of statistical predictions (with discussion). *Journal of the Royal Statistical Society B, 36*, 111–147.

Thybo, A. K., Kuhn, B. F., & Martens, H. (2002). Explaining Danish children's preferences for apples using instrumental, sensory and demographic/behavioural data. *Food Quality and Preference, 15*, 53–63.

Wold, S., Martens, H., & Wold, H. (1983). The multivariate calibration problem in chemistry solved by the PLS method. In A. Ruhe, & B. Kågstrom (Eds.), *Proceedings of the Conference on Matrix Pencils* (pp. 286–293), Lecture Notes in Mathematics. Heidelberg: Springer.

Wold, S., Hellberg, S., Lundstedt, T., Sjostrom, M., & Wold, H. (1987). PLS modelling with latent variables in two or more dimensions. In *Proceedings PLS Model Building. Theory and Applications*. Symposium Frankfurt am Main, Sep 1987.

Chapter 17
Regression Modelling Analysis on Compositional Data

Huiwen Wang, Jie Meng, and Michel Tenenhaus

Abstract In data analysis of social, economic and technical fields, compositional data is widely used in problems of proportions to the whole. This paper develops regression modelling methods of compositional data, discussing the relationships of one compositional data to one or more than one compositional data and the interrelationship of multiple compositional data. By combining centered logratio transformation proposed by Aitchison (The Statistical Analysis of Compositional Data, Chapman and Hall, 1986) with Partial Least Squares (PLS) related techniques, that is PLS regression, hierarchical PLS and PLS path modelling, respectively, particular difficulties in compositional data regression modelling such as sum to unit constraint, high multicollinearity of the transformed compositional data and hierarchical relationships of multiple compositional data, are all successfully resolved; moreover, the modelling results rightly satisfies the theoretical requirement of log-contrast. Accordingly, case studies of employment structure analysis of Beijing's three industries also illustrate high goodness-of-fit and powerful explainability of the models.

17.1 Introduction

In data analysis of social, economic and technical fields, compositional data is widely used in problems of proportions to the whole, such as investment structure, industrial structure, consumption structure, etc.

H. Wang
School of Economics and Management, Beihang University, 37 Xueyuan Road, Haidian District, Beijing 100083, China
e-mail: wanghw@vip.sina.com

J. Meng
School of Statistics, Central University of Finance and Economics, Beijing 100081, China
e-mail: mengjie517@126.com

M. Tenenhaus
Department of SIAD, HEC School of Management, 1 rue de la Libération, 78351 Jouy-en-Josas, France
e-mail: tenenhaus@hec.fr

V. Esposito Vinzi et al. (eds.), *Handbook of Partial Least Squares*, Springer Handbooks of Computational Statistics, DOI 10.1007/978-3-540-32827-8_18,

According to the definition, a compositional data with p components refers to a non-negative vector $X = (x_1, x_2, \ldots, x_p)$, of which $x_j (j = 1, 2, \ldots, p)$ satisfy

$$\sum_{j=1}^{p} x_j = 1 \tag{17.1}$$

Equation (17.1) is also called "sum to unity" constraint which is the basic characteristic of compositional data. The concept of compositional data originally came from research by Ferrers (1866). Pearson (1897) indicated in an article about spurious correlation that in practice of compositional data analysis, the sum to unity constraint was often intentionally or unintentionally ignored and some statistical methods designed for data without constraint were frequently misused, which sometimes led to disastrous results. Aitchison (1986) published the first systemic work on compositional data – *The Statistical Analysis of Compositional Data*, which presented logratio transformation, discussed the theory of logistic-normal distributions and introduced some related statistical models of compositional data. Hinkle and Rayens (1994) proposed logcontrast partial least squares (LCPLS) regression. Zhang (2000) discussed ordinary least squares regression modelling on compositional data.

Based on the above research, this paper further develops regression modelling methods on compositional data, including PLS regression models of one dependent compositional variable on one or more than one independent compositional variables, and PLS path modelling on compositional data.

Confined by the sum to unity constraint, the following problems may arise when using classic linear regression methods in the modelling process: (1) more than one dependent variables should be considered in the modelling process; (2) the sum to unity constraint of compositional data should always be satisfied throughout the modelling; (3) the components of compositional data rang within (0,1), that makes trouble in prediction of the dependent variable; (4) considering each compositional data, which is composed of multiple components and represents a thematic meaning, the hierarchical relationship in terms of multiblock variables should be mainly investigated for the purpose of interpreting.

Studies in this paper show that these problems could be well resolved by combining centered logratio transformation with PLS regression, hierarchical PLS and PLS path modelling, respectively. The remainder of the paper is organized as follows: in the next section, some basic related knowledge of compositional data is introduced; in Sect. 17.3, simple linear regression model of compositional data is established by integrating centered logratio transformation with "standard" PLS regression method, which is then used to analyze the relationship of employment structure on GDP structure in Beijing's three industries ; afterwards, multiple linear regression model on compositional data is built in Sect. 17.4 by adopting hierarchical PLS regression method, and the investment structure in Beijing's three industries is further added in the case study as the second independent compositional variable; in Sect. 17.5, PLS path modelling is employed to explore the direct and indirect connections of multiple compositional data, which is then used to analyze the causal relationships of the investment, GDP and employment structures in Beijing's three industries; based on the above studies, a comparison of the three models is given in Sect. 17.6; finally, Sect. 17.7 is a conclusion of the contribution.

17.2 Overview of Compositional Data

In this section, definitions and properties of logratio transformation, compositional covariance matrix and logcontrast combination are introduced as an overview of compositional data.

17.2.1 Logratio Transformation

Logratio transformation is first introduced by Aitchison (1986), which can be easily calculated and has some proper mathematic properties.

Definition 17.2.1 *The logratio transformation of a p-part composition is the* $(p-1)$- *dimensional vector given by*

$$\hat{x}_j = \log \frac{x_j}{x_p}, \quad j = 1, 2, \ldots, p-1 \tag{17.2}$$

Obviously, logratio transformation could overcome the "sum to unity" constraint and it is much easier to model on $\hat{x}_j (j = 1, 2, \ldots, p-1)$ ranging within $(-\infty, +\infty)$. However, the model has a disadvantage in interpreting, because the asymmetrically transformed variables cannot rightly match to the original variables. As a result, it is still difficult to be applied in practice.

Definition 17.2.2 *The centered logratio transformation of a p-part composition is the p-dimensional vector given by*

$$\tilde{x}_j = \log \frac{x_j}{\sqrt[p]{\prod_{i=1}^{p} x_i}}, \quad j = 1, 2, \ldots, p \tag{17.3}$$

Denote $\tilde{X} = (\tilde{x}_1, \tilde{x}_2, \ldots, \tilde{x}_p)$, obviously we have $\tilde{x}_j \in (-\infty, +\infty)$.
And the reverse transformation of (3) is given by

$$\begin{aligned} v_j &= \tilde{x}_j - \tilde{x}_p, \quad j = 1, 2, \ldots, p-1 \\ x_j &= \frac{e^{v_j}}{1 + \sum_{i=1}^{p-1} e^{v_i}}, \quad j = 1, 2, \ldots, p-1 \\ x_p &= \frac{1}{1 + \sum_{i=1}^{p-1} e^{v_i}} \end{aligned} \tag{17.4}$$

In centered logratio transformation, every element in $\tilde{X} = (\tilde{x}_1, \tilde{x}_2, \ldots, \tilde{x}_p)$ is symmetrical to the components in compositional data $X = (x_1, x_2, \ldots, x_p)$, thus it can properly represent every original component and has more powerful ability in explaining the result. However, there is still a problem that the centered transformed variables are completely correlated, since

$$
\begin{aligned}
\tilde{X}\mathbf{1}_{p\times 1} &= \sum_{j=1}^{p} \log \frac{x_j}{\sqrt[p]{\prod_{i=1}^{p} x_i}} \\
&= \sum_{j=1}^{p} \log x_j - p \log \sqrt[p]{\prod_{i=1}^{p} x_i} \\
&= \sum_{j=1}^{p} \log x_j - \log \prod_{i=1}^{p} x_i \\
&= 0
\end{aligned}
\tag{17.5}
$$

17.2.2 Compositional Covariance Matrix

Based on the above logratio transformations, Aitchison presented the following corresponding covariance matrices.

Definition 17.2.3 *The covariance matrix of $\hat{X}$(Definition 17.2.1) is termed the logratio covariance matrix*

$$
\Sigma = \left[\sigma_{ij}\right] = \left[\operatorname{cov}\left(\log \frac{x_i}{x_p}, \log \frac{x_j}{x_p}\right)\right] \quad i, j = 1, \ldots, p-1
$$

Definition 17.2.4 *The covariance matrix of $\tilde{X}$(Definition 17.2.2) is termed the centered logratio covariance matrix*

$$
\Gamma = \left[\gamma_{ij}\right] = \left[\operatorname{cov}\left(\log \frac{x_i}{g(X)}, \log \frac{x_j}{g(X)}\right)\right] \quad i, j = 1, \ldots, p
$$

where $g(X)$ is the geometric mean of the p components of X.

and it should be noticed that elements in every row of Γ sum to zero.

$$
\begin{aligned}
\sum_{j=1}^{p} \gamma_{ij} &= \sum_{j=1}^{p} \operatorname{cov}(\tilde{x}_i, \tilde{x}_j) \\
&= \operatorname{cov}(\tilde{x}_i, \sum_{j=1}^{p} \tilde{x}_j) \\
&= 0
\end{aligned}
$$

Finally, we should emphasize that compositional covariance matrix is the foundation in multiple statistical analysis of compositional data. Both Principal component analysis on compositional data (Aitchison 1986) and LCPLS (Hinkle and Rayens 1994) were established by using the centered logratio covariance matrix. Similarly, regression models presented below will also follow this way of investigation.

17.2.3 Linear Combination and Logcontrast

As a linear combination to unconstrained Euclidean space, Aitchison proposed a "logcontrast" to the simplex.

Definition 17.2.5 *A logcontrast for a p-part composition X is any loglinear combination*

$$\sum_{j=1}^{p} a_j \log x_j, \quad \textit{where} \sum_{j=1}^{p} a_j = 0 \tag{17.6}$$

Actually, the logcontrast combination of compositional data comes from the following connection with the linear combination of logratio transformation

$$\begin{aligned}\sum_{j=1}^{p-1} a_j \log \frac{x_j}{x_p} &= \sum_{j=1}^{p-1} a_j \left(\log x_j - \log x_p\right) \\ &= \sum_{j=1}^{p-1} a_j \log x_j + a_p \log x_p \\ &= \sum_{j=1}^{p} a_j \log x_j\end{aligned}$$

where $a_p = -\sum_{j=1}^{p-1} a_j$. Thus a linear combination of the variates in $\hat{X}$ can be viewed as a logcontrast.

Moreover, a logcontrast is the same as a contrast in the centered logratio transformed composition

$$\begin{aligned}\sum_{j=1}^{p} a_j \log \frac{x_j}{g(X)} &= \sum_{j=1}^{p} a_j \left(\log x_j - \log g(X)\right) \\ &= \sum_{j=1}^{p} a_j \log x_j - \log g(X) \sum_{j=1}^{p} a_j \\ &= \sum_{j=1}^{p} a_j \log x_j\end{aligned} \tag{17.7}$$

17.3 Simple Linear Regression Model of Compositional Data

There has been some research on the regression model of compositional data. Aitchison (1986) first made some study in the situation of one dependent compositional variable on several independent ordinary variables. By combining compositional data analysis with PLS regression, Hinkle and Rayens (1994) proposed LCPLS with one dependent ordinary variable on one independent compositional variable. Zhang (2000) discussed ordinary least squares regression of dependent ordinary variable on independent compositional variable. In this paper, the simple linear regression model of compositional data involves in studying the relationship of one dependent compositional variable on one independent compositional variable. Developed from LCPLS (Hinkle and Rayens 1994), the definition and theorem for conducting the modelling are presented below.

17.3.1 PLS Algorithm Based on Covariance Matrix

In PLS regression, variables are usually standardized firstly to overcome the problem of large scale variance of them, and the modelling process is actually based on the correlation matrix of the original data. However, as to the compositional data, on one hand, the components of composition have the same scale of percentage; on the other hand, as mentioned in Sect. 17.2.2, studies show it is more reasonable to adopt compositional covariance matrix and the result can meet the logcontrast condition. Therefore, in this part we specially present a PLS algorithm which is directly based on covariance matrix, so that it'll be much easier to expand to the analysis of compositional data.

Let $X = (x_1, x_2, \ldots, x_p)$ be an $(n \times p)$ matrix containing n rows of observations on p explanatory variables and $Y = (y_1, y_2, \ldots, y_q)$ an $(n \times q)$ matrix containing n rows of corresponding observations on q response variables. The covariance structure of $\{X, Y\}$ is given by

$$\begin{aligned} S_X &= \text{cov}(X) = (X - \bar{X})^{\mathrm{T}}(X - \bar{X}) \\ s &= \text{cov}(X, Y) = (X - \bar{X})^{\mathrm{T}}(Y - \bar{Y}) \end{aligned}$$

PLS produces factors $(t_1, \ldots, t_A)$,$(u_1, \ldots, u_A)$ of X and Y, respectively, given by

$$\begin{aligned} t_k &= X w_k \\ u_k &= Y c_k \end{aligned} \qquad k = 1, \ldots, A \tag{17.8}$$

where $W = [w_1, \ldots, w_A]_{p \times A}$ and $C = [c_1, \ldots, c_A]_{q \times A}$ are weight matrices.

Here, A is the number of components needed to adequately model the data $\{X, Y\}$ based on some minimization or stopping rule. The conditions that ensure uniqueness of the weight matrix are inherent in the following definition.

Definition 17.3.1 *For the data* $\{X, Y\}$*, the PLS factors of* X *and* Y *are given by the matrix equation in (17.8) with*

$$(w_k, c_k) = \arg\max \left\{ \mathrm{cov}(Xw, Yc) : \right. \tag{17.9}$$
$$\left. w^T w = 1, c^T c = 1, \{w^T S_X w_j = 0\}_{j=1}^{k-1} \right\}, k = 1, \ldots, A$$

where arg max *stands for the argument of the maximum, that is to say, the value of the given argument for which the value of the given expression attains its maximum value.*

Theorem 17.3.1 *For the PLS factors given in Definition 17.3.1, the vector solutions of (17.9) are:*

1. $w_{k+1} (k = 0, \ldots, A-1)$ *is the eigenvector associated with the largest eigenvalue of matrix* $H_k s s^T$*, where* $H_0 = I$*,* $H_k = I - S_X W_k [W_k^T S_X^2 W_k]^{-1} W_k^T S_X$*;*
2. $c_{k+1} = \dfrac{s^T w_{k+1}}{\|s^T w_{k+1}\|}$

Proof. Suppose that we have the first k solution vectors $W_k = [w_1, \ldots, w_k]$ of (17.9), then by using the Lagrange multiplier technique let

$$\phi_{k+1}(w, c) = \mathrm{cov}(Xw, Yc) - \lambda_1(w^T w - 1) - \lambda_2(c^T c - 1) - w^T S_X W_k \theta$$

where λ_1,λ_2 and $\theta = [\theta_1, \ldots, \theta_k]^T$ are Lagrange multipliers corresponding to the constraints in (17.9). Since $\mathrm{cov}(Xw, Yc) = w^T sc$, we have the vector of partial derivatives of ϕ_{k+1} with respect to the elements of w,c set equal to zero

$$\frac{\partial \phi_{k+1}(w, c)}{\partial w} = sc - 2\lambda_1 w - S_X W_k \theta = 0 \tag{17.10}$$

$$\frac{\partial \phi_{k+1}(w, c)}{\partial c} = s^T w - 2\lambda_2 c = 0 \tag{17.11}$$

Premultiplication of (17.10), (17.11) by w^T, c^T, respectively, and solving for λ_1, λ_2 gives

$$\begin{gathered} 2\lambda_1 = 2\lambda_2 = w^T sc = c^T s^T w = \mathrm{cov}(Xw, Yc) = 2\lambda \\ c = \tfrac{1}{2\lambda} s^T w \end{gathered} \tag{17.12}$$

and premultiplication of (17.10) by $W_k^T S_X$ and solving for θ gives

$$\theta = [W_k^T S_X^2 W_k]^{-1} W_k^T S_X sc \tag{17.13}$$

Using (17.12) and (17.13) to simplify (17.10), results in the eigenvector problem

$$H_k s s^T w = (2\lambda)^2 w$$

Since $\mathrm{cov}(Xw, Yc) = 2\lambda$ is the objective function of being max, w is the eigenvector associated with the maximal eigenvalue of matrix $H_k ss^{\mathrm{T}}$; and $c = \frac{s^{\mathrm{T}} w_{k+1}}{\|s^{\mathrm{T}} w_{k+1}\|}$ is obtained according to (17.12).

17.3.2 Simple Linear Regression Model of Compositional Data

Following both Aitchison's contention that the linear combinations of X should be replaced with logcontrasts and Hinkle and Rayens's LCPLS, the following developed definition of simple linear regression model of one dependent compositional variable on one independent compositional variable is suggested.

Definition 17.3.2 *For the data $\{X, Y\}$, where X is a p-part composition and Y is a q-part composition, the logcontrast PLS (LCPLS) factors of X and Y are given by*

$$\begin{aligned} t_k &= \log X w_k \\ u_k &= \log Y c_k \end{aligned} \qquad k = 1, \ldots, A$$

where $\log X = (\log x_1, \ldots, \log x_p)$, $\log Y = (\log y_1, \ldots, \log y_q)$

$$\begin{aligned} (w_k, c_k) = \arg\max & \Big\{ \mathrm{cov}(\log Xw, \log Yc) : \\ & w^{\mathrm{T}} w = 1, c^{\mathrm{T}} c = 1, \mathbf{1}^{\mathrm{T}} w = 0, \mathbf{1}^{\mathrm{T}} c = 0, \{w^{\mathrm{T}} S_{\log X} w_j = 0\}_{j=1}^{k-1} \Big\}, \\ & k = 1, \ldots, A \end{aligned} \tag{17.14}$$

Theorem 17.3.2 *The logcontrast PLS factors defined in Definition 17.3.2 can be formed by constructing the PLS factors (Definition 17.3.1) of the centered logratio transformation of the composition $\{X, Y\}$.*

Proof. The logratio transformations are $\tilde{X}, \tilde{Y}$, as defined in Definition 17.2.2. To compute the A PLS factors of $\tilde{X}, \tilde{Y}$ given by

$$\begin{aligned} t_k &= \tilde{X} w_k \\ u_k &= \tilde{Y} c_k \end{aligned} \qquad k = 1, \ldots, A$$

we will use the covariance structures of the data $\{\tilde{X}, \tilde{Y}\}$. These are

$$\begin{aligned} \Gamma_X &= \mathrm{cov}(\tilde{X}) = (\tilde{X} - \bar{\tilde{X}})^{\mathrm{T}} (\tilde{X} - \bar{\tilde{X}}) \\ \gamma &= \mathrm{cov}(\tilde{X}, \tilde{Y}) = (\tilde{X} - \bar{\tilde{X}})^{\mathrm{T}} (\tilde{Y} - \bar{\tilde{Y}}) \end{aligned}$$

From Theorem 17.3.1 the weight vectors defining the factors are given by

$$\begin{aligned} H_k \gamma \gamma^{\mathrm{T}} w_{k+1} &= \lambda_{max} w_{k+1} \\ c_{k+1} &= \frac{\gamma^{\mathrm{T}} w_{k+1}}{\|\gamma^{\mathrm{T}} w_{k+1}\|} \end{aligned} \qquad k = 0, \ldots, A-1 \tag{17.15}$$

where $H_0 = I$, $H_k = I - \Gamma_X W_k [W_k^{\mathrm{T}} \Gamma_X^2 W_k]^{-1} W_k^{\mathrm{T}} \Gamma_X$.

Now to see how computing the above weights and factors of $\tilde{X}, \tilde{Y}$ is equivalent to doing LCPLS.

Notice from (17.5) that $\tilde{X}\mathbf{1}_{p\times 1} = \tilde{Y}\mathbf{1}_{q\times 1} = \mathbf{0}_{n\times 1}$. This implies

$$\mathbf{0} = \mathrm{cov}(\tilde{X}\mathbf{1}, \tilde{Y}) = \mathbf{1}^{\mathrm{T}}\mathrm{cov}(\tilde{X}, \tilde{Y}) = \mathbf{1}^{\mathrm{T}}\gamma$$
$$\mathbf{0} = \mathrm{cov}(\tilde{X}, \tilde{Y}\mathbf{1}) = \mathrm{cov}(\tilde{X}, \tilde{Y})\mathbf{1} = \gamma\mathbf{1}$$

and from (17.15), $\mathbf{1}^{\mathrm{T}} w_k = \mathbf{1}^{\mathrm{T}} c_k = 0 (k = 1, \ldots, A)$, since $\mathbf{1}^{\mathrm{T}}\Gamma_X = 0$. Thus the weight vectors resulting from standard PLS on $\{\tilde{X}, \tilde{Y}\}$ are contrasts; that is, they each sum to zero. Using this result and (17.7) we have the following relations

$$\max_{\substack{w^{\mathrm{T}}w=1, c^{\mathrm{T}}c=1 \\ \mathbf{1}^{\mathrm{T}}w=0, \mathbf{1}^{\mathrm{T}}c=0}} \mathrm{cov}(\log Xw, \log Yc) = \max_{\substack{w^{\mathrm{T}}w=1, c^{\mathrm{T}}c=1 \\ \mathbf{1}^{\mathrm{T}}w=0, \mathbf{1}^{\mathrm{T}}c=0}} \mathrm{cov}(\tilde{X}w, \tilde{Y}c) \tag{17.16}$$

and

$$\max_{\hat{w}^{\mathrm{T}}\hat{w}=1, \hat{c}^{\mathrm{T}}\hat{c}=1} \mathrm{cov}(\tilde{X}\hat{w}, \tilde{Y}\hat{c}) \geq \max_{\substack{w^{\mathrm{T}}w=1, c^{\mathrm{T}}c=1 \\ \mathbf{1}^{\mathrm{T}}w=0, \mathbf{1}^{\mathrm{T}}c=0}} \mathrm{cov}(\tilde{X}w, \tilde{Y}c) \tag{17.17}$$

But the maximizing vectors of the left side of (17.17), subject to the PLS constraints, are simply the weight vectors given by (17.15). Thus (17.17) is an equality and hence the weights and factors of LCPLS are exactly the weights and factors computed above.

In conclusion, the algorithm of simple linear regression model of compositional data can be summarized as follows:

1. Take centered logratio transformation on both dependent and independent compositional variables
2. Apply PLS on the transformed variables and analyze the regression relationships of them
3. Take reverse transformations as equation (4) on the estimations from the built model and the predictions of compositional data can be obtained resultingly

17.3.3 Case Study on Simple Linear Regression Model of Compositional Data

In this part, regression model of employment structure on GDP structure in Beijing's three industries is built to illustrate the modelling process of simple linear regression on compositional data.

The trendlines of GDP and employment structures in Beijing's three industries from 1990 to 2003 are shown in Fig. 17.1, where the solid lines with solid scatters denote the employment proportions of the three industries and the dash lines with hollow scatters denote GDP proportions of the three industries.

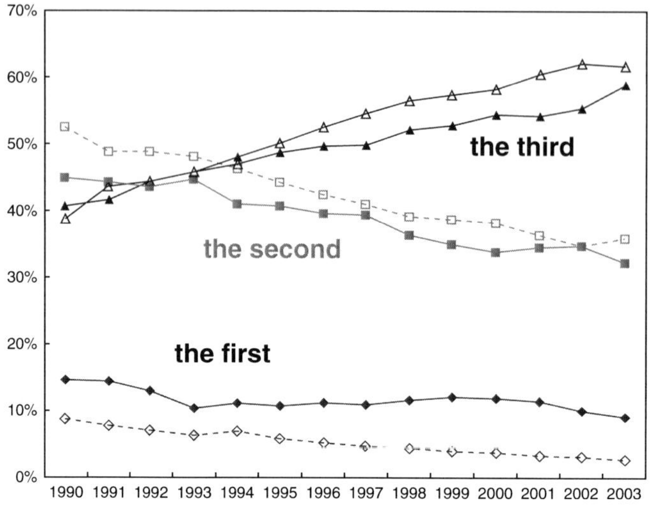

Fig. 17.1 Trendlines of GDP and employment proportions of Beijing's three industries in 1990–2003

Table 17.1 Weights of LCPLS of employment structure on GDP structure

Independent variables	w_1	w_2	Dependent variables	c_1	c_2
$\tilde{GDP}_1$	−0.689	−0.439	$e\tilde{m}p_1$	−0.349	−0.675
$\tilde{GDP}_2$	−0.036	0.818	$e\tilde{m}p_2$	−0.465	0.735
$\tilde{GDP}_3$	0.725	−0.379	$e\tilde{m}p_3$	0.814	−0.060
RdX	72%	28%	RdY	69%	0.8%

Here what we called GDP or employment structure refers to the proportions of each industry to the three total value, both of which are obviously compositional data according to the definition. Now we establish regression model of employment structure on GDP structure by applying simple linear regression model of compositional data in Sect. 17.3.2.

Denote the compositional data GDP and employment structures of the three industries as $GDP = (GDP_1, GDP_2, GDP_3)$ and $emp. = (emp_1, emp_2, emp_3)$; and their centered logratio transformations as $\tilde{GDP} = (\tilde{GDP}_1, \tilde{GDP}_2, \tilde{GDP}_3)$ and $e\tilde{m}p. = (e\tilde{m}p_1, e\tilde{m}p_2, e\tilde{m}p_3)$.

Apply PLS regression on the transformed variables. Two weight vectors of extracted PLS factors are recorded in Table 17.1, satisfying the logcontrast condition. The percentages in the bottom row of the table represent the explainablity of PLS factors to the independent and dependent variable sets, showing a satisfying result.

Meanwhile, the loading plot of the first two PLS factors in Fig. 17.2 also displays the relationship between the dependent and independent variables. It is visible that

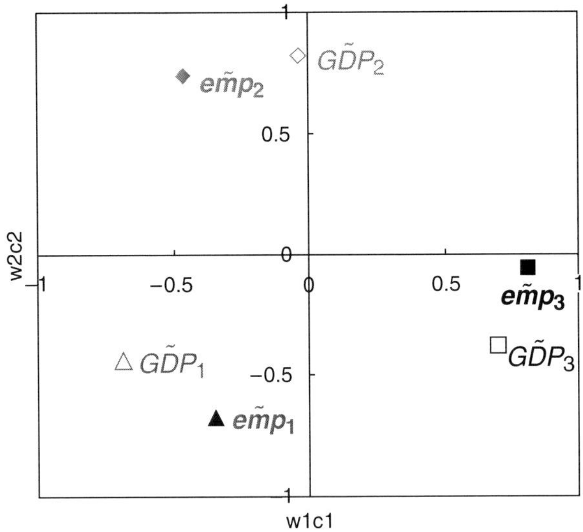

Fig. 17.2 Loading plot of the first two PLS factors

Table 17.2 Coefficients of LCPLS of employment structure on GDP structure

Independent variables	$e\tilde{m}p_1$	$e\tilde{m}p_2$	$e\tilde{m}p_3$
$G\tilde{D}P_1$	0.235	0.020	−0.254
$G\tilde{D}P_2$	−0.220	0.254	−0.034
$G\tilde{D}P_3$	−0.015	−0.274	0.288
R^2	47%	93%	95%

employment and GDP proportions of each industry have a high correlation with each other.

Furthermore, LCPLS regression coefficients of employment structure on GDP structure are calculated in Table 17.2, meeting logcontrasts as well. The R^2 in the bottom row of the table indicate good fitness of the model.

Finally, reverse transformations are implemented to get the fitted values of the original employment proportions of the three industries in Fig. 17.3, where the solid scatters are observed values and the dash lines denote the fitted values from the LCPLS model.

17.4 Multiple Linear Regression Model of Compositional Data

Multiple linear regression model of compositional data is developed from simple linear regression model. As it has been mentioned in Sect. 17.1, the hierarchical relationship should be especially considered when there are more than one composition

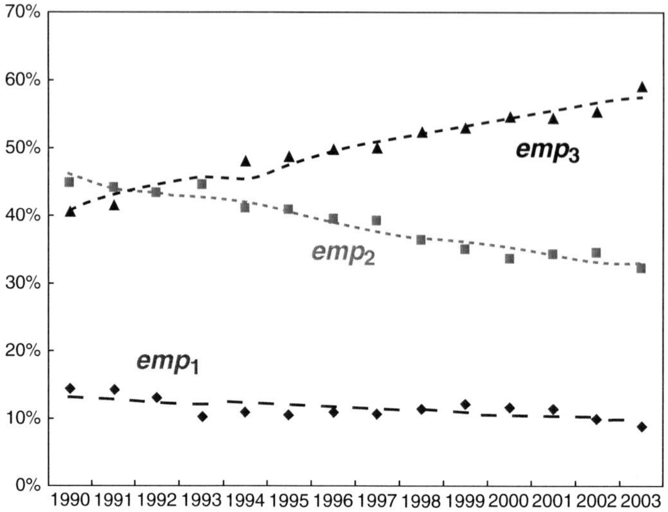

Fig. 17.3 Observed and fitted values of employment structure in Beijing's three industries

as independent variables. Therefore, hierarchical PLS regression method could be a choice for solving the problem.

In this section, we first introduce hierarchical PLS regression. And then multiple linear regression on compositional data is presented. Finally, we add the investment structure as another independent compositional variable to the model established in Sect. 17.3.3 to analyze the effect of both GDP and investment structure to the employment structure in Beijing's three industries.

17.4.1 Introduction to Hierarchical PLS Regression

Hierarchical PLS regression was first introduced by Wold (1996), which is a very efficient technic in dealing with large numbers of variables in the field of complex data analysis. Although the "standard" PLS regression is still feasible in the modelling on too many variables, the results, such as plots and lists of loadings, weights, coefficients, VIP, etc., become messy and difficult to interpret. To solve the problem, researchers attempt to find out ways to simplify the variable set. However, modelling on a subclass of the former set is somewhat unreasonable, for excessively deleting variables may increase the risk of losing information, receiving fake answers and misleading interpretation.

A better alternative is to divide the variables into conceptually meaningful blocks, which leads to two model levels: the upper level where the relationships between blocks are modelled, and the lower level showing the details of each block. On each level, PLS regression, principal component analysis (PCA) and other standard models could be used. Therefore, the hierarchical modelling method can obtain a compactly integrated model (top model) which provides an overview of the whole and several sub models (base models) which allows "zooming" onto interesting subsets of data.

Let $X = (X_1, \ldots, X_p)$ be a p-block of explanatory variable set, where $X_j (j = 1, \ldots, p)$ is a sub block with p_j variables, that is $X_j = (x_{j1}, \ldots, x_{jp_j})$; while $Y = (y_1, \ldots, y_q)$ is a q corresponding response variable set.

Definition 17.4.1 *The PLS factors in hierarchical PLS regression are given by*

$$t_j = X_j W_j, j = 1, \ldots, p$$
$$u^{top} = Y C^{top}$$

where $t_j = (t_{j1}, \ldots, t_{jA_j})$, $W_j = (w_{j1}, \ldots, w_{jA_j})$ *are PLS factors and weights of* X_j *in each base model of* Y *on* X_j*; and* $u^{top} = (u_1^{top}, \ldots, u_A^{top})$, $C^{top} = (C_1^{top}, \ldots, C_A^{top})$ *are factors and weights of* Y *in the top model of* Y *on* $(t_1, \ldots, t_p)$.

Figure 17.4 gives a visualization of hierarchical PLS regression modelling process. Seen from the modelling process, hierarchical PLS regression is very powerful in integrating and explaining information of the data set. Compared to the standard PLS regression, it is more effective and practical in managing high dimensional data.

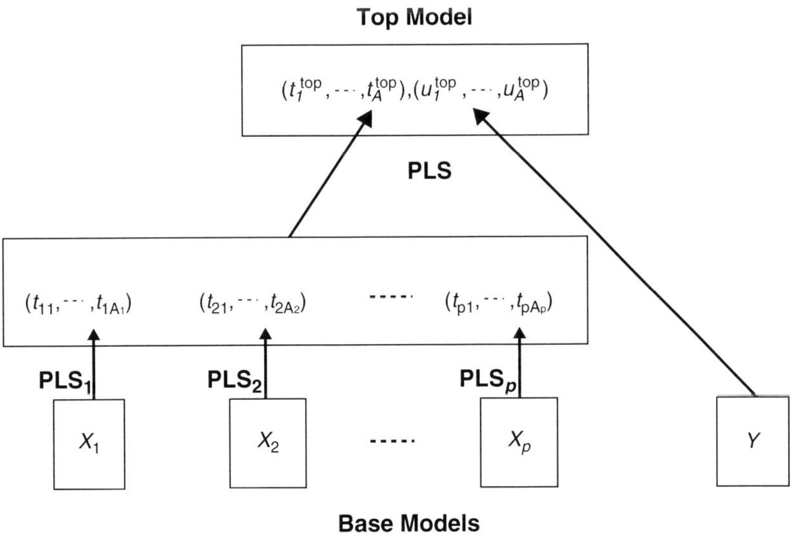

Fig. 17.4 Hierarchical PLS modelling

17.4.2 Multiple Linear Regression Model of Compositional Data

In this part, we study on the linear regression modelling method of one dependent compositional variable on more than one independent compositional variables. In the multiple model, except the "sum to unity" constraint, the hierarchical relationship of multiple compositions with multiple components should be mainly discussed to enhance the explainablity of the model. For this reason, we could first summarize each independent composition and extract its corresponding thematic variables; then regress the dependent composition on these thematic variables.

One way to carry out the above idea is to combine centered logratio transformation with hierarchical PLS regression.

Let $X = (X_1, \ldots, X_p)$ be an independent variable set of p compositions, where $X_j (j = 1, \ldots, p)$ is a composition with p_j components, that is $X_j = (x_{j1}, \ldots, x_{jp_j})$; while $Y = (y_1, \ldots, y_q)$ is a composition with q components of corresponding dependent variable set.

Definition 17.4.2 *The logcontrast PLS factors in hierarchical PLS regression are given by*

$$\begin{aligned} t_j &= \log X_j W_j, j = 1, \ldots, p \\ u^{top} &= \log Y C^{top} \end{aligned}$$

where $\log X_j = (\log x_{j1}, \ldots, \log x_{jp_j})$, $\log Y = (\log y_1, \ldots, \log y_q)$; t_j, W_j *are logcontrasts PLS factors and weights of* X_j *in each base model;* u, C *are logcontrasts factors and weights of composition* Y *in the top model.*

Theorem 17.4.1 *The logcontrast PLS factors defined in Definition 17.4.2 can be formed by constructing the PLS factors (Definition 17.4.1) of the centered logratio transformation of the multiple compositions* $\{(X_1, \ldots, X_p), Y\}$ *in hierarchical PLS regression.*

Proof. The logratio transformations are $(\tilde{X}_1, \ldots, \tilde{X}_p), \tilde{Y}$. As Definition 17.4.1, we have

$$\begin{aligned} t_j &= \tilde{X}_j W_j, j = 1, \ldots, p \\ u^{top} &= \tilde{Y} C^{top} \end{aligned}$$

According to Theorem 17.3.2, t_j, W_j are logcontrast factors and weights of composition X_j in each base model.

Similar to (17.15), $c_j^{top} = \dfrac{\mathrm{cov}\big((t_1, \ldots, t_p), \tilde{Y}\big)^{\mathrm{T}} w_j^{top}}{\|\mathrm{cov}\big((t_1, \ldots, t_p), \tilde{Y}\big)^{\mathrm{T}} w_j^{top}\|}$ and $1^{\mathrm{T}} c_j^{top} = 0$, since $1^{\mathrm{T}}\mathrm{cov}\big((t_1, \ldots, t_p), \tilde{Y}\big)^{\mathrm{T}} = \mathrm{cov}\big((t_1, \ldots, t_p), \tilde{Y}\big)1 = \mathrm{cov}\big((t_1, \ldots, t_p), \tilde{Y}1\big) = 0$.

Using this result and (17.7) we have $u^{top} = \tilde{Y} C^{top} = \log Y C^{top}$.

Therefore, the logcontrast factors and weighs of compositional data in hierarchical PLS regression are exactly those computed in hierarchical PLS regression of the centered logratio transformation of composition data.

In conclusion, the algorithm of multiple linear regression model of compositional data can be summarized as follows:

1. Take centered logratio transformation on both dependent and independent compositional variables
2. Apply hierarchical PLS on the transformed variables and analyze the regression relationships of them
3. Take reverse transformations as (4) on the estimations from the built model and the predictions of compositional data can be obtained resultingly

17.4.3 Multiple Linear Regression Model for Predicting the Employment structure

In this part, on the basis of simple LCPLS regression model built in Sect. 17.3.3, we further bring in the investment structure as the second composition of independent variable, analyze the effects of proportions of investment and GDP to the proportions of employment in Beijing's three industries, and illustrate the multiple PLS regression modelling process on compositional data and its validity.

Denote the investment structure of the three industries as $inv. = (inv_1, inv_2, inv_3)$ and its centered logratio transformation as $\tilde{inv}. = (\tilde{inv}_1, \tilde{inv}_2, \tilde{inv}_3)$; while the notations of GDP and employment structures are same as those in Sect. 17.3.3.

Conduct hierarchical PLS regression on the transformed data. There are two base models of $\tilde{emp}.$ on $\tilde{GDP}$ and $\tilde{inv}.$, and one top model of $\tilde{emp}.$ on the factors extracted in the base models. The weight vectors and their corresponding percents of explainablity are recorded in Table 17.3.

Meanwhile, the loading plots are shown in Fig. 17.5, from which we could see that the employment proportion of the third industry has high correlations with its GDP and investment proportions in the direction of the first factor, and so does it in the first industry; while the correlations in the second industry is weak.

According to the above built hierarchical PLS model, the fitted values of employment proportions of the first, second and third industries could also be calculated, and the R^2 of them are 52%, 95% and 92%, respectively. Figure 17.6 shows a good fitness of the model, where the solid scatters are observed values and the dash lines denote the fitted values.

Table 17.3 Weights of base models and top models

Base model 1	w_{11}	w_{12}	Base model 2	w_{21}	w_{22}	Top model	c_1	c_2
$\tilde{GDP}_1$	−0.689	−0.439	$\tilde{inv}_1$	−0.769	0.192	$\tilde{emp}_1$	−0.374	−0.251
$\tilde{GDP}_2$	−0.036	0.818	$\tilde{inv}_2$	0.146	−0.783	$\tilde{emp}_2$	−0.441	0.798
$\tilde{GDP}_3$	0.725	−0.379	$\tilde{inv}_3$	0.623	0.591	$\tilde{emp}_3$	0.815	−0.547
RdX_1	72%	28%	RdX_2	77%	23%	RdY	61%	12%

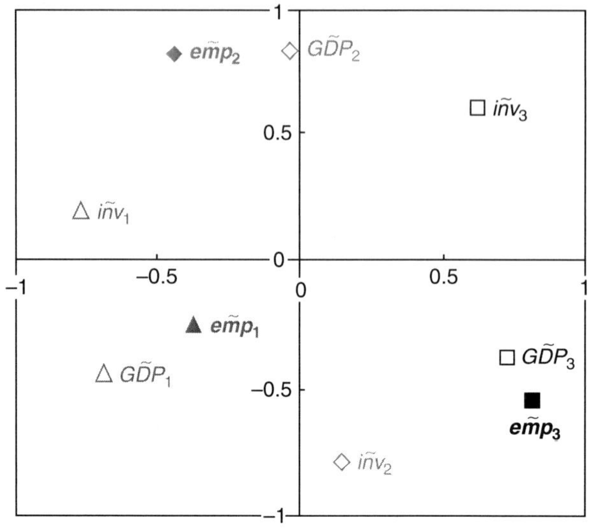

Fig. 17.5 Loading plots of the hierarchical PLS regression

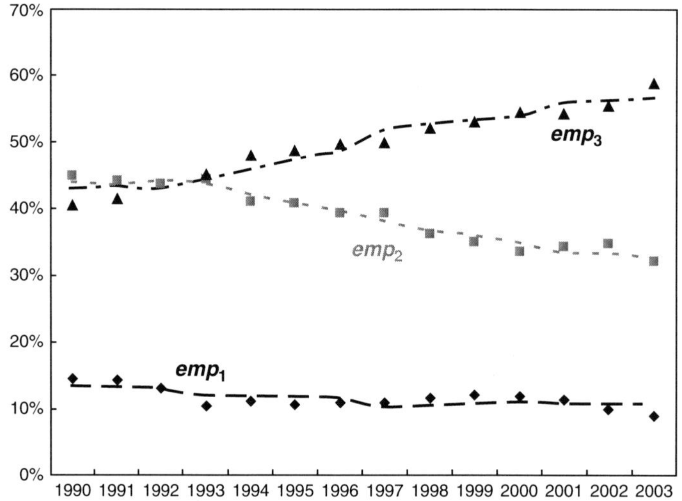

Fig. 17.6 Observed and fitted values of employment structure in Beijing's three industries

17.5 PLS Path Modelling on Compositional Data

In this section, by employing PLS path modelling, we make further development in analyzing the more complex interrelationships of multiple compositions. First, PLS path modelling is introduced. And then PLS path modelling on compositional data is presented. Finally, as the case study used in Sect. 17.4.3, we apply the model to analyze the causal correlations of the investment, GDP and employment structures in Beijing's three industries.

17.5.1 Introduction to PLS Path Modelling

PLS path modelling is originally presented by the founder of PLS regression, Wold (1985), and further developed by Lohmoller (1989), who also exploited the software of it. The objective of PLS path modelling is similar to LISREL (LInear Structural RELations) by Joreskog (1970), which is mainly used to analyze the relationship of several aggregations. However, different with LISREL in parameter estimation, PLS path modelling is more practical with fewer assumptions. In fact, there are many restrictions hardly satisfied in LISREL. For example, LISREL modelling is based on the covariance matrix of the observation variables, variables should obey normal distribution assumption, and the number of observations should be large enough. Besides, problems of getting an unexplainable results or computing nonconvergence may also arise in practice, which severely restrict its application fields. Compared to LISREL, PLS path modelling implements an iterative algorithm with a series of simple or multiple linear regressions, which is verified more practical and effective.

In PLS path model, a latent variable (LV) is an unobservable variable (or construct) indirectly described by a block of observable variables which are called manifest variables (MV) or indicators; and the causality model leads to linear equations relating the LVs between them. Consider p-block of variables $X_1, \dots, X_p$, where $X_j = (x_{j1}, \dots, x_{jp_j})(j = 1, \dots p)$ is a block of MVs and ξ_j is the related LV of X_j.

17.5.1.1 The PLS Path Model

A PLS path model is described by two models: (1) a measurement model (also called outer model) relating the MVs to their own LV and (2) a structural model (also called inner model) relating some endogenous LVs to other LVs. (An LV is an exogenous variable if it never appears as a dependent variable, or else an endogenous variable.)

In the measurement model, there are two ways to relate the MVs to their LVs.

(1) The reflective way
In the model of the reflective way, each MV reflects its LV, which can be expressed by a simple regression

$$x_{jh} = \pi_{jh0} + \pi_{jh}\xi_j + \varepsilon_{jh} \tag{17.18}$$

where ε_{jh} is the residual term with the hypothesis of a zero mean and uncorrelated with the LV ξ_j.

(2) The formative way

In the formative way, it is supposed that the LV ξ_j is generated by its own MVs, expressed by a linear combination of its MVs

$$\xi_j = \sum_{h=1}^{p_j} w_{jh} x_{jh} + \delta_j \tag{17.19}$$

where δ_j is the residual term with the hypothesis of a zero mean and uncorrelated with the MV x_{jh}.

In the structural model, there are a series of linear equations of LVs to describe their causality relationships

$$\xi_j = \beta_{j0} + \sum_{i \neq j} \beta_{ji} \xi_i + \zeta_j \tag{17.20}$$

where ζ_j is the residual term satisfying the predictor specification hypothesis of a zero mean and uncorrelated with LVs $\xi_j (i \neq j)$.

The causality model must be a causal chain, that is there is no loop in the causality model. Besides, a structural model can be summarized by a 0/1 square matrix called the inner design matrix, where rows and columns represent the LVs and a cell (i, j) is filled with a 1 if LV ξ_j explains LV ξ_i, and 0 otherwise.

17.5.1.2 The Algorithm of PLS Path Modelling

The algorithm of PLS path modelling is composed of LVs estimation and estimation of the structural equations.

In the LVs estimation, PLS path modelling iteratively computes LVs and weights according to the following three steps until convergence.

(1) Outer estimation Y_j

The standardized LVs are estimated as linear combinations of their MVs

$$Y_j = X_j w_j \tag{17.21}$$

where $w_j = [w_{j1}, \ldots, w_{jp_j}]^{\mathrm{T}}$ is outer weight vector.

(2) Inner estimation Z_j

The inner estimation is defined by

$$Z_j = \sum_{i:\beta_{ji} \neq 0} e_{ji} Y_i \tag{17.22}$$

where β_{ji} are the coefficients in formula (17.20) and $e_{ji} = \mathrm{sign}\big(r(Y_j, Y_i)\big)$ are the inner weights.

(3) Weights estimation w_j

There are two ways to estimate weight vector w_j: mode A and mode B.

In mode A, w_{jh} is the regression coefficient of Z_j in the simple regression of x_{jh} on the inner estimation Z_j

$$w_{jh} = \text{cov}(x_{jh}, Z_j) \tag{17.23}$$

In mode B, the weight vector w_j is the regression coefficient vector in the multiple regression of Z_j on the MVs

$$w_j = (X_j^{\mathrm{T}} X_j)^{-1} X_j^{\mathrm{T}} Z_j \tag{17.24}$$

Mode A is appropriate for a block with a reflective measurement model and mode B for a formative one.

In estimation of the structural equations, the structural equations (17.20) are estimated by individual OLS multiple regressions where the LVs ξ_j are replaced by their estimations $\hat{\xi}_j$.

Actually, the above algorithm could be summarized by an optimality criterion.

Definition 17.5.1 *For the multiblock data, the LVs in PLS path model are given by*

$$Y_j = X_j w_j, j = 1, \ldots, p$$

with $w_j = \arg\max \left\{\text{cov}(X_j w, \sum_{i=1, i\neq j}^{p_j} e_{ji} Y_i) : w^{\mathrm{T}} w = 1\right\}$.

17.5.2 PLS Path Modelling on Compositional Data

Followed above, it is clear to see that PLS path modelling is usable in extracting the thematic variables of multiblocks of variables and analyze their path correlations. Therefore, it is adoptable in exploring the interrelationships of multiple compositions.

Let $X = (X_1, \ldots, X_p)$ be a group of p compositions, where $X_j (j = 1, \ldots, p)$ is a composition with p_j components, that is $X_j = (x_{j1}, \ldots, x_{jp_j})$. Here we mainly discuss their reflective way in the measurement model.

Definition 17.5.2 *For the multiple compositional data, the logcontrast LVs in PLS path model of multiple compositions are given by*

$$Y_j = \log X_j w_j, j = 1, \ldots, p$$

where $\log X_j = (\log x_{j1}, \ldots, \log x_{jp_j})$ $w_j = \arg\max \left\{\text{cov}(\log X_j w, \sum_{i=1, i\neq j}^{p_j} e_{ji} Y_i) : w^{\mathrm{T}} w = 1, 1^{\mathrm{T}} w = 0\right\}$.

Theorem 17.5.1 *The logcontrast LVs of compositional data in PLS path modelling defined in Definition 17.5.2 can be formed by constructing LVs in PLS path model of the centered logratio transformation of the multiple compositions* $(X_1, \ldots, X_p)$.

Proof. The logratio transformations are $(\tilde{X}_1, \ldots, \tilde{X}_p)$.

As the algorithm of PLS path modelling, the outer weight is calculated as mode A $w_j = \text{cov}(\tilde{X}_j, Z_j)$. And

$$1^{\mathrm{T}} w_j = 1^{\mathrm{T}} \text{cov}(\tilde{X}_j, Z_j) = \text{cov}(\tilde{X}_j 1, Z_j) = 0$$

Thus the outer weights resulting from standard PLS path modelling on $(\tilde{X}_1, \ldots, \tilde{X}_p)$ are contrasts.

Using this result and (17.7) we have the following relations

$$\begin{aligned} Y_j &= \log X_j w_j = \tilde{X}_j w_j \\ &\max_{w^{\mathrm{T}} w=1, 1^{\mathrm{T}} w=0} \text{cov}\Big(\log X_j w, \sum_{i=1, i\neq j}^{p_j} e_{ji} Y_i\Big) \\ &= \max_{w^{\mathrm{T}} w=1, 1^{\mathrm{T}} w=0} \text{cov}\Big(\tilde{X}_j w, \sum_{i=1, i\neq j}^{p_j} e_{ji} Y_i\Big) \\ &= \max_{w^{\mathrm{T}} w=1} \text{cov}\Big(\tilde{X}_j w, \sum_{i=1, i\neq j}^{p_j} e_{ji} Y_i\Big). \end{aligned}$$

Thus the weights and LVs in PLS path modelling on compositional data defined in Definition 17.5.2 are exactly the weights and LVs computed in the PLS path modelling on the centered logratio transformation of composition data.

In conclusion, the algorithm of PLS path modelling on compositional data can be summarized as follows:

(1) Take centered logratio transformation on multiple compositional variables
(2) Apply PLS path modelling on the transformed variables and analyze the interrelationships of them

17.5.3 PLS Path Modelling for Analyzing the Employment structure

Here we apply the above model to analyze the interrelationships of investment, GDP and employment structures in Beijing's three industries. The data set and variable notations are same as those in Sect. 17.4.3. Let investment, GDP, employment be LVs and their proportions of the three industries be the corresponding MVs (seen in Table 17.4).

According to the rule of macroeconomic theory, investment directly affects GDP and employment, and indirectly affects employment through GDP. Therefore, investment can be regarded as an exogenous LV, while GDP and employment as endogenous LVs. As a result, the initial inner design matrix is shown in Table 17.5 and the corresponding path model graph in Fig. 17.7.

Table 17.4 Definitions of LVs and MVs

LV	Investment structure ($\tilde{inv}.$)	GDP structure ($G\tilde{D}P$)	Employment structure ($e\tilde{m}p.$)
MVs	$i\tilde{n}v_1$	$G\tilde{D}P_1$	$e\tilde{m}p_1$
	$i\tilde{n}v_2$	$G\tilde{D}P_2$	$e\tilde{m}p_2$
	$i\tilde{n}v_3$	$G\tilde{D}P_3$	$e\tilde{m}p_3$

Table 17.5 Inner design matrix of the structural model (1)

	$i\tilde{n}v.$	$G\tilde{D}P$	$e\tilde{m}p.$
$i\tilde{n}v.$	0	0	0
$G\tilde{D}P$	1	0	0
$e\tilde{m}p.$	1	1	0

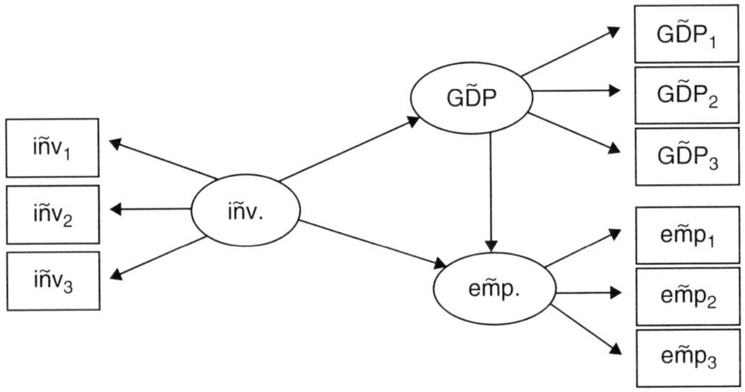

Fig. 17.7 Path model of investment, GDP and employment structures

Table 17.6 is the centered logratio transformed data of the proportions of investment, GDP and employment in Beijing's three industries.

According to the model in Fig. 17.7, apply PLS Path model to the data in Table 17.6 and obtain the inner weights of LVs in Table 17.7.

Because the direct effect of $i\tilde{n}v.$ to $e\tilde{m}p.$ (0.102) is not significant, this path could be removed; then the developed inner design matrix are given in Table 17.8.

Reconduct PLS path modelling based on the modified design matrix; and the inner weights of LVs are calculated in Table 17.9.

Meanwhile, the outer weights of LVs on their corresponding MVs are recorded in Table 17.10, which rightly meet the logcontrast condition.

Resultingly, the PLS path model could be summarized in Fig. 17.8.

From the above results, we could get the following conclusions.

(1) According to the structural model, the investment and GDP structures in Beijing's three industries greatly influence it's employment structure: the direct effect of $i\tilde{n}v.$ to $G\tilde{D}P$ is 0.846, the indirect effect of $i\tilde{n}v.$ to $e\tilde{m}p.$ through $G\tilde{D}P$ is $0.846 \times 0.965 = 0.816$, and $G\tilde{D}P$ to $e\tilde{m}p.$ is 0.965.

(2) From the measurement model, the employment proportion of the third industry plays a great role in the employment structure, and it is actually driven by its corresponding investment and GDP proportions. Besides, the first industry also

Table 17.6 Transformed data of the proportions of investment, GDP and employment

Year	$\tilde{inv}.$			$\tilde{GDP}$			$\tilde{emp}.$		
	$\tilde{inv}_1$	$\tilde{inv}_2$	$\tilde{inv}_3$	$\tilde{GDP}_1$	$\tilde{GDP}_2$	$\tilde{GDP}_3$	$\tilde{emp}_1$	$\tilde{emp}_2$	$\tilde{emp}_3$
1990	−2.172	0.721	1.451	−1.093	0.696	0.397	−0.722	0.411	0.311
1991	−2.152	0.793	1.359	−1.202	0.655	0.547	−0.730	0.395	0.335
1992	−2.145	0.912	1.232	−1.276	0.685	0.590	−0.804	0.399	0.405
1993	−2.880	1.392	1.488	−1.348	0.698	0.650	−0.976	0.481	0.495
1994	−2.790	1.126	1.665	−1.273	0.627	0.646	−0.928	0.385	0.543
1995	−2.808	1.096	1.712	−1.390	0.632	0.758	−0.956	0.389	0.567
1996	−2.958	1.177	1.780	−1.474	0.628	0.846	−0.929	0.349	0.580
1997	−3.818	1.591	2.228	−1.539	0.625	0.914	−0.940	0.349	0.591
1998	−3.556	1.345	2.211	−1.594	0.613	0.982	−0.888	0.263	0.625
1999	−3.427	1.232	2.196	−1.640	0.623	1.017	−0.849	0.216	0.633
2000	−3.032	0.927	2.105	−1.709	0.641	1.068	−0.861	0.188	0.673
2001	−3.305	0.981	2.324	−1.774	0.630	1.144	−0.893	0.217	0.676
2002	−3.352	1.052	2.300	−1.816	0.617	1.199	−0.988	0.259	0.729
2003	−3.314	1.037	2.277	−1.927	0.692	1.234	−1.057	0.224	0.833

Table 17.7 Inner weights of LVs (1)

	$\tilde{inv}.$	$\tilde{GDP}$	$\tilde{emp}.$
$\tilde{inv}.$	0	0	0
$\tilde{GDP}$	0.843	0	0
$\tilde{emp}.$	<u>0.102</u>	0.879	0

Table 17.8 Inner design matrix of the structural model (2)

	$\tilde{inv}.$	$\tilde{GDP}$	$\tilde{emp}.$
$\tilde{inv}.$	0	0	0
$\tilde{GDP}$	1	0	0
$\tilde{emp}.$	0	1	0

Table 17.9 Inner weights of LVs (2)

	$\tilde{inv}.$	$\tilde{GDP}$	$\tilde{emp}.$
$\tilde{inv}.$	0	0	0
$\tilde{GDP}$	0.846	0	0
$\tilde{emp}.$	0	0.965	0

Table 17.10 Outer weights of LVs

$\tilde{inv}.$		$\tilde{GDP}$		$\tilde{emp}.$	
$\tilde{inv}_1$	1.18	$\tilde{GDP}_1$	1.95	$\tilde{emp}_1$	2.04
$\tilde{inv}_2$	−0.17	$\tilde{GDP}_2$	0.13	$\tilde{emp}_2$	2.70
$\tilde{inv}_3$	−1.01	$\tilde{GDP}_3$	−2.08	$\tilde{emp}_3$	−4.74

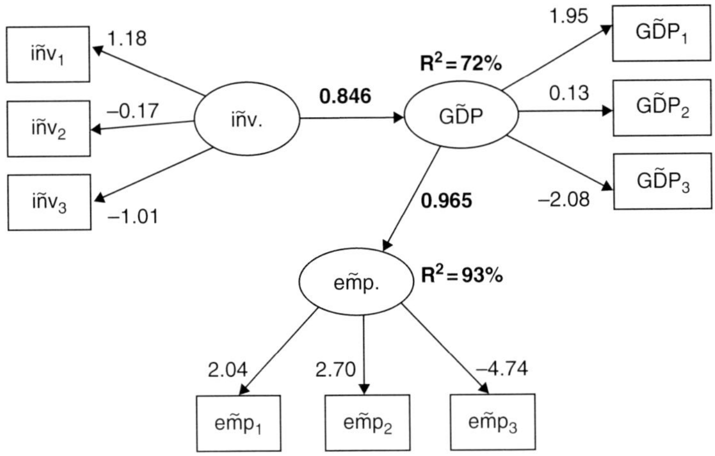

Fig. 17.8 Results of PLS path modelling of investment, GDP and employment structures

has a similar relationship between its investment, GDP and employment proportions. However, the outer weights of investment and GDP proportions of the second industry are much smaller.

(3) The high values of R^2 verify the good fitness and validity of the built model.

17.6 Comparison of the Three Models of Compositional Data

In Sects. 17.3–17.5, regression modelling analysis on simple or multiple compositional data has been introduced by adopting PLS regression, hierarchical PLS and PLS path modelling, respectively. Meanwhile, they are all theoretically proved satisfying the logcontrast property and could be easily calculated by conducting standard methods on the centered logratio transformed compositional data. As the case study, these three models are applied to investigate and fit the employment structure of Beijing's three industries. Seen from the results, they are all proved successful in modelling on compositional data, which have high precisions of fitness and powerful ability in interpretation to the data set.

In terms of the modelling process of these three methods, simple and multiple linear regression models of compositional data hold something basically alike; hierarchical PLS regression on multiple compositional data is just developed and upgraded from standard PLS regression. In the case study of simple and multiple regressions on employment structure, it is shown that multiple model, which introduces another explanatory factor (investment structure), has a more comprehensive exposition of the data and a high level of accuracy.

As to the multiple model, hierarchical PLS and PLS path modelling can both be used in regression analysis on multiblock variables. However, they still hold different characteristics in aspects of modelling objective, process and results.

(1) They have different modelling assumption. In PLS path modelling, the reflective way in the measurement model requires unidimensionality of each block of MVs while hierarchical PLS not.
(2) They have different complexity of computation. PLS path modelling adopts an iterative way to estimate the path coefficients of direct and indirect relationships of variables, which seems a little complicated; while hierarchical PLS is easy to be conducted for its straightforward modelling process involving twice standard PLS regression.
(3) They present different analyzing results. PLS path modelling particularly emphasizes on the summarizing of each LV to its corresponding MVs and the direct or indirect effects of LVs; while hierarchical PLS only focuses on the direct impact of multiblock independent variables to dependent variables except the interrelations between independent variables themselves. Hence hierarchical PLS is unavailable in exploring embedded interactions of multiblocks of variables.

Because of those features mentioned above, we should choose different models for different purpose in practice. Seen from the case study, hierarchical PLS on compositional data mainly gives a linear regression model of employment structure to investment and GDP structures; moreover, it can also detailedly investigate the extent of investment and GDP proportions of each industry in boosting its employment proportion, respectively. It is more practical from the view of forecasting. On the other hand, PLS path modelling on compositional data illustrates upper connections of summarized variables from a higher level, indicating how the investment structure indirectly affects employment structure through GDP structure.

17.7 Conclusion

This paper proposes simple, multiple regressions and path modelling on compositional data by combining centered logratio transformation with PLS related technics.

By introducing centered logratio transformation, compositional data can then be scaled up to a broader range of $(-\infty, +\infty)$. On one hand, it avoids the "sum to unity" constraint; on the other hand, it rightly represents the characteristic of every original component and is much more convenient and reasonable in explaining the result.

By employing PLS related technics to the transformed compositional data, the evil consequence of complete correlation can be solved effectively. In the simple regression modelling analysis on compositional data, standard PLS regression method is conducted on one-to-one composition variables. In the multiple regression modelling analysis, hierarchical meanings of the variables should be mainly considered. One solution is to adopting hierarchical PLS regression: extracting thematic variables in each independent composition by the base models; then regressing dependent composition on the thematic variables by the top model, which can be easily managed by twice using standard PLS regression. Finally, PLS path modelling can especially analyze the direct and indirect path links of the thematic variables (LVs).

What's more, consistent with theories of compositional data, the three models are all rightly satisfying the logcontrast condition, which verifies their validity and rationality from a theoretical point of view.

For the case studies, regression modelling procedures of compositional data are illustrated by analyzing employment structure of Beijing's three industries from 1990 to 2003. A simple regression model on compositional data of employment structure to GDP structure in Beijing's three industries is firstly established; afterwards, a multiple regression model is built by further bringing in the second independent composition – investment structure; finally, PLS path modelling on compositional data of investment, GDP and employment structures is presented to summarize an upper interrelations of them. The results of these three cases verify the validity and rationality of the models discussed in this paper, which have a high level of accuracy in prediction applications and are quite powerful in explaining the relationships of multiple compositional variables.

Acknowledgements This research was supported by the National Natural Science Foundation of China (NSFC) under grant number 70125003, 70371007, 70531010 and 70521001.

References

Aitchison, J. (1986). The statistical analysis of compositional data. London: Chapman and Hall.

Bayol, M. P., Delafoye, A., Tellier, C., & Tenenhaus, M. (2000). Use of PLS path modeling to estimate the European consumer satisfaction index (ECSI) model. *Statistica Applicata - Italian Journal of Applied Statistics, 12*(3), 361–375.

Chin, W. W. (1998). The partial least squares approach for structural equation modeling. In: G. A. Marcoulides (Ed.), Modern methods for business research. NJ, USA: Lawrence Erlbaum Associates.

Chin, W. W., & Newsted, P. R. (1999). Structural equation modeling analysis with small samples using partial least squares. In: R. Hoyle (Ed.), Statistical strategies for small sample research. Thousand Oaks, CA, USA: Sage.

Eriksson, L., Johansson, E., Wold, N., & Wold, S. (2001). Multi- and megavariate data analysis–principles and applications. Umea, Sweden: Umetrics AB.

Ferrers, N. M. (1866). An elementary treatise on trilinear coordinates. London: Macmillan.

Guinot, C., Latreille, J., & Tenenhaus, M. (2001). PLS path modellind and multiple table analysis. Application to the cosmetic habits of women in Ile-de-France. *Chemometrics and Intelligent Laboratory Systems, 58*, 247–259.

Johan, A. W., Theodora, K., & John, F. M. (1998). Analysis of Multiblock and Hierarchical PCA and PLS models. *Journal of Chemometrics, 12*, 301–321.

Joreskog, K. G. (1970). A general method for analysis of covariance structure. *Biometrika, 57*, 239–251.

Lohmoller, J. B. (1984). LVPLS Program Manual Version 1.6. Zentralarchiv fr Empirische Sozialforschung. Koln: Universitat zu Koln.

Lohmoller, J. B. (1989). Latent variables path modeling with partial least squares. Heidelberg: Physica.

Noonan, R., & Wold, H. (1982). PLS path modeling with indirectly observed variables: a comparison of alternative estimates for the latent variable. In: K. G. Joreskog, & H. Wold (Eds.), Systems under indirect observation. Amsterdam: North-Holland.

Pages, J., & Tenenhaus, M. (2001). Multiple factor analysis combined with PLS path modelling. Application to the analysis of relationships between physicochemical variables, sensory profiles and hedonic judgements. *Chemometrics and Intelligent Laboratory Systems, 58*, 261–273.

Pearson, K. (1897). On a form of spurious correlation which may arise when indices are used in the measurement of organs. In: M. Emery (Ed.), Mathematical contributions to the theory of evolution. London: Proceedings of the Royal Society.

Wang, H. W. (1999). Partial least-squares regression method and application. Beijing: National Defence Industry Press.

Wold, H. (1982). Soft modeling: the basic design and some extensions. In: K. G. Joreskog, & H. Wold (Eds.), Systems under indirect observation, Part 2. Amsterdam: North-Holland.

Wold, H. (1985). Partial least squares. In: S. Kotz, & N. L. Johnson (Eds.), Encyclopedia of statistical sciences. New York: Wiley.

Wold, S., Wold, K., & Tjessem, K. (1996). Hierarchical Multiblock PLS and PC Models for easier model interpretation and as an alternative to variable Selection. *Journal of Chemometrics, 10*, 463–482.

Wold, S., Trygg, J., Berglund, A., & Antti, H. (2001). Some recent developments in PLS modeling. *Chemometrics and Intelligent Laboratory Systems, 58*, 131–150.

Zhang, Y. T. (2000). The statistical analysis generality of compositional data. Beijing: Science.

Part II
Applications to Marketing and Related Areas

Chapter 18
PLS and Success Factor Studies in Marketing

Sönke Albers

Abstract While in consumer research the "Cronbach's α – LISREL"-paradigm has emerged for a better separation of measurement errors and structural relationships, it is shown here that studies involving an evaluation of the effectiveness of marketing or organizational strategies based on structural relationships require the application of PLS. This is because we no longer distinguish between constructs and their reflecting measures but rather between abstract marketing policies (constructs) and their forming detailed marketing instruments (indicators). It is shown with the help of examples from literature that many studies of this type applying LISREL have been misspecified and would have better made use of the PLS approach. I also demonstrate the appropriate use of PLS in a study of success factors for e-businesses. I conclude with recommendations on the appropriate design of success factor studies, including the use of higher-order constructs and the validation of such studies.

18.1 Introduction

Based on research primarily in the area of salesperson behavior, Churchill (1979), in an influential article, has advocated a better measurement approach for empirical studies in marketing. He stresses that complex constructs like role conflict or role ambiguity cannot be measured with a single item because each measure has an idiosyncratic error and will not give a reliable measure. Rather, it is better to work with multiple measures. This allows the researcher to separate the relationships between various constructs from their measurement errors. According to classical test theory as developed in psychology, Churchill promoted the evaluation of the internal consistency of the items with the help of Cronbach's α (Cronbach 1951). Furthermore, relationships between constructs can

S. Albers
Institute of Innovation Research, Christian-Albrechts-University at Kiel, Westring 425, 24098 Kiel, Germany
e-mail: albers@bwl.uni-kiel.de

V. Esposito Vinzi et al. (eds.), *Handbook of Partial Least Squares*, Springer Handbooks of Computational Statistics, DOI 10.1007/978-3-540-32827-8_19,

be modeled by structural equation approaches that can be estimated with the help of variance–covariance-based approaches like LISREL. This has been the standard in marketing for many years.

While many studies using this approach were at first in the field of consumer behavior research, later studies used the same approach for studies of organizational effectiveness. Typical examples are studies on market orientation by, e.g., Homburg and Pflesser (2000), Matsuno et al. (2002), and Zhou et al. (2005). While the approach by Churchill is mostly applicable to psychological constructs, this no longer holds for organizational constructs. The reason is that the measurement approach advocated by Churchill (1979) works with the assumption that constructs may be operationalized through indicators that reflect the construct. However, if one is interested in the area of organizational and marketing effectiveness in the factors that drive success, then the constructs have to be operationalized as different aspects and thus form the construct. Unfortunately, many articles have not paid attention to this important distinction and have estimated their models under the assumption of reflective indicators (Jarvis et al. 2003; Fassott 2006). This misspecification does not only have consequences with respect to the estimation of parameters, but also to the selection of the right indicators and to the derivation of the implications of the results.

I, therefore, describe typical misspecifications in structural equation modeling approaches applied in empirical marketing studies in Sect. 18.2. In Sect. 18.3, I derive the consequences of these misspecifications, which require a paradigm shift from the so-called "Cronbach's α – LISREL" approach for mostly psychological constructs to the derivation of success factor constructs based on content validity grounds and to estimate the structural equations with the help of Partial Least Squares (PLS). Section 18.4 describes a prototypical application with recommendations on how to report results. Based on these experiences, I formulate some recommendations for the design of success factor studies in marketing in Sect. 18.5 and, in Sect. 18.6, present a conclusion.

18.2 Misspecification of Marketing Studies

According to Jarvis et al. (2003), 28% of all structural equation modeling articles in marketing top-A-Journals, especially in Journal of Marketing, use misspecified models. This refers to the fact that studies assume reflective indicators while they are in fact different facets and, hence, must be formative. The same is true for articles in the leading German journals. Fassott (2006) reports that ca. 33% of those articles use misspecified models.

The authors of these articles follow a common scheme that can be deduced, for example, from the articles by Steenkamp and van Trijp (1991), Baumgartner and Homburg (1996), and Steenkamp and Baumgartner (2000). In Germany an article by Homburg and Giering (1996) was very influential and was later misused as a kind of recipe for empirical work. The basic premise is that the success of marketing is due to complex influences like market orientation and so on. These influences cannot be measured error-free. Rather, it is advisable to measure the construct with the help of indicators. According to classical test theory, one should use indicators or items that are reliable and should discriminate between different constructs. Therefore, tests are done whether the constructs and its operationalizations are supported by Cronbach's α (Cronbach 1951) and other reliability measures and confirmatory factor analysis. If the constructs do not comply with the tests, they are purified in the sense that indicators that do not correlate sufficiently high with the other indicators of a construct are deleted. Based on these modified operationalizations of constructs, the statistical analysis is carried out with the help of LISREL, which tries to fit the variance–covariance matrix as best as possible. The advantage is that LISREL is readily available and provides many test statistics to assess the overall model fit. In the very end, the overall fit was used to support or reject theories. This approach represented the ruling paradigm for many years, so that Homburg and Baumgartner (1995, p. 1093) conclude that one cannot get papers accepted that do not follow these rules.

Only later on, it was pointed out by Diamantopoulos and Winklhofer (2001) and Rossiter (2002) that this approach may be misleading if a researcher wants to investigate the drivers of success. In this case, the constructs have to be operationalized by formative rather than reflective indicators. The misspecification of structural equation models by implicitly assuming reflective indicators, although they are actually formative, can mostly be observed when the articles deal with organizational constructs like market orientation, customer orientation, and service orientation of companies, salesforces or employees.

In the following, I want to make a clear distinction between reflective and formative indicators. Figure 18.1 shows that, in principle, a construct can be operationalized in both ways (Albers and Hildebrandt 2006). The right side of Fig. 18.1 shows reflective indicators. In this case, the causal direction is that a construct is reflected by indicators and therefore the causal relationship goes from the construct to the indicators. This might be appropriate when a researcher wants to test theories with respect to satisfaction. However, in managerially oriented business studies we want to find out what are the most important drivers of satisfaction that ultimately lead to the retention of a customer. In this case, we need as many facets of satisfaction as required for a success factor study and the causal relationship goes from the indicators to the constructs.

In the case of formative indicators, the classical test theory no longer applies. The items are no longer replaceable and very often do not correlate enough. As a consequence, researchers who struggled with the requirement of 0.7 for Cronbach's α, but found that their constructs did not meet this criterion, relaxed this requirement with the excuse that weaker requirements should be applied to new constructs (Petersen 1994, p. 382). However, formative constructs need not be correlated and could therefore not be tested with test theory at all. If researchers either tried to delete unreliable items (false purification) or assumed the wrong direction of the relationship, this could lead to misspecified models (Jarvis et al. 2003).

This misspecification is still present in many articles in top-A-journals. I want to demonstrate this with the help of some examples that are typical for this area.

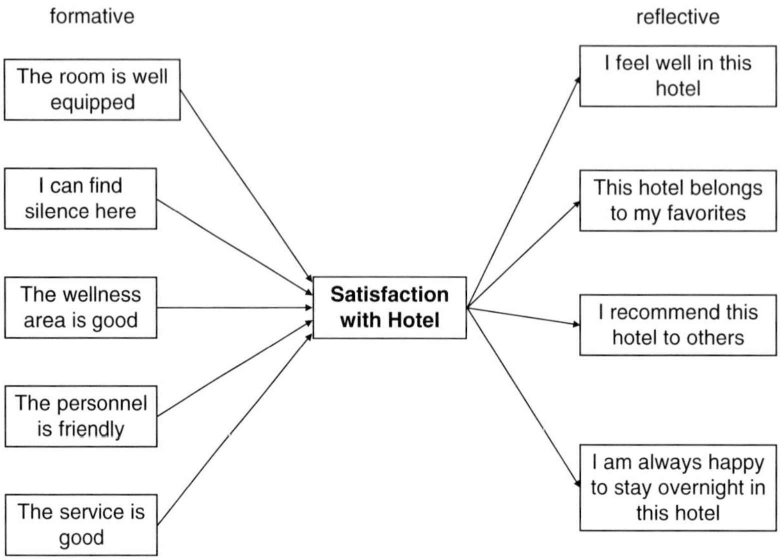

Fig. 18.1 Satisfaction as a formative and reflective construct
Source: Albers and Hildebrandt (2006)

However, it is not my purpose to single out certain authors but rather to discuss the way the particular methodology applied in these articles was misspecified:

In a study by Homburg and Pflesser (2000), they investigated whether market orientation is based on an organizational culture that has shared basic values influencing norms, and these in turn influence artifacts and behaviors, and these in turn again finally impact market and financial performance. They operationalized their constructs with the help of 23 aspects that are themselves first-order constructs, which are in turn operationalized by 78 indicators or items. They applied the usual techniques of Cronbach's α and confirmatory factor analysis (CFA) to purify the scales of the 23 first-order constructs. As they considered their model to be too complex, the authors tested simplified alternative measurement models and selected the one with the best CAIC (consistent Akaike information criterion). Based on a 5-factor model with only one dimension of shared values (out of eight aspects), one dimension of norms (out of eight aspects), one dimension of market-oriented behaviors (20 original items were reduced to 12 items), and two dimensions of artifacts (out of six), they finally tested their hypothesized structure. Although this implies a reflective philosophy of construct operationalization, Homburg and Pflesser (2000) report different dimensions (aspects) of their constructs, which is only possible if one assumes that the first-order factors form the second-order constructs. If one deletes different aspects, then the overall meaning of the constructs has changed and the results no longer hold for this construct in general but only for this particular operationalization. While this procedure allows the application of statistical criteria for construct validation and the final application of LISREL with reflective indicators, this would have strongly altered the originally postulated model. If we really have different aspects, then they are not interchangeable and could therefore not be deleted. Deletion of items is only possible if all items stem from a universe of alternative but interchangeable measures that reflect the whole construct but not special facets or aspects (Rossiter 2002).

The study by Matsuno et al. (2002) investigates whether market orientation and entrepreneurial proclivity have an influence on business performance. The authors develop a structural equation model (SEM) in which entrepreneurial proclivity influences market orientation via constructs describing organizational structure (formalization, centralization, and departmentalization) while market orientation finally impacts business performance. All the constructs have been operationalized with the help of item batteries that have been purified according to Cronbach's α and CFA. The model itself has been estimated by LISREL, assuming reflective relationships between the constructs and indicators.

A closer look at the operationalization of the construct market orientation shows that nearly all the indicators represent different (formative) aspects of market orientation but are not total reflections of the construct. Indeed, on the basis of a factor analysis Matsuno et al. (2002) find three sets of indicators to be sufficiently correlated. They, therefore, argue that this implies that market orientation is a second-order factor with the sets of correlated indicators as first-order factors. These first-order factors are termed intelligence generation, intelligence dissemination, and responsiveness and are in turn representations of a total of 22 indicators. In order to run LISREL, the authors determined the values of the three first-order factors by calculating the unweighted mean of the respective indicators. This means that the model is based on market-orientation with its first-order factors as indicators. As these first-order factors represent different aspects (otherwise they could not emerge as different factors in a confirmatory factor analysis), it would have been mandatory to handle them as formative indicators. Unfortunately, they were erroneously considered to be reflecting indicators of market orientation. In the same way, entrepreneurial proclivity was operationalized by the unweighted means of the indicators of the three first-order factors innovativeness, risk taking, and proactiveness. The authors derive as managerial implication that both constructs positively affect business performance. However, because of the reflective nature of their indicators, they could not give an indication of what drives business performance the most. Therefore, the value of this investigation is limited.

Zhou et al. (2005) analyze whether market orientation impedes breakthrough innovations. They work with constructs like market orientation, technology orientation, and entrepreneurial orientation for strategic orientation as well as demand uncertainty, market turbulence, and competitive intensity for market force. With the help of covariance-based structural equation modeling, they analyze whether these factors exert an influence on either technology-based or market-based innovations and, finally, on performance. These factors are constructs and therefore are operationalized by a total of 54 indicators. Although many of these indicators represent different aspects, the authors have validated them with the help of confirmatory factor analysis, leading to the dropping of a number of indicators. As some of their constructs are second-order factors, they replaced the first-order factors by the summated scores of their indicators. In the final model, they treated all constructs as reflected by indicators. This is of course not possible given the nature of the indicators representing aspects or drivers but not reflections. It is surprising to see that despite all the articles already published, such as the ones by Diamantopoulos and Winklhofer (2001), Rossiter (2002), and Jarvis et al. (2003), authors and reviewers of top-A-journals apparently do not know about this problem of misspecification.

The review by Jarvis et al. (2003) also makes it clear that many researchers like to make a compromise by considering different aspects or drivers with the help of several first-order factors forming one second-order factor while staying in the tradition of classical test theory by operationalizing the first-order factors with the help of multiple reflecting measures. Very often this is the outcome of a process of an exploratory factor analysis and a purification following a confirmatory factor analysis. Owing to the difficulties of handling second-order factors in LISREL as well as other SEM approaches, authors frequently calculate unweighted means as measures for the first-order factors and continue to work with those by mostly using regression analysis.

18.3 Consequences of Misspecification

The discussion of several applications in top-A-journals in marketing makes clear that the models suffer from manifold misspecifications:

(a) Items have been deleted despite their relevance for the construct.
(b) Many constructs are not measured in a general way but only represent the meaning of their sample of indicators.
(c) The estimation of misspecified models leads to biased estimates.
(d) In the case of unweighted linear combinations in order to run linear regressions, the relationships are underestimated because stronger relationships can be found with a weighting of the indicators.
(e) With the assumption of reflective indicators, it is only possible to derive results for the constructs but not for the differential effect of the indicators. This is especially a problem in success factor studies where learning that market orientation has a positive impact on market results (this is a highly plausible conclusion) is of less concern than which drivers (indicators) are mostly responsible for the success.

Ad (a) The recommendation by Churchill (1979) to use better measures in marketing by using multiple items only holds for constructs that are measured by reflective indicators. As each indicator has its idiosyncratic error, it will not give a reliable measure. Rather, it is better to work with multiple items. However, this also implies that all multiple measures must come from a universe of equally suitable items and are drawn randomly from it. Therefore, with the help of classical test theory it can be checked whether the selected items show internal consistency, e.g., by calculating Cronbach's α and testing for one-dimensionality. While this procedure very often makes sense in the field of consumer behavior research, this is generally no longer true for studies of organizational effectiveness. Here, we are interested, for example, in finding out which drivers of the organizational structure and culture that lead to market orientation have an impact on success. Only this information provides recommendations on concrete actions that improve business, while the information that market orientation has a positive impact on success does not tell us what to do. Therefore, researchers have strived to operationalize their constructs with as many aspects as extractable from expert interviews. Now, if the so called "Cronbach's α – LISREL" paradigm (which may also include other reliability measures) is applied, researchers have found that these indicators are no longer internally consistent or sufficiently intercorrelated. They have, therefore, deleted all items that showed a low reliability. Authors frequently report that up to 50% of items were deleted.

Ad (b) It can seen from our two examples that authors have either deleted items that do not show reliability during the purification process (see a) or have selected only some aspects. However, if these items do not represent interchangeable items drawn randomly, then any selection of indicators alters the meaning of a construct. Therefore, the findings can only refer to the special operationalization of this construct in this study and do not allow for any kind of generalization. In essence, under reflective assumptions, one selects the set of maximally intercorrelated items while, in a formative approach, one tries to avoid intercorrelated items. This means that the operationalization can differ as much as the sets of items, as both approaches are distinct. Diamantopoulos and Siguaw (2002) re-analyzed an already published study by Cadogan et al. (1999) in which reflective indicators were erroneously assumed and, therefore, some of them that did not provide internal consistency were deleted. In contrast, Diamantopoulos and Siguaw (2002) assumed that the indicators of their construct export coordination were formative. This allows for the inclusion of many facets but resulted in multicollinearity. Therefore, the authors eliminated some intercorrelated indicators. Now, it is no surprise that only 2 indicators out of a pool of 30 indicators are the same according to both methods. This implies that the meaning of the constructs, even if they have the same name, is drastically altered. In addition, the authors find that the relationship to 14 different export success measures can better be explained with the help of the remaining 5 uncorrelated formative indicators than with the correlated 16 reflective indicators.

Ad (c) If one attempts to estimate a model with the help of covariance-based structural equation modeling approaches like LISREL (Jöreskog and Sörbom 1996) or AMOS (Arbuckle 1999), which is based on reflective indicators that are actually formative, then one obtains biased estimators (Jarvis et al. 2003). The purification process may result in model structures that give totally different results compared to true models with formative indicators estimated with the help of PLS (Wold 1985). The results of a simulation study show that the coefficients explaining the influence of the various constructs (inner model) are positively biased. With respect to the level of the coefficients, there is a significantly negative relationship with respect to the level of the intercorrelation of the indicators of the constructs. However, this result only holds for maximum-likelihood estimations of variance–covariance models (Jarvis et al. 2003). When comparing LISREL with

PLS, Albers and Hildebrandt (2006) found that the coefficients are surprisingly robust if the models are specified correctly, while formative models estimated by PLS and, alternatively, by LISREL under the assumption of reflective indicators lead to completely different conclusions.

Ad (d) Very often the approach of using multiple items for measurement leads to complex models with large numbers of indicators. As estimation procedures require a sufficient number of degrees of freedom (five observations per parameter is an often-used rule), authors have to work with a smaller number of indicators. Rather than specifying a full model and estimating it with the help of LISREL, these authors only evaluate the measurement model with the help of confirmatory factor analysis and then work with indices that comprise the items as unweighted means. This allows them to apply simple OLS regression to estimate relationships. As the weights are equal, this means that the explanatory power of the construct is less than in the weighted case and the structural relationships may be underestimated. We have also seen in the study by Matsuno et al. (2002) that these authors handle formative aspects by defining second-order constructs that have different aspects or facets as first-order constructs, which are in turn measured by reflective multiple items. If one collapses one level of this second order construct relationship by forming indices of unweighted means of indicators and works with the indices as indicators reflecting the construct, then the indicators are not sufficiently intercorrelated and therefore we arrive at a bad model fit.

Ad (e) By assuming reflective indicators, it is implied that the indicators are interchangeable representations of the construct. Therefore, these studies only argue with the effects of the constructs. In the case of studies on the influence of market orientation, one has no knowledge of how to achieve this market orientation. If the indicators are formative and modeled as such, then one can determine the influence of the indicators on the construct. However, if reflective indicators are used, no such interpretation is possible. However, it might be argued that we can use reflective indicators if we are only interested in the influence of a holistic strategy with highly intercorrelated strategy elements. If we understand the strategy as all indicators have to be altered in case the construct is altered then the direction of causality is not crucial. However, the explanation power of the model is limited because we can only investigate the influence of a complete strategy and not that of its components and we cannot be sure whether the strategy has been operationalized completely. Moreover, in the case of such strategies, there is no need for a separation of the measurement model and the structural model because the operationalization of a strategy cannot involve a measurement error according to classical test theory. Rather, it can only be incomplete, which would determine the meaning of the strategy. Insofar, indicators can serve as reflective effects of a construct as well as formative aspects of a construct, depending on the purpose of the study.

Hence, contrary to the intention of Jarvis et al. (2003, p. 203) to provide decision rules that allow the either reflective or formative character of indicators to be determined unequivocally, it is argued in the preceding paragraph that it is not possible to assess whether a construct operationalization approach is correct or not. Rather, the approach (working with holistic strategies or components determining a strategy) limits the kind of results obtainable and may lead to an inappropriate model. In the case of reflective indicators, the model might be correct but only allows an analysis of whether changing all indicators at the same time will lead to more success or not. Whether single indicators have more importance cannot be the goal of such a study. In the same way, one cannot evaluate the validity of constructs with formative indicators. Rather, the chosen indicators determine the meaning of the construct and, thereby, the explanatory power of the model (Albers and Hildebrandt 2006).

18.4 Success Factor Models in Marketing

Success factor studies should concentrate on the impact of success drivers. Insofar, hypothesis testing takes second place to identifying the differential impact of the various factors. As success in marketing is driven by many factors, one first faces the problem of selecting the relevant factors. According to Rossiter (2002), this should be done on the basis of expert interviews and a thorough literature review. If some of these factors belong to the same domain, they are subsumed under more abstract constructs that allow for a more aggregate discussion. I only consider studies that have such structural relationships. In a second step, multicollinearity has to be removed. In success factor studies, intercorrelated factors do not imply that indicators reflect a construct, but rather that they are the result of applying certain holistic strategies in practice. As the multicollinearity of indicators within a construct inflates the standard error, it is advisable to either remove correlated indicators or to aggregate them within a single index.

In this way, Albers (2003) investigated which marketing strategies, business models, characteristics of founders, financial incentives, job characteristics, organizational culture, and kind of IT solution have the highest positive impact on business performance as measured by market share, revenue and profitability. In this study, the author elicited measures that represent different stages of the success chain: satisfaction with the achieved level of market share and with its development over the last 12 months and the same for revenue. Both are, of course, only a prerequisite of profitability, which is operationalized by cash-flow and ROI considerations in order to capture absolute as well as relative effects. In addition, the company's judgment of achieved customer satisfaction and the employee fluctuation rate were questioned. The assumed causal relationships are visualized in Fig. 18.2.

Based on expert interviews and a literature study, 42 indicators were aggregated to 10 constructs. In the majority, the indicators were measured on 7-point Likert scales. In addition, the model takes some dummy-variables into account. Details of the measurement model are given in Albers (2003). On this basis, the following relationships are assumed: Market share is the heart of all activities and therefore influenced by the marketing concept, the communication strategy, and the non-imitability. In addition, the founders' network and experience, job attractiveness, financial incentives, and corporate culture influence the effort of the employees to fight for market share. In the very end, all these variables explain 30% of the mar-

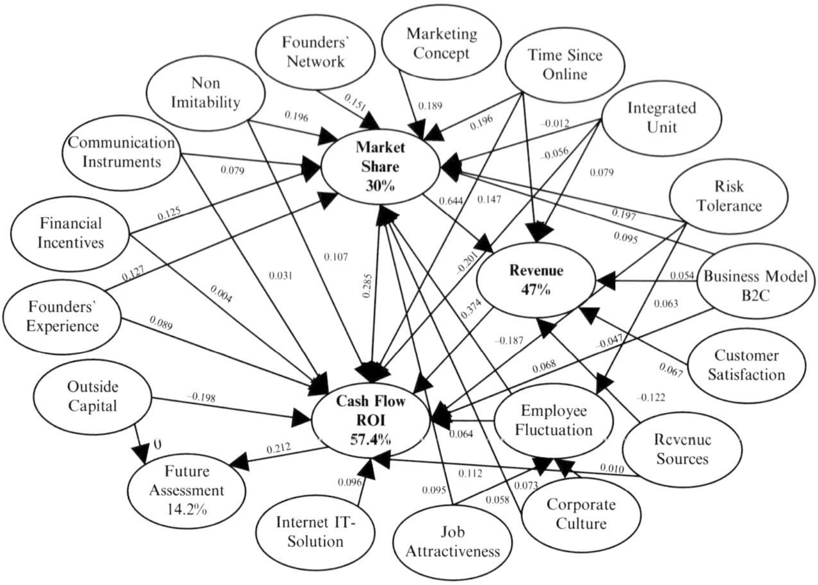

Fig. 18.2 Structure and estimation results of structural equation model

ket share. Market share determines revenue, which is also directly influenced by the revenue sources and some other variables, which also influence market share. 47% of the variance of revenue can, thus, be explained. Revenue, then, determines profitability. Profitability is also directly influenced by marketing variables, if they determine the margins that can be realized, and by variables that have cost consequences like the IT solution and the financial incentives. Based on this model, the explained variance with respect to profitability is 57.4%. Finally, profitability explains 14.2% of the variance of the future assessment of the operations given continuing the operations and/or increasing investments.

If we want to investigate data from a broad cross-section of companies, it is advisable to include covariates as additional explanatory variables in order to absorb the heterogeneity of the sample. Since large differences are observed between B2C and B2B operations, a respective dummy variable is included. Companies had already reached their break-even if they started earlier, so the time since they first went online was included. In addition, it was taken into account whether the operation was financed by outside capital, which may impose different expectations on profitability. Finally, a dummy variable, which distinguishes between start-ups and integrated units, was included to capture residual variance that can not be explained by the constructs. The relationships described so far represent a structural equation model given in Fig. 18.2.

In order to empirically test the relative importance of the various success factors, Albers (2003) distributed a questionnaire via e-mail (in a few cases also by fax and surface mail) to 590 companies. The addressees of the companies were obtained with the help of an online search for phrases like e-commerce, start-up, online-shop, etc. Moreover, a systematic search was done of Web newspapers and shopping indices. The questionnaire was addressed to members of the board, chief executive officers as well as managers of the e-business operations. A total of 191 companies responded, which resulted in a response rate of 32%. Unfortunately, 21 out of 191 companies refused to fill in the necessary information on business success. In addition, we excluded another 23 companies because they had more than 4 missing values in the success factors. In the other cases, missing values were replaced by the mean values of the respective variables. As a result, we can base our analysis on 147 complete questionnaires. The sample of responses does not claim to be representative but appears to be typical.

Based on 147 complete questionnaires, Albers estimated a PLS model despite the unfavorable ratio of observations to parameters of less than 4. This is because PLS partially estimates parameters per construct so that the degree of freedoms $147 - 10 - 1 = 136$ was still satisfactory.

The study did not reveal any surprising relationships with respect to the sign of the regression coefficients. Rather, the derivation of the impact of the various indicators was the goal of this investigation. This means that, first of all, the effects of the various indicators via all paths on the endogenous variable ROI had to be calculated. By adding up all single effects one obtains the total effect. Table 18.1 presents the results of the impact of the various success factors on success in terms of standardized regression coefficients. Even better would be to sort the variables in a matrix with different classes of overall parameter values as well t-values for the total effect. However, this requires a Monte-Carlo simulation to determine the distribution of the parameter value over several paths. This has still to be implemented by current software.

In a similar way, Thies and Albers (2010) investigated the success of drivers of cooperation strategies between content providers and ecommerce companies.

18.5 Recommendations on the Use of PLS for Success Factor Studies in Marketing

On the basis of my criticism as well as the description of typical applications of success factor studies (with constructs and indicators) in marketing, the following recommendations can be made:

(a) Indicators of success factor studies should be actionable and therefore need to be formative.
(b) Indicators can only be evaluated by means of content validity.
(c) The application of SEM estimated by PLS is superior to working with indices and running simple regressions or applying LISREL to indices.
(d) There is no need for second-order constructs.

Table 18.1 Total effects of single indicators for ROI (Albers 2003)

Classes	Positive total effects		Negative total effects	
>0.15	Time since online	0.2456	Venture capital funded	−0.1980
	Brand advantage	0.1872		
	Prior experience in same industry	0.1634		
0.05–0.15	Individually developed IT-solution	0.1040		
	Employee fluctuation	0.0998		
	Network for data on market and competition	0.0983		
	Secure jobs	0.0690		
	Wide assortment	0.0678		
	Selling to target group	0.0615		
	Prior consulting experience	0.0604		
0.03–0.05	Career opportunities	0.0571	Margins on sold products	−0.0562
	Time advantage	0.0512	Prior entrepreneur experience	−0.0439
	Commercially available IT-solution	0.0501	Public relations	−0.0413
	Stock options and shares	0.0480	Prior experience in IT-technology	−0.0405
	Online advertising communication	0.0457	Network for access to suppliers and customers	−0.0342
	TV communication	0.0449	IT-solution tested upfront by customers	−0.0310
	Open source software or freeware	0.0434		
	Above average salary	0.0389		
	Attractive price	0.0382		
	IT-solution influenced by lead-user	0.0329		

<0.03	Customer satisfaction	0.0251	Network for acceptance and references	−0.0288
	Business model (B2C)	0.0232	Revenue from online-advertising	−0.0276
	Commissions	0.0192	Print communication	−0.0255
	Revenue fixed fees	0.0120	Technology and know-how network	−0.0236
	Hierarchical decision making	0.0096	Created new brand	−0.0214
	Autonomous unit	0.0070	Prior experience in finance and accounting	−0.0150
	Risk tolerance	0.0029	Interesting tasks	−0.0115
	One-to-one marketing	0.0003	Selling brands	−0.0109
			Radio communication	−0.0107
			Prior experience in marketing and sales	−0.0088
			Continuous education	−0.0078
			Supply of own ideas	−0.0068
			Complaint management	−0.0057

(e) Rather than reporting significant coefficients, the impact of indicators should be reported (standardized b versus t-values).
(f) Owing to structural equations, the total impact of exogenous indicators on endogenous indicators should be evaluated by counting all paths.
(g) We need finite mixture programs for the PLS framework to capture unobserved heterogeneity.

Ad (a) With respect to success factor studies we are not so much interested in supporting hypotheses of the type that a construct such as market orientation has a positive impact on business performance. Such a relationship is highly plausible. Valuable information is only generated for the business community if we know the level of impact that the various drivers of market orientation have. The indicators should be actionable, which implies that they must form a construct and not reflect it.

Ad (b) A set of formative indicators should cover all aspects or facets of a construct. Such indicators can therefore not be drawn randomly from a universe of interchangeable indicators. This implies that we cannot apply statistical criteria for the validation of the measurement of a construct. Rather, we can only test content validity through appropriate reasoning (Rossiter 2002). Although some authors (Diamantopoulos 2005; Finn and Kayande 2005) have argued that this is unsatisfactory and should be accompanied by appropriate statistical tests Rossiter (2005) argues against it. Even the tetrad-test is only a test whether indicators are truly reflective but not a test to prove the contrary (Gudergan et al. 2008).

Ad (c) In order to handle formative indices, we frequently found that authors aggregate indicators to indices in order to run LISREL or simple OLS regressions. This implies that the indices have been aggregated by computing the unweighted mean of all indicators forming a construct. Unfortunately, this has the consequence that one cannot determine the different impacts of different indicators as drivers. In addition, equal weights will underestimate the relationship between the construct and a final endogenous construct. Besides this, current programs like LISREL or AMOS also enable the user to include formative indicators but one can estimate many more parameters with PLS because the degree of freedoms in PLS is determined on the basis of the maximum number of indicators or relationships per construct.

Ad (d) In the literature, we find the second-order constructs approach. Its use has become popular because it allows for the operationalization on the basis of aspects or facets (first-order factors forming the second-order factor) and at the same time allows working with multiple items that reflect a first-order factor and to evaluating their measurement according to the "Cronbach's α – CFA" paradigm. Unfortunately, a second-order factor can only be handled in SEMs if it is itself reflected by some indicators. One proposal is to use the unweighted sum of all indicators as a reflecting indicator (Jarvis et al. 2003). This practice is questionable because it means that the construct will be explained by just one indicator and thereby does not allow for different weights of the different aspects. It is therefore better to refrain from using the highly abstract second-order construct and work with all the first-order factors as constructs. This will give richer information on the impact of the various constructs. Even in this case, it would be better to work with indicators forming first-order factors because this gives actionable results.

Ad (e) The purpose of studies with reflective or formative indicators is different. In the first case, the test of theories has been the dominant research goal of studies. We have seen that a test whether market orientation has a positive impact on business performance provides limited insights because the relationship is highly plausible and the result of a significance test heavily depends on the number of investigated cases or other non-controlled effects. Therefore, we get richer information if we determine the level of impact that different drivers have on business performance. Insofar, we advocate that significance testing is not the main purpose of success factor studies but that the determination of the parameter levels is. To better visualize the different impacts, it is proposed to present a table with indicators classified according to different intervals of importance (total effects) and standard errors.

Ad (f) In the application, it has been proposed to determine not only the direct effect but also an indicator's total effect on the endogenous construct via all indirect paths. However, the standard errors of the total effect are not as yet given by programs like PLS-graph (Chin 1998). Rather, one has to determine via simulation what the standard error is of the sum of paths with different standard errors per connection in the graph.

Ad (g) In marketing, it is observed that regression results are heavily distorted because of heterogeneity across cases. When investigating success and its drivers, one has to concede that decision units combine drivers in different ways, which makes it impossible to determine just one uniform relationship. Rather, it will be found that at least segments of units (cases) behave in a similar way. It is therefore advisable, to simultaneously determine segments of cases and regression equations per segment. This is done with the help of finite mixture regressions. Unfortunately, there is no program available for PLS that can perform this kind of estimation. So far, only the program FIMIX has been developed to determine the regression equations once the weights of the various indicators have been determined beforehand (Hahn et al. 2002).

18.6 Conclusion

Success factor studies in marketing have traditionally been analyzed with the help of an approach that not only determines the structural relationships but also the measurement error of complex constructs. Researchers implied reflective indicators that could be validated and estimated according to the traditional "Cronbach's α – LISREL" paradigm. Unfortunately, reflective indicators in a structural equation model do not allow for actionable results. Rather, success factors should consider all facets of a construct and be treated as formative indicators. If researchers worked with facets as indicators but treated them as reflective misspecifications are the result. The item purification process may lead to the deletion of important aspects and the estimation might result in substantial biases.

This article proposes a new paradigm for success factor studies in marketing. In such studies, the significance of highly plausible relationships is no longer of interest. Rather, the differential impact of the various variables in the model as a whole is of interest. PLS is the most suitable model for such applications, as it allows for quantifying the total effects of success factors.

References

Albers, S. (2003). Relative advantages of e-business start-ups versus integrated units of bricks-and-moritar companies. In G. Fandel et al. (Eds.), Modern Concepts of the Theory of the Firm-Managing Enterprises of the New Economy, Springer, Berlin (pp. 33–57).

Albers, S., & Hildebrandt, L. (2006). Methodological problems in success factor studies – measurement error, formative versus reflective indicators and the choice of the structural equation model. *Zeitschrift für betriebswirtschaftliche Forschung, 58*(2), 2–33.

Arbuckle, J. L. (1999). AMOS user's guide, version 4.0.

Baumgartner, H., & Homburg, C. (1996). Applications of structural equation modeling in marketing and consumer research: A review. *International Journal of Research in Marketing, 22*(1), 1–9.

Cadogan, J. W., Diamantopoulos, A., & Pahud de Mortanges, C. (1999). A measure of export market orientation: Scale development and cross-cultural validation. *Journal of International Business Studies, 30*, 689–707.

Chin, W. W. (1998). The partial least squares approach to structural equation modelling. In G. A. Marcoulides (Ed.), *Modern methods for business research* (pp. 295–336). New Jersey: Lawrence Erlbaum.

Churchill, G. A., Jr. (1979). A paradigm for developing better measures of marketing constructs. *Journal of Marketing, 16*(1), 64–73.

Cronbach, L. J. (1951). Coefficient alpha and the internal structure of tests. *Psychometrika, 16*(3), 297–324.

Diamantopoulos, A. (2005). The C-OAR-SE procedure for scale development in marketing: A comment. *International Journal of Research in Marketing, 22*(1), 1–9.

Diamantopoulos, A., & Winklhofer, H. M. (2001). Index construction with formative indicators: An alternative to scale development. *Journal of Marketing Research, 38*, 269–277.

Diamantopoulos, A., & Siguaw, J. A. (2002). *Formative vs. reflective indicators in measure development: Does the choice of indicators matter?* Cornell School of Hotel Administration, Working paper.

Fassott, G. (2006). Operationalization of latent variables in structural equation models: Current status. *Zeitschrift für betriebswirtschafliche Forschung, 58*(2), 67–88.

Finn, A., & Kayande, U. (2005). How fine is C-OAR-SE? A generalizability theory perspective on Rossiter's procedure. *International Journal of Research in Marketing, 22*(1), 11–21.

Gudergan, S. (2008). Confirmatory tetrad analysis in PLS path modeling. In S. P. Gudergan, C. M. Ringle, S. Wende & A. Will (Eds.). *Journal of Business Research*, **61**(12), 1238–1249. Elsevier.

Hahn, C., Johnson, M. D., Herrmann, A., & Huber, F. (2002). Capturing customer heterogeneity using a finite mixture PLS approach. *Schmalenbach Business Review, 54*, 243–269.

Homburg, C., & Baumgartner, H. (1995). Die Kausalanalyse als Instrument der Marketingforschung. Eine Bestandsaufnahme. *Zeitschrift für Betriebswirtschaft, 65*, 1091–1108.

Homburg, C., & Giering, A. (1996). Konzeptualisierung und Operationalisierung komplexer Konstrukte. *Marketing ZFP, 18*(1), 5–24.

Homburg, C., & Pflesser, C. (2000). A multiple-layer model of market-oriented organizational culture: Measurement issues and performance outcomes. *Journal of Marketing Research, 37*, 449–462.

Jarvis, C. B., MacKenzie, S. B., & Podsakoff, P. M. (2003). A critical review of construct indicators and measurement model misspecifications in marketing and consumer research. *Journal of Consumer Research, 30*, 199–218.

Jöreskog, K., & Sörbom, D. (1996). *LISREL 8: User's reference guide*. Scientific Software International Inc.

Matsuno, K., Mentzer, J. T., & Özsomer, A. (2002). The effects of entrepreneurial proclivity and market orientation on business performance. *Journal of Marketing, 66*(3), 18–32.

Petersen, R. A. (1994). A meta-analysis of Cronbach's coefficient alpha. *Journal of Consumer Research, 21*, 381–391.

Rossiter, J. (2002). The C-OAR-SE procedure for scale development in marketing. *International Journal of Research in Marketing, 19*, 305–335.

Rossiter, J. (2005). Reminder: A horse is a horse. *International Journal of Research in Marketing*, 22(1), 23–25.

Steenkamp, J. B., & Baumgartner, H. (2000). On the use of structural equation modeling for marketing modeling. *International Journal of Research in Marketing, 17*, 195–202.

Steenkamp, J. B., & van Trijp, H. C. M. (1991). The use of LISREL in validating marketing constructs. *International Journal of Research in Marketing, 8*, 283–299.

Thies, S., & Albers, S. (2010). Application of PLS in Marketing: Content strategies in the internet. In V. Esposito. Vinzi, W. W. Chin, J. Henseler, & H. Wang (Eds.), *Handbook Partial Least Squares*. Heidelberg: Springer.

Wold, H. (1985). Partial least squares. In S. Kotz & N. L. Johnson (Eds.), *Encyclopedia of statistical sciences* (pp. 581–591). New York: Wiley.

Zhou, K. Z., Yim, C. K., & Tse, D. K. (2005). The effects of strategic orientations on technology- and market-based breakthrough innovations. *Journal of Marketing, 69*(2), 42–60.

Chapter 19
Applying Maximum Likelihood and PLS on Different Sample Sizes: Studies on SERVQUAL Model and Employee Behavior Model

Carmen Barroso, Gabriel Cepeda Carrión, and José L. Roldán

Abstract Structural equation modelling (SEM) has been increasingly utilized in marketing and management areas. This increasing deployment of SEM suggests that a comparison should be made of the different SEM approaches. This would help researchers choose the SEM approach that is most appropriate for their studies. After a brief review of the SEM theoretical background, this study analyzes two models with different sample sizes by applying two different SEM techniques to the same set of data. The two SEM techniques compared are: Covariance-based SEM (CBSEM) – specifically, maximum likelihood (ML) estimation – and Partial Least Squares (PLS). After presenting the study findings, the paper provides insights regarding when researchers should analyze models with CBSEM and when with PLS. Finally, practical suggestions concerning PLS use are presented and we discuss whether researcher considered these.

19.1 Introduction

Marketing and management research has been increasing and adding more sophisticated methodological tools. Owing to a higher elaboration level, marketing and management researchers have been able to design and test more complex models to explain reality. Among these methodological tools, structural equation modeling (SEM, hereafter) is a way to run multiple regressions between variables and latent variables. LISREL and AMOS are the most popular of the SEM software packages. In fact, many researchers believe these programs *are* SEM. Nevertheless, not all SEM is covariance-based and factorial analysis. Recently, the use of Partial Least

C. Barroso and J. L. Roldán
Management and Marketing Department, University of Seville, Ramón y Cajal, 1, 41018 Sevilla, Spain
e-mail: barroso@us.es, jlroldan@us.es

G. C. Carrión
Departamento de Administración de Empresas y Marketing, Universidad de Sevilla, Ramón y Cajal, 1, 41018 Sevilla, Spain
e-mail: gabi@us.es

V. Esposito Vinzi et al. (eds.), *Handbook of Partial Least Squares*, Springer Handbooks of Computational Statistics, DOI 10.1007/978-3-540-32827-8_20,

Squares (PLS) has increased. The PLS objective, unlike that of covariance based SEM (CBSEM), is latent variable prediction and the method is not covariance-based but variance-based. PLS tries to maximize the variance explained of the dependent variables. Many management and marketing researchers have applied and popularized PLS (e.g., Fornell and Cha 1994; Hulland 1999) since, compared to CBSEM, it offers many benefits with respect to distribution requirements, type of variables, sample size and the complexity of the model to be tested. Nevertheless, the main drawbacks of PLS are its predictive and exploratory nature and that it only achieves consistency at large (McDonald 1996). However, few papers have been written about how to be rigorous when applying PLS and how to solve all methodological problems that could arise in models when PLS is applied. While there are various papers and books that explain the PLS theoretical background (Tenenhaus and Esposito 2005), almost no papers explain and analyze PLS' real application in management research. Many relevant authors in the SEM field call for concrete suggestions regarding specific issues such as the deployment of various techniques and a guideline of when to apply one technique or the other (Vinzi et al. 2006). Other specific issues to be addressed by SEM researchers are: (1) identifying the most adequate technique when several are applicable; (2) analyzing the similarities and divergences between CBSEM and PLS regarding sample size as a possible cause for concern.

Following these research suggestions, we address a results comparison estimating two well-known models of the marketing and management literature, both of which apply CBSEM-ML (maximum likelihood) and PLS with the goal of showing the similarities and divergences between the two techniques. These two models provide one more concern regarding the comparison: sample size. The comparative analysis conclusions permit the clarifying of issues such as: when is PLS more pertinent than CBSEM-ML? When does this not matter? We begin with the theoretical context and an outline of the main SEM concepts on which we anchor this chapter. Thereafter, we present a theoretical comparison between CBSEM-ML and PLS and a description of how to analyze models applying SEM. Subsequently, we describe the two models and their results applying both techniques. Finally, we provide the implications and conclusions for researchers.

19.2 Structural Equation Modeling

Structural equation modeling (SEM) emerges as a result of the conjunction of two traditions (Chin 1998a; Goldberger 1971). On the one hand, an econometric perspective focused on prediction, on the other, a psychometric approach that models concepts as latent (unobserved) variables that are indirectly inferred from multiple observed measures (indicators or manifest variables).

Compared to the first generation of multivariate methods,[1] SEM approaches, as a second generation of multivariate analysis (Bagozzi and Fornell 1982), allow to

[1] For example, linear regression, principal components analysis, factor analysis, LOGIT, ANOVA, and MANOVA.

(Chin 1998a; Fornell 1982; Haenlein and Kaplan 2004): (1) explicitly model measurement error for observed variables; (2) incorporate abstract and unobservable constructs (latent variables) measured by indicators (also called items, manifest variables, or observed measures); (3) simultaneously model relationships among multiple predictor (independent or exogenous) and criterion (dependent or endogenous) variables; and (4) combine and test a priori knowledge and hypotheses with empirical data. In this regard, SEM tends to be confirmatory rather than exploratory.

In a single, systematic, and comprehensive analysis, SEM evaluates (Diamantopoulos 1994; Gefen et al. 2000): (1) The measurement model, i.e., loadings of observed items (indicators or measures) on their expected constructs (latent variables). The measurement part describes how each of the latent variables is operationalized via the manifest variables and provides information about the validities and reliabilities of the latter. (2) The structural model, i.e. the assumed causation in a set of dependent and independent latent variables. These relationships between the latent variables reflect substantive hypotheses based on theoretical considerations. Furthermore, the structural model shows the amount of unexplained variance.

SEM permits complicated variable relationships to be expressed through hierarchical or non-hierarchical, recursive or non-recursive structural equations to present a more complete picture of the entire model (Bullock et al. 1994; Hanushek and Jackson 1977). The intricate causal networks enabled by SEM characterize real-world processes better than simple correlation-based models. Therefore, SEM is more suited for the mathematical modeling of complex processes to serve both theory (Bollen 1989) and practice (Dubin 1976).

The holistic analysis that SEM is capable of performing is carried out via one of two distinct statistical techniques (Gefen et al. 2000): (1) Covariance-based SEM (CBSEM) and (2) a variance-based (or components-based) method, i.e. Partial Least Squares (PLS). The two types of SEM differ in the objectives of their analyses, the statistical assumptions on which they are based, and the nature of the fit statistics they produce.

19.3 PLS or Covariance-Based SEM?

PLS and Covariance-Based SEM (CBSEM) have been designed to achieve different objectives. CBSEM attempts to estimate the parameters of the model (i.e., loadings and path values) in order to minimize the difference between the sample covariances and those predicted by the theoretical model. Thus, the parameter estimation process tries to reproduce the covariance matrix of the observed measures' (Chin and Newsted 1999) overall goodness-of-fit measures to see how well the hypothesized model fits the data (Barclay et al. 1995). CBSEM emphasizes the overall model fit; that is, this approach is oriented towards testing a strong theory. Therefore, CBSEM is best suited for confirmatory research (Gefen et al. 2000).

PLS path modeling focuses on the prediction of the dependent variables (both latent and manifest). This objective is achieved by maximizing the explained variance (R^2) of the dependent variables. Thus, parameter estimates are obtained based on the ability to minimize the residual variances of dependent variables. Compared to CBSEM, PLS is more suited for predictive applications and theory building (exploratory analysis), although PLS can be also used for theory confirmation (confirmatory analysis).

The decision between these approaches is whether to use SEM for theory testing and development or for predictive applications (Anderson and Gerbing 1988). In situations where prior theory is strong and further testing and development are the goal, covariance-based full-information estimation methods (i.e. Maximum Likelihood (ML) or Generalized Least Squares (GLS)) are more appropriate. However, for application and prediction, a PLS approach is often more suitable. Indeed, Wold (1979) states that PLS is primarily intended for causal-predictive analysis, where the problems explored are complex (i.e. models with a large number of variables, indicators and relationships), and prior theoretical knowledge is scarce. Barclay et al. (1995, p. 288) conclude: (1) PLS is generally recommended for predictive research models where the emphasis may be more on theory development. (2) CBSEM is more suited for confirmatory testing of how well a theoretical model fits observed data, requiring much stronger theory than PLS.

Chin (1998b, p. 299) makes three basic distinctions for choosing between PLS and CBSEM: (1) the constructs are modeled as indeterminate or determinate (defined),[2] (2) the extent to which, in the theoretical model and auxiliary theory, the researcher links measures to constructs, and (3) the researcher is parameter-oriented or prediction-oriented. If the second option applies best to each question, the PLS approach is likely to be more suitable.

Certainly, PLS can be a powerful method of analysis because of the minimal demands on measurement scales,[3] sample size, and residual distributions (Fornell and Bookstein 1982). With reference to CBSEM, PLS avoids two serious problems which often interfere with meaningful modeling: improper, i.e. inadmissible solutions[4] (Fornell and Bookstein 1982), and factor indeterminacy.[5] As a consequence of the use of an iterative algorithm that consists of a series of ordinary

[2] According to Fornell (1982, p. 5), a determinate or defined construct is a composite (often called a component or a derived variable) of its indicators (manifest variables). An indeterminate construct (often called factor) is a composite of its indicators plus an error term. Defined constructs sacrifice the theoretical desirability of allowing for imprecise measurement for the practical advantage of construct estimation and direct calculation of component scores. A determinate construct is completely determined by its indicators and assumes that the combined effect of the indicators is free from measurement error.

[3] Nominal, ordinal, and interval scaled variables are permissible in PLS (Falk and Miller 1992, p. 32; Wold 1985, p. 234). In this respect, nominal variables should be replaced by a set of Boolean variables or dummy-coded variables to be admissible in a PLS model (Falk and Miller 1992, p. 67; Lohmöller 1989, p. 143).

[4] For example, negative estimates of variance and standardized loadings greater than 1. One possible cause of improper solutions might be failure of the model to fit the data.

[5] Factor indeterminacy occurs when case values for the latent variables can not be obtained in the estimation process. PLS avoids factor indeterminacy by explicitly defining the unobservable

least squares (OLS) analyses, neither is identification a problem for recursive nor does PLS require the measured variables to follow any particular distribution (Chin 1998b).

PLS is a technique designed to reflect the theoretical and empirical conditions present in the behavioral and social sciences, where these are habitual situations with no solid theories and scarce knowledge. This kind of modeling is called soft modeling (Wold 1980). Mathematical and statistical procedures underlying the system are rigorous and robust[6] (Wold 1979); however, the mathematical model is soft in the sense that it makes no measurement, distributional, or sample size assumptions. The goal to be achieved is milder than hard modeling (i.e. CBSEM, particularly using maximum-likelihood estimation procedures). In soft modeling, the concept of causation must be abandoned and replaced by the concept for predictability. While causation guarantees the ability to control events, predictability allows only a limited degree of control (Falk and Miller 1992). In CBSEM, each established causal relationship should be due to a justification based on a substantial theory, and the proposed causality could be simple, circular or complex (Bullock et al. 1994; Hair et al. 1998). In fact, establishing causation is difficult in research. According to Cook and Campbell (1979), establishing causation requires the demonstrating of: association, temporal precedence, and isolation. Therefore, statistical analysis alone can not prove causation, because it does not establish isolation or temporal ordering (Bollen 1989; Bullock et al. 1994). Besides, this problem is more pronounced in SEM because of the complexity of the structural models and the potential existence of equivalent models. Given these reasons, SEM methods should be used as a confirmatory and not as an exploratory method, particularly in the covariance-based techniques (Bullock et al. 1994; Bollen 1989). Taking into account the nature of epistemic relationships,[7] it should be pointed out that CBSEM was originally designed to operate with reflective indicators (Fornell 1982). In this case, the latent variable is thought to give rise to what is observed – indicators – (e.g., personality traits and attitudes). On the other hand, there are so-called formative indicators, which are manifest variables giving rise to an unobserved theoretical construct (e.g., the social status construct could be defined as produced by occupation, income, location of residence, etc.). It should be highlighted that PLS allows working with both types of measures (Fornell and Bookstein 1982). In contrast, any attempts to model formative indicators in CBSEM can lead to identification problems, implied covariances of zero among some indicators, and/or the existence of equivalent models (MacCullum and Browne 1993). Therefore, authors such as Diamantopoulos and Winklhofer (2001) suggest the use of PLS as an alternative for incorporating formative measurement models.

---

variables. In this way, PLS produces latent variable scores that can be used to predict its own indicators or other latent variables scores.

[6] Monte Carlo simulations show that the PLS method is quite robust against (Cassel et al. 1999): (1) skew instead of symmetric distributions of manifest variables, (2) multi-collinearity within blocks of manifest variables and between latent variables; and (3) misspecification of the structural model (omission of regressors).

[7] An epistemic relationship describes the link between theory and data.

Concerning the directional relationships among constructs, they can be both recursive (unidirectional) and nonrecursive (bidirectional). CBSEM allows both, whereas PLS currently only works with recursive.[8]

Finally, according to Wold (1985), CBSEM and PLS should be considered as complementary rather than competitive methods, and both have a rigorous rationale of their own. As Jöreskog and Wold – parents of LISREL and PLS, respectively – state: "ML is theory-oriented, and emphasizes the transition from exploratory to confirmatory analysis. PLS is primarily intended for causal-predictive analysis in situations of high complexity but low theoretical information" (Jöreskog and Wold 1982). Subsequently, Wold distinguished a division of labor between LISREL and PLS: "LISREL is at a premium in small models where each parameter has operative significance, and accurate parameter estimation is important. PLS comes to the fore in larger models, where the importance shifts from individual variables and parameters to packages of variables and aggregate parameters" (Wold 1985).

19.4 Relevant SEM Analysis Characteristics

The specific literature indicates two stages of the SEM analysis (Hair et al. 1998): measurement model and structural model assessment. The measurement model defines the latent variables that the model will use, and assigns observed variables (indicators) to each. It attempts to analyze whether the theoretical constructs are correctly measured by the manifest variables. This analysis is carried out with reference to reliability and validity attributes. The structural model defines the causal relationships between the latent variables. The structural model is assessed according to the meaningfulness and significance of the hypothesized relationships between the constructs.

The basic terms used are the following (Diamantopoulos 1994; Falk and Miller 1992; Wold 1985; Barclay et al. 1995): (1) The theoretical construct or latent variable (graphically represented by a circle), which makes a distinction between the exogenous constructs (ξ) that act as predictor or causal variables of the endogenous constructs (η). (2) Indicators, measures, manifest or observable variables (graphically symbolized by squares).

19.4.1 Measurement Model Assessment

The measurement model is evaluated by examining individual item reliability, internal consistency or construct reliability, average variance extracted analysis, and discriminant validity.

[8] Hui (1978, 1982) developed a fixed-point PLS method to model nonrecursive relations. However, this algorithm has not been implemented in the present PLS software applications.

In PLS, individual item reliability is assessed by inspecting the loadings (λ), or simple correlations of the indicators with their respective latent variable. A widely accepted rule of thumb has been proposed by Carmines and Zeller (1979). They indicate that to accept an indicator as a constituent of a construct, the manifest variable should have a loading of 0.707 or more. This implies more shared variance between the construct and its measures than error variance. Nonetheless, several researchers think this rule of thumb should not be as rigid at the early stages of scale development (Chin 1998b) and when scales are applied across different contexts (Barclay et al. 1995). In contrast to covariance-based SEM, where including additional poor indicators will lead to a worse fit, in the case of PLS, the inclusion of weak items will help to extract what useful information is available in the indicator to create a better construct score. It should not be forgotten that PLS works with determinate constructs; consequently, worse indicators are factored in by lower weights (Chin 2002).

Nonetheless, in the case of a construct with formative indicators, the loadings are misleading because the intraset correlations for each block are never taken into account in the estimation process that this technique follows to obtain the construct parameters. Therefore, it makes no sense to compare loadings among manifest variables within a block. The interpretation of a construct with formative indicators should be based on the weights (Chin 1998b). Like the canonical correlation analysis, the weights allow us to understand the make-up of each emergent construct. That is to say, these provide information on how each dimension or indicator (formative) contributes to the respective construct. However, a concern related to using formative measures deals with the potential multicollinearity among the formative items. This would produce instable estimates, and would make it difficult to separate the distinct effect of the indicators on the emergent construct (Diamantopoulos and Winklhofer 2001; Mathieson et al. 2001).

In CBSEM, the item reliability shows the variance rate that such an item and the construct share, which is equivalent to communality in the exploratory factor analysis. An indicator should have at least 50% of its variance in common with the latent variable, establishing a value of 0.5 as an acceptation limit (Sharma 1996).

The construct reliability assessment allows the evaluation of the extent to which a variable or set of variables is consistent in what it intends to measure (Straub et al. 2004). As a measure of internal consistency, the composite reliability (ρ_c) developed by Jöreskog (1974) fulfills the same task as Cronbach's alpha. The interpretation of both indexes is similar. Nunnally (1978) suggests 0.7 as a benchmark for "modest" reliability applicable in early stages of research, and a more strict 0.8 value for basic research. Nevertheless, as measures of internal consistency, both composite reliability (ρ_c) and Cronbach's alpha, are only applicable to latent variables with reflective indicators (Chin 1998b). However, in an emergent construct with formative manifest variables, indicators need not covary with one another (Jarvis et al. 2003). Thus, such measures are not necessarily correlated and, consequently, traditional reliability and validity assessment have been argued as inappropriate and illogical for this type of constructs when refering to its indicators (Bollen 1989).

Another measure of reliability is the average variance extracted (AVE, Fornell and Larcker 1981). This measure quantifies the amount of variance that a construct captures from its manifest variables or indicators relative to the amount due to measurement error (Chin 1998b). This ratio tends to be more conservative than composite reliability (ρ_c). AVE values should be greater than 0.50. This means that 50% or more of the indicator variance should be accounted for. Moreover, as in the previous case, this measure is only appropriate for constructs with reflective indicators.

Finally, discriminant validity indicates the extent to which a given construct differs from other constructs. To assess discriminant validity, (Fornell and Larcker 1981) suggest that the AVE should be greater than the variance between the construct and other constructs in the model (i.e., the squared correlation between two constructs).

19.4.2 Structural Model Assessment

In CBSEM, the first step consists of analyzing the significance achieved by the coefficient estimates ($t > 1.96$). A non-significant parameter indicates the necessity to re-formulate such a model, taking into account the theoretical basis. Subsequently, the researcher should carefully analyze the overall model fit measures (Hair et al. 1998). For an adequate evaluation of the structural model in PLS, there are two key indexes: the explained variance in the endogenous variables (R^2) and the path coefficients (β).

When asked the key question – where are the goodness-of-fit measures? – regarding a PLS analysis in any SEM-based study, the answer should be that it is impossible to offer this information. The reason for this answer is based on the fact that the existing goodness-of-fit measures are related to the model's ability to account for the sample covariances and therefore assume that all measures are reflective. Nevertheless, PLS does not have any explicit objective function and allows for formative indicators, therefore, it is, by design, unable to provide such fit indexes (Chin 1998a).

In order to estimate the precision of the PLS estimates, nonparametric techniques of re-sampling should be used. Consequently, jackknifing and bootstrapping [9] are two approaches commonly used in PLS analysis. Both methods provide the standard errors and t-statistics of the parameters.

Together with these resampling techniques, the Q^2 test, developed by Geisser (1975) and Stone (1974), is used to assess the predictive relevance of the endogenous constructs. This test is an indicator of how well observed values are reproduced by the model and its parameter estimates. Two types of Q^2 can be obtained,

[9] See Efron (1982), Efron and Gong (1983), Efron and Tibshirani (1993), and Chapter 3 of this book for further details.

depending on the form of prediction: cross-validated communality and cross-validated redundancy (Fornell and Cha 1994). Chin (1998b) suggests using the latter to examine the predictive relevance of the theoretical/structural model. A Q^2 greater than 0 implies that the model has predictive relevance, whereas a Q^2 less than 0 suggests that the model lacks predictive relevance.

19.5 PLS or Covariance-Based SEM?

In order to carry out the proposed comparisons, we strive to achieve the following objectives:

1. To compare results achieved by means of the same model using covariance-based SEM (CBSEM-ML) and PLS.
2. To present two very different models with respect to sample size and number of indicators.

We provide a SERVQUAL model, which possesses many indicators and uses a large sample, and an employee behavior model, which has less indicators and uses a smaller sample. These two models are widely referenced in marketing and management literature.

19.5.1 Comparing Data in a Big Simple: The SERVQUAL Model and Customer Satisfaction

The service marketing literature devoted much attention to the relationship between perceived quality service and customer satisfaction as a "loyalty chain" component (e.g., Beerli et al. 2004; Bitner and Hubbert 1994; Caruana 2002; Cronin and Taylor 1992; Spreng and Mackoy 1996; Falk and Miller 1992; Sureshchandar et al. 2002; Tam 2004; Yi 2004; Zeithaml and Berry 1996). These contributions have illustrated that service quality should be considered an attitude that is highly related to satisfaction, but not equivalent (Spreng and Mackoy 1996; Taylor and Baker 1994). Service quality can be defined as the degree and direction of the discrepancies between service delivery perceptions and customers' previous expectations (Parasuraman et al. 1988). Nowadays, most researchers agree that service quality is an antecedent of customer satisfaction (Cronin and Taylor 1992; Zeithaml and Berry 1996; Bitner and Hubbert 1994). The most used scale of perceived quality service has been developed by Parasuraman et al. (1985). These authors suggest an instrument called SERVQUAL, which is applied in numerous studies despite various criticisms (Teas 1993; Cronin and Taylor 1994). In our study, we use a modified version of this proposed scale (Parasuraman et al. 1988). This scale contains 22 items belonging to five underlying factors: "tangibles" (four items); "reliability" (five items), "responsiveness" (four items); "security" (four items); "empathy" (five items). Customer satisfaction is measured by Maloles's scale (Maloles 1997). Our sample consists of 3,624 bank industry customers.

We present the outcomes obtained after applying both techniques (ML estimation and PLS).

Measurement Model Analysis.

We start our analysis with a confirmatory factor analysis (CFA). As can be observed (Table 19.1), indicators composing both measurement models (CBSEM-ML and PLS) are exactly alike in this model. In fact, the cs19 indicator is dropped in both estimations. According to the concrete values, such as those described in the theoretical section, it can be observed that the CFA values for CBSEM-ML estimation are slightly lower than those obtained with PLS. The average of the ratio λ (ML)/λ (PLS) is 0.9064 for service quality and 0.9929 for customer satisfaction. Notwithstanding, the indicator hierarchy is very similar in both techniques.

The Composite reliability coefficient (ρ_c) is used to address construct reliability on both SEM analyses. It is possible to assess internal consistency through Cronbach's alpha (Werts et al. 1974), but we choose composite reliability following Barclay et al. (1995) and Fornell and Larcker's suggestions (Fornell and Larcker 1981), since composite reliability is not influenced by existent items number in each scale and uses item loadings extracted from the causal model analyzed. Composite

Table 19.1 Individual item loadgings

Customer satisfaction			Service quality			Service quality		
Items	Factor loadings		Items	Factor loadings		Items	Factor loadings	
	CBSEM	PLS		CBSEM	PLS		CBSEM	PLS
s1	0.838	0.8382	Tangibles	0.610	0.7243	Responsiveness	0.767	0.8854
s2	0.791	0.7811	cs1	0.545	0.6900	cs10	0.649	0.7266
s3	0.714	0.6366	cs2	0.587	0.7330	cs11	0.624	0.7733
s4	0.826	0.7506	cs3	0.695	0.7686	cs12	0.785	0.8237
s5	0.672	0.8421	cs4	0.578	0.7075	cs13	0.516	0.6783
s6	0.672	0.7170	Reliability	0.795	0.8562	Security	0.850	0.8756
s7	0.833	0.8346	cs5	0.703	0.7742	cs14	0.836	0.8770
s8	0.814	0.8378	cs6	0.744	0.8024	cs15	0.792	0.8502
s9	0.871	0.8736	cs7	0.784	0.8320	cs16	0.763	0.8259
			cs8	0.802	0.8471	cs17	0.672	0.7555
			cs9	0.546	0.6230	Empathy	0.781	0.8355
						cs18	0.818	0.8226
						cs19	–	–
						cs20	0.852	0.8474
						cs21	0.642	0.7879
						cs22	0.700	0.8188

Table 19.2 Composite reliability and AVE coefficients

Construct	Composite reliability (ρ_c)		AVE	
	CBSEM	PLS	CBSEM	PLS
Tangibles	0.6949	0.8160	0.4010	0.5262
Reliability	0.8498	0.8848	0.5101	0.6082
Responsiveness	0.7096	0.8385	0.4703	0.5661
Security	0.8511	0.8971	0.5896	0.6862
Empathy	0.8420	0.8910	0.5343	0.6715
Service quality	0.8307	0.9382	0.5849	0.7013
Customer satisfaction	0.9295	0.9211	0.6241	0.6296

Table 19.3 CBSEM-ML correlation matrix. Diagonal elements (values in parentheses) are the square root of the AVE

	TANG	RELIAB	RESPONS	SECU	EMPAT	SERVQ	CUSTSAT
TANG	(0.633)	–	–	–	–	–	–
RELIAB	0.647	(0.714)	–	–	–	–	–
RESPONS	0.721	0.798	(0.685)	–	–	–	–
SECU	0.724	0.773	0.855	(0.767)	–	–	–
EMPAT	0.657	0.723	0.746	0.734	(0.730)	–	–
SERVQ	–	–	–	–	–	(0.764)	–
CUSTSAT	–	–	–	–	–	0.81	(0.79)

reliability values are appropriated for both approaches (CBSEM-ML and PLS). All values are above or very close to 0.7 (Table 19.2). As can be appreciated, values for PLS are greater than for CBSEM-ML due to the greater values of the indicators estimations in PLS.

Regarding the Average Variance Extracted (AVE) for each construct, all values (except for the tangibles construct with CBSEM-ML) are above 0.5 (Table 19.2), which means variance explained by indicators exceeds variance explained by error. Again, the values for PLS are greater than the values for CBSEM-ML. To address discriminant validity, we compare whether the average variance extracted is greater than the square correlations between the construct and each of the other constructs in the model (Fornell and Larcker 1981). This highlights that one construct differs from the others.

To make the calculation process agile, we carry out a reverse procedure. That is, to determine construct discriminant validity, we calculate the square root of the AVE, and it should be greater than each of the construct correlations. These values are shown in the following tables: (one for CBSEM-ML (Table 19.3) and another for PLS (Table 19.4), where diagonal elements (values in parentheses) represent the square root of the AVE.

All constructs satisfy that condition for PLS, whereas the CBSEM-ML estimation presents a problem between the tangibles and reliability dimensions of service

Table 19.4 PLS correlation matrix. Diagonal elements (values in parentheses) are the square root of the AVE

	TANG	RELIAB	RESPONS	SECU	EMPAT	SERVQ	CUSTSAT
TANG	(0.725)	–	–	–	–	–	–
RELIAB	0.502	(0.780)	–	–	–	–	–
RESPONS	0.549	0.712	(0.752)	–	–	–	–
SECU	0.568	0.668	0.751	(0.828)	–	–	–
EMPAT	0.646	0.646	0.658	0.655	(0.819)	–	–
SERVQ	–	–	–	–	–	(0.837)	–
CUSTSAT	–	–	–	–	–	0.814	(0.793)

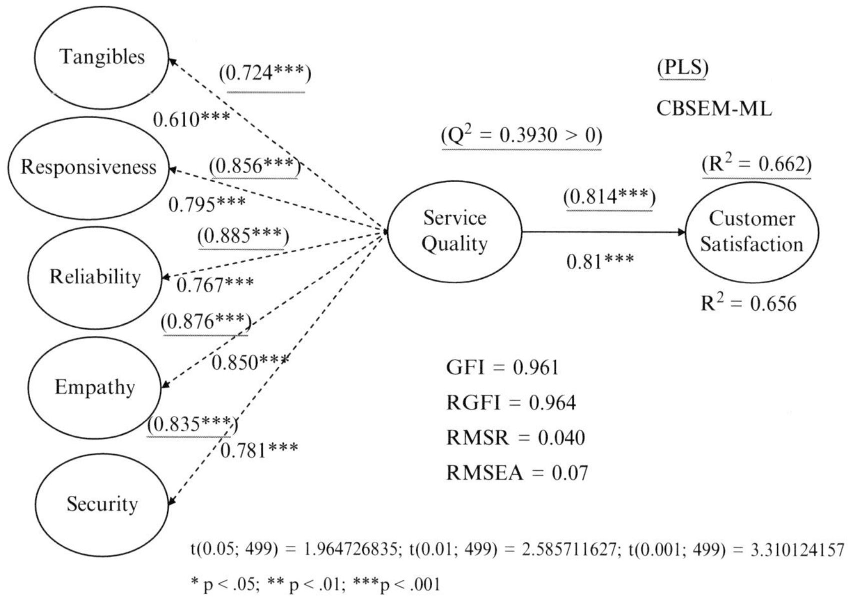

Fig. 19.1 Structural model analysis

quality. Nonetheless, this is a well-known problem in the service quality literature (Teas 1993; Cronin and Taylor 1994). In our case, we applied an ANOVA analysis to demonstrate the discriminant validity between both dimensions (see Barroso et al. 2004, for further details).

Structural Model Analysis

The structural model (Fig. 19.1) shows the existent relationships between the constructs in both CBSEM-ML and PLS. The goodness-of-fit measures for CBSEM (excluding χ^2) indicate that the data fits the model. Thus, GFI is 0.961, above the desired 0.9 value. This likewise applies to the RGFI value, which rates 0.964, also above 0.9. Another fit ratio, such as RMSE, reaches 0.07, above the 0.05 threshold. Correlations estimated between the latent variables are 0.81, and SERVQUAL dimensions values vary between 0.610 and 0.850: all significant. The standardized loadings for every SERVQUAL dimension can be interpreted as the square root of the composite reliability of the associated dimension. Thus, for example, 72% of the empathy dimension variation is associated with SERVQUAL, whether or not the model is considered correct.

From the point of view of the PLS technique (Chin et al. 2003), the model shows a good predictability, reaching an explained variance (R^2) of the dependent variable of 0.662. Furthermore, the predictive measure for the endogenous construct also achieves a value higher than 0 ($Q^2 = 0.3930$), pointing out that the model has predictive relevance. Anyhow, we would like to highlight how very close the path coefficients achieved by the two techniques are: these are clearly significant.

It can be seen that the PLS values for dimension loadings are higher than those of CBSEM-ML. Hence, in our model, latent variables are something better measured by PLS than by CBSEM-ML. Another difference in the results that is not very well noted here, is that the construct correlations are lower for PLS than for CBSEM-ML. Probably due to sample size, those values are almost coincident in our case. Both differences justify Chin's statements (Chin 1995) when results are compared in the same model using CBSEM-ML and PLS. The extreme similarity between correlations in such a big sample explains Herman Wold's (father of PLS) words when he states that "The PLS estimates are consistent at large in the sense that they tend to the true values when there is indefinite increase not only in the number of observed cases, but also in the number of indicators for each latent variable" (Wold 1985).

19.5.2 Comparing Data in a Small Sample: The Employee Behavior Model

The employee behavior literature states that both job conflict and job ambiguity are employee satisfaction antecedents (Babin and Boles 1998; Hartline and Ferrell 1996; Mackenzie et al. 1998; Singh 1998). We define job conflict as the degree to which expectations and requirements in a job are incompatible for two or more employees. Job ambiguity is defined as the uncertainty level about activities and tasks that shape a particular job (Rizzo et al. 1970). Both elements constitute what the literature calls job stress. The literature states that there is a trade-off between employee satisfaction with a job and both job conflict and job ambiguity (Hartline and Ferrell 1996; Mackenzie et al. 1998). Furthermore, job ambiguity has a positive relationship with the likelihood of turnover in a organization, and job conflict has a negative link with employee commitment (Brown and Peterson 1993). The job conflict scale is shaped by six items and job ambiguity by five items. We use Rizzo,

Table 19.5 Individual item loadings

Job conflict	Factor loadings		Job ambiguity	Factor loadings		Employee satisfaction	Factor loadings	
Items	CBSEM	PLS	Items	CBSEM	PLS	Items	CBSEM	PLS
c1	0.688	0.7174	a1	–	–	s1	0.821	0.8633
c2	0.760	0.7146	a2	–	0.7465	s2	0.696	0.7589
c3	0.421	0.7330	a3	0.720	0.7795	s3	0.630	0.7402
c4	–	–	a4	0.726	0.7791	s4	–	–
c5	0.747	0.7207	a5	–	0.7005	s5	0.671	0.7365
c6	–	0.7323				s6	–	–
						s7	0.633	0.7140
						s8	–	–
						s9	–	–

Table 19.6 Composite reliability and AVE coefficients

Construct	Composite reliability (ρ_c)		AVE	
	CBSEM-ML	PLS	CBSEM-ML	PLS
Job conflict	0.771	0.846	0.483	0.524
Job ambiguity	0.892	0.839	0.512	0.566
Employee satisfaction	0.882	0.875	0.556	0.584

House and Lirtzman's scale (Rizzo et al. 1970). The scale of employee satisfaction with a job has been adapted from Babin and Boles (1998). Our sample comprises 176 bank industry employees.

We present the outcomes obtained after applying both techniques (CBSEM-ML estimation and PLS).

Measurement Model Analysis

We start our analysis with a confirmatory factor analysis. With reference to individual item reliability (Table 19.5), we can observe that the final list of indicators included in the measurement model differs in the two methods. After an item trimming process, CBSEM-ML deletes indicators that are maintained by PLS (e.g., c6, a2 and a5). Another observable difference is the increase in the distance between the loading estimates as developed by both techniques. It seems that the differences are more intense when the sample decreases. The average ratio λ (ML)/λ (PLS) is 0.9132. To compare reliability measures, we use the composite reliability coefficient (ρ_c) again. In respect of both methods, all latent variables seem to satisfy the conditions imposed for composite reliability, as all are above 0.7 (Table 19.6). As in the previous study, PLS also generates greater measures than the CBSEM-ML estimation.

Table 19.7 CBSEM-ML correlation matrix. Diagonal elements (values in parentheses) are the square root of the AVE

	Job conflict	Job ambiguity	Employee satisfaction
Job conflict	(0.694)		
Job ambiguity	0.346	(0.715)	
Employee satisfaction	−0.432	−0.508	(0.745)

Table 19.8 PLS Correlation Matrix. Diagonal elements (values in parentheses) are the square root of the AVE

	Job conflict	Job ambiguity	Employee satisfaction
Job conflict	(0.768)		
Job ambiguity	0.194	(0.872)	
Employee satisfaction	−0.232	−0.331	(0.764)

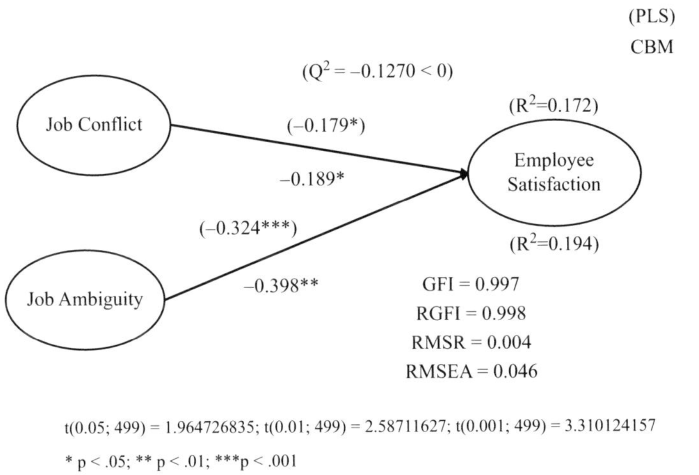

Fig. 19.2 Structural model analysis

In both methods, all the latent variables seem to satisfy the conditions imposed for the AVE indexes. As in the previous study, PLS also generates greater measures than ML. Finally, according to the estimations developed by both the methods, all the constructs achieve discriminant validity (see Tables 19.7 and 19.8).

Structural Model Analysis

The structural model (Fig. 19.2) shows the existent relationships between the constructs in both CBSEM-ML and PLS. Again, according to the CBSEM-ML estimation, the goodness-of-fit indexes (excluding χ^2) suggests that the data fits the model. Thus, GFI achieves 0.997, above the desired 0.9 value. This also applies to the RGFI value. It achieves 0.998, also above 0.9. Another fit ratio, the RMSE, reaches 0.046, very close to the 0.05 threshold.

The correlations estimated between the latent variables for the CBSEM-ML approach are −0.189 and −0.398. The PLS values are slightly lower but significant too: these are −0.179 and −0.324 (Fig. 19.2). On the other hand, PLS provides a R^2 value very close to CBSEM-ML, whereas the Q^2 index ($-0.1270 < 0$) indicates no predictive relevance of the model (Fornell and Cha 1994; Sellin 1989). This outcome is comprehensible due to the low level of R^2 on the dependent variable. Therefore, we would need new variables and relationships for an increase in the explained variance of the employee satisfaction construct. The standardized coefficients are significant for CBSEM-ML and PLS. Nevertheless, compared to CBSEM-ML, PLS underestimates the path coefficients.

It can be observed that PLS loading values are higher than those of CBSEM-ML. Again, in this model, latent variables are something better measured by PLS than by CBSEM-ML. Another clear difference between results is construct correlations are lower for PLS than for ML. Both differences justify Haenlain and Kaplan's statements (Haenlein and Kaplan 2004) regarding comparing the same model using CBSEM-ML and PLS. It seems that when the case number decreases, the regular differences found return.

19.6 Conclusions and Implications for Researchers

The growing interest in SEM analysis among social researchers leads to the necessity to make comparisons between various SEM techniques. Marketing and management researchers are not only interested in the main characteristics of each technique, but they also want to know when the use of a particular technique is more appropriate (Vinzi et al. 2010). The main goal of this paper is to focus on addressing the similarities and differences between CBSEM (ML estimation) and PLS. The final objective is to begin a research stream that will help researchers with their empirical studies. Therefore, our study compares two well-known models in the management and marketing literature whose main divergences are their indicators numbers and sample size. Obviously, this study constitutes a first approximation to this issue. New studies such as Reinartz et al. (2009) have emerged using Monte Carlo simulations to compare the behaviour of both techniques (CBSEM vs. PLS). This stream continues the early contributions by Jarvis et al. (2003) and Mackenzie et al. (2005).

Our findings lead to the following conclusions: first, as many SEM researchers state, the objectives of the two techniques differ. The main objective of PLS is prediction, while the CBSEM objective is more confirmative. CBSEM considers the analysis of covariance, while PLS takes the observed variances of dependent as point of departure. Second, in PLS, hypotheses are derived from a general theory that does not recognize all relevant variables. Thus, the theory is less sound. The CBSEM models are based on solid theories that they try to confirm. In sum, taking into account the predictive or explanatory character of the model to be tested, and the soundness of the theoretical background, researcher should choose either one approach or the other.

Third, findings achieved in our study suggest that when various sample sizes are utilized, the differences between these two approaches are become evident (Reinartz et al. 2009). Thus, PLS increases its consistency when the sample size and number of indicators included in the model are increased. In this case, the outcomes are very close. Hence, PLS and CBSEM-ML tend to converge in models with many indicators and large sample sizes. In turn, when the sample size and number of indicators decrease, PLS' consistency is reduced and there is a bigger gap between the outcomes in the two techniques. In conclusion, after our findings we deduce that CBSEM-ML is more exigent with data in order to adjust them to the theory utilized. However, PLS does not discard anything that SEM-ML models assume, both in the indicators level and in the values of the relationships between the latent variables. PLS is even more conservative than the CBSEM-ML models in this regard. In comparison, PLS tends to increase the factor loadings but to decrease the path coefficient values.

These study findings encourage us to continue this research stream. Hence, as future research streams, we propose analyzing the other recognized differences between these techniques not covered by this study. First, the inclusion of reflective or formative indicators and their possible influence on the outcomes of the final model fit. Up to now, both PLS and some software packages with ML estimation (EQS) allow formative indicators. It should be taken into account that in the management and marketing field, there are a relevant number of constructs whose indicators are formative and not reflective. An unsuitable use of techniques generates misspecification problems in models (Jarvis et al. 2003).

Therefore, we consider studying the outcomes of both techniques when the indicators that link latent variables are formative and not only reflective, as very interesting. Second, considering the complexity level of models. CBSEM and PLS' different views (confirmatory vs. exploratory) sometimes determines that PLS models need many constructs and indicators, while CBSEM models (due to their confirmatory nature) require more parsimony. Thus, previous studies suggest that CBSEM models tend to be less complex (number of variables implied and analyzed) than SEM-PLS models. A relevant empirical study could shed light on this field. Finally, our research clarifies one of the most discussed questions in social sciences: which is the more suitable technique for an empirical study?

References

Anderson, J. C., & Gerbing, D. W. (1988). Structural equation modeling in practice: a review and recommended two-step approach. *Psychological Bulletin, 103*, 411–423.

Babin, B. J., & Boles, J. S. (1998). Employee behaviour in a service environment: a model and test of potential differences between men and women. *Journal of Marketing, 62*, 77–91.

Bagozzi, R. P., & Fornell, C. (1982). Theoretical concepts, measurement, and meaning. In C. Fornell (Ed.), *A second generation of multivariate analysis*, Vol. 2 (pp. 5–23). New York: Praeger.

Barclay, D., Higgins, C., & Thompson, R. (1995). The partial least squares (PLS) approach to causal modelling: Personal computer adoption and use as an illustration. *Technology Studies (Special Issue on Research Methodology)*, 2(2), 285–309.

Barroso, C., Martin, E. M., & Martin, D. (2004). The influence of employee organizational citizenship behavior on customer loyalty. *International Journal of Service Industry Management, 15*(1), 27–53.

Beerli, A., Martn, J., & Quintana, A. (2004). A model of customer loyalty in the retail banking market. *European Journal of Marketing, 38*, 253–275.

Bitner, M. J., & Hubbert, A. (1994). Encounter satisfaction versus overall satisfaction versus quality. In R. T. Rust, & R. L. Oliver (Eds.), *Service quality: new directions in theory and practice* (pp. 72–94). Thousand Oaks, CA: Sage.

Bollen, K. A. (1989). Structural equations with latent variables. New York: Wiley.

Brown, S., & Peterson, A. (1993). Antecedents and consequences of salesperson job satisfaction: a meta-analysis and assessment of causal effects. *Journal of Marketing, 30*, 63–77.

Bullock, H. E., Harlow, L. L., & Mulaik, S. A. (1994). Causation issues in structural equation modeling research. *Structured Equation Modeling, 1*(3), 253–267.

Carmines, E. G., & Zeller, R. A. (1979). Reliability and validity assessment, N. 07-017, *Sage University Paper Series on Quantitative Applications in the Social Sciences*. Beverly Hills, CA: Sage.

Caruana, A. (2002). Service loyalty: the effects of service quality and the mediating role of customer satisfaction. *European Journal of Marketing, 36*, 811–822.

Cassel, C., Hackl, P., & Westlund, A. H. (1999). Robustness of partial least-squares method for estimating latent variable quality structures. *Journal of Applied Statistics, 26*(4), 435–446.

Chin, W. W. (1995). Partial least squares is to LISREL as principal components analysis is to common factor analysis. *Technology Studies, 2*(2), 315–319.

Chin, W. W. (1998a). Issues and opinion on structural equation modelling. *MIS Quarterly, 22*(1), 7–15.

Chin, W. W. (1998b). The partial least squares approach to structural equation modelling. In G. A. Marcoulides (Ed.), *Modern methods for business research* (pp. 295–336). Mahwah, NJ: Lawrence Erlbaum.

Chin, W. W. (2002). *Exploring some FAQs regarding PLS including to report results.* I Workshop on partial least squares methodology. Seville, Spain: University of Seville.

Chin, W. W. (2003). PLS Graph Version 3.0. Soft Modeling Inc.

Chin, W. W., & Newsted, P. R. (1999). Structural equation modeling analysis with small samples using partial least squares. In R. Hoyle (Ed.), *Statistical strategies for small samples research* (pp. 307–341). Thousand Oaks, CA: Sage.

Chin, W. W., Marcolin, B. L., & Newsted, P. R. (2003). A partial least squares latent variable modeling approach for measuring interaction effects: Results from a Monte Carlo simulation study and an electronic-mail emotion/adoption study. *Information Systems Research, 14*(2), 189–217.

Cook, T. D., & Campbell, D. T. (1979). *Quasi experimentation: design and analytical issues for field settings*. Chicago: Rand McNally College.

Cronin, J. J., & Taylor, S. A. (1992). Measuring service quality: a reexamination and extension. *Journal of Marketing, 56*, 55–68.

Cronin, J. A., & Taylor, S. A. (1994). SERVPERF versus SERVQUAL: Reconciling performance-based and perceptions-minus-expectations measurement of service quality. *Journal of Marketing, 58*, 125–131.

Diamantopoulos, A. (1994). Modelling with LISREL: A guide for the uninitiated. *Journal of Marketing Management, 10*, 105136.

Diamontopoulos, A., & Winklhofer, H. M. (2001). Index construction with formative indicators: an alternative to scale development. *Journal of Marketing Research, 38*, 269–277.

Dubin, R. (1976). Theory building in applied areas. In M. D. Dunnette (Ed.), *Handbook of industrial and organizational psychology* (pp. 17–26). Chicago: Rand McNally College.

Efron, B. (1982). *The Jackknife, the bootstrap and other resampling plans*. Philadelphia: Society for Industrial and Applied Mathematics.

Efron, B., & Gong, G. (1983). A leisurely look at the bootstrap, the jackknife, and cross-validation. The American Statistician, 37, 36–48.

Efron, B., & Tibshirani, R. J. (1993). *An introduction to the bootstrap*. No: 57, Monographs on Statistics and Applied Probability. New York: Chapman and Hall.

Falk, R. F. & Miller, N. B. (1992). *A primer for soft modeling*. Akron, Ohio: The University of Akron.

Fornell, C. (1982). A second generation of multivariate analysis: an overview. In C. Fornell (Ed.), *A second generation of multivariate analysis*, Vol. 1 (pp. 1–21). New York: Praeger.

Fornell, C., & Bookstein, F. L. (1982). A comparative analysis of two structural equation models: Lisrel and PLS applied to market data. In C. Fornell (Ed.), *A second generation of multivariate analysis*, Vol. 1 (pp. 289–324). New York: Praeger.

Fornell, C., & Cha, J. (1994). Partial least squares. In R. Bagozzi (Ed.), *Advanced methods of marketing* (pp. 52–78). Cambridge: Blackwell.

Fornell, C., & Larcker, D. F. (1981). Evaluating structural equation models with unobservable variables and measurement error. *Journal of Marketing Research, 18*, 39–50.

Gefen, D., Straub, D., & Boudreau, M. (2000). Structural equation modeling and regression: guidelines for research practice. *Communications of the Association for Information Systems, 7*(7), 1–78.

Geisser, S. (1975). The predictive sample reuse method with applications. *Journal of the American Statistical Association, 70*(350), 320–328.

Goldberger, A. S. (1971). Econometrics and psychometrics: a survey of communalities. *Psychometrika, 36*, 83–107.

Haenlein, M., & Kaplan, A. M. (2004). A beginners guide to partial least squares analysis. *Understanding Statistics, 3*(4), 283–297.

Hair, J. F., Jr., Anderson, R. E., Tatham, R. L., & Black, W. C. (1998). *Multivariate data analysis with readings*, 5th Edn. Englewood Cliffs, NJ: Prentice Hall.

Hanushek, E. A., & Jackson, J. E. (1977). Statistical methods for social scientists. Orlando, FL: Academic.

Hartline, M. D., & Ferrell, O. C. (1996). The management of customer-contact service employees: an empirical investigation. *Journal of Marketing, 60*, 52–70.

Hui, B. S. (1982). On building partial least squares models with interdependent inner relations. In K. G. Jöreskog, & H. Wold (Eds.), *Systems under indirect observations: causality, structure, prediction*, Vol. 2 (pp. 249–271). Amsterdam: North-Holland.

Hui, B. S. (1978). The partial least squares approach to path models of indirectly observed variables with multiple indicators. Unpublished Ph.D. dissertation, Graduate Group in Managerial Science and Applied Economics, University of Pennsylvania.

Hulland, J. (1999). Use of partial least squares (PLS) in strategic management research: A review of four recent studies. *Strategic Management Journal, 20*(2), 195–204.

Jarvis, C. B., MacKenzie, S. B., & Podsakoff, P. M. (2003). Critical review of construct indicators and measurement model misspecification in marketing research. *Journal of Consumer Research, 30*, 199–218.

Jöreskog, K. G., & Wold, H. (1982). *Systems under indirect observation - causality structure prediction*. Amsterdam: North Holland.

Lohmöller, J. B. (1989). *Latent variables path modeling with partial least squares*. Heidelberg: Physica.

MacCallum, R. C., & Browne, M. W. (1993). The use of causal indicators in covariance structure models: some practical issues. *Psychological Bulletin, 114*, 533–541.

MacKenzie, S. B., Podsakoff, P. M., & Ahearne, M. (1998). Some possible antecedents and consequences of in-role and extra-role salesperson performance. *Journal of Marketing, 62*, 87–98.

MacKenzie, S. B., Podsakoff, P. M., & Jarvis, C. B. (2005). The problem of measurement model misspecification in behavioural and organizational research and some recommended solutions. *Journal of Applied Psychology, 90*, 710–730.

Maloles, C. M. (1997). The Determinants of Customer Retention. Doctoral Thesis, The City of University of New York.

Mathieson, K., Peacock, E., & Chin, W. W. (2001). Extending the technology acceptance model: The influence of perceived user resources. *Data Base for Advances in Information Systems, 32*(3), 86–112.

McDonald, R. P. (1996). Path analysis with composite variables. *Multivariate Behavioral Research, 31*(2), 239–270.

Nunnally, J. (1978). *Psychometric theory*, 2nd Edn. New York: McGraw-Hill.

Parasuraman, A., Berry, L. L., & Zeithaml, V. A. (1991). Understanding customer expectations of service. *Sloan Management Review, 32*(3), 39–48.

Parasuraman, A., Zeithaml, V. A., & Berry, L. L. (1985). A conceptual model of service quality and its implications for future research. *Journal of Marketing, 49*, 41–50.

Parasuraman, A., Zeithaml, V. A., & Berry, L. L. (1988). SERVQUAL: a multiple item scale for measuring consumer perceptions of service quality. *Journal of Retailing, 64*(1), 12–40.

Reinartz, W., Haenlein, M., & Henseler, J. (2009). An empirical comparison of the efficacy of covariance-based and variance-based (SEM). *International Journal of Research in Marketing, 26*(4) 332–344.

Rizzo, J. R., House, R. J., & Lirtzman, S. I. (1970). Role conflict and ambiguity in complex organizations. *Administrative Science Quarterly, 15*, 150–163.

Sellin, N. (1989). Partial least square modeling in research on educational achievement. In W. Bos, & R. H. Lehmann (Eds.), *Reflections on educational achievement*, Papers in Honour of T. Neville Postlethwaite (pp. 256–267). New York: Waxmann Munster.

Sharma, S. (1996). Applied multivariate techniques. New York: Wiley.

Singh, J. (1998). Striking a balance in boundary-spanning positions: an investigation of some unconventional influences of role stressors and job characteristics on job outcomes of salesperson. *Journal of Marketing, 62*, 69–86.

Spreng, R. A., & Mackoy, R. D. (1996). An empirical examination of a model of perceived service quality and satisfaction. *Journal of Retailing, 72*(2), 201–214.

Stone, M. (1974). Cross-validatory choice and assessment of statistical predictions. *Journal of the Royal Statistical Society, 36*(2), 111–147.

Straub, D., Boudreau, M. C., & Gefen, D. (2004). Validation guidelines for IS positivist research. *Communications of the Association for Information Systems, 14*, 380–426.

Sureshchandar G. S., Rajendran C., & Anantharaman R. N (2002). Determinants of customer-perceived service quality: a confirmatory factor analysis approach. *Journal of Services Marketing, 16*(1), 9–34.

Tam, J. L. M. (20 04). Customer satisfaction, service quality and perceived value. *Journal of Marketing Management, 20*(7/8), 21–35.

Taylor, S. T., & Baker, T. L. (1994). An assessment of the relationship between service quality and customer satisfaction in the formation of consumer purchase intentions. *Journal of Retailing, 70*(2), 163–178.

Teas, R. K. (1993). Expectations, performance evaluation, and consumers perceptions of quality. *Journal of Marketing, 57*, 18–34.

Tenenhaus, M., Vinzi, V. E., Chatelin, Y.-M., & Lauro, C. (2005). PLS path modeling. Computational Statistics & Data Analysis, *48*(1), 159-205.

Vinzi, V. E., Tenenhaus, M., Lauro, C., & Amato, S. (2006). *PLS path modeling: recent developments and open issues for model assessment and improvement*, Chap. 2.

Vinzi, V. E., Trinchera, L., & Amato, S. (2010). PLS path modeling: from foundations to recent developments and open issues for model assessment and improvement. In: V. Esposito Vinzi, W. Chin, J. Hensler, and H. Wold (Eds.,) *Handbook of partial least squares*, (pp. 47–82). Heidelberg: Springer.

Werts, C. E., Linn, R. L., & Jöreskog, K. G. (1974). Interclass reliability estimates: testing structural assumptions. *Educational and Psychological Measurement, 34*, 25–33.

Wold, H. (1979). *Model construction and evaluation when theoretical knowledge is scarce: an example of the use of partial least squares*. Cahiers du Dpartement Dconomtrie. Genve: Facult des Sciences conomiques et Sociales, Universit de Genve.

Wold, H. (1980). Soft modeling: intermediate between traditional model building and data analysis. *Mathematical Statistics, 6*, 333–346.

Wold, H. (1985). Systems analysis by partial least squares. In P. Nijkamp, H. Leitner, & N. Wrigley (Eds.), *Measuring the unmeasurable* (pp. 221–251). Dordrecht: Martinus Nijhoff.

Yi, Y., & La, S. (2004). What influences the relationship between customer satisfaction and repurchase intention? Investigating the effects of adjusted expectations and customer loyalty. *Psychology and Marketing, 21*, 351–373.

Zeithaml, V. A., Berry, L. L., & Parasuraman, A. (1996). The behavioral consequences of service quality. *Journal of Marketing, 60*, 31–46.

Chapter 20
A PLS Model to Study Brand Preference: An Application to the Mobile Phone Market

Paulo Alexandre O. Duarte and Mário Lino B. Raposo

Abstract Brands play an important role in consumers' daily life and can represent a big asset for companies owning them. Owing to the very close relationship between brands and consumers, and the specific nature of branded products as an element of consumer life style, the branded goods industry needs to extend its knowledge of the process of brand preference formation in order to enhance brand equity.

This chapter show how Partial Least Squares (PLS) modeling can be used to successfully test complex models where other approaches would fail due to the high number of relationships, constructs and indicators. Here, PLS modeling is applied to brand preference formation regarding mobile phones.

With a wider set of explanatory factors than prior studies, this one explores the factors that contribute to the formation of brand preference using a PLS model to understand the relationship between those and consumer preference for mobile phone brands.

Despite the exploratory nature of the study, the results reveal that brand identity, personality and image, together with self-image congruence have the highest impact on brand preference. Some other factors linked to the consumer and the situation also affect preference, but to a lesser degree.

20.1 Introduction

Owing to their massive presence in today's market and the huge diversity of products, brands play an important role in the consumer decision process. Brands are used to differentiate sellers' offers, and function as a sign of guarantee for consumers.

Brands are composed of many different elements, both tangible and intangible (Gardner and Levy 1955; Levy 1959a, b; Broadbent and Cooper 1987; Keller 2003). They exist in customers' minds as a sum of those elements and deliver a variety

P.A.O. Duarte and M.L.B. Raposo
Departamento de Gestão e Economia, Universidade da Beira Interior, Estrada do Sineiro, 6200-209 Covilhã, Portugal
e-mail: pduarte@ubi.pt, mraposo@ubi.pt

V. Esposito Vinzi et al. (eds.), *Handbook of Partial Least Squares*, Springer Handbooks of Computational Statistics, DOI 10.1007/978-3-540-32827-8_21,

of benefits, which can be classified as satisfying a buyer's rational and emotional needs (De Chernatony and McDonald 2001). The importance of brand preference is emphasized by Zajonc (1980) when he points out that the affective component can continue to exist, even after the cognitive basis has been erased from a consumer's memory.

However, as pointed by Creyer and Ross (1997) and Muthukrishnan and Kardes (2001), much remains unknown about the formation of preference, especially how and according to which factors consumers develop preference for one brand. The majority of research on brand preference is partial by nature, as it is mostly confined to measuring the impact of one single factor on brand preference, which is an obvious limitation (Stafford 1966; Hawkins 1970; Ross 1971; Monroe 1976; Dunn and Murphy 1986; Bushman 1993; Schmitt and Shultz 1995; Sengupta and Fitzsimons 2000; Jamal and Goode 2001; Niedrich and Swain 2003).

This research has three main objectives. First, it attempts to identify and compile the various factors reported in the literature that can influence brand preference. Second, it develops a model to study brand preference formation to improve our understanding of the interaction of the elements involved in the emergence of preference and which potentially affect the decision choice process. Third, it show that PLS can be successfully used to test complex models with a large number of constructs and indicators.

20.2 Theoretical Background

20.2.1 Brand Preference Formation

Over the years several attempts have been made to explain the development of brand preference, some of them have been labeled as models of consumer behavior. The Howard and Sheth (1969) model is one example of those models, as it seeks to explain brand choice behavior.

The literature review of brand preference formation reveals two main theoretical perspectives, labeled as "archeological" and "architectural" (Payne et al. 1999). The first assumes that there is a well-defined preference and therefore the task of the researcher is just to uncover or reveal this. On the other hand, the second believes that preference is formed when the consumer needs to choose, and is produced using stable values associated with the object being evaluated, and a situation-specific component that represents the joint effect of the task and context contingencies. This second perspective believes that the situation-specific component is a major determinant of judgment responses (Payne et al. 1992, 1999).

However, noting that consumers do not always behave in a consistent way when choosing a brand, a probabilistic perspective of preference emerged (see, e.g., Bass 1974; Srinivasan 1975; Bass and Pilon 1980; Blin and Dodson 1980; Sharma 1981; DeSarbo and Rao 1984, 1986; Currim and Sarin 1984; Carroll et al. 1990; Russel and Kamakura 1997).

Aware of the complexity of preference, Nowlis and Simonson (1997) state that there is no single path to brand preference formation. Trying to integrate the various approaches, Shocker and Srinivasan (1979) stressed that it makes sense to treat choice as a stochastic process and relate it to a determinist measure of preference. We believe, just like various other authors (Lehmann 1972; Bettman and Jones 1972), that the two perspectives are complementary rather than substitutes.

In spite of the discussion, some general stages and elements that appear in every model can be identified. The process seems to start with stimuli which are selected, absorbed and codified by the consumer, combined with information retrieved from their memory. This package of data is then processed, a representation is formed and brand preference is developed and stored in a consumer's memory. Regardless of how we look at the process, it is essential to know what those stimuli are and how they interact with other factors to form brand preference.

To identify the major influences on brand preference an exhaustive review of the literature between 1942 and 2005 was conducted to gather information about current knowledge and, to provide the framework for the brand preference formation model proposed in this chapter.

For the literature review, the factors were divided into three groups (consumer, product/brand and situation) following Woodside and Trappey's (1992) and Belk's (1974, 1975a, b) indications that consumer behavior is conditioned by the characteristics of the consumer himself, by the situation, and the object. We assume that is also true for preference; consequently, the determinants of preference identified by the literature review were classified into one of the three groups previously mentioned.

20.2.2 Consumer-Oriented Factors

Consumer characteristics are the first main group of factors of interest for this study. This group should reflect the most important characteristics and dominant influences present in individuals and are expected to be responsible for guiding their brand preference.

For example, Schmitt and Shultz (1995) suggest the existence of an ideal consumer for every brand, based on their characteristics. Relying on this assumption, we expect to find a set of characteristics common to consumers who prefer one specific brand.

Following this same thought, several researchers have tried to identify meaningful relationships between demographic characteristics and consumer behavior (Bass and Talarzyk 1972; Fennell et al. 2003; Jamal and Goode 2001). Practically all those studies only reveal weak effects of demographic characteristics on consumer behavior (Rossi et al. 1996; Bucklin et al. 1995). Such a case is the influence of consumers' age and gender on brand perception (Elliot 1994; Sethuraman and Cole 1999). Likewise, Lin (2002) shows that consumers' values change with age, gender, education, and social class. Some other factors correlated with preference, like satisfaction or need for cognition also seems to be linked to the demographic profile of consumers (Bryant and Cha 1996; Mittal and Kamakura 2001; Jamal and Goode 2001; Lin 2002).

We feel that demographic variables are important for this, and despite the discussion about their importance, they should be considered when modeling preference.

But it is not only the demographic characteristics that have caught the attention of researchers. Several authors have been looking for a way to predict preference and behavior from personality. Unfortunately, the conclusions of those studies are conflicting, and lack consensus about the true power of personality to predict consumer behavior (Evans 1959; Westfall 1962; Birdwell 1968; Kaponin 1960; Shank and Langmeyer 1994; Alpert 1972; Kassarjian 1971; Horton 1974; Kassarjian 1979). In any case, in the face of the evidence of the existence of an association between a consumer's personality self-concept and brand values, namely brand identity and personality, we cannot exclude the existence of a possible influence (Graeff 1996; Fournier 1998; Aaker 1997, 1999).

Other studies explore the relationship between involvement and preference, showing that involvement plays an important role in defining how consumers receive and process information (Bolfing 1988; Zhang and Markman 2001; Chernev 2001; Muthukrishnan and Kardes 2001). For instance, high levels of involvement lead to different levels of the need for cognition and motivation to search for information (Witt and Bruce 1972; Celsi and Olson 1988; Maheswaran and Mackie 1992), and the way it is used and interpreted (Bettman et al. 1975; Jain and Maheswaran 2000).

The predisposition to process information also depends on the need for cognition. This concept by Cacioppo and Petty (1982) refers to the individual's tendency to engage in and enjoy effortful cognitive endeavors. Research on the need for cognition suggests that this characteristic is predictive of the way in which people deal with tasks and social information and subsequently influences the way individuals develop their preference.

A final element is the memory and the capacity to store and recall information. The way information is stored and retrieved from memory also seems to play some part in generating preference (Costley and Brucks 1992; Haley and Case 1979; Hutchinson et al. 1994). Brands that are easily remembered seem to be preferred over brands that are difficult to memorize.

To summarize, we think that is very unlikely, if not impossible, that a single preference model based on the characteristics of consumers can fit all consumers and products, in order to be universally applicable. Instead, we feel that the appropriateness of a preference model is likely to vary across individuals and products. In our opinion, despite all the difficulties and discussions, the identification of the relevant influences of consumer-related factors on preference, either directly or through other variables, can be useful and, therefore, those effects should not be ignored.

20.2.3 Brand-Related Factors

The second group specifically addresses the factors related to the object, i.e. the product and brand attributes. As previously mentioned, products and brands have a special and personal value for consumers that exceeds the functional value and is capable of expressing social identities and symbolizing class and status (Bristow and Asquith 1999).

Prior research suggests that product and brand-related factors, such as brand name (Zinkhan and Martin 1987; Klink 2001), can affect how consumers look at brands and the inferences made about quality (Sappington and Wernerfelt 1985). Perceived quality impacts preference (Morton 1994; Dickerson 1982; Hugstad and Durr 1986; Stephen et al. 1985; Wall and Heslop 1989; Olsen 2002; Hellier et al. 2003) and is also influenced by price (Peterson 1970; Zeithaml 1988; Lichtenstein and Burton 1989; Lichtenstein et al. 1993; Chapman and Wahlers 1999), which influences preference too (Monroe 1976; Rao and Monroe 1988; Venkataraman 1981), and by country of origin (Han and Terpstra 1988; Khachaturian and Morganosky 1990; Powers and Nooh 1999; Tse and Gorn 1993; Thakor and Katsanis 1997) which additionally seems to impact perceived value (Ahmed and D'Astous 1993) and preference (Papadopoulos et al. 1990; Peris et al. 1993; Kim 1995).

Another important factor is brand identity, personality, and image. Our theoretical research reveals that this variable seems to interact with self-image congruence and the preference showed by consumers (Sirgy 1982; Phau and Lau 2001; Jamal and Goode 2001).

All those factors, together with product attributes (Urban and Hauser 1993), perceived value (Hellier et al. 2003), package (Keller 2003), and familiarity (Meyers-Levy 1989), appear in the literature on preference.

20.2.4 Situational Factors

This group of factors was the most challenging for three reasons. The first was the difficulty experienced with classifying one factor as situational. Second, the extremely high number of potential situational variables and, finally, the limited support found in the literature. However, Belk (1974) stresses that, situational factors are essential to predict consumer behavior, while Payne et al. (1999) believe that this component of situational factors has a large impact on preference.

To classify one factor as situational, we use Belk's (1974, 1975a, b) definition that situational factors are those present at a precise moment and place, which do not result from the consumer or object of choice, but which can, beyond any doubt, affect consumer behavior.

Owing to the large number of situational factors, and to the difficulty in classifying some factors as situational, as was previously mentioned, only a few were used in this study, specifically those that appeared the most important in previous studies.

As a result of those limitations, only five situational factors (communication, social environment, risk perception, pioneering advantage, and product visibility), which had proved to be related to preference, were used. For example, several authors report that a higher level of communication (namely advertising) induces high levels of preference (Paivio 1971; Shepard 1978; Mitchell and Olson 1981; Woodside and Wilson 1985; Carrol et al. 1990).

The impact of the social environment is supported by the works of Sheth (1968), Hawkins and Coney (1974) and Keillor et al. (1996). Product visibility is somehow related to this last factor. Graeff (1997), Dickson (1982) and Becherer et al. (1982) reported an association between it, the consumption context and the preference for one brand.

Another factor that emerged from the literature review was risk perception. The relationship between risk perception and preference appears in the studies of Peter and Ryan (1976), Pras and Summers (1978), Campbell and Goodstein (2001) and Hellier et al. (2003).

Finally, the pioneering advantage factor is based on the work by Carpenter and Nakamoto (1989), which suggests that the first brand in the market tends to build a standard for preference which influences the following brands. These authors' basic idea was confirmed by recent studies by Zhang and Markman (1998), Alpert et al. (2001), Rettie et al. (2002), Niedrich and Swain (2003), and Desai and Ratneshwar (2003).

Additionally, a construct which reflects the information search, acquisition and processing was included in the model due to the various references to it in the literature.

20.3 Theoretical Model

The theoretical model was developed by searching in the available literature for variables reportedly related to brand preference. The review of the literature on brand preference between 1942 and 2004 reveals a final set of 22 principal factors (constructs), and a total of 54 relationships that may be significant for the development of brand preference as modeled. The proposed model, with 23 constructs and 106 indicators, incorporates many of the factors and relations that the review indicates as directly and individually contributing to explain brand preference. Table 20.1 summarizes the most relevant studies supporting the selection of variables and relations used in the formulation of the model presented in Fig. 20.1. The inclusion of a construct or relation in the model was based on its relevance for the study, the degree of differentiation, and its effective operationalization. Nevertheless, due to the complexity of the process of brand preference formation, it is assumed that not all the factors and relations were included, which could be seen as a limitation.

Table 20.1 Studies supporting the variables and relations

Path			Studies
Demographic profile	→	Self-concept	Lin (2002)
Demographic profile	→	Satisfaction	Bryant and Cha (1996); Mittal and Kamakura (2001); Olsen (2002)
Demographic profile	→	Need for cognition	Elliot (1994)
Demographic profile	→	Communication	Ginter and Bass (1972)
Demographic profile	→	Preference	Jamal and Goode (2001); Sethuraman and Cole (1999); Bass and Talarzyk (1972)
Demographic profile	→	Information search	Mandrik (1996)
Self-concept	→	Preference	Landon (1974); Sirgy (1982, 1985); Hughes (1976)
Self-concept	→	Self-image congruence	Gardner and Levy (1955); Levy (1959); Sirgy (1982, 1985)
Satisfaction	→	Preference	Taylor and Baker (1994); Hellier et al. (2003); Jamal and Goode (2001)
Need for cognition	→	Self-concept	Malhotra (1988); Sadowski and Cogburn (1997)
Need for cognition	→	Social environment	Cacioppo et al. (1996)
Need for cognition	→	Preference	Garbarino and Edell (1997)
Need for cognition	→	Information search	Mandrik (1996); Bloch and Richins (1983); Zaichkowsky (1985); Celsi and Olson (1988)
Need for cognition	→	Self-image congruence	Sadowski and Cogburn (1997); McCrea and John (1992)
Memory	→	Preference	Hutchinson et al. (1994); Nedungadi (1990); Ettenson (1993); Fisher et al. (1999)
Involvement	→	Need for cognition	Antil (1984); Celsi and Olson (1988)
Involvement	→	Preference	Zhang and Markman (2001)
Involvement	→	Information search	Witt and Bruce (1972); Celsi and Olson (1988); Maheswaran and Mackie (1992); Bolfing (1988); Jain and Maheswaran (2000)
Communication	→	Need for cognition	Zhang and Buda (1999)
Communication	→	Memory	Rheingold (1985); Fisher et al. (1999); Macklin (1996); Alreck and Settle (1999)
Communication	→	Preference	Paivio (1971); Shepard (1978); Mitchell and Olson (1981); Woodside and Wilson (1985); Carrol et al. (1990); D'Souza and Rao (1995); Alreck and Settle (1999)
Communication	→	Familiarity	Bogart and Lehman (1973); Cobb-Walgren et al. (1995); Alreck and Settle (1999); Lin et al. (2000)
Communication	→	Information search	Harris and Monaco (1978); Gruenfeld and Wyer (1992); Creyer and Ross (1997); Garbarino and Edell (1997)
Social environment	→	Preference	Sheth (1968); Stafford (1966); Hawkins and Coney (1974); Schmitt and Shultz (1995); Keillor et al. (1996); Yang et al. (2002); Ji (2002)

(*continued*)

Table 20.1 (continued)

Path		Studies
Risk	→ Preference	Peter and Ryan (1976); Pras and Summers (1978); Campbell and Goodstein (2001); Hellier et al. (2003); Muthukrishnan and Kardes (2001)
Visibility	→ Preference	Belk (1975); Miller and Ginter (1979); Dickson (1982); Becherer et al. (1982); Graeff (1997)
Visibility	→ Information search	Mandrik (1996)
Familiarity	→ Memory	Meyer-Levy (1989a, b)
Familiarity	→ Preference	Monroe (1976); Moreland and Zajonc (1982); Rheingold (1985)
Familiarity	→ Information search	Mandrik (1996)
Brand indent/ pers/image	→ Preference	Birdwell (1968); Ross (1971); Sirgy (1982, 1985); Graeff (1997); Phau and Lau (2001)
Brand indent/ pers/image	→ Self-image congruence	Fournier (1998); Helman and De Chernatony (1999); Sheth, Newman and Gross (1991)
Brand name	→ Preference	Klink (2001); Bristow et al. (2002); Venkataraman (1981); Woodside and Wilson (1985)
Brand name	→ Quality	Zinkhan and Martin (1987); Zeithaml (1988); Zaichkowsky and Vipat (1993); Rao et al. (1999); Srinivasan and Till (2002); Sappington and Wernerfelt (1985); Jacoby et al. (1977); Rigaux Bricmont (1981); Zeithaml (1988); Dick et al. (1996)
Brand origin	→ Preference	Papadopoulos et al. (1990); Peris et al. (1993); Kim (1995); Thorelli et al. (1989)
Brand origin	Perceived value	Ahmed and D'Astous (1993)
Brand origin	→ Quality	Han and Terpstra (1988); Khachaturian and Morganosky (1990); Powers and Nooh (1999); Tse and Gorn (1993); Thakor and Katsanis (1997)
Perceived value	→ Preference	Hellier et al. (2003); Morton (1994)
Quality	→ Preference	Morton (1994); Dickerson (1982); Hugstad and Durr (1986); Stephen et al. (1985); Wall and Heslop (1989); Olsen (2002); Hellier et al. (2003)
Quality	→ Perceived value	Morton (1994); Agarwal and Teas (2001); Hellier et al. (2003); Snoj et al. (2004)
Quality	Information search	Mandrik (1996)
Price	→ Preference	Monroe (1976); Rao and Monroe (1988); Wheatley et al. (1977)
Price	→ Perceived value	Sivakumar (1996); Chapman and Wahlers (1999)
Price	→ Quality	Peterson (1970); Zeithaml (1988); Lichtenstein and Burton (1989); Lichtenstein et al. (1993); Chapman and Wahlers (1999)
Price	→ Information search	Mandrik (1996)
Product attributes	→ Preference	Urban and Hauser (1993); Fisher et al. (1999); Carpenter et al. (1994); Dhar et al. (1999); Chernev (2001); Zhang and Markman (2001)
Product attributes	→ Quality	Kirmani and Zeithaml (1993); Richardson et al. (1994); Dick et al. (1996)

(continued)

Table 20.1 Table 1 (continued)

Product attributes	→ Information search	Allison and Uhl (1964); Russo et al. (1998)
Package	→ Preference	Alsop (1984); Banks (1950); Krugman (1962) Keller (2003)
Package	→ Quality	Riezebos (2003); Alsop (1984); Rigaux Bricmont (1981)
Information search	→ Preference	Fisher et al. (1999)
Pioneering	→ Preference	Carpenter and Nakamoto (1989); Zhang and Markman (1998); Alpert et al. (2001); Rettie et al. (2002); Niedrich and Swain (2003); Desai and Ratneshwar (2003).
Self-image congruence	→ Satisfaction	Jamal and Goode (2001); Aaker (1997); Moutinho and Goode (1995)
Self-image congruence	→ Preference	Belk et al. (1982); Onkvisit and Shaw (1987); Belk (1988); Richins (1994a, b); Hong and Zinkhan (1995); Ericksen (1996); Aaker (1999); Jamal and Goode (2001); Sirgy (1982)

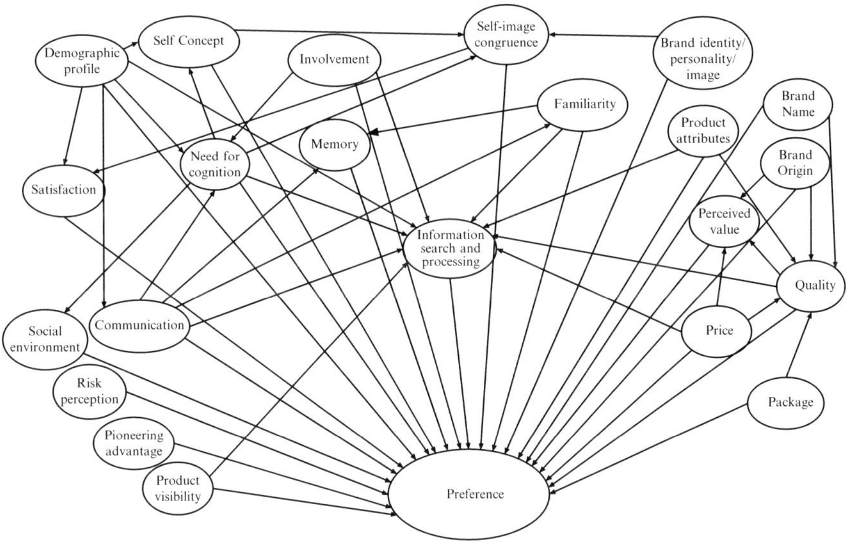

Fig. 20.1 Theoretical model of brand preference

20.4 Design and Methodology

To select the product class for the empirical research, a small questionnaire was conducted in a sample of 50 university students, using the brand dependence and brand disparity scales from Bristow et al.'s (2002) study. The data was analyzed, the results were interpreted, and mobile phones proved to be the best product class, of the ones tested, to study brand preference.

The empirical data was obtained from a sample of Portuguese students studying between the 9th grade of secondary school and the last year of university, all of whom study at state schools throughout the country. Those students were asked to state their preference regarding the brand of mobile phone to buy, and to evaluate the various factors identified in the literature reviewed, using the multi-item Likert-type scales, previously selected, adapted and pre-tested for the current context.

A balance was sought between covering the maximum and most important indicators and the extent of the questionnaire. Where several measures were available, preference was given to those judged most easily read, and those with strong predictive power. Finally, a set of 106 indicators was selected from existing questionnaires and the handbook of marketing scales. The questions were adapted for readability prior to pre-testing.

Table 20.2 presents a summary of the studies reviewed to identify the indicators used to measure the constructs of the model (the full list of measures is available from the authors).

Table 20.2 Constructs, number of indicators and studies

Construct	Studies
Demographic profile	Sethuraman and Cole (1999); Jamal and Goode (2001)
Self-concept	Malhotra (1981); Sirgy et al. (1997); Lau and Lee (1999)
Involvement	Traylor (1981); Zaichkowsky (1985); Zinkhan and Martin (1987); Rodgers and Schneider (1993); Zaichkowsky (1994); D'Astous and Gargouri (2001)
Need for cognition	Cacioppo et al. (1984)
Memory	Lange and Dahlén (2003)
Brand name	Mandrik (1996); Kohli and LaBahn (1997)
Brand identity, personality and image	Lewis and Stubbs (1999); Del Río et al. (2001)
Price and perceived value	Petroshius and Monroe (1987); Schmitt and Shultz (1995); Agarwal and Teas (2001); D'Astous and Gargouri (2001); Del Río et al. (2001); Quester and Lim (2003)
Quality	Dodds et al. (1991); Schmitt and Shultz (1995); Burton et al. (1998); Chapman and Wahlers (1999); Agarwal and Teas (2001); Ballester and Alemán (2002)
Familiarity	Low and Lamb (2000); D'Astous and Gargouri (2001); Mackay (2001); Lange and Dahlén (2003)
Satisfaction	Lau and Lee (1999); Jamal and Goode (2001)
Self image congruence	Lau and Lee (1999)
Social environment	Lau and Lee (1999); Del Río et al. (2001)
Risk	Mitchell (1992); Agarwal and Teas (2001)
Information search and processing	Srinivasan and Ratchford (1991)
Preference	Moschis (1981); Duncan and Nelson (1985); Stayman and Aaker (1988); Petroshius and Crocker (1989); Costley and Brucks (1992); Sirgy et al. (1997); Jamal and Goode (2001); Mackay (2001); Quester and Lim (2003); Hellier et al. (2003).

Table 20.3 Sample characterization

Education level	Age				Total
	<15	15–18	19–25	≥26	
9th grade	39	97	1	0	137
10th grade	2	63	0	0	65
11th grade	1	62	3	0	66
12th grade	0	38	12	0	50
University students	1	13	132	31	177
Bachelor's degree	0	1	19	4	24
University degree	0	1	1	5	7
Total	43	275	168	40	526

Note: Gender is missing for two subjects

The following indicators were used to evaluate: demographic profile, satisfaction, self-concept, need for cognition (Cacioppo and Petty 1982), involvement, memory, self-image congruence, communication, social environment, risk perception, pioneering advantage (Carpenter and Nakamoto 1989), product visibility, information search, familiarity, brand identity/personality and image, product attributes, brand name, brand origin, price, quality, perceived value, and package. Using the guidelines proposed by Jarvis et al. (2003), two constructs (demographic profile and self-concept) were modeled as formative and the remaining as reflective.

The sample was stratified according to the number of students in each grade. A total of 700 questionnaires were mailed and 542 were received. Of those, 14 were eliminated, for various reasons, resulting in a valid sample of 528 subjects. Table 20.3 presents the participants' distribution by education level.

To evaluate the strength of brand in the consumer mind, a top-of-mind analysis (TOMA) was made. A TOMA allows the investigator to explore people's perceptions and immediate associations with a particular issue. It works by asking: what is the first brand that comes to mind when the product class is mentioned? The results of the TOMA can somehow be regarded as an indicator of brand preference. It is conceivable that consumers will automatically think of their preferred brand when a given product category is mentioned.

The TOMA performed in this study reveals that Nokia is the winner by far, followed by Siemens, as can be seen in Table 20.4.

When looking at the subjects' first brand of mobile phone and their actual brand an interesting point emerged. Alcatel was the first brand for 21.3% of the respondents, but is the actual brand for only 2.8% (see Table 20.5).

Inversely, the preference for Nokia and Siemens seems to increase as they have more actual users who had first bought another brand. These findings can be an especially interesting starting point for Alcatel to try to find why they lose so much market share and cannot retain consumer preference over time.

To assess the predictive power of our theoretical model, a structural equation modeling (SEM), specifically Partial Least Squares (PLS) (using PLS-Graph Version 3.0 by Wynne Chin), was used to evaluate the relationships between the constructs, and to estimate both the measurement and structural parameters in

Table 20.4 Top-of-mind analysis

Brand	Order of response			Total
	1st	2nd	3rd	
Alcatel	11	49	95	155
Mitsubishi	1		1	2
Motorola	14	52	97	163
Panasonic	1	2	3	6
Philips	1	2	3	6
Nokia	409	83	20	512
Samsung	11	54	91	156
Sharp		7	9	16
Siemens	61	207	115	383
Sony Ericsson	15	58	77	150
Sendo	1	2	7	10
Telit		1		1
Maxon		1		1
Sagem	1	3	4	8
Trium		4	1	5
Audiovox	2			2
Total	528	525	523	

Note: Some respondents didn't mention a second or third brand name

Table 20.5 Comparison between first and actual brand

	Actual	%	First	%
Alcatel	15	2.84	108	20.7
Mitsubishi	4	0.76	4	0.8
Motorola	24	4.55	57	10.9
Panasonic	1	0.19	6	1.1
Philips	2	0.38	15	2.9
Nokia	312	59.09	147	28.1
Samsung	21	3.98	14	2.7
Sharp	3	0.57	0	0.0
Siemens	111	21.02	87	16.6
Sony Ericsson	23	4.36	40	7.6
Sendo	2	0.38	1	0.2
Maxon	1	0.19		
Sagem	1	0.19	11	2.1
Trium	7	1.33	12	2.3
Bosh			8	1.5
Aeg	1	0.19	9	1.7
Audiovox			3	0.6
Nec			1	0.2
Total	528		523	

the proposed structural equation model. The choice of PLS is due to the nature of the study and the size and complexity of the model. Furthermore, the model has two constructs measured with formative indicators and PLS is appropriate for the analyses of measurement models with both formative and reflective items (Diamantopoulos and Winklhofer 2001).

20.5 PLS Analyses

The Partial Least Squares (PLS) was used to evaluate the proposed theoretical model. PLS is a structural equation modeling (SEM) technique that can simultaneously test the measurement model (relationships between indicators or manifest variables and their corresponding constructs or latent variables) and the structural model (relationships between constructs). Additionally, PLS has the capacity to deal with very complex models with a high number of constructs, indicators, and relationships (Garthwaite 1994; Barclay et al. 1995), what makes it ideal to our study.

The PLS algorithm generates loadings between reflective constructs and their indicators and weights between formative constructs and their indicators. It also produces standardized regression coefficients between constructs, and coefficients of multiple determination (R^2) for all endogenous constructs in the model.

In PLS, the relationship between a construct and its indicators can be modeled as either formative or reflective, which is an advantage compared to the covariance-based methods. In addition, PLS allows working with small sample sizes and makes less strict assumptions about the distribution of the data (Chin and Newsted 1999).

However, rather than being viewed as competitive models, PLS and covariance-based SEM techniques should be viewed as complementary. They differ regarding the objective (prediction for PLS and theory testing for covariance-based SEM) and the approach (variance for PLS and covariance for covariance-based SEM) (Chin and Newsted 1999).

According to Jöreskog and Wold (1982), "ML is theory-oriented, and emphasizes the transition from exploratory to confirmatory analysis. PLS is primarily intended for causal-predictive analysis in situations of high complexity but low theoretical information."

Certain conditions are required to evaluate the appropriateness of PLS compared to its covariance-based counterpart, which can be classified into four groups (Falk and Miller 1992): theoretical conditions, measurement conditions, distributional conditions, and practical conditions. According to these authors, PLS could be used when there is no strong existing theory, and hypotheses are derived from a macro-level theory in which all relevant variables are not known, relationships between constructs are conjectural, some of the manifest variables are categorical and they may have some degree of unreliability, distribution of the data may not be normal, sample size is very large or small, and a large number of manifest and latent variables are modeled. After a systematic review of all these conditions, it was decided that PLS was the most appropriate technique for this study.

20.5.1 Measurement Model

In PLS, the relationship between a construct and its indicators can be modeled as either formative or reflective. Formative indicators are also known as cause or induced indicators, while reflective indicators are also known as effect indicators. Our study uses both kinds of indicators.

In a PLS analysis, reflective and formative indicators must be treated differently. For constructs with reflective measures (i.e., latent constructs), it's necessary to examine the loadings, which can be interpreted in the same manner as the loadings in a principal component analysis. For constructs using formative measures (i.e., emergent constructs), it's necessary to look at the weights, as they provide information about the composition and relative importance of each indicator in the creation/formation of the construct. Since the construct is viewed as an effect rather than a cause of the item responses, no interdependencies can be assumed among the formative items. As a result, traditional reliability and validity assessments have been argued as inappropriate and illogical for this type of factor, referring to its dimensions (Bollen 1989). Their interpretation is similar to the canonical correlation analysis (Sambamurthy and Chin 1994).

The measurement model for constructs with reflective measures is assessed by looking at: individual item reliability, internal consistency and discriminant validity. The individual item reliability is evaluated by examining the loadings of the measures with the construct they intend to measure.

Using the rule of thumbs of accepting items with loadings of 0.707 or more, we notice that 18 indicators of the 106 did not reach the level of acceptable reliability. However, as pointed by Chin (1998) and Barclay et al. (1995), loadings of at least 0.5 might be acceptable if other questions measuring the same construct had high reliability scores. Falk and Miller (1992) propose as a rule of thumb retaining manifest variables with loadings that exceed 0.55, i.e. 30% of the variance of the manifest variable is related to the component. Upon examination of the cross-loadings (available from the authors) of our model six indicators were eliminated as they presented loadings lower than 0.5 and some presented higher loadings in other constructs than in the one they were intended to measure. In the whole model, only two indicators present loadings between 0.5 and 0.55 (COGN1, COGN3), so we decide to keep them.

The internal consistency was examined using the composite reliability index by Fornell and Larcker (1981). In our model the composite reliability index for all constructs exceed the minimum acceptable value of 0.7 (Hair et al. 1998), with need for cognition presenting the lowest (0.736) and package the maximum (0.938).

The next step was evaluating discriminant validity. Discriminant validity indicates the extent to which a given construct is different from other latent constructs. As a means of evaluating discriminant validity, Fornell and Larcker (1981) suggest the use of the Average Variance Extracted (AVE).

A score of 0.5 for the AVE indicates an acceptable level. (Fornell and Larcker 1981). Table 20.6 shows that the average variances extracted by our measures range from 0.536 to 0.791 above the acceptable value, except for the need for cognition construct which has a value of 0.361.

This value may be an effect of tailoring the scale. However, looking at the composite reliability index, the discriminant validity of the constructs (Table 20.7), and the cross-loading, we decide to keep the construct in the model, as we believed that it actually measures the respondents' degree of need for cognition.

Table 20.7 compares the square root of the AVE (diagonal values) with the correlations among the reflective constructs. All constructs were more strongly correlated with their own measures than with any other of the constructs, suggesting good convergent and discriminant validity.

For adequate discriminant validity, this measure should be greater than the variance shared between the construct and other constructs in the model. This, according to Chin (1998), can also be accomplished by examining the loadings and cross-loadings matrix. In our model the assessment of discriminant validity does not reveal any problem, as all indicators showed higher loadings with their respective construct than with any other reflective construct.

As formative indicators are not expected to correlate with one another and therefore traditional measures of validity are not appropriate, Chin (1998) suggests the evaluation of the Variance Inflation Factor and condition index to assess multicollinearity, and the significance of the weights (Table 20.8).

Using four conservative criteria by Olmo and Jamilena (2000), we see that the measures of demographic profile and self-concept components present VIF values lower than the limit specified, indicating the absence of multicollinearity.

The condition index confirms the absence of multicollinearity, as its value for every dimension never exceeds 30.

For formative items, the magnitude and significance of the weight indicate the importance of the contribution of the associated latent variable. The education level is by far the most important variable in forming the demographic profile. For the self-concept construct, the level of formality (PERS8) seems to be the most important variable.

The significance of the weight was assessed using the bootstrap procedure. The results of 500 resamples indicate that several indicators were not significant even at the 0.1 level, but given the exploratory nature of the study and following Chin's (1998) recommendation, those items were retained in the model to assess the strength of the demographic profile.

Table 20.6 Weights, loadings, composite reliability and average variance extracted

Constructs and indicators	Type	Weight	Loading	Composite reliability ρ_c	Average variance extracted AVE
Demographic profile	F			n.a.	n.a.
Age		0.219	0.703		
Education		0.761	0.836		
Gender		0.262	0.289		
Family_DIM		−0.281	−0.394		
Marit_status		−0.389	−0.113		
Resid		0.170	−0.012		
Self-concept	F			n.a.	n.a.
PERS1		0.143	0.101		
PERS2		0.206	0.282		
PERS3		0.094	0.207		
PERS4		−0.223	−0.319		
PERS5		0.353	0.166		
PERS6		0.170	0.168		
PERS7		0.205	0.277		
PERS8		−0.800	−0.737		
PERS9		0.288	0.360		
Satisfaction	R			0.900	0.693
SATGLOB		0.220	0.707		
SAT1		0.342	0.858		
SAT2		0.303	0.875		
SAT3		0.325	0.879		
Need for cognition	R			0.736	0.361
COGN2		0.369	0.601		
COGN4		0.448	0.710		
COGN5		0.307	0.625		
COGN1		0.300	0.523		
COGN3		0.213	0.525		
Memory	R			0.780	0.546
MEM1		0.439	0.750		
MEM2		0.542	0.840		
MEM3		0.354	0.609		
Involvement	R			0.874	0.538
ENV1		0.225	0.768		
ENV2		0.203	0.768		
ENV3		0.284	0.702		
ENV4		0.210	0.593		
ENV5		0.197	0.707		
ENV6		0.249	0.840		
Communication	R			0.913	0.601
COM1		0.230	0.816		
COM2		0.213	0.836		
COM3		0.170	0.785		
COM4		0.209	0.814		

(*continued*)

Table 20.6 (continued)

COM5		0.166	0.774		
COM6		0.162	0.742		
COM7		0.126	0.643		
Social environment	R			0.859	0.607
SOC1		0.273	0.759		
SOC2		0.191	0.637		
SOC3		0.398	0.865		
SOC4		0.392	0.835		
Perceived risk	R			0.899	0.643
RSC1		0.187	0.679		
RSC2		0.251	0.818		
RSC3		0.275	0.867		
RSC4		0.263	0.842		
RSC5		0.264	0.791		
Product visibility	R			0.840	0.724
VIS1		0.513	0.811		
VIS2		0.657	0.889		
Preference	R			0.828	0.616
PREF1		0.455	0.776		
PREF2		0.420	0.825		
PREF3		0.400	0.752		
Familiar	R			0.883	0.659
FAM1		0.313	0.872		
FAM2		0.351	0.876		
FAM3		0.317	0.858		
FAM4		0.242	0.610		
Brand identity. image	R			0.908	0.587
IPI1		0.189	0.775		
IPI2		0.165	0.750		
IPI3		0.178	0.787		
IPI4		0.166	0.812		
IPI5		0.203	0.719		
IPI6		0.173	0.793		
IPI7		0.238	0.722		
Brand name	R			0.825	0.545
NOM1		0.329	0.786		
NOM2		0.365	0.834		
NOM3		0.348	0.738		
NOM6		0.318	0.568		
Brand origin	R			0.835	0.629
ORIG1		0.470	0.830		
ORIG2		0.412	0.826		
ORIG3		0.376	0.718		
Perceived value	R			0.858	0.606
VLP1		0.221	0.573		
VLP2		0.310	0.808		
VLP3		0.364	0.877		

(continued)

Table 20.6 (continued)

Constructs and indicators	Type	Weight	Loading	Composite reliability ρ_c	Average variance extracted AVE
Demographic profile	F			n.a.	n.a.
VLP4		0.370	0.822		
Quality	R			0.922	0.747
QLD1		0.259	0.819		
QLD2		0.303	0.882		
QLD3		0.295	0.911		
QLD4		0.300	0.842		
Price	R			0.808	0.584
PRC3		0.379	0.754		
PRC4		0.443	0.717		
PRC5		0.485	0.819		
Product attributes	R			0.852	0.536
ATB1		0.275	0.715		
ATB2		0.270	0.750		
ATB3		0.255	0.703		
ATB4		0.345	0.812		
ATB5		0.211	0.674		
Package	R			0.938	0.791
EMB1		0.248	0.865		
EMB2		0.293	0.909		
EMB3		0.311	0.916		
EMB4		0.271	0.866		
Information search	R			0.882	0.656
INF1		0.343	0.866		
INF2		0.316	0.841		
INF3		0.327	0.883		
INF4		0.239	0.623		
Pioneering advantage	R			n.a.	n.a.
PRIMMC		1.000	1.000		
Self-image congruence	R			0.857	0.667
CNS1		0.456	0.830		
CNS3		0.393	0.824		
CNS5		0.375	0.796		

Notes: Type: *R* reflective, *F* formative, *n.a.* not applicable

20.5.2 Structural Model

The structural model represents the relationships between constructs or latent variables that were hypothesized in the research model. Since the primary objective of PLS is prediction, the goodness of a theoretical model is established by the strength of each structural path and the combined predictiveness (R^2) of its exogenous constructs (Chin 1998). Falk and Miller (1992) suggest that the variance explained, or

Table 20.7 Discriminant validity coefficients

	Satis	Need	Mem	Invol	Comm	Soc En	Risk	Visibi
Satis	0.832							
Need	−0.120	0.601						
Mem	0.292	−0.273	0.739					
Invol	0.239	−0.286	0.214	0.733				
Comm	0.344	−0.258	0.455	0.487	0.775			
Soc Env	0.365	−0.377	0.252	0.378	0.433	0.779		
Risk	0.537	−0.301	0.319	0.398	0.562	0.452	0.802	
Visibi	0.177	−0.401	0.204	0.398	0.343	0.547	0.359	0.851
Prefer	0.302	−0.312	0.235	0.463	0.443	0.544	0.489	0.458
Famil	0.273	−0.270	0.330	0.252	0.341	0.247	0.335	0.224
Br.Iden	0.433	−0.364	0.338	0.388	0.526	0.593	0.499	0.356
Br.Name	0.532	−0.273	0.327	0.445	0.580	0.532	0.557	0.360
Br.Orig	0.184	−0.416	0.155	0.373	0.327	0.361	0.370	0.449
Value	0.642	−0.347	0.273	0.346	0.444	0.467	0.687	0.346
Quality	0.792	−0.223	0.351	0.338	0.454	0.498	0.705	0.285
Price	−0.253	0.329	−0.232	−0.361	−0.421	−0.445	−0.525	−0.390
Prod At	0.276	−0.268	0.389	0.435	0.502	0.337	0.431	0.307
Package	0.076	−0.326	0.150	0.437	0.333	0.356	0.282	0.404
Inform	0.214	−0.338	0.288	0.439	0.382	0.296	0.363	0.299
Pion.	−0.007	−0.039	−0.015	0.000	0.002	0.031	−0.003	0.062
Congr.	0.221	−0.391	0.254	0.412	0.436	0.462	0.380	0.450

	Prefer	Famil	Br.Iden	Br.Name	Br.Orig	Value	Quality	Price
Prefer	0.785							
Famil	0.300	0.812						
Br.Iden	0.648	0.460	0.766					
Br.Name	0.564	0.412	0.655	0.738				
Br.Orig	0.425	0.317	0.420	0.426	0.793			
Value	0.423	0.294	0.504	0.549	0.280	0.778		
Quality	0.432	0.312	0.513	0.587	0.274	0.755	0.864	
Price	−0.406	−0.253	−0.434	−0.398	−0.341	−0.420	−0.395	0.764
Prod At	0.427	0.528	0.481	0.488	0.470	0.346	0.355	−0.347
Package	0.415	0.171	0.349	0.401	0.502	0.168	0.161	−0.271
Inform	0.429	0.649	0.494	0.435	0.432	0.315	0.304	−0.329
Pion.	0.033	0.003	0.024	−0.002	0.020	−0.009	−0.004	−0.068
Congr.	0.590	0.270	0.563	0.538	0.530	0.322	0.331	−0.352

	Prod At	Package	Inform	Pion.	Congruence
Prod At	0.732				
Package	0.376	0.889			
Inform	0.560	0.331	0.810		
Pion.	0.056	−0.013	0.073	1.000	
Congr.	0.453	0.515	0.434	0.045	0.817

Notes: Diagonal elements are the square root of average variance extracted (AVE) between the constructs and their measures. Off-diagonal elements are correlations between constructs. For discriminant validity, diagonal elements should be larger than off-diagonal elements in the same row and column

Table 20.8 Multicollinearity statistics

Component	Indicator	Tolerance	VIF	t-Statistic
Demographic profile	Age	0.378	2.648	1.232
	Education	0.376	2.656	4.082***
	Gender	0.929	1.076	1.391
	Family_dim	0.952	1.050	2.053*
	Marit_status	0.906	1.104	4.254***
	Residence	0.887	1.127	1.302
Self-concept	PERS1	0.862	1.160	0.763
	PERS2	0.943	1.061	1.538
	PERS3	0.859	1.165	0.658
	PERS4	0.850	1.176	1.423
	PERS5	0.916	1.091	1.633
	PERS6	0.953	1.049	0.864
	PERS7	0.878	1.139	1.667
	PERS8	0.938	1.066	4.439***
	PERS9	0.898	1.113	2.031*

Notes: *** $p < 0.001$, ** $p < 0.01$, * $p < 0.05$ (based on $t_{(499)}$, two-tailed test)

Table 20.9 Variance explained

Dependent construct	R^2
Satisfaction	0.059
Communication	0.049
Need for cognition	0.138
Self-concept	0.096
Social environment	0.143
Memory	0.242
Self-image congruence	0.358
Information search	0.539
Familiarity	0.116
Perceived value	0.589
Quality	0.392
Preference	0.574

R^2s, for endogenous variables should be greater than 0.1. The variance explained for each dependent construct is showed in Table 20.9.

As can be seen, three of the 12 endogenous constructs do not meet Falk and Miller's (1992) rule of 0.1. In this study, the final dependent construct (preference) has an R^2 value of 0.574, which can be considered satisfactory, taking into account the complexity of the model. Other constructs in the model also present acceptable levels of explained variance above the 0.1 level.

After computing the path estimates in the structural model, a bootstrap analysis was performed to assess the statistical significance of the path coefficients. From the initial set of paths, five were revealed as significant at 0.95, six at the 0.99 level, and the remaining 18 were significant at the 0.999 level, as shown in Table 20.10.

Table 20.10 Path coefficient

Path			Path coefficient	T statistic	Sign
Demographic profile	→	Satisfaction	−0.104	2.052	*
Demographic profile	→	Need for cognition	0.205	3.447	***
Demographic profile	→	Communication	−0.221	4.345	***
Demographic profile	→	Preference	0.099	2.667	**
Need for cognition	→	Self-concept	−0.169	2.103	*
Need for cognition	→	Social environment	−0.377	9.187	***
Need for cognition	→	Information search	−0.088	2.406	*
Need for cognition	→	Self-image congruence	−0.211	4.624	***
Involvement	→	Need for cognition	−0.163	3.148	**
Involvement	→	Preference	0.119	2.821	**
Involvement	→	Information search	0.201	5.380	***
Communication	→	Need for cognition	−0.133	2.853	**
Communication	→	Memory	0.388	9.957	***
Communication	→	Familiarity	0.341	7.491	***
Social environment	→	Preference	0.099	1.991	*
Risk	→	Preference	0.129	2.169	*
Visibility	→	Preference	0.118	2.901	**
Familiarity	→	Memory	0.198	4.412	***
Familiarity	→	Information search	0.463	11.951	***
Brand indent/pers/image	→	Preference	0.331	6.190	***
Brand indent/pers/image	→	Self-image congruence	0.480	12.971	***
Brand name	→	Quality	0.521	10.568	***
Quality	→	Perceived value	0.690	23.289	***
Price	→	Perceived value	−0.131	3.822	***
Price	→	Quality	−0.192	4.101	***
Product attributes	→	Information search	0.194	4.890	***
Package	→	Quality	−0.139	3.265	**
Self-image congruence	→	Satisfaction	0.201	4.247	***
Self-image congruence	→	Preference	0.195	4.532	***

Notes: *** $p < 0.001$, ** $p < 0.01$, * $p < 0.05$; (based on $t_{(499)}$, two-tailed test)

Figure 20.2 shows the significant paths (at the minimum level of 0.05) for our model. As can be seen, of the initial 22 constructs, only 7 seem to have a direct and statistical significant impact on brand preference, with brand identity, personality and image and self image congruence constructs having the strongest influence.

In PLS, no global criterion is optimized and, consequently, there is no that allows us to evaluate the overall model. Trying to surpass this problem, Tenenhaus et al. (2004) propose a global criterion of goodness-of-fit (*GoF*) that represents an operational solution for this gap, and can be seen as an index for validating the PLS model globally. This *GoF* measure is the geometric mean of the average communality and the average R^2. The average communality is computed as a weight average of the different communalities with the number of manifest variables or indicators of every construct as weights. It is worth noting that single indicator constructs should not be used for the computation of the average communality, because they lead to communalities equal to 1 (Tenenhaus et al. 2005).

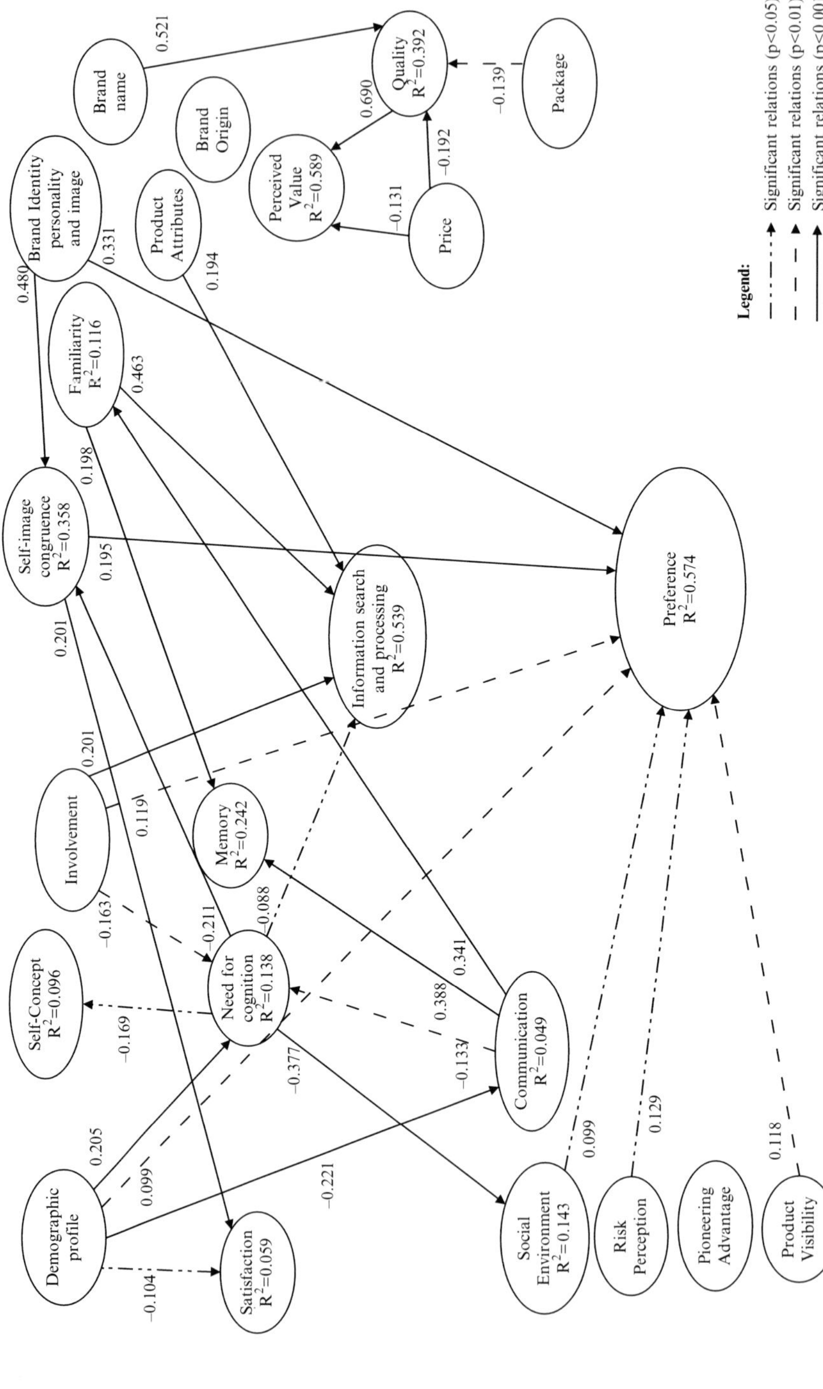

Fig. 20.2 Statistical significance of path coefficients

Table 20.11 Communality, redundancy and *GoF*

Block	R^2	Average communality	Average redundancy	Manifest variables (MV)	AvComm. × MV
Demog. profile (*)		0.2411		6.0000	0.930
Self-concept (*)	0.0957	0.1178	0.0113	9.0000	1.026
Satisfaction	0.0593	0.6941	0.0412	4.0000	2.773
Need for cognition	0.1369	0.3610	0.0494	5.0000	1.804
Memory	0.2421	0.5458	0.1321	3.0000	1.637
Involvement		0.5379		6.0000	3.229
Communication	0.0489	0.6012	0.0294	7.0000	4.206
Social environment	0.1402	0.6102	0.0856	4.0000	2.428
Risk perception		0.6427		5.0000	3.216
Visibility		0.7248		2.0000	1.449
Preference	0.6159	0.4997	0.3077	3.0000	1.847
Familiar	0.1159	0.6588	0.0763	4.0000	2.636
Brand iden/ier/imag		0.5879		7.0000	4.107
Brand name		0.5442		4.0000	2.182
Brand origin		0.6288		3.0000	1.886
Perceived value	0.5892	0.6064	0.3573	4.0000	2.425
Quality	0.3960	0.7466	0.2957	4.0000	2.986
Price		0.5844		3.0000	1.753
Product		0.5360		5.0000	2.680
Package		0.7911		4.0000	3.165
Information search	0.5393	0.6559	0.3537	4.0000	2.624
Pioneering advant.		1.0000		1.0000	
Self-image congru.	0.3543	0.6665	0.2361	3.0000	2.001
Average	0.2778			100.00	0.529887
				GoF	*0.3814*

Note: (*) For latent variables (LVs) measured with formative indicators the communalities were replaced with the R^2 obtained through the multiple regression of the LVs scores from internal estimation, over its own formative manifest variables (MVs)

For our model, the Amato et al. (2004) *GoF* was 0.3814, as can be seen in Table 20.11.

Another test applied in PLS models is the Stone-Geisser test of predictive relevance. This test can be used as an additional assessment of model fit in PLS analysis (Stone 1974, Geisser 1975). The Q^2 statistic is a jackknife version of the R^2 statistic. According to Chin (1998), the "Q^2 represents a measure of how well observed values are reconstructed by the model and its parameter estimates." Models with Q^2 greater than zero are considered to have predictive relevance. Models with higher positive Q^2 values are considered to have more predictive relevance.

The procedure to calculate the Q^2 involves omitting or "blindfolding" one case at a time and reestimating the model parameters based on the remaining cases, and predicting the omitted case values on the basis of the remaining parameters (Sellin 1989). The procedure results in the Q^2 test statistic.

The Stone-Geisser Q^2 can be obtained through the underlying latent variable score case from which the cross-validated communality is obtained, or through those latent variables that predict the block in question from which the cross-validated redundancy is obtained.

The cv-communality measures the capacity of the path model to predict the manifest variables or data points from their own latent variable score, and serves as an indicator of the quality of the measurement model. The cv-redundancy measures the capacity of the model to predict the endogenous manifest variables using the latent variables that predict the block in question, and serve as a sign of the quality of the structural model (Tenenhaus et al. 2005).

We compute measures of cross-validation to evaluate both the measurement model (cv-communality H^2) and the structural model (cv-redundancy F^2). For our model, blindfolding has been carried out using $G = 30$. According to Wold (1982), the omission distance should be an integer between the number of indicators and cases. Chin (1998) indicates that values between 5 and 10 are feasible but, considering the complexity of the model, we believe that a larger number is preferable. The results are in Table 20.12.

As can be seen, several blocks do not present an acceptable cross-validated redundancy index. More, due to blindfolding procedure, the cv-communality and

Table 20.12 Blindfolding results: cv-communality and cv-redundancy

Block	Cv-communality H^2	Cv-redundancy F^2
Demographic profile	0.0049	
Self-concept	−0.0768	−0.0532
Satisfaction	0.4862	−0.3575
Need for cognition	0.0660	−0.1023
Memory	0.1460	−0.0123
Involvement	0.3579	
Communication	0.4708	−0.3451
Social environment	0.3628	−0.1539
Risk perception	0.4657	
Visibility	0.2037	
Preference	0.2474	0.2884
Familiar	0.4416	−0.2067
Brand iden/per/imag	0.4462	
Brand name	0.2624	
Brand origin	0.2750	
Perceived value	0.3650	0.3194
Quality	0.5613	0.1861
Price	0.1962	
Product	0.3124	
Package	0.6289	
Information search	0.4376	0.3017
Pioneering advantage		
Self-image congruence	0.3358	0.1244

the cv-redundancy measures may be negative, which happens in this study and, according to Tenenhaus et al. (2005), implies that the corresponding latent variable has been badly estimated. These results may be attributed to the size and complexity of the theoretical model proposed.

20.6 Discussion

In keeping with the evidences retrieved from the literature review (Rossi et al. 1996; Bucklin et al. 1995), the demographic profile in this study shows a small, but statistical significant, impact on brand preference. This impact can be even higher, as this construct represents several other effects on other components of the model, and consequently, we think that demographic variables should not be ignored in brand preference studies.

The need for cognition construct presents a rich set of significant relations with other elements, but these results should be carefully considered given the AVE value obtained in the measurement model evaluation. Nevertheless, it can be observed that all the paths, starting with the need for cognition, have negative signs, suggesting that consumers with a high level of need for cognition, i.e. who appreciate the effort of thinking over things, tend to pay little attention and assign little importance to, and rely less on other factors. In line with the indications by Zhang and Buda (1999) and Sadowski and Cogburn (1997), these results show that the level of need for cognition has the capability of influencing the way consumers look at the environment and the stimulus received.

The need for cognition is also influenced by the importance placed on communication, suggesting that consumers who place higher importance on communication are less likely to engage in complex mental processes. On the other hand, communication shows a positive impact on memory and familiarity, which is consistent with previous studies. The absence of a direct impact on the preference confirms Hawkins (1970) and Higie and Sewall's (1991) doubts about the existence of a direct link between communication and preference and reinforces the indication by D'Souza and Rao (1995) that communication itself is not sufficient to increase brand preference. Nevertheless, communication has a significant impact on memory, as pointed out by Ettenson (1993), on familiarity according to Bogart and Lehman (1973), Cobb-Walgren et al. (1995), Alreck and Settle (1999), Lin et al. (2000) and Riezebos (2003), and on the need for cognition, but none of those links directly to preference, only through other constructs.

Our findings also suggest that familiarity enhances memory, but contrary to the observations by Haley and Case (1979) and Hutchinson et al. (1994), memory has a negative impact (non significant) on brand preference, suggesting that preference can be negatively affected by memory capacity, perhaps because consumers with better memories are able to retain more data and produce more complex comparisons.

More consistent with the evidence from the literature reviewed, namely Witt and Bruce (1972), Celsi and Olson (1988) and Maheswaran and Mackie (1992), is the effect of involvement, which exhibits a positive impact on information search, thus pointing to a high level of involvement inducing a more extensive information search. Also, the degree of familiarity and the importance placed on product attributes display a positive influence on information search, suggesting that consumers more familiar with the class and those who weighted product attributes more heavily, tend to place more importance on information search and processing. Conversely, consumers with a high need for cognition are less willing to engage in information search, which could be explained by the confidence they have in on their own mental skills.

Looking at the attributes related to the brand, we notice that only brand identity, personality, and image components exhibit a significant relation with preference, suggesting that most consumers use brands as a way of expressing themselves or their lifestyle and, consequently, they tend to prefer brands whose identity, personality, and image are closer to them, pointing out that a consumer's relationship with brands becomes increasingly symbiotic.

Companies have long stimulated consumers to identify with products or brands and their identity/personality. Brands becomes extremely attractive to consumers, and so become new friends, who over time become old friends. Consumers prefer brands with a strong identity, personality and image (Sirgy 1982; Phau and Lau 2001), especially those that reinforce their self-concept. Fournier (1998) has even identified a total of 15 types of consumer/brand relationships.

Consequently, the congruence between brand identity, personality and image, and consumer self-image, called self-image congruence seems to be very important for brand preference. Many studies (Belk et al. 1982; Onkvisit and Shaw 1987; Belk 1988; Richins 1994a,b; Hong and Zinkhan 1995; Ericksen 1996; Aaker 1999; Jamal and Goode 2001) have confirmed the importance of self-image congruence, which our study now confirms. If we look at the path coefficients we notice that brand identity, personality and image, and self-image congruence have the strongest relations with preference, stressing the importance of those constructs in the development of brand preference.

Other constructs related to the brand show strong and significant relations, especially brand name/quality and quality/perceived value, but none have a significant impact on preference.

Finally, of the situational factors, only social environment and product visibility exhibit a significant positive influence on preference. These findings suggest that consumers try to match the brand of their mobile phone with the brands of their friends and family. A product with social visibility also seems to have a positive impact on preference, which was previously noted by Graeff (1997).

Several other constructs also show strong relations, namely: communication, familiarity, brand name, quality, need for cognition, product attributes and demographic profile, but, as was anticipated, a large number does not exhibit a statistical significant relation with preference. The explanation for this contradiction, in our opinion, may result from two conditions. First, the product class used in this investigation has different characteristics from the products used in the studies reviewed, most of which were consumer goods. Second, as was anticipated in the introduction, we think that the interaction between factors plays a crucial role in the development of preference. This is, perhaps, an issue that could explain the results found, because previous studies focus only on the impact of one or a very limited set of factors on preference.

Consequently, we cannot say that our results are contrary to those found in the literature; rather, they should serve as a new starting point for investigators to consider, revise, and extend upon.

In conclusion, these results show that the social environment and the context in which the product will be used influence the brand preference for mobile phones. Further, the results stress the importance of brand identity and its relationship with the self-image of the consumer for the formation of brand preference and, therefore, reinforce the conviction of several authors that consumers tend to prefer brands that are closer to their self-image.

20.7 Summary, Conclusions, and Limitations

The goal of our research was to uncover factors that lead to the formation of brand preference and improve our understanding of the interaction of those factors. At the same time, we hope to show that PLS can be successfully used to test big and complex models, where other statistical techniques would fail.

From the analysis, we were be able to show that several factors contribute to brand preference, specially those related to brand identity, personality and image and their congruence with consumer self-image. The findings of this study are partially supported by the literature, and the estimation model validates 29 of the 54 relationships hypothesized in our conceptual model at the 0.05 significance level. The R-square for the model was 0.574, which we think can be considered very satisfactory, taking into account its complexity.

In the light of the controversy about the nature of brand preference and consumer behavior, the results of this investigation support Best's (1978) vision of a pattern of preference, which can result in a buying pattern or a pattern of choice behavior. Nevertheless, we cannot ignore or underestimate the power of situational factors in determining consumer preference. Consequently, in our opinion, the results of this research reinforce the conciliatory perspective by Lehmann (1972), Bettman and Jones (1972), and Shocker and Srinivasan (1979), which points to the integration of the deterministic and probabilistic approaches.

The main direct effects on brand preference are the self-image congruence and the identity/personality and image of the brand. In addition to those, the level of involvement, social environment, risk perception, demographic profile, and product visibility also show a positive influence on brand preference. Several other constructs present indirect, but significant and robust, contributions to explain the development of brand preference.

On the other side, 15 constructs in this research do not exhibit a direct influence on brand preference. Of those, the pioneering advantage and brand origin are the only ones that do not show a single significant relationship with any other construct in the model.

The results of the demographic variables, as previously noted, follow the evidence from previous studies (e.g., Jamal and Goode 2001); that is to say, present a small but significant relation with preference construct and, consequently, should not be ignored in future investigations.

In conclusion, our findings suggest that brand preference formation is a complex process, in which factors should not be considered independently because interaction plays a determinant role.

These findings must, of course, be interpreted with extreme caution; moreover, the model needs to be tested with improved and more objective measures for some constructs to solve methodological problems associated with the statistical significance of those measures. In addition, the model clearly does not include all the relevant variables. The possible inclusion of more situational, brand-related or other consumer-related variables to further extend the proposed model should be actively pursued by future research. Additionally, other relationships currently not supported by other studies, may be included in the model, for example, the relationship between brand identity, personality, and image and the perceived value or perceived quality.

Finally, we believe that this study is important to show how PLS path modeling can be used to successfully assess complex models and, in our case, provide some explanation of the relationships between the selected factors and brand preference formation. Furthermore, it shows that factors that are individually significant, can lose their power when assessed together with other factors due to the interaction effect. In our opinion, the new insight into the interaction effect provides important and usable information to managers. Nevertheless, this study needs to be replicated with new samples of consumers and different products and be improved with the introduction of new and relevant variables and perhaps the refinement of the scales used to measure some of the constructs.

References

Aaker, J. L. (1997). Dimensions of brand personality. *Journal of Marketing Research, 34*(3), 347–356.

Aaker, J. L. (1999). The malleable self: The role of self-expression in persuasion. *Journal of Marketing Research, 36*(1), 45–57.

Agarwal, S., & Teas, R. K. (2001). Perceived value: Mediating role of perceived risk. *Journal of Marketing Theory and Practice, 9*(4), 1–14.

Ahmed, S. A., & D'Astous, A. (1993). Cross-national evaluation of made-in concept using multiple cues. *European Journal of Marketing, 27*(7), 39–52.

Allison, R. I., & Uhl, K. P. (1964). Influence of beer brand identification on taste perception. *Journal of Marketing Research, 1*(3), 36–39.

Alpert, F., Kamins, M., Sakano, T., Onzo, N., & Graham, J. (2001). Retail buyer beliefs, attitude and behavior toward pioneer and me-too follower brands: A comparative study of Japan and the USA. *International Marketing Review, 18*(2), 160–187.

Alpert, M. I. (1972). Personality and the determinants of product choice. *Journal of Marketing Research, 9*(1), 89–92.

Alreck, P. L., & Settle, R. B. (1999). Strategies for building consumer brand preference. *Journal of Product and Brand Management, 8*(2), 130–144.

Alsop, R. (1984, November 29). Color grows more important in catching consumers' eyes. *Wall Street Journal (Eastern edition)*, 37.

Antil, J. H. (1984). Conceptualization and operationalization of involvement. *Advances in Consumer Research, 11*, 203–209.

Ballester, M. E. D., & Alemán, J. L. M. (2002). Construcción de un índice de medición con indicadores formativos. *Investigación y Marketing, 75*, 16–20.

Banks, S. (1950). The measurement of the effect of a new packaging material upon preference and sales. *Journal of Business, 23*(2), 71–80.

Barclay, D., Thompson, R., & Higgins, C. (1995). The partial least squares (PLS) approach to causal modeling: Personal computer adoption and use as an illustration. *Technology Studies, 2*(2), 285–309.

Bass, F. M. (1974). The theory of stochastic preference and brand switching. *Journal of Marketing Research, 11*(1), 1–20.

Bass, F. M., & Pilon, T. (1980). A stochastic brand choice framework for econometric modeling of time series market share behavior. *Journal of Marketing Research, 17*(4), 486–497.

Bass, F. M., & Talarzyk, W. (1972). An attitude model for the study of brand preference. *Journal of Marketing Research, 9*(1), 93–96.

Becherer, R. C., Morgan, F. W., & Richard, L. M. (1982). Informal group influence among situationally/dispositionally-oriented consumers. *Journal of the Academy of Marketing Science, 10*, (3), 269–281.

Belk, R. W. (1974). An exploratory assessment of situational effects in buyer behavior. *Journal of Marketing Research, 11*(2), 156–163.

Belk, R. W. (1988). Possessions and the extended self. *Journal of Consumer Research, 15*(2), 139–168.

Belk, R. W. (1975a). Situational variables and consumer behavior. *Journal of Consumer Research, 2*(3), 157–164.

Belk, R. W. (1975b). The objective situation as a determinant of consumer behavior. *Advances in Consumer Research, 2*, 427–438.

Belk, R. W., Bahn, K. D., & Mayer, R. N. (1982). Developmental recognition of consumption symbolism. *Journal of Consumer Research, 9*(1), 4–17.

Best, R. J. (1978). Validity and reliability of criterion-based preferences. *Journal of Marketing Research, 15*(1), 154–160.

Bettman, J. R., & Jones, M. J. (1972). Formal models of consumer behavior: A conceptual overview. *Journal of Business, 45*(4), 544–562.

Bettman, J. R., Capon, N., & Lutz, R. J. (1975). Cognitive algebra in multi-attribute attitude models. *Journal of Marketing Research, 12*(2), 151 164.

Birdwell, A. E. (1968). A study of the influence of image congruence on consumer choice. *Journal of Business, 41*(1), 76–88.

Blin, J.-M., & Dodson, J. A. (1980). The relationship between attributes, brand preference, and choice: A stochastic view. *Management Science, 26*(6), 606–619.

Bloch, P. H., & Richins, M. L. (1983). A theoretical model for the study of product importance perceptions. *Journal of Marketing, 47*(3), 69–81.

Bogart, L., & Lehman, C. (1973). What makes a brand name familiar? *Journal of Marketing Research, 10*(1), 17–22.

Bolfing, C. P. (1988). Integrating consumer involvement and product perceptions with market segmentation and positioning strategies. *Journal of Consumer Marketing, 5*(2), 49–57.

Bollen, K. A. (1989). *Structural equations with latent variables*. New York: Wiley.

Bristow, D. N., & Asquith, J. A. L. (1999). What's in a name? An intracultural investigation of Hispanic and Anglo consumer preferences and the importance of brand name. *Journal of Product and Brand Management, 8*(3), 185–203.

Bristow, D. N., Schneider, K. C., & Schuler, D. K. (2002). The brand dependence scale: Measuring consumers' use of brand name to differentiate among product alternatives. *Journal of Product and Brand Management, 11*(6), 343–356.

Broadbent, K., & Cooper, P. (1987). Research is good for you. *Marketing Intelligence and Planning, 5*(1), 3–9.

Bryant, B. E., & Cha, J. (1996). Crossing the threshold. *Marketing Research, 8*(4), 20–28.

Bucklin, R. E., Gupta, S., & Han, S. (1995). A brand's eye view of response segmentation in consumer brand choice behavior. *Journal of Marketing Research, 32*(1), 66–74.

Burton, S., Lichtenstein, D. R., Netemeyer, R. G., & Garretson, J. A. (1998). A scale for measuring attitude toward private label products and an examination of its psychological and behavioral correlates. *Journal of the Academy of Marketing Science, 26*(4), 293–306.

Bushman, B. J. (1993). What's in a name? The moderating role of public self-consciousness on the relation between brand label and brand preference. *Journal of Applied Psychology, 78*(5), 857–861.

Cacioppo, J. T., & Petty, R. E. (1982). The need for cognition. *Journal of Personality and Social Psychology, 42*, 116–131.

Cacioppo, J. T., Petty, R. E., & Kao, C. F. (1984). The efficient assessment of need for cognition. *Journal of Personality Assessment, 48*, 306–307.

Cacioppo, J. T., Petty, R. E., Feinstein, J. A., Jarvis, W., & Blair, G. (1996). Dispositional differences in cognitive motivation: The life and times of individuals varying in need for cognition. *Psychological Bulletin, 119*(2), 197–253.

Campbell, M. C., & Goodstein, R. C. (2001). The moderating effect of perceived risk on consumers' evaluations of product incongruity: Preference form the norm. *Journal of Consumer Research, 21*(3), 439–449.

Carpenter, G. S., & Nakamoto, K. (1989). Consumer preference formation and pioneering advantage. *Journal of Marketing Research, 25*(3), 285–298.

Carpenter, G. S., Glazer, R., & Nakamoto, K. (1994). Meaningful brands from meaningless differentiation: The dependence on irrelevant attributes. *Journal of Marketing Research, 31*(3), 339–350.

Carroll, J. D., De Soete G., & DeSarbo, W. S. (1990). Two stochastic multidimensional choice models for marketing research. *Decision Sciences, 21*(2), 337–356.

Celsi, R. L., & Olson, J. C. (1988). The role of involvement in attention and comprehension processes. *Journal of Consumer Research, 15*(2), 210–224.

Chapman, J., & Wahlers, R. (1999). A revision and empirical test of the extended price-perceived quality model. *Journal of Marketing Theory and Practice, 7*(3), 53–64.

Chernev, A. (2001). The impact of common features on consumer preferences: A case of confirmatory reasoning. *Journal of Consumer Research, 27*, 475–488.

Chin, W. W. (1998). The partial least squares approach to structural equation modeling. In G. A. Marcoulides (Ed.), *Modern methods for business research* (pp. 295–336). New Jersey: Lawrence Erlbaum.

Chin, W., & Newsted, P. R. (1999). Structural equation modeling analysis with small samples using partial least squares. In R. Hoyle (Ed.), *Statistical strategies for small sample research* (pp. 307–341). Thousand Oaks: Sage.

Cobb-Walgren, C., Ruble, C. A., & Donthu, N. (1995). Brand equity, brand preference, and purchase intent. *Journal of Advertising, 24*(3), 25–40.

Costley, C. L., & Brucks, M. (1992). Selective recall and information use in consumer preferences. *Journal of Consumer Research, 18*(4), 464–474.

Creyer, E. H., & Ross, W. T. (1997). Tradeoffs between price and quality: How a value index affects preference formation. *Journal of Consumer Credit Management, 31*(2), 280–302.

Currim, I. S., & Sarin, R. K. (1984). A comparative evaluation of multiattribute consumer preference models. *Management Science, 30*(5), 543–561.

D'Astous, A., & Gargouri, E. (2001). Consumer evaluations of brand imitations. *European Journal of Marketing, 35*(1/2), 153–167.

De Chernatony, L., & McDonald, M. H. B. (2001). *Creating powerful brands*. Oxford: Butterworth-Heinemann.

Del Río, A., Vázquez, R., & Iglesias, V. (2001). The role of the brand name in obtaining differential advantages. *Journal of Product and Brand Management, 10*(7), 452–465.

Desai, K. K., & Ratneshwar, S. (2003). Consumer perceptions of product variants positioned on atypical attributes. *Journal of the Academy of Marketing Science, 34*(1), 22–35.

DeSarbo, W. S., & Rao, V. R. (1984). GENFOLD2: A set of models and algorithms for the general unfolding analysis of preference/dominance data. *Journal of Classification, 1*, 147–186.

DeSarbo, W. S., & Rao, V. R. (1986). A constrained unfolding methodology for product positioning. *Marketing Science, 5*(1), 1–19.

Dhar, R., Nowlis, S. M., & Sherman, S. J. (1999). Comparison effects on preference construction. *Journal of Consumer Research, 26*, 293–306.

Diamantopoulos, A., & Winklhofer, H. M. (2001). Index construction with formative indicators: An alternative to scale development. *Journal of Marketing Research, 38*(2), 269–277.

Dick, A., Jain, A., & Richardson, P. (1996). How consumers evaluate store brands. *Journal of Product and Brand Management, 5*(2), 19–28.

Dickerson, K. G. (1982). Imported versus U.S. produced apparel: Consumer views and buying patterns. *Home Economic Research Journal, 10*(3), 241–253.

Dickson, P. R. (1982). Person-situation: Segmentation's missing link. *Journal of Marketing, 46*(4), 56–64.

Dodds, W. B., Monroe, K. B., & Grewal, D. (1991). Effects of price brand and store information on buyers' product evaluations. *Journal of Marketing Research, 28*(3), 307–319.

D'Souza, G., & Rao, R. C. (1995). Can repeating an advertisement more frequently than the competition affect brand preference in mature market. *Journal of Marketing, 59*, 32–42.

Duncan, C. P., & Nelson, J. E. (1985). Effect of humor in a radio advertising experiment. *Journal of Advertising, 14*(2), 33–40.

Dunn, M. G., & Murphy, P. E. (1986). Research note: The influence of perceived risk on brand preference for supermarket products. *Journal of Retailing, 62*(2), 204–216.

Elliott, R. (1994). Exploring the symbolic meaning of brands. *British Food Journal, 5*(Special), s13–s19.

Ericksen, M. K. (1996). Using self-congruity and ideal congruity to predict purchase intention: A European perspective. *Journal of Euromarketing, 6*(1), 41–56.

Ettenson, R. (1993). Brand names and country-of-origin effects in emerging market economies of Russia, Poland, and Hungary. *International Marketing Review, 10*(5), 14–36.

Evans, F. B. (1959). Psychological and objective factors in the prediction of brand choice. *Journal of Business, 32*, 340–369.

Falk, R. F., & Miller, N. B. (1992). *A primer for soft modeling*. Ohio: The University of Akron Press.

Fennell, G., Allenby, G. M., Yang, S., & Edwards, Y. (2003). The effectiveness of demographic and psychographic variables for explaining brand and product category use. *Quantitative Marketing and Economics, 1*(2), 223–244.

Fisher, G. W., Carmon, Z., Ariely, D., & Zauberman, G. (1999). Goal-based construction of preferences: Task goals and the prominence effect. *Management Science, 45*(8), 1057–1075.

Fornell, C., & Larcker, D. (1981). Evaluating structural equation models with unobservable variables and measurement error. *Journal of Marketing Research, 18*, 39–50.

Fournier, S. (1998). Consumers and their brands: Developing relationship theory in consumer research. *Journal of Consumer Research, 24*(3), 343–373.

Garbarino, E. C., & Edell, J. A. (1997). Cognitive effort, affect and choice. *Journal of Consumer Research, 24*(2), 147–158.

Gardner, B. B., & Levy, S. J. (1955). The product and the brand. *Harvard Business Review, 33*, 33–39.

Garthwaite, P. H. (1994). An interpretation of partial least squares. (425), 122–127.

Geisser, S. (1975). The predictive sample reuse method with applications. *Journal of the American Statistical Association, 70*(350), 320–328.
Ginter, J. L., & Bass, F. M. (1972). An experimental study of attitude change, advertising, and usage in new product introduction. *Journal of Advertising, 1*(1), 33–39.
Graeff, T. R. (1996). Using promotional messages to manage the effects of brand and self-image on brand evaluations. *The Journal of Consumer Marketing, 13*(3), 4–18
Graeff, T. R. (1997). Consumption situations and the effects of brand image on consumers' brand evaluations. *Psychology and Marketing, 14*(1), 49–70.
Gruenfeld, D. H., & Wyer, R. S., Jr. (1992). Semantics and pragmatics of social influence: How affirmations and denials affect beliefs in referent propositions. *Journal of Personality and Social Psychology, 62*(1), 38–49.
Hair, J. F., Jr., Anderson, R. E., Tatham, R. L., & Black, W. C. (1998). *Multivariate data analysis* (5th ed.). London: Prentice Hall International.
Haley, R. I., & Case, P. B. (1979). Testing thirteen attitude scales for agreement and brand discrimination. *Journal of Marketing, 43*(4), 20–32.
Han, C. M., & Terpstra, V. (1988). Country-of-origin effects for uni-national and bi-national. *Journal of International Business Studies, 19*(2), 235–255.
Harris, R. J., & Monaco, G. E. (1978). Psychology of pragmatic implications: Information processing between the lines. *Journal of Experimental Psychology, 107*(1), 1–27.
Hawkins, D. I. (1970). The effects of subliminal stimulation on drive level and brand preference. *Journal of Marketing Research, 7*(3), 322–326.
Hawkins, D. I., & Coney, K. A. (1974). Peer group influences on children's product preferences. *Academy of Marketing Science Review Online, 2*(2), 322–331.
Hellier, P. K., Guersen, G. M., Carr, R. A., & Rickard, J. A. (2003). Customer repurchase intention. A general structural equation model. *European Journal of Marketing, 37*(11/12), 1762–1800.
Helman, D., & De Chernatony, L. (1999). Exploring the development of lifestyle retail brands. *The Service Industries Journal, 19*(2), 49–68.
Higie, R. A., & Sewall, M. A. (1991). Using recall and brand preference to evaluate advertising effectiveness. *Journal of Advertising Research, 31*(2), 56–63.
Hong, J. W., & Zinkhan, G. M. (1995). Self-concept and advertising effectiveness: The influence of congruency conspicuousness and response mode. *Psychology and Marketing, 12*(1), 53–77.
Horton, R. L. (1974). The Edwards personal preference schedule and consumer personality research. *Journal of Marketing Research, 11*(3), 335–337.
Howard, J. A., & Sheth, J. N. (1969). The theory of buyer behavior. New York: Wiley.
Hughes, R. E. (1976). Self-concept and brand preference: A partial replication. *Journal of Business, 49*(4), 530–540.
Hugstad, P. S., & Durr, M. (1986). A study of country of manufacturer impact on consumer perceptions. *Journal of the Academy of Marketing Science, 9*, 115–119.
Hutchinson, J. W., Raman, K., & Mantrala, M. K. (1994). Finding choice alternatives in memory: Probability models of brand name recall. *Journal of Marketing Research, 31*(4), 441–461.
Jacoby, J., Szybillo, G. J., & Busato-Schach, J. (1977). Information acquisition behavior in brand choice situations. *Journal of Consumer Research, 3*(4), 209–216.
Jain, S. P., & Maheswaran, D. (2000). Motivated reasoning: A depth-of-processing perspective. *Journal of Consumer Research, 26*(4), 358–371.
Jamal, A., & Goode, M. M. H. (2001). Consumers and brands: A study of the impact of self-image congruence on brand preference and satisfaction. *Marketing Intelligence and Planning, 19*(7), 482–492.
Jarvis, C. B., Mackenzie, S. B., & Podsakoff, P. M. (2003). A critical review of construct indicators and measurement model misspecification in marketing and consumer research. *Journal of Consumer Research, 30*(2), 199–218.
Ji, M. F. (2002). Children's relationships with brands: True love or one-night stand? *Psychology and Marketing, 19*(4), 369–387.
Jöreskog, K. G., & Wold, H. (1982). The ML and PLS techniques for modeling with latent variables: Historical and comparative aspects. In H. Wold & K. Jöreskog (Eds.), *Systems*

under indirect observation: Causality, structure, prediction (Vol. 1, pp. 263–270). Amsterdam: North-Holland.

Kaponin, A. (1960). Personality characteristics of purchasers. *Journal of Advertising Research, 1*(1), 6–12.

Kassarjian, H. H. (1971). Personality and consumer behavior: A review. *Journal of Marketing Research, 8*(4), 409–418.

Kassarjian, H. H. (1979). Personality: The longest fad. *Advances in Consumer Research, 6*, 122–124.

Keillor, B. D., Parker, R. S., & Schaefer, A. (1996). Influences on adolescent brand preferences in the United States and Mexico. *Journal of Advertising Research, 36*(3), 47–56.

Keller, K. L. (2003). *Strategic brand management – building, measuring and managing brand equity* (2nd ed.). New Jersey: Prentice Hall.

Khachaturian, J. L., & Morganosky, M. A. (1990). Quality perceptions by country of origin. *International Journal of Retail and Distribution Management, 18*(5), 21–30.

Kim, C. K. (1995). Brand popularity and country image in global competition: Managerial implications. *Journal of Product and Brand Management, 4*(5), 21–33.

Kirmani, A., & Zeithaml, V. (1993). Brand equity and advertising: Advertising's role in building strong brands. In D. A. Aaker & A. L. Biel (Eds.), *Brand equity and advertising: Advertising's role in building strong brands* (pp. 143–162). New Jersey: Lawrence Erlbaum.

Klink, R. R. (2001). Creating meaningful new brand names: A study of semantics and sound symbolism. *Journal of Marketing Theory and Practice, 9*(2), 27–34.

Kohli, C., & Labahn, D. W. (1997). Creating effective brand names: A study of the naming process. *Journal of Advertising Research, 37*(1), 67–75.

Krugman, H. E. (1962). The learning of consumer preference. *Journal of Marketing, 26*(2), 31–33.

Landon, L. E., Jr. (1974). Self concept, ideal self concept, and consumer purchase intentions. *Journal of Consumer Research, 1*(2), 44–51.

Lange, F., & Dahlén, M. (2003). Let's be strange: Brand familiarity and ad-brand incongruency. *Journal of Product and Brand Management, 12*(7), 449–461.

Lau, G. T., & Lee, S. H. (1999). Consumers' trust in a brand and the link to brand loyalty. *Journal of Market Focused Management, 4*, 341–370.

Lehmann, D. R. (1972). Judged similarity and brand-switching data as similarity measures. *Journal of Marketing Research, 9*(3), 331–334.

Levy, S. J. (1959a). Symbols by which we buy. In M. J. Baker (Ed.) (2001), *Marketing-critical perspectives on business and management, II* (pp. 432–438). London: Routledge.

Levy, S. J. (1959b). Symbols for sale. *Harvard Business Review, 37*, 117–124.

Lewis, C., & Stubbs, S. (1999). National expansion of British regional brands: Parallels with internationalization. *Journal of Product and Brand Management, 8*(5), 369–386.

Lichtenstein, D. R., & Burton, S. (1989). The relationship between perceived and objective price-quality. *Journal of Marketing Research, 26*(4), 429–443.

Lichtenstein, D. R., Ridgway, N. M., & Netemeyer, R. G. (1993). Price perceptions and consumer shopping behavior: A field study. *Journal of Marketing Research, 30*(2), 234–245.

Lin, C.-F. (2002). Segmenting customer brand preference: Demographic or psychographic. *Journal of Product and Brand Management, 11*(4), 249–268.

Lin, C., Wu, W.-Y., Wang, Z.-F. (2000). A study of market structure: A brand loyalty and brand switching behaviors for durable household appliances. *International Journal of Market Research, 42*(3), 277–300.

Low, G. S., & Lamb, C. W., Jr. (2000). The measurement and dimensionality of brand associations. *Journal of Product and Brand Management, 9*(6), 350–368.

Mackay, M. M. (2001). Evaluation of brand equity measures: Further empirical results. *Journal of Product and Brand Management, 10*(1), 38–51.

Macklin, M. C. (1996). Preschoolers' learning of brand names from visual cues. *Journal of Consumer Research, 23*, 251–261.

Maheswaran, D., & Mackie, D. M. (1992). Brand name as a heuristic cue: The effects of task importance and expectancy confirmation on consumer judgments. *Journal of Consumer Psychology, 1*(4), 317–336.

Malhotra, N. K. (1981). A scale to measure self-concepts, person concepts, and product concepts. *Journal of Marketing Research, 18*(4), 456–464.

Malhotra, N. K. (1988). Self concept and product choice: An integrated perspective. *Journal of Economic Psychology, 9*(1), 1–28.

Mandrik, C. A. (1996). Consumer heuristics: The tradeoff between processing effort and value in brand choice. *Advances in Consumer Research, 23*, 301–307.

McCrae, R. R., & John, O. P. (1992). An introduction to the five-factor model and its application. *Journal of Personality, 60*, 175–215.

Meyers-Levy, J. (1989a). The influence of a brand name's association set size and word frequency on brand memory. *Journal of Consumer Research, 16*(2), 197–207.

Meyers-Levy, J. (1989b). Investigating dimensions of brand name that influence the perceived familiarity of brands. *Advances in Consumer Research, 16*, 258–263.

Miller, K. E., & Ginter, J. L. (1979). An investigation of situational variation in brand choice behavior and attitude. *Journal of Marketing Research, 16*(1), 111–123.

Mitchell, A. A. (1982). Models of memory: Implications for measuring knowledge structures. *Advances in Consumer Research, 9*, 45–51.

Mitchell, A. A., & Olson, J. C. (1981). Are product attribute beliefs the only mediator of advertising effects on brand attitude? *Journal of Marketing* Research, *18*(3), 318–332.

Mittal, V., & Kamakura, W. A. (2001). Satisfaction, repurchase intent, and repurchase behavior: Investigating the moderating effect of customer characteristics. *Journal of Marketing Research, 38*(1), 131–142.

Monroe, K. B. (1976). The influence of price differences and brand familiarity on brand preferences. *Journal of Consumer Research, 3*(1), 42–49.

Moreland, R. L., & Zajonc, R. B. (1982). Exposure effects in person perception: Familiarity, similarity, and attraction. *Journal of Experimental Social Psychology, 18*, 395–415.

Morton, J. (1994). Predicting brand preference. *Marketing Management, 2*(4), 32–44.

Moschis, G. P. (1981). Patterns of consumer learning. *Journal of the Academy of Marketing Science, 9*(2), 110–126.

Moutinho, L., & Goode, M. (1995). Gender effects to the formation of overall product satisfaction: A multivariate approach. *Journal of International Consumer Marketing, 8*(1), 71–91.

Muthukrishnan, A. V., & Kardes, F. R. (2001). Persistent preferences for product attributes: The effects of the initial choice context and uninformative experience. *Journal of Consumer Research, 28*(1), 89–104.

Nedungadi, P. (1990). Recall and consumer consideration sets: Influencing choice without altering brand evaluations. *Journal of Consumer Research, 17*(3), 263–276.

Niedrich, R. W., & Swain, S. D. (2003). The influence of pioneer status and experience order on consumer brand preference: A mediated effects model. *Journal of the Academy of Marketing Science, 31*(4), 468–480.

Nowlis, S. M., & Simonson, I. (1997). Attribute-task compatibility as a determinant of consumer preference reversals. *Journal of Marketing Research, 34*(2), 205–218.

Olmo, J. M. C., & Jamilena, D. M. F. (2000). Regresión lineal. In T. L. Martínez (coord.), *Técnicas de análisis de datos en investigación de mercados* (pp. 247–280). Madrid: Editorial Pirámide.

Olsen, S. O. (2002). Comparative evaluation and the relationship between quality, satisfaction, and repurchase loyalty. *Journal of the Academy of Marketing Science, 30*(3), 240–249.

Onkvisit, S., & Shaw, J. (1987). Self-concept and image congruence: Some research and managerial implications. *Journal of Consumer Marketing, 4*(1), 13–23.

Paivio, A. (1971). *Imagery and verbal processes*. New York: Holt, Rinehart, and Winston.

Papadopoulos, N., Heslop, L. A., & Bamossy, G. (1990). A comparative image analysis of domestic versus imported products. *International Journal of Research in Marketing, 7*(4), 283–294.

Payne, J. W., Bettman, J. R., & Johnson, E. J. (1992). Behavioral decision research: A constructive processing perspective. *Annual Review of Psychology, 43*, 87–131.

Payne, J. W., Bettman, J. R., & Schkade, D. A. (1999). Measuring constructed preferences: Towards a building code. *Journal of Risk and Uncertainty, 19*(3), 243–270.
Peris, S. M., Newman, K., Bigne, E., & Chansarkar, B. (1993). Aspects of Anglo-Spanish perceptions arising from country-of-origin image. *International Journal of Advertising, 12*(12), 131–142.
Peter, J. P., & Ryan, M. J. (1976). An investigation of perceived risk at the brand level. *Journal of Marketing Research, 13*(2), 184–188.
Peterson, R. A. (1970). The price-perceived quality relationship: Experimental evidence. *Journal of Marketing Research, 7*(4), 525–528.
Petroshius, S. M., & Monroe, K. B. (1987). Effect of product-line pricing characteristics on product evaluations. *Journal of Consumer Research, 13*(4), 511–519.
Petroshius, S. M., & Crocker, K. E. (1989). An empirical analysis of spokesperson characteristics on advertisement and product evaluations. *Journal of the Academy of Marketing Science, 17*(3), 217–225.
Phau, I., & Lau, K. C. (2001). Brand personality and consumer self-expression: Single or dual carriageway? *Brand Management, 8*(6), 428–444.
Powers, T., & Nooh, S., Md. (1999). The impact of country-of-origin on product choice: A developing country perspective. *Journal of Practical Global Business, 1*(1), 18–40.
Pras, B., & Summers, J. O. (1978). Perceived risk and composition models for multiattribute decisions. *Journal of Marketing Research, 15*(3), 429–437.
Quester, P., & Lim, A. L. (2003). Product involvement/brand loyalty: Is there a link? *Journal of Product and Brand Management, 12*(1), 22–38.
Rao, A. R., & Monroe, K. B. (1988). The moderating effect of prior knowledge on cue utilization in product evaluations. *Journal of Consumer Research, 15*(2), 253–264.
Rao, A. R., Qu, L., & Ruekert, R. W. (1999). Signaling unobservable product quality through a brand ally. *Journal of Marketing Research, 36*(2), 258–268.
Rettie, R., Hilliar, S., & Alpert, F. (2002). Pioneer brand advantage with UK consumers. *European Journal of Marketing, 36*(7/8), 895–913.
Rheingold, H. L. (1985). Development as the acquisition of familiarity. *Annual Review of Psychology, 36*, 1–17.
Richardson, P. S., Dick, A. S., & Jain, A. K. (1994). Extrinsic and intrinsic cue effects on perceptions of store brand quality. *Journal of Marketing, 58*(4), 28–36.
Richins, M. L. (1994a). Valuing things: The public and private meanings of possessions. *Journal of Consumer Research, 21*(3), 504–521.
Richins, M. L. (1994b). Special possessions and the expression of material values. *Journal of Consumer Research, 21*(3), 522–533.
Riezebos, R. (2003). *Brand management. A theoretical and practical approach.* Essex: Prentice Hall/Financial Times.
Rigaux-Bricmont, B. (1981). Influences of brand name and packaging on perceived quality. *Advances in Consumer Research, 9*, 472–477.
Rodgers, W. C., & Schneider, K. C. (1993). An empirical evaluation of the kapferer-laurent consumer involvement profile scale. *Psychology and Marketing, 10*(1), 333–345.
Ross, I. (1971). Self-concept and brand preference. *Journal of Business, 44*(1), 38–50.
Rossi, P. E., McCulloch, R. E., & Allenby, G. M. (1996). The value of purchase history data in target marketing. *Marketing Science, 15*(4), 321–340.
Russel, G. J., & Kamakura, W. A. (1997). Modeling multiple category brand preference with household basket data. *Journal of Retailing, 73*(4), 439–461.
Russo, J. E., Meloy, M. G., & Husted-Medvec, V. (1998). Predecisional distortion of product information. *Journal of Marketing Research, 35*(4), 438–452.
Sadowski, C. J., & Cogburn, H. E. (1997). Need for cognition in the big-five factor structure. *Journal of Psychology, 131*(3), 307–312.
Sambamurthy, V., & Chin, W. W. (1994). The effects of group attitudes toward alternative GDSS designs on the decision-making performance of computer-supported groups. *Decision Sciences, 25*(2), 215–241.

Sappington, D. E. M., & Wernerfelt, B. (1985). To brand or not to brand? A theoretical and empirical question. *Journal of Business, 58*(3), 279–293.
Schmitt, B. H., & Shultz, II, C. J. (1995). Situational effects on brand preference for image products. *Psychology and Marketing, 12*(5), 433–446.
Sellin, N. (1989). *PLSPath version 3.01 application manual*. Germany: Hamburg.
Sengupta, J., Fitzsimons, G. J.(2000). The effects of analysing reasons for brand preferences: Disruption or reinforcement. *Journal of Marketing Research, 37*(3), 318–330.
Sethuraman, R. & Cole, C. (1999). Factors influencing the price premiums that consumers pay for national brands over store brands. *Journal of Product and Brand Management, 8*(4), 350–351.
Shank, M. D., & Langmeyer, L. (1994). Does personality influence brand image? *Journal of Psychology, 128*(2), 157–164.
Sharma, S. (1981). The theory of stochastic preferences: Further comments and clarifications. *Journal of Marketing Research, 18*(3), 364–369.
Shepard, R. N. (1978). The mental image. *American Psychologist, 33*(2), 125–137.
Sheth, J. N. (1968). How adults learn brand preference. *Journal of Advertising Research, 8*(3), 25–36.
Sheth, J. N., Newman, B. I., & Gross, B. L. (1991). Why we buy what we buy: A theory of consumption values. *Journal of Business Research, 22*(2), 159–170.
Shocker, A. D., & Srinivasan, V. (1979). Multiattribute approaches for product concept evaluation and generation: A critical review. *Journal of Marketing Research, 16*(2), 159–180.
Sirgy, M. J. (1982). Self-concept in consumer behavior: A critical review. *Journal of Consumer Research, 9*(3), 287–300.
Sirgy, M. J. (1985). Using self-congruity and ideal congruity to predict purchase motivation. *Journal of business Research, 13*(3), 195–206.
Sirgy, M. J., Grewal, D., Mangleburg, T. F., Park, J.-O., Chon, K.-S., Claiborne, C. B., et al. (1997). Assessing the predictive validity of two methods of measuring self-image congruence. *Journal of the Academy of Marketing Science, 25*(3), 229–241.
Sivakumar, K. (1996). An empirical investigation of the two roles of price on brand choice. *Pricing Strategy and Practice, 4*(2), 15–22.
Snoj, B., Korda, A. P., & Mumel, D. (2004). The relationships among perceived quality, perceived risk and perceived product value. *The Journal of Product and Brand Management, 13*(2/3), 156–167.
Srinivasan, N., & Ratchford, B. T. (1991). An empirical test of a model of external search for automobiles. *Journal of Consumer Research, 18*, 233–242.
Srinivasan, S. S., & Till, B. D. (2002). Evaluation of search, experience and credence attributes: Role of brand name and product trial. *Journal of Product and Brand Management, 11*(7), 417–431.
Srinivasan, V. (1975). A general procedure for estimating consumer preference distributions. *Journal of Marketing Research, 12*(4), 377–389.
Stafford, J. E. (1966). Effects of group influences on consumer brand preferences. *Journal of Marketing Research, 3*, 68–75.
Stayman, D. M., & Aaker, D. A. (1988). Are all the effects of ad-induced feelings mediated by attitude toward the Ad? *Journal of Consumer Research, 15*(3), 368–373.
Stephen, K. T., Fox, H. W., & Leonard, M. J. (1985). A comparison of preferences concerning the purchase of domestics products versus imports: United States and Jamaica. *Journal of the Academy of Marketing Science, 8*, 100–104.
Stone, M. (1974). Cross-validatory choice and assessment of statistical predictions. *Journal of the Royal Statistical Society, Series B (Methodological), 36*(2), 111–147.
Taylor, S. A., & Baker, T. L. (1994). An assessment of the relationship between service quality and customer satisfaction in the formation of consumers' purchase intentions. *Journal of Retailing, 70*(2), 163–178.
Tenenhaus, M., Vinzi, V. E., Chatelin, Y.-M., & Lauro, C. (2005). PLS path modeling. *Computational Statistics and Data Analysis, 48*, 159–205.

Tenenhaus, M., Amato, S., and Esposito Vinzi, V. (2004). A global goodness-of-fit index for PLS structural equation modelling. *Proceedings of the XLII SIS Scientific Meeting*, Vol. Contributed Papers, CLEUP, Padova, pp. 739–742.

Thakor, M. V., & Katsanis, L. P. (1997). A model of brand and country effects on quality dimensions: Issues and implications. *Journal of International Consumer Marketing, 9*(3), 79–100.

Thorelli, H. B., Lim, J.-S., & Ye, J. (1989). Relative importance of country of origin, warranty and retail store image on product evaluations. *International Marketing Review, 6*(1), 35–46.

Traylor, M. B. (1981). Product involvement and brand commitment. *Journal of Advertising Research, 21*(6), 51–56.

Tse, D. K., & Gorn, G. J. (1993). An experiment on the salience of country-of-origin in the era of global brands. *Journal of International Marketing, 1*(1), 57–76.

Urban, G. L., & Hauser, J. R. (1993). *Design and marketing of new products*. Englewood Cliffs, NJ: Prentice-Hall.

Venkataraman, V. K. (1981). The price-quality relationship in an experimental setting. *Journal of Advertising Research, 21*(4), 49–52.

Wall, M., & Heslop, L. A. (1989). Consumer attitudes toward the quality of domestic and imported apparel and footwear. *Journal of Consumer Studies and Home Economics, 13*, 337–358.

Westfall, R. (1962). Psychological factors in predicting product choice. *Journal of Marketing, 26*(2), 34–40.

Wheatley, J. J., Walton, R. G., & Chiu, J. S. Y. (1977). The influence of prior product experience, price and brand on quality perception. *Advances in Consumer Research, 4*, 72–77.

Witt, R. E., & Bruce, G. D. (1972). Group influence and brand choice congruence. *Journal of Marketing Research, 9*(4), 440–443.

Wold, H. (1982), Soft modeling, the basic design and some extensions. In K. G. Jreskog & H. Wold (Eds.), *Systems under indirect observation*, I–II. Amsterdam: North-Holland.

Woodside, A., & Trappey, III, R. J. (1992). Finding out why customers shop your store and buy your brand: Automatic cognitive processing models of primary choice. *Journal of Advertising Research, 32*(6), 59–78.

Woodside, A. G., & Wilson, E. J. (1985). Effects of consumer awareness of brand advertising on preference. *Journal of Advertising Research, 25*(4), 41–47.

Yang, S., Allenby, G. M., & Fennell, G. (2002). Modelling variation in brand preference: The roles of objective environment and motivating conditions. *Marketing Science, 21*(1), 14–31.

Zaichkowsky, J. L. (1985). Measuring the involvement construct. *Journal of Consumer Research, 12*(3), 341–352.

Zaichkowsky, J. L. (1994). The personal involvement inventory: Reduction, revision, and application to advertising. *Journal of Advertising, 23*(4), 59–70.

Zaichkowsky, J. L., & Vipat, P. (1993). Inferences from brand names. *European Advances in Consumer Research, 1*, 534–540.

Zajonc, R. B. (1980). Feeling and thinking: Preferences need no inferences. *American Psychologist, 35*, 151–175.

Zeithaml, V. A. (1988). Consumer perceptions of price, quality, and value: A means-end model and synthesis of evidence. *Journal of Marketing, 52*, 2–22.

Zhang, S., & Markman, A. B. (1998). Overcoming the early entrant advantage: The role of alignable and nonalignable differences. *Journal of Marketing Research, 35*(4), 413–426.

Zhang, S., & Markman, A. B. (2001). Processing product unique features: Alignability and involvement in preference construction. *Journal of Consumer Psychology, 11*(1), 13–27.

Zhang, Y., & Buda, R. (1999). Moderating effects of need for cognition on responses to positively versus negatively framed advertising messages. *Journal of Advertising, 28*(2), 1–14.

Zinkhan, G. M., & Martin, C. R., Jr. (1987). New brand names and inferential beliefs: Some insights on naming new products. *Journal of Business Research, 15*(2), 157–172.

Chapter 21
An Application of PLS in Multi-Group Analysis: The Need for Differentiated Corporate-Level Marketing in the Mobile Communications Industry

Markus Eberl

Abstract The paper focuses on the application of a very common research issue in marketing: the analysis of the differences between groups' structural relations. Although PLS path modeling has some advantages over covariance-based structural equation modeling (CBSEM) regarding this type of research issue – especially in the presence of formative indicators – few publications employ this method. This paper therefore presents an exemplary model that examines the effects of corporate-level marketing activities on corporate reputation as a mediating construct and, finally, on customer loyalty. PLS multi-group analysis is used to empirically test for differences between stakeholder groups in a sample from Germany's mobile communications industry.

21.1 Motivation

The escalating competition in global markets has compelled companies throughout all industries to analyze their (potential) customer base. Subsequently, the application of differentiation strategies has emerged as an extremely successful possibility in saturated markets (Markwick and Fill 1997). Customer segmentation approaches have inspired quantitative marketing research to develop methods with which to identify customer segments. Simultaneously, a growing stream of research has aimed at broadening the understanding of product- and customer-driven organizational success factors (Hall 1992; Markwick and Fill 1997; Wilson 1985; Weigelt and Camerer 1988). Assets that are intangible by nature can, from a resource-based view, be a strategic success factor for companies, since they cannot be easily imitated by competitors. For various reasons, which are described later, a company's reputation is one of its most interesting intangible assets. Empirical research into the consequences of intangibles – and especially reputation – has been relatively scarce

M. Eberl
Senior Consultant Models and Methods, TNS Infratest Forschung GmbH,
Landsberger Straße 338, 80687 München, Germany
e-mail: markus.eberl@tns-infratest.com

V. Esposito Vinzi et al. (eds.), *Handbook of Partial Least Squares*, Springer Handbooks of Computational Statistics, DOI 10.1007/978-3-540-32827-8_22,

and is only slowly providing empirical evidence of the theoretically postulated effects of a "fine" corporate reputation. The aim of this paper is to show that PLS path modeling can be a helpful tool when the question is whether one should also segment the relevant interested parties in respect of reputation management.

From a methodological point of view, the problem analyzed in this study is quite popular, as it boils down to the analysis of differences in structural relations (e.g., the effect of customer satisfaction on retention) between groups (i.e. subsamples). Although PLS path modeling has advantages (e.g., the softer distributional assumptions, or the possibility to deal with large numbers of formative indicators) over covariance-based structural equation modeling (hereafter referred to as CBSEM) regarding this type of research question, few publications employ this method. This paper therefore presents a typical PLS application in marketing research. Research into intangible resources' outcomes and corporations' success factors generally refers to the question of control levers, i.e. which activities management should preferably undertake in order to achieve a sustainable competitive advantage from this intangible. These drivers are often formative indicators that rule out CBSEM approaches in many cases (MacCullum and Browne 1993; Bollen 1989; Eberl 2006).

The model proposed in the following sections deals with the effects of corporate-level marketing activities on corporate reputation (as an exemplary intangible resource) and, finally, on customer loyalty. Reputation will be modeled with two constructs that mediate the activities' effect on customer loyalty. (Note that according to Baron and Kenny (1986), a moderator is defined as a qualitative (e.g., sex, race, class) or quantitative (e.g., level of reward) variable that affects the direction and/or strength of an independent – or predictor – variable and a dependent – or criterion – variable's relationship, while a variable functions as a mediator to the extent that it accounts for the relation between the predictor and the criterion.) We further introduce a stakeholder group (i.e. a subsample) as a moderator variable into our model in order to explore the possible need for differentiated marketing activities in order to gain a high reputation and customer loyalty in different subgroups. This part of the research is more exploratory in nature and expands the theoretical knowledge to be gained from the hypothesized relationships between the model constructs in the model structure. Owing to the presence of many formative indicators, PLS path modeling is the only simultaneous method to quantify all relationships, including the latent variables in this model. The model is to be tested empirically with a sample from Germany's mobile communications industry. PLS multi-group analysis, with parametric t-testing based on the PLS re-sampling technique (Chin 2000), will be used to empirically test for differences between stakeholder groups in a sample.

The paper is organized as follows: in the next section, the theoretical aspects of the model are presented, which include the concept of corporate reputation and the levers for corporate-level marketing. Thereafter, the measures are presented that were used to capture the latent variables in the model. As mentioned above, PLS path modeling is the only viable methodology with which to estimate these. Section 21.4 briefly discusses the methodology of parametric multi-group comparisons of PLS estimates, which will later be employed to explore stakeholder groups' moderating influence. The empirical results are presented in Sect. 21.5, while the final section concludes the paper and presents some implications for corporate behavior.

21.2 Corporate-Level Marketing and Reputation

21.2.1 Reputation's Consequences

Companies are becoming increasingly aware that a purely shareholder-oriented approach to doing business can be problematic. A firm's long-term goals are often not purely financially oriented, thus affecting a broader set of stakeholders. Besides, sustainable competitive advantages can be more easily obtained from intangible assets than from more product-related sources, as they are much harder to imitate. A company's reputation is one of those intangibles that are extremely hard to imitate (Hunt and Morgan 1995). The literature ascribes many potential company benefits to a "good" reputation: With regard to consumers (Shapiro 1983; Zeithaml 1988), reputation functions as a risk-reduction mechanism (Kotha et al. 2001), leads to higher product satisfaction (Aaker 1991), and ultimately increases loyalty (Rogerson 1983). But one has to acknowledge that although the cited authors agree on the more or less theory-based fact that reputation is a source of competitive advantage, there has been relatively weak empirical evidence of the consequences of a "good" reputation (Roberts and Dowling 2002) as well as the marketing levers that can be used in reputation management.

21.2.2 The Concept of Corporate Reputation

Initially, research into reputation revealed great dissent in respect of the construct's definition. Although many authors have published on the subject, this problem is still, to some degree, present (Fombrun and van Riel 1997; Gotsi and Wilson 2001). Consequently, current research is faced with a large number of different definitions as well as operationalizations of reputation.

21.2.2.1 Definitions

The discussion of definitions – which has been part of reputation research from the start – has not led to an integrative conceptualization that can be used in all research areas in which the term "reputation" is relevant (e.g., sociology, signaling theory, or corporate level marketing). Nevertheless, there are a number of useful definitions of "reputation" that are, to some extent at least, based on scientific work. They differ with regard to the various interested parties' point of view as well as regarding the distinction between "corporate image" and "corporate reputation" (Eberl 2006). Since the discussion of definitions is not integral to this paper, only one definition will be presented (cf. Gotsi and Wilson, 2001; Eberl, 2006). Although there are still certain differences, various authors have tried to provide an integrative definition of recent conceptualizations. In their cross-disciplinary literature review, Gotsi and Wilson (2001) defined corporate reputation as "(..) a stakeholder's overall evaluation of a company over time. This evaluation is based on the stakeholder's direct experiences with the company, any other form of communication and symbolism that provides information about the firm's actions and/or a comparison with the actions of other leading rivals" (Gotsi and Wilson 2001).

21.2.2.2 The Dimensionality of Reputation

Many "reputation indexes" with which to quantify reputation have, however, not been developed according to scientific operationalization procedures. Among these are rankings such as Fortune Magazine's "America's/Global Most Admired Companies" indexes (Hutton 1986) (henceforth referred to as AMAC and GMAC), and a large number of European magazines' indices like Germany's "Manager Magazin Imageprofile". Eidson and Master (2000) as well as Schwaiger (2004) provide an overview of the various measurement concepts, all of which have been criticized to some extent.

An important validity problem in prior reputation research has always been that reputation's multidimensionality has not been in accordance with the relevant conceptualization. This critique is especially valid regarding the Fortune "Most Admired" indices as formulated by Fryxell and Wang (1994). Fortune presents two indices: the AMAC (America's Most Admired Companies) and the GMAC (Global Most Admired Companies). In the AMAC study, an overall reputation score is achieved as the mean of eight attributes rated by experts from within the company's industry on 11-point scales (Hutton 1986). While AMAC only incorporates American companies, the GMAC features the 500 largest companies worldwide. The GMAC overall score is computed from the eight AMAC categories plus one item that refers to the company's international activities. An important problem with these measures is that there is no clear definition of the concept "reputation" (Sobol et al. 1992). Fombrun and Shanley (1990) analyzed the reputation measures' scores and items and concluded that the Fortune scales are problematic because of their unidimensionality (Brown and Perry 1994), and because financial criteria uniformly determine the Fortune data. While the Fortune surveys marked the kick-off of reputation research, most authors publishing on the topic agree that there are other criteria according to which a reputation should be assessed (Dunbar and Schwalbach 2001; Fombrun and Shanley 1990; Herremans et al. 1993; Weiss et al. 1999; Benjamin et al. 1999; Shenkar and Yuchtman-Yaar 1997; Shamsie 2003). Balmer (2001a,b, 2003) makes a strong point for corporate branding as a major research area in twenty-first century marketing. A strong corporate brand cannot be easily assessed, but past financial performance is a prerequisite for a company to be held in high esteem, i.e. to be highly reputed (Balmer 2003). A unidimensional construct would also contradict the definitional framework given above. Therefore, a conceptual broadening of the "reputation" construct, as evaluated with the Fortune data, is necessary.

This broadening is accomplished as follows: for our study, we draw on a definition of reputation as a concept similar to attitudes. Common knowledge from attitude theory provides us with reputation's two-dimensionality, which comprises a cognitive as well as an affective component (Schwaiger 2004). In his empirical study, Schwaiger (2004) likewise conceptualized reputation: Based on a definition of corporate reputation as an attitude-related construct – which is consistent with recent definitions – he modeled reputation with two dimensions, using a cognitive and an affective component. Twenty-one explanatory variables that formed antecedents of corporate reputation were gained from open-ended expert interviews. After pretesting, a large-scale representative data set (3,300 judgements on the 21 driver items) from Germany, the United Kingdom, and the United States was split in half. The first half of the sample was used to explore the strength of these drivers' influence on corporate reputation. Cross-validation with the rest of the sample yielded satisfying results. The model proved to be reliable and valid in explaining the drivers of reputation. The structure of four constructs that drive reputation has been shown to be robust across different data sets, countries and industries (e.g. Eberl and Schwaiger 2004, 2005; Eberl 2006).

This paper suggests that it should also be taken into account that it is possible – through communication – to substitute individuals' direct experiences with surrogate experiences and thus allow a reputation to exist within the overall public (Mahon 2002; Dozier 1993). The term "surrogate experiences" defines the communications of other stakeholders' direct experiences (as customers, employees, media etc.) with a company to third-party stakeholders. This view is in accordance with Fishbein's view of attitudes being ultimately "obtained from direct experiences with objects and from communications about them received from other sources" (Loudon and Della Bitta 1993). Note that the concept explicitly allows reputation to vary within different stakeholder groups.

21.2.3 Antecedents and Consequences of Reputation

While the different concepts of reputation have been thoroughly discussed in literature, recommendations on how a reputation can actually be managed are scarce. Some evidence has been provided to show that corporate-level marketing's activities (comprising product quality as well as corporate communications and corporate social responsibility, etc.) actually influence reputational judgements. Nevertheless, it is not known whether these judgments affect a company's customer-specific marketing objectives (e.g., customer satisfaction and loyalty). Since this paper does not endeavor to present a complete model for all possible outcomes of reputation, the focus on satisfaction and loyalty are deemed sufficient at this stage.

This paper argues that isolated corporate-level marketing activities do not directly influence consumers' loyalty decisions. It is far more plausible that individuals process the perception of a single company activity in the light of existing evaluations stored in their minds, which will eventually lead to a confirmation or contradiction of the existing evaluation called reputation. Hence, reputation is an important mediator in the analysis of corporate-level marketing activities' impact on customer loyalty. It is conceptualized in a two-dimensional way: (1) a dimension comprising all of the stakeholders' cognitive evaluations of the company (which can be labeled "competence") and (2) a dimension capturing all of the stakeholders' affective judgments (which can be labeled "likeability"). It is hypothesized that both dimensions influence customer satisfaction directly, while the emotional dimension also influences loyalty directly. Previous research into the "drivers of reputation" (Schwaiger 2004; Eberl and Schwaiger 2005), i.e. a firm's corporate-level marketing instruments, has identified four formative constructs that aggregate the relevant corporate levers ("quality", "performance", "attractiveness" and "corporate social responsibility"). Figure 21.1 displays the relationships taken into account. The supposition that the two dimensions of reputation are mediators implies the need to test for their mediating influence, which a later section describes.

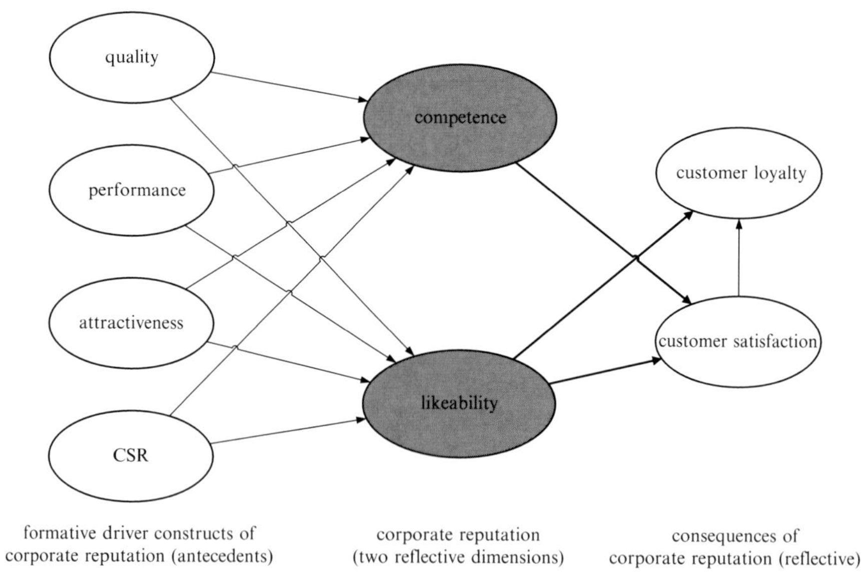

Fig. 21.1 Research model

21.2.4 Stakeholder-Specific Reputation Management

The analysis of the path coefficients in the proposed model allows for a detailed prioritization of marketing activities' levers on the four driver constructs' aggregate level as well as on the more detailed level of the formative indicators used to operationalize the constructs. Since it is in the very nature of stakeholder groups to have differing interests in respect of a company, it is plausible that some of those stakeholder groups will tend to weigh the various aspect of a company's reputation and behavior differently, which will lead to marketing activities influencing reputation to various degrees. Moreover, it is possible that by assessing companies professionally, and thus on a more cognitive basis, the cognitive components of reputational aspects will have a relative dominating effect on customer satisfaction and loyalty. The relative importance of the influence of reputation's cognitive dimension compared to that of its affective dimension is key to the model's interpretation. This is due to this influence's importance, inducing a choice of some marketing levers over others in reverse: When choosing which corporate-level marketing activity to emphasize, it will be this activity's contribution (i.e. the path coefficient) to competence and likeability that will lead to the company taking concrete measures. And the relative importance of competence and likeability for satisfaction and loyalty (i.e. the specific path coefficient) will provide the answer to whether competence or likeability will have greater influence on satisfaction and loyalty. If stakeholder groups react differently to corporate-level activities, this implies that a company has to act in a differentiated and segment-specific way. Consequently, the stakeholder group should be interpreted as a discrete moderator variable. This moderator may account for differences in reputation levers' strength. The analysis of group differences undertaken in this study therefore seeks to answer two questions of a more exploratory nature:

(1) Is there such a moderating effect *at all*, i.e. do drivers of reputation and satisfaction differ depending on the stakeholders?
(2) Is this an issue for all stakeholder groups and all paths in the hypothesized model, or are there some levers for reputation management that should be employed in subgroups only?

21.3 Operationalization and Measures

The operationalization of the four exogenous constructs that drive reputation ("quality", "performance", "attractiveness", and "corporate social responsibility") is based on previous research by Schwaiger (2004) as well as Eberl and Schwaiger (2005), in which a procedure similar to Rossiter's C-OAR-SE approach has been applied (Rossiter 2002). If one ignores this approach's dogmatic rejection of statistical measures in the item validation process (cf. the recent critique by Diamantopoulos 2005), it is a good guideline for operationalization. In interviews with experts from various industries, corporate reputation was briefly defined and discussed to ensure a common understanding of reputation and the constructs quality, performance, attractiveness, and CSR. The experts were then asked to take a broad view of their organizational environment and think of aspects that could possibly drive their company's reputation in the four fields of organizational quality, the performance aspects, the company's attractiveness, and its responsible behavior. After gaining 21 items for the four constructs (presented in the appendix), the causal direction between each indicator and its respective construct was inspected (Chin 1998b; Jarvis et al. 2003; Eberl 2004). It was found that the 21 indicators have to be specified as formative. This is not surprising, since the aim of the expert interviews was to identify drivers, i.e. antecedents of reputation. This is also in accordance with the intention to model quality, performance, attractiveness, and social responsibility as driver constructs for corporate reputation and, ultimately, being able to identify important levers (i.e. the indicators). The measures thus capture the stakeholders' assessment of 21 levers for corporate-level marketing activities and can be used as input variables in respect of reputation management and controlling. For example, corporate social responsibility is captured by the aspects

- "I have the feeling that [company] is not only concerned about profit"
- "[company] is concerned about the preservation of the environment"
- "[company] behaves in a socially conscious way"
- "I have the impression that [company] is forthright in giving information to the public" and
- "I have the impression that [company] has a fair attitude towards competitors"

Intuitively, it is possible for a company to be forthright in giving information to the public, while simultaneously not necessarily behaving in a socially conscious way (in the eyes of the stakeholders). Therefore, these indicators need not necessarily correlate from a theoretical point of view. The same argument applies to the rest of the indicators presented here and in the appendix. These aspects represent target variables for marketing activities. The extent of their effects on customer-specific target variables such as customer satisfaction and loyalty is therefore crucial for the prioritization of such activities.

The three indicators gained in respect of competence as well as the three indicators of likeability were identified as being exchangeable indications of their underlying constructs and were treated as reflective (Schwaiger 2004). Likeability was operationalized by the following items:

- "[company] is a company that I can better identify with than with other companies"
- "[company] is a company that I would more regret not having if it no longer existed than I would other companies", and
- "I regard [company] as a likeable company"

The measures of the cognitive dimension of reputation were:

- "[company] is a top competitor in its market"
- "As far as I know, [company] is recognized world-wide" and
- "I believe that [company] performs at a premium level"

Finally, the constructs of customer satisfaction and loyalty were operationalized with reflective measures that are well known in empirical marketing studies (Zeithaml and Berry 1996): overall satisfaction, intention to repurchase, propensity for recommendation, and intention to remain a customer in the long-run.

All the items in the study were measured with 7-point rating scales. However, a reassessment of the specification of the constructs likeability, competence, customer satisfaction, and loyalty via expert interviews could not verify that the measures have to be specified in a formative fashion.

21.4 PLS Path Modeling in Multiple Groups

The proposed model presents an application of PLS path modeling in corporate-level marketing. An interpretation of the path coefficients that determine the four formative constructs allows corporate-level marketing measures, as represented in the formative indicators, to be prioritized (MacCullum and Browne 1993). In fact, the same model would not be identified in a CBSEM environment.

"Stakeholder group" can be described as a moderator variable in this model. As such, it is hypothesized as influencing the other main effects' strength in the model (Baron and Kenny 1986) (i.e. the effects of the four exogenous driver constructs as well as the impact of the two reputational dimensions on customer satisfaction and loyalty). There are several ways of including moderating effects within PLS path models.

21.4.1 *Moderating Influences Within Structural Models*

One way would be to include an exogenous interaction term within the model. The model would then not only comprise the main effect under consideration (a) and the moderator variable's main effect on the endogenous variable (b), but also an interaction variable's effect (c) (predictor × moderator). Proof of moderation would be provided if path c was found to differ significantly from zero (Baron and Kenny 1986). This approach is especially appealing for continuous (and latent) moderator variables, but cannot be used in path modeling with covariance-based techniques (CBSEM). These models assume that the correlation between latent variables' error terms equals zero. These assumptions would, of course, be violated by the very way in which the interaction term is constructed. PLS, conversely, has no such restriction, so that the interaction technique is a feasible alternative for testing moderation in PLS models. Chin et al. (2003) point out that due to PLS' bias, it is actually superior to traditional OLS regression in respect of the same model: "While problematic if not accounted for within covariance-based modeling software such as LISREL, these correlations may actually help provide a more accurate estimation of the interaction effect when using PLS" (Chin et al. 2003).

The construction of the interaction term ($a \times b$) is accomplished by formulating a latent interaction variable. The cross-product of the predictor and moderator variables' indicators yields the indicators of the latent interaction variable (Chin et al. 2003). This approach can be applied without any drawbacks if both the predictor and moderator construct are modeled as having reflective indicators. If, however, at least one of the two constructs is operationalized in a formative fashion, the cross-product of the indicators must not be applied: "Since formative indicators are not assumed to reflect the same underlying construct (i.e. can be independent of one another and measuring different factors), the product indicators between two sets of formative indicators will not necessarily tap into the same underlying interaction effect" (Chin et al. 2003). It is therefore recommended that in respect of a formative predictor and/or moderator variable, the latent variable scores of one or both should, as a first step, be estimated in a main effects model and that the latent interaction variable should then be constructed as a single-indicator construct via the product variable of the two latent score variables. One drawback of this approach is, however, that it is not possible to interpret the moderator variable's impact on the predictor variable's weights (and/or loadings). This is a drawback when conducting driver analysis

21.4.2 Multiple Group Analysis

The second approach to the analysis of moderating effects in path models is multiple group analysis, which is especially useful for discrete moderator variables (e.g., sex, customer status [yes/no], stakeholder group). Group comparisons are also used in CBSEM environments (Jöreskog 1971), but can also be applied in PLS (Chin 2000; Keil et al 2000). Basically, a discrete moderator variable can be interpreted as dividing the data into groups of subsamples. The same PLS path model can then be estimated in each of the distinct subsamples. CBSEM models usually report having used different measures for global fit (based on their hard distributional assumptions), which allows for a statistical assessment of the group differences in terms of the structural invariance between the groups. This approach is an easy-to-apply instrument for testing discrete moderators. Nevertheless, the approach does have some drawbacks. One of the prerequisites of this parametric testing procedure is that – as in any t-test – the data is largely normal. This is a huge problem in many applications, since violation of the assumption may lead to biased results. Recent publications have, however, developed alternatives. Dibbern and Chin (2005) proposed an alternative distribution-free approach by using a random permutation procedure in accordance with Edgington (1987) and Good (2000). This rather new approach to PLS path modeling is a very interesting alternative for further research on this topic. For more information, see also the paper by Chin in this handbook. As the approach requires a huge number of simulation runs, and has not, for practical reasons, been used in practical research to date, it was not adopted in this study either.

It is not possible to compare groups in PLS by using a global criterion. However, there is a possibility to compare the path coefficients between two groups at a time, which allows an interpretation of the differences in effects between groups. In the context of this study, the different mechanisms concerning impact on loyalty can be revealed by comparing models' resulting path estimators across groups. According to Chin, these structural differences can, furthermore, be tested for significance with pair-wise t-tests (Chin 2000). This approach merely requires, that (1) every model considered has to be acceptable in terms of goodness of fit (not necessarily equal goodness of fit), (2) the data should not be too non-normal, and (3) there should be measurement invariance (Chin 2000). The approach uses the re-sampling estimates for the standard errors of the structural paths in two samples under consideration gained from the bootstrapping procedure usually used for model evaluation (Chin 1998a). Differences between the path estimators are tested for significance with a t-test. The approach's test statistic has to be constructed according to the fact whether the standard errors of the path estimators in the two subgroups are equal or not. If they are equal, the test statistic is computed as follows (Chin 2000):

$$t = \frac{Path_{sample1} - Path_{sample2}}{\sqrt{\frac{(m-1)^2}{(m+n-2)} \cdot s.e._{sample1}^{2} + \frac{(n-1)^2}{(m+n-2)} \cdot s.e._{sample2}^{2}} \cdot \sqrt{\frac{1}{m} + \frac{1}{n}}} \sim t_{m+n-2} \tag{21.1}$$

with

$Path_{sample1/2}$	original sample estimate for the path coefficient in both subsamples respectively
m	number of cases in sample 1
n	number of cases in sample 2
$s.e._{sample1/2}$	standard error of the path coefficient in both subsamples respectively (gained from the re-sampling procedure implemented in PLS)

Should there be evidence of the standard errors' inequality in the two groups, the test statistic can be computed as (Chin 2000):

$$t = \frac{Path_{sample1} - Path_{sample2}}{\sqrt{s.e._{sample1}^{2} + s.e._{sample2}^{2}}} \tag{21.2}$$

Further, the t-test's degrees of freedom (df) would then have to be computed as follows:

$$df = \frac{\left(s.e._{sample1}^{2} + s.e._{sample2}^{2}\right)^2}{\left(\frac{s.e._{sample1}^{2}}{m+1} + \frac{s.e._{sample2}^{2}}{n+1}\right)} - 2 \tag{21.3}$$

The groups can be compared pair-wise. If there are moderators with more than two realizations, the groups have to be compared pairwise before an overall interpretation of the results is undertaken.

21.5 Data and Results

21.5.1 Germany's Mobile Communications Market

The model was tested in Germany's mobile communications market on four major service providers. Together, they have more than 71 million customers with a market penetration rate in 2004 of approximately 82%. The market is, therefore, close to saturation. Consequently, there has been a steep decline in prices since the beginning of 2003. The market's increasing competitiveness means that customer loyalty is an important issue in the industry. It is also difficult to maintain product innovations and product-based competitive advantages. Thus, corporate-level activities are an important possibility for differentiation. At the same time, the corporations are facing various stakeholder groups with very different demands (e.g., environmental issues in respect of the discussion of electromagnetic radiation vs. customers' concerns regarding availability). Companies in the market therefore face the question of how the different activities that they could undertake could affect the different stakeholder groups.

21.5.2 Sample Demographics

Together with experts from the mobile communications industry, four stakeholder groups were identified as being most important for the industry: customers, media representatives and opinion leaders, politicians, and the financial community (which includes analysts and other opinion leaders in the financial industry).

Data were collected by means of CATI interviews in February 2005. The subjects rated the indicators of reputation (i.e. competence and likeability) and the driver constructs of the four service provider companies on 7-point Likert scales. Each interviewee was asked about his satisfaction with and loyalty regarding his own service provider. Customers were randomly selected from the general public, while the other stakeholders were randomly selected from industry databases. Since it was difficult to rule out the possibility of a person belonging to various stakeholder groups at a time, all interviewees belonging to the stakeholder group "customers" had to be described as being "customers only." It was believed that in the other groups, each group's specific characteristics would provide politicians or members of the financial community or media with another perspective of looking at the firm. Simultaneously, the results would not be distorted by the effects of not having personal experience with the company.

The dataset comprises a total of $N = 352$ persons representing the four most important stakeholder groups identified: representatives of the media ($n = 34, 9.7\%$), politics ($n = 58, 16.5\%$), the financial community ($n = 50, 14.2\%$), and randomly selected persons representing the general public and customers ($n = 210, 59.7\%$). The aggregate sample's demographic characteristics correspond to the distribution in the total overall population. The demographics of the two subsamples "politics" and "media" deviate slightly from the aggregate sample's demographic characteristics, which is not surprising.

21.5.3 Mediation in the Model

PLS' estimation of the model and bootstrapping was performed with SmartPLS (Hansmann and Ringle 2004), employing the centroid weighting scheme and the construct-level sign change option in the bootstrapping procedure (for a discussion of weighting schemes applied in the PLS algorithm cf. Lohmöller 1989 and Chin and Newsted, 1999).

Before one can assess a stakeholder group's moderating effects, it is necessary to clarify whether the two reputational dimensions can at all be justified as mediators beyond theoretical aspects. The research model proposed in this paper may be interpreted as suggesting that corporate reputation functions as a mediator variable of the four driver constructs quality, performance, attractiveness, and corporate social responsibility. While the theoretical point differs somewhat, the mediating structure presented here is clear. It raises the issue of how reputation can be justified as a variable in the model if the latter does not assume that the four drivers have direct effects on both customer satisfaction and loyalty.

However, the concept of reputation as used in this paper is a rather elaborate concept that captures a construct and simultaneously allows the analysis of its drivers. On the other hand, reputation is covered by its two dimensions, and the drivers used actually capture the aspects that drive reputation. One can interpret the model as a type of second-order formed construct as defined by Rossiter (2002): the four constructs quality, performance, attractiveness, and CSR are antecedents and thus "form" the two reputational constructs competence and likeability. Of course, in technical terms, reputation is not a second-order construct, as it is also operationalized by means of unique indicators.

Thus, one surmises that without the "mediator" reputation, the four constructs quality, performance, attractiveness, and corporate social responsibility could explain customer satisfaction and loyalty well, i.e. that the two constructs competence and likeability are unnecessary and do not contribute to variance explanation beyond theoretical aspects. Our model's derivation does not allow us to share this point of view: starting off with reputation as our focal concept, we incorporated the four constructs on the left-hand side of our model as antecedents, and customer satisfaction as an external criterion in order to validate the outcomes of reputation. Nevertheless, this would not generally rule out the possibility of also taking the four driver constructs' direct effects on satisfaction and loyalty into account. Our theoretical assumption therefore aims at perfect mediation. Note that in this case, a direct relationship between two variables a and b is not significantly different from zero in a model that also takes a mediator variable into account (and, consequently, the paths $a \rightarrow b$ and $c \rightarrow b$) (Baron and Kenny 1986). An example of such a relationship can be found in the stimulus-organism-response (SOR) models that have found the mediating constructs of the individual to mediate the relationships between the observable input stimuli and output responses.

To underpin our theoretical assumption of perfect mediation with empirical results, we therefore tested the mediating effect of corporate reputation's two dimensions with an alternative model. This alternative model also comprised the driver constructs' four direct links to customer satisfaction. Generally, mediation in path models can be assessed by examining the relationship of the direct link between two latent variables (c) and the indirect link via the potential mediator variable (path a from the predictor to the mediator and path b from the mediator to the endogenous variable). Mediation can be assumed if $H_0 : a \times b = 0$ can be rejected. The asymptotical normally distributed $z = \frac{a \cdot b}{\sqrt{b^2 \cdot s_a^2 + a^2 \cdot s_b^2}}$ (Sobel 1982) can be used as a test statistic.

VAF (variance accounted for) can be used as a means of assessing the size of the effect: $\text{VAF} = \frac{a \cdot b}{a \cdot b + c}$ (Shrout and Bolger 2002). The z-test for the overall alternative model proposed in this paper ($n = 352$) yielded a significant ($p < 0.1$) mediation effect in respect of likeability as a mediator for attractiveness, quality and CSR's influence on satisfaction. For each of the four driver constructs, at least one of the two reputational dimensions showed values > 25.89% (even in the case of performance, although this was not significant). These results are very consistent with the results of the overall model to be presented later and provide the theoretical aspects discussed in this section with further evidence from a purely data-oriented perspective.

From this sections' results, we conclude that the model structure provided is appropriate, and these results therefore uphold our rejection of there being direct effects between the drivers (quality, performance, attractiveness, CSR) and the outcomes (satisfaction and loyalty). In the next step, we attempt to answer our actual research question regarding the existence of stakeholder groups' moderating effects.

21.5.4 The Overall Model and the Moderating Effect of Stakeholder Groups

Table 21.1 provides a brief overview of the path coefficients in the overall sample as well as the different stakeholder groups. While all four groups were compared pairwise, the following section will present a more thorough discussion of the differences between the stakeholder groups that were found to be significant, or of certain interest to the mobile communications industry.

Table 21.1 PLS path estimators for the complete sample and the stakeholder subsamples

n	all stakeholders 352		media 34		politicians 58		financial community 50		general public 210	
	coeff.	t	coeff.	t	coeff.	t	coeff.	t	coeff.	t
quality → competence	**0.453**	**2.496**	**0.769**	**5.633**	**0.431**	**2.350**	**0.472**	**2.443**	**0.467**	**2.746**
performance → competence	**0.293**	**1.853**	0.067	0.490	**0.306**	**1.743**	0.142	0.735	**0.314**	**2.053**
attractiveness → competence	0.095	0.674	0.207	1.472	0.131	1.137	0.220	1.424	0.037	0.288
csr → competence	0.021	0.156	−0.181	1.585	0.076	0.521	−0.042	0.259	0.043	0.304
quality → likeability	**0.393**	**2.205**	0.172	1.055	**0.569**	**3.202**	**0.355**	**1.884**	**0.410**	**2.514**
performance → likeability	0.124	0.808	**0.667**	**3.966**	0.138	0.785	0.111	0.680	0.056	0.324
attractiveness → likeability	0.158	1.019	−0.177	1.124	0.107	0.739	**0.352**	**2.278**	0.164	1.039
csr → likeability	0.160	1.211	0.152	1.014	0.093	0.712	0.078	0.663	0.220	1.638
competence → satisfaction	0.187	1.071	0.148	0.795	0.078	0.411	−0.023	0.157	0.187	1.039
likeability → satisfaction	**0.363**	**2.497**	**0.521**	**4.453**	**0.473**	**3.006**	**0.433**	**2.937**	**0.416**	**2.404**
satisfaction → loyalty	**0.524**	**4.400**	**0.650**	**6.563**	**0.405**	**3.831**	**0.480**	**4.780**	**0.523**	**4.834**
likeability → loyalty	**0.325**	**2.731**	0.154	1.444	**0.479**	**5.010**	**0.307**	**2.938**	**0.344**	**2.976**

21.5.4.1 Results for the Overall Model

The R^2 values of the endogenous reflective construct customer loyalty are very acceptable in respect of the overall model (0.545) as well as regarding each subsample (0.56 for media, 0.59 for politicians and general public and 0.45 for the financial community model). Table 21.2 displays the complete list of all endogenous constructs and submodels' R^2 values in the discussion of differences between subgroups. The results show that a corporate reputation's dimensions are actually good predictors of the latent variable customer loyalty. Figure 21.2 presents the results of the overall path model.

In contrast to other industries (Schwaiger 2004; Eberl and Schwaiger 2005, 2004), the affective dimension clearly dominates. The t-values of the paths of competence in respect of satisfaction are relatively small compared to those of likeability in each subsample. This implies that in the mobile communications market, investments in a favorable assessment of corporate competence do not necessarily pay

Table 21.2 Goodness of fit for endogenous constructs

	overall model		general public		media		politics		financial c.	
	R^2	α	R^2	α	R^2	α	R^2	α	R^2	α
Competence	0.631	0.777	0.628	0.750	0.788	0.879	0.779	0.815	0.522	0.727
Likeability	0.546	0.824	0.560	0.803	0.576	0.836	0.733	0.844	0.587	0.851
Cust.loyalty	0.545	0.833	0.585	0.850	0.557	0.815	0.587	0.773	0.448	0.840
Satisfaction	0.252	n/a	0.308	n/a	0.372	n/a	0.292	n/a	0.177	n/a

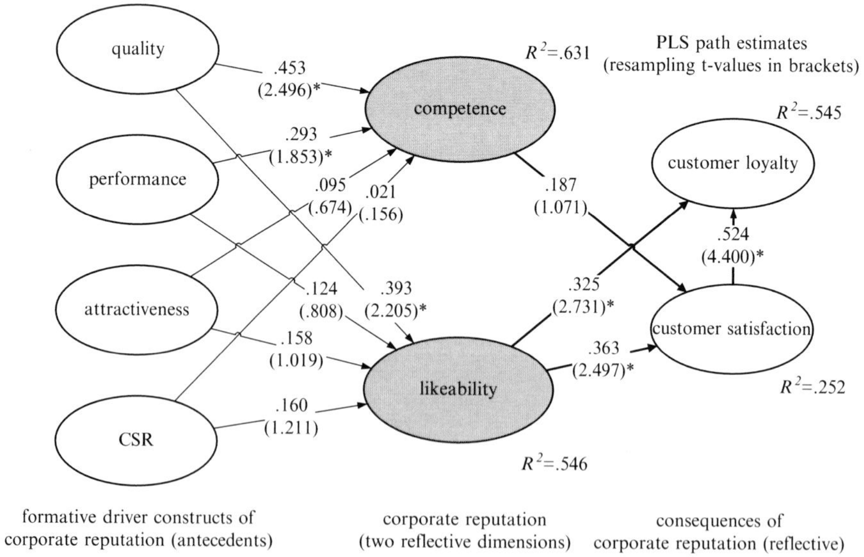

Fig. 21.2 Results of the overall model (aggregated dataset) (*: $p < 0, 1$)

off in terms of an increase in customer satisfaction (and, consequently, in loyalty). Being regarded as a competent firm does not pay off in terms of consumer satisfaction. This is consistent with the characteristics of the market's products: as discussed, it is very difficult to gain product-based competitive advantages because the consumer finds the product "mobile telecommunications" – a rather intangible product – very difficult to distinguish. The data therefore support the notion that the emotional component is the driver that the industry should preferably use as a target variable when conducting corporate-level marketing. This effect can be found when looking at the aggregate sample of all the stakeholders as well as at that of the four individual subsamples. Hence, this seems to be an industry effect rather than a stakeholder group effect. Note that the strength of the relationship between competence and satisfaction is based on a cross-sectional analysis. It could be hypothesized that competence may be a penalty factor in the industry, eventually leading to dissatisfaction if the company falls short of a certain threshold. Since this could also be an effect of the multicollinearity of reputation's two dimensions, we also inspected the correlation between competence and likeability, but found that the correlation of 0.431 was not a source of interpretational bias. Further, the model only considers the firm's customer-specific goals. It is plausible that the stakeholders' good assessment of the cognitive component will have positive consequences in other aspects of corporate governance. These effects would have to be investigated in a separate research model. The results of this model suggest that it is more promising for companies to conduct corporate-level marketing activities with the aim of maximizing likeability instead of competence.

The results of the impact of the four driver constructs quality, performance, attractiveness, and corporate social responsibility differ quite substantially in the four stakeholder groups. A discussion of the aggregate sample's level will therefore be postponed in favor of a more thorough discussion of the differences between the stakeholder groups in the next subsection.

21.5.4.2 Differences Between Stakeholder Groups

Prerequisites for employing multiple t-tests for group comparisons

As discussed in Sect. 21.4.1, the procedure of comparing multiple groups with pairwise t-tests as performed in this paper is subject to several assumptions about the data and the model: (1) the data should not be too non-normal, (2) each submodel considered has to achieve an acceptable goodness of fit, and (3) there should be measurement invariance (Chin 2000).

We visually inspected normality by means of QQ-plots, which, for brevity's sake, are not presented in this paper. Visual inspection of normality is the normal way of checking distributional assumptions when dealing with quasimetric scales – such as the symmetric 7-point rating scale that we employed (Bromley 2002). The author carried out the visual inspection. A later validation by an expert who was unfamiliar with the aim of this study did not alter the results. None of the 31 variables that were used in the analysis were found to deviate strongly from the distributional assumption. We employed consecutive F-tests to decide on the equality of the standard errors gained from the resampling procedure implemented in PLS. The test's null hypothesis of variance homogeneity was rejected at the 0.05 level in only 4 out of 186 tests; therefore, formula (21.1) (p. 497) was employed in respect of all comparisons.

To check that each submodel considered achieved acceptable fit, we relied on the R^2 values realized in respect of the endogenous constructs in each subgroup, since there is no other overall parametric criterion in PLS. Table 21.2 shows the R^2 values of all endogenous constructs in all subgroups. All values are very acceptable within the usual boundaries of interpretation. Note that it is not surprising that the proportion of explained variance for the construct customer satisfaction is lower than for the other endogenous constructs. A brief look at the literature dealing with customer satisfaction reveals a huge number of possible determinants of satisfaction, only one of which refers to intangible assets like corporate reputation. In fact, we were pleasantly surprised that reputation could explain satisfaction of approximately 20–30% – quite a large percentage. Table 21.2 also displays the reliability values (coefficient α) of the reflective constructs. They are also well within the boundaries usually required for acceptance (≥ 0.7).

The final prerequisite for group comparisons to be made is measurement invariance, i.e. the loadings and weights of the eight constructs' measurement models must not differ significantly within the model. This is to ensure that the paths compared in the test are comparable in terms of the causal relationships that they represent. In this study, the measurement invariance of the constructs is also compared with pair-wise t-tests, i.e. the same procedure of pair-wise comparisons was followed for all measurement models as well as for the path coefficients later on. At the 5% level, no difference between any subsample was significant, but at the 10% level, the following differences were found to be significantly different: (1) three indicators of quality between the models for media and politics ($p = 0.0533, 0.0692$ and 0.0958 respectively), (2) one indicator of quality between the models for media and financial community ($p = 0.0829$), and (3) one indicator of CSR between the models for media and politics. We conclude that, in terms of our test procedure, structural invariance is given for two reasons: (1) the number of significant differences found between the groups is only a very small fraction of all $6 \times 31 = 186$ tests performed and (2) the differences appear mainly in respect of the construct quality, which uses a relatively large number of indicators for operationalization. We conclude that, between different groups, this formative construct's content is not heavily biased by this result. Consequently, we can proceed to the interpretation of the results of the subgroups.

Interpreting the group differences

Table 21.1 shows the estimated values of the structural relations within the subsamples. Besides the cognitive reputation dimension's minor effect – which is found in every subsample –, the four stakeholder groups seem, first of all, to have rather different levers for the construction of a "good" reputation: Table 21.1 reveals that the relative importance of quality, performance, attractiveness, and CSR differs quite substantially within the four subsamples. Hence, one can see the different mechanisms at work in the various stakeholder groups from the path coefficients' absolute values and the t-values reported. The pair-wise t-tests performed to test for structural invariance in keeping with 21.4.1, allow these differences to be analyzed with respect to significance.

As this part of the study is more exploratory in nature, we do not formulate separate hypotheses for the 12 (paths) $\times$ 6 (group comparisons) $=$ 72 differences between groups, but stick to the general rationale, as established in Sect. 21.2.4, that stakeholders may react differently to corporate-level marketing activities (i.e. changes in the drivers of reputation). If the path coefficients in two subsamples are not significantly different, one could conclude that the strength of the influence between the two constructs involved is generalizable (with respect of the two groups involved). This conclusion could also be interesting for a company, as this aspect does not have to be treated separately in respect of each of the stakeholder groups. Of course, if a null-hypothesis cannot be rejected, this does not imply proof of it in a statistical sense, due to the possibility of beta errors. In terms of this section, paths should therefore only be cautiously interpreted as "generalizable between stakeholder groups." Nevertheless, because structural equivalence between groups cannot be rejected, this will also be relevant information for the industry. This is also valuable information for reputation management. In our study, structural invariance could not be rejected for a number of paths in pair-wise comparisons, therefore, for brevity's sake, we will concentrate on some interesting significant differences in this section. Interpretation of the non-significant differences will certainly be an important aspect of the marketing implications discussion in the next section.

Two stakeholder groups can be considered "professional judges of corporations": the media and the financial community. In these two subsamples, performance's influence on reputation's cognitive dimension is very small. On the other hand, corporate "quality" aspects are the predominant drivers of "competence" in these two groups and not in the other stakeholder groups. This is indicative of these two groups' wider view of the company.

The importance of the media

Interestingly, the media turn out to be the most interesting subgroup, as more structural invariance hypotheses can be significantly rejected due to this group than other subsamples. The other stakeholder groups seem a little more homogenous with respect to implications for corporate behavior. We will consequently discuss the differences between the media and the other subsamples in a little more depth, beginning with a look at the resulting path coefficients estimated for the subgroup. Besides, the media are anyway a very important stakeholder group due to their

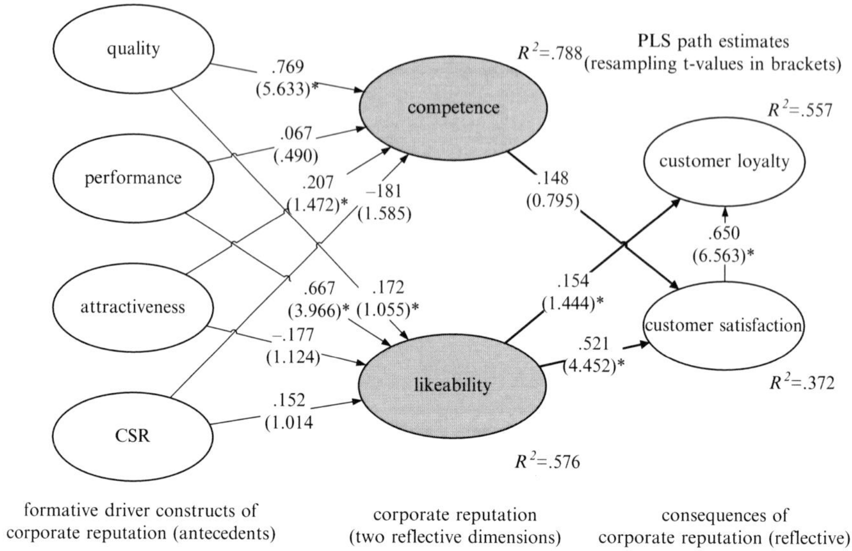

Fig. 21.3 Results of the model in respect of the subsample media (*: $p < 0, 1$)

function as opinion leaders (because we conceptualized reputation as also being based on communicated messages of interactions with the company, e.g., via the media). Figure 21.3 presents the estimated path coefficients of this subsample.

Testing the differences between the media and the other subgroups

The media are a very interesting target for the mobile communications firms' stakeholder management, since the same marketing levers in terms of corporate reputation have quite different consequences when compared to the consequences in other stakeholder groups: The negative effect of perceived CSR activities seems to express a critical perspective. This is underpinned by the fact that likeability has a relatively small influence on customer loyalty.

This perspective is further strengthened when examining the results of the pair-wise t-tests that tested the differences between the path estimators in the four groups. Table 21.3 shows the total differences between the media subsample's path estimators in comparison to those of the other stakeholder groups as well as the results of the pair-wise t-tests. Negative differences imply that the specific path is larger in the media subsample. The differences are very large in many constellations, even though only some pairs prove to be significant. It is very possible that this is an effect of sample size and the test's resulting degrees of freedom. For example, the impact of performance on likeability is significantly larger in the media group than in the politician group (difference = 0.440), or the customers selected from the general public (difference = −0.525). Yet, only the first difference is significant. As can be

Table 21.3 Differences in path estimators between the "media" group and other groups

differences of path coefficients	**politicians**			**financial community**			**general public**		
of subsample **media** to subsample..	diff.	pooled s.e.	p	diff.	pooled s.e.	p	diff.	pooled s.e.	p
quality → competence	−0.286	1.197	0.272	0.274	1.156	0.289	−0.234	2.302	0.583
performance → competence	0.225	1.154	0.369	−0.140	1.162	0.589	0.222	2.076	0.563
attractiveness → competence	−0.072	0.847	0.695	0.037	0.984	0.866	−0.174	1.746	0.590
csr → competence	0.198	0.963	0.344	−0.130	0.960	0.544	0.201	1.910	0.570
quality → likeability	0.423	1.210	0.109	−0.213	1.183	0.420	0.271	2.217	0.509
performance → likeability	−0.440	1.208	**0.095**	0.471	1.078	**0.053**	−0.525	2.351	0.228
attractiveness → likeability	0.180	1.028	0.420	−0.420	1.011	**0.065**	0.276	2.149	0.488
csr → likeability	−0.074	0.944	0.718	0.054	0.836	0.772	0.032	1.828	0.925
competence → satisfaction	−0.038	1.312	0.894	0.147	1.037	0.525	0.042	2.450	0.926
likeability → satisfaction	−0.077	1.027	0.729	0.096	0.903	0.634	−0.109	2.337	0.801
satisfaction → loyalty	0.329	0.686	**0.029**	−0.156	0.689	0.311	0.172	1.575	0.555
likeability → loyalty	−0.250	0.724	0.113	0.171	0.650	0.240	−0.121	1.466	0.656

derived from the formula (21.1) on p. 497, the pooled standard error used to weigh the path differences is a function of the respective sample sizes. The large sample size difference thus leads to an increase in the pooled standard errors. Consequently, absolute path differences of about 0.5 turn out to be significant in a comparison of the media and financial community and not significant in a comparison of the media and general public. It is obvious that the path estimators in the media subsample are very distinct from the paths in the other stakeholder groups.

Differences between the subsample media's path coefficients and other subgroups

When comparing the results of the path estimators in Table 21.3, one finds that in the media sample, product and corporate quality aspects covered by the formative driver "quality" explain an overwhelming proportion of the competence assessment in the group, while performance is a driver of likeability rather than competence. Interestingly, performance has a larger impact on the assessment of likeability than corporate social responsibility. The company's communication and reputation management should therefore stress performance drivers rather than corporate social responsibility when communicating with the media. While this may seem counterintuitive at first, industry experts provided a possible explanation: as the mobile communications industry is a relatively young industry, the rise and fall of the new economy has led to the media regarding mobile communications firms quite skeptically. High organizational performance may therefore be regarded as a sign of the company's sustainability, therefore minimizing fears that the company could be lost.

Furthermore, corporate social responsibility (CSR) as perceived by stakeholders has a negative effect on the judgment of competence in the media and financial analysts sample, while it only has a positive effect on the emotional component likeability. This is indicative of a very rationalist perspective of the company: media stakeholders apparently include the cost of CSR activities in their calculations of corporate competence and disregard their somewhat financial assessment of competence due to corporate social behavior. Conversely, modest CSR may lead to the assessment of competence increasing; this effect can be observed in stock markets' favorable evaluation of rationalization programs. This effect also holds true for the financial community stakeholder group as can be derived from Table 21.1 (p. 501).

Although the t-values are relatively small, an investment in CSR activities and the communication of those activities to all stakeholders alike would present the firm with a trade-off: whilst CSR activities foster the assessment of likeability, they negatively influence the assessment of competence. But since competence is less important for customer satisfaction in the industry, the positive effects of CSR activities on likeability outweigh the negative effects on loyalty through the media and financial communities' decreased competence judgment.

When, in an undifferentiated fashion, conducting similar corporate-level marketing activities as a means of reputation management in respect of all stakeholder groups, the consequences would differ greatly in respect of each stakeholder group. Although a positive change in competence would result in a positive change in respect of customer satisfaction in the media sample, the same initial change in competence assessment would result in a notably larger change in the financial community subsample.

Further, Table 21.3 (p. 507) reveals that the payoff of a good reputation is lower for the media than for politicians, which is expressed as a lower path coefficient for all the relations that have been taken into account. Yet, in the media subgroup, satisfaction seems to have a significantly higher impact on customer loyalty. On the other hand, the media seem to be a more promising target group for reputation management than the financial community (differences in the path estimators are greater than zero for all the considered consequences of reputation).

21.6 Marketing Implications

In the first place, the overall model's results make a strong case for corporate reputation management's effectiveness in terms of customer-specific goals: corporate reputation does have a positive impact on customer satisfaction and loyalty. Intangible assets thus play a key role in differentiation, besides mere product-attributed differentiation. We can therefore underpin the theoretical assumptions regarding the effects of a positive reputation with these empirical findings. We further find that the mobile communications industry is to a greater degree subject to reputation's affective aspects. This is not surprising, as differentiation via product features is quite hard to accomplish in this market. This positive link can also be regarded as a potential threat for companies in the industry: a drop in likeability during communicational crises (e.g., product harm crises, boycotts or scandals) may eventually lead to a drop in customer satisfaction and loyalty. A strong reputation may provide the means with which to attenuate the reputational effects of a crisis. Companies looking for control levers need to know which levers can be used to achieve this. The present study can answer this question through the example of the mobile communications industry. Companies should pay more attention to the affective reputational dimension by investing in their likeability. The driver constructs model can also be used to derive tangible drivers regarding how likeability can be influenced. This can be accomplished by interpreting the total effect sizes (through the multiplication of path coefficients) from the formative driver indicators of likeability (cf. Table 21.4 in the appendix for the weights of the formative indicators on their respective constructs in the overall model). One can, consequently, identify those levers that will show the most impact on likeability (and thus, satisfaction and loyalty).

But the results also advise companies not to act undifferentiatedly regarding reputation management, and reveal that for corporate-level marketing to be successful, management has to prioritize the stakeholder groups in which reputation management is to be employed, as well as having to monitor which actions are effective in which groups. In our study, we found a relatively large number of paths that are generalizable across stakeholder groups. Hence, companies must be very careful when choosing activities, since some activities will influence different stakeholder groups in a different way. We have explored the media subgroup in a little more detail, as it is not only an important stakeholder group due to its effect on others via mass communication, but also as this specific subgroup has a distinct way of interpreting a company's activities in terms of reputation. We have found that the rational antecedental mechanisms of reputation can be generalized between subsamples, i.e. the influence of the four drivers quality, performance, attractiveness, and CSR on the competence dimension does not differ significantly between any subgroup. When examining the antecedents of the affective dimension likeability, we simply find that the impact of CSR activities (as interpreted by the stakeholders) on likeability can be generalized across all the stakeholders. The other driver constructs have different impacts on likeability. This finding is of much importance for the industry, since we have found that reputation's likeability dimension is of greater importance with regard to the generation of customer satisfaction and loyalty. Note again that we have shown that satisfaction and loyalty effects in this model are really effects of reputation, and not merely direct consequences of the drivers.

Furthermore, it can be observed that in the industry researched, the stakeholder groups employ different degrees of rationality when constructing decisions regarding satisfaction. Irrationalities like the overall evaluation of a company are more likely to influence customers who are not members of other stakeholder groups.

Nevertheless, the results do not suggest that "professional" stakeholder groups like the media or the financial community are more rational regarding all aspects, they are merely influenced by other aspects of the company's reputation. Whether a company should stress their reputation management's cognitive or emotional components in respect of the various stakeholder groups can be easily deduced from the path coefficients of the endogenous constructs' two dimensions in the model. On examining the coefficients in the measurement model, a company can further learn which corporate-level marketing action will lead to an increase or decrease in the two reputational components in which stakeholder group. The model consequently allows the prediction of the different levers' consequences in each and every stakeholder group.

PLS path modeling has been demonstrated as a very powerful and reliable tool for this kind of research question and it allows an analysis to be made of the differences between groups in even relatively small subsamples.

21.7 Appendix

Table 21.4 Indicators for the exogenous constructs

	Indicator	Outer weight in overall model
	The products / services offered by . . . are of high quality.	0.144
	In my opinion . . . tends to be an innovator, rather than an imitator with respect to mobile communications	0.135
	I think that . . .'s products / services offer good value for money.	0.006
quality	The services . . . offers are good.	0.071
	Customer concerns are held in high regards at	−0.017
	. . . seems to be a reliable partner for customers.	0.100
	I have the impression that . . . is forthright in giving information to the public.	0.086
	I regard . . . as a trustworthy company.	0.237
	I have a lot of respect for	0.142
	. . . is a very well managed company.	0.338
performance	. . . is an economically stable company.	0.136
	I assess the business risk for . . . as modest compared to its competitors.	0.123
	I think that . . . has growth potential.	0.236
	. . . has a clear vision about the future of the company.	0.147
attractiveness	In my opinion . . . is successful in attracting high-quality employees.	0.287
	I could see myself working at	0.099
	I like the physical appearance of . . . (company buildings, shops etc..)	0.452
CSR	I have the feeling that . . . is not only concerned about the profit.	0.062
	. . . behaves in a socially conscious way.	0.364
	. . . is concerned about the preservation of the environment.	0.099
	I have the impression that . . . has a fair attitude towards competitors.	0.310

References

Aaker, D. A. (1991). Managing brand equity: capitalizing on the value of a brand name. New York: Free.

Balmer, J. M. T. (2001a). Corporate identity, corporate branding and corporate marketing: seeing through the fog. *European Journal of Marketing, 35*(3–4), 248–291.

Balmer, J. M. T. (2001b). The three virtues and seven deadly sins of corporate branding. *Journal of General Management, 27*(1), 1–17.

Balmer, J. M. T. (2003). Corporate brands: what are they? What of them? *European Journal of Marketing, 37*(7–8), 972–997.

Baron, R. M., & Kenny, D. A. (1986). The moderator-mediator variable distinction in social psychological research: conceptual, strategic, and statistical considerations. *Journal of Personality and Social Psychology, 51*(6), 1173–1182.

Benjamin, B. A., & Podolny, J. M. (1999). Status, quality, and social order in the California wine industry. *Administrative Science Quarterly, 44*(3), 563–589.

Bollen, K. A. (1989). Structural equations with latent variables. Free: New York.

Brosius, F. (2002). SPSS 11, Bonn.

Brown, B., & Perry, S. (1994). Removing the financial performance halo from fortune's most admired companies. *Academy of Management Journal, 37*(5), 1347–1359.

Chin, W. W. (1998a). The partial least squares approach for structural equation modeling. In: G. A. Marcoulides (Ed.), *Modern methods for business research* (pp. 295–336). Mahwah.

Chin, W. W. (1998b). Issues and opinion on structural equation Modelling. *MIS Quarterly, 22*(1), 7–16.

Chin, W. W. (2000). Frequently asked questions – partial least squares and PLS-graph. http://disc-nt.cba.uh.edu/chin/plsfaq.htm (accessed 6-10-2005).

Chin, W. W., Marcolin, B. L., & Newsted, P. R. (2003). A partial least squares latent variable modeling approach for measuring interaction Effects: Results from a Monte Carlo Simulation study and an electronic-mail emotion/adoption study. *Information Systems Research, 14*(2), 189–217.

Chin, W. W., & Newsted, P. R. (1999). Structural equation modeling analysis with small samples using partial least squares. In: R. H. Hoyle (Ed.), *Statistical strategies for small sample research* (pp. 307–341). Thousand Oaks: Sage.

Diamantopoulos, A. (2005). How fine is C-OAR-SE? A generalizability theory perspective on Rossiter's procedure. *International Journal of Research in Marketing, 22*(1), 1–9.

Dibbern, J., & Chin, W. W. (2005). Multi-group comparison: testing a PLS-model on the sourcing of application software services across Germany and the U.S.A. using a permutation based algorithm. In: F. W. Bliemel, A. Eggert, G. Fassott, J. Henseler (Eds.), *Handbuch PLS Pfadmodellierung: Methode, Anwendung, Praxisbeispiele* (pp. 135–159). Stuttgart: Schffer-Poeschel.

Dozier, D. M. (1993). Image, reputation and mass communication effects. In: W. Armbrecht, H. Avenarius, & U. Zabel (Eds.), Image and PR. Opladen: Westdt. Verlag.

Dunbar, R. L. M., & Schwalbach, J. (2001). Corporate reputation and performance in Germany. *Corporate Reputation Review, 3*(2), 115–123.

Eberl, M. (2004). Formative und reflektive Indikatoren im Forschungsprozess: Entscheidungsregeln und die Dominanz des reflektiven Modells. *Schriften zur Empirischen Forschung und Quantitativen Unternehmensplanung, 5* Vol. 19.

Eberl, M. (2006). Unternehmensreputation und Kaufverhalten - Methodische Aspekte komplexer Strukturmodelle. Wiesbaden: DUV.

Eberl, M., & Schwaiger, M. (2004). Struktur und Parameter der Corporate Reputation: Die Branche als Determinante. Ein internationaler kausalanalytischer Modellvergleich. In: P. Albrecht, E. Lorenz, B. Rudolph (Eds.), Risikoforschung und Versicherung (pp. 623–654). Karslruhe: VVW.

Eberl, M., & Schwaiger, M. (2005). Corporate reputation – Disentangling the effects on financial performance. *European Journal of Marketing, 39*(7–8), 838–854.

Edgington, E. S. (1987). *Randomization tests*, 2nd Edn. New York: Marcel Dekker.

Eidson, C., & Master, M. (2000). Top Ten ... Most admired ... most respected: who makes the call? *Across the Board 37*(3), 16–22.

Fombrun, C. J., & Shanley, M. (1990). What's in a Name? Reputation building and corporate strategy. *Academy of Management Journal, 33*(2), 233–258.

Fombrun, C. J., & van Riel, C. B. M. (1997). The reputational landscape. *Corporate Reputation Review, 1*(1), 5–13.

Fryxell, G. E., & Wang, J. (1994). The Fortune corporate reputation index: reputation for what? *Journal of Management, 20*(1), 1–14.

Gardberg, N. A., Fombrun, C. J. (2002). The global reputation quotient project: first steps towards a cross-nationally valid measure of corporate reputation. *Corporate Reputation Review, 4*(4), 303–307.

Good, P. (2000). *Permutation tests, a practical guide to resampling methods for testing hypotheses*, 2nd edn. Springer: New York.

Gotsi, M., & Wilson, A. M. (2001). Corporate reputation: seeking a definition. *Corporate Communications, 6*(1), 24–30.

Hall, R. (1992). The strategic analysis of intangible resources. *Strategic Management Journal, 13*(2), 135–144.

Hansmann, K.-W., & Ringle, C. M. (2004). Smart PLS user's guide. Hamburg: Universitt Hamburg.

Herremans, I. M., Akathaporn, P., McInnes, M. (1993). An investigation of corporate social responsibility, reputation and economic performance. *Accounting, Organizations & Society, 18*(7–8), 587–604.

Hunt, S. D., & Morgan, R. M. (1995). The comparative advantage theory of competition. *Journal of Marketing, 59*(2), 1–15.

Hutton, C. (1986). America's most admired corporations. *Fortune, 113*(1), 16–22.

Jarvis, C. B., MacKenzie, S. B., & Podsakoff, P. M. (2003). A critical review of construct indicators and measurement model misspecification in marketing and consumer research. *Journal of Consumer Research, 30*(3), 199–218.

Jöreskog, K. G. (1971). Simultaneous Factor analysis in several populations. *Psychometrika, 36*(4), 409–426.

Jöreskog, K. G., & Goldberger, A. S. (1975). Estimation of a model with multiple indicators and multiple causes of a single latent variable. *Journal of the American Statistical Association, 70*(351), 631–639.

Keil, M., Tan, B. C. Y., Wei, K.-K., Saarinen, T., Tuunainen, V., & Wasenarr, A. (2000). A cross-cultural study on escalation of commitment behavior in software projects. *MIS Quarterly, 24*(2), 299–325.

Kotha, S., Rajgopal, S., & Rindova, V. (2001). Reputation building and performance: an empirical analysis of the Top-50 pure internet firms. *European Management Journal, 19*(6), 571–586.

Lohmöller, J.-B. (1989). Latent variables path modeling with partial least squares. Heidelberg: Springer.

MacCallum, R. C., & Browne, M. W. (1993). The use of causal indicators in covariance structure models: some practical issues. *Psychological Bulletin, 114*(3), 533–541.

Mahon, J. F. (2002). Corporate reputation: a research agenda using strategy and stakeholder literature. *Business and Society, 41*(4), 415–445.

Marwick, N., & Fill, C. (1997). Towards a framework for managing corporate identity. *European Journal of Marketing, 31*(5–6), 396–409.

Loudon, D. L., & Della Bitta, A. J. (1993). *Consumer behavior: concepts and applications*, 4th Edn. New York: McGraw-Hill.

Roberts, P. W., & Dowling, G. R. (2002). Corporate reputation and sustained superior financial performance. *Strategic Management Journal, 23*(12), 1077–1094.

Rogerson, W. P. (1983). Reputation and product quality. *The Bell Journal of Economics, 14*(2), 508–516.

Rossiter, J. R. (2002). The C-OAR-SE procedure for scale development in marketing. *International Journal of Research in Marketing, 19*(4), 305–335.

Schwaiger, M. (2004). Components and parameters of corporate reputation – an empirical study. *Schmalenbach Business Review, 56*(1), 46–71.

Shamsie, J. (2003). The context of dominance: an industry-driven framework for exploiting reputation. *Strategic Management Journal, 24*(3), 199–215.

Shapiro, C. (1983). Premiums for high quality products as returns to reputations. *The Quarterly Journal of Economics, 98*(4), 659–679.

Shrout, P. E., & Bolger, N. (2002). Mediation in experimental and nonexperimental studies: new recommendations. *Psychological Methods, 7*(4), 422–445.

Shenkar, O., & Yuchtman-Yaar, E. (1997). Reputation, image, prestige, and goodwill: an interdisciplinary approach to orgnaizational standing. *Human Relations, 50*(11), 1361–1381.

Sobel, M. E. (1982). Asymptotic confidence intervals for indirect effects in structural equation models. *Sociological Methodology, 13*(1), 290–312.

Sobol, M. G., Farelly, G. E., & Taper, J. S. (1992). Shaping the corporate image – an analytical guide for executive decision makers. New York: Quorum Books.

Weiss, A. M., Anderson, E., MacInnes, D. J. (1999). Reputation management as a motivation for sales structure decisions. *Journal of Marketing, 63*(4), 74–89.

Weigelt, K., & Camerer, C. (1988). (1988). Reputation and corporate strategy: a review of recent theory and applications. *Strategic Management Journal, 9*(5), 443–454.

Wilson, R. (1985). Reputations in games and markets. In: A. E. Roth (Ed.), *Game-theoretic models of bargaining*, (pp. 27–62). Cambridge: Cambridge University Press.

Zeithaml, V. A. (1988). Consumer perceptions of price, quality and value. *Journal of Marketing, 52*(3), 2–22.

Zeithaml, V. A., Berry, L. L., & Parasuraman, A. (1996). The behavioral consequences of service quality. *Journal of Marketing, 60*(2), 31–46.

Chapter 22
Modeling the Impact of Corporate Reputation on Customer Satisfaction and Loyalty Using Partial Least Squares

Sabrina Helm, Andreas Eggert, and Ina Garnefeld

Abstract Reputation is one of the most important intangible assets of a firm. For the most part, recent articles have investigated its impact on firm profitability whereas its effects on individual customers have been neglected. Using data from consumers of an international consumer goods producer, this paper (1) focuses on measuring and discussing the relationships between corporate reputation, consumer satisfaction, and consumer loyalty and (2) examines possible moderating and mediating effects among the constructs. We find that reputation is an antecedent of satisfaction and loyalty that has hitherto been neglected by management. Furthermore, we find that more than half of the effect of reputation onto loyalty is mediated by satisfaction. This means that reputation can only partially be considered a substitute for a consumer's own experiences with a firm. In order to achieve consumer loyalty, organizations need to create both, a good reputation and high satisfaction.

22.1 Introduction

Marketing research relies on hypothetical constructs to explain the behavior of market actors. This paper focuses on the construct of corporate reputation which is deemed an important intangible asset and competitive advantage of the firm (Fombrun 1996). It may be defined as stakeholders' overall evaluation of a company over time (Gotsi and Wilson 2001; Fombrun 1996). Reputation serves as a point of reference when judging the firm's contribution to stakeholders' own and

S. Helm
University of Arizona, John and Doris Norton School of Family and Consumer Sciences, 650 N. Park Ave, PO Box 210078, Tucson, AZ 85721-0078, USA
e-mail: helm@email.arizona.edu

A. Eggert and I. Garnefeld
University of Paderborn, Warburger Str. 100, 33098 Paderborn, Germany
e-mail: andreas.eggert@notes.upb.de, ina.garnefeld@notes.upb.de

V. Esposito Vinzi et al. (eds.), *Handbook of Partial Least Squares*, Springer Handbooks of Computational Statistics, DOI 10.1007/978-3-540-32827-8_23,

the public's welfare. Therefore, it is decisive for stakeholders' contributions to the firm (Lewis 2001).

In the past, research on reputation mainly focused on reputation as an indispensable condition of market exchange because customers evaluate a firm's reputation before entering into a business relationship. How reputation affects already established relationships, has not been investigated in detail, though.

In order to manage reputation, it seems important to analyze its behavioral effects. Customers are believed to be more loyal to the products of firms with a good reputation (Morley 2002). However, empirical evidence on the effect of reputation in the formation of customer satisfaction and loyalty is scarce (Andreassen and Lindestad 1999) and led to divergent findings (Anderson and Sullivan 1993; Yoon et al. 1993; Abdullah et al. 2000; Andreassen and Lindestad 1999). Loyalty management could profit from an investigation of another determinant of the construct. Therefore, we take this gap in the literature as a starting point.

Additionally, we want to focus the mediating and moderating effects among these constructs. From a methodological point of view, the paper at hand aims at illustrating how to quantify mediating and moderating effects in structural equation models with latent variables. Mediating effects are often hypothesized in structural equation modeling, but rarely explicitly tested. This leads to a validity problem regarding the model as a whole and regarding the managerial implications. Moderating effects are fundamental to the marketing discipline because complex phenomena are typically subject to contingencies. Identifying and quantifying these contingencies is an important challenge within marketing research. While the literature frequently proclaims the importance of contingencies, empirical research is rather limited.

Also, covariance-based measurement approaches such as LISREL fall short of quantifying moderating effects due to their inherent assumptions. Partial Least Squares (PLS) is a competing estimation approach for structural equation models. Using PLS, one can directly assess the strength of latent moderating variables.

Based on a customer survey of an international consumer goods producer, the paper investigates the interplay of corporate reputation, consumer satisfaction, and loyalty. The research objectives of the paper are

1. to discuss and assess three hypotheses on the relationships between the three constructs using partial least squares and
2. to examine possible mediating and moderating effects among the constructs.

In order to address these objectives, the paper is structured as follows: In the next section, the relevant literature on reputation, satisfaction, loyalty, and their measures is reviewed. The third section describes the process of scale development. The fourth section deals with the empirical research design, followed by an overview of the major research findings. We discuss the findings and implications in a sixth section and conclude by presenting limitations of the present research in a final section.

22.2 Literature Review and Conceptual Model

22.2.1 *Reputation, Satisfaction, and Loyalty*

The growing body of literature has led to an abundance of definitions of corporate reputation. Fombrun (1996, p. 72) defines the construct as "a perceptual representation of a company's past actions and future prospects that describes the firm's overall appeal to all of its key constituents when compared with other leading rivals". Reputation is a socially shared impression, a consensus about how a firm will behave in any given situation (Bromley 2002; Sandberg 2002). Morley (2002, p. 8) explains that "corporate reputation – or image as advertising professionals prefer to term it – is based on how the company conducts or is perceived as conducting its business". While he uses the terms image and reputation synonymously, other authors differentiate the constructs. Middleton and Hanson (2002, p. 4), for example, provide a cogent summarization of the image construct and define it to consist of "attitudes and beliefs about the company held by the company's stakeholders shaped by the organisation's own communication processes". Markwick and Fill (1997), as well as Nguyen and Leblanc (2001a), explain that corporate image represents a variable portrait of a firm and its products/brands in the mind of a consumer that is mostly influenced by the firm's promotion efforts that may be altered relatively quickly, whereas reputation reflects the degree of trust in a firm's ability and willingness to meet consumers' expectations continuously.

In the context of this paper, we focus on the construct of reputation and define it as a stakeholder's overall evaluation of a firm over time in respect to its handling of stakeholder relationships (Fombrun 1996). Reputation is a perceptual collective construct (Wartick 2002) as it relies on an individual's perception of the public's impression about a firm. In discerning it from the image construct, we follow Balmer and Gray (1999), who suggest that image is an immediate mental picture that individuals conceive of an organization. In contrast, reputation is "formed over time; based on what the organization has done and how it has behaved" (Balmer and Greyser 2003, p. 177), meaning that it evolves as a result of consistent behavior that created trust.

The growing interest in reputation has led to the development of a variety of different construct measures. Rankings of companies are the most common approach to measure reputation. They are usually based on a cluster of different corporate associations that represent different stakeholders' expectations regarding the activities of a firm. Examples for such social expectations are the delivery of high-quality products, treating employees fairly, and delivering a good financial performance. Indicators used to measure corporate reputation usually represent one facet of these expectations. The set of indicators is then aggregated to make up the construct of reputation in the sense of an index. Examples are Fortune's annual study on the Most Admired Companies and the Reputation Institute's Reputation Quotient^{sm} (RQ^{sm}). Among others, Fombrun (1998), Lewis (2001), and Wartick (2002) have reviewed the existing measurement approaches, highlighting the Fortunes annual

"Most Admired Companies" and the RQ^{sm} as the most frequently used and discussed data sets.

The loyalty construct has gained vast attention in marketing research leading to several different definitions and conceptualizations of the construct with varying levels of complexity. According to Oliver (1997, p. 392), customer loyalty is "a deeply held commitment to re-buy or re-patronize a preferred product or service consistently in the future, despite situational influences and marketing efforts having the potential to cause switching behavior". Dick and Basu (1994, p. 102) understand it to be the "favorable correspondence between relative attitude and repeat patronage", and thus, attitude and repetitive behavior are reflected in consumer loyalty. Because individuals usually act according to an attitudinal predisposition, in modeling the loyalty construct, we integrate an emotional predisposition of the consumer as well as a behavioral intention to maintain an ongoing relationship with a firm (Oliver 1999).

Furthermore, we propose that satisfaction is a main determinant of loyalty. This means that episodic experiences as a main part of satisfaction are linked to relational connotations, a notion that is supported by attitude-behavior consistency arguments (Oliver 1997; Singh and Sirdeshmukh 2000). Consumer satisfaction results from a favorable correspondence between a consumer's expectations and his/her experiences with a firm or its products and services (Churchill and Surprenant 1982). Due to the importance of satisfaction in explaining loyalty, we include the construct in our analysis. We focus only on the experience-part of satisfaction though, as expectations can be developed partly on the basis of reputational information about a firm which might lead to an overlap of reputation and an expectancy-based satisfaction construct.

22.2.2 The Relationship Between Reputation, Satisfaction, and Loyalty

A review of the literature on reputation and loyalty shows inconsistent findings concerning the causal relationships between both constructs. Fombrun (1996, p. 78) points out that "reputation breeds customer loyalty". Nguyen and Leblanc (2001b), as well as Gray (1986), interpret reputation as an important determinant of loyalty. According to Anderson and Weitz (1989), a highly reputable firm that was able to build trust will have more loyal customers than less reputable firms. Finally, Anderson and Sullivan (1993, p. 132) claim that reputation "determines customers' sensitivity to short-run deviations in product quality and satisfaction", indicating that reputation may compensate for a consumer's bad experiences or dissatisfaction. A good reputation guarantees that the firm will soon return to producing the high quality products its reputation was built upon. This signalling function of reputation has been investigated in the literature on the economics of information (Shapiro 1982; Herbig and Milewicz 1994).

Besides these conceptual analysis, empirical evidence regarding the relationships between corporate reputation, satisfaction, and loyalty has been established. Yoon et al. (1993) find a positive relationship between corporate reputation and the intention to buy a firm's products. Abdullah et al. (2000) show that, compared to satisfaction, there is a relatively large impact of a firm's image on consumer loyalty; these authors do not differentiate between image and reputation. Nguyen and Leblanc (2001b) find a significant relationship between reputation and loyalty. Anderson and Weitz (1992), however, only find a partial effect of reputation on loyalty and commitment in supplier–retailer relationships. Finally, Andreassen and Lindestad (1999) failed to support the hypothesized direct impact of corporate image (interpreted synonymously to reputation) on loyalty.

Against the background of these divergent findings, different causalities appear to be reasonable. In the context of our research, we posit a positive effect of satisfaction and reputation on loyalty. The positive relationship between consumer satisfaction and loyalty has been investigated in a number of empirical studies (Oliver 1999; Anderson and Sullivan 1993; Rust and Zahorik 2003). An explanation for the impact of satisfaction can be found in social exchange theory (Thibaut and Kelley 1959). Perceived satisfaction is a stimulus or reinforcement that an individual repeatedly wants to achieve and which therefore leads to loyalty. A positive relationship between reputation and loyalty can be explained based on the economics of information. The consumer is uncertain as to the question whether staying in a certain business relationship is more profitable than establishing a new one. The reputation of a firm serves as a signal to the consumer that is used to reduce his/her uncertainty. "In a context of imperfect information, the customer has tendency to use corporate reputation to infer the quality of a specific product or service offered by a firm or to predict its future action" (Nguyen and Leblanc 2001a, p. 233).

We further posit that satisfaction is positively influenced by corporate reputation. The existence of a positive relationship between reputation and satisfaction can be explained by self-perception theory (Bem 1967) and the motivation for self-affirmation, as well as by Festinger's (1957) theory of cognitive dissonance. A poor public reputation might influence a consumer's perceptions of his own experiences with a firm's products and services, urging him/her to reconsider his/her perception and possibly leading to a negative impact on satisfaction. A positive public reputation, on the other hand, confirms and reinforces the consumer's own experiences and satisfaction scores.

Therefore, we hypothesize:

H.1: *Consumers' loyalty to a firm is positively influenced by*

- a) *the degree of consumer satisfaction (i.e., consumer experiences)*
- b) *the degree of corporate reputation as perceived by the consumer.*

H.2: *Consumers' satisfaction as reflected in their experiences with a firm's offerings is positively influenced by the degree of reputation as perceived by the consumer.*

As satisfaction is based on consumers' own experiences with a firm's offerings, its link to loyalty should be stronger than the link from reputation to loyalty.

Consequently, reputation might also be considered as a moderator of the relationship between satisfaction and consumer loyalty. According to self-perception theories, this moderating effect should be positive: The consumer's own experiences are reinforced by the reputation of the firm which is based on the experiences of a multitude of consumers. In an information economic sense, the moderating effect could be negative, though: good reputation guarantees that even after a lapse in quality, the firm will soon return to producing the high quality products its reputation was built upon. As Anderson and Sullivan (1993) claim, reputation influences a customer's sensitivity to deviations in product quality and satisfaction indicating that reputation may compensate a consumer's bad experiences and therefore serve as a negative moderator of the relationship between satisfaction and loyalty. Therefore, we hypothesize:

H.3: *Reputation moderates the relationship between satisfaction and loyalty.*

Furthermore, satisfaction – especially the consumer's experiences as a part of the satisfaction construct – might function as a mediator in the relationship between reputation and loyalty so that (nearly) no direct effect of reputation on loyalty becomes evident. Consumer's own experiences are more viable and important in determining loyalty than experiences communicated by others (i.e., reputation). If this were the case, reputation could not compensate consumer dissatisfaction. Although the theoretical foundation remains thin, we hypothesize:

H.4: *Satisfaction mediates the effect of reputation on loyalty.*

The structural equation model visualizing these relationships is shown in Fig. 22.1.

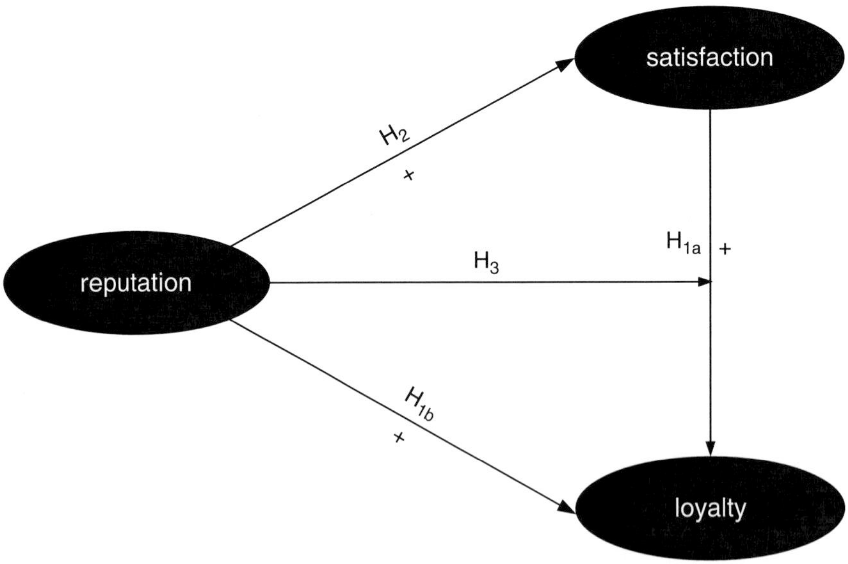

Fig. 22.1 Basic structural model

22.3 Scale Development

A novel measure for reputation was developed due to a lack of consensus on valid scales (Nguyen and Leblanc 2001a). As the epistemic nature of reputation is thought to be formative, the methodology proposed by Diamantopoulos and Winklhofer (2001) for building formative construct measures was followed. It contains four steps: content and indicator specification, test for indicator collinearity, and test for external validity.

The first step includes content specification. Drawing on existing scales and the literature on reputation, satisfaction, and loyalty, we conducted two focus group interviews with fellow researchers with different academic backgrounds, and forty individual, in-depths interviews with consumers. In this way, corporate reputation could be defined taking into account input by the potential respondents (Berens and van Riel 2004). Potential indicators for the measures were identified by looking for items commonly used in prior measurement models and by integrating the results of the interviews.

In a second step, an item-sorting task (Anderson and Gerbing 1991) showed how well the items tapped their underlying constructs. Participants consisting of 12 fellow academics were told about the basic research design, the definition of reputation and two other constructs, i.e., customer satisfaction and loyalty. They were asked to assign individual items to what they believed to be the correct construct out of the set of three. Two indices proposed by Anderson and Gerbing (1991) – the proportion of substantive agreement (p_{sa}) and the substantive-validity coefficient (c_{sv}) – were computed for each item to identify those that were difficult to assign to the corresponding construct. The equations for these calculations are:

$$p_{sa} = \frac{n_c}{N}$$

$$c_{sv} = \frac{n_c - n_o}{N}$$

N number of participants
n_c number of correctly allocated items
n_o highest number of assignments of the item to another construct

For reputation, this led to a reduced item set of 10 as only items with a p_{sa} of above 0.75 and a c_{sv} above 0.5 were kept in the measure.

Finally, questionnaires were administered to 20 consumers in a "think aloud" answer mode. The ten remaining reputation indicators, four indicators for satisfaction (i.e., experiences), and eight indicators for loyalty were then included in the final survey. In accordance with Rossiter (2002), no statement-based approach was used. Instead, bipolar, entirely verbalized seven-point scales were used as shown in Table A.1 in the appendix.

Corporate reputation and satisfaction were modeled with formative indicators, while loyalty was conceptualized as a reflective construct. Conceptualizing reputation as a formative construct means that it is an aggregation of all its indicators such as treatment of employees, commitment to protecting the environment, etc. (Bollen and Lennox 1991; Jarvis et al. 2003). This implies that because it treats its employees right, a firm has a good reputation; because it protects the environment, it has a good reputation. The same applies to the measure of satisfaction, which is also seen as a summation of a firm's performances such as the quality of products, customer orientation, etc. As common measures for customer satisfaction are not conceptualized formatively but reflectively (Bettencourt 1997; Mano and Oliver 1993; Westbrook and Oliver 1981), we also derive a new measure for this construct.

In accordance with the literature (Andreassen 1994; Oliver 1997), loyalty was conceptualized as a reflective construct. Increasing loyalty of a consumer will usually result in a variety of different attitudinal and behavioral consequences. The more loyal, the more often the consumer might refer the products to others, re-buy products of the same firm, etc. He or she will show the entirety of the possible characteristics of loyal customers, including a positive personal disposition towards the firm. Loyalty therefore leads to the behavioral indicators which characterizes a reflective construct structure.

22.4 Empirical Study

22.4.1 Research Design and Sample Structure

The procedure and results described below were part of a larger research project that focused the reputation of an international consumer goods producer (fast-moving consumer goods such as detergent or cosmetics) and its effects on different stakeholders. Here, we only discuss the results of the study conducted in the German consumer sample.

Interviewers of a leading research institute contacted 1,681 consumers following a random-route design. Personal, computer-aided interviews took place at consumers' households at 210 sample points all across Germany. In 729 cases, the household or targeted person refused to take part in the interview, leading to a response rate of 56.6% (952 cases). Respondents had to be knowledgeable about the firm's reputation and to have actual experience with the firm as customers, and were therefore identified by two filter questions. This led to an effective sample size of 45.3% and 762 usable questionnaires.

22.4.2 Data Analysis

Partial least squares (PLS) analysis was used because the model contains formative and reflective constructs; for an overview and a discussion of the features of PLS see Fornell and Bookstein (1982). Specifying formative indicators poses problems with software for covariance structure analysis such as LISREL (MacCullum and Browne 1993), as covariance-based methods often lead to improper and uninterpretable solutions when formative measurement models are involved (Fornell and Bookstein 1982). The software package employed was SPAD-PLS.

22.5 Results

22.5.1 Measurement Model

In a first step, we analyze the formative and reflective measurement models. The results are shown in Table A.1 in the appendix. As no indicator of the satisfaction construct has a weight below 0.1, there is no need for scale purification (Chin 1998; Baumgartner and Homburg 1996).

Concerning the reflective latent variable "consumer loyalty", all items were subjected to an exploratory factor analysis with varimax rotation (Hair et al. 1998), resulting in a KMO-value of 0.94 and a one-factor solution. Although the indicators used to measure loyalty contain behavioral and affective aspects, the construct is uni-dimensional. Individual item reliability, factor loadings, t-values, and average variance extracted were compared against established standards (Baumgartner and Homburg 1996; Bagozzi et al. 1991). The average variance extracted amounts to 65.3% (see Table A.1).

Five indicators of the latent variable reputation have a weight below .1 and one has a negative sign (weights that are not significant at $p = 0.5$ are printed in italics in Table A.1). Seltin and Keeves (1994) claim such indicators to be "trivial" and call for their removal in order to build parsimonious models. Concerning formative variables, however, indicator deletion is problematic as "omitting an indicator is omitting a part of the construct" (Bollen and Lennox 1991, p. 305). Facets of the reputation construct would be removed resulting in the formation of a new construct. Therefore, Rossiter (2002, p. 315) claims that "Item selection to increase the 'reliability' of the formed scale is definitely not appropriate". In our case, reputation would be reduced to product quality, environmental issues, customer orientation, credibility of advertising claims, and value for money. These might well be the most important facets of reputation from a consumer's point of view, but if the aim of the researcher is to build a reputation measure applicable to different stakeholder groups, an elimination of items would reduce the capacity of the measurement model to cover other stakeholder groups as well. As we aimed at building such a stakeholder-oriented measurement model, the whole set of reputation indicators are contained in the reputation measure.

Multi-collinearity might pose a relevant problem as the formative measurement model is based on multiple regression (Diamantopoulos and Winklhofer 2001). In the data set, the highest value for the variance inflation factor (VIF) was 3.09 for reputation, which is far below the common cut-off threshold of 10 (Kleinbaum et al. 1998). Therefore, multi-collinearity does not represent a serious problem.

22.5.2 *Structural Relationships*

In a second step, the inner model is considered. The results are depicted in Fig. 22.2

All path coefficients are significant at $p = 0.01$. The strongest effect shows path a linking reputation and satisfaction (hypothesis 2). There is also a strong effect from satisfaction onto loyalty. This is consistent with hypothesis 1a. The direct path c from reputation to loyalty (hypothesis 1b) is weaker than the direct effect of satisfaction onto loyalty (path b). Although the effect is not very strong, it is not negligible either. This shows that loyalty is not only caused by customer satisfaction but also by corporate reputation.

After having tested the direct effects, the moderating effect is tested (see Chin et al. 2003 for details regarding the methodology). The effect structure of reputation on the relationship between satisfaction and loyalty is shown in Fig. 22.3.

"In general terms, a moderator is a qualitative (e.g., sex, race, class) or quantitative (e.g., level of reward) variable that affects the direction and/or strength of

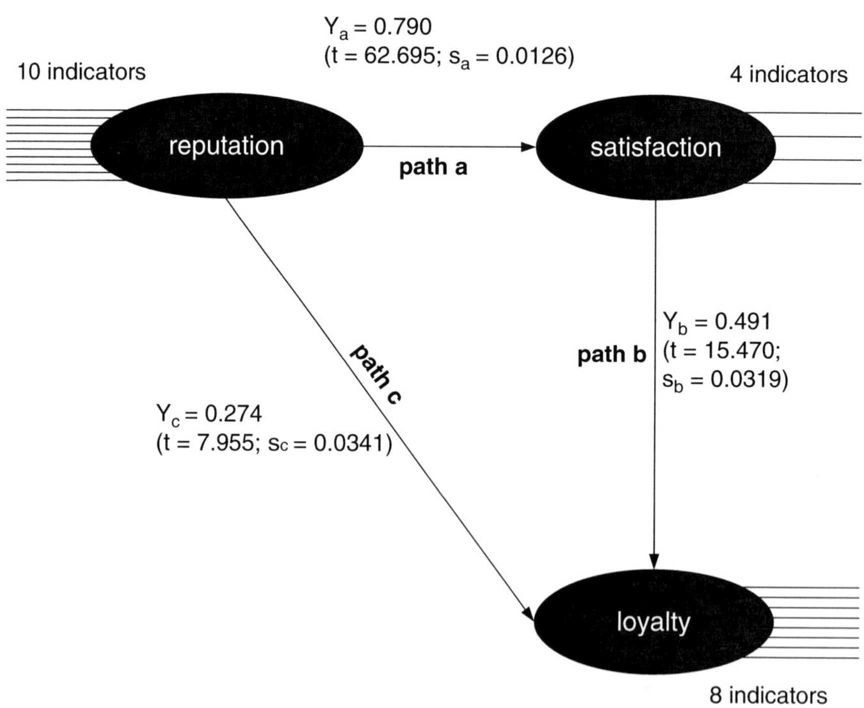

Fig. 22.2 Information on the structural model

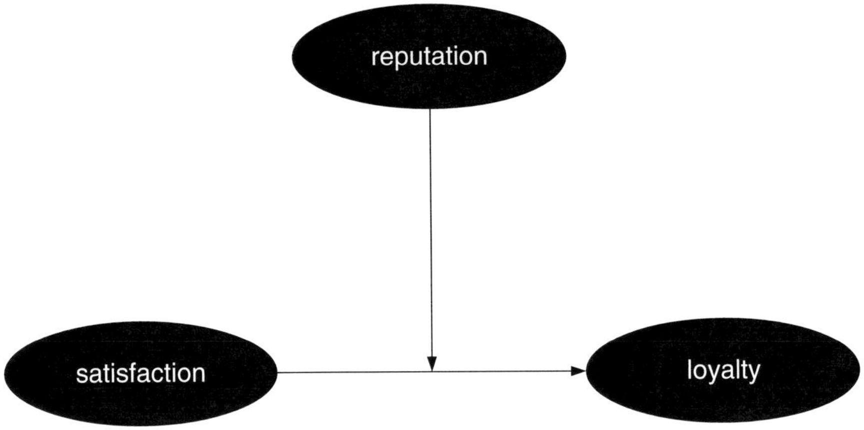

Fig. 22.3 The moderating effect of reputation on the satisfaction-loyalty link

the relation between an independent or predictor variable and a dependent or criterion variable" (Baron and Kenny 1986, p. 1174). Moderator variables are of high relevance as complex relationships are normally subject to contingencies. However, they are rarely tested within the context of structural equation modeling.

To test the moderating effect, the influence of the exogenous variable on the endogenous variable, the direct effect of the moderating variable on the endogenous variable and the influence of the interaction variable on the endogenous variable are estimated (see Fig. 22.4). The moderator hypothesis is confirmed if the interaction effect (i.e., path c) is significant, independently of the magnitude of the path coefficients a and b (Baron and Kenny 1986).

$$f^2 = \frac{R^2_{\text{model with moderator}} - R^2_{\text{model without moderator}}}{1 - R^2_{\text{model without moderator}}}$$

Considering the contradictory statements found in the literature, we did not hypothesize on the direction of the effect, i.e., whether reputation enhances or diminishes the satisfaction-loyalty link. Therefore, a two-sided test of significance is applied. As reputation and satisfaction both are measured using formative scales, the interaction variable is formed by multiplying the construct coefficients of reputation and satisfaction (see Fig. 22.5).

A moderating effect can be confirmed if path c is significant, independent of the magnitude of pathes a and b. In our model, we estimate a standardized path coefficient of −0.039. However, the effect is not significant ($p < 0.05$). The effect size is calculated as follows (Chin et al. 2003):

$$f^2 = \frac{0.546 - 0.545}{1 - 0.546} = 0.0002$$

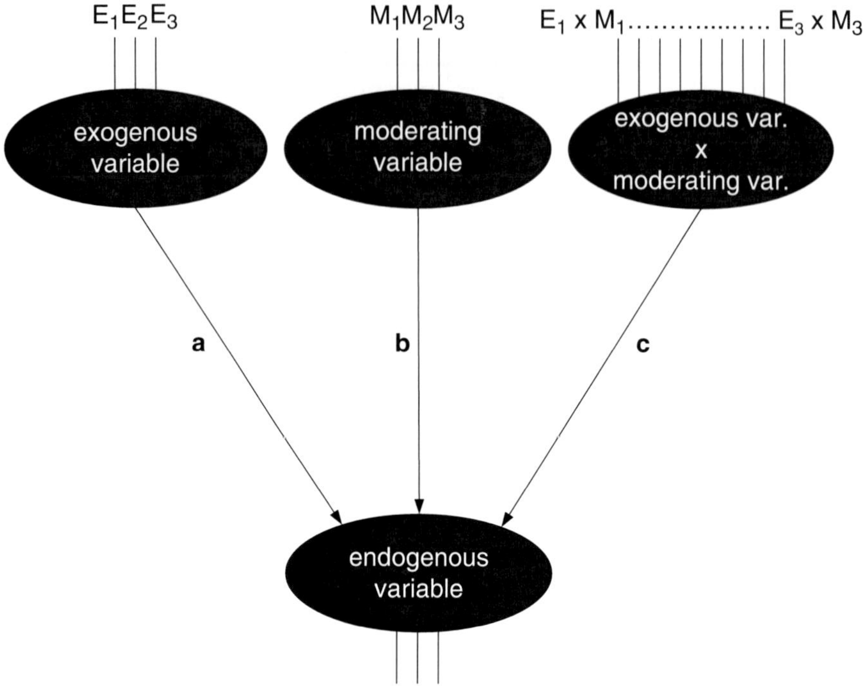

Fig. 22.4 Moderator model

Consequently, reputation cannot be considered a moderating variable; hypothesis 3 has to be rejected.

To address hypothesis 4, the mediating effect of satisfaction is analyzed. While the traditional approach, following Hoyle and Kenny's (1999) suggestions, recommends a two-step approach, Iacobucci and Duhachek (2003) argue for the superiority of a simultaneous assessment of the mediating effect as shown in Fig. 22.6.

To establish the mediating effect, the indirect effect $a \times b$ has to be significant. To test for significance, the z-statistic (Sobel 1982) is applied. If the z-value exceeds 1.96 (at $p < 0.05$) the null hypothesis can be rejected, i.e., there is no indirect effect of reputation on loyalty via the construct of satisfaction. The z-value is formally defined as follows:

$$z = \frac{a \times b}{\sqrt{b^2 \times s_a^2 + a^2 \times s_b^2 + s_a^2 \times s_b^2}}$$

As shown in Fig. 22.2, there is a significant effect of reputation onto satisfaction (0.790, $p < 0.001$) as well as of satisfaction onto loyalty (0.491, $p < 0.001$). As there is also a significant direct relationship between reputation and loyalty (0.274, $p < 0.001$), satisfaction is established as a partial mediator. This mediating effect is

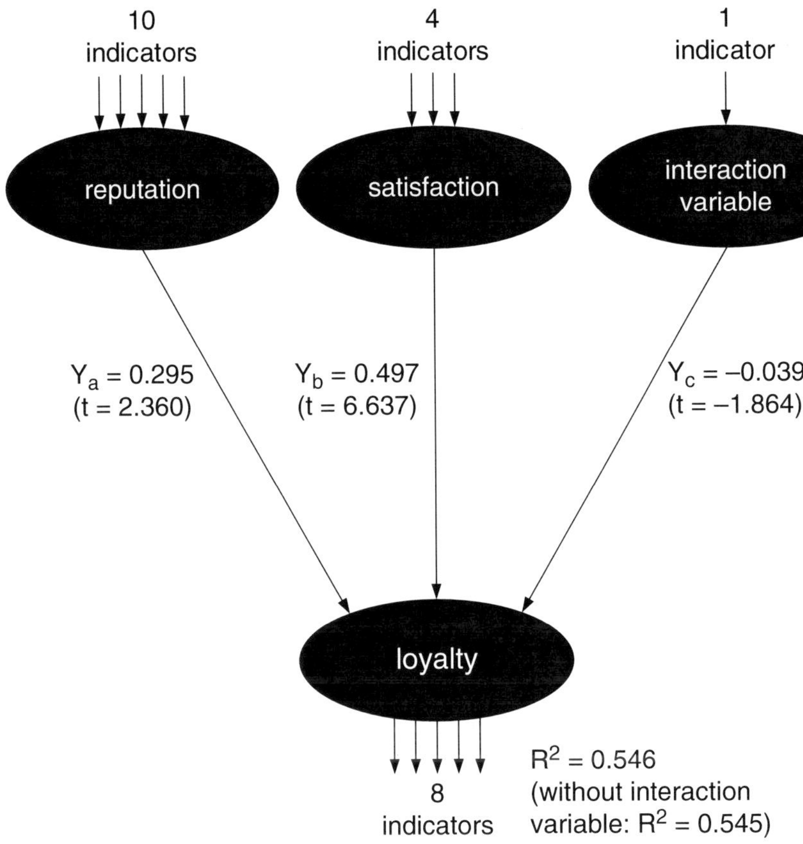

Fig. 22.5 Information on the moderator model

confirmed by the z-statistic (Sobel 1982):

$$z = \frac{0.790 \times 0.491}{\sqrt{(0.491)^2 \times (0.013)^2 + (0.790)^2 \times (0.032)^2 + (0.013)^2 \times (0.032)^2}} = 14.87$$

The result shows that reputation has a direct effect on loyalty as well as an indirect effect via the satisfaction construct

To estimate the magnitude of the indirect effect Iacobucci and Duhachek (2003) use the VAF (Variance Accounted For) value, which represents the ratio of the indirect effect to the total effect.

$$VAF = \frac{a \times b}{a \times b + c}$$

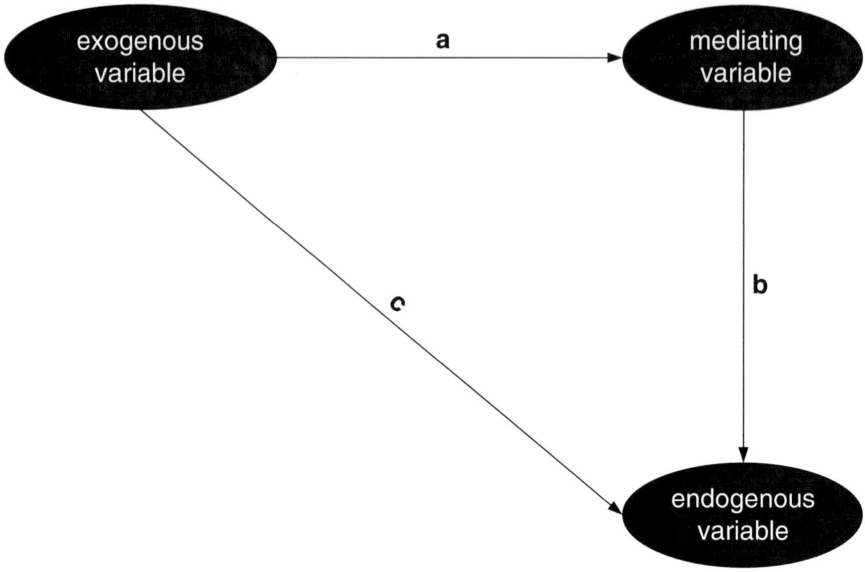

Fig. 22.6 Illustration of the mediating effect

The magnitude of the indirect effect of reputation via satisfaction is illustrated by the high VAF value:

$$VAF = \frac{0.79 \times 0.491}{0.79 \times 0.491 + 0.274} = 0.586$$

A VAF value of 58.6% indicates that more than half of the total effect of reputation onto loyalty is explained by the indirect effect.

22.6 Discussion

In the past, literature on reputation has been dominated by conceptual research. The inclusion of reputation in structural equation modeling to examine its interplay with other focal constructs has remained an exception. However, in order to make use of reputation as a strategic asset (Fombrun 1996; Lewis 2001), its link to other important marketing variables needs to be understood. The present study contributes to a better knowledge of the interplay between perceived corporate reputation, satisfaction of consumers as manifested in their experiences with the firm, and loyalty.

We found that reputation not only influences a consumer's own experiences with the products of the firm. It also determines consumers' loyalty. Investing in reputation should therefore have positive effects on the bonding of customers although such investments usually aim at fostering the esteem the firm is held in by the general public, not at improving individual stakeholder relationships. This lever to improve loyalty has hitherto been neglected and is largely ignored in loyalty management.

Understanding reputation as a formative construct allows for identifying the drivers of reputation. Hence, our results cannot only show that loyalty can be improved by a good reputation but also how a good reputation can be established. We found value for money and quality of the products as well as the credibility of advertising claims to be the most important aspects of reputation. Therefore, a firm that aims at managing its reputation in order to achieve higher loyalty should especially concentrate on these key drivers of reputation.

The results also show that more than half of the effect of reputation on loyalty is mediated by satisfaction. In the absence of satisfaction, even the best reputation lacks most of its effect on loyalty. Therefore, firms need to create both, a good reputation and high satisfaction rates. This is also important considering the acquisition of new customers, a task that has not been covered within the scope of this research. Lacking own experiences, new customers rely on reputational information and word-of-mouth from satisfied customers to make a first-time purchase decision.

Reputation was expected to be a moderator for the satisfaction-loyalty link. However, we found no empirical evidence for this moderating influence of reputation. This finding could possibly be due to the study's limitation to fast-moving consumer goods. Different findings in other industry settings may occur. It is expected that reputation will play an important role in products or services that lead to higher levels of perceived risk. Furthermore, other firms within the same industry and other stakeholders should be investigated.

The present study focused on the importance of reputation in determining consumer satisfaction and loyalty. Reputation has been found to be a determinant of both constructs, having the strongest effect on loyalty via the satisfaction construct. This leads to a more thorough understanding of the interplay between the three constructs.

From a methodological standpoint, the study shows the importance of including moderator and mediator variables into structural equation modeling. These variables are often discussed in conceptual literature but empirical testing remains scarce. One explanation could be the lack of knowledge in identifying and assessing mediators and moderators. To close this gap, this paper aimed at showing a step-by-step framework for systematically integrating these variables into PLS-path models.

22.7 Limitation

As in any empirical research, the results of the present study cannot be interpreted without taking into account its limitations. Furthermore, this study generates a set of researchable issues that might be addressed in future projects.

With respect to the hypothesized causality between the constructs investigated in our study, the suggested directionality needs to be determined by theoretical arguments. Satisfaction has been interpreted as the experience with a firm's offerings made by the individual respondents. Therefore, reputation should be viewed as an antecedent to this, not an outcome. Experiences with a firm's products cannot immediately affect a firm's overall reputation. Only if a number of customers experience deteriorating quality and disseminate this information in the market, reputation will deteriorate over time. As reputation is built by word-of-mouth communication (Yoon et al. 1993; Fombrun 1996), satisfaction levels among consumers and other stakeholders will eventually impact a firm's reputation (besides possible impacts of the media). This effect may only be studied in a longitudinal design. The same reasoning applies to the causality flow between reputation and loyalty. In the long run, improvements in loyalty (i.e., increases in favorable word-of-mouth, in resources for investments in product quality due to re- and cross-buying, etc.) will positively affect reputation, but this effect could not be investigated in this study. The expected positive long-term effects of satisfaction and loyalty on reputation further strengthen the importance of a combined reputation and satisfaction management.

The sample was reduced to consumers who were questioned in their role as customers of one specific firm. This randomized sample contained interviewees who were representative of German consumers. As previously pointed out, a cross-sectional and even cross-cultural study could provide important insights. Also, it would be interesting to distinguish between products and services. It is to be expected that reputation will play an important role in establishing cooperative relationships with consumers in service settings that are characterized by experience and credence qualities. This role might be less important in transaction-oriented market settings, meaning that relationship-orientation or quality might also moderate the effects of reputation.

Appendix A

Table A.1 Information on the measurement models

Indicator	Description	Weight/ Loading	t-value
	Reputation[a]		
x1	Quality of products	0.2733	51.129
x2	Commitment to protecting the environment	0.2204	31.633
x3	Corporate success	*0.0183*	*0.3245*
x4	Treatment of employees	*−0.0243*	*0.2915*
x5	Customer orientation	0.1287	27.766
x6	Commitment to charitable and social issues	*0.0987*	*14.297*
x7	Value for money of products	0.3029	61.554
x8	Financial performance	*0.0218*	*0.3988*
x9	Qualification of management	*0.0308*	*0.5454*
x10	Credibility of advertising claims	0.3023	57.450
	Satisfaction[b] **(experience concerning attributes)**		
y1	Quality of products	0.3330	223.232
y2	Value for money of products	0.2947	188.717
y3	Customer orientation	0.3007	233.442
y4	Adherence to advertising claims	0.3057	232.657
	Loyalty[c] **(y5 to y8: affective loyalty/y9 to y12: behavioral loyalty) Average variance extracted = 0.653; Composite reliability = 0.938**		
y5	To what extent do you feel bonded to x?	0.7807	181.585
y6	To what extent would you regret if products made by x were no longer available?	0.8098	192.441
y7	To what extent are products made by x part of your everyday life?	0.8430	200.218
y8	To what extent are you loyal to products made by x?	0.8577	189.218
y9	When shopping next time, are you going to buy products made by x?	0.8103	192.146
y10	Would you refer products made by x to your family and friends?	0.7829	177.206
y11	Do you prefer products made by x to products of competitors?	0.8376	193.390
y12	Are you going to try new products made by x?	0.7368	158.072

[a] Question: "Concerning the following attributes, does company x have a good or bad reputation in the public?" Scale: 1 = "very good reputation", 7 = "very bad reputation"; the scale was entirely verbalized.
[b] Question: "How would you rate your experiences with x concerning the following attributes?" Scale: 1 = "very good experiences", 7 = "very bad experiences" the scale was entirely verbalized.
[c] Scale for item y5 to y8: 1= "to a very high extent", 7 = "not at all"; the scale was entirely verbalized. Scale for items y9 to y12: 1= "yes, very likely", 7 = "no, not likely at all"; the scale was entirely verbalized.

References

Abdullah, M., Al-Nasser, A., & Husain, N. (2000). Evaluating functional relationship between image, customer satisfaction and customer loyalty using general maximum entropy. *Total Quality Management, 11*, 826–829.

Anderson, E., & Sullivan, M. (1993). The antecedents and consequences of customer satisfaction for firms. *Marketing Science, 12*, 125–143.

Anderson, E., & Weitz, B. (1989). Determinants of continuity in conventional industrial channel dyads. *Marketing Science, 8*(2), 310–323.

Anderson, E., & Weitz, B. (1992). The use of pledges to build and sustain commitment in distribution channels. *Journal of Marketing Research, 24*(1), 18–34.

Anderson, J., & Gerbing, D. (1991). Predicting the performance of measures in a confirmatory factor analysis with a pretest assessment of their substantive validities. *Journal of Applied Psychology, 76*(5), 732–740.

Andreassen, T. (1994). Satisfaction, loyalty and reputation as indicators of customer orientation in the public sector. *Internatinal Journal of Public Sector Management, 7*(2), 16–34.

Andreassen, T., & Lindestad, B. (1999). The effect of corporate image in the formation of customer loyalty. *Journal of Service Research, 1*, 82–92.

Bagozzi, R., Yi, Y., & Phillips, L. (1991). Assessing construct validity in organizational research. *Administrative Science Quarterly, 36*(3), 421–458.

Balmer, J., & Gray, E. (1999). Corporate identity and corporate coummunications: Creating a competitive advantage. *Corporate Commincations, 4*(4), 171–176.

Balmer, J., & Greyser, S. (2003). *Revealing the Corporation*. London/New York: Routledge.

Baron, R., & Kenny, D. (1986). The moderator-mediator variable distinction in social psychological research: Conceptual, strategic, and statistical considerations. *Journal of Personality and Social Psychology, 51*(6), 1173–1182.

Baumgartner, H., & Homburg, C. (1996). Applications of structural equation modeling in marketing and consumer research: A review. *International Journal of Research in Marketing, 13*(2), 139–161.

Bem, D. (1967). Self-perception: An alternative interpretation of cognitive dissonance phenomena. *Psychological Review, 74*(3), 183–200.

Berens, G., & van Riel, C. (2004). Corporate associations in the academic literature: Three main streams of thought in the reputation measurement literature. *Corporate Reputation Review, 7*(2), 161–178.

Bettencourt, L. (1997). Customer voluntary performance. customers as partners in service delivery. *Journal of Retailing, 3*, 383–406.

Bollen, K., & Lennox, R. (1991). Conventional wisdom on measurement: A structural equation perspective. *Psychological Bulletin, 110*(2), 305–314.

Bromley, D. (2002). Comparing corporate reputations: League tables, quotients, benchmarks, or case studies. *Corporate Reputation Review, 5*(1), 35–50.

Chin, W. (1998). The partial least squares approach to structural equation modeling. In G. Marcoulides (Ed.), *Modern Methods for Business Research* (pp. 295–336). Mahwah/London: Lawrence Erlbaum.

Chin, W., Marcolin, B., & Newsted, P. (2003). A partial least squares latent variable modeling approach for measuring interaction effects: Results from a monte carlo simulation study and an electronic-mail emotion/adoption study. *Information Systems Research, 14*(2), 189–217.

Churchill, G., & Surprenant, C. (1982). An investigation into the determinants of customer satisfaction. *Journal of Marketing Research, 19*(10), 491–504.

Diamantopoulos, A., & Winklhofer, H. (2001). Index construction with formative indicators: An alternative to scale development. *Journal of Marketing Research, 38*, 269–277.

Dick, A., & Basu, K. (1994). Customer loyalty: Toward an integrated conceptual framework. *Journal of the Academy of Marketing Science, 22*(2), 99–113.

Festinger, L. (1957). *A Theory of Cognitive Dissonance*. Stanford CA: Stanford University Press.

Fombrun, C. (1996). *Reputation*. Boston: Harvard Business School Press.
Fombrun, C. (1998). Indices of corporate reputation: an analysis of media rankings and social monitors ratings. *Corporate Reputation Review*, *1*, 327–340.
Fornell, C., & Bookstein, F. (1982). A comparison analysis of two structural equation models: Lisrel and pls applied to market data. In C. Fornell (Ed.), *A Second Generation of Multivariate Analysis* (pp. 290–312). New York: Greenwood.
Gotsi, M., & Wilson, A. (2001). Corporate reputation: Seeking a definition. *Corporate Communications*, *6*, 24–30.
Gray, J. (1986). *Managing the Corporate Image*. Westport/London: Greenwood Press.
Hair, J., Anderson, R., Tatham, R., & Black, W. (1998). *Multivariate Data Analysis* (5th ed.). Englewood Cliffs: Prentice Hall.
Herbig, P., & Milewicz, J. (1994). Marketing signals in service industries. *Journal of Services Marketing*, *8*(2), 19–35.
Hoyle, R. H., & Kenny, D. A. (1999). Sample size, reliability, and tests of mediation. In R. H., Hoyle (Ed.), *Statistical strategies for small sample research* (pp. 195–222). Thousand Oaks: Sage.
Iacobucci, D., & Duhachek, A. (2003). *Mediation analysis – round table acr 2003*. Presentation at the round table of the ACR Conference, Toronto.
Jarvis, C., Mackenzie, S., & Podsakoff, P. (2003). A critical review of construct indicators and measurement model misspecification in marketing and consumer research. *Journal of Consumer Research*, *30*, 199–218.
Kleinbaum, D., Kupper, L., Muller, K., & Nizam, A. (1998). *Applied Regression Analysis and Other Multivariable Methods* (3rd ed.). Pacific Grove: Duxbury.
Lewis, S. (2001). Measuring corporate reputation. *Corporate Communications*, *6*, 31–35.
MacCullum, R., & Browne, M. (1993). The use of causal indicators in covariance structure models: Some practical issues. *Psychological Bulletin*, *114*(1), 533–541.
Mano, H., & Oliver, R. (1993). Assessing the dimensionality and structure of the consumption experience: Evaluation, feeling, and satisfaction. *Journal of Consumer Research*, *20*, 451–466.
Markwick, N., & Fill, C. (1997). Towards a framework for managing corporate identity. *European Journal of Marketing*, *31*(5/6), 396–409.
Middleton, S., & Hanson, D. (2002). Reputation, research, and the resource bases view: Determining a suitable methodolgy. In R. H. Hoyle (Ed.), *Proceedings of the 6th International Conference on Corporate Reputation, Identity and Competitiveness*. Boston.
Morley, M. (2002). *How to Manage Your Global Reputation* (2nd ed.). New York: New York University Press.
Nguyen, N., & Leblanc, G. (2001a). Corporate image and corporate reputation in customers' retention decision in services. *Journal of Retailing and Consumer Services*, *8*, 227–236.
Nguyen, N., & Leblanc, G. (2001b). Image and reputation of higher education institutions in students' retention decisions. *The International Journal of Educational Management*, *15*(6), 303–311.
Oliver, R. (1997). *Satisfaction*. New York: McGraw-Hill.
Oliver, R. (1999). Whence consumer loyalty? *Journal of Marketing*, *63*, 33–44.
Rossiter, J. (2002). The c oar se procedure for scale development in marketing. *International Journal of Research in Marketing*, *19*(4), 305–335.
Rust, R. T., & Zahorik, A. J. (2003). Customer satisfaction, customer retention,and market share. *Journal of Retailing*, *69*, 193–215.
Sandberg, K. (2002). Kicking the tires of corporate reputation. *Harvard Management Communication Letter*, *5*(1), 3–4.
Seltin, N., & Keeves, J. (1994). Path analysis with latent variables. In T. Husn & T. Postlethwaite (Eds.), *The International Encyclopedia of Education* (2nd ed., pp. 4352–4359). Oxford: Pergamon.
Shapiro, C. (1982). Consumer information, product quality, and seller reputation. *Bell Journal of Economics*, *13*(1), 20–35.

Singh, J., & Sirdeshmukh, D. (2000). Agency and trust mechanisms in consumer statisfaction and loyalty judgments. *Journal of the Academy of Marketing Science*, *28*(1), 150–167.

Sobel, M. (1982). Asymptotic confidence intervals for indirect effects on structural equation models. In S. Leinhardt (Ed.), *Sociological Methodology* (pp. 290–312). Jossy-Bass.

Thibaut, J., & Kelley, H. (1959). *The Social Psychology of Groups*. New York: Wiley.

Wartick, S. (2002). Measuring corporate reputation. *Business & Society*, *41*(4), 371–392.

Westbrook, R., & Oliver, R. (1981). Developing better measures of consumer satisfaction. some preliminary results. *Journal of Consumer Research*, *18*, 84–91.

Yoon, E., Guffey, H., & Kijewski, V. (1993). The effects of information and company reputation on intentions to buy a business service. *Journal of Business Research*, *27*, 215–228.

Chapter 23
Reframing Customer Value in a Service-Based Paradigm: An Evaluation of a Formative Measure in a Multi-industry, Cross-cultural Context

David Martín Ruiz, Dwayne D. Gremler, Judith H. Washburn, and Gabriel Cepeda Carrión

Abstract Customer value has received much attention in the recent marketing literature, but relatively little research has specifically focused on inclusion of service components when defining and operationalizing customer value. The purpose of this study is to gain a deeper understanding of customer value by examining several service elements, namely service quality, service equity, and relational benefits, as well as perceived sacrifice, in customer assessments of value. A multiple industry, cross-cultural setting is used to substantiate our inclusion of service components and to examine whether customer value is best modeled using formative or reflective measures. Our results suggest conceptualizing customer value with service components can be supported empirically, the use of formative components of service value can be supported both theoretically and empirically and is superior to a reflective operationalization of the construct, and that our measure is a robust one that works well across multiple service contexts and cultures.

D.M. Ruiz
Escuela Universitaria de Estudios Empresariales, 41012 Sevilla, Spain
e-mail: dmartin@us.es

D.D. Gremler
Department of Marketing, College of Business Administration, Bowling Green State University, Bowling Green, OH 43403, USA
e-mail: gremler@cba.bgsu.edu

J.H. Washburn
John H. Sykes College of Business, University of Tampa, 401 W. Kennedy Blvd., UT Box 48F, Tampa, FL 33606, USA
e-mail: jwashburn@ut.edu

G.C. Carrión
Departamento de Administración de Empresas y Marketing, Universidad de Sevilla, Ramón y Cajal, 1, 41018 Sevilla, Spain
e-mail: gabi@us.es

V. Esposito Vinzi et al. (eds.), *Handbook of Partial Least Squares*, Springer Handbooks of Computational Statistics, DOI 10.1007/978-3-540-32827-8_24,

23.1 Introduction

Companies have recognized the strategic relevance of maintaining a solid base of loyal customers for survival, growth, and financial performance (Arnett et al. 2003). Scholars and successful firms have highlighted the delivery of customer value as a key strategy for achieving customer loyalty and reducing defection rates (Parasuraman and Grewal 2000). In some sense, customer value creation has emerged as a new paradigm that is a more comprehensive approach than the focus on service quality and customer satisfaction in creating and sustaining a competitive advantage (Stewart 2002; Vargo and Lusch 2004; Woodall 2003). Gale (1997) notes, "the customer value paradigm is newer, includes many of the elements of the customer satisfaction paradigm, plus additional features, and is more widely adopted." Similarly, Holbrook (1994) points out that "customer value is the fundamental basis for all marketing activity." Customer value research is viewed as being in its early stages and still underdeveloped to the extent that its definition remains confusing (Flint et al. 2002).

Customer value has been addressed in the marketing literature for some time, but only recently has consideration been given to understanding value in the context of service delivery. It is widely held that customer value leads to competitive advantage (Woodruff 1997) and that value is typically seen as a tradeoff between what customers receive versus what they give up (e.g., Monroe 1990; Zeithaml's 1988. Zeithaml's (1988) definition of product value, "consumers' overall assessment of the utility of a product based on perceptions of what is received and what is given," is representative of how value has been described in tangible goods contexts. However, relatively little research has specifically focused on the inclusion of service components when defining and operationalizing customer value. Indeed, researchers have traditionally implied that service value should be conceived as a special case of customer value that could lead to a competitive advantage for service providers (e.g., Parasuraman and Grewal 2000). More recently, however, Vargo and Lusch (2004) have proposed that the traditional goods-based marketing paradigm is evolving into a service-based paradigm. Following this paradigm shift, we suggest that the conceptualization of customer value should be reframed and extended to include service elements.

The conceptualization and measurement of customer value has been approached in different ways in the marketing literature. The unidimensional approach describes customer value in a global fashion and often operationalizes the construct directly through single measures of utility or value for money (e.g., Bolton and Drew 1991; Cronin et al. 1997; Hartline and Jones 1996) or multiple items (e.g., Teas and Agarwal 2000). However, in conceptualizing customer value in this way, researchers lose the conceptual richness of the construct. Alternatively, the multidimensional approach considers customer value as a highly complex concept with many components (e.g., de Ruyter et al. 1997; Sheth et al. 1991). Recent studies addressing customer value have suggested that the construct is too complex to be operationalized as unidimensional (Lam et al. 2004; Rust et al. 2000; Wang et al. 2004; Woodall 2003). A question that arises when taking a multidimensional approach is, whether customer value should be modeled as consisting of reflective or formative indicators. Indeed, understanding the underlying essence of the construct, whether it is reflective (i.e., changes in the underlying construct cause changes in the indicators) or formative (i.e., indicators impact or cause the underlying construct), is an essential first step in modeling its structure (Jarvis et al. 2003). However, no prior study has examined whether customer value is better modeled with reflective or formative indicators.

The purpose of our study is to gain a deeper understanding of the customer value construct by looking at service components, to analyze how customer value is best measured, and to investigate this conceptualization across contexts and cultures. Specifically, we examine several service elements, namely service quality, service equity, and relational benefits (both social and confidence benefits) to see what role they play in customers' assessments of value. We conduct our study in a multiple-industry, multiple-culture setting to validate and generalize the proposed conceptualization of customer value. Our analysis also examines how customer value should be modeled by comparing a multidimensional, formative approach with a unidimensional, reflective approach.

23.2 Literature Review

23.2.1 Previous Conceptualizations of Customer Value

Early research on customer value is based in the pricing literature (Dodds and Monroe 1985), where perceived quality and sacrifice are the main components in determining the perceived value of a product, and extrinsic and intrinsic attributes are the determinants of quality and sacrifice. The widely held view is that "buyers' perceptions of value represent a tradeoff between the quality or benefits they perceive in the product relative to the sacrifice they perceive by paying the price" (Monroe 1990, p. 46). Zeithaml's (1988) customer value model, one of the first to appear in the literature, has been empirically assessed in a variety of different product categories and with numerous attribute cues (e.g., Dodds et al. 1991; Grewal et al. 1998; Kerin et al. 1992; Naylor and Frank 2000; Sweeny and Soutar 2001; Sweeny et al. 1999; Teas and Agarwal 2000; Yang and Peterson 2004). These studies, which all conceptualize customer value in a unidimensional manner, have identified how different product attributes (e.g., country of origin, perceived risk, price, perceived quality) relate to customer perceived value and behavioral intentions.

Other scholars have conceptualized customer value as multidimensional. As we indicate in Table 23.1, many studies have adopted Zeithaml's (1988) approach (i.e. tradeoff model) by arguing that customer value consists of various benefits and sacrifices (e.g., Lapierre 2000; Lin et al. 2005). Other frameworks have also been proposed. For example, Woodruff (1997, p. 142) proposes that customer value "incorporates both desired and received value and emphasizes that value stems from customers' learned perceptions, preferences, and evaluations." This view depicts customer value as a hierarchy or means-end chain that begins with customers thinking about desired attributes and performance and builds to customers' goal-directed

Table 23.1 Recent multidimensional approaches used to examine customer value empirically

Author(s) / Context	Type of components	Components of customer value(items)	
de Ruyter et al. (1997) *Hotelservice*	Reflective	Benefits components emotional value (5), practical value (5), logical value (5)	Sacrifice components
Grewal et al. (1998) *Bicycles*	Reflective	perceived acquisition value (9)	perceived transaction value (3)
Lapierre (2000) *ICE Information, communication, entertainment), distribution, and finance services*	Reflective	alternative solutions (3), product quality (4), product customization (4), responsiveness (3), flexibility (4), reliability (5), technical competence (5), supplier's image (2), trust (5), solidarity (4)	price (5), time/effort/energy (5), conflict (3)
Mathwick et al. (2001) *Internet and catalog shopping*	Reflective	aesthetics (6), playfulness (5), service excellence (2), customer ROI (6)	
Sweeny and Soutar (2001) *Durables*	Reflective	emotional value (5), social value (4), performance/quality (6)	price (4)
Petrick (2002) *Fast food restaurant service*	Reflective	quality (4), emotional response (5), reputation (5)	monetary price (6), behavioral price (5)
Lam et al. (2004) *Courier services (business-to-business)*	Reflective	service quality (5)	price competitiveness (5)
Heinonen (2004) *Online bill payment service*	Reflective	technical value (1), functional value (1), temporal value (1), spatial value (1)[a]	technical value (1), functional value (1), temporal value (1), spatial value (1)
Wang et al. (2004) *Security firms*	Reflective	functional value (4), social value (3), emotional value (5)	perceived sacrifice (6)
Liu et al. (2005) *Financial staffing services*	Reflective	core service (3), support service (4)	economic value (3)
Pura (2005) *Directory services*	Reflective[b]	social value (3), emotional value (2), epistemic value (3), conditional value (2)	monetary value (3), convenience value (4)

(*continued*)

Table 23.1 (continued)

Author(s) / Context	Type of components	Components of customer value (items)	
Lin et al. (2005) *Web services*	Reflective and formative	web site design (5), fulfillment/reliability (3), security/privacy (3), customer service (3)	monetary sacrifice (2)

[a]The value components were each assumed to include an assessment of benefits and sacrifices.
[b]Six value components were investigated independently; the discussion does not suggest a formative conceptualization.

and purposeful behavior or their satisfaction with the received value; only a handful of studies have followed this approach, including those by Flint et al. (2002), Overby et al. (2004), and Woodruff and Gardial (1996). Sheth et al. (1991) propose five dimensions of customer value-epistemic, social, functional, emotional, and conditional dimensions of consumption; and their study serves as a framework for research conducted by de Ruyter et al. (1997) and Sweeny and Soutar (2001). Finally, Holbrook's (1994) multidimensional conceptualization suggests that value not only serves as the basis for a purchase decision, but is also the result of a particular consumption experience. He proposes a value typology based on three criteria – extrinsic/intrinsic value, reactive/passive value, and internal/external orientation – that has been tested by other researchers (e.g., Mathwick et al. 2001). However, of these alternative conceptualizations of value, the most commonly used framework remains Zeithaml's (1988) tradeoff model. We adopt her approach and conceptualize customer value in service contexts as consisting of various benefits and sacrifices.

23.2.2 *Service Value*

The call for more of a service focus in marketing research has recently been made in the literature. For example, Vargo and Lusch (2004, p. 2) argue that "the traditional dominant, goods-centered view of marketing not only may hinder a full appreciation for the role of services but also may partially block a complete understanding of marketing in general." The service view of marketing is customer-centric, suggesting that value is defined by and cocreated with the customer rather than embedded in the output (Sheth et al. 2000). Similarly, Grönroos (2000, pp. 24–25) states that "value for customers is created throughout the relationship by the customer, partly in interactions between the customer and the supplier or service provider. The focus is on the customers' value-creating processes where value emerges for customers and is perceived by them."

Following these arguments, and consistent with Vargo and Lusch's (2004) suggested service-dominant paradigm, we focus on better understanding customer value by examining service-related issues. Thus, in this study, we are interested in examining *the customer's perception of quality and benefits weighed against sacrifices in the context of service delivery*. From this point forward, we will use the term service value as a synonym for customer value since our focus is on demonstrating the role various service components can have in shaping customers' perceptions of value. In the next section, we identify major components of service value – in terms of benefits and sacrifices – present in the service delivery process.

23.3 Toward a Conceptualization of Service Value

23.3.1 Service Value Components

In multidimensional approaches, value has been described as depending on a combination of monetary and non-monetary sacrifice, quality, performance, and disconfirmation experiences that represent a "richer, more comprehensive measure of customers' overall evaluation of a service than service quality" (Bolton and Drew 1991, p. 383). We contend that service value is primarily a cognitive consumer response since most of its components are assessed rationally. Our review of the literature suggests that customers consider several issues when making cognitive assessments of service value including service quality, service equity, relational benefits, and perceived sacrifice. The following paragraphs briefly discuss each of these components and argue why, based on our review, they should be considered salient components of service value.

Service Quality. The delivery of a high-value service offering is generally expected to be based on customer perceptions of quality (Berry 1995; Gremler and Brown 1996; Gronroos 1995). If a company's service delivery is built on a core physical product (e.g., a cellular phone in wireless communication services), product quality will be a component of perceived value for the customer (Rust and Oliver 1994). However, independent of where an offering stands on the goods-services continuum, perceived service quality is considered to be an essential pillar of value (Gronroos 1995). Service quality is difficult for competitors to imitate (Parasuraman and Grewal 2000), and it therefore represents a basis for differentiation (Berry 1995) and competitive advantage (Reichheld and Earl Sasser 1990) in building service value.

Service Equity. We suggest that service equity, which is also referred to as service image or service brand equity, should be considered as a second component of service value. Berry and Parasuraman (1991) contend that service image can be a source of customer value creation as company communications and customer experiences with the service define perceptions of the brand. A strong brand can create feelings of proximity, affection, and trust, and thus contribute significantly to customer perceptions of value. Cultivating brand equity in services is especially important given the intangible nature of the "invisible purchase" that a service represents for the customer (Berry 2000). As a consequence, service equity plays the role of a signaling indicator for the customer in a wide number of service settings (Singh and Sirdeshmukh 2000). Therefore, service equity is likely to be a salient dimension of perceived customer value in services, and a path to value creation for the customer.

Relational Benefits. The benefits derived from an ongoing relationship with the service provider represent another value component that should be considered in evaluations of the service delivery process. Grönroos (1997) argued that a relationship has a value of its own, acting as a softener in the case of discrete service failures, since the relational customer judges the relationship with the provider as a whole. Building on the early work of Barnes (1994), Bendapudi and Berry (1997), and Berry (1995), Gwinner et al. (1998) developed, and empirically supported, a typology of three relational benefits: confidence benefits, social benefits, and special treatment benefits. These are all benefits that exist above and beyond the core service being delivered (Hennig-Thurau 2002). Confidence benefits refer to customer feelings of trust and anxiety reduction. As customers engage in relational behavior and accumulate service encounter experiences, their level of uncertainty decreases as their knowledge of the service provider increases. Social benefits refer to the friendship, recognition, and fraternization that might arise between the customer and the service provider; they pertain to the emotional part of the relationship and are characterized by personal recognition of customers by employees, the customer's familiarity with employees, and the creation of friendships between customers and employees. Because service encounters are mostly social encounters (Czepiel 1990), Gwinner et al. (1998) found such benefits are often highly valued by customers. Finally, special treatment refers to functional benefits such as "... the customer's perception of preferential treatment, extra attention or personal recognition, and special service not available to other customers" (Gwinner et al. 1998, p. 105). A number of authors have found that these benefits significantly affect customer assessments of the service provider (cf. Bolton et al. 2000; Hennig-Thurau 2002; Price and Arnould 1999; Reynolds and Beatty 1999). Therefore, we contend that relational benefits are part of service value – at least for those customers who actively participate in an ongoing relationship – since these customers are able to evaluate such benefits as their experience with the service provider accumulates.

Perceived Sacrifice. Finally, customers may face a number of sacrifices, which involve both monetary and non-monetary costs, to obtain a service. The price paid for the service is the obvious monetary sacrifice, which is clearly a component of service value (Voss et al. 1998). Indeed, price or sacrifices have been empirically tested as either the antecedents or dimensions of value in both product and service settings (Cronin et al. 1997; Teas and Agarwal 2000). However, although customers do not always want low prices, they do consistently want the service to be worth the money expended. For some customers or in some specific situations, non-monetary sacrifices (e.g., convenience with respect to time, effort, and energy) might be even more important than monetary sacrifices when making choices. For example, time-constrained consumers patronize convenience stores and increasingly shop online to save time and effort. In this regard, time spent on making the buying decision and time spent waiting to access, receive, and complete the service are all relevant (Berry et al. 2002). In conclusion, the literature suggests perceived sacrifice – including both price and non-monetary sacrifices – should also be included in a conceptualization of service value.

23.3.2 Operationalizing Service Value

Because of the multidimensional conceptualization of service value, we propose that the construct is best operationalized as a formative index. The calculation of such an index requires the use of formative rather than reflective indicators (Arnett et al. 2003). When reflective indicators are used, the latent construct is assumed to cause the observed indicators; that is, with reflective indicators the observed variables "reflect" the changes in the latent construct (Bollen 1989). In comparison, when a latent construct is measured using formative indicators, the observed indicators are assumed to cause or "form" the latent construct. As such, omitting one or more formative indicators in effect omits part of the construct. The literature suggests that *each* of the service value components discussed earlier, should be essential to customer perceptions of value. Thus, our index is comprised of measures that influence the underlying latent construct rather than being influenced by it. Although the use of reflectively measured latent constructs dominates much of the research in marketing (Diamantopoulos and Winklhofer 2001), formative indexes have a long and rich tradition in social science research (e.g., Cronbach and Glesser 1953; Warner et al. 1949). Examples of formative indexes used in marketing research include the American Customer Satisfaction Index (Fornell et al. 1996), the Swedish Customer Satisfaction Barometer (Fornell 1992), the Deutsche Kundenbarometer (Meyer 1994), the job descriptive scale (Futrell 1979), and the retailer equity index (Arnett et al. 2003).

In this study, we conceptualize and measure service value as an index formed by the following components: service equity, service quality, relational benefits, and perceived sacrifice. It is appropriate to conceptualize service value as an index since changes in any of these dimensions would cause a change in the service value index. Furthermore, a change in one of the observed variables is not necessarily accompanied by changes in any of the other observed dimensions. For example, devoting more time to reach the dentist's office because of the longer distance to the office from the patient's home than other such offices (an indicator of perceived sacrifice) would not necessarily be accompanied by a change in service quality, service equity, or relational benefits displayed by the service provider. Therefore, the measurement of service value is modeled as having formative components that cause changes in the latent construct service value index (see Fig. 23.1).

23.4 Methodology

In view of the earlier discussion, the intent of the present study is threefold: (1) to identify components expected to be strong indicators of service value – namely, service quality, service equity, relational benefits, and perceived sacrifice; (2) to compare this multidimensional conceptualization of service value with a direct (reflective) conceptualization of the construct; and (3) to generalize this conceptualization by examining its robustness across differing services and across two cultures.

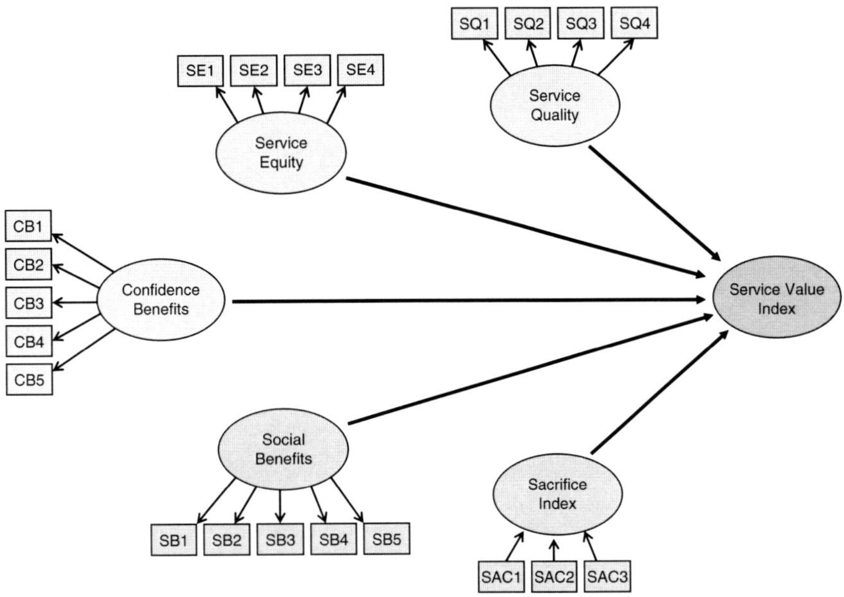

Fig. 23.1 Service value components

To examine the robustness of the conceptualization across various types of services, we grouped service organizations into three categories – following Bowen's (1990) classification of service industries – based on: the degree to which the offering is directed to the person or the person's property; whether the service has high, moderate, or low levels of customer contact; and the extent to which the service is highly customized, moderately customized, or standardized. To examine this conceptualization of service value across cultures, we conducted studies of both U.S. and Spanish consumers.

23.4.1 Measures and Data Collection

A self report questionnaire that examines relationships with service providers was administered to 800 respondents (500 U.S. and 300 Spanish consumers). Respondents in both countries completed one of three questionnaire forms representing the three categories of service providers suggested by Bowen (1990): Group 1 – high contact, customized, personalized services (e.g., medical care, barber shop); Group 2 – moderate contact, semi-custom, non-personal services (e.g., dry cleaning, auto repair); and Group 3 – moderate contact/standardized services (e.g., health club, fast-food restaurant). Each respondent was asked to report on a service provider with whom he or she perceived having a strong, established relationship (cf. Gwinner et al. 1998).

The service value components under evaluation consist of a collection of 23 items that measure each of the components previously described: service quality (five items), service equity (five items), relational benefits – specifically confidence benefits (five items) and social benefits (five items),[1] and perceived sacrifice (three items). All items were taken directly or modified slightly from previously validated measures in the literature. Specifically, the service quality scale was adopted from Taylor and Baker (1994) and Gremler and Brown (1996); service equity items were taken from Yoo and Donthu (2001) and Ha (1996); relational benefits (specifically, confidence benefits and social benefits) items were taken from Gwinner et al. (1998); and the perceived sacrifice measures were from Sweeney and Soutar (2001) and Blackwell et al. (1999). The scales, presented in the Appendix, are seven-point Likert scales with anchors "strongly disagree" and "strongly agree."[2]

Both reflective and formative measures can be associated with a particular construct (Fornell 1982). As indicated earlier, of the service value components we considered, only perceived sacrifice is considered to be a formative construct (formed by price, time, and effort indicators). Our perceived sacrifice index combines both monetary and non-monetary sacrifices measures in a formative way since monetary sacrifices (e.g., price) and non-monetary sacrifices (e.g., time) are not necessarily positively correlated and, in fact, may sometimes be negatively correlated. The remaining components – service quality, service equity, confidence benefits, and social benefits – are first-order latent constructs measured by reflective indicators.

Finally, three other sets of measures were included in the study. To compare our index with a reflective operationalization of the construct, seven items were included as a direct reflective measure of value (Grewal et al. 1998; Sweeny and Soutar 2001). Two constructs were also included to provide an external validity assessment, including customer satisfaction – measured with six items based on Taylor and Baker (1994) and Oliver (1980), and repurchase intentions – with three items based on Zeithaml et al. (1996) and Taylor and Baker (1994).

23.4.2 Respondent Samples

U.S. Sample. Students served as data collectors for this sample, a technique that has been successfully used in a variety of services marketing studies (e.g., Bitner et al. 1990; Gwinner et al. 1998; Keaveney 1995). A total of 100 undergraduate

[1] We chose to focus on only two of the three relational benefits delineated by Gwinner et al. (1998), namely confidence benefits and social benefits. This decision was based on the necessity for parsimony and the desire to avoid weighting the service value construct too heavily on the dimension of relational benefits.

[2] Measures were pretested in both the U.S. (56 respondents) and Spain (66 respondents), following a double translation procedure (from English to Spanish and then back to English). As a consequence of the pretest results, two items were slightly reworded. In general, items and measurement scales in the pretest worked properly, displaying good reliability with Cronbach's alphas all above 0.80.

students from a public university in the midwestern U.S. participated as data collectors as part of a class assignment; a total of 500 questionnaires were distributed to U.S. customers. Each student distributed five questionnaires among their network of acquaintances from each of five age ranges (i.e., 19–29, 30–39, 40–49, 50–59, and over 60) and was instructed to collect data from at least two respondents of each gender. Three versions of the questionnaire, representing each of Bowen's (1990) three industry groups, were randomly distributed within each data collector's set of five. All questionnaires were collected within 14 days of distribution. Of the 500 questionnaires, six were not usable as they did not contain a complete set of responses; thus, 494 responses were usable (170, 158, and 166 per service Industry Groups 1, 2, and 3, respectively).

Spanish Sample. In Spain, two doctoral students trained in field research at a public university in Spain distributed 300 questionnaires to customers, with 254 of the responses deemed usable (55, 107, and 92 per Industry Groups 1, 2, and 3, respectively). As with the U.S. sample, data collectors followed age and gender quotas to prevent response bias. The industry group quota was not strictly followed, as it turned out to be difficult for the researchers to identify customers within the Spanish sample who perceived they had a strong, established relationship with a service provider from Industry Group 1 – only 55 usable responses were collected for this group.

In total, we obtained 748 valid questionnaires (225 from Industry Group 1, 265 from Industry Group 2, and 258 from Industry Group 3). The U.S. respondents averaged 45.0 years of age and 56.6% were female; Spanish respondents averaged 30.8 years of age and 57.0% were female. The average length of the customer/service provider relationship was 10.1 years in the U.S. sample and 5.1 years in the Spanish sample.

23.4.3 Data Analysis

Data analysis was performed using Partial Least Squares (PLS), a structural equation modeling technique that uses a principal-component-based estimation approach (Chin 1998). The use of PLS has certain advantages: (1) it does not suffer from indeterminacy problems like other causal modeling techniques using EQS or LISREL; (2) it is a nonparametric technique and, therefore, does not assume normality of the data; (3) it does not require as large a sample size as other causal modeling techniques; and (4) it can be used to estimate models that use both formative and reflective indicators. Research suggests the characteristics of PLS analysis make it an especially useful tool for index construction (Arnett et al. 2003; Diamantopoulos and Winklhofer 2001; Fornell et al. 1996).

For index development testing using PLS, Chin (1995,1998) recommends two procedures: the bootstrapping procedure and the Stone-Geisser test. In bootstrapping, a large number of random samples – Chin (1998) suggests 500 samples generated from the original dataset by sampling with replacement (Efron and Tibshirani 1993). Path coefficients are estimated with each random sample, and mean parameter estimates and standard errors are computed across the total number of samples.

In addition, the Stone-Geisser test of predictive relevance is used to assess model fit (Geisser 1975; Stone 1974); predictive relevance can be considered a type of model fit indicator as PLS does not provide assessment of causal relationships. The Stone-Geisser test, which does not require assumptions about the distribution of residuals, involves omitting or "blindfolding" one case at a time, re-estimating the model parameters based on the remaining cases, and predicting the omitted case values on the basis of the remaining parameters (Sellin 1995). The procedure results in the Q^2 test statistic, a measure representing how well observed values are reconstructed by the model and its parameter estimates (Chin 1998). If $Q^2 > 0$, the model has predictive relevance. Conversely, if $Q^2 \leq 0$, the model lacks predictive relevance.

In PLS, results are presented in two stages: the measurement model, which includes an assessment of the reliability and validity of the measures, and the structural model, which tests: (1) the amount of variance explained, (2) the significance of the relationships, and (3) the model's predictive relevance (Barclay et al. 1995). In this study, we assess the external validity of the index by evaluating the relationship between the service value index and measures of customer satisfaction and repurchase intentions.

23.5 Results

23.5.1 Measurement Model Analysis

The measurement model in PLS is assessed in terms of inter-construct correlations, item-to-construct correlations, Cronbach's alphas, composite reliabilities, and the average variance extracted for each construct. As indicated in Fig. 23.1, we model the service value index as a second-order formative construct with the five components independent from one another. Each of the scales for service equity (SE), service quality (SQ), confidence benefits (CB), and social benefits (SB) consist of reflective items, while the scale for perceived sacrifice (SAC) is formed by formative items. In the following paragraphs, we assess measure reliability, internal consistency, and discriminant validity for each of the service value components and the other measures included in the study. Table 23.2 displays factor loadings of the reflectively formed components of service value and the weights of the formative component (perceived sacrifice); Table 23.3 includes descriptive statistics and their correlations.

In order to assess *measure reliability* of each service value component, as well as the other measures in the study, we examined how each item relates to the latent constructs.[3] When assessing measures associated with a particular construct, the type

[3] In assessing formative indicators, it is important to keep in mind that they may be completely uncorrelated and, therefore, internal consistency across components is not appropriate. According to Diamantopoulos and Winklhofer (2001), the correlation among formative indicators is not

Table 23.2 Assessment of reflective and formative constructs

(A) Reflective constructs: factor loadings

	Service Equity	Service Quality	Confidence (Relational) Benefits	Social (Relational) Benefits	Customer Value (reflective measure)	Customer Satisfaction	Repurchase Intentions
SE1	**0.83**	0.23	0.11	0.22	0.30	0.20	0.03
SE2	**0.91**	0.23	0.20	0.18	0.26	0.22	0.03
SE3	**0.92**	0.26	0.24	0.14	0.26	0.23	0.03
SE4	**0.82**	0.15	0.20	0.12	0.18	0.19	0.01
SQ1	0.21	**0.88**	0.15	0.31	0.27	0.31	0.03
SQ2	0.22	**0.88**	0.13	0.13	0.31	0.29	0.05
SQ3	0.27	**0.90**	0.14	0.21	0.28	0.27	0.04
SQ4	0.26	**0.86**	0.13	0.15	0.25	0.28	0.03
CB1	0.11	0.11	**0.88**	0.16	0.09	0.06	0.02
CB2	0.13	0.03	**0.88**	0.18	0.12	0.14	0.00
CB3	0.19	0.10	**0.92**	0.18	0.17	0.11	0.00
CB4	0.12	0.09	**0.89**	0.29	0.15	0.14	0.00
CB5	0.10	0.12	**0.86**	0.08	0.12	0.17	0.01
SB1	0.22	0.21	0.52	**0.89**	0.21	0.15	0.06
SB2	0.23	0.21	0.43	**0.90**	0.18	0.28	0.03
SB3	0.26	0.21	0.32	**0.89**	0.20	0.19	0.01
SB4	0.08	0.16	0.37	**0.84**	0.21	0.34	0.02
SB5	0.22	0.25	0.34	**0.89**	0.28	0.31	0.03
CV1	0.31	0.27	0.19	0.21	**0.79**	0.42	0.07
CV2	0.11	0.10	0.15	0.04	**0.82**	0.19	0.02
CV3	0.23	0.26	0.14	0.19	**0.88**	0.23	0.04
CV4	0.13	0.13	0.13	0.09	**0.86**	0.19	0.02
CV5	0.24	0.20	0.18	0.19	**0.88**	0.28	0.03
CV6	0.27	0.18	0.10	0.26	**0.78**	0.24	0.05
CV7	0.22	0.30	0.14	0.16	**0.87**	0.22	0.02
SAT1	0.21	0.23	0.24	0.22	0.32	**0.90**	0.08
SAT2	0.21	0.32	0.20	0.24	0.39	**0.95**	0.05
SAT3	0.23	0.30	0.21	0.23	0.37	**0.92**	0.03
SAT4	0.32	0.30	0.19	0.25	0.36	**0.93**	0.05
SAT5	0.26	0.34	0.15	0.23	0.39	**0.96**	0.06
SAT6	0.31	0.32	0.19	0.22	0.38	**0.95**	0.07
RP1	0.02	0.04	0.01	0.04	0.04	0.06	**0.92**
RP2	0.02	0.02	0.00	−0.01	0.02	0.02	**0.90**
RP3	0.03	0.03	0.02	0.04	0.05	0.04	**0.84**

(B) Formative constructs: component weights

	Component Weights
SAC1	**0.51**
SAC2	**0.57**
SAC3	**0.12**

Table 23.3 Descriptive statistics and correlation matrix

	Mean[a]	SD	CA	CR	AVE	1	2	3	4	5	6	7	8	9
1. Service equity	5.72	1.06	0.89	0.92	0.76	**(0.87)**								
2. Service quality	5.26	1.00	0.90	0.93	0.77	0.62	**(0.88)**							
3. Social benefits	4.30	0.99	0.93	0.95	0.79	0.42	0.41	**(0.89)**						
4. Confidence benefits	5.34	1.00	0.93	0.94	0.76	0.58	0.66	0.69	**(0.87)**					
5. Sacrifice index[b]	3.22	1.12	n.a.	n.a.	n.a.	−0.07[c]	−0.21	−0.16	−0.19	n.a.				
6. Service value index[b]	4.77	1.02	n.a.	n.a.	n.a.	0.76	0.73	0.58	0.85	−0.31	n.a.			
7. Customer value[d]	5.31	1.42	0.93	0.94	0.69	0.62	0.70	0.43	0.63	−0.37	0.78	**(0.83)**		
8. Customer satisfaction	5.72	1.32	0.96	0.97	0.83	0.65	0.80	0.46	0.73	−0.26	0.86	0.80	**(0.91)**	
9. Repurchase intentions	4.94	1.46	0.87	0.91	0.72	0.62	0.64	0.40	0.61	−0.28	0.70	0.65	0.76	**(0.85)**

Notes:
[a]Mean = the average score for all of the items included in this measure; S.D. = Standard Deviation; CA = Cronbach's Alpha; CR = Composite Reliability; AVE = Average Variance Extracted; n.a. = not applicable. The bold numbers on the diagonal are the square root of the Average Variance Extracted. Off-diagonal elements are correlations among constructs
[b]Formative construct
[c]For this correlation, $p < 0.05$; for all other correlations in the table, $p < 0.01$
[d]This construct is formulated using seven reflective indicators

of measure dictates whether one looks at the weights when examining formative measures, or factor loadings when examining reflective measures (Mathwick et al. 2001). Table 23.2 shows construct-to-item loadings and cross-loadings of the reflective service value measures. All of the loadings exceed 0.82 for these items and load more highly on their own construct than on others. The loadings for the direct reflective measures of customer value, as well as for customer satisfaction and repurchase intentions, are also as expected (i.e., all above 0.70). These results provide strong support for the reliability of the reflective measures.

explained by the measurement model but is exogenously determined. Therefore, internal consistency across components is of minimal importance since two components that might even be negatively related could both serve as meaningful indicators. As a result, "conventional procedures used to assess the validity and reliability of scales composed of reflective indicators are not appropriate for indexes with formative indicators" (Diamantopoulos and Winklhofer 2001, p. 271). In contrast to formative indicators, reflective indicators are essentially interchangeable because they mirror or reflect the latent construct. Omitting a single reflective measure will not compromise the essential nature of the construct. Reflective indicators should be internally consistent and changes in the latent construct cause changes in the reflective variable(s). Thus, we examine the internal consistency within each reflective service value component and the other reflective constructs in the study, but not across the service value components.

In the case of formative measures, instead of examining the factor loadings, one examines factor weights – which represent a canonical correlation analysis and provide information about how each indicator contributes to the respective construct (Mathwick et al. 2001). As indicated in Table 23.2, all three formative items for perceived sacrifice significantly contribute to the measure ($p < 0.01$), with time (weight = 0.57) and money (weight = 0.51) being the major contributors to the sacrifice index, followed distantly by effort (weight = 0.12). A concern with formative measures is the potential multicollinearity among the items (Mathwick et al. 2001), which could produce unstable estimates. Thus, we performed a collinearity test; the results showed minimal collinearity with the variance inflation factor (VIF) of all items ranging between 1.30 and 1.80, far below the common cut-off threshold of 5 to 10. These results suggest that the three items are salient contributors to the perceived sacrifice index.

Internal consistency is assessed using two measures: Cronbach's alpha and composite reliability. Nunnally (1978) suggests 0.70 as a benchmark for a "modest" reliability applicable in early stages of research and 0.80 as a more "strict" reliability applicable in basic research. As shown in Table 23.3, both the alpha and composite reliability of each set of reflective measures for each component of the service value index, as well as each of the other measures included in the study, exceeds 0.89. Additionally, the factor loadings for each of the components of the service value index are all greater than 0.82, and for all of the other constructs examined, the loadings are greater than 0.78, suggesting all of the items are good indicators of their respective components.

Discriminant validity was assessed in two ways. First, we examined the Average Variance Extracted (AVE) – which indicates the amount of variance that is captured by the construct in relation to the variance due to measurement error. Values for AVE should exceed 0.50 (BAR95). As the statistics presented in Table 23.3 indicate, all AVE values are greater than 0.69. Second, we compared the square root of the AVE (i.e. the diagonal in Table 23.3) with the correlations among constructs (i.e. the off-diagonal elements in Table 23.3). In Table 23.3, the square root of AVE for all of the reflective constructs exceeds 0.83 and each is greater than the correlation between the constructs; in order to demonstrate discriminant validity, diagonal elements should be greater than off-diagonal elements (Fornell and Larcker 1981). These statistics suggest that each construct relates more strongly to its own measures than to measures of other constructs; that is, all constructs share more variance with their own measures than with the others. These two sets of findings provide strong evidence of discriminant validity among the constructs.

Collectively, these results provide support for the overall quality of our measures. In particular, the statistics suggest our component measures are reliable, are internally consistent, and have discriminant validity.

Finally, we assessed the service value index as a formative *second-order factor*. The previous discussion provides support for the quality of the measures of the various service value components. Also of interest are the weights of the five service value components. The statistics for all but one of the components were as expected. As indicated in Table 23.4, the weights for service quality (weight = 0.46), service

Table 23.4 Service value statistics across contexts

	Entire sample	Industry group 1	Industry group 2	Industry group 3	U.S. sample	Spanish sample
Service value index weights[a]						
Service quality (SQ) component	0.46	0.46	0.46	0.42	0.44	0.55
Service equity (SE) component	0.34	0.28	0.39	0.28	0.31	0.36
Confidence benefits (CB) component	0.23	0.33	0.16	0.22	0.30	0.13
Social benefits (SB) component	0.00	0.05	−0.04	0.00	−0.03	0.03
Sacrifice (SAC) component	−0.30	−0.25	−0.30	−0.43	−0.29	−0.37
MIMIC model:						
Structural path						
SV index → CV (reflective measure)	0.79	0.73	0.83	0.77	0.80	0.71
Standard error[b]	0.01	0.03	0.02	0.02	0.02	0.03
R2	0.63	0.54	0.69	0.60	0.64	0.51
Q2	0.56	0.56	0.60	0.52	0.60	0.55
External validity model:						
Structural path						
SV index → SAT	0.88	0.86	0.88	0.81	0.88	0.77
Standard error[b]	0.01	0.02	0.02	0.02	0.01	0.02
R2	0.78	0.74	0.78	0.66	0.78	0.60
Q2	0.74	0.75	0.73	0.73	0.80	0.63
Structural path						
SV index → RP	0.72	0.68	0.69	0.68	0.69	0.58
Standard error	0.02	0.05	0.04	0.04	0.04	0.05
R2	0.51	0.46	0.48	0.46	0.48	0.34
Q2	0.53	0.42	0.56	0.47	0.53	0.41

[a] All weights are standardized
[b] Standard error values are estimated using a bootstrapping procedure
Industry Group 1: (High Contact/Customized/Personalized Services) Nice Restaurants, Beauty Salon, Medical Care Services, Barber Shop, Dental Care, Legal Services, Investment Brokerage Firms, Financial Consulting/Accounting Services
Industry Group 2: (Moderate Contact/Semi-customized/Non-personal Services) Photo Finishing Services, Shoe Repair, Laundry and Dry Cleaning Services, Computer Repair, Auto Repair, Veterinarian Care, Banking Services, Cellular/Mobile Phone Service
Industry Group 3: (Moderate Contact/Standardized Services) Health Club, Airlines, Movie Theater, Grocery Store, Express Mail Services, Copying/Printing Services, Retail Clothing Store, Fast Food Restaurant

equity (weight = 0.34), confidence benefits (weight = 0.23), and sacrifice (weight = −0.30) suggest they are major determinants of service value. Surprisingly, the weight for social benefits was essentially zero (weight = 0.004). We performed a collinearity test on the index; the results showed minimal collinearity among the five components, with the variance inflation factor (VIF) of all items ranging between 1.06 and 3.00, far below the common cut-off threshold of 5 to 10. Thus, the five service value components are independent from one another. Overall, these results suggest four of the five components are salient contributors to the service value index. In the discussion section, we discuss this finding further.

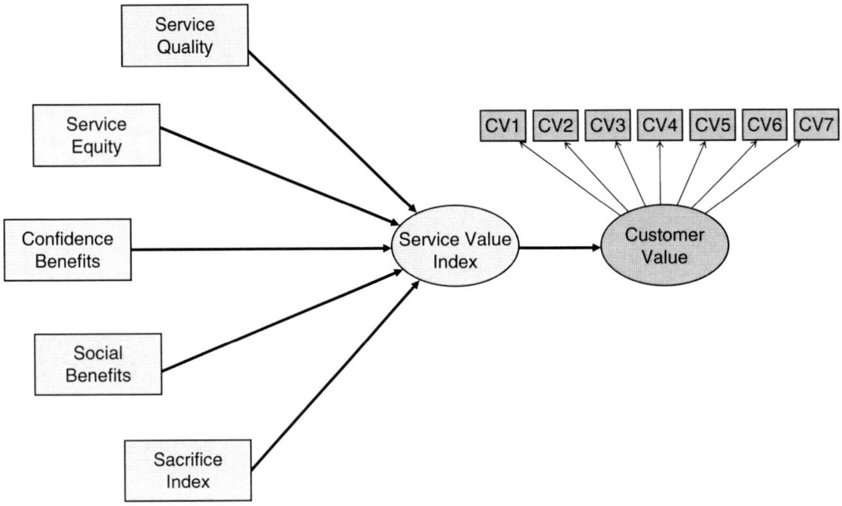

Fig. 23.2 MIMIC model for PLS analysis of the service value index

23.5.2 Structural Model Assessment

A model estimated through PLS algorithms can only be analyzed if it is placed within a larger model that incorporates consequences of the latent variable in question. In our case, we examine several models: (1) a multiple indicators and multiple causes (MIMIC) model, where the dependent variable is a direct measure of customer value; and (2) two models with other theoretically related dependent variables included for external validity assessment.

A MIMIC model approach (Jöreskog and Goldberger 1975) can be used to assess the appropriateness of a set of formative indicators (Diamantopoulos and Winklhofer 2001). To test the validity of our five-component service value index, our MIMIC model (see Fig. 23.2) includes a reflective seven-item measure of customer value as an external criterion variable that is explained by the service value index. (See the Appendix for a list of the items included in this measure.) According to the MIMIC model statistics, our index explains a relatively large amount of variance in this seven-item measure of value; the model's R^2 value, the main criteria by which model fit is assessed in PLS analysis (Chin 1998), is 0.63. In addition, the Stone-Geisser statistic (Q^2) is 0.56; values greater than zero indicate that the model has predictive relevance. Furthermore, the path from the service value index to the seven-item customer value measure is positive and significant ($\beta = 0.79$, $p < 0.001$) and the standard error is low (SE $= 0.01$), indicating the service value index adequately captures the construct being measured by the reflective indicators. In sum, the data provide support for the proposed formative model of service value.

To provide evidence of *external validity*, the service value index should be significantly correlated to other constructs that theory suggests should be associated

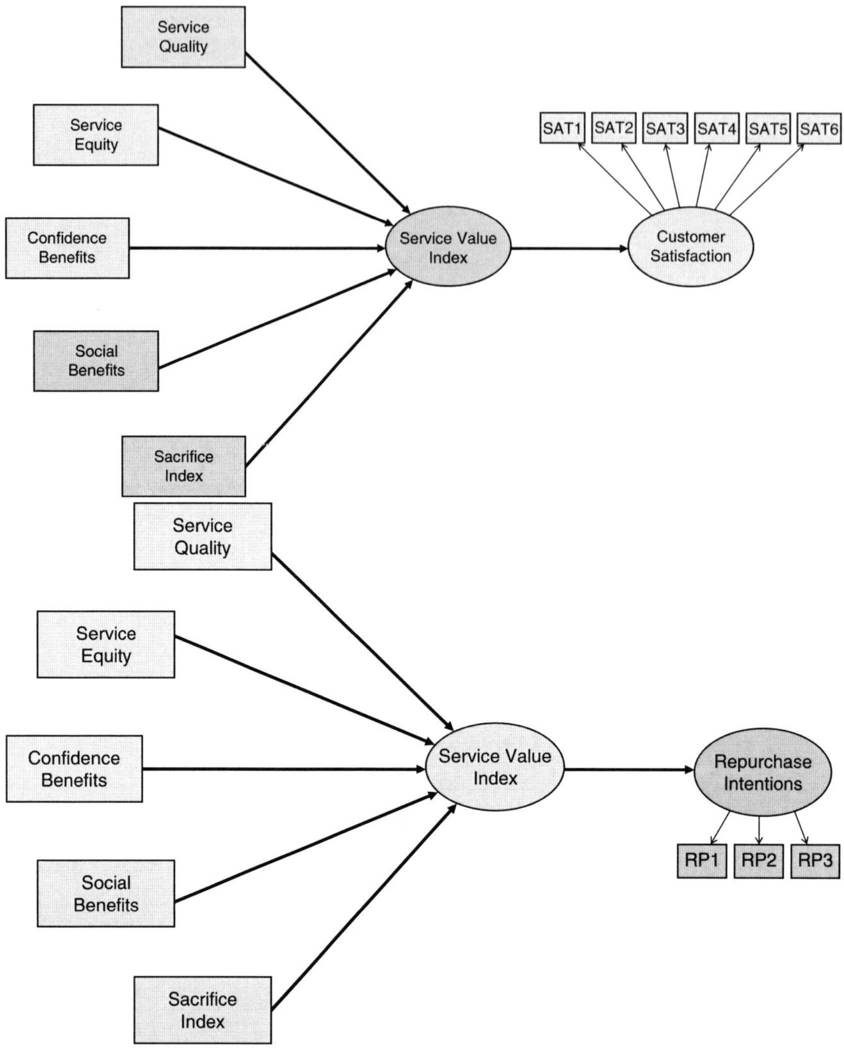

Fig. 23.3 External validation models for PLS analysis of the service value index

with the construct (Bagozzi 1994). As indicated earlier and depicted in Fig. 23.3, we included two constructs in the study – namely, customer satisfaction and repurchase intentions – that theory suggests should be related to service value. Consistent with the services literature (e.g., Cronin et al. 2000), we estimated two models in which the service value index serves as an antecedent for these two constructs (see Fig. 23.3). The resulting statistics suggest each model fits the data well: for customer satisfaction, $R^2 = 0.74$, and for repurchase intentions, $R^2 = 0.51$. We also estimated these models using the reflective seven-item measure of value. The service value index outperforms the reflective measure as the R^2 values are smaller when value is modeled using reflective indicators: for customer satisfaction, $R^2 = 0.64$,

Table 23.5 Comparison of formative and reflective measures of customer value

	Service value index (Formative measure)	Customer value (Reflective measure)
Customer satisfaction		
Structural path		
Service value → Customer satisfaction (SAT)	0.86	0.80
Standard error[a]	0.01	0.01
R^2	0.74	0.64
Repurchase intentions		
Structural path		
Service value → Repurchase intentions (RP)	0.72	0.65
Standard error[a]	0.02	0.02
R^2	0.51	0.43

[a] Standard error values are estimated using a bootstrapping procedure

and for repurchase intentions, $R^2 = 0.43$. We conducted an f^2 analysis to compare the R^2 values in the two external validity models for both value measures. The f^2 statistic for a comparison of the customer satisfaction R^2 values is 0.64 and for a comparison of the repurchase intentions R^2 values it is 0.43; both f^2 statistics are greater than 0.35, the level that suggests a substantial difference between each pair of R^2 values (Chin 1998), indicating that the service value index is a substantially better predictor of these two constructs than the reflective measure.

We also examined the path coefficients between the service value index and the two constructs, using the bootstrapping test mentioned earlier with 500 subsamples (Chin 1998). As reported in Table 23.5, the coefficients are significant ($p < 0.001$; $SE = 0.01$) in each relationship: for customer satisfaction $\gamma = 0.86$, and for repurchase intentions $\gamma = 0.71$. These coefficients are greater than those that result from using a model with a reflective measure of value: for customer satisfaction $\gamma = 0.80$, and for repurchase intentions, $\gamma = 0.65$. As we did with the external validity models mentioned in the previous paragraph, we conducted an f^2 analysis to compare the path coefficients in the external validity model for both value measures (formative and reflective). The f^2 statistic for a comparison of the customer satisfaction coefficients is 0.64 and for the repurchase intentions is 0.43; and, as before, both values are greater than 0.35, the level that suggests a substantial difference between the path coefficients (Chin 1998), indicating that the service value index is a substantially better predictor of these two constructs.

Overall, statistics from the MIMIC model and the external validation models provide evidence in support of the external validity of the service value index. The external validity results also suggest the superiority of the formative service value measure compared to the reflective measure of the construct, as the R^2 values and path coefficients are all significantly greater when using the (formative) service value index than when using the reflective seven-item value measure.

23.5.3 Salience of Service Value Components across Contexts

To assess the salience of the various service value components across service contexts, we split the data into three sets corresponding to the three industry groups described earlier. As displayed in Table 23.4, the relative importance of the service value components is very consistent and varies minimally across industry contexts. In particular, the salient role of service quality is not dependent on the context, as the weight of this component in the index is similar across industry groups. That is, across the three industry groups, service quality consistently emerges as the most salient component of service value, with weights ranging from 0.42 to 0.46.

Service equity, perceived sacrifice, and confidence benefits also have relatively consistent weights across the three industry groups. In particular, the range of the service equity weights, although slightly larger than the range of weights for service quality, is relatively small; the component weight for semi-customized non-personal services (Industry Group 2) (weight = 0.39) is a little more than it is for both high contact (Industry Group 1) (weight = 0.28) or standardized services (Industry Group 30) (weight = 0.28). For perceived sacrifice, the range of the weights is a little greater. As the level of personalization and interpersonal contact decreases (i.e., going from Industry Group 1 to Industry Group 3), the relative importance of perceived sacrifice increases (with weights of –0.25, –0.30, and –0.43 for Industry Groups 1, 2, and 3, respectively). Confidence benefits also make a similar contribution to the service value index across all three industry groups (with weights ranging from 0.16 to 0.33).

As mentioned earlier, the weight for social benefits is essentially zero when the entire data set is analyzed. This is also true when looking at the contribution of social benefits to the service value index across contexts. In general, the weights of the five service value components (displayed in Table 23.4) suggest the contributions of each are relatively consistent – both in terms of the magnitude and the relative order – across service contexts.

Although the importance of the various components is fairly consistent across the three industry groups, there is some variation. For example, in standardized services (Industry Group 3), the weight of perceived sacrifice is the largest component of the service value index (weight = –0.43), matching the contribution of service quality (weight = 0.42); however, for moderate contact, semi-customized services (Industry Group 2), the relative weight of perceived sacrifice decreases (weight = –0.30), reaching its lowest level (weight = –0.25) for personalized high-contact services (Industry Group 1). However, the pattern of weights is, in general, consistent across contexts.

23.5.4 Salience of Service Value Components Across Cultures

In addition to investigating the service value components across contexts, we also examined the components across cultures by comparing the U.S. sample with the Spanish sample. In general, as was the case in looking across the industry groups, the importance of the service value components is also relatively consistent across the two cultures. That is, the weights displayed in Table 23.4 suggest the largest contribution to the service value index is made by service quality, followed by service equity, perceived sacrifice, and confidence benefits. The magnitude of the weights are fairly similar for each component across cultures, except that confidence benefits appear to be more important in the U.S. (weight = 0.30) than in Spain (weight = 0.13).

23.6 Discussion

Our review of the literature suggests three salient issues arise when considering customers' perceptions of value: whether customer value should be conceptualized as unidimensional or multidimensional, whether the components of customer value should be modeled as reflective or formative, and whether service components should be included in conceptualizations of the construct. This study contributes to the literature by addressing these issues. In particular, our study (1) identifies service components expected to be strong indicators of customer value – namely, service quality, service equity, relational benefits (including confidence benefits and social benefits), and perceived sacrifice; (2) demonstrates the superiority of this multidimensional conceptualization of customer value to a direct (reflective) conceptualization of the construct; and (3) provides evidence in support of the robustness of this conceptualization by assessing it across differing service contexts and cultures.

23.6.1 Unidimensional Versus Multidimensional Conceptualization of Customer Value

The conceptualization of customer value has been approached in different ways in the marketing literature. The *unidimensional* approach describes customer value in a global fashion; using this approach, the construct is often measured directly by reflective items attempting to capture the concept of utility or value for money. However, this conceptualization of customer value prevents researchers from capturing the conceptual richness of the construct. Alternatively, the *multidimensional* approach considers customer value as a highly complex concept with many components. We contend, as do many recent studies, that the customer value construct is too complex to be conceptualized as unidimensional and should be considered multidimensional.

In support of our claim, we compare a unidimensional conceptualization of the construct with a multidimensional approach. Following Arnett et al. (2003), we construct a MIMIC model, which includes a reflective seven-item measure of customer value as an external criterion variable, to test the validity of our multidimensional service value construct. The resulting statistics indicate that the service value index adequately captures the construct being measured by the reflective indicators, providing support for our multidimensional conceptualization of service value.

23.6.2 Usage of Reflective or Formative Components in Operationalizing Customer Value

A question that arises when taking a multidimensional approach is whether customer value should be modeled as consisting of reflective or formative components. A reflective approach would suggest that each dimension is (or should be) highly correlated with the others because *changes in the underlying construct cause changes in the dimensions*; a formative approach suggests the various dimensions may be independent of each other as *they cause the underlying construct*. The fundamental essence of any construct, whether it is reflective or formative, is crucial in modeling the construct's structure (Jarvis et al. 2003). However, we are not aware of any prior study that has examined customer value using a formative approach or has addressed whether the construct is better modeled with reflective or formative components.

To address this gap in the literature, we proposed a formative index of customer value to capture a more complete portrayal of the construct and compared this to an operationalization of the construct using reflective measures. We found our formative index significantly outperforms a reflective measure. In particular, the variance explained (measured via R^2) for customer satisfaction and repurchase intentions is significantly greater when using our index and the magnitude of the path coefficients between the two customer value measures and each of these two constructs is significantly greater with our index. These results suggest that formative index of customer value is a significantly better predictor of these two constructs than a reflective measure of the construct.

23.6.3 Inclusion of Service Components in Conceptualizing Customer Value

Customer value has received much attention in recent marketing literature, but relatively little attention has been given to the inclusion of service components when defining and operationalizing customer value. That is, most conceptualizations of customer value tend to have a product focus, a likely consequence of the traditional goods-based marketing paradigm that has dominated thought for the past few decades (Vargo and Lusch 2004). Since service components are generally not considered in conceptualizations of customer value, we believe the discipline's conceptualization of the construct is incomplete. Following Vargo and Lusch's call to shift to a more service-based paradigm, we have argued in this study that the conceptualization of customer value should be reframed to include service elements, including service quality, service equity, and relational benefits.

Our conceptualization of customer value was tested across a variety of service settings and in two countries (the U.S. and Spain); the results are fairly uniform across contexts and cultures. First, service quality consistently emerges as the major determinant of service value across both cultures and three industry groups, supporting previous literature suggesting quality is an essential pillar of the value creation process. That is, the evidence confirmed the essential role that service quality plays in the value perception of service as a major source of competitive advantage for companies. Second, we found that service equity is also a significant component of service value, especially for moderate-contact, semi-customized services. While the literature supports the importance of branding in services, to our knowledge, this is the first empirical exploration of the relevance of service equity in the global context of value. Importantly, this research shows that service quality and service equity are the consistently significant drivers of service value.

Perceived sacrifice, the third major component of service value we examined, generally has a relative weight close to that of service equity. However, the influence of perceived sacrifice appears to be context-dependent; the importance of sacrifice (weight = –0.43) increases when the service is standardized and nonpersonal in nature (Industry Group 3), suggesting customers are more sacrifice-conscious when they have fewer interactions with the provider. On the other hand, the relevance of sacrifice for service value decreases when it comes to high-contact, customized services (weight = –0.25). Perceived sacrifice appears to be less important when the customer has more direct contact with the service provider.

One type of relational benefit we included in our study, confidence benefits, appears to be relatively more important when the service is more personal in nature and with a higher level of customer-employee contact (Industry Group 1). Customers apparently value feelings of confidence in, and reduced anxiety with, a service provider when the service is more complex. This finding is consistent with the key role that trust plays in high-contact, customized services (such as dental services, legal services, and financial consulting).

One unanticipated finding is the negligible contribution social benefits appear to make to the service value index. Although the respondents were asked to evaluate a service provider with whom they had a strong relationship, they apparently did not identify service providers where they have a strong interpersonal relationship with their employees. That is, most respondents did not report having a particularly strong social connection with the service provider – the average social benefits score of 4.30 is just above the midpoint on the 1 to 7 scale. However, the fact that social benefits had no impact even for respondents from Industry Group 1 was very surprising since these customers used services that tend to have significantly more interactions with employees than the other two industry groups. In standardized services (Industry Group 3), one could perhaps argue that customers are not interested in developing close interpersonal relationships, which would explain why social benefits are irrelevant in this context. Clearly, the insignificant contribution of social benefits to customer value needs further investigation.

23.7 Implications

23.7.1 Managerial Implications

This study highlights issues that are directly relevant to managers responsible for creating or measuring customer value. Consistent with the emergent thinking on competing through service, our study supports the notion that competitive advantage is achieved by focusing on the service elements of customer value. In an environment that is increasingly competitive on a global scale, management efforts directed toward a better understanding of and measuring customer value, and, in particular, service value, will improve an organization's competitive position. Results from this study can influence managerial decisions in at least three areas: 1) customer value measurement, 2) customer value perceptions for global companies, and 3) company performance on elements of customer value.

Measuring Customer Value. Managers should reexamine current customer value measures to ensure these tools capture the richness of this multidimensional construct. Our findings clearly suggest that a simple, direct measure is inadequate for capturing the complexities of customer value. Our development of a service value index implies that, for the measure to be comprehensive, it should contain several service components; omitting these aspects of customer value prevents a complete understanding of the construct. In addition, we confirm that service value is strongly correlated with such critical outcomes as customer satisfaction and repurchase intention.

Customer Value for Global Companies. Global managers can similarly measure customer value across cultures with confidence. Our study indicates that the value model is robust across the U.S. and Spanish cultures. While complete generalization requires further validation, managers can begin to develop improved programs and measurement instruments with the expectation that customers in different markets may define value in similar ways.

Performance on Customer Value Elements. Our study suggests customer perceptions of value are influenced by service elements; therefore, service should be an integral part of any customer value strategy. Our model clearly demonstrates that service quality is consistently the strongest driver of service value, across cultures and across industries. This finding suggests service quality is the key to improving customer value perceptions and should be emphasized in all customer encounters.

Managers should also take note of the importance of service equity and begin to incorporate this component in measures and programs. Service equity elements are particularly relevant for such service providers as dry cleaners or auto repair shops (i.e., Industry Group 2 – moderate contact, non-personal services) where service equity rivals service quality as the most important component of customer value. The image the company portrays through its communications and customer interactions plays heavily into customers' value perceptions. The auto repair shop that projects an image of integrity, efficiency, and professionalism at each customer contact point will increase its customer value proposition.

Managers must also recognize that the level of importance customers attach to what they perceive to be sacrifices in purchasing and/or using a service is likely to vary across industries. Our study shows that customers are more "sacrifice-conscious" when consuming impersonal, standardized services and become less so as the service becomes more personalized. To increase customers' value perceptions, managers – especially those in standardized, moderate-contact industries – should attempt to reduce customers' perceptions of sacrifice. Movie theater managers, for example, might allow customers to pre-purchase tickets online, thereby reducing the sacrifice of standing in a long ticket line.

Our findings on the importance of relational benefits were mixed. Confidence benefits (e.g., trust, anxiety reduction) are consistently important but the level of importance varies across industries and cultures. Confidence benefits are more important when the service is highly personal and involves high contact and, interestingly, in the U.S. in comparison to Spain. Therefore, confidence benefits should be emphasized for service providers such as doctors, lawyers, and financial consultants and should be considered especially vital in the U.S. Visual cues that inspire trust (e.g., sedate dcor in a lawyer's office) may be more influential in improving perceived customer value for the lawyer than for the dry cleaner. On the other hand, our study suggests social benefits may not contribute to customers' value perceptions in the manner previously suggested by the literature. Rather, our findings suggest companies might consider carefully examining the effectiveness of programs designed to increase customers' social benefits (e.g., building friendships or familiarity with employees).

23.7.2 Research Implications

At least three research implications arise from our study. First, researchers should avoid unidimensional conceptualizations of customer value whenever possible. Scholars who attempt to capture the essence of customer value by defining it as a single dimension are likely to have an incomplete portrayal of the construct, limiting the understanding of a customer's perceptions of value as well as its drivers and consequences.

Second, scholars who conceptualize customer value as multidimensional but operationalize it by including reflective dimensions are likely to incorrectly specify the construct. For example, there is no reason why the "what I receive" components of customer value (such as service quality) should necessarily be correlated with the "what I give up" components (such as perceived sacrifice). Yet, this assumption is normally made when the components are considered to be reflective. By using reflective measures, previous models of customer value may have been misspecified; these misspecifications can affect the conclusions and evidence drawn from empirical research (Jarvis et al. 2003). In future studies, we recommend that researchers who intend including multiple dimensions of customer value consider using a formative approach unless a convincing argument can be made for a reflective approach being appropriate.

Third, given the influence that the service components of a product's offering can have on a customer's experience, scholars would be well advised to include elements of service when conceptualizing customer value. Ignoring the service dimensions of customer value may mean that an important domain of customer value construct is not being captured.

Our study has provided a framework for conceptualizing customer value to provide guidance to future researchers in terms of each of these implications. That is, we have developed a robust, formative index of customer value that (1) is superior to a reflective measure of value, (2) includes relevant service components, and (3) works well across contexts and cultures.

23.7.3 Limitations and Future Research

We acknowledge certain limitations in this study and suggest some directions for future research. First, our list of service components may not be exhaustive. In this study, a primary objective was to find a salient group of service components that is consistent across contexts and consumers. However, other service components of customer value may be salient in specific situations or for some types of customers. For example, special treatment, another of Gwinner et al.'s (1998) relational benefits, may be meaningful in those contexts where a strong service relationship exists between the provider and customer. Similarly, our division of services into three groups may have prevented us from looking at individual elements pertaining to single service industries. Thus, exploring a single context more deeply may identify some specific components that have been overlooked. And, as mentioned earlier, the insignificant contribution that social benefits – a concept well supported in the literature – makes to the customer value index needs further investigation.

Second, we did not thoroughly analyze customer value differences across contexts. Future study is needed to understand the extent to which value differs not only among service industries, but also among cultures and customer types. For example, future studies should examine the extent to which the relative weights of the various service components differ across cultures. Also, although the importance of the various service value components is fairly consistent across the three industry groups, there is some variation. These variations should be explored in future research.

Third, we did not explore the extent to which customer-related variables might account for differences in the weights of the various value components. Perhaps some customer characteristics (demographics, psychographics, experience with the service, etc.) influence which component of service value is more important. For example, are some value components more important to female customers, to older customers, or to customers with extensive experience with a particular type of service?

Finally, the relative impact of each service value dimension on outcomes of interest to marketers (e.g., customer loyalty, future purchase intentions, word-of-mouth communication) should be assessed. We examined the relationship between service value and two such outcomes (customer satisfaction and repurchase intentions), but only as part of a validity test of the index. Although a positive relationship between unidimensional conceptualizations of customer value and customer loyalty has been established (e.g., Cronin et al. 2000), future research should determine the extent to which the relationship holds when using a multidimensional conceptualization of value. Other research might explore the relative impact that each service value dimension has on these marketing outcomes.

APPENDIX

Measurement Items

SQ: Service Quality
SQ1. In general, this company's service is reliable and consistent.
SQ2. My experience with this company is always excellent.
SQ3. I would say that this company provides superior service.
SQ4. Overall, I think this company provides good service.

SE: Service Equity
SE1. It makes sense to buy this company's services compared to others, even if they are the same.
SE2. Even if another company offers the same service, I would still prefer this company.
SE3. If another company offers services as good as this company's, I would still prefer this company.
SE4. If another company is not different from this company in any way, it still seems smarter to purchase this company's services.

CB: Confidence (Relational) Benefits
CB1. I have more confidence the service will be performed correctly.
CB2. I have less anxiety when I buy/use the services of this company.
CB3. I believe there is less risk that something will go wrong.
CB4. I know what to expect when I go to this company.
CB5. I feel I can trust this company.
SB: Social (Relational) Benefits
SB1. I am recognized by certain employees.
SB2. I enjoy certain social aspects of the relationship.
SB3. I have developed a friendship with the service provider.
SB4. I am familiar with the employee(s) that perform(s) the service.
SB5. At this company, they know my name.

SAC: Perceived Sacrifice
SAC1. The price charged to get this company's services is high.
SAC2. The time required to receive this company's services is high.
SAC3. The effort I expend to receive this company's services is high.

CV: Customer Value (reflective measure)
CV1: The value I receive from this company's services is worth the time, effort and money I have invested
CV2. This company's services are reasonably priced.
CV3. This company offers good services for the price.
CV4. I am happy with the price of this company's services.
CV5. This company makes me feel that I am getting my money's worth.
CV6: The value of this company's services compares favorably to other service providers.
CV7. This company offers good value for the price I pay.

SAT: Customer Satisfaction
SAT1. I am happy with this company's services.
SAT2. Overall, I am pleased when I purchase this company's services.
SAT3. Using this company's services is a satisfying experience.
SAT4. My choice to use this company was a wise one.
SAT5. Overall, I am satisfied with this company.
SAT6. I think I did the right thing in deciding to use this company for my service needs.

RP: Repurchase Intentions
RP1. I intend to continue doing business with this company in the future.
RP2. As long as the present service continues, I doubt that I would switch companies.
RP3. I will choose this company the next time I need this service.

Note: All items used seven-point Likert scales with anchors 1 ("strongly disagree") and 7 ("strongly agree").

References

Arnett, D. B., Laverie D. A., & Meiers, A. (2003). Developing parsimonious retailer equity indexes using partial least squares analysis: a method and applications. *Journal of Retailing, 79*(Fall), 161–170.

Bagozzi, R. P. (1994). *Advanced methods of marketing research.* Cambridge, MA: Blackwell Business.

Barclay, D. W., Higgins, C. A., & Thompson, R. L. (1995). The partial least squares (PLS) approach to causal modeling: personal computer adaptation and use as an illustration. *Technology Studies, 2*(2), 285–324.
Barnes, J. G. (1994). Close to the customer: but is it really a relationship? *Journal of Marketing Management, 10*(7), 561–570.
Bendapudi, N., & Berry, L. L. (1997). Customers' motivations for maintaining relationships with service providers. *Journal of Retailing, 73*(Spring), 15–37.
Berry, L. L. (1995). *On great service: a framework for action*. New York: Free.
Berry, L. L. (2000). Cultivating service brand equity. *Journal of the Academy of Marketing Science, 28*(Winter), 128–137.
Berry, L. L., & Parasuraman, A. (1991). *Marketing services: competing through quality*. New York: Free.
Berry, L. L., Seiders, K., & Grewal, D. (2002). Understanding service convenience. *Journal of Marketing, 66*(July), 1–17.
Bitner, M. J., Booms, B. H., & Tetreault, M. S. (1990). The service encounter: diagnosing favorable and unfavorable incidents. *Journal of Marketing, 54*(January), 71–84.
Blackwell, S. A., Szeinbach, S. I., Barnes, J. H., Garner, D. W., & B. Victoria. (1999). The antecedents of customer loyalty: an empirical investigation of the role of personal and situational aspects on repurchase decisions. *Journal of Service Research, 1*(May), 362–375.
Bollen, K. A. (1989). *Structural equations with latent variables*. New York: Wiley.
Bolton, R., & Drew, J. (1991). A multistage model of customer assessment of service quality and value. *Journal of Consumer Research, 17*(March), 375–384.
Bolton, R. N., Kannan, P. K., & Bramlett, M. D. (2000). Implications of loyalty program membership and service experiences for customer retention and value. *Journal of the Academy of Marketing Science, 28*(Winter), 95–108.
Bowen, J. (1990). Development of a taxonomy of services to gain strategic marketing insights. *Journal of the Academy of Marketing Science, 18*(Winter), 43–49.
Chin, W. (1995). Partial least squares is to LISREL as principal components analysis is to common factor analysis. *Technology Studies, 2*, 315–319.
Chin, W. (1998). The partial least squares approach to structural equation modeling. In G. A. Marcoulides (Ed.), *Modern methods for business research* (pp. 295–336). London: Lawrence Erlbaum Associates.
Cronbach, L. J., & Glesser, G. C. (1953). Assessing similarity between profiles. *Psychological Bulletin, 50*(6), 456–473.
Cronin, J. J., Jr., Brady, M. K., & Hult, G. T. M. (2000). Assessing the effects of quality, value, and customer satisfaction on consumer behavioral intentions in service environments. *Journal of Retailing, 76*(Summer), 193–218.
Cronin, J. J., Jr., Brady, M. K., Brand, R. R., Hightower, R., Jr., & Shemwell, D. J. (1997). A cross-sectional test of the effect and conceptualization of service value. *Journal of Services Marketing, 11*(6), 375–391.
Cudeck, R., & Browne, M. W. (1983). Cross-validation of covariance structures. *Multivariate Behavioral Research, 18*, 147–167.
Czepiel, J. A. (1990). Service encounters and service relationships. implications for research. *Journal of Business Research, 20*(January), 13–21.
de Ruyter, K., Lemmink, J., Wetzels, M., & Mattson, J. (1997). Carry-over effects in the formation of satisfaction: the role of value in a hotel service delivery process. In T. A. Swartz, D. E. Bowen, & D. Iacobucci (Eds.), *Advances in services marketing and management* (pp. 61–78), Vol. 6. Greenwich, CT: JAI.
Diamantopoulos, A., & Winklhofer, H. M. (2001). Index construction with formative indicators: an alternative to scale development. *Journal of Marketing Research, 43*(May), 269–277.
Dodds, W. B., & Monroe, K. B. (1985). The effect of brand and price information on subjective product evaluations. In E. Hirschman, & M. Holbrook (Eds.), *Advances in Consumer Research* (pp. 85–90), Vol. 12. Provo, UT: Association for Consumer Research.

Dodds, W. B., Monroe, K. B., & Grewal, D. (1991). Effects on price, brand, and store information on buyers' product evaluations. *Journal of Marketing Research, 28*(August), 307–319.

Efron, B., & Tibshirani, R. J. (1993). *An introduction to the bootstrap*. New York: Chapman and Hall.

Flint, D. J., Woodruff, R. B., & Gardial, S. F. (2002). Exploring the phenomenon of customers' desired value change in a business-to-business context. *Journal of Marketing, 86*(October), 102–117.

Fornell, C. (1982). *A second generation of multivariate analysis*. New York: Praeger.

Fornell, C. (1992). A national customer satisfaction barometer: the Swedish experience. *Journal of Marketing, 56*(January), 6–21.

Fornell, C., & Larcker, D. F. (1981). Evaluating structural models with unobservable variables and measurement error. *Journal of Marketing Research, 28*(February), 39–50.

Fornell, C., Johnson, M. D., Anderson, E. W., Cha, J., & Bryant, B. E. (1996). The American customer satisfaction index: nature, purpose, and findings. *Journal of Marketing, 60*(October), 7–18.

Futrell, C. M. (1979). Measurement of salespeople's job satisfaction: convergent and discriminant validity of corresponding INSCALES and job descriptive index scales. *Journal of Marketing Research, 16*(November), 594–597.

Gale, B. (1997). Satisfaction is not enough. *Marketing News, 31*(22), 18.

Geisser, S. (1975). The predictive sample reuse method with applications. *Journal of the American Statistical Association, 70*, 320–328.

Gremler, D. D., & Brown, S. W. (1996). Service loyalty: its nature, importance, and implications. In B. Edvardsson, S. W. Brown, R. Johnston, et al. (Eds.), *Advancing service quality: a global perspective* (pp. 171–180). USA: International Service Quality Association.

Grewal, D., Monroe, K. B., & Krishnan, R. (1998). The effects of price-comparison advertising on buyers' perceptions of acquisition value, transaction value, and behavioral intentions. *Journal of Marketing, 62*(April), 46–59.

Gronroos, C. (1995). Relationship marketing: the strategy continuum. *Journal of the Academy of Marketing Science, 23*(Fall), 252–254.

Grönroos, C. (1997). Relationship marketing: interaction, dialogue, and value. Working Paper No. 344, Swedish School of Economics and Business Administration, Finland.

Grönroos, C. (2000). *Service management and marketing. A customer relationship management approach*. New York: Wiley.

Gwinner, K., Gremler, D. D., & Bitner, M. J. (1998). Relational benefits in services industries: the customer's perspective. *Journal of the Academy of Marketing Science, 26*(Spring), 101–114.

Ha, L. (1996). Observations: advertising clutter in consumer magazines: dimensions and effects. *Journal of Advertising Research, 36*(July–August), 76–85.

Hartline, M. D., & Jones, K. C. (1996). Employee performance cues in a hotel environment: influence on perceived service quality, value, and word-of-mouth intentions. *Journal of Business Research, 35*(3), 207–215.

Heinonen, K. (2004). Reconceptualizing customer perceived value: the value of time and place. *Managing Service Quality, 13*(2/3), 205–215.

Hennig-Thurau, T., Gwinner, K. P., & Gremler, D. D. (2002). Understanding relationship marketing outcomes: an integration of relational benefits and relationship quality. *Journal of Service Research, 4*(February), 230–247.

Holbrook, M. B. (1994). The nature of customer value: an axiology of services in the consumption experience. In R. T. Rust, & R. L. Oliver (Eds.), *Service quality: new directions in theory and practice* (pp. 21–71). Sage: Thousand Oaks.

Jarvis, C. B., Mackenzie, S. B., & Podsakoff, P. M. (2003). A critical review of construct indicators and measurement model misspecification in marketing and consumer research. *Journal of Consumer Research, 30*(September), 199–218.

Jöreskog, K. G., & Goldberger, A. S. (1975). Estimation of a model with multiple indicators and multiple causes of a single latent variable. *Journal of the American Statistical Association, 70*(351), 631–639.

Keaveney, S. M. (1995). Customer switching behavior in service industries: an exploratory study. *Journal of Marketing, 59*(April), 71–82.

Kerin, R. A., Jain, A., & Howard, D. J. (1992). Store shopping experience and consumer price-quality-value perceptions. *Journal of Retailing, 68*(Winter), 376–397.

Lam, S. Y., Shankar, V., Krishna Erramilli, M., & Murthy, B. (2004). Customer value, satisfaction, loyalty, and switching costs: an illustration from a business-to-business service context. *Journal of the Academy of Marketing Science, 32*(Summer), 293–311.

Lapierre, J. (2000). Customer-perceived value in industrial contexts. *Journal of Business and Industrial Marketing, 15*(2–3), 122–140.

Lin, C.-H., Sher, P. J., & Shih, H.-Y. (2005). Past progress and future directions in conceptualizing customer perceived value. *International Journal of Service Industry Management, 16*(4), 318–336.

Mathieson, K., Peacock, E., & Chin, W. W. (2001). Extending the technology acceptance model: the influence of perceived user resources. *The Data Base for Advances in Information Systems, 32*(Summer), 86–112.

Mathwick, C., Malhotra, N., & Rigdon, E. (2001). Experiential value: conceptualization, measurement, and application in the catalog and internet shopping environment. *Journal of Retailing, 77*(Spring), 39–54.

McDougall, G. H. G., & Levesque, T. (2000). Customer satisfaction with services: putting perceived value into the equation. *Journal of Services Marketing, 14*(5), 392–410.

Meyer, A. (1994). *Das Deutsche Kundenbarometer*. Mnchen, Germany: Ludwig-Maximilians-Universitt Mnchen.

Monroe, K. (1990). Pricing: making profitable decisions. New York: McGraw Hill.

Naylor, G. & Frank, K. E. (2000). The impact of retail sales force responsiveness on consumers' perceptions of value. *Journal of Services Marketing, 14*(4), 310–322.

Nunnally, J. C. (1978). *Psychometric theory*. New York: McGraw-Hill.

Oliver, R. L. (1980). A cognitive model of the antecedents and consequences of satisfaction decisions. *Journal of Marketing Research, 17*(November), 460–469.

Overby, J. W., Gardial, S. F., & Woodruff, R. B. (2004). French versus American consumers' attachment of value to a product in a common consumption context: a cross-national comparison. *Journal of the Academy of Marketing Science, 32*(4), 437–460.

Parasuraman, A., & Grewal, D. (2000). Serving customers and consumers effectively in the twenty-first century: a conceptual framework and overview. *Journal of the Academy of Marketing Science, 28*(Winter), 9–16.

Petrick, J. F. (2002). Development of a multi-dimensional scale for measuring the perceived value of a service. *Journal of Leisure Research, 34*(2), 119–134.

Price, L. L., & Arnould, E. J. (1999). Commercial friendships: service provider-client relationships in context. *Journal of Marketing, 63*(October), 38–56.

Pura, M. (2005). Linking perceived value and loyalty in location-based mobile services. *Managing Service Quality, 15*(6), 509–538.

Reichheld, F. F., & Earl Sasser, W., Jr. (1990). Zero defections: quality comes to services. *Harvard Business Review, (September–October)*, 105–111.

Reynolds, K. E., & Beatty, S. E. (1999). Customer benefits and company consequences of customer-salesperson relationships in retailing. *Journal of Retailing, 75*(Spring), 11–32.

Rust, R. T., & Oliver, R. L. (1994). Service quality: insights and managerial implications from the frontier. In R. T. Rust, & R. L. Oliver (Eds.), *Service quality: new directions in theory and practice* (pp. 1–19). Thousand Oaks: Sage.

Rust, R. T., Lemon, K. N., & Zeithaml, V. A. (2004). Return on marketing: using customer equity to focus marketing strategy. *Journal of Marketing, 68*(January), 109–127.

Rust, R. T., Zeithaml, V. A., & Lemon, K. N. (2000). *Driving customer equity*. New York: Free.

Sellin, N. (1995). Partial least squares modeling in research on education achievement. In W. Boss, & R. H. Lehmann (Eds.), *Reflections on education achievement* (pp. 256–267). Münster: Waxmann Verlag.

Sheth, J. N., Newman, B. I., & Gross, B. L. (1991). Why we buy what we buy: a theory of consumption values. *Journal of Business Research, 22*(March), 159–170.
Sheth, J. N., Sisodia, R., & Sharma, A. (2000). The antecedents and consequences of customer-centric marketing. *Journal of the Academy of Marketing Science, 28*(Winter), 55–66.
Singh, J., & Sirdeshmukh, D. (2000). Agency and trust mechanisms in consumer satisfaction and loyalty judgments. *Journal of the Academy of Marketing Science, 28*(Winter), 150–167.
Sirdeshmukh, D., Singh, J., & Sabol, B. (2002). Consumer trust, value, and loyalty in relational exchanges. *Journal of Marketing, 66*(January), 15–37.
Sirohi, N., McLaughlin, E. W., & Wittink, D. R. (1998). A model of consumer perceptions and store loyalty intentions for a supermarket retailer. *Journal of Retailing, 74*(Summer), 223–245.
Stewart, D. W. (2002). Getting published: reflections of an old editor. *Journal of Marketing, 66*(October), 1–6.
Stone, M. (1974). Cross-validatory choice and assessment of statistical predictions. *Journal of the Royal Statistical Society, Series B, 36*(2), 111–147.
Sweeney, J. C., & Soutar, G. N. (2001). Consumer perceived value: the development of a multiple item scale. *Journal of Retailing, 77*(Summer), 203–220.
Sweeney, J. C., Soutar, G. N., & Johnson, L. W. (1999). The role of perceived risk in the quality-value relationship: a study in a retail environment. *Journal of Retailing, 75*(Spring), 77–105.
Taylor, S. A., & Baker, T. L. (1994). An assessment of the relationship between service quality and customer satisfaction in the formation of consumers' purchase intentions. *Journal of Retailing, 70*(Summer), 163–178.
Teas, R. K., & Agarwal, S. (2000). The effects of extrinsic product cues on consumers' perceptions of quality, sacrifice, and value. *Journal of the Academy of Marketing Science, 28*(Spring), 278–290.
Vargo, S. L., & Lusch, R. F. (2004). Evolving to a new dominant logic for marketing. *Journal of Marketing, 68*(January), 1–17.
Voss, G. B., A. Parasuraman, & Grewal, D. (1998). The roles of price, performance, and expectations in determining satisfaction in service exchanges. *Journal of Marketing, 62*(October), 46–61.
Wang, Y., Lo, H. P., Chi, R., & Yang, Y. (2004). An integrated framework for customer value and customer-relationship-management performance: a customer-based perspective from China. *Managing Service Quality, 14*(2–3), 169–182.
Warner, W. L., Meeker, M., & Eells, K. (1949). *Social class in America: a dynamic study of a system that powerfully influences the lives of all Americans*. Chicago: Science Research Associates.
Woodall, T. (2003). Conceptualising 'value for the customer': an attributional, structural, and dispositional analysis. *Academy of Marketing Science Review [Online]* (12), available: www.amsreview.org/articles/woodall12-2003.pdf.
Woodruff, R. B., & Gardial, S. F. (1996). *Know your customer: new approaches to customer value and satisfaction*. Cambridge, MA: Blackwell.
Woodruff, R. (1997). Customer value: the next source for competitive advantage. *Journal of the Academy of Marketing Science, 25*(Spring), 139–153.
Yang, Z., & Peterson, R. T. (2004). Customer perceived value, satisfaction, and loyalty: the role of switching costs. *Psychology and Marketing, 21*(10), 799–822.
Yoo, B., & Donthu, N. (2001). Developing and validating a multidimensional consumer-based brand equity scale. *Journal of Business Research, 52*(April), 1–15.
Zeithaml, V. A. (1988). Consumer perceptions of price, quality, and value: a means-end model and synthesis of evidence. *Journal of Marketing, 52*(July), 2–22.
Zeithaml, V. A., Berry, L. L., & Parasuraman, A. (1996). The behavioral consequences of service quality. *Journal of Marketing, 60*(April), 31–46.

Chapter 24
Analyzing Factorial Data Using PLS: Application in an Online Complaining Context

Sandra Streukens, Martin Wetzels, Ahmad Daryanto, and Ko de Ruyter

Abstract Structural equation modeling (SEM) can be employed to emulate more traditional analysis techniques, such as MANOVA, discriminant analysis, and canonical correlation analysis. Recently, it has been realized that this emulation is not restricted to covariance-based SEM, but can easily be extended to components-based SEM, or partials least squares (PLS) path analysis (Guinot et al. 2001; Tenenhaus et al. 2005; Wetzels et al. 2005). In this paper, we will apply PLS path analysis to a fixed-effects, between-subjects factorial design in an online complaint-handling context. The results of our empirical study reveal that satisfaction with online recovery is determined by the level of both procedural and distributive justice. Furthermore, customers' satisfaction with the way their complaints are handled has a positive influence on the customers' intentions to repurchase and to spread positive word of mouth. Taking into account the entire chain of effects, we find that the influence of justice perceptions on behavioral intentions is almost fully mediated by satisfaction. From a managerial perspective, the results of our study provide insight into how to design effective complaint-handling strategies in order to maintain a satisfied and loyal customer base.

24.1 Introduction

Structural equation modeling (SEM) has the potential to fundamentally improve experimental research in social sciences (MacKenzie 2001). Compared to traditional approaches (i.e., (M)AN(C)OVA) used to analyze data from factorial experimental

S. Streukens
Department of Marketing and Strategy, Universiteit Hasselt, Campus Diepenbeek, Agoralaan Gebouw DBE 3590 Diepenbeek, Belgium
e-mail: sandra.streukens@uhasselt.be

M. Wetzels and Ko de Ruyter
Department of Marketing and Marketing Research, Maastricht University, P.O. Box 616 6200 MD, The Netherlands
e-mail: m.wetzels@mw.unimaas.nl, k.deruyter@mw.unimaas.nl

A. Daryanto
Department of Business Analysis, Systems and Information Management, Newcastle Business School, City Campus East, Northumbria University, Newcastle upon Tyne, NE1 8ST, UK
e-mail: ahmad.daryanto@northumbria.ac.uk

V. Esposito Vinzi et al. (eds.), *Handbook of Partial Least Squares*, Springer Handbooks of Computational Statistics, DOI 10.1007/978-3-540-32827-8_25,

designs, the use of SEM offers the following advantages: ability to control for measurement error and enhanced testing of nomological webs among multiple dependent variables (cf. MacKenzie 2001). Despite these fundamental strengths, it appears that the proposed covariance-based SEM approaches to analyzing experimental data perform rather poorly in small sample conditions under non-normality and do not have the ability to handle complex models (e.g., Bagozzi et al. 1991; McDonald et al. 2002). Given the fundamental properties of PLS estimation, it has the potential to offer a method for analyzing data from factorial experimental designs that offers many of the abovementioned advantages of SEM-based analysis but overcomes the often-encountered drawbacks. Thus, a PLS-based approach to experimental designs offers a strong methodological tool that can be applied in many circumstances. In this paper, we show how PLS can be used to analyze data from factorial experimental designs.

In this chapter, we will apply the proposed PLS approach to data obtained from a factorial experimental design in an online service recovery context. The significance of this application and the relevant literature will be discussed in Sect. 2. In Sect. 3, we will demonstrate how PLS can be used to analyze factorial data and how to interpret the accompanying output. We will end this chapter with a discussion and conclusion.

24.2 Online Service Recovery: Significance and Literature Review

Several empirical studies indicate that organized service recovery policies are an important tool in order to maintain satisfied and loyal customers (Blodgett et al. 1997; Maxham and Netemeyer 2002; Tax et al. 1998). In contrast to complaint-handling in traditional (i.e., offline) services, only limited attention has been paid to the antecedents and consequences of satisfaction in complaint-handling in online settings despite the great differences that exist between online and offline settings and, therefore, the way complaint management procedures are perceived by customers in both settings. First of all, effective complaint management is particularly important for e-services, as customers can terminate their relationship with the service provider by just a simple mouse click (Holloway and Beatty 2003). Second, Holloway and Beatty (2003) state that satisfaction with complaint recovery is especially crucial for online service providers as poor service online may hurt online as well as offline sales. Third, in an online environment, customers cannot directly see and touch the product, nor can they directly bring it home after buying it (Reichheld and Schefter 2000). Fourth, the formation of customer evaluative judgments is different in online settings (Shankar et al. 2003). Fifth, the types of service failures experienced may be different for the online and offline environment and customers tend to complain more online than in traditional marketplace (Holloway and Beatty 2003). Finally, given the lack of human interaction in e-services, we cannot

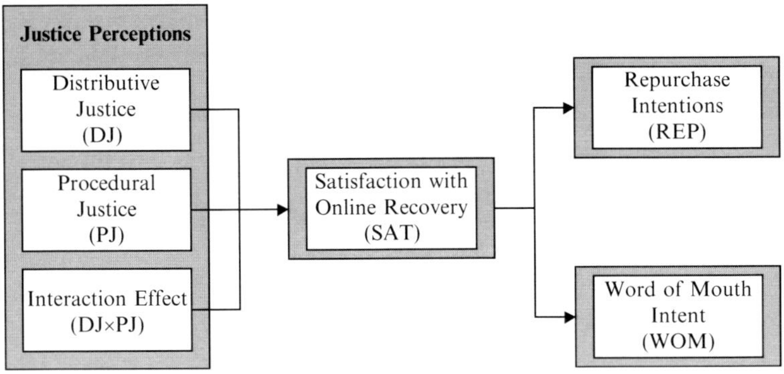

Fig. 24.1 Conceptual model

simply extrapolate the empirical findings concerning complaint-handling that were established in offline/regular services (Reichheld and Schefter 2000).

The research objectives guiding our work are formulated as follows:

1. To examine how justice perceptions of complaint-handling procedures influence key customer evaluative judgments in an online setting.
2. To show how PLS path modeling can be used to analyze factorial design (i.e., data from experimental studies).

Figure 24.1 provides an overview of the conceptual model underlying our study. The relevant literature underlying our conceptual framework will be summarized below.

Concerning traditional offline service delivery formats, equity or justice theory has been proven to be a powerful approach to understand and explain customers' perceptions regarding companies' service recovery efforts (e.g., Smith et al. 1999; Blodgett et al. 1997; Maxham and Netemeyer 2002; Tax et al. 1998). In the literature two reasons can be distinguished that clarify the significant explanatory power of justice perceptions in understanding customer's perceptions of service recovery strategies. First of all, Maxham and Netemeyer (2002) state that implicit promises of fairness are salient because it is often difficult for customers to evaluate service before, and sometimes after, the transaction has been made. This is especially true for (online) complaint management procedures as these are characterized by high degree of experience quality, meaning that a customer can only evaluate the service in retrospection (Brush and Artz 1999; Klein 1998). Second, as complaint-handling can be considered a process (Tax et al. 1998), justice theory provides researchers with a comprehensive framework to understand customer evaluations as each part of the complaint-handling process is subject to fairness considerations and each aspect of a complaint resolution creates a justice episode (Bies 1987; Tax et al. 1998). As these characteristics apply to online service delivery formats as well, in our opinion, justice theory will also very likely be a strong approach to explain customer's post-recovery attitudes and behaviors in an online context.

Building on the principals of equity theory, we believe that the evaluation of an online recovery process is a function of the recovery process itself (referred to as procedural justice) and the outcomes of the recovery process (referred to as distributive justice). The suggested impact of procedural and distributive justice on online service evaluations is supported by the work of Zeithaml et al. (2000) who state that customer evaluative judgments in an online service context are based on what customers receive as an outcome as well as on how the process of service delivery takes place.

Procedural justice can be defined as the perceived fairness of the way the complaint is handled (Netemeyer and Maxham 2002). According to Tax et al. (1998) procedural justice is meaningful because it aims to resolve conflicts in ways that encourage the continuation of a relationship even when outcomes are not satisfactory to one/both parties. Flexibility, speed of recovery, accessibility of complaint procedure, the freedom of the complainant in rejecting or accepting the refund offered and the extent to which a complainant is free to express their view of the complaint-handling procedure are important factors in the formation of procedural justice perceptions (Tax et al. 1998; Blodgett et al. 1997). Although the complaint-handling in online settings may be different in form, the positive effect of procedural justice on recovery satisfaction may still hold (Janda et al. 2002; Montoya-Weiss et al. 2003). Consequently, we hypothesize:

H_1 *Procedural justice positively affects satisfaction with the online complaint recovery.*

Distributive justice relates to the outcome of the complaint-handling effort. The degree to which a customer perceives the outcome of complaint-handling as fair in terms of distributive justice depends on the benefits received and the costs associated with the experienced service failure (Netemeyer and Maxham 2002). It is reasonable to assume that the outcome of complaint-handling efforts is itself independent of the channel through which the service is provided. Based on this assumption, we believe that the positive relationship between perceived distributive justice and satisfaction with complaint-handling as empirically supported in offline service settings can be extended to an online setting. Therefore, we hypothesize:

H_2 *Distributive justice positively affects satisfaction with the online complaint recovery.*

It has been empirically demonstrated (e.g., Sparks and McColl-Kennedy 2001) that in a service recovery context, outcomes and procedures work together to create a sense of justice. Following the principle of referent cognition theory, Tax et al. (1998) state that the value of a service recovery outcome may be enhanced or comprised by the procedures by which the outcome is established. We extend this finding to an online service context. The underlying premise is that human-computer interaction is fundamentally social and that individuals respond to computers in much the same way that they respond to human beings (cf. Reeves and Nash 1996). Hence, we posit:

H_3 *Perceptions of procedural justice affect the nature of the positive relationship between distributive justice and satisfaction with the online complaint recovery.*

This study examines the effects of procedural and distributive justice on three types of customer outcomes: satisfaction, loyalty intentions, and word of mouth intentions. Ample empirical evidence is available concerning the relevance of these three outcome variables in a complaint management context (e.g., Maxham and Netemeyer 2002; Blodgett et al. 1997). In brief, these customer outcomes can be described as follows: satisfaction is the customer's overall affective psychological response based on subjective evaluations of the overall service performance after organizational recovery efforts (Hess et al. 2003). Word of mouth intent can be defined as the likelihood that one would favorably recommend doing business with a certain firm after a failure and recovery effort, while purchase intent refers to the degree to which customers intend to purchase a firm's products/services in the future (Netemeyer and Maxham 2002).

Although both satisfaction and behavioral intentions are key constructs in studying the effectiveness of service recovery efforts, consideration of the nomological web that exists among them is crucial to obtain valid and unbiased estimates of the effects justice perceptions have on these outcome variables.

Our previously formulated hypotheses state that justice perceptions only have a direct impact on the formation of satisfaction. This is congruent with the existing literature (e.g., Maxham and Netemeyer 2002; Wirtz and Mattila 2004) on service recovery, which states that satisfaction mediates the positive impact of justice perceptions on repurchase intentions and the intention to engage in word of mouth. Finally, it should be noted that similar to traditional services, the relationship between satisfaction and behavioral intentions is also evidenced in e-services (Anderson and Srinivasan 2003, Holloway et al. 2005). Overall, the literature cited above leads to the formulation of the following hypotheses:

H_4 *Satisfaction with service recovery positively affects repurchase intentions.*

H_5 *Satisfaction with service recovery positively affects the intention to engage in word of mouth.*

H_6 *Satisfaction with service recovery mediates the relationship between justice perceptions and (a) repurchase intentions and (b) word of mouth intentions.*

24.3 Method

24.3.1 Study Design

In order to test the hypotheses outlined above, a 2 × 2 between-groups quasi-experimental design was employed using written scenarios. Subjects were randomly assigned to the various treatments and were ask asked to read a scenario in which a customer was dissatisfied with a product (a pair of sports shoes starting to fall apart after only limited use) that s/he bought online and sought to redress from the online retailer via the website. Sports shoes were chosen as it is a product with which most subjects are familiar and have at least some experience of purchasing them (cf. Blodgett et al. 1997).

Manipulations were conducted as follows: under the high distributive justice condition, the customer received a full refund, whereas under the low distributive justice condition, the customer was offered a 15% discount on a new pair of shoes. In respect of procedural justice, we manipulated the scenarios with regard to when the complainant receives a response from the company and the level of effort the customer has to exert to obtain this response. Under the high procedural justice condition the customer received a response within 24 hours of his/her first email, whereas under the low procedural justice condition the customer received an answer from the company only after five working days after having sent a second email.

After having read one of the four scenarios, the respondents were asked to fill out a questionnaire containing the following measures: to assess whether manipulations indeed achieved the desired effect, we included the items of Blodgett et al.'s (1997) scale on procedural (3 items) and distributive justice (3 items). Furthermore, we included measures to assess customer satisfaction (Maxham and Netemeyer 2002; 3 items), repurchase intentions (Blodgett et al. 1997; 3 items) and word of mouth intent (Maxham and Netemeyer 2002; 3 items). For all constructs we used seven-point Likert scales, with higher scores reflecting a more favorable attitude. Table 24.1 provides an overview of the items used to measure customer satisfaction, repurchase intentions, and word of mouth intent. The items used for the manipulation checks are presented in the appendix A to this chapter.

24.3.2 Sampling Procedure and Sample Characteristics

All respondents ($n = 147$) were students participating in a business research course at a European university. They were asked to take part in the study and filled out the questionnaire during the last 15–20 min of their classes. Participation in the study was rewarded with a candy bar.

The mean age of the respondents was 23.12 years with a standard deviation of 2.88 years. Furthermore, the proportion of males and females in the sample was equal (i.e., 49.7% male; 50.3% female). As a results of the international orientation of the university at which we collected the data, various nationalities are represented in the sample: Dutch (51.0%), German (35.4%), Belgian (4.1%), and 9.5% of the respondents were non-European.

24.3.3 Analytical Results

Unless mentioned otherwise, we used PLS-GRAPH version 3.0 to estimate the parameters in our model, with the number of bootstrap samples J equaling 1,000 and all containing 147 cases. Below, we describe the empirical results pertaining to our study. First, we assess the measurement properties of the scales used in our study. More specifically, we assess whether the multiple-items scales used possess

Table 24.1 Measurement properties

	Coefficient	t-value	p-value
Satisfication			
$\lambda_1 = 2.570$ $\lambda_2 = 2.570$			
$\lambda_3 = 2.570$ $\alpha = 0.95$			
$ave = 0.86$			
1 Company provided a satisfactory resolution to problem	0.95	115.79	<0.0001
2 Not satisfied with company's problem handling (−)	0.90	32.41	<0.0001
3 Regarding the problem resolution satisfied with company	0.90	49.74	<0.0001
Word of mouth			
$\lambda_1 = 2.746$ $\lambda_2 = 0.161$			
$\lambda_3 = 0.093$ $\alpha = 0.97$			
$ave = 0.92$			
1 Likelihood to spread positive word-of-mouth about company	0.96	153.04	<0.0001
2 Recommend company to others	0.94	70.97	<0.0001
If asked for advice, recommend company	0.97	115.83	<0.0001
Repurchase intent			
$\lambda_1 = 2.552$ $\lambda_2 = 0.271$			
$\lambda_3 = 0.177$ $\alpha = 0.95$			
$ave = 0.85$			
1 Likelihood to shop at this online retail store in the future	0.92	62.86	<0.0001
2 If this situation happened, would never shop there again (−)	0.91	37.10	<0.0001
3 If this situation happened, would still shop there in the future	0.94	60.02	<0.0001

Satisfaction 1 = totally disagree; 7 = totally agree
Word of mouth and Repurchase intent 1 = very unlikely; 7 = very likely

favorable psychometric properties in terms of unidimensionality, reliability, convergent and discriminant validity. Second, we discuss how PLS can be used to analyze factorial data and its relative advantage of existing methods, and apply the suggested approach to our data.

24.3.3.1 Measurement Properties

In order to assess the psychometric properties of the multiple item scales used in our study, we follow the procedures suggested by Tenenhaus et al. (2005). The empirical results related to the analysis of the scale's measurement properties are summarized in Table 24.1.

Starting with the assessment of unidimensionality, we conducted a principle component analysis (using SAS v8) for each of the three scales. For all three scales, unidimensionality is evidenced as the first eigenvalue (λ_1) of the block of variables exceeds one and the second eigenvalue (λ_2) is smaller than one (see also Table 24.1).

The internal consistency of the measurement scales under study is evidenced by the fact that the composite reliability values, indicated by α, all exceed the recommended cut-off values of 0.70 (Nunnally and Bernstein 1994).

Having substantiated the existence of unidimensionality and the reliability of the scales used in this study, we proceed by examining whether the scales possess a substantial degree of within-method convergent validity and discriminant validity. Within-method convergent validity is evidenced by the large (>0.50) and significant item loadings on their respective constructs (cf. Anderson and Gerbing 1988). Finally, discriminant validity is established as the square root value of average trait extracted is greater than the correlation coefficient between the two relevant constructs. Figures regarding the evidence of discriminant validity are provided in Table 24.2. Furthermore, Table 24.2 provides key descriptive statistics of the scales used in our study, as well as the correlations and covariances among all pairs of variables.

Structural Model

The effects of our factorial design are captured by dichotomous variables. As the number of respondents per cell is not equal, we opted for dummy coding rather than effects coding the justice manipulations used in our study (cf. Pedhazur 1997).

Prior to the actual analysis of our conceptual model, we first need to examine whether the intended justice manipulations achieved the desired effect. Although manipulation checks are typically conducted by means of a series of one-way ANOVAs, they can also be directly performed in PLS by estimating a model that connects the dichotomous manipulations to the variables intended to measure the effect of the manipulation as well. For the situation at hand, the model to conduct manipulation checks in PLS is graphically displayed in Fig. 24.2.

In Fig. 24.2, the variables $D(PJ)$ and $D(DJ)$ represent the dummy coded manipulations for procedural and distributive justice respectively, and are formative indicators of a latent construct representing the *actual* manipulation used in the study. The constructs "PJ Manipulation Check" and "DJ Manipulation Check" assess the respondents' perceptions regarding the manipulations of procedural and distributive justice. These latter constructs are both assessed by multi-item scales (see appendix A for details of the scales). The significant values of ρ_1 ($t = 25.071$; $p < 0.0001$) and ρ_2 ($t = 20.359$; $p < 0.0001$) indicate that the procedural justice and distributive justice manipulations achieved the desired effects.

Below, different types of models are outlined in order to clearly and convincingly demonstrate the added value of PLS over other methods (i.e., (M)ANOVA and covariance-based SEM) in analyzing data from factorial designs.

Table 24.2 Correlations, covariance, and descriptive statistics

	SAT	*WOM*	*REP*
SAT	0.93[a]	2.67[c]	2.56
WOM	0.82[b]	0.96	2.73
REP	0.88	0.84	0.92
Complete sample ($n = 147$)			
Mean	3.93	3.98	4.12
SD	1.63	1.81	1.79
Skewness ($SE = 0.200$)	−0.13	0.07	−0.19
Kurtosis ($SE = 0.397$)	−1.20	−1.14	−1.11
$LP\text{–}LD(n = 37)$			
Mean	2.32	5.60	2.37
SD	0.94	1.21	1.23
Skewness ($SE = 0.388$)	0.61	−0.79	0.84
Kurtosis ($SE = 0.759$)	−0.41	−0.21	−0.08
$LP\text{–}HD(n = 36)$			
Mean	4.34	3.61	4.74
SD	1.21	1.30	1.43
Skewness ($SE = 0.393$)	−0.40	−0.08	−0.51
Kurtosis ($SE = 0.768$)	−0.85	−0.98	−0.45
$HP\text{–}LD(n = 38)$			
Mean	3.53	4.46	3.53
SD	1.48	1.70	1.36
Skewness ($SE = 0.383$)	0.05	−0.11	−0.06
Kurtosis ($SE = 0.750$)	−1.25	−0.98	−0.31
$HP\text{–}HD(n = 36)$			
Mean	5.60	2.18	5.91
SD	0.71	0.98	0.75
Skewness ($SE = 0.393$)	0.38	0.55	−0.10
Kurtosis ($SE = 0.768$)	−0.57	−0.99	−0.91

[a]Square root of average variance extracted values are on the diagonal of the matrix.
[b]Correlation coefficients are placed in the lower triangle of the matrix.
[c]Covariances are placed in the upper triangle of the matrix. A correlation/covariance matrix as well descriptive statistics at the item level of the constructs can be obtained from the first author.

The first model is a PLS model that exactly replicates a (M)ANOVA estimation approach (see also Fig. 24.3).[1] To achieve this, we propose a path model containing only latent variables with a single indicator. To capture the design effects,

[1] The model in Fig. 24.3 represents a MANOVA approach as typically used in the marketing literature (e.g. Blodgett et al. 1997): the experimental effects are hypothesized to influence all outcome variables and there are no effects hypothesized among the outcome variables. To exactly assess the hypotheses outlined in this paper following a (M)ANOVA approach one would actually need separate models: one ANOVA model with satisfaction as an outcome variable and two regression models to estimate the effects of satisfaction on repurchase intentions and word of mouth intentions, respectively.

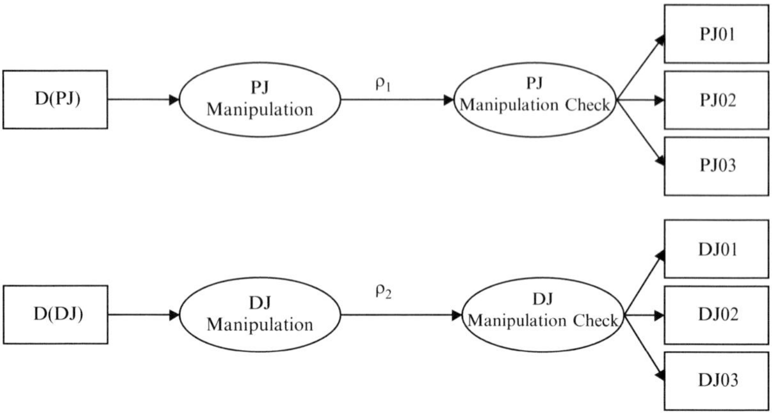

Fig. 24.2 Conducting manipulation checks in PLS

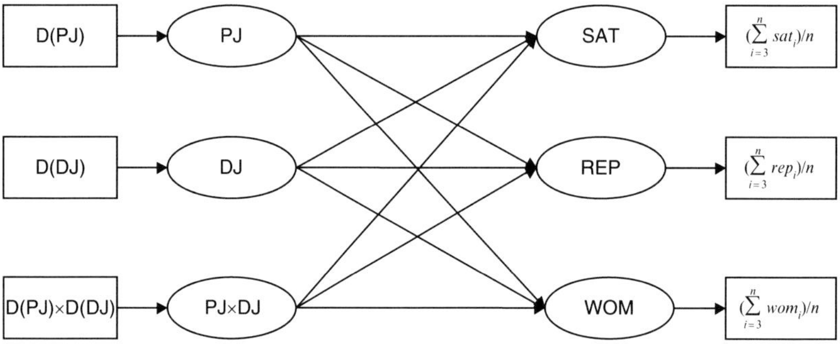

Fig. 24.3 (M)ANOVA using PLS

formative indicators are used, whereas each outcome variable is represented by a latent variable for which the (reflective) indicator is formed by the sum of its items.

The added value of PLS analysis over traditional MANOVA is that one can allow for structural paths among the various outcome variables, thereby substantially diminishing the effects of omitted variable bias. The introduction of covariance-based SEM approaches to modeling factorial data (Bagozzi and Yi 1989) was a giant leap forward in analyzing factorial data, as structural paths among dependent variables can be taken into account whilst controlling for measurement error. However, the methodology cannot always be feasibly used in empirical research as it requires multivariate normal data, large sample sizes and cannot be used for complex models (Bagozzi et al. 1991). Compared to covariance-based SEM models, the PLS approach offers the following advantages to analyzing factorial data: first of all, PLS poses less stringent assumptions regarding the distributional characteristics of the data. Second, its ability to model both reflective and formative indicators, whereas

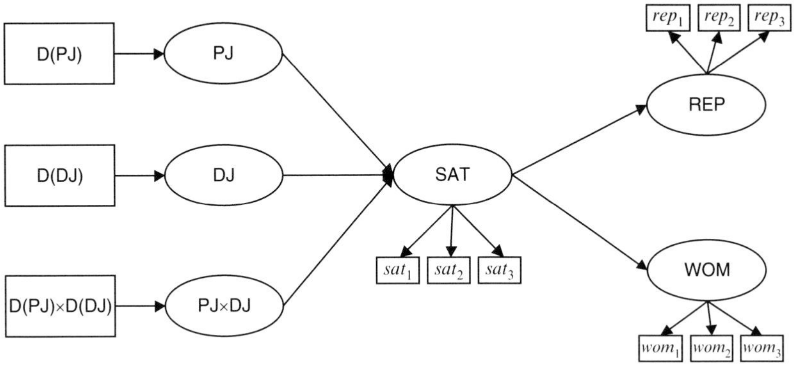

Fig. 24.4 A PLS approach to modeling factorial data

covariance-based SEM approaches can typically handle only reflective indicators. Third, PLS can be used well in case of small and medium sized samples. Fourth, PLS can handle more complex designs.

In Fig. 24.4 we outline a PLS model to model factorial data, which also allows for structural relationships among the outcome variables as outlined in our conceptual model (see also Fig. 24.1).

Regarding the model presented in Fig. 24.4, the experimental manipulations are modeled as latent variables with dummy variables as their formative indicators and the outcome variables are modeled as latent variables with multiple items as their reflective[2] indicators. As the model presented in Fig. 24.4 provides us with the most valid representation of the situation at hand, we will only discuss the empirical results pertaining to this model. Although in the majority of cases that build on the principles of Theory of Reasoned Action (TRA) developed by Fishbein and Ajzen (1975), the effects of beliefs (i.e., justice) on behavioral intentions (i.e., repurchase intent and word of mouth) are fully mediated by attitude (i.e., satisfaction with complaint recovery), Bagozzi (1982) provides empirical support for a model in which attitude only partially mediates the relationship between beliefs on behavioral intentions. Thus, to increase the validity of our findings regarding the mediating role of satisfaction with complaint recovery in our conceptual model, we estimate a model that contains both indirect and direct effects between the justice manipulations and behavioral intentions.

To assess H_6, which states that satisfaction with online recovery mediates the effect of justice perceptions on behavioral intentions, we use the procedure outlined

[2] In respect of the outcome variables, the choice of using reflective indicators is guided by the work of Jarvis et al. (2003). If the guidelines presented by Jarvis et al. (2003) on the specification of indicators suggest the use of formative indicators, this can be readily applied in our suggested PLS approach to analyze factorial data.

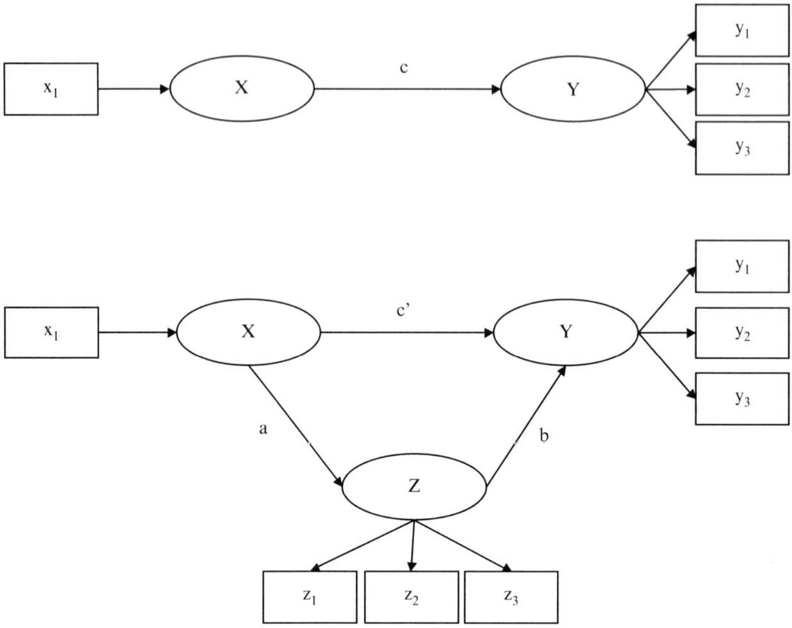

Fig. 24.5 Hoyle and Kenny's (1999) mediation test

by Hoyle and Kenny (1999). In summary, the Hoyle and Kenny[3] approach requires the estimation of the two types of models presented in Fig. 24.5.

In terms of the labels used in Fig. 24.5, "X" denotes one of the justice perceptions, "Y" the respondent's behavioral intentions (either repurchase intent or word of mouth intent), and "Z" reflects the possible mediator, in this case satisfaction with the online recovery.[4]

Statistical evidence of mediation in a structural equation modeling context requires the following (cf. Hoyle and Kenny 1999): first, evidence of a causal influence of X on Y ($c \neq 0$). Second, a significant indirect effect of X on Y ($ab \neq 0$), indicative of a decline in the direct effect of X on Y when the mediator is accounted for (please note that $ab = c - c'$). If $ab \neq 0$ and $c' \neq 0$, M only partially mediates

[3] For situations in which the independent variable(s), mediator variable, and/or dependent variable(s) are embedded in a larger nomological network (i.e., have their own additional antecedents or consequences), the approach by Iacobucci et al. (2007) is preferred over the Hoyle and Kenny (1999) approach.

[4] The form and number of indicators used in the models presented in Fig. 24.5 are chosen to reflect the situation of our study. The Hoyle and Kenny (1999) approach also applies to other forms and numbers of indicators.

the relationship between X and Y. If $ab \neq 0$ and $c' = 0$, M fully mediates the effect of X on Y.

To assess the significance of the various effects, we employed a bootstrap procedure, ($J = 1{,}000$ with $n = 147$). Based on the outcomes of the bootstrap procedure, we constructed several 95% confidence intervals. The bootstrap percentile confidence interval is preferred over the standard normal confidence interval for small sample sizes ($n < 400$), which are often characterized by a skewed and leptokurtic sample distribution of the indirect effect ab (Preacher and Hayes 2006; Shrout and Bolger 2002; Bollen and Stine 1990). A further improvement came from Efron and Tibshirani (1998), who proposed a bias-corrected bootstrap percentile confidence interval, which corrects for the bias in the central tendency of the estimate. A simulation study by MacKinnon et al. (2004) shows that the bias-corrected version of the bootstrap percentile method outperforms the regular bootstrap percentile method in terms of statistical power and accuracy of the confidence intervals. Computational details on how to construct (bias corrected) bootstrap percentile confidence interval are presented in appendix B. The accompanying estimation results of the structural model are presented in Table 24.3.

Inspection of the estimation results of the structural model reveals the following. First of all, we can conclude that our conceptual model is well supported by the data as indicated by the R-squared values ($R^2_{SAT} = 0.54 (p < 0.0001)$; $R^2_{REP} = 0.77 (p < 0.0001)$; $R^2_{WOM} = 0.67\ (p < 0.0001)$). Turning to the individual effects, we see that both distributive and procedural justice have a significant[5] influence on the formation of satisfaction with service recovery in an online setting. Hence, H_1 and H_2 are supported. However, we fail to find a significant interaction effect of procedural and distributive justice in the development of satisfaction. Consequently, H_3 is not supported. The crucial role of satisfaction with recovery in shaping both customers' repurchase intentions and customers' intentions to spread word of mouth is also reflected in the data, thereby providing support for H_4 and H_5. In addition to the hypothesized direct effects, our analysis also reveals a direct influence of distributive justice on repurchase intent.

Based on the empirical results we can conclude that the effect of procedural justice on behavioral intentions is fully mediated by satisfaction, whereas the effect of distributive justice on behavioral intentions is only partially mediated (41%) by satisfaction with online recovery. Overall, H_6 is fully supported for procedural justice and only partly for distributive justice. Please note that the mediation analysis does not apply to the interaction effect as there is no effect of $PJ \times DJ$ on SAT (i.e., $a = 0$).

[5] Although the three types of confidence intervals are very consistent for the effects found is this study, we base our hypothesis testing on the bias-corrected bootstrap percentile confidence interval given its superior performance as demonstrated by MacKinnon et al. (2004).

Table 24.3 Estimation results structural model

Effects	Estimate	SE	Standard normal	95% CI		Mean	Bootstrap SD
				Bootstrap e percentile	Bias corrected bootstrap perc.		
Direct effects							
Proc→sat	0.37	0.08	(0.21;0.53)	(0.21;0.54)	(0.21;0.55)	0.37	0.09
Proc→rep	0.06	0.07	(−0.09;0.20)	(−0.08;0.18)	(−0.06;0.22)	0.05	0.07
Proc→wom	0.07	0.09	(−0.11;0.25)	(−0.07;0.29)	(−0.12;0.23)	0.11	0.09
Dist→sat	0.63	0.07	(−0.48;0.77)	(0.48;0.77)	(0.46;0.76)	0.63	0.07
Dist→rep	0.21	0.07	(0.07;0.35)	(0.06;0.36)	(0.09;0.39)	0.20	0.07
Dist→wom	0.14	0.07	(−0.01;0.28)	(0.00;0.29)	(0.00;0.28)	0.14	0.07
Proc ∗ dist → sat	0.01	0.09	(−0.17;0.19)	(−0.18;0.19)	(−0.06;0.33)	0.01	0.09
Proc ∗ dist → rep	0.01	0.07	(−0.14;0.15)	(−0.12;0.14)	(−0.12;0.14)	0.01	0.06
Proc ∗ dist → wom	0.06	0.08	(−0.10;0.22)	(−0.19;0.14)	(−0.02;0.22)	0.03	0.08
Sat→rep	0.73	0.06	(0.61;0.85)	(0.62;0.85)	(0.60;0.83)	0.73	0.06
Sat→wom	0.67	0.08	(0.51;0.83)	(0.50;0.79)	(0.52;0.81)	0.65	0.07
Indirect effects							
Proc→sat→rep	0.27	0.06	(0.14;0.39)	(0.15;0.41)	(0.15;0.41)	0.27	0.07
Proc→sat→wom	0.25	0.06	(0.13;0.37)	(0.14;0.37)	(0.15;0.39)	0.24	0.06
Dist→sat→rep	0.45	0.06	(0.33;0.58)	(0.34;0.59)	(0.35;0.56)	0.46	0.06
Dist→sat→wom	0.42	0.07	(0.28;0.56)	(0.28;0.55)	(0.30;0.58′)	0.41	0.07
Total effects							
(dist → rep)+ (dist → sat → rep)	0.66	0.08	(0.50;0.82)	41% mediation			

24.4 Discussion and Conclusion

The use of factorial experimental design is ubiquitous in social sciences. Although traditional analysis techniques, such as (M)AN(C)OVA, can be considered powerful for this type of study under certain conditions, they fail to meet some often-encountered modeling circumstances such as structural dependency among the outcome variables, non-normal data, and small samples. Although considerable research has been devoted to developing covariance-based models to overcome the limitations of these traditional estimation approaches, only limited effort has been made to show how component-based techniques such as PLS can be used to estimate these more realistic, but more complex, models of factorial experimental data.

In this paper we showed how PLS can be used to analyze data of factorial designs. First, we indicated how PLS is related to traditional MANOVA. Compared to traditional estimation approaches (i.e., MANOVA) the PLS model provides a more accurate and insightful picture of the phenomenon under study as it allows researchers to take into account the nomological web that may exist among the dependent variables. Compared to covariance-based SEM approaches to analyzing factorial data, the PLS approach offers a much greater practical applicability as it requires no distributional assumptions regarding the data, can be used well in small and medium sample sizes, can incorporate both reflective and formative indicators, and does not run into trouble when estimating complex models.

As choosing the best technique for the research design at hand is a critical step in conducting sound research, it is also important to acknowledge that there are circumstances in which covariance-based SEM approaches to modeling factorial data are preferred over PLS path modeling. Based on a Monte Carlo simulation conducted by Hoyle and Kenny (1999), it can be concluded that the bias in parameter estimates is inversely related to the reliability of the constructs. As covariance-based SEM techniques allow correcting parameter estimates for measurement error, it is favored in situations in which the reliability of the measures is less optimal.

Balancing the relative (dis)advantages of covariance-based SEM and PLS, we can nevertheless state that PLS has the potential to fundamentally improve the analysis of experimental designs in social sciences.

From a marketing perspective, our work offers the following insights: in contrast to studies conducted in offline service settings, it appears that distributive and procedural justice have independent positive effects on satisfaction with online recovery. A possible explanation for this finding could be due to the inherent differences between electronic services and traditional services. Owing to the lack of human interaction both with employees and other customers, e-service customers may produce less strong and clear perceptions regarding the procedures in complaint recovery situations. As such, the prediction based on referent cognitions theory (cf. Folger 1984; Tax et al. 1998) that perceived procedural injustice will exacerbate feelings of distributive injustice when customers believe that a better outcome could have been achieved with a fairer procedure, may not hold.

Taking a look at the individual effects of procedural and distributive justice, we see that distributive justice has a larger positive impact on the formation of satisfaction with online recovery than procedural justice. This finding is in contrast to empirical results obtained by various researchers (e.g., Maxham and Netemeyer 2002; Tax et al. 1998) in offline service settings. Again, the difference in the nature of the interaction between offline and online service contexts may play a key role in explaining this finding. In an offline context, the costs involved in the actual complaint recovery procedure may be substantially higher compared to online service delivery formats (e.g., traveling to the store, waiting in line). Consequently, customers may be more likely to form more negative perceptions of procedural justice in an offline service delivery format. Drawing on prospect theory (cf. Mittal et al. 1998), more negative evaluations are weighted more heavily, thereby explaining the larger effect of procedural justice in traditional service delivery formats. From a different angle, distributive justice in online service complaint-handling may be easier for customers to evaluate than procedural justice. As a result, customers may place more weight on the evaluation of distributive justice in developing their post-recovery attitudes and behaviors. From a practical perspective, the finding that customers place more value on distributive justice that on procedural justice provides managers with insights into setting priorities when developing effective online recovery strategies.

In line with research conducted in offline complaint-handling situations, we also find support for positive associations between satisfaction with recovery efforts and the intent of the customer to again do business with the company. This relationship is relevant as loyalty intentions are a significant antecedent of actual behavior, which is crucial to a firm's long-term survival. In a similar vein, the significant positive relationship between satisfaction with the online recovery and customer's intent to engage in word of mouth entails good news for the company, as satisfied customers may persuade others to do business with the company.

Finally, various limitations of the current study need to be recognized, which, it is hoped, will provide fruitful directions for further research efforts. First of all, our results relate to a single setting. Although, on the one hand, this allows us to control for cross-industry difference, on the other hand, it would be interesting to examine the generalizability of our findings. Second, in terms of measurement a cross-sectional approach was pursued. Related work in offline service settings demonstrates interesting longitudinal effects (e.g., Maxham and Netemeyer 2002), which have remained unexplored in online service contexts. Third, our chain of effects ends with behavioral intentions. Extending this chain with actual behavior or financial measures would allow managers to make an economically justified analysis of the value and design of effective recovery strategies.

Appendix A

Overview of the items used in the manipulation checks. All items are based on the work of Blodgett et al. (1997). Conform the work of Blodgett et al. (1997) and other researchers who employed the scale, the items were modeled as reflective indicators. See Tables 24.4 and 24.5

Table 24.4 Scales and psychometric properties

	Coefficient	t-value	p-value
Distributive justice $\lambda_1 = 3.564$ $\lambda_2 = 0.213$ $\lambda_3 = 0.094$ $\alpha = 0.97$ $ave = 0.92$			
1. Taking everything into consideration the company's refund offer was quite fair	0.97	139.54	<0.0001
2. Regarding the refund the customer did not get what s/he deserved (−)	0.95	47.88	<0.0001
3. Given the circumstances, I feel that the company offered adequate compensation	0.96	107.76	<0.0001
Procedural justice $\lambda_1 = 2.669$ $\lambda_2 = 0.212$ $\lambda_3 = 0.119$ $\alpha = 0.96$ $ave = 0.92$			
1. The customer's complaint was handled in a very timely manner	0.93	38.16	<0.0001
2. The customer's complaint was not resolved as quickly as it should have been (−)	0.96	86.86	<0.0001
3. The customer had to write too many e-mails in order to resolve the problem	0.95	104.29	<0.0001

Scale anchors: 1 = totally disagree; 7 = totally agree

Table 24.5 Descriptive statistics

	LD–LP	HD–LP	LD–HP	HD–HP	Overall
N	37	36	38	36	147
Mean DJ	2.34	5.81	3.00	6.52	4.38
SD DJ	0.89	0.93	1.34	0.61	2.03
Skewness DJ	0.36	−0.40	0.63	−1.49	−0.13
Skewness DJ SE	0.39	0.39	0.38	0.39	0.20
Kurtosis DJ	−0.37	−0.86	−0.27	1.84	−1.47
Kurtosis DJ SE	0.76	0.77	0.70	0.77	0.40
Mean PJ	2.14	2.71	6.04	6.79	4.43
SD PJ	0.79	1.13	1.13	0.34	2.22
Skewness PJ	0.56	0.88	−1.37	−1.49	−0.09
Skewness PJ SE	0.39	0.39	0.38	0.39	0.20
Kurtosis PJ	−0.62	1.05	0.97	0.98	−1.68
Kurtosis PJ SE	0.76	0.77	0.70	0.77	0.40

LD Low distributive justice; *HD* High distributive justice; *LP* Low procedural justice; *HP* High procedural justice. Data on item level as well as correlation/covariance matrices can be obtained from the first author.

Appendix B

Constructing a Bootstrap Percentile Confidence Interval

The bootstrap percentile interval for parameter β (regardless whether it is a direct of indirect effect) is constructed by the following steps (Shrout and Bolger 2002; Bollen and Stine 1990):

1. Using the original data set as a population reservoir, create J bootstrap samples of N subjects by randomly sampling observations with replacement from the data set. Parameters J and N can be set in PLSGRAPH via *options* > *resampling*.
2. For each bootstrap sample, estimate parameter $\hat{\beta}$ and save the result. The possibility to save bootstrap estimates can also be found under *options* > *resampling* in PLSGRAPH. To proceed with the following step we pasted the bootstrap results produced by PLSGRAPH into Excel® (SPSS® is also a good option).
3. Examine the distribution of the bootstrap estimates and determine the $(\alpha/2) \times 100\%$ and $(1-\alpha/2) \times 100\%$ percentiles of the distribution. These percentile represent, respectively, the lower and upper bound of the confidence interval.

Constructing a Bias Corrected Bootstrap Confidence Interval

1. Define Z_{lower} and Z_{upper} as the corresponding z-scores in a standard normal distribution.
2. Define Z'_{lower} and Z'_{upper} as the z-scores that define the percentile for the bias-corrected bootstrap confidence interval. Equations B1 and B2 summarizes how to determine Z'_{lower} and Z'_{upper}.

$$Z'_{lower} = Z_0 + \frac{Z_0 + Z_{lower}}{1 - \hat{a}(Z_0 + Z_{lower})} \tag{B1}$$

$$Z'_{upper} = Z_0 + \frac{Z_0 + Z_{upper}}{1 - \hat{a}(Z_0 + Z_{upper})} \tag{B2}$$

where Z_0 is the z-score corresponding to the percentage of the q bootstrap estimates that are less than the original sample estimate. To determine Z_0 the following website offer very helpful calculator: http://davidmlane.com/hyperstat/z_table.html.

Furthermore, coefficient $\hat{a}$ is the acceleration constant as is defined as:

$$\hat{a} = \frac{\sum_{i=1}^{n} (\overline{\theta} - \theta_i)^3}{6\left[\sum_{i=1}^{n} (\overline{\theta} - \theta_i)^2\right]^{3/2}} \tag{B3}$$

where θ_i is the ith jackknife estimate of the parameter computed after deleting case i, and $\overline{\theta}$ is the average value of the n jackknife estimates.

3. After having computed Z'_{lower} and Z'_{upper}, determine the proportion of the normal distribution to the left of Z'_{lower} and Z'_{upper} respectively. Again, a handy calculator can be found on http://davidmlane.com/hyperstat/z_table.html.

Assume that the proportion of the normal distribution to the left of Z'_{lower} and Z'_{upper} is respectively π_{lower} and π_{upper}, then the limits of the confidence interval are determined as follows (with J denoting the number of bootstrap samples).

The lower bound is the $(\pi_a * J)^{th}$ estimate in the sorted distribution of bootstrap estimates and the upper bound is the $(\pi_b * J)^{th}$ estimate in the sorted distribution of bootstrap estimates.

We conducted the calculations needed to construct the bias corrected bootstrap interval in Excel®. For more details on the construction of bias corrected bootstrap confidence intervals see Preacher and Hayes (2006) and Efron and Tibshirani (1998).

References

Anderson, J. C. & Gerbing, D. W. (1988). Structural equation modeling in practice: a review and recommended two-step approach. *Psychological Bulletin, 103*, 411–423.

Anderson, R. E. & Srinivasan, S. S. (2003). E-satisfaction and e-loyalty: a contingency framework. *Psychology and Marketing, 20*(2), 123–138.

Bagozzi, R. P. (1982). A field investigation of causal relations among cognition, affect, intentions, and behavior. *Journal of Marketing Research, 19*(4), 562–583.

Bagozzi, R. P., & Yi, Y. (1989). On the use of structural equation models in experimental designs. *Journal of Marketing Research, 26*(3), 271–284.

Bagozzi, R. P., Yi, Y. & Singh, S. (1991). On the use of structural equation models in experimental designs: two extensions. *International Journal of Research in Marketing, 8*(2), 125–140.

Bies, R. J. (1987). The predicament of injustice: The management of moral outrage. *Research in Organizational Behavior, 9*, 289–319.

Blodgett, J. G., Hill, D. & Tax, S. S. (1997). The effects of distributive, procedural, and interactional justice on postcomplaint behavior. *Journal of Retailing, 73*(2), 185–210.

Brush, T. H. & Artz, K. W. (1999). Toward a contingent resource-based theory: the impact of information asymmetry on the value of capabilities in veterinary medicine. *Strategic Management Journal, 20*(3), 223–250.

Bollen, K. A. & Stine, R. (1990). Direct and indirect effects: classical and bootstrap estimates of variability. In C. C., Clogg (Ed.), *Sociological methods 1990 Vol. 20*, (pp. 115–140), Oxford: Basil Blackwell.

Efron, B., & Tibshirani, R. J. (1998). *An introduction to the bootstrap*, Boca Raton, FL: Chapman and Hall.

Fishbein, M., & Ajzen, I. (1975). *Belief, attitude, intention and behavior: an introduction to theory and research.* Reading, MA: Addison-Wesley.

Folger, R. (1984), Emerging issues in the social psychology of justice. In R. Folger (Eds.), *The sense of injustice: social psychology perspectives* (pp. 4–23), New York: Plenum Press.

Guinot, C., Latreille, J., & Tenenhaus, M. (2001). PLS path modeling and multiple table analysis: Application to the cosmetic habits of women in Ile-de-France. *Chemometrics and Intelligent Laboratory Systems, 58*(2), 247–259.

Hess, J. R., Jr., Ganesan, S., & Klein, N.M. (2003). Service failure and recovery: The impact of relationship factors on customer satisfaction. *Journal of the Academy of Marketing Science 31*(2), 127–145.

Holloway, B. B., & Beatty, S. H. (2003). Service failures in online retailing: a recovery opportunity. *Journal of Service Research, 6*(1), 92–105.

Holloway, B. B., Wang, S., & Parish, J. T. (2005). The role of cumulative online purchasing experience in service recovery management. *Journal of Interactive Marketing 19*(3), 54–66.

Hoyle, R. H., & Kenny, D. A. (1999). Sample size, reliability, and tests of statistical mediation. In R. H., Hoyle (Ed.), *Statistical Strategies for Small Sample Research* (pp. 197–223), Thousand Oaks, CA: Sage Publications.

Iacobucci, D., Saldanha N., & Deng, X. (2007). "A meditation on mediation: Evidence that structural equations models perform better than regressions" *Journal of Consumer Psychology 17*(2), 139–153.

Janda, S., Trocchia, P. J., & Gwinner, K. (2002). Customer perceptions of Internet retail service quality. *International Journal of Service Industry Management, 13*(5), 412–431.

Jarvis, C. B., MacKenzie, S. B., & Podsakoff, P. M. (2003). A critical review of construct indicators and measurement model misspecification in marketing and consumer research. *Journal of Consumer Research, 30*(2), 199–218.

Klein, L. R. (1998). Evaluating the potential of interactive media through a new lens: search versus experience goods. *Journal of Business Research, 41*(3), 195–203

MacKenzie, S. B. (2001), Opportunities for improving consumer research through latent variable structural equation modeling. *Journal of Consumer Research, 28*(1), 159–166.

MacKinnon, D. P., Lockwood, C. M., & Williams, J. (2004). Confidence limits for the indirect effect: Distribution of the product and resampling methods. *Multivariate Behavioral Research, 39*(1), 99–128.

McDonald, R. A., Seifert, C. F., Lozeret, S. J., Givens, S., & Jaccard, J. (2002). The effectiveness of methods for analyzing multivariate factorial data. *Organizational Research Methods, 5*(3), 255–274.

Maxham, J. G., & Netemeyer, R. G. (2002). Modeling customer perceptions of complaint handling over time: The effects of perceived justice on satisfaction and intent. *Journal of Retailing, 78*(4), 239–252.

Mittal, V., Ross, W. T., & Baldasare, P. M. (1998). The asymmetric impact of negative and positive attribute level performance on overall satisfaction and repurchase intentions. *Journal of Marketing, 62*(1), 33–47.

Montoya-Weiss, M. M., Voss, G. B., & Grewal, D. (2003). Determinants of online channel use and overall satisfaction with a relational, multichannel service provider. *Journal of the Academy of Marketing Science, 31*(4), 448–458.

Nunnally, J. C., & Bernstein, I. H. (1994). *Psychometric theory*, New York: McGraw-Hill.

Pedhazur, E.J. (1997), *Multiple regression in behavioral research: explanation and prediction.* Singapore: Wadsworth/Thomson Learning.

Preacher, K. J., & Hayes, A. F. (2006). Asymptotic and resampling strategies for assessing and comparing indirect effects in simple and multiple mediater models. *Working Paper.*

Reichheld, F. F., & Schefter, P. (2000). E-Loyalty. *Harvard Business Review, 78*(4), 105–113.

Reeves, B., & Nass, C. I. (1996). *The media equation: how people treat computers, television, and new media like real people and places.* Cambridge, UK: Cambridge University Press.

Shrout, P. E., & Bolger, N. (2002). Mediation in experimental and nonexperimental studies: New procedures and recommendations. *Psychological Methods 7*(4), 422–445.

Smith, A. K., Bolton, R. N., & Wagner, J. (1999). A model of customer satisfaction with service encounters involving failure and recovery. *Journal of Marketing Research 36*(3), 356–372.

Sparks, B. A., & McColl-Kennedy, J. (2001). Justice strategy options for increased customer satisfaction in a services recovery setting. *Journal of Business Research 54*(3), 209–218.

Shankar, V., Smith, A. K., & Rangaswamy, A. (2003). Customer satisfaction and loyalty in online and offline environments. *International Journal of Research in Marketing, 20*(2), 153–175.

Tax, S. S., Brown, S. W., & Chandrashekaran, M. (1998). Customer evaluations of service complaint experiences: Implications for relationship marketing. *Journal of Marketing 62*(2), 516–533.

Tenenhaus, M., Esposito Vinzi, V., Chatelin, Y., & Lauro, C. (2005). PLS path modeling. *Computational Statistics and Data Analysis 48*(1), 15–205.

Wirtz, J., & Mattila, A. S. (2004). Consumer responses to compensation, speed of recovery and apology after a service failure. *International Journal of Service Industry Management, 15*(2), 150–166.

Wetzels, M. G. M., Lindgreen, A., Ruyter, K. de & Wouters, J. (2005). The effect of corporate image and service delivery on customer evaluative judgments in service organizations: Analyzing an experimental study using partial least squares. In Bliemel, F., Eggert, A., Fassott, G. and J. Henseler (Eds.), *Handbuch PLS-pfadmodellerung: Methode, anwendung, praxisbeispiele* (pp. 225–240), Stuttgart: Schäffer-Poeschel Verlag.

Zeithaml, V. A., Parasuraman, A., & Malhotra, A. (2000). *A conceptual framework for understanding e-service quality: implications for future research and managerial practice.* (Working Paper 00–115) Cambridge, MA.

Chapter 25
Application of PLS in Marketing: Content Strategies on the Internet

Silvia Boßow-Thies and Sönke Albers

Abstract In an empirical study the strategies are investigated that content providers follow in their compensation policy with respect to their customers. The choice of the policy can be explained by the resource based view and may serve as recommendations. We illustrate how a strategy study in marketing can be analyzed with the help of PLS thereby providing more detailed and actionable results. First, complex measures have to be operationalized by more specific indicators, marketing instruments in our case, which proved to be formative in most cases. Only by using PLS it was possible to extract the influence of every single formative indicator on the final constructs, i.e., the monetary form of the partnerships. Second, PLS allows for more degrees of freedom so that a complex model could be estimated with a number of cases that would not be sufficient for ML-LISREL. Third, PLS does not work with distributional assumptions while significance tests can still be carried out with the help of bootstrapping. We recommend the use of PLS for future strategy studies in marketing because it is possible to extract the drivers at the indicator level so that detailed recommendations can be given for managing marketing instruments.

25.1 Introduction

Although a high proportion of the population uses the Internet for information and communication content providers still struggle with the question of how to manage their product in the most profitable way. It is still very difficult to overcome the "content for free mentality" of users and to introduce paid content models to the Internet. Another way to market content is to syndicate different content bundles to other players in the market who need interesting content to increase the number and

S. Boßow-Thies
Capgemini Telecom Media and Networks Deutschland GmbH, Neues Kanzler Eck 21, 10719 Berlin, Germany
e-mail: silvia.bossow-thies@capgemini.com

S. Albers
Institute of Innovation Research, Christian-Albrechts-University at Kiel, Westring 425, 24098 Kiel, Germany
e-mail: albers@bwl.uni-kiel.de

V. Esposito Vinzi et al. (eds.), *Handbook of Partial Least Squares*, Springer Handbooks of Computational Statistics, DOI 10.1007/978-3-540-32827-8_26,

duration of visits on their website (Werbach 2000). Generally, content providers can follow two different marketing strategies in this context: On the one hand, a provider can follow a "sales strategy." In this case, he aims at generating direct profits through content licensing and regards the Internet as an additional distribution channel. On the other hand, it might be more effective for some providers to follow a "promoting strategy." Here, the provider strives to increase his own traffic or own brand awareness and image with the transfer of his content. According to the structure-follows-strategy paradigm one can assume that a provider who follows a sales strategy will primarily be directly paid by his subscribers according to classical licensing arrangements. Other actors who follow a promoting strategy will only be rewarded in an indirect way by content branding or the integration of a link leading to the own website. Here, the transfer of content can be seen as a more cost-effective alternative to banner advertising. Although syndication is widely used in practice, it is still not obvious which marketing strategy and compensation policy is accepted by the different providers.

This article investigates which strategies the players follow and how the choice of the strategy depends on its antecedents. As the providers differ in several characteristics, it is obvious to presume that the particular content-relevant resources of the providers might have an impact on their strategies and, consequently, on the compensation policy. This leads to the theory of the resource based view of a firm (RBV). The RBV focuses specifically on the question of how different resource endowments determine corporate strategies and, ultimately, on the characteristics of interorganizational relationships (Penrose 1959; Prahalad and Hamel 1990; Wernerfelt 1984).

On the basis of the RBV, several hypotheses are deduced in the second section of this chapter and an explanatory model is built. In the third section, we illustrate that PLS can be regarded as an adequate statistical method. Our model contains the abstract constructs "content relevant resources," "marketing strategies," and "compensation policy", which needs to be operationalized by detailed indicators – here, marketing instruments. The indicators cause the constructs and can therefore be seen as formative measures, which can be handled only by PLS in a simple manner. In this section, we also illustrate the standard procedures which should be undertaken by using PLS as the method of analysis. Additionally we demonstrate which further statistical methods should be used to increase the explanatory power of the PLS analysis. The results in Sect. 4 show that the marketing strategies and compensation policies should be implemented on the basis of the intangible resource endowments of the content providers. Only by using PLS were we able to determine the individual influence of every formative indicator, i.e., the marketing instruments in our case. We close with a concluding section and an outlook for further research.

25.2 Resource-Strategy-Relationship Model

Although most providers syndicate their content on the Internet, the question of which marketing strategy and compensation have been accepted has not been solved. As previous analyses in Table 25.1 show, providers who follow a sales

Table 25.1 Descriptive illustration of the strategy-structure relationship

	Monetary relationships (%)	Non-monetary relationships (%)
Sales strategy $n = 65$	82.5	17.5
Promoting strategy $n = 35$	23.9	76.1

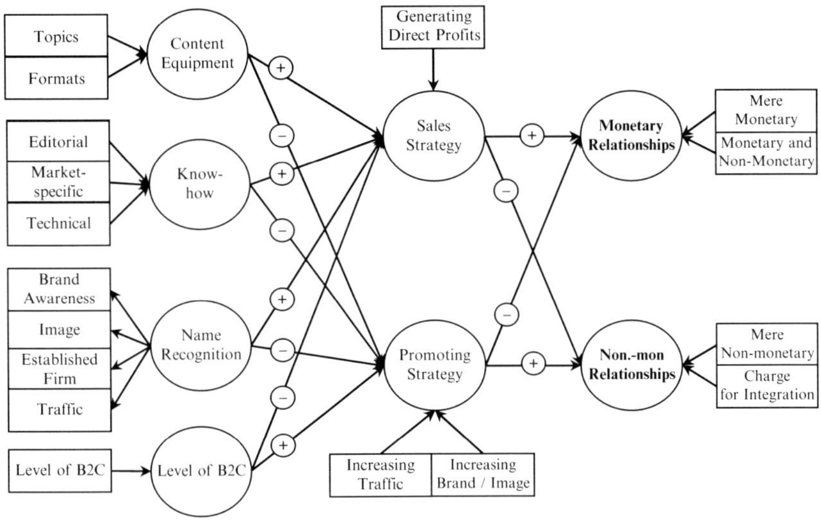

Fig. 25.1 Resource-strategy-relationship model

strategy also work under non-monetary relationships and vice versa (Thies 2005). Therefore the question is: which antecedents influence the providers' marketing strategy and compensation policy?

The model of how certain antecedents influence the choice of a content provider's strategy is visualized in Fig. 25.1, which shows two separate submodels. The first one deals with the relationship between different *marketing strategies* and *compensation policies*. For these, we assume a good fit according to the *structure-follows-strategy paradigm*: A provider with a sales strategy should primarily maintain monetary partnerships while others who follow a promoting strategy should mainly have relationships with non-monetary rewards. In this setting, the *structure* is represented by the formation of the interorganizational relationship (monetary versus non-monetary).

The second submodel contains the effects of the antecedents on the strategies of the content providers. As the providers differ in several characteristics, one can presume that the particular content-relevant resources of the providers might have an impact on their strategies and, consequently, on the compensation policy. This leads to the *resource-based view of a firm (RBV)*. The resource-based view assumes that a business combines various resources on the basis of which strategies for gaining competitive advantages are implemented (Wernerfelt 1984).

Therefore, several content-relevant resources were identified in the explanatory stage of the project by conducting *focus interviews*. As a result of the qualitative research, the content-relevant resources could be classified into physical, organisational, financial and intangible ones (Bamberger and Wrona 1996). According to the RBV, resources must be proved to be valuable, rare, imperfectly imitable and non-substitutable to have the potential to create a comparative advantage. Only the *intangible resources*, namely the content equipment, content-specific know-how and name recognition, fulfilled all the mentioned criteria above. Therefore, they need to be considered in implementing the marketing strategy in the following.

As the mentioned resources are abstract dimensions, they need to be operationalized by more detailed indicators. These can be extracted with the help of expert interviews in a next step (Rossiter 2002).

The content equipment of a provider contains different kinds of content (text, graphs, pictures, etc.) the provider can syndicate to other players in the market. Regarding the *content equipment* of the provider, one can assume that the more extensive the content equipment suitable for the syndication process, the higher the possibility to gain a competitive advantage by syndicating the content. If the content equipment is extensive, the set-up of an additional distributional channel is worthwhile and the provider will follow a sales strategy. However, if a provider only owns little content that he can syndicate on the Internet, e.g., due to rights of disposal problems, he would do this to increase his own traffic or brand awareness as well as his image. Therefore, he would rather follow a promoting strategy. As the expert interviews showed, the abstract construct "content equipment" can be operationalized by the amount of syndicated topics and formats. These indicators are independent of each other since a provider can syndicate different content topics just as texts and does not necessarily have to provide the content in various formats. The indicators form the construct "content equipment", which thus acts as an *index*.

Moreover, the *content-specific know-how* might be of importance for the implementation of the providers' marketing strategy. The content-specific know-how incorporates all knowledge areas which are essential to run the syndication business. Here, it is essential to cover the single steps of the value production process. The higher the know-how, the more reasonable it is to concentrate on an additional distribution channel and to follow a sales strategy. If the level of know-how is low, the content may only be transferred to a slight extent and the special requirements have to be handled by the subscriber. Under these circumstances the provider would rather follow a promoting strategy. As the construct "content-specific know-how" is still quite abstract, it also needs to be more specified by its indicators. To provide content of high quality and therefore create a competitive advantage, personnel trained in journalisms are necessary. Additionally, the technical know-how can be seen as a limiting factor for syndication activities on the Internet. The personnel must know how to configure the content from possibly different data formats, maintain and transfer the offered product. Moreover, one can assume that it is necessary to have a deep understanding of the market, the underlying trends and the relevant players. The outlined indicators are independent from one another, which means that a provider with a high level of editorial know-how can also have a high level of technical or market-specific know-how, but he does not necessarily have to have this. Consequently, the indicators form the construct, which means that the construct "content-specific know-how" can be regarded as an index.

Furthermore, it can be assumed that the *name recognition*, the positive awareness, of a provider might be an important factor in the outlined context, which is reflected by his brand awareness and image, the amount of traffic and the extent to which he is established in the market. The lower the values of these indicators, the more the provider depends on an increase in his name recognition. As a result, he requires increasing traffic, growing brand awareness as well as image and will follow a promoting strategy. The higher the name recognition, is on the other hand, the less his dependence and the higher the provider's preference for generating direct profits. In this case he will follow a sales strategy. Here the indicators are representations of the underlying factor and thus represent reflective indicators of the construct "name recognition."

In addition to the mentioned intangible resources, the provider's *level of B2C* should be taken into account as a control variable referring to the overall strategy of the provider. The higher the level of B2C, the more the provider will focus on advertising revenue or profits from online or accordant offline deals and therefore follow a promoting strategy. Otherwise, the sales strategy will be preferred. As previous analysis showed, the level of B2C can be measured in a direct way and can be operationalized as a single item. Figure 25.1 and Table 25.2 give an overview of the hypotheses.

Table 25.2 Resource-strategy-relationship model hypotheses

	Overview of hypotheses
H1	The more extensive the *content equipment* of the content provider the more likely he will follow a *sales strategy*.
H2	The less extensive the *content equipment* of the content provider the more likely he will follow a *promoting strategy*.
H3	The higher the *content-specific know-how* of the content provider the more likely he will follow a *sales strategy*.
H4	The less the *content-specific know-how* of the content provider the more likely he will follow a *promoting strategy*.
H5	The higher the *name recognition* of the content provider the more likely he will follow a *sales strategy*.
H6	The less the *name recognition* of the content provider the more likely he will follow a *promoting strategy*.
H7	The higher the *level of B2C* of the content provider the more likely he will follow a *promoting strategy*.
H8	The less the *level of B2C* of the content provider the more likely he will follow a *sales strategy*.
H9	The more the content provider follows a *sales strategy* the more *monetary relationships* he will maintain.
H10	The more the content provider follows a *sales strategy* the less *non-monetary relationships* he will maintain.
H11	The more the content provider follows a *promoting strategy* the less *monetary relationships* he will maintain.
H12	The more the content provider follows a *promoting strategy* the more *non-monetary relationships* he will maintain.

To test the model, a *survey* was conducted. A standardized questionnaire was sent via email to nearly all content providers in the German-speaking market. The respondents had the opportunity to complete the questionnaires directly on the screen and send it back via email or to fill out a printed version of the questionnaire and to fax or post it. Overall 309 companies had been identified as suitable participants in an investigation of the Internet and were finally contacted. The respective informants were reminded twice: via email and by phone. Twenty-one indicated that they would not transfer any content to a partner, which reduced the number of possible answers to 288. A total of 136 firms took part in the survey, which led to a response rate of 47.22 %.

Preliminary analysis showed no bias between early and late respondents. Owing to the low level of missing values per item, all the indicators could remain in the analysis. The small number of missing values was replaced by their means.

25.3 Analysis with Partial Least Squares (PLS)

Based on the hypotheses derived from the RBV, the explanatory model was analysed in a next step. As Fig. 25.1 and Table 25.2 show we hypothesized the relationship between various constructs. To reduce the complexity and enhance the explanatory power of the model we operationalized the abstract constructs with more specific indicators. As *structural equation modelling (SEM)* deals with multilevel relationships between latent variables measured by multiple manifest items it seems to be the appropriate method for statistical analysis of the case at hand.

The procedures for estimating SEM can principally be separated into variance-covariance-based procedures such as *ML-LISREL* or *AMOS* and principal components-regression-approaches such as *PLS*. In this case we chose PLS to test the relationships in the model because it works with less restrictive requirements.

The most widely used variance-covariance-based procedure, *ML-LISREL*, uses the *maximum likelihood estimation method* and therefore several assumptions, have to be fulfilled. Especially in situations of high complexity but low level of information, some of the assumptions might be violated (Dupacavá and Wold 1982, p. 293). Firstly, *ML-LISREL* needs large sample sizes ($N > 200$) and relatively few indicators and constructs for the algorithm to converge (Hair et al. 1998, p. 605). PLS, however, is applicable to relatively small sample sizes and complex models (Fornell and Bookstein 1982, p. 450; Wold 1985, p. 590). Regarding our survey, the number of cases might not be sufficient for *ML-LISREL* to obtain proper results if the complexity of the model is borne in mind. Secondly, while covariance-based methods depend on a multivariate normal distribution of the data, PLS makes no distributional assumptions. Therefore, PLS is also applicable in situations with an explorative character like our analysis where a multivariate normal distribution of the data cannot be ensured. Thirdly, formative indicators can be handled much simpler by PLS. This means that the measurement model in PLS may not only include *reflective indicators*, which are caused by an underlying construct (Mode A), but also *formative* ones, which form the construct (Mode B) and, hence, act as an index (Diamantopoulos and Winklhofer 2001, p. 269). The PLS algorithm can deal with both kinds of indicators which leads to mode C if modes A and B are both integrated into one explanatory model. While the estimation of indicators in mode A follows a series of single regressions with the indicators as the dependent variables, the estimation of mode B is based on a multiple regression, treating the indicators as independent variables. The distinction between formative and reflective indicators has often been neglected in the literature leading to misspecified models and poor results (Albers 2010; Jarvis et al. 2003; Rossiter 2002). Therefore, one really has to prove whether the change of the direction of one item will necessarily result in an alteration of the other items in the same direction. If this is not the case, the indicators cannot be regarded as reflective. In our case, most of the variables, e.g., the "content-specific know-how", are formative measures. Here, the indicators are independent from one another: A provider with a high level of editorial know-how, can also have a high level of technical or market-specific know-how, but it is not a compulsory relationship. Only the construct "name recognition" is measured in a reflective way. This implies that the values of the indicators brand awareness, image, established firm and traffic should co-vary with one another.

Having ensured a theoretically based model with appropriate specifications, the PLS analysis can be conducted. The empirical PLS analysis and interpretation of the results are presented in *two steps*. In a first step, the quality of the measurement model is assessed. Only in the case of *reliable* and *valid* measures of the latent variables can a valuable analysis of the structural model and interpretation be undertaken (Anderson and Gerbing 1988, p. 417; Hulland 1999, p. 198). As PLS makes no distributional assumptions, only *non-parametric tests* can be used to evaluate the explanatory model (Chin 1998, p. 316).

The quality of *reflective measures* can be assessed by the *individual reliability* of the items as well as by the *convergent validity* and the *discriminant validity* of the latent variables (Hulland 1999, p. 198ff.). As formative indicators cause their constructs, they do not have to be highly correlated with one another. Therefore, formative indicators have to be evaluated according to their *content validity* (Chin 1998, p. 367; Hulland 1999, p. 201).

With name recognition, we have only one reflective construct (see Fig. 25.1) for which the usual tests are applied. Regarding the *reliability of the items*, Table 25.3 shows that all loadings exceed the threshold level of 0.707, indicating that more than 50% of the variance in the observed variable is due to the construct (Hulland 1999). Furthermore, a *bootstrap test* shows high significance levels for all loadings. With respect to the *convergent validity* of a construct, *Cronbach's alpha* (Cronbach 1951) and the *internal consistency measure (IC)*, developed by Werts, Linn und Jöreskog (Werts et al. 1974), should be used. Both measures differ in that the IC takes individual loadings into account, whereas Cronbach's alpha assumes a priori that each indicator contributes equally to its construct (Barclay et al. 1995, p. 297). Nevertheless, the interpretation of the measures is similar and 0.707 should be exceeded in both cases (Hulland 1999, p. 199).

Table 25.3 Results of the outer model

	Proposed effect	Loadings, weights	Observed t-value	Signif.-level 1-tail
Content equipment (formative)				
Topics	+	0.708	4.274	0.000
Formats	+	0.537	2.542	0.006
Know-how (formative)				
Editorial	+	0.664	1.755	0.041
Market-specific	+	−0.254	0.572	0.284
Technical	+	0.767	1.977	0.025
Name recognition (reflective)				
Brand awareness	+	0.834	3.778	0.000
Strong image	+	0.853	4.153	0.000
Established firm	+	0.858	3.328	0.000
Sufficient traffic	+	0.722	3.179	0.001
Promoting strategy (formative)				
Increasing own traffic	+	0.823	7.338	0.000
Index: increasing brand/image	+	0.307	2.072	0.020
Monetary relationships (form.)				
Mere monetary compensation	+	0.997	34.124	0.000
Monetary and non-monetary compensation	+	0.403	3.295	0.000
Non-monetary relations (form.)				
Mere non-monetary compensation	+	0.992	53.650	0.000
Charge for integration	+	0.125	2.467	0.008

Furthermore, the *AVE* measure developed by Fornell and Larcker (1981) should be considered. It measures the amount of variance of the indicator which is accounted for by the construct relative to the amount due to the measurement error. Therefore, the AVE should exceed 0.5, indicating that more than 50% of the indicators' variance can be captured by the construct. In our case, Cronbach's alpha is 0.841 while the internal consistency measure (IC) is 0.890. Hence, both values meet the respective marginal values. The same is true for the average variance extracted (AVE) value of 0.670, which exceeds the required 0.5.

The *discriminant validity* is the traditional counterpart of the convergent validity. To evaluate to which extent measures of a given construct differ from other indicators of the latent variables, the AVE-value can be used again. Overall, the average shared variance of a construct and its indicators should exceed the shared variance with every other construct of the model. Therefore, the square root of AVE should surpass the correlation coefficient of the construct with every other construct of the model, which is the case in the outlined model. Furthermore, as a reflective indicator should load higher on its corresponding construct than on the other ones, the cross-loadings should be examined. Additionally, all indicators of the construct in question should have a higher loading than the indicators of further constructs. As there is only one reflective construct in the outlined model, the examination of the cross-loadings is not appropriate in this case.

The other constructs of the model are caused by *formative indicators*. As formative indicators do not have to be highly correlated with each other, the application of the mentioned measures is inappropriate. Rather, in order to investigate the quality of the formative indicators, their *content validity* has to be evaluated (Albers 2010; Diamantopoulos and Winklhofer 2001; Rossiter 2002, 2005). Diamantopoulos (2005) and Finn and Kayande (2005), however, plead for generalized measures in this context. Hence, the effects and the weights resulting from a bootstrapping should be considered.

Table 25.3 presents significant values for the proposed effects and adequate weightings according to the conducted expert interviews (Chin 1998; Hulland 1999; Rossiter 2002). Only the indicator "market-specific know-how" shows no significance. An investigation of multicollinearity demonstrates that the formative indicators "increasing own brand awareness" and "increasing own image" are correlated too much. Therefore, an index was created by the means of these items. We had no further problems with multicollinearity as the Variance Inflation Factors (VIF) were shown to be less than 2.0 in each case. Table 25.4 gives an overview of the VIFs of formative indicators.

Furthermore, the correlations between the exogenous variables showed to be relatively low: *r* (content equipment, know-how) = 0.04; r (content equipment, name recognition) = 0.19; r (know-how, name recognition) = 0.17. Nevertheless, in the very end the achieved explained variance (R^2) of the endogenous constructs determines whether a theoretically sound exogenous construct is operationalized appropriately.

Based on a sound measurement model, the *structural model* is estimated. To evaluate the inner model and test the hypotheses, the path coefficient of the inner model as well as the R^2 and R^2_{adj} of the endogenous latent variables have to be

Table 25.4 Overview of VIFs

	VIF
Content equipment (formative)	
Topics	1.083
Formats	1.083
Know-how (formative)	
Editorial	1.018
Market-specific	1.374
Technical	1.391
Promoting strategy (formative)	
Increasing own traffic	1.257
Index: increasing brand/image	1.257
Monetary relationships (form.)	
Mere monetary compensation	1.040
Monetary and non-monetary compensation	1.040
Non-monetary relations (form.)	
Mere non-monetary compensation	1.000
Charge for integration	1.000

inspected. Although PLS provides a relatively unbiased estimation of path coefficients, the method follows no distributional assumptions and does not present significance levels. Therefore, a bootstrap, with $N = 100$ samples, will be run, providing t-values and 1-tail significance levels (Efron and Gong 1983; Efron and Tibshirani 1993; Hinkley 1988). Table 25.5 presents the R^2 and $R^2{}_{\text{adj}}$ of the tested model, which have to be evaluated at first.

The results show that a substantial part of the variance of the latent constructs can be explained, which also refers to a sound measurement of the model. Consequently, the different strategies and monetary form of the relationships are explained to a comparable extent. Given that regressions with cross-sectional data arrive at an explained variance of between 30 % and 40 %, the nomological validity of the model is satisfactory.

A next step can now examine which hypotheses are supported by the analysis. Table 25.6 presents the path coefficient of the inner model along with the results of the conducted bootstrap. Figure 25.2 shows the results of the inner model graphically.

Nine of the 12 hypotheses are supported while three show no significance. Every significant relationship is characterized by a path coefficient > 0.1 and can therefore

Table 25.5 R^2 und $R^2{}_{\text{adj}}$ of the endogenous variables

	R^2	$R^2{}_{\text{adj}}$
Sales strategy	0.300	0.275
Promoting strategy	0.307	0.282
Monetary relationships	0.531	0.523
Non-monetary relationships	0.526	0.518

Table 25.6 Results of the inner model

	Hypothesized effect	Path coefficient	Observed t-value	Significance level 1-tail
Sales strategy				
Content equipment	+	0.294	0.275	0.000
Know how	+	0.176	1.536	0.064
Name recognition	+	*0.090*	0.845	0.200
Level of B2C	−	−0.355	4.642	0.000
Promoting strategy				
Content equipment	−	−0.310	−0.317	0.000
Know-how	−	*−0.056*	0.486	0.314
Name recognition	−	*0.006*	−0.025	0.483
Level of B2C	+	0.438	4.410	0.000
Monetary relationships				
Sales strategy	+	0.524	4.936	0.000
Promoting strategy	−	−0.292	3.064	0.001
Non-mon. relationships				
Sales strategy	−	−0.538	5.748	0.000
Promoting strategy	+	0.272	3.561	0.000

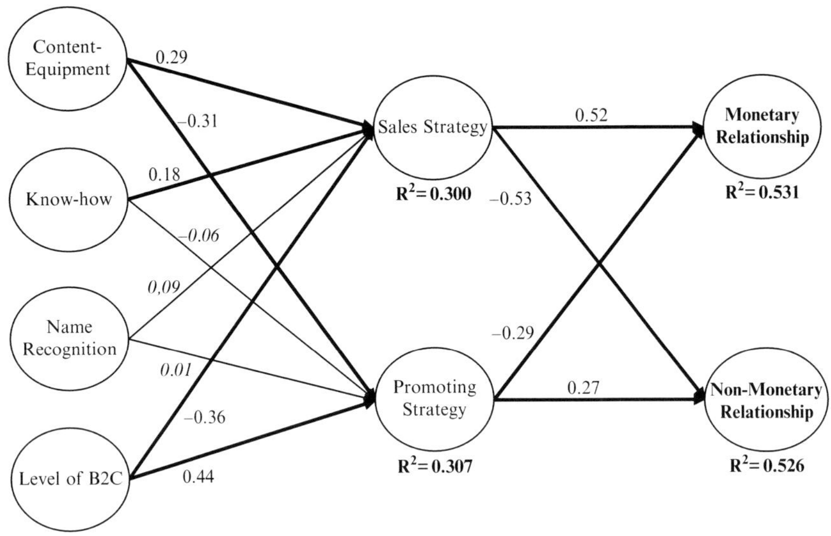

Fig. 25.2 Results of the inner model

not be neglected (Seltin and Keeves 1994, p. 4356). The sales strategy of the provider is influenced by the level of B2C, the content equipment, and know-how of the provider in descending order. The same holds for the promoting strategy with reversed sign, although there is no significant relationship between the know-how and this strategy. Additionally, the name recognition of the provider has no influence on either strategy.

For further *cross-validation* of the model, the data-splitting approach is applied as simultaneous methods like the Stone-Geisser approach are only applicable for mode A models (reflective constructs). The sample was randomly split into an estimation sample and a hold-out sample. According to the recommendations of Steckel and Vanhonacker (Steckel and Vanhonacker 1993), 75 % of the cases were used for the estimation sample, while 25% created the hold-out sample. As we are consequently implicitly testing the *predictive validity* of the model, it is advisable to include only the relationships with an observed t-value > 1. In this case, only relationships that have a higher information value than white noise are considered (Hansen 1987; Chin 2006). Table 25.7 gives an overview of the results.

High correlations (r) between the calculated and observed values of the hold-out sample (0.443–0.852) indicate a good predictive validity of the model and the generality of the results. The same is shown by the small difference between the calculated r^2 and the R^2 of the hold-out sample.

For managerial purposes it is not so much the significance that counts but the differential effects of the variables. One way is to assess whether a predictor variable has a substantive influence on the dependent variable, which can be explored through the *effect size* f^2:

Table 25.7 R^2 und R^2_{adj} of the endogenous variables

	Sales strategy	Promoting strategy	Monetary relationships	Non-monetary relationships
R	0.443	0.515	0.852	0.841
r^2	0.197	0.265	0.726	0.708
R^2	0.300	0.307	0.531	0.526

Table 25.8 Effect sizes of the latent variables

	f^2	Rating
Sales strategy		
Content equipment	0.120	Small
Know-how	0.041	Small
Name recognition	0.011	
Level of B2C	0.174	Middle
Promoting strategy		
Content equipment	0.128	Small
Know-how	0.004	
Name recognition	0.001	
Level of B2C	0.268	Middle
Monetary relationships		
Sales strategy	0.371	Large
Promoting strategy	0.115	Small
Non-monetary relationships		
Sales strategy	0.416	Large
Promoting strategy	0.105	Small

$$f^2 = \frac{R^2{}_{\text{included}} - R^2{}_{\text{excluded}}}{1 - R^2{}_{\text{included}}}.$$

$R^2{}_{\text{included}}$ or $R^2{}_{\text{excluded}}$ indicates the R^2 of the dependent variable when the independent variable is included or excluded as a predictor of the dependent variable. The higher f^2 the greater the influence of the independent construct whereby values of 0.02, 0.15 and 0.35 can be respectively regarded as small, medium or large (Chin 1998, p. 317). The results are given in Table 25.8.

Another way is to calculate the *total effects* that single indicators have on the determination of either a monetary or non-monetary compensation policy. Table 25.9 shows the results. Of course, the total effects of indicator antecedents can only be computed for formative indicators.

The results show that the providers actually follow two different strategies, namely a sales and promoting strategy. Both can be regarded as antipodal to each other as most of the respective path coefficients have roughly the same value, but with reversed signs. Hence, the providers either aim at an increase in direct profits and regard the Internet as a further distribution-channel of their content or intend to improve the own traffic, image and/or name recognition. Here the results of the measurement model show that a growth of traffic is more important than the other goals.

Both strategies are explained by the proposed content relevant resources of the providers which influence the strategies to different extents. As the resources have

Table 25.9 Importance of the indicator antecedents on the compensation policy

Importance interval	Positive total effects		Negative total effects	
Monetary relationships				
>0.20	Generating direct profits	0.520	Level of B2C	−0.315
			Increasing own traffic	−0.239
0.20	Content topics	0.170		
−0.11	Content formats	0.129		
0.10	Technical know-how	0.085	Increasing brand/image	−0.089
−0.05	Editorial know-how	0.074		
<0.05	Name recognition	0.044	Market-specific know-how	−0.028
Non-monetary relationships				
>0.20	Level of B2C	0.310	Generating direct profits	−0.530
	Increasing own traffic	0.222		
0.20–0.11			Content topics	−0.168
			Content formats	−0.128
0.10–0.05	Increasing brand/image	0.083	Technical know-how	−0.086
			Editorial know-how	−0.074
<0.05	Market-specific know-how	0.028	Name recognition	−0.045

an impact on the strategies they also influence the form of the provider's remuneration. Overall the resourced-based view of a firm can therefore be regarded as a suitable theory in our case.

25.4 Discussions of the Results

Table 25.9 summarizes the importance of single indicators (if formative) and constructs (if reflective) on the choice of the monetary or non-monetary relationship. Indicators and constructs are sorted with respect to a positive or a negative influence. Furthermore, they are classified into importance intervals according to their total effects. The results illustrate that the level of B2C has the biggest impact on the choice of either using the content for increasing direct profits (sales strategy) or for raising own traffic, brand awareness and image (promoting strategy) and finally on the monetary form of the relationships. As a result, the management of the content has to be tightly coupled with the further activities of the provider. Consequently, the content can be seen as a suitable alternative form of advertising. Hence, one has to compare the earnings, which could be gained with a special kind of content, with the cost reduction in advertising when content is used as an alternative.

Furthermore, it was demonstrated that the content equipment also has a high influence on the strategies and the compensation policies negotiated for the partnerships although it is lower than the B2C level. The higher the amount of content suitable for syndication, e.g., without the limitation of property rights, the more the provider will seek a sales strategy. As indicated in Table 25.9, the content topics are much more important than the formats, as text is still the most syndicated and demanded kind of content. Like the level of B2C, the content equipment has a similar impact on both strategies.

The results show that the content-specific know-how has an impact on the sales strategy. The more knowledge the provider possesses the more the provider follows a sales strategy, whereas no relationship could be confirmed between the know-how and the promotion strategy. As the results also show the impact of the market-specific know-how on the strategies and the compensation of the provider can be ignored while the technical know-how is more important than the editorial know-how.

Finally we can maintain that name recognition has a negligible impact on the choice of strategy and compensation policy. As a result, it has not been determined that firms with high name recognition, like publishing houses, follow a sales strategy, whereas start-ups with a low level of name recognition primarily embark on a promoting strategy. To sum up, the choice of strategy is sufficiently explained by its antecedents.

Regarding the fit-problem between the marketing strategies and compensation policies, it can be shown that the strategies explain a substantial part of the outcomes "monetary" and "non-monetary partnerships." Nevertheless, the results illustrate no definite link between the strategies and the form of the partnerships. Providers who follow a sales strategy, also work under non-monetary partnerships and vice versa. As the results show, the sales strategy has a higher impact on both forms of the relationship. More providers agree to form a monetary relationship although they more often follow a promoting strategy than the other way round. This implies that additional factors might have an impact on the monetary form of the relationship. Here, one might assume that not only providers, but also the content subscribers influence the financial form of the relationship and, therefore, their resources and strategies might also have an impact in this context. This has to be determined in further investigations.

25.5 Conclusion and Outlook

This article illustrates how a strategy study in marketing can be analyzed with the help of PLS, thereby providing more detailed and actionable results. We also discuss additional methods with which the explanatory power of the analysis can be increased. In our case, PLS turned out to be the adequate statistical method. First, complex measures had to be operationalized with more specific indicators – marketing instruments in our case –, which proved to be formative in the most cases. Only by using PLS was it possible to extract the influence of every single formative indicator on the final constructs, i.e., the monetary form of the partnerships. Second,

PLS allows for more degrees of freedom, so that a complex model could be estimated with a number of cases that would not have been sufficient for ML-LISREL. Third, PLS does not work with distributional assumptions, while significance tests can still be carried out with the help of bootstrapping. To sum up, we recommend the use of PLS for future strategy studies in marketing because it is possible to extract the drivers at the indicator level, allowing detailed recommendations to be made regarding managing marketing instruments.

Our analysis shows that content providers follow certain strategies in their compensation policy with respect to their customers. The choice of the policy can be explained by the resource-based view and may serve as recommendations. While this is based on sales results only, further studies should also include the cost side. Finally, this analysis is carried out from the viewpoint of the content provider while the compensation contract is also influenced by the situation of the content subscriber, which also has to be taken into account.

References

Albers, S. (2010). PLS and success factor studies in marketing. In V. E. Vinzi, W. W. Chin, J. Henseler, & H. Wang (Eds.), *Handbook of Partial Least Squares*. Heidelberg: Springer.

Anderson, J., & Gerbing, D. W. (1988). Structural equation modelling in practice: A review and recommended two-step approach. *Psychological Review, 103* (3), 411–423.

Bamberger, I., & Wrona, T. (1996). The resource based view of a firm and its importance for strategic management (in German). *Schmalenbachs Zeitschrift für betriebswirtschaftliche Forschung, 48*(2), 130–153.

Barclay, D., Higgins, S., & Thompson, R. (1995). The partial least squares approach to causal modelling personal computer adoption and use as an illustration. *Technology Studies, 2*(2), 285–374.

Chin, W. W. (1998). The partial least squares approach to structural equation modelling. In: G. A. Marcoulides (Ed.), *Modern methods for business research* (pp. 295–336). Mahwah, NJ: Lawrence Erlbaum.

Chin, W. W. (2010). Bootstrap cross-validation indices for PLS path model assessment. In V. E. Vinzi, W. W. Chin, J. Henseler, & H. Wang (Eds.), *Handbook of Partial Least Squares*. Heidelberg: Springer.

Cronbach, L. J. (1951). Coefficient alpha and the internal structure of tests. *Psychometrika, 16*(3), 297–324.

Diamantopoulos, A. (2005). The C-OAR-SE procedure for scale development in marketing: A comment. *International Journal of Research in Marketing, 22*(1), 1–9.

Diamantopoulos, A., & Winklhofer, H. M. (2001). Index construction with formative indicators: An alternative to scale development. *Journal of Marketing Research, 38*, 269–277.

Dupacavá, J., & Wold, H. (1982). On some identification problems in ML modelling of systems with indirect observations. In K. G. Jöreskog & H. Wold (Eds.), *Systems under indirect observations: Causality, structure, prediction, part 2* (pp. 293–315). New York: Amsterdam.

Efron, B., & Gong, G. (1983). A leisurely look at the bootstrap, the jackknife, and cross validation. *The American Statistican, 37*(1), 36–48.

Efron, B., & Tibshirani, R. J. (1993). *An introduction to the bootstrap – monographs on statistics and applied probability*. New York: Chapman and Hall.

Finn, A., & Kayande, U. (2005). How fine is C-OAR-SE? A generalizability theory perspective on rossiter's procedure. *International Journal of Research in Marketing, 22*(1), 11–21.

Fornell, C., & Bookstein, F. L. (1982). Two structural equation models: LISREL and PLS applied to consumer exit-voice theory. *Journal of Marketing Research, 19*, 440–452.

Fornell, C., & Larcker, D. F. (1981). Evaluating structural equation models with unobservable variables and measurement error. *Journal of Marketing Research, 18*, 39–50.

Hair, J. F., Jr., Anderson, R. E., Tatham, R. L., & Black, W. C. (1998). *Multivariate data analysis* (5th ed.). Upper Saddle River, NJ: Prentice Hall.

Hansen, G. (1987). Forecasting error and multicollinearty (in German). *Jahrbuch für Nationalökonomie und Statistik, 203*(5–6), 517–531.

Hinkley, D. V. (1988). bootstrapping methods. *Journal of the Royal Statistical Society, Series B, 50*(3), 321–337.

Hulland, J. (1999). Use of partial least squares (PLS) in strategic management research: A review of four recent studies. *Strategic Management Journal, 20*, 195–204.

Jarvis, C. B., MacKenzie, S. B., & Podsakoff, P. M. (2003). A critical review of construct indicators and measurement model misspecifications in marketing and consumer research. *Journal of Consumer Research, 30*, 199–218.

Penrose, E. T. (1959). *The theory of the growth of the firm*. Oxford: Oxford University Press.

Prahalad, C. K., & Hamel, G. (1990). The core competence of the corporation. *Harvard Business Review, 68*, 79–91.

Rossiter, J. (2002). The C-OAR-SE procedure for scale development in marketing. *International Journal of Research in Marketing, 19*, 305–335.

Rossiter, J. (2005). Reminder: A horse is a horse. *International Journal of Research in Marketing, 22*(1), 23–25.

Seltin, N., & Keeves, J. P. (1994). Path analysis with latent variables. In T. Husen & Postlethwaite, T. N. (Ed.), *International encyclopedia of education*. (2nd ed., pp. 4352–4359). Oxford: Pergamon.

Steckel, J. H., & Vanhonacker, W. R. (1993). Cross-validation regression models in marketing research. *Marketing Science, 12*(4), 415–427.

Thies, S. (2005). Content-Interaktionsbeziehungen in Internet: Ausgestaltung und Erfolg (engl.: Content-cooperations on the Internet – forms and performance). DUV: Wiesbaden.

Werbach, K. (2000). Syndication – the emerging model for business in the internet Era. *Harvard Business Review, 78*, 85–93.

Wernerfelt, B. (1984). A resource-based view of the firm. *Strategic Management Journal, 5*(2), 171–180.

Werts, C. E., Linn, R. L., & Jöreskog, K. G. (1974). Interclass reliability estimates: Testing structural assumptions. *Educational and Psychological Measurement, 34*, 24–33.

Wold, H. (1985). Partial least squares.In S. Kotz & N. L. Johnson (Eds.), *Encyclopedia of statistical sciences* (pp. 581–591). New York: Wiley.

Chapter 26
Use of Partial Least Squares (PLS) in TQM Research: TQM Practices and Business Performance in SMEs

Ali Turkyilmaz, Ekrem Tatoglu, Selim Zaim, and Coskun Ozkan

Abstract Advances in structural equation modeling (SEM) techniques have made it possible for management researchers to simultaneously examine theory and measures. When using sophisticated SEM techniques such as covariance-based structural equation modeling (CBSEM) and partial least squares (PLS), researchers must be aware of their underlying assumptions and limitations. SEM models such as PLS can help total quality management (TQM) researchers achieve new insights. Researchers in the area of TQM need to apply this technique properly in order to better understand the complex relationships proposed in their models. This paper attempts to apply PLS in the area of TQM research. Consequently, special emphasis is placed on identifying the relationships between the most prominent TQM constructs and business performance based on a sample of SMEs operating in the Turkish textile industry. The analysis of PLS results indicate that a good deal of support is found for the proposed model where a satisfactory percentage of the variance in the dependent constructs is explained by the independent constructs.

A. Turkyilmaz and S. Zaim
Department of Industrial Engineering, Fatih University, Buyukcekmece, 34500 Istanbul, Turkey
e-mail: aturkyilmaz@fatih.edu.tr, szaim@fatih.edu.tr

E. Tatoglu
Faculty of Economics and Administrative Sciences, Chair of International Trade and Business, Bahcesehir University, Besiktas, Istanbul, Turkey
e-mail: ekremt@bahcesehir.edu.tr

C. Ozkan
Department of Industrial Engineering, Kocaeli University, Veziroglu Yerleskesi, 41040 Kocaeli, Turkey
e-mail: coskun_ozkan@yahoo.com

V. Esposito Vinzi et al. (eds.), *Handbook of Partial Least Squares*, Springer Handbooks of Computational Statistics, DOI 10.1007/978-3-540-32827-8_27,

26.1 Introduction

Advances in structural equation modeling (SEM) techniques have made it possible for management researchers to simultaneously examine theory and measures. SEM is a comprehensive statistical approach to testing hypotheses about relationships between observed and latent variables. It combines features of factor analysis and multiple regression to study both the measurement and the structural properties of theoretical models. Such techniques are considered superior to more traditional statistical techniques such as multiple regression, factor analysis, and multidimensional scaling. However, researchers should apply these new techniques appropriately. They must be aware of the underlying assumptions and limitations of SEM techniques.

SEM is formally defined by two sets of linear equations called the inner and outer model. The inner model specifies the relationships between unobserved or latent variables (LVs), while the outer model specifies the relationships between LVs and their associated observed or manifest variables (MVs).

There are two common statistical approaches for structural model estimation. The most prominent SEM technique is the maximum likelihood (ML) based covariance structure analysis method the so-called CBSEM (Bollen 1989; Jöreskog 1970; Rigdon 1998). The second approach is the Partial Least Squares (PLS)-based variance analysis developed by Wold (1975, 1982, 1985). These two distinct methods of SEM differ in terms of their objectives, statistical assumptions and the nature of the fit statistics they produce (Gefen et al. 2000). The main concern of PLS is, in general, related to the explanatory power of the path model along with the significance level of standardized regression weights. In contrast, the objective of CBSEM is to show that the complete set of paths as specified in the model is reasonable, and that the operationalization of the theory is corroborated and not disconfirmed by the sample data. These two methods also differ with respect to the type of relationship they support between the observed variables and their associated latent constructs (i.e. outer model). PLS supports two types of relationships, formative and reflective, whereas CBSEM supports only reflective indicators (Fornell and Bookstein 1982).

Although CBSEM has been widely adopted as a powerful approach and has been used for parameter estimation in most applications of structural modeling, there are some situations where PLS approach is superior to CBSEM. CBSEM is poorly suited to deal with small data samples and can provide nonunique or otherwise improper solutions in some cases (Hulland 1999). Moreover, data from management research often do not satisfy the requirements of multinormality and interval scaling for maximum likelihood estimation. More fundamentally, two serious problems often interfere with meaningful covariance structure analysis: inadmissible solutions and factor indeterminacy (Fornell and Cha 1994; Wold 1985).

PLS is a general method for the estimation of path models involving latent constructs indirectly measured by multiple indicators (Wold 1982). This tool is primarily intended for causal-predictive analysis in which the problems explored are complex and theoretical knowledge is scarce. PLS is an appropriate technique to use in a theory development situation (Wold 1979). This technique uses a component-based approach to estimation. Consequently, it places minimal demands on sample size and residual distributions (Lohmöller 1989).

While SEM techniques such as CBSEM and PLS can enhance existing methodological approaches to conducting quality management research, they should be applied properly. Most quality management researcher are very familiar with the fundamentals of covariance-based-type SEM models, whereas current familiarity with PLS is relatively low in the field of quality management, making it difficult for researchers to properly evaluate its use.

Employing the PLS approach, Cassel et al. (2000) measured the European Customer Satisfaction Index (ECSI). In this study, they also used the Monte Carlo simulation method to evaluate the robustness of partial least squares. The authors noted that PLS is reasonably robust against multicollinearity, skew response distributions, and various types of model misspecifications (Cassel et al. 2000, 1999; Cassel 2000). In another survey, Kanji (1998) also employed the PLS approach to develop the Business Excellence Index model that simultaneously measures customers', employers' and shareholders' satisfaction within an organization in order to obtain a comprehensive evaluation of the organizational performance (Kanji 1998; Kanji and Wallace 2000).

The purpose of this study is to help shape application of PLS in the area of total quality management (TQM). In doing this special emphasis is placed on investigating the relationships between TQM practices and the business performance of small and medium-sized enterprises (SMEs) in Turkey. The rest of this study is organized as follows: The next section provides a brief review of the theoretical background of TQM. The third section presents the methodology of the study, followed by the results. A discussion and conclusions are provided in the final section.

26.2 Theoretical Background

Any organization, regardless of its nature, is advised to adopt TQM practices to generate high quality products or services and to meet the challenge of global competition. Total quality management (TQM) is an integrated management philosophy aiming at continuous improvement in all functions of an organization to produce and deliver commodities or services in line with customers' needs or requirements, and it covers many important aspects, ranging from customer satisfaction, meeting customers' requirements, and reducing rework and waste to increased employee involvement, process management and supplier relations.

TQM helps firms establish an organizational culture committed to customer satisfaction through continuous improvement. This culture varies from one country to another and between different industries, but has certain essential principles, which can be implemented to secure greater market share, increased profits, and reduced costs (Kanji and Wallace 2000). A review of extant literature on TQM and continuous improvement programs identifies a number of common aspects, which include committed leadership, closer customer relationships, benchmarking, supplier relations, increased training, employee empowerment, zero defects mentality, flexible manufacturing, process improvement and measurement (Saraph et al. 1989; Flynn et al. 1995; Anderson et al. 1994; Black and Porter 1996; Demirbag et al. 2006). Furthermore, to determine the critical factors of TQM, various studies were undertaken and different instruments were developed by individual researchers and institutions such as the Malcolm Baldrige Award, the EFQM (European Foundation for Quality Management), and the Deming Prize criteria Based on these studies, a wide range of management issues, approaches, and systematic empirical investigations have been generated.

Measuring business performance is crucial for the effective management of an organization. Therefore, to improve business performance, one needs to determine the extent of TQM implementation and measure its impact on business performance (Gadenne and Sharma 2002). Traditionally, business performance has been measured by using financial indicators, which may include inter alia profit, market share, earnings, and growth rate. Kaplan and Norton (1996) emphasized that financial indicators would measure only past performance. Therefore, in order to overcome the potential shortcomings of traditional business performance systems they added non-financial categories to the traditional performance measurement system.

There is a relatively large body of empirical studies that measures business performance by means of TQM criteria (see, e.g., Benson et al. 1991; Samson and Terziovski 1998; Flynn et al. 1995; Wilson and Collier 2000; Fynes and Voss 2001; Montes et al. 2003). These studies explore a variety of theoretical and empirical issues. If the TQM plan is implemented properly, it has an impact on a wide range of areas, including better process management, understanding customers' needs, improved customer satisfaction, improved internal communication, better problem solving, and fewer errors.

Large-size firms have recently had a greater tendency to focus on their core business areas and have therefore extensively relied on outsourcing. As the quality of products and services depends extensively on the quality of suppliers' products and services, large firms encourage the application of TQM practices by their suppliers, the majority of which are small and medium-sized enterprises (SMEs). Despite some attempts to investigate the relationships between TQM practices and the business performance of SMEs (Ahire and Golhar 1996; McAdam and McKeown 1999; Yusof and Aspinwall 2000; Sun and Cheng 2002; Lee 2004; Demirbag et al. 2006), there is a lack of systematic empirical evidence regarding the level of TQM implementation and its effect on the business performance of SMEs.

26.3 Variables and the Model

Based on a review and classification of the relevant empirical literature, the following TQM factors were, in a broad sense, identified as the most appropriate TQM constructs within the context of SMEs, which include customer focus, top management involvement, process management, supplier management, and employee relations (Lee 2004; Demirbag et al. 2006).

Customer focus (CF), which is considered the major "driver" of TQM practices, addresses how and how well the organization determines current and emerging customer requirements and expectations; provides effective customer relationship management, and determines customer satisfaction (Kaynak 1995). In this study, we measure the CF construct using the following indicators: in-house market research activities of the firm, survey of customer choices, and reviewing the business environment from the customers' point of view.

Top management involvement (TMI) is an important factor in TQM implementation as it improves business performance by influencing other TQM practices (Saraph et al. 1989; Ahire et al. 1996; Anderson et al. 1994; Flynn et al. 1995; Wilson and Collier 2000). In SMEs, the success of TQM applications depends on a strong leadership that must be initiated by the top management. Quality improvement plans proposed by various gurus primarily emphasize the commitment of top management. The top management of the firm determines an appropriate organizational culture, vision, and quality policy. Managers of organizations should determine objectives, and set specific measurable goals to satisfy customer expectations, and improve their organizations' performance. They must also provide adequate resources for the implementation of quality efforts.

The employee relations (ER) construct investigated in this study includes a variety of organizational development (OD) techniques to facilitate changes within the organization such as employee participation in decisions, employee recognition, teamwork and the use of effective communications to create an awareness of organizational goals. These OD techniques are generally considered the most relevant human resource practices in organizations that make effective use of TQM techniques.

Process Management (PM) is mainly concerned with how the organization designs and introduces products and services, and integrates production and delivery requirements (Kaynak 1995). It is therefore vital to the success of an organization. The PM construct is composed of the following items: availability of quality data, the extent to which quality data are used as tools for managing quality and the extent to which quality data and control charts are displayed for the production processes in the firm.

Suppliers play a well-recognized key role in quality management and have a significant impact on several quality dimensions. Once it is recognized that the materials and the components purchased are the main cause of quality problems and that the blame for this can often be placed on incorrect relations with suppliers, the logical conclusion is that, in order to achieve adequate quality control of critical inputs, companies must invest more in forging relations with their suppliers. The construct supplier relations (SR) is measured by the following items in this study: supplier selection criteria, longer term relationships, clarity of specifications, and reliance on a sufficiently small number of dependable suppliers.

Business performance (BF) is the final construct used in this study and represents the performance measure in the latent variable model. All six constructs used in this study are listed in Appendix A along with their associated indicators.

Drawing on a sample of SMEs in the Turkish apparel industry, we utilize the PLS method to evaluate the proposed relationships as indicated in Fig. 26.1.

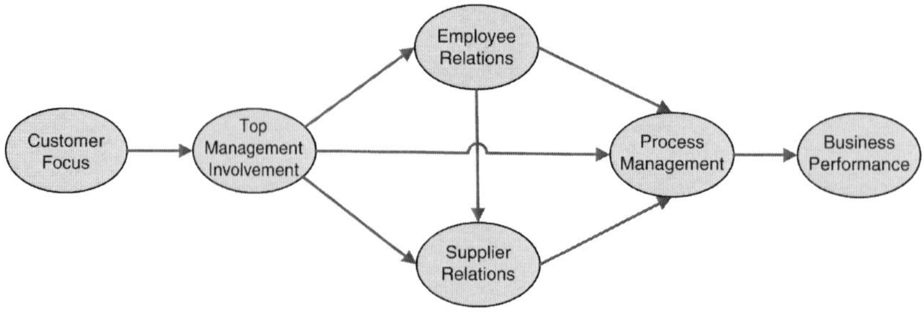

Fig. 26.1 The structural equation model

26.4 Research Methodology

26.4.1 Survey Instrument and Data Collection

The survey instrument used in this study was largely derived from the work of Saraph et al. (1989) with the purpose of identifying critical factors of TQM in a business unit.

The original version of the questionnaire was in English. This questionnaire was translated into the local language (Turkish). The local version was back translated until a panel of experts agreed that the two versions were comparable. Each item was rated on a five-point Likert scale, ranging from "very low" to "very high." The questionnaire was pre-tested several times to ensure that the wording, format, and sequencing of questions were appropriate.

The study focused on the textile industry, including textile mill products and apparel (SIC codes 22 and 23), since it has been a leader in implementing progressive quality management practices in Turkey. A self-administered questionnaire was distributed to 500 SMEs in the textile industry in the city of Istanbul, selected randomly from the database of Turkish Small Business Administration (KOSGEB). It was requested that the questionnaire be completed by a senior officer/executive in charge of quality management. The responses indicated that a majority of the respondents completing the questionnaire were, in fact, members of the top management. After one follow-up, 138 useable questionnaires were returned, giving a response rate of 28 percent, which was considered satisfactory for subsequent analysis. A comparison of the annual sales volume, number of employees and sub-industry variation revealed no significant differences between the responding and non-responding firms ($p > 0.1$). Thus, the responses adequately represented the total sample group.

26.5 Results

26.5.1 PLS Estimation of the Structural Equation Model

Before starting to analyze the path model, the unidimensionality of each construct in the proposed model was checked, using principal component analysis, Cronbach's alpha, and Dillon-Goldstein's ρ (Tenenhaus et al. 2005). According to the test results, all six constructs were found to be reliable for the path analysis, as shown in Table 26.1.

In this model, the manifest variables do not differ in nature with respect to their latent variables, and they should be utilized and improved simultaneously. Thus, reflective representation is more appropriate than a formative one for each construct. Table 26.2 presents the latent and manifest variables along with the inner and outer model equations.

26.5.2 Discriminant Validity

Discriminant validity refers to the degree to which measures of different dimensions of TQM are unique from one another. According to Venkatraman (1989), "this is achieved when measures of each dimension converge on their corresponding true scores (which is unique from other dimensions) and can be tested that the correlations between pairs of dimensions are significantly different from unity." Table 26.3 reports the results of 15 pair-wise tests conducted for discriminant validity. 13 of the 15 tests indicated strong support for the discriminant validity, while two tests failed to satisfy the criterion for discriminant validity. Therefore, it is necessary to evaluate if there is evidence to conclude that the dimensions are identical or not. Venkataraman (1989) states "since the conceptual domains of these dimensions do not overlap significantly and they exhibit different patterns of relationships with other dimensions, it is possible to accept the distinctive characteristics of these dimensions." We can conclude that the discriminant validity criterion is satisfied by these dimensions, as two of the 15 tests did not satisfy this criterion.

PLS procedure uses two-stage estimation algorithms to obtain the weights, loadings and path estimates. In the first stage an iterative scheme of simple and/or

Table 26.1 Reliability and validity of the constructs

Construct	Number of indicators	Cronbach's Alpha	Dillon-Goldstein's rho	First eigenvalue	Second eigenvalue
CF	3	0.750	0.860	2.008	0.536
TMI	7	0.902	0.923	4.423	0.705
ER	4	0.826	0.896	2.227	0.446
SR	4	0.818	0.885	2.595	0.622
PM	3	0.835	0.891	2.688	0.503
BP	5	0.862	0.902	3.223	0.578

Table 26.2 Model variables, parameters and relations

Latent variables and Inner model equations	Manifest variables	Outer model equations
ξ_1 Customer focus	x_{11} CF1	$x_{1i} = \lambda_{1i}\xi_1 + \delta_{1i}$
	x_{12} CF2	
	x_{13} CF3	
η_1 Top management involvement	y_{11} TMI1	$y_{1i} = \lambda_{1i}\eta_1 + \varepsilon_{1i}$
$\eta_1 = \gamma_{11}\xi_1 + \zeta_1$	y_{12} TMI2	
	y_{13} TMI3	
	y_{14} TMI4	
	y_{15} TMI5	
	y_{16} TMI6	
	y_{17} TMI7	
η_2Employee relations	y_{21} ER1	$y_{2i} = \lambda_{2i}\eta_1 + \varepsilon_{2i}$
$\eta_2 = \beta_{21}\eta_1 + \zeta_2$	y_{22} ER2	
	y_{23} ER3	
	y_{24} ER4	
η_3Supplier relations	y_{31} SM1	$y_{3i} = \lambda_{3i}\eta_1 + \varepsilon_{3i}$
$\eta_3 = \beta_{31}\eta_1 + \beta_{32}\eta_2 + \zeta_3$	y_{32} SM2	
	y_{33} SM3	
	y_{34} SM4	
η_4Process management	y_{41} PM1	$y_{4i} = \lambda_{4i}\eta_1 + \varepsilon_{4i}$
$\eta_4 = \beta_{41}\eta_1 + \beta_{42}\eta_2 + \beta_{43}\eta_3 + \zeta_4$	y_{42} PM2	
	y_{43} PM3	
η_5 Business performance	y_{51} BP1	$y_{5i} = \lambda_{5i}\eta_1 + \varepsilon_{5i}$
$\eta_5 = \beta_{54}\eta_4 + \zeta_5$	y_{52} BP2	
	y_{53} BP3	
	y_{54} BP4	
	y_{54} BP5	

multiple regressions is performed until a solution converges on a set of weights used for estimating the latent variables scores. The second stage involves the non-iterative application of PLS regression for obtaining loadings, path coefficients, mean scores, and location parameters for the latent and manifest variables (Fornell and Cha 1994; Chin 1998; Tenenhaus et al. 2005).

26.5.3 *Outer Model Estimation*

PLS results are estimated after 5 iterations using Decisia Spad software. The findings of the study were divided into outer and inner model estimations. Table 26.4 presents the estimation results of the outer model including outer weights, correlation between a manifest and its latent variable, communality and redundancy measures.

As shown in Table 26.4, the correlation values between the manifest variables and their respective latent variables were found to be very satisfactory. The communality

Table 26.3 Assessment of discriminant validity

Test ♯	Description	Chi-squared Constrained Model	Chi-squared Unconstrained Model	Difference
1	Top management – customer focus	29.171	20.221	8.950**
2	Top management – process management	42.7	31.6	11.1**
3	Top management – employee relations	52.48	49.1	3.38*
4	Top management – supplier relations	69.61	64.9	4.71*
5	Top management – business performance	67.65	57.58	10.030**
6	Customer focus – employee relations	27.3	18.84	8.46**
7	Customer focus – supplier relations	28.16	18.0	10.16**
8	Customer focus – process management	22.66	13.2	9.46**
9	Customer focus – business performance	22.22	12.3	9.92**
10	Employee relations – supplier relations	23.2	15.0	8.2**
11	Employee relations – process management	29.459	28.5	0.959
12	Employee relations – business performance	31.4	29.5	1.9
13	Supplier relations – process management	33.61	13.66	19.95**
14	Supplier relations – business performance	45.46	23.7	21.76**
15	Process management – business performance	29.91	24.9	5.01*

$*p < 0.01$; $**\ p < 0.001$

measure, which might be considered the R-square value, is the squared correlation between the manifest variable and its own related latent variable. It measures the capacity of the manifest variable to describe the related latent variable (Tenenhaus et al. 2005). A communality measure is expected to be higher than 0.60 for each manifest variable. In this application, the communality measures of all the manifest variables were found to be satisfactory, with most of them above the threshold value of 0.60.

For an endogenous latent variable, redundancy is the capacity of the model to predict its manifest variables from the indirectly connected latent variable (Tenenhaus et al. 2005). For such a complex model, the redundancy results are also satisfactory.

26.5.4 Inner Model Estimation

Once the outer weights of the latent variables have been identified, the path model or inner model is estimated by individual OLS multiple regressions (Fornell and Cha 1994; Chin 1998; Tenenhaus et al. 2005).

The full set of 26 variables comprising all six constructs loads significantly on their respective corresponding factors. In addition, all of the hypothesized paths are significant, as shown in Table 26.5. The standardized coefficients of these paths and the R-square values of each construct are shown in Fig. 26.2. Analysis of the PLS results indicates that a good deal of support has been found for all of the proposed

Table 26.4 Outer model estimation results

Latent variable	Manifest variable	Outer weight	Correlation	Communality	Redundancy
Customer focus	CF1	0.3634	0.8466	0.7167	
	CF2	0.3472	0.8531	0.7277	
	CF3	0.2719	0.7441	0.5537	
Top management involvement	TMI1	0.1329	0.7875	0.6202	0.1961
	TMI2	0.1320	0.7710	0.5945	0.1880
	TMI3	0.1833	0.8243	0.6794	0.2149
	TMI4	0.1499	0.7946	0.6313	0.1997
	TMI5	0.1516	0.8007	0.6411	0.2028
	TMI6	0.1530	0.8060	0.6496	0.2055
	TMI7	0.1565	0.7724	0.5966	0.1887
Employee relations	ER1	0.2064	0.7623	0.5812	0.3256
	ER2	0.2115	0.7750	0.6007	0.3366
	ER3	0.2427	0.8731	0.7624	0.4272
	ER4	0.2673	0.8573	0.7350	0.4118
Supplier relations	SR1	0.2386	0.7079	0.5011	0.1935
	SR2	0.2209	0.7624	0.5812	0.2244
	SR3	0.2346	0.8225	0.6766	0.2613
	SR4	0.3291	0.9084	0.8253	0.3187
Process management	PM1	0.2225	0.8542	0.7296	0.1469
	PM2	0.2716	0.8683	0.7540	0.1518
	PM3	0.2934	0.8593	0.7383	0.1487
Business performance	BP1	0.2642	0.8780	0.7708	0.2518
	BP2	0.1936	0.7637	0.5832	0.1905
	BP3	0.2177	0.8248	0.6803	0.2223
	BP4	0.1381	0.7108	0.5052	0.1651
	BP5	0.2186	0.8177	0.6687	0.2185

relationships in the model. The relationships between the TQM constructs, in addition to the relationship between process management and business performance, were all found to be positive and significant ($p<0.01$). The R-square values of the path model are satisfactory, ranging from 0.32 to 0.56, indicating that a satisfactory percentage of the variance in the dependent constructs is explained by the independent constructs. While Fig. 26.2 shows only the significant direct effects between the constructs, Table 26.4 provides the decomposition of these effects between the constructs.

26.6 Discussion and Conclusions

When using sophisticated SEM techniques such as CBSEM and PLS, researchers must be aware of their underlying assumptions and limitations. While most researchers have a good basic understanding of CBSEM-type models, their familiarity

Table 26.5 Inner (Path) model estimation results

Construct	Factor	Regression coefficient	Standard deviation	t-value
TMI	Intercept	1.5892		
$R^2 = 0.32$	CF	0.5624	0.0725	7.75**
ER	Intercept	0.7564		
$R^2 = 0.38$	TMI	0.4487	0.0784	5.72**
SR	Intercept	1.0912		
$R^2 = 0.39$	TMI	0.5143	0.0772	6.66**
	ER	0.1874	0.0772	2.43*
PM	Intercept	−0.2171		
$R^2 = 0.56$	TMI	0.3124	0.0760	4.11**
	ER	0.2784	0.0671	4.15**
	SR	0.3285	0.0748	4.39**
BP	Intercept	1.9390		
$R^2 = 0.33$	PM	0.5716	0.0720	7.94**

$*p < 0.05$; $**\ p < 0.01$

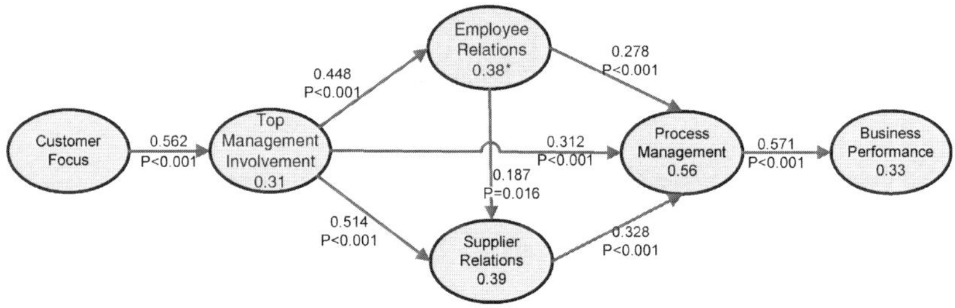

Fig. 26.2 Path model results

with PLS in the area of quality management slight. SEM models such as PLS can help TQM researchers achieve new insights. PLS requires a higher level of rigor and clarity than more traditional methodological approaches. Researchers in the area of quality management need to master this technique properly in order to better understand the complex relationships proposed in their models.

By applying PLS in the area of TQM research, this study has sought to investigate the relationships between TQM practices and to identify the direct and indirect effects of TQM practices on business performance. Consequently, special emphasis was placed on identifying the relationships between the most prominent TQM constructs and business performance based on a sample of SMEs operating in the Turkish textile industry. The analysis of PLS results indicated that a good deal of support has been found for the proposed model where a satisfactory percentage of the variance in the dependent constructs is explained by the independent constructs.

The findings show that TQM practices start with customer focus. There is a strong and positive relationship between customer focus and top management involvement. Under increasing competitive pressure, the purpose of companies is to retain their customers. Therefore, determining and meeting customer requirements are a necessary step to create a better business performance. Delivering quality to customers in the competitive marketplace emphasizes the need to continually enhance customer's satisfaction, which in turn leads many companies to adopt a more customer-oriented approach. Having a customer focus has now become a key concern for every company intent on increasing the value of its customer assets and boosting its business performance.

Top management involvement is necessary when the effectiveness of TQM implementation is investigated. Effective leadership by top management also indirectly affects firm performance through the mediating effects of process management. In fact, the success of TQM applications hinges on strong leadership that must be initiated by the top management. Quality improvement plans proposed by various quality gurus strongly emphasize the top management commitment. The top management of the organization is directly responsible for determining an appropriate organizational culture, vision, and quality policy. Top managers should also determine objectives, and develop specific and measurable goals to satisfy customer expectations and improve their organizations' performance. In order to enhance their business performance, managers must convey their priorities and expectations to their employees. In this study, management leadership has been found to have a direct and positive relationship with employee relations, supplier relations and process management.

Supplier relations are another important underlying dimension of TQM practices to improve business performance. Traditionally, vendors are selected from among many suppliers due to their ability to meet the quality requirements, delivery schedule, and the price offered. In this approach, suppliers compete aggressively with one another. The relationship between the buyer and the seller is usually adversarial. This traditional purchasing approach places special emphasis on the commercial transaction between the supplier and the customer. The main purchasing objective in this approach is to obtain the lowest possible price by creating strong competition between the suppliers, and negotiating with them. However, in the modern business world, many firms prefer the strategy of few suppliers. The few supplier strategy implies that a buyer wants to have a long-term relationship and the cooperation of a few dedicated suppliers. Using few suppliers can create value for the buyer and yield both lower transaction and production costs. The relationship between the buyer and the supplier includes specified work-flow, sharing information through electronic data interchange and the Internet, and joint planning and other mechanisms that allow a just in time (JIT) system and TQM in the company.

Based on the survey results, a strong and positive relationship between the top management involvement and employee relations has been noted. Building quality awareness among employees, recognition of employees for superior quality performance, employee-involvement-type programs, and feedback about their performance are very important to achieve successful employee relations. Firm must develop formal reward and recognition systems to encourage employee involvement, and support teamwork.

Process management, which includes such sub-factors as the availability of quality data, the extent to which quality data are used as a tool to manage quality, and the extent to which data and control charts are displayed in work areas, has been found to have a strong impact on business performance. This might be explained by the low level of personnel compliance with the implicit and explicit norms and rules of the workplace. Under such circumstances, the marginal contribution of the inputs used for process management (inspection, supervision etc.) purposes to the total quality would be high. This could explain the relatively high value of the process management-coefficient in the model.

The TQM approach places a great deal of emphasis on the maintenance of process control; in other words, it ensures that these processes do not only behave as expected, but also that the behavior of these processes does not create problems for the future. Thus, greater attention is paid to controlling the behavior of the processes that generate the products than to product conformity control. To achieve this objective, statistical instruments are used (e.g., the control sheet) in order to determine whether the machinery and the various production processes are under control. These instruments are weak when only used by quality control specialists, but they become extremely powerful when the whole staff learns how to use them and apply them to their own activities. Consequently, the production personnel receive timely and visible feedback on quality, i.e. information on the level of quality such as the percentage of defective items and the frequency of mechanical breakdowns. In particular, the process data gathered through quality controls are supplied in both a visible and timely way.

Another important conclusion and a managerial implication of this study is that SMEs should focus more on reducing variation in the production process to improve business performance. To improve process performance, top management involvement, supplier relations, and employee relations must be ensured. For SMEs in Turkey, customer focus or orientation is the most important quality practice for top management involvement.

APPENDIX

Top Management Involvement

1. Extent to which top executives assume responsibility for quality performance (current practice)
2. Acceptance of responsibility for quality by major department heads
3. Degree to which top management is evaluated for quality performance
4. Extent to which the top management has objectives for quality performance
5. Degree to which top management considers quality improvement as a way to increase profits
6. Extent to which top management has developed and communicated a vision
7. Quality is emphasized throughout the company by the senior management

Supplier Relations

1. Extent to which suppliers are selected based on quality rather than price or delivery schedule
2. Extent to which longer term relationships are offered to suppliers
3. Clarity of specifications provided to suppliers
4. Extent to which suppliers are selected based on quality rather than price or delivery schedule

Process Management

1. Availability of quality data
2. Extent to which quality data are used as tools to manage quality
3. Extent to which quality data, control charts are displayed in work areas

Employee Relations

1. Effectiveness of quality teams or employee involvement type program in company
2. Amount of feedback provided to the employees on their quality performance
3. Extent to which quality awareness building among employees is on-going
4. Extent to which employees are recognized for superior quality performance

Customer Focus

1. We do a lot in-house market research
2. We often talk with or survey those who can influence our customer's choices
3. We periodically review the likely effect of changes in our business environment

Business Performance

1. Investments in R&D aimed at new innovations
2. Capacity to develop a unique competitive profile
3. New product/service development
4. Market development
5. Market orientation

References

Ahire, S. L., Golhar, D. Y., & Waller, M. A. (1996). Development and validation of TQM implementation constructs. *Decision Sciences, 27*(1), 23–56.

Anderson, J. C., Rungtusanatham, M., & Schroeder, R. G. (1994). A Theory of quality management underlying the deming management method. *Academy of Management Review, 19*(3), 472–509.

Benson, G. P., Saraph, J. V., & Schroeder, R. G. (1991). The effects of organizational context on quality management: an empirical investigation. *Management Science, 37*(9), 1107–1124.

Black, S. E., & Porter, L. J. (1996). Identification of the critical factors of TQM. *Decision Sciences, 27*(1), 1–21.

Bollen, K. A. (1989). *Structural equations with latent variables*. New York, NY: Wiley.

Cassel C. M. (2000). Measuring customer satisfaction on a national level using a super-population approach. *Total Quality Management, 11*(7), 909–915.

Cassel C. M., Hackl, P., & Westlund, A. H. (1999). Robustness of partial least squares method for estimating latent variable quality structures. *Journal of Applied Statistics, 26*(4), 435–446.

Cassel C. M., Hackl, P., & Westlund, A. H. (2000). On measurement of intangible assets: a study of robustness of partial least squares. *Total Quality Management, 11*(7), 897–907.

Chin, W. W. (1998). The partial least squares approach for structural equation modeling. In G. A. Marcoulides (Ed.), *Modern methods for business research* (pp. 295–336). London: Lawrence Erlbaum Associates.

Demirbag, M., Tatoglu, E., Tekinkus, M., & Zaim, S. (2006). An analysis of the relationship between TQM implementation and organizational performance. *Journal of Manufacturing Technology Management, 17*(6), 829–847.

Flyyn, B. B., Schroder, R. G., & Sakakibara, S. (1994). A framework for quality management research and an associated measurement instrument. *Journal of Operations Management, 11*, 339–366.

Flyyn, B. B., Schroeder, R. G., & Sakakibara, S. (1995). The impact of quality management practices on performance and competitive advantage. *Decision Sciences, 26*(5), 659–691.

Fornell, C., & Cha, J. (1994). Partial least squares. In R. P. Bagozzi (Ed.), *Advanced methods in marketing research* (pp. 52–78). Cambridge: Basil Blackwell.

Fornell, C., & Bookstein, F. L. (1982). Two structural equation models: LISREL and PLS applied to consumer exit-voice theory. *Journal of Marketing Research, 19*(4), 440–452.

Fynes, B., & Voss, C. (2001). A path analytic model of quality practices, quality performance and business performance. *Production and Operations Management, 10*(4), 494–513.

Gadenne, D., & Sharma, B. (2002). An inter-industry comparison of quality management practices and performance. *Managing Service Quality, 12*(6), 394–404.

Gefen D., Straub W. D., & Boudreau M. C. (2000). Structural equation modeling and regression: guidelines for research practice. *Communications of the Association for Information Systems, 4*(7).

Hulland, J. (1999). Use of partial least squares (PLS) in strategic management research: a review of four recent studies. *Strategic Management Journal, 20*, 195–204.

Jöreskog, K. G. (1970). A general method for analysis of covariance structure. *Biometrika, 57*, 239–251.

Kanji, G. K. (1998). Measurement of business excellence. *Total Quality Management, 9*(7), 633–643.

Kanji, G. K., & Wallace, W. (2000). Business excellence through customer satisfaction. *Total Quality Management, 11*(7), 979–998.

Kaplan, R. S., & Norton, D. P. (1996). *The balanced scorecard: translating strategy into action*. Boston: Harvard Business School Press.

Kaynak, H. (2003). The relationship between total quality management practices and their effects on firm performance. *Journal of Operations Management, 21*, 405–435.

Lee, C. Y. (2004). Perception and development of total quality management in small manufacturers: an exploratory study in China. *Journal of Small Business Management, 42*(1), 102–115.

McAdam, R., & McKeown, M. (1999). Life after ISO 9000: an analysis of the impact of ISO 9000 and total quality management on small businesses in Northern Ireland. *Total Quality Management, 10*(2), 229–241.

Montes, F. J. L. M., Jover, A. V., & Fernandez, L. M. M. (2003). Factors affecting the relationship between total quality management and organizational performance. *International Journal of Quality and Reliability Management, 20*(2), 189–209.

Ridgon, E. E. (1998). Structural equations modeling. In G. A. Marcoulides (Ed.), *Modern methods for business research* (pp. 251–294). Mahwah, NJ: Lawrence Erlbaum Associates.

Samson, D., & Terziovski, M. (1998). The relationship between total quality management practices and operational performance. *Journal of Operations Management, 17*, 393–409.

Saraph, J. V., Benson, G. P., & Schroder, R. G. (1989). An instrument for measuring the critical factors of quality management. *Decision Sciences, 20*, 810–829.

Sun, H., & Cheng, T. K. (2002). Comparing reasons, practices and effects of ISO 9000 certification and TQM implementation in Norwegian SMEs and large firms. *International Small Business Journal, 2*(4), 421–440.

Tenenhaus, M., Vinzi, V. E., Chatelin, Y. M., & Lauro, C. (2005). PLS path modeling. *Computational Statistics & Data Analysis, 48*, 159–205.

Wilson, D. D., & Collier, D. A. (2000). An empirical investigation of the Malcolm Baldrige National Quality Award causal model. *Decision Sciences, 31*(2), 361–390.

Wold, H. (1975). Soft modeling by latent variables: the non-linear iterative partial least squares (NIPALS) approach. In J. Gani (Ed.), *Perspectives in probability and statistics: Papers, in Honour of M.S. Bartlett on the occasion of his 68th Birthday* (pp. 117–142). London: Applied Probability Trust, Academic.

Wold, H. (1982). Soft modeling: the basic design and some extensions. In K.G. Jöreskog, & H. Wold (Eds.), *Systems under indirect observation*, Part 2 (pp. 1–54). Amsterdam: North-Holland.

Wold, H. (1985). Partial least squares. In S. Kotz, N. L. Johnson (Eds.), *Encyclopedia of statistical sciences*, Vol. 6 (pp. 581–591). New York: Wiley.

Yusof, S. M., & Aspinwall, E. (2000). Total quality management implementation frameworks: comparison and review. *Total Quality Management, 1*(3), 281–294.

Chapter 27
Using PLS to Investigate Interaction Effects Between Higher Order Branding Constructs

Bradley Wilson

Abstract This chapter illustrates how PLS can be used when investigating causal models with moderators at a higher level of abstraction. This is accomplished with the presentation of a marketing example. This example specifically investigates the influence of brand personality on brand relationship quality with involvement being a moderator. The literature is reviewed on how to analyze moderational hypotheses with PLS. Considerable work is devoted to the process undertaken to analyze higher order structures. The results indicate that involvement does moderate the main effects relationship between brand personality and brand relationship quality. This chapter makes a unique contribution and applied researchers will appreciate the descriptive way it is written with regards to analytical process.

27.1 Chapter Overview

Many models in the social sciences have posited the existence of a moderating variable(s)[1] impacting relations between independent and dependent latent variables. Recent advances in methodology and available software have resulted in many new approaches for assessing the effect of moderators[2] within structural models (Cortina et al. 2001). This chapter builds on the theoretical contribution in this monograph of Henseler and Fassott (2010) by applying Partial Least Squares (PLS) (Chin et al. 1996, 2003) to investigating a marketing example with interaction terms. The use of PLS in modeling interactions allows more complex models to be investigated, lowers the sample size required and allows the researcher to flexibly deal with data that violates distributional assumptions. Given that interaction modeling

---

[1]A moderator is defined as, "a qualitative or quantitative variable that affects the direction and/or strength of the relation between an independent and dependent or criterion variable" (Baron and Kenny, 1986, p. 1174).

[2] For this chapter the term of moderation and interaction effects are used interchangeably.

B. Wilson
School of Media and Communication, RMIT University, 124 LaTrobe Street,
GPO Box 2476V, Melbourne, Victoria 3000, Australia
e-mail: brad.wilson@rmit.edu.au

V. Esposito Vinzi et al. (eds.), *Handbook of Partial Least Squares*, Springer Handbooks of Computational Statistics, DOI 10.1007/978-3-540-32827-8_28,

introduces distributional problems when creating cross product terms and that model complexity is naturally exacerbated when adding these interaction terms, PLS provides a flexible means for addressing these concerns. It is not the intention of this chapter to illustrate further the well known advantages of PLS and its' algorithm or utility for investigating moderating effects as these have been discussed previously by others and within other contributions in this monograph. The contribution of this work is that it illustrates the use of the Chin et al. (2003) technique to investigate interaction effects between three second order constructs. Specifically, this chapter investigates the role of category involvement (CIP) (Laurent and Kapferer 1985) moderating the main effects relationship between Brand Personality (BP) (Aaker 1997) and Brand Relationship Quality (BRQ) (Fournier 1994). This work builds on the call for research by Fournier (1994) indicating that the influence of Brand Personality on Brand Relationship Quality needs to be investigated.[3] The majority of studies testing interaction effects with PLS have been limited to other disciplines outside of marketing (Denham et al. 2003; Khalifa and Cheng 2002; Kwong and Lee 2002).

27.2 Introduction

The primary goal of this chapter is to illustrate how PLS can be used when investigating models with moderators at a higher level of abstraction. This is accomplished with the presentation of a marketing example. Firstly, there is a brief outline of the conceptual development of the theoretical model under investigation. The main hypotheses are presented. Secondly, the methodology is outlined. Thirdly, the majority of this chapter presentation is devoted to the process undertaken in obtaining the results. Before outlining the results, some problematic results initially obtained whilst analyzing the measurement models with Covariance-Based Structural Equation Modeling (CBSEM) methods are pragmatically discussed. Fourthly, PLS results are illustrated with the main effects model compared with the interactions model to assess the utility of the interaction effect. Finally, study limitations and suggestions for future research of a methodological and theoretical nature are outlined.

27.3 Literature Review and Model Development

It has been an interest and priority of academics and practitioners alike to investigate how the softer attributes of a brand's image (such as brand personality) influence and relate to brand loyalty (Aaker and Biel 1993). A brief explanation is given for the main constructs of interest and a model is subsequently developed. This

[3] The author would like to thank Professor Fournier for inspiring this research avenue. A majority of the work highlighted in the literature review stems from ideas developed in her seminal dissertation and subsequent journal publications.

work is unique in drawing together two relatively important contributions within the marketing literature.

27.3.1 Brand Personality

In order to give a richer picture into the conceptual groundings of the model to be investigated it was decided to selectively investigate the personal relationship, psychology and marketing literatures. Davis and Todd (1982, p. 93) defined a relationship as: "a particular state of affairs- one which conveys information about how two or more persons or objects are connected." This definition acknowledges the ability of people to form relations with objects. Marketers' have long been trying to embed their brands with personalities to encourage some degree of person–brand personality congruence (Sirgy 1982). They use a range of tools to create brand images and brand personalities. Aakers' (1995) seminal work on brand personality has spawned a revisitation of the whole concept. Brand personality as defined by Aaker (1997, p. 347) is "the set of human characteristics associated with a brand." For instance, the use of a celebrity endorser and/or animated characters may have a personality trait "rub off" effect into the brand (Callcott and Lee 1994). This may happen by association through an image transfer process (McCracken 1988). The country of origin, manufacture or ownership all contributes to the creation of brand personality (Thakor and Kohli 1996). Other "strategies used by advertisers to imbue a brand with personality traits include: anthropomorphization, personification, and the creation of user imagery (Aaker 1997, p. 347)." It could be argued that a firm's processes and how it is distributed impacts on its' brand personality. For instance, Dell computers may be considered more innovative and leading edge as opposed to its other competitors by the way it practises direct marketing. The product being often stylishly black in colour communicates elements of sophistication.

Aaker and Fournier (1995, p. 394) emphasize that, "personality, is used differently in the context of brands (consumer behavior) than in the context of persons (psychology). For example, while a person's personality is determined by multidimensional factors (e.g., appearance, traits and behavior), a brand, by its nature of being an inanimate object, has a personality that is determined by different factors (e.g., attributes, benefits, price, user imagery)."

Jennifer Aaker refined a Brand Personality Scale (BPS) initially refined from "The Big Five" of human personality in her dissertation (Aaker 1995). Her work based on both exploratory and confirmatory factor analyzes on large brand sets (37 brands and 20 brands) and large samples ($n = 637$ and $n = 180$) identified that brand personality was a second order reflective representation with five first order factors: Sincerity (Down-to Earth, Honest, Wholesome, Cheerful), Excitement (Daring, Spirited, Imaginative, Up-to-date), Competence (Reliable, Intelligent, Successful), Sophistication (Upper Class, Charming) and Ruggedness (Outdoorsy,

Tough). Under each construct there are facets and these are represented in the brackets.

The brand personality constructs are considered to be the independent constructs in this research. This study conceptualizes brand personality as a higher order construct. This is in keeping with Aaker (1995) whom established that brand personality was a reflective second order construct using CBSEM methods with a validation sample.

It is beyond the scope of this presentation to review all of the other research on brand personality. Numerous studies (Aaker et al. 2001; Bhat and Reddy 1998; Capara et al. 1998, 2001) have attempted cross-cultural replications, whilst other studies have focused on applying brand personality to different contexts: corporate personality (Bromley 2000); non-profit entity personality (Venables et al. 2003) and sport sponsorship (Deane et al. 2003). It is notable that they also treated brand personality with reflective measures.

27.3.2 Brand Relationship Quality

The second main domain of interest for this study was developed by Fournier (1994, 1995, 1998). Through the use of grounded theory methods and ethnographic techniques she qualitatively illustrated that people in fact do have relationships with brands.

Her qualitative analysis has also been very convincing in reinforcing the belief that brands are given animate qualities by their users. This provides further validity to the notion of people sustaining dyadic relationships with brands and that they project animate human-like qualities onto inanimate branded objects. Further, qualitative work by Andreou (1994), Hanby (1999), and Hess (1998) have argued that consumers are able to form active and reciprocal consumer–brand relationships, supporting the validity of the brand-relational metaphor. Understanding consumers and the relationships they form with brands provides knowledge about the enduring bonds that develop between a consumer and brand. Some of these relationship and brand loyalty affiliations are developed from childhood (Ji 2002).

At the conclusion of her ethnographic qualitative work, Fournier (1994) in her dissertation continues to develop an item battery to measure the quality of the person–brand bond. She termed this Brand Relationship Quality. "Brand relationship quality (BRQ) is best thought of as a customer-based indicator of the strength and depth of the person–brand relationship. It reflects the intensity and viability of the enduring association between a consumer and a brand (Fournier 1994, p. 124)." Fournier (1994) considers the multi-faceted measure of brand relationship quality to be "a refined articulation of the brand loyalty notion." David Aaker (1996, p. 167) reemphasizes this point by stating that, "the dimensions can be viewed as variants of brand loyalty." The notion of what BRQ is (and what it is not) is best expressed by a direct quotation from Fournier's (1994) seminal dissertation work:

"Several fundamental principles apply to the brand relationship quality construct which also, serve to differentiate it from existing marketing constructs (such as brand loyalty, satisfaction, etc):

(1) *BRQ is a property of the relationship between a person and a brand.* BRQ is not a characteristic of either the individual or the brand per se, but rather reflects an aspect of the intersection or joining of the two parties.

(2) *BRQ is dynamic;* it changes as a function of time in line with evolution in relationship partners and in response to specific behaviors enacted by them in the context of the relationship. Static measures of BRQ identify characteristics of the relationship at a given point in time. This research measures the person–brand relationship at one point in time. The results presented represent a cross-sectional measurement of the person–brand relationship. However, it must be acknowledged that this person–brand relationship is continually evolving and developing over time.

(3) *BRQ is defined as perceived by the individual in the relationship;* it is reflected in the thoughts, feelings, and behaviors exhibited by the person toward a particular brand and is not an objective characteristic of the brand relationship (as with statistical quality control measures of product performance, for example) (Fournier 1994, p. 125)."

Using confirmatory factor analysis techniques on a calibration ($n = 270$) and validation sample ($n = 209$), brand relationship quality was revealed to be a second order construct with seven reflective first order constructs. Subsequent work reported by Fournier (1998) revealed that the constructs were: brand partner quality, love and passion, intimacy, self-connection, nostalgic connection, interdependence and commitment. BRQ in this study is also treated as a reflective second order construct based on the validation work of Fournier (1994).

Until recently, researchers have not had the requisite theory and measures to adequately explore the contribution of softer, intangible, emotional drivers such as brand personality on brand loyalty. The work of Aaker and Fournier allows these links to be explored further. Their work in reinvigorating the areas of brand personality and placing a new perspective on the old notion of brand loyalty has created numerous new research opportunities.

Fournier (1998) has specifically called for research investigating the relations between brand personality and brand relationship quality. Others have echoed this sentiment, "It is unclear whether brand personality affects some Consumer Based Brand Equity facets (Netemeyer et al. 2004, p. 222)." David Aaker (1996, p. 165) makes a strong case for investigating specific relations between brand personality and brand relationships when he states, "brand behavior and imputed motivations, in addition to affecting brand personality, can also directly affect the brand–customer relationship." This study addresses this important call for research.

27.3.3 Product Class Involvement

Product class involvement is the third main construct of interest within this study. Product class involvement has been used in many marketing studies. Research on involvement has been prolific over a 30 year period in marketing. Numerous definitions and measures for involvement have been constructed. Involvement is often viewed as a "property of the relationship between a person and a product category, rather than a specific possession (Ball and Tasaki 1992, p. 159)." It is generally accepted that the level of involvement is associated with the level of perceived personal relevance or importance of a specific product category to the customer (Zaichkowsky 1985). Involvement has both intrinsic (enduring) and extrinsic (situational) elements (Richins and Bloch 1986). Enduring involvement pertains to the accumulation of knowledge in long term memory compared with situational involvement which is much more temporal and influenced by the purchase situation (Richins and Bloch 1986). The Consumer Involvement Profile (CIP) encompasses some aspects of both enduring involvement and situational involvement. The CIP was the scale utilized in this work to measure involvement (Laurent and Kapferer 1985; Kapferer and Laurent 1986).

The original CIP (Laurent and Kapferer 1985) included 19 items (four constructs, not five constructs). Initial investigations using three samples with data collected via in-home interviewing and analyzed using reliability and exploratory factor analysis found that perceived risk/importance and probability of mispurchase were not distinct constructs.

Discriminant validity was adequately demonstrated with low between construct intercorrelations. This was deemed satisfactory. The interest construct was not investigated in the initial 1985 study and was added after further research. The four constructs in the Laurent and Kapferer (1985) article for the 14 product categories under investigation were presented as averages out of 100.

Further studies (Kapferer and Laurent 1985; Kapferer and Laurent 1986) refined the CIP by including the interest construct. This new structure was examined for validity and reliability with a sample of 1,568 including some 20 product categories. Nomological validity was supported by investigation of relationships with several dependent measures such as: level of extensive decision making, brand commitment, and reading articles (Bearden et al. 1993).

The final CIP was developed to be a multidimensional measure of involvement. This was in keeping with previous involvement studies that conceptualized involvement as being multidimensional in nature (Arora 1982). The final CIP is a collection of 16 items that measure five first order constructs namely: product risk/importance, symbolic value, hedonic value, probability of mispurchase, and enduring interest. Some authors believe that the final five CIP constructs represent antecedents of involvement (Day et al. 1995; Zaichkowsky 1994). Many scaling methods have been used with the CIP. Jain and Srinivasan (1990) transformed the original Likert scale into a semantic differential format. However, the 5-point Likert version of the scale is implemented in this study. The final CIP is a collection of 16 items that measure five first order reflective constructs. This involvement measure was chosen as it provides a richer description tapping the full involvement domain. It is also a second order representation like the independent and dependent constructs within the structural model. Therefore, the analysis is undertaken at the same level of abstraction.

27.3.4 Hypotheses to be Tested

The need for research for the main effects model of brand personality and brand relationship quality has previously been outlined. The next major consideration is establishing the role of product class involvement in this relationship. It has been shown that consumers are more likely to relate positively to relationship marketing tactics when consumers are involved in a product category (Gordon and van der Sprong 1998). Involvement has been a moderator in numerous marketing studies (Homburg and Giering 2001; Low and Mohr 2001; Suh and Yi 2006) and is also treated as a moderator within this study. Product class involvement was also treated as a moderator when investigating BRQ (Fournier 1994). A conceptualized structural model demonstrating the moderating role of involvement is represented diagrammatically in Fig. 27.1.

A review of the literature has resulted in the following hypotheses:

H_1: *There is a positive relationship between brand personality and overall brand relationship quality.*

H_2: *The relationship between brand personality and overall brand relationship quality is moderated by the level of consumer product class involvement: that is, the relationship is weaker under conditions of low product class involvement and stronger under conditions of high product class involvement.*

In Fig. 27.1, the dependent variable (Y) would be brand relationship quality with the predictor (X) and moderator (Z) variables being brand personality dimensions and product class involvement, respectively. The necessity to investigate the relationships at the higher level of abstraction is to ascertain the respective contribution at the global or macro level. This is deemed appropriate so as to remain consistent with past contributions and theory derivations. It maintains a level of continuity with each original individual conceptualization as second order representations. It is also worthy to address the numerous calls for research. Previously, such issues could not be explored due to the available methods. This represents a first contribution and exploration using PLS.

The methodology section is presented next.

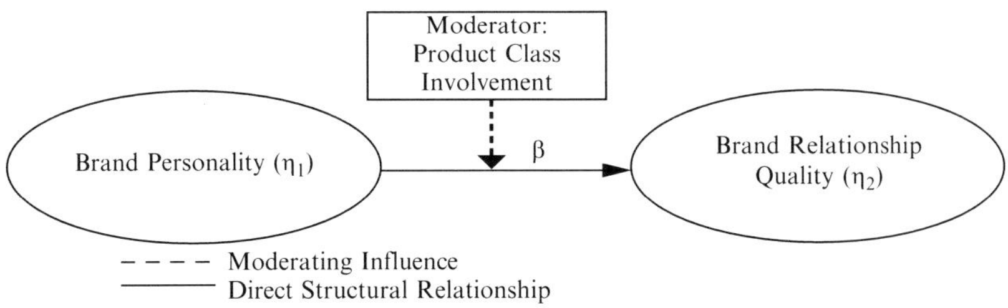

Fig. 27.1 Structural model to be tested

27.4 Methodology

27.4.1 The Sample and Data Collection

The data was collected via a national random mail self-completion questionnaire using Marketing Pro (a national white pages directory with addresses) as the sample frame. Marketing Pro is a CD ROM directory consolidating some seven million residential listings, Australia wide. The product categories were selected after four pretests were completed involving expert opinion, two studies of undergraduate student product class mentions and another study analyzing awareness and equity scores from the Australia Scan national survey (Callaghan and Wilson 1998). The final categories (and brands) chosen also considered such issues as: the product class having national distribution, product class familiarity, and whether the final product classes selected would provide a mix of different involvement levels. The selection was also mindful of previous brands studied (Fournier 1994).

27.4.2 The Brands Studied

The product classes (and brands) studied included: Cola Soft Drinks (Coca-Cola and Pepsi), Film (Kodak and Fuji), Airlines (Qantas and Ansett), Credit Cards (Mastercard and Visa), Cars (Ford and Holden), and Athletics Shoes (Nike and Reebok). Respondents filled in the questionnaire for two brands, thus violating the assumption of independence between observations that techniques like CBSEM require. The brands are all large and familiar brands with extensive product ranges. The brands in this investigation have received years of advertising and marketing support. The brands are all well established. Therefore, it was believed, they would have an established brand personality.

It must be noted that each brand personality, given the scope and range of the brands studied, would be multi-faceted. That is, through years of advertising campaigns and innovations the brands' chosen have come to stand for multiple brand personality traits for its' respective target markets. For instance, if we were to consider a brand such as Levi's which has product variants directed at very diverse demographics and various product styles and distribution points, it is easy to consider the brand possessing many personalities. Others have echoed that underlying the brand personality may be notions of masculinity, ruggedness, rebelliousness and individualism, however, "one execution cannot necessarily reflect every brand value (Fuller 1995)." In fact, each separate communication may only concentrate on communicating and enhancing a limited range of brand personality traits at a time. Only a number of limited messages can be communicated within a 15 or 30 second television execution. This idea would have to be revised in the face of a targeted marketing campaign containing a series of executions. For example, the Engineered Levi's range is very distinct from the range of Levi's 501's directed at those over 40 years of age. Therefore, it is the collection of traits delivered across multiple media executions to various target audiences and segments focusing on many product ranges and distributed accordingly that represents the brands' holistic personality. This is how large brands develop multiple personalities and this is considered to represent brand personality strength in this study. This is captured in the higher level construct developed by Aaker (1995) when modeled as a second order construct. Niche brand personalities are not represented in this study as only nationally distributed familiar brands with adequate penetration were sought. This brand selection was necessary to facilitate respondent recruitment and participation.

27.4.3 The Measures

There were three main item batteries (Brand Personality Scale: 47 items, Brand Relationship Quality Scale: 62 items, Consumer Involvement Profile: 16 items) used in this study.[4]

The Brand Personality Scale of Aaker (1995, 1997) was adapted slightly. In deciding to add items, a panel of experts consisting of three marketing academics reviewed the items to determine their relevance to the Australian culture. The panel was briefed on each construct representing brand personality and discussed the nuance of each trait descriptor with reference to its' suitability to the Australian culture. Two items were added to the scale as it was deemed that Australian respondents would not take out the same meaning from items such as: western and small town. There was also consensus that the sophistication trait descriptors were not clear enough. Based on this two items were added to the established 45 items scale. The items added were: sophisticated and outback. The Brand Personality Scale was measured on a 5-point modified semantic differential scale (not at all descriptive – extremely descriptive).

The same panel of experts was also used to assess the potential for item misinterpretation with the other scales. There were no underlying concerns with the CIP item battery. A 5-point Likert scale was considered appropriate in keeping with the previous research.

The Brand Relationship Quality Scale implemented within this study was slightly different to the original scale developed within Fournier's (1994) dissertation work. Fournier supplied an extended version that was being subjected to further scale validation in ongoing research.[5] This version of the BRQ scale was conceptually

[4] There are a total of 125 possible items (ignoring item deletion) in the main effects model. This in itself is a complex model. Conventional sample size rules of a minimum of at least five observations per item is often recommended (e.g., Tabachnick and Fidell (1996)), and a ratio of ten or greater is preferred. This would make the required sample size for CBSEM to be very large. When the interactions terms are added this model (and required sample size) becomes more complex again. When such data is non-normal and necessitates ADF estimation the required sample size becomes unpractically large.

[5] The author would like to acknowledge and thank Professor Fournier for her initial support, inspiration and for supplying the most up to date BRQ scale for investigation.

discussed in Fournier (1998) and is used in this work. There were no significant changes between the two versions. Some constructs had benefited from the rewording of items and the introduction of a few new items. The items used a 7-point scale which was increased from the 5-point scale in Fourniers' (1994) original BRQ scale to allow greater discrimination. The scale was a modified semantic differential (Does not describe my feelings toward the brand at all – Does describe my feeling toward the brand very well).

All measures in this study were treated as being reflective in keeping with the initial mode they were specified. Fornell and Bookstein (1982, p. 292) believe that "constructs such as 'personality' or 'attitude' are typically viewed as underlying factors that give rise to something that is observed. Their indicators tend to be realized, then as reflective." Although this statement is referring to human personality its' applicability can be transferred to brand personality measurement. Similarly, items within the brand relationship quality and involvement construct are all "attitudinal-style" items. The panel of experts were in agreement with the items being reflective when briefed on the individual items representing constructs. The work of Jarvis et al. (2003) was consulted post hoc to confirm whether reflective or formative operationalizations should be applied and all agreed that the constructs should follow their originally developed conceptualizations. Bollen and Ting (2000) would suggest the implementation of Confirmatory Tetrad analysis as a quantitative test that is more data driven. This test was not used due to the strong support above.

27.4.4 Profile of Respondents

Data was collected from around Australia. A final sample size of 1,290 was obtained. The final response rate was 25.8%. A lottery (similar to Aaker 1995) and a small incentive (movie ticket) was utilized to encourage response. The questionnaire was mailed out with two reminder letters. Reminder letters were sent out when the responses received had reached a plateau and were starting to decline.

An analysis of the sample characteristics indicates that the sample is representative of the Australian population. A distribution of the age of respondents revealed that: 22% were aged 15–34, 51% were aged 35–54 and 27% were aged 55–75. This is similar to the age distribution within the general populace. The gender split was 53% males and 47% females. Again this mirrored roughly the breakdown within Australian society. Around half of the sample indicated having at least a high school education, with a further quarter having undergraduate and another quarter having completed postgraduate study.

People indicated they had a high level of knowledge and familiarity of the brands for which they were responding to. Around 43% of total respondents had purchased the brands under study in the past year and 68% had at least bought and used the brand at some time in their life. The sample was deemed adequate for further analysis.

The next section describes the main choices made before analyzing the data. This will assist the interested reader in understanding the process behind each stage in the analysis.

27.4.5 PLS Interactions Approaches

There are two main approaches that can be utilized when using PLS to investigate interactions. The reader should consult the contribution of Henseler and Fassott (2010) in this monograph for complete details for the two approaches. They are briefly outlined below.

The first approach deals with (psuedo-) continuous interaction terms [(eg., Numerical scales, Likert scales, etc) (Chin et al. 2003)]. In this approach each item representing the independent construct (X) is multiplied with each item representing the moderating construct (Z) to create interaction terms (X.Z). PLS is capable of explaining complex relationships (Fornell and Bookstein 1982). This is important with continuous variable PLS interactions modeling as the number of indicators for the interaction construct is the multiple of the number of indicators for the predictor and moderator constructs (If the independent (X) construct is measured by 8 indicator variables and the moderator (Z) has 8 then $Z * X = 64$ interaction variables would be introduced). A large number of interaction items result.

Figure 27.2 presents a graphical model of how you would set up your analyzes in the available PLS software. This model is a replication of the one presented in Chin et al. (1996). It must be remembered that prior to creating all the interactions terms that all predictive and moderator variables be mean centered or standardized (Chin et al. 1996; Low and Mohr 2001; Ping 1996a, b). This helps minimize multicollinearity that develops when creating the product terms. The main effects model is the specified relationship between the independent (X) and dependent constructs (Y). The interaction model features the introduction of the moderator and interaction terms (Z and X.Z) into the model.

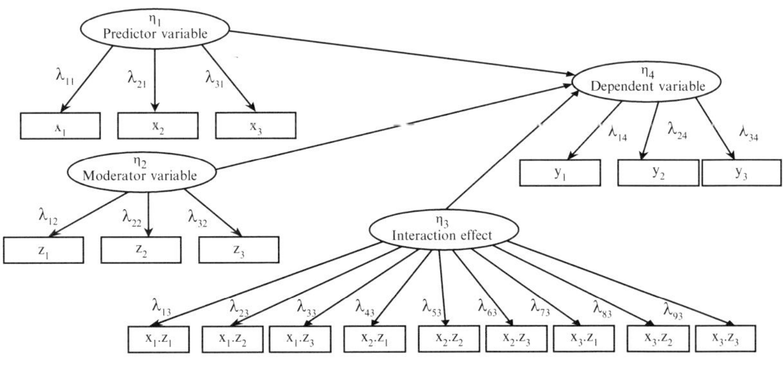

Fig. 27.2 PLS model with interactions effect. *Source*: Chin et al. (2003, p. 198)

The second approach involves implementing a multiple group PLS model. This approach uses dichotomous variables (like occupation, etc) or creates two or more groups artificially from continuous variables (Chin 2002). Researchers are often interested in discrete variables like gender or occupation moderating relations between constructs of interest (Dick and Sauer 1992).[6] The effect of gender as a moderator has also been explored in email usage behavior (Gefen and Straub 1997). Gender could also be worthy of investigation when investigating relations between brand personality and brand relationship quality dimensions. It could be argued that females have "a people-centered approach" (Rigg and Sparrow 1994, p. 9). It is believed that females are more nurturing and caring. Fournier (1998) in her qualitative ethnographic study on brand relationships purposively chose this group. Females may be better able to accept the notion of brands having human-like personality traits and better affiliate with Fournier's "brand as a relationship partner" notion. Although not explicitly demonstrated within this paper the multiple group PLS model could also have applicability to the research domain. Past research by Chin et al. (1996, 2003) would advise against the use of multiple group models when researchers have continuous moderator variables at their disposal as it could result in inadequate power to detect the moderator/interaction effect. It is this author's contention that the prevalent use of the two group method in CBSEM may be a function of the extensive level of expertise required to implement some of the continuous interaction modeling approaches (see Cortina et al. 2001 for a review). Implementation of CBSEM when dealing with continuous interactions is very specialized and often beyond the level of competence for all but the most advanced covariance modeling users.

Both PLS interactions methods have received scant application in the marketing literature. The multiple group approach has been applied in marketing within a retailing context investigating how different retail store formats influence purchase intentions (Grace and O'Cass 2005). The author is not aware of any further applications within the marketing domain. Both PLS interaction modeling approaches have numerous operational advantages. When using PLS, "models consisting of over 200 indicators can be easily executed.... LISREL, conversely, will tend to reject the model (based on covariance fit). This rejection occurs, in part, because the model needs to account for more covariances. As the number of indicators increases and as sample size increases, the power to detect even minor model misspecifications increase (Chin 1998, p. 332)." The use of PLS is often mandatory in such situations. When all of the interaction terms are included within the model it becomes very complex which further justifies the use of PLS. For the current study, both approaches could have been implemented. This study only implements the continuous interaction term approach.

Furthermore, it must be remembered that the relative newness of this research area makes PLS applicable. The theoretical model is not well formed and represents

[6] The intention of this chapter is not to discuss moderation approaches that utilize the two group approach through dichotomous (sex) or items that have been artificially dichotomised (eg., median/mean splits etc).

a first attempt at consolidating both the brand personality and brand relationship quality theories (with small to moderate sample sizes). Barclay et al. (1995) suggest that PLS is suitable for "research models where the emphasis may be more on theory development." In fact, the presumption that the application of CBSEM to the above mentioned theories at such an early stage of theoretical development may have been premature. This is the first time the main effects relationship has been investigated within Australia.

The data was analyzed with PRELIS, PLS-Graph 3.0 and SPAD 6.0. Zumastat 3.1 (an SPSS add-in) was used to create the interaction terms. It must be acknowledged that the author initially utilized CBSEM (Lisrel 8.50) to analyze the involvement construct measurement models (see the results Sect. 27.5.3 for a cursory discussion).

The next section outlines the results.

27.5 Results

27.5.1 *Preliminary Data Analysis*

The preliminary data analysis involved missing value analysis and descriptives analysis. Firstly, missing data patterns were visually inspected. There appeared to be no one item suffering from extensive missingness. A test to determine the randomness of missing data was conducted (Hair et al. 1995). The data set was recoded with missing values being coded zero and non-missing values being coded one. A correlation matrix was then run with low correlations indicating a low association between the missing data process for pairs of variables. All correlations were suitably low to suggest the missing data imputation could be considered appropriate.

The choice of missing data imputation was carefully considered. A process of EM imputation was undertaken due to many reasons. Given that this research involves interaction modeling with higher order constructs, replacing with the EM estimated value is believed to have a minor effect on variables undergoing further analysis in the structural model. Mean replacement was also considered as mean centering is recommended when dealing with interaction terms to avoid multicollinearity between interaction terms (X.Z) however, given that the interaction terms are created from derived standardized factor scores (after the hierarchical components measurement models are estimated, see Sect. 27.5.4) this was not considered a problem (Aiken and West 1991).

Variable distributions were then inspected and statistics calculated to test normality. Statistics indicated that the normality assumption is violated. There was a positive skew and leptokurtic distribution to the data (Byrne and Campbell 1999). This is not uncommon with social science data. For the sake of conserving presentation space, descriptive statistics at the item level, such as: the mean, standard deviation, skewness, and kurtosis figures are omitted from this chapter. However, the non-normality of the data provides further support for utilizing PLS.

27.5.2 Common Method Bias

Common method bias can arise when using similar scales with the same number of response options. A similar source can introduce spurious relationships among the variables. Common method bias could be exacerbated as higher order constructs for the main measurement models are represented by components measured in a similar format. Each question is obviously different as are the constructs and all measurement models. A factor analysis (ex post one-factor test) was run to demonstrate that there is no common factor loading on all measures. This is the same as Harmon's one factor test. The results revealed that there was no common factor loading on all measures (Podsakoff and Organ 1986). Therefore common method bias was considered not to be a problem with this dataset.

27.5.3 Initial Results Obtained with Covariance-Based Methods

It was the authors' original intention to investigate the full structural model with CBSEM methods. All theoretical constructs had previously undergone what was believed to be quite solid psychometric testing within their respective countries of development. There was an adequate level of cross-validation which had also been completed. Limited work had been done within an Australian context. Due to the level of theoretical development it was initially deemed prudent to apply CBSEM methods to the data. The three major domains had also utilized CBSEM methods in their development or subsequent validation studies. The involvement measurement model was the first theory exposed to CBSEM modeling. This was because it was the simplest in structure, containing the smallest number of items and constructs.

Given that the data was non-normal a suitable CBSEM estimator was selected. One of the key assumptions of maximum likelihood estimation in CBSEM is that the variables in the model need to be multivariate normal (Cortina et al. 2001). Some authors suggest that maximum likelihood estimation is relatively robust against violations of normality (Boomsma 1983; Gerbing and Anderson 1985) whilst others believe asymptotic distribution-free estimation (ADF, WLS) (Browne 1984) should be implemented. Using ADF estimation is much more computationally intensive requiring larger sample sizes. In this case, the sample size is large enough under conventional rules (Holmes-Smith and Rowe 1994; Steenkamp and van Trijp 1991).[7] The LISREL analyzes were run using both ML and ADF estimators. The polychoric correlation matrix (ML and ADF) with asymptotic covariance matrix (ADF only) was used as the data input, as is typical when using these estimators (Rigdon and Ferguson 1991). The involvement measurement model was run as a Single second order factor model (five uncorrelated first order factors reflecting one

[7] Minimum required sample size for use of ADF estimation [$1.5q(q + 1)$ if $q > 12$], where q is the number of items. So if q = 16, [$1.5 \times 16(17) = 408$ required sample size.

second order involvement factor) and also as a Saturated model (five correlated first order factors). An inspection of all of the results revealed the presence of negative unique error variances (Heywood Cases) in all solutions. Despite all other fit statistics being in acceptable ranges (Hoyle 1995) the four solutions could not be utilized further without purification. These results are not surprising as CBSEM models are often affected by many factors such as: Heywood cases, an inability to converge to a solution, parameters that are outside reasonable limits, large standard errors of parameter estimates, and large correlations among parameter estimates (Rindskopf 1984). There were two offending items out of the 16 items producing negative error estimates. Rindskopf (1984, p. 118) states that, "negative error variance estimates are often the result of an attempt to compensate for large factor loadings." The results revealed this trend with some loadings in the 0.90 range. Negative unique variance estimates are frequently encountered (Jöreskog 1967) in CBSEM. "It is well-known . . . , one third of the data yield one or more nonpositive estimates of the unique variances (Lee 1980, p. 313)" as modified from (Dillon et al. 1987, p. 127). To help rectify these problems, the strategies of Dillon et al. (1987) were followed such as constraining the error variances to zero, a small positive value, and model reparameterization through item deletion. All strategies did not solve the problem satisfactorily (often resulting in non convergence). Other fixes such as using the generalized least-squares estimator was also implemented with the same results. Model respecification was attempted via merging factors together into a single involvement construct and a four-factor representation, however, similar results were obtained.

Fornell and Bookstein (1982, p. 444) believe that poor LISREL estimates "suggest several possibilities: (1) the theory is wrong, (2) the data are inaccurate, (3) the sample size is too small, or (4) covariance structure analysis is not appropriate for this analysis task." Previous replication studies would suggest that (1) is unlikely (Laurent and Kapferer 1985). It is a possibility that the data is inaccurate (2) and technically it should be tested on split half samples or with a validation sample. The sample size (3) analyzed was deemed adequate by conventional standards. It is believed that that CBSEM was not satisfactory in this case due to data distribution problems. PLS was chosen to overcome the problems experienced above and that it reflects the exploratory nature of this investigation being an investigation of higher order interactions.

It is heartening that other academics have experienced similar problems, although, the reporting of such problems is relatively scarce within the literature.[8] In a study on exit-voice theory Fornell and Bookstein (1982) firstly utilize CBSEM methods and after coming to improper solutions (negative error variances and standardized loadings greater than 1), they choose to finish their analysis with PLS where the method subsequently converges to a solution. With a PLS solution they then highlight their structural model results. More recently, whilst studying mobile phone data for the European Customer Satisfaction Index (ECSI)

[8] It is the author's contention that these CBSEM issues are not as readily reported due to it drawing a negative connotation to the research in general and editorial reluctance to accept articles with negative results.

Tenenhaus et al. (2005) analyze a reduced form of the full ECSI model (with less constructs included) to compare LISREL and PLS estimates due to LISREL non-convergence of the full ECSI model.[9] It appears that CBSEM methods may have limitations when researchers are investigating complex models with sample size constraints (Chin and Newsted 1999). PLS practically always converges (Wold 1981). PLS is also robust against deviations from the normal distribution (Cassel et al. 1999) and does not rest on the assumption of observation independence[10] (Falk and Miller 1992). PLS also circumvents "inadmissible solutions in the form of negative variances and out-of-range covariances common in CBSEM" (Chin and Newsted 1999, p. 309). PLS was deemed most appropriate for the analysis.

27.5.4 Data Analysis Strategy

The modeling strategy employed is partially related to the two-step approach advocated by Anderson and Gerbing (1988). The measurement models are assessed for adequate validity and unidimensionality prior to commencing the structural main effects and interactions modeling.[11] For this study, the two-step approach involves: (1) a detailed assessment of the measurement models at the item level and higher-order level, and (2) includes an analysis of the posited structural relationships. Reliability and validity was verified at each stage.

The internal consistency of the measures, i.e., their unidimensionality and reliability, were the first properties to be assessed. The indicators used to measure a construct (or latent variable) must be unidimensional. Convergent validity for the measures was assessed by running a separate factor analysis for each construct under

[9] It must be noted that author's of such articles evidencing non-convergence problems do not outline the specific causes for non-convergence. This may be due to the numerous possible causes of such problems including: "(1) sampling fluctuations, (2) model misspecification to the extent that no factor analysis model will fit the data, and (3) "indefiniteness" (underidentification) of the model, (4) empirical underidentification (Rindskopf, 1984) and (5) outliers/influential cases. (as modified from Chen et al. (2001, p. 470))."

[10] Marketers often carry out attitudinal studies and run the same item battery sometimes many times for the same individual. Data is often then stacked for analysis. For example, each respondent rates five brands on the same scale (Aaker, 1995). This violates the independence of case assumption in CBSEM. However, Aaker may have minimized such problems by focusing on the brand as the unit of analysis (not case).

[11] In principle, the goal is the same to establish adequate construct validity and unidimensionality, however, the process is slightly different when undertaking a two-step approach with PLS. With covariance-based methods the goal is to create adequate congeneric measurement models with the aim of reducing the number of indicators (item purification) and then to create composite single indicators proportionately weighted by each item factor score. The second stage in covariance based methods involves evaluating the structural model taking these composite single indicators and fixing paths and error variances by formula (to allow model identification). The researcher then can estimate the coefficients for the posited structural relations and discuss in relation to hypotheses.

investigation. This follows the procedure followed by many researchers utilizing PLS in recent times (Bontis 1998; Grace and O'Cass 2003, 2005). The analysis is undertaken to confirm that one dimension represents each reflective construct. This determines if each construct can be regarded as unitary.

As the research involves exploring relationships at a higher level of abstraction each second order measurement model (three measurement models; brand personality, involvement, and brand relationship quality) are then estimated separately using the repeated indicators approach, also known as the hierarchical components model suggested by Wold (Lohmöller 1989, p. 130–133; Chin et al. 2003). "In essence, a second order factor is directly measured by observed variables for all the first order factors. While this approach repeats the number of manifest variables used, the model can be estimated by the standard PLS algorithm (Reinartz et al. 2003, p. 19)." Standardized latent scores (representing the first order constructs) are saved during this stage of the analysis. The standardized scores are automatically computed in the PLS analysis. These scores are copied into the PLS data file for further analysis. These scores subsequently become the observed variables representing the first order constructs in the structural model. Factor scores are frequently estimated and used as input for further statistical calculations (Field 2005; Hair et al. 1995). Other researchers have used the PLS repeated indicators approach and utilized latent construct scores in further analyzes within models in recent times (Reinartz et al. 2004; Venaik 1999; Venaik et al. 2001, 2005; Zhang et al. 2006). The hierarchical components model is diagrammatically represented in Fig. 27.3 below. The abovementioned authors' did not implement interactions modeling at the second order level of abstraction.

It is notable that Ping's (1995, 1996a, b) CBSEM interaction modeling method has simplified the creation of interaction terms through the use of composites. Recently, his work has explored (Ping 2005) the implementation of interactions modeling at the second order level. He follows his standard approach of creating composite constructs but does this at the second order level using alternative specifications to represent the higher order latent construct (second order latent variable, summed indicator first order latent variable, and factor scored first order latent variable). His work shows there is little difference in overall results between the composite methods. The Jöreskog and Yang (1996) CBSEM method also uses derived factor scores to simplify the process of interactions analysis (Yang Jonsson 1998).

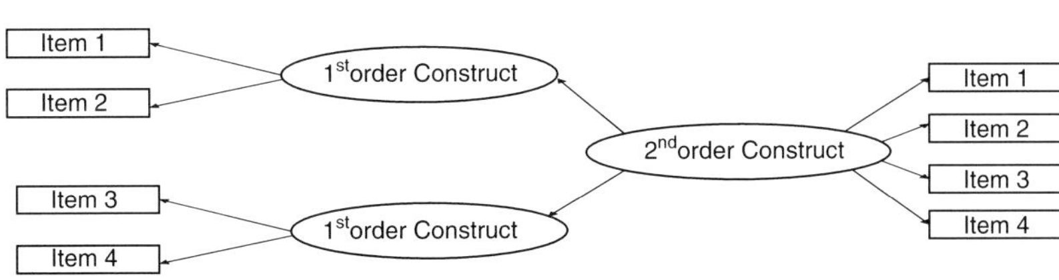

Fig. 27.3 Conceptual representation of hierarchical components model

Convergent and discriminant validity tests are in essence repeated twice. The first time, is at the item-level when the measurement models for brand personality, involvement and brand relationship quality are separately investigated. Secondly, the process is undertaken again at a higher level of abstraction when construct scores representing first order constructs are substituted into the model. This representation then allows relationships between higher level constructs to be investigated.

Because the derived construct (factor) scores are standardized by nature this helps avoid computational errors by lowering the correlations between the interaction terms and their individual components (multicollinearity) (Chin et al. 2003). Interaction terms for all X and Z variables were then created using ZumaStat. Finally, the main effects and moderating relationships (interactions model) are estimated.

Aside from the usual statistics [R^2, Average Variance Accounted For (AVA)], there are some useful statistics to help assist the researcher assess the contribution the introduction of interaction terms has above and beyond the main effects relationship. "In formulating and testing for interaction effects using PLS, one needs to follow a hierarchical process similar to multiple regression where you compare the results of two models (i.e., one with and one without the interaction construct) (Limayem et al. 2001, p. 282)." The main effects and interaction models (with all cross-products variables) are subsequently modeled with the effect size (f^2) of the main effects and interactions model being assessed by the Cohen effect size formula (Cohen 1988):

$$f^2 = \frac{R^2_{\text{model with moderator}} - R^2_{\text{model without moderator}}}{1 - R^2_{\text{model with moderator}}} \tag{27.1}$$

The difference in R-squares between the main effects model and interaction model is calculated to assess the overall effect size f^2 for the interaction where 0.02, 0.15, and 0.35 has been suggested as small, moderate, and large effects respectively (Cohen 1988). The effect size and significance of interaction terms determines the utility of the interaction model over the main effects model. The critical ratios to determine structural parameter significance were estimated via bootstrapping (Efron and Tibshirani 1993). The number of samples in the bootstrap procedure was set to 500 exceeding the recommendation of 200 by Chin (1998, p. 320). The following section outlines results for the measurement models.

27.5.5 Measurement Model Results

In the first step, exploratory principal components analysis (PCA) and reliability analysis (RA) to assess the validity of the model measures for each construct are completed. In the second step, the measurement model estimated in step one is used for simultaneously estimating three separate hierarchical measurement models. Each construct was explored via PCA with varimax rotation. Others have used this approach (Bontis 1998; Grace and O'Cass 2003). PCA is used extensively within consumer research (Baumgartner and Homburg 1996). The initial analysis indicated that the items of each construct were loading appropriately. Falk and Miller (1992) suggest that loadings of indicators on constructs need to be greater than 0.55. They believe this level is adequate to establish item reliability. Chin (1998, p. 325) believes, "loadings of 0.5 and 0.6 may still be acceptable if there exists other indicators in the block for comparison." The Chin (1998) recommendation is adhered to here as each construct has multiple measures. Most of the loadings (item reliability) exceeded the more stringent cut-off threshold (0.707) which implies that more than 50% (0.707^2) of the variance in the observed variable is shared with the construct (Barclay et al. 1995). The remaining loadings satisfied the Chin (1998) requirement of being greater than 0.6. Three items were eliminated in this process. Correlations between the construct where also inspected and illustrated that items correlated more highly with like items. The item-level principal components results and correlation matrix is not presented so as to preserve space.

Table 27.1 presents key statistics such as: Cronbach's Alpha (Cronbach 1951), Composite Reliability [often referred to as Internal Consistency (IC) statistic or Dillon–Goldstein statistics] (Werts et al. 1974) and the Average Variance Extracted (AVE) (Fornell and Larcker 1981) for each construct.

Cronbach's Alpha is only reported as a matter of convention and should be not be given much credence as it is the lower bound estimate of reliability (Raykov 2001). The composite reliability statistic is considered to be a better indicator of the unidimensionality of a block than the Cronbach's alpha (Chin 1998, p. 320). All composite reliabilities were high ranging between 0.8029 and 0.9422. These reliabilities provide evidence of unidimensionality and illustrate that the constructs are suitable for further analysis (Hattie 1985). The calculated values are all above conventional cut offs for reliability > 0.70 (Nunnally and Bernstein 1994). The AVE illustrates the amount of variance the items share with the construct it purports to measure (Fornell and Larcker 1981). It is important that the items share more variance with its' measures than with other constructs in a given model. This is the case with AVE's ranging between 0.5249 and 0.8116. The results demonstrate adequate convergent validity and unidimensionality. Convergent validity was therefore satisfied. The hierarchical measurement models could now be estimated.

There were three separate measurement models (one each for brand personality, consumer involvement profile, brand relationship quality) estimated using the hierarchical components method. This tests whether the first order constructs loaded onto their posited second order constructs.

All loadings and path coefficients between the first order and second order constructs were inspected and significance was assessed via 500 bootstrapped iterations. Standardized factor scores (latent variable scores in this case as they come to represent the construct in the structural modeling later) were saved during this stage of the analysis. All loadings were again above 0.6 as recommended by Chin (1998). Having computed the latent variable scores an assessment of discriminant validity was initiated. Discriminant validity was satisfied with all correlations between composite constructs (latent variable scores) being lower than their respective reliability estimates (Gaski 1984; Gaski and Nevin 1985; Grace and O'Cass 2003, 2005;

Table 27.1 Reflective measurement model constructs

Construct	Original No. of Ind[a]	After No. of Ind[b]	Item Loading (λ) range[c]	Alpha[d] (α)	Comp Rel[e] ($\rho_{\xi X}$)	AVE[f]
Brand personality						
Sincerity (SIN)	12	11	0.699 → 0.806	0.9309	0.9410	0.5946
Excitement (EXC)	12	12	0.657 → 0.809	0.9259	0.9369	0.5538
Competence (COMP)	9	9	0.645 → 0.837	0.8986	0.9181	0.5554
Sophistication (SOP)	7	7	0.710 → 0.843	0.8977	0.9200	0.6202
Ruggedness (RUG)	7	7	0.602 → 0.821	0.8460	0.8840	0.5249
Brand relationship quality						
Partner quality (PQUAL)	11	11	0.699 → 0.826	0.9316	0.9419	0.5962
Love and passion (LOV)	9	9	0.677 → 0.858	0.9288	0.9408	0.6426
Intimacy (INTM)	11	9	0.604 → 0.822	0.8875	0.9108	0.5313
Self-connection (SCON)	7	7	0.781 → 0.852	0.9227	0.9388	0.6859
Nostalgic connection (NCON)	7	7	0.640 → 0.822	0.8917	0.9170	0.6081
Commitment (COMM)	9	9	0.704 → 0.847	0.9304	0.9422	0.6468
Interdependence (INTD)	8	8	0.658 → 0.849	0.9213	0.9360	0.6541
Consumer involvement profile						
Product risk/importance (RIS)	3	3	0.646 → 0.820	0.6313	0.8029	0.5775
Symbolic value (SYMV)	3	3	0.849 → 0.914	0.8527	0.9109	0.7726
Hedonic value (HEDV)	3	3	0.898 → 0.907	0.8840	0.9286	0.8116
Probability of mispurchase (PMIS)	4	4	0.717 → 0.844	0.8138	0.8789	0.6319
Interest (INT)	3	3	0.636 → 0.907	0.7378	0.8521	0.6693

[a] Original Number of indicators; [b] Number of indicators after deletion ; [c] Highest and lowest loading after deletion; [d] Cronbach's alpha composite; [e] Reliability composite Reliability ; [f] Average variance extracted (AVE)

Table 27.2 Hierarchical measurement model results

Higher order construct name	Component name	Loading (λ_i)	Significance[a]
Brand personality	Sincerity (SIN)	0.9216	***
$\rho_{\xi X} = 0.9761$	Excitement (EXC)	0.9302	***
AVE = 0.8277	Competence (COMP)	0.9193	***
	Sophistication (SOP)	0.8941	***
	Ruggedness (RUG)	0.8785	***
Brand Relationship Quality	Partner quality (PQUAL)	0.8944	***
$\rho_{\xi X} = 0.9860$	Love and passion (LOV)	0.9589	***
AVE = 0.8788	Intimacy (INTM)	0.9188	***
	Self-connection (SCON)	0.9459	***
	Nostalgic connection (NCON)	0.9403	***
	Commitment (COMM)	0.9601	***
	Interdependence (INTD)	0.9404	***
Consumer involvement profile	Product risk/importance (RIS)	0.5753	***
$\rho_{\xi X} = 0.8256$	Symbolic value (SYMV)	0.5947	***
AVE = 0.4830	Hedonic value (HEDV)	0.8074	***
	Probability of mispurchase (PMIS)	0.2961	***
	Interest (INT)	0.7928	***
Consumer Involvement Profile	Product risk/importance (RIS)	0.5202	***
Resestimated $\rho_{\xi X} = 0.8674$	Symbolic value (SYMV)	0.5977	***
AVE = 0.5095	Hedonic value (HEDV)	0.8552	***
	Interest (INT)	0.8358	***

[a] Bootstrapping results (n = 500)*** $p < 0.001$** $p < 0.01$* $p < 0.05$ n.s = not significant
$\rho_{\xi X}$ – composite reliability
AVE – Average variance extracted

O'Cass and Pecotich 2005). Parameter results and significance levels are presented in Table 27.2. Please note that although these results are presented together each higher order construct domain was estimated as three separate hierarchical measurement models. All higher level construct composite reliabilities and AVE's were in the acceptable range.

27.5.6 Structural Model Results

The same process that was undertaken before when determining the reliability and validity for the item level measurement models was applied again at the higher order of abstraction. The modeling occurs now with latent variable scores which effectively become observed indicants representing the first order constructs. PCA analysis was firstly undertaken and this confirmed brand personality, and brand relationship quality as being unitary constructs. Loadings for the components representing the BPS and BRQ construct ranged between 0.879<–>0.930 and 0.894<–>0.960, respectively. The CIP revealed a two component representation with the component Probability of Mispurchase loading highly by itself. This was not surprising as the earlier results showed that this construct had the lowest path coefficient of 0.2961. This was by far the lowest value compared with the other four paths reflecting the involvement construct. Earlier studies by Laurent and Kapferer (1985) also found problems with this construct. This variable was deleted. Another PCA was reestimated and loadings now ranged between 0.546 and 0.849. The construct is now unitary. This hierarchical measurement model was re-run without the Probability of Mispurchase construct and factor scores were again derived. Table 27.2 shows the results without this construct and it is evident that this deletion improves composite reliability. Also the AVE is now in an acceptable range. The latent variable scores derived for the three models could now be used further in the structural modeling. The composite reliability estimates were all acceptable (BPS 0.9761, CIP 0.8674, and BRQ 0.9860). Discriminant validity is also satisfied because in no case is the correlation between any variable and another as high as its reliability coefficient. The results in relation to the outer components of structural model are all adequate and the critical ratios are all significant ($p < 0.05$). The two structural models (main effects and interaction model) can now be estimated using PLS.

The structural model coefficient results for Model 1 (Main Effects Model) and Model 2 (Interaction Model) specifying relationships between the latent variables (brand personality, involvement, brand relationship quality and interaction term) are reported in Table 27.3. These coefficients are interpreted just like standardized regression coefficients (Fornell and Cha 1994).

The results in Table 27.3 demonstrate that the standardized beta coefficients for brand personality and involvement (0.5571 and 0.1709) with an R^2 of 0.3875 for brand relationship quality. The inclusion of the interaction term shows a smaller beta of 0.1277 increasing the R-square to 0.4027. These results imply that one standard deviation increase in brand personality will impact brand relationship quality directly by 0.5571. The R-squares are both high indicating the predictive capacity of the model. The contribution to R-square illustrates the importance of each construct and its' relative contribution to overall R-squared. These results satisfy the requirement of Falk and Miller (1992) whom state that the variance in the endogenous construct explained by any one individual path must exceed 1.5%.

Table 27.3 Structural model results

Structural relation	Model 1 (Main effects)		Model 2 (Interaction model)	
	Path Coeff	Sig. Cont R^2	Path Coeff	Sig. Cont R^2
Brand personality → BRQ	0.5571	*** 86.2%	0.5683	*** 84.5%
CIP → BRQ	0.1709	*** 13.8%	0.1658	*** 12.8%
Interaction construct/term			0.1277	*** 2.7%
R^2		0.3880		0.4037
R^2 adjusted		0.3875		0.4027

Bootstrapping results (n = 500)*** $p < 0.001$ ** $p < 0.01$ * $p < 0.05$
Path Coeff = Path coefficient. Cont R^2 = Contribution to R^2

To determine the merit of the interaction term being added into the model the effect size was calculated. This is assessed by inputting R^2s of the main effects and interactions model using the Cohen (1988) effect size formula (27.1):

$$f^2 = [0.4027 - 0.3875]/[1 - 0.4027] = 0.0254.$$

The effect size (f^2) when an interaction is 0.02, 0.15, and 0.35 has been suggested as small, moderate, and large respectively (Cohen 1988). It is important to understand that a small f^2 does not necessarily imply an unimportant effect. "If there is a likelihood of occurrence for the extreme moderating conditions and the resulting beta changes are meaningful, then it is important to take these situations into account (Limayem et al. 2001, p. 281)."

In this particular study the effect size has been deemed to be small (0.0254). Product Class Involvement does have a role to play. Involvement has a strong positive impact on BRQ. The interaction between brand personality and level of involvement shows a significant positive effect on the strength of BRQ. This would suggest that these constructs should be modeled together in future studies. The next section features a discussion and outlines opportunities for further research.

27.6 Discussion

These results are quite positive and show the flexibility of using PLS for interactions modeling of higher order constructs to overcome CBSEM estimation difficulties. Both hypotheses are supported. It should be noted that a weakness of PLS interactions modeling is that it has a tendency to underestimate structural parameter estimates. PLS performs better with more indicants per construct. "This implies caution against putting too much emphasis on PLS loadings when there are few indicators (Chin et al. 2003, p. 205)." The same situation has been highlighted in CBSEM interactions modeling with derived factor scores. Yang-Wallentin et al. (2004, p. 147) results complement Chin et al. (2003) when they state that, the more indicators a latent variable had, the better the estimated (factor) scores will be. The above stated problems have been negated somewhat due to the use of an adequate number of indicators represented in both the initial measurement models and in the final structural model.

This study was quite ambitious in trying to model relationships at a higher order of abstraction. It represents a first attempt. One that will be refined with future validation studies. It provides some preliminary evidence to illustrate (at least for the brands in this study) that the softer image drivers of brand loyalty do indeed matter. Some authors' argue that a brand's true source of differentiation lies not in the superiority of functional (physical) attributes of the product. Rather, sustainable differentiation is now focused on "softer" intangible issues such as; emotions, feelings, images, personality, and relationships with the brand (Carpenter et al. 1994). This research provides valuable support to this viewpoint and is a call to action for brand custodians to invest further and manage carefully the "softer" emotional elements of branding. The importance of brand personality is central in this process. Product class involvement has also been shown to have a significant moderating effect. This supports others that believe the moderating role of product class involvement is important (Homburg and Giering 2001).

Other moderators could obviously be explored within the main effects model. These were not covered in this chapter so as to give a richer description of the process undertaken within the analysis. This should assist others in replicating the process with their own data. Implementation of multiple group PLS analysis as demonstrated in (Chin 2002; Grace and O'Cass 2005; Henseler and Fassott 2010; Lee 2000) using dichotomous (sex, age) or artificially dichotimized variables (interpersonal orientation (Swap and Rubin 1983; Wilson et al. 2003; personal attachment styles (Paulssen and Fournier 2005)) could also provide valuable theoretical advances and insights.[12]

27.6.1 Study Limitations and Suggestions for Further Research

There are some study limitations and opportunities for future research worth mentioning. The usual caveats concerning the use of single informants and self-reported data apply to this study and, consequently, some caution is advised when generalizing the findings. Although not presented in this chapter, it may also be of interest to adopt a more fine-grained approach in concert with the higher level analysis undertaken here to examine other relationships not explored between first order constructs. For instance, future research may wish to examine whether the individual independent brand personality constructs at the first order level interrelate with brand relationship quality and its' individual first order constructs. This could be implemented using regression analysis. Unfortunately, the sample sizes are too small at the brand level to investigate with confidence.

There are study limitations concerning the scope (12 brands) and the number of product categories (6 product classes) studied. It is recommended that future research consider other brands within the product category, extending past the two most familiar brands and incorporating lower-level brands (or even niche brands) within the category. As the sample size to number of variables ratio was too low, contrasts could not be made between brands. A larger longitudinal study is suggested. This could profile the evolving and dynamic changes occurring within a brand's personality and help track brand relationships over time. To achieve maximum faith in the findings, future researchers should employ an experimental design to more suitably test causation and eliminate any extrageneous influences. This study has been very useful in demonstrating that exploring relations between higher order constructs is possible. Future researchers are also encouraged to be ambitious in their plans to investigate such abstract constructs. It may be possible (designs

[12] This idea was initially suggested by Professor Fournier.

permitting) in the future to complete growth curve modeling for there underlying these constructs with the use of CBSEM or PLS methods. This would be exciting in allowing insight into understanding brand relationship trajectories.

Further methodological studies with PLS continuous moderators need to implement Monte Carlo designs to investigate the stability of coefficients estimated and determine adequate indicator-to-sample size ratios. Chin et al. (2003) have provided some initial results, but given the substantive literature that exists for CBSEM methods, such work in the PLS realm is in its' infancy. This could encompass many model types of varying strength and complexity. Nonlinear and quadratic terms could also be explored with PLS. This important task will give PLS methods greater legitimacy and allow researchers to more confidently understand when (and when not) to use it.

Chin et al. (2003) have shown that using PLS in interactions modeling leads to increased power to detect relationships (thus, further minimizing Type I errors) and allows the researcher to flexibly deal with data that violates distributional assumptions. This is notable for this study as the results indicated the interaction effect size was small according to Cohen (1988) and although no substantive comparison could be implemented with CBSEM interactions methods (due to CBSEM non-convergence) it is believed that possibly this would not have been unearthed without the use of the PLS interactions modeling method. These well known advantages make it very suitable for many social research studies. Researchers also need to be aware that a major weakness of PLS is that being a limited information method the bias and consistency of parameter estimates are less than optimal. The estimates will be asymptotically correct under the joint conditions of large sample size and large number of indicators per latent variable (Chin 1998; Lohmöller 1989). This is the consistency at large assumption (Wold 1980). The one Monte Carlo study with this technique has shown that PLS also has a tendency to underestimate structural parameter estimates and inflate interaction estimates ($X * Z$). "This implies caution against putting too much emphasis on PLS loadings when there are few indicators (i.e., < 8) (Chin et al. 2003, p. 205)." The number of indicants used in this study was consistent with this recommendation. One of the greatest benefits of using PLS to analyze psychosocial data is that it can deal with complex models. The illustrated example would not have been able to be explored with other techniques. Again, researchers are urged to run more Monte Carlo PLS studies to gain a better understanding of the above issues. The Chin et al. (2003) study provides the only limited guidance available at this time. Researchers are best to err on the side of caution when interpreting results. There has been over 20 years of Monte Carlo studies in the CBSEM domain. CBSEM studies have investigated: structural coefficient stability, standard error accuracy and goodness-of-fit statistic behavior for various population model types. To place the level of development in perspective, PLS researchers are now just starting to develop their own goodness-of-fit statistics (Amato et al. 2004). There are many research opportunities.

PLS also has numerous managerial advantages in that variables can be left in the model (pending adequate reliability and validity is demonstrated) to more aptly explain the drivers of complex relationships between constructs. Creating managerial useful benchmarks has been illustrated within a retailing context (Arnett et al. 2003). Herein, lies one of the strengths of PLS for practitioners. When marketers are investigating such areas as: brand personality, brand associations and other image dimensions often the utility of such research from the practitioner's perspective is based around whether the research allows a rich enough distinction to determine a "point of difference." The question a manager would often like answered is: "How can I make my brand/s different or distinctive?" This is much more powerful than, "How are my brands similar to other brands?" When statistical methods inhibit (do not converge or produce nonsense solutions) or reduce the full flavour of the qualitative nuance underlining such concepts they become less useful for managers. This provides a strong argument for practitioners preferring the use of qualitative research methodologies in such substantive domains. Blackston (1992) in support of qualitative methods when investigating the two-sided nature of brand relationships concluded, "the use of factor analysis eliminated outlying statements so that what remains are image statements which represent a sort of lowest common denominator." He goes on to state, "we most often discard the very things that would allow us to see what makes a brand really different or unique (Blackston 1992, p. 232)." It is acknowledged that the method chosen must address the substantive research questions under study. However, it is believed that use of the PLS method in addressing some of the issues mentioned herein allows a balance between greater substantive insight and the application of standard psychometric principles. This is most appropriate to most forms of psychosocial research undertaken. It is believed the use of PLS allows the richer descriptive dimensions to remain relatively intact, thus, allowing managers a solid baseline from which to implement decisions at the tactical level. This becomes more pronounced when the researcher decides to use formative indicants within their models.

27.6.2 Concluding Remarks

This work is the first of its' kind employing a new method (PLS interactions modeling with higher order constructs) within an important research area. After overcoming initial difficulties with CBSEM methods a solution was obtained with PLS. This approach was more suitable from a philosophical perspective in that the theory was relatively new, the model was complex and the work was being undertaken within a different culture. The PLS approach was better able to deal with inherent data distribution problems. An exciting era is here. I have highlighted the importance for further Monte Carlo studies to be undertaken. PLS will also be increasingly chosen when researchers use formative measures. Malhotra (1996), in a meta-analytic study on statistical methods implemented in major marketing journals, illustrated that the use of structural equation methods had increased dramatically. His study did not distinguish between PLS and CBSEM methods. A recent examination of top marketing journals (Journal of Marketing, Journal of Marketing Research and Journal of Consumer Research) between 2000 and 2003 (inclusive) indicated that only one article had featured the use of PLS (Goodhue et al. 2006). It is my contention that PLS will be increasingly used within the marketing domain. In fact, both CBSEM and PLS modeling methods will increasingly be implemented in marketing studies well into the future. Understanding when each technique "could and should" be used will be the key.

References

Aaker, D. A., & Biel, A. L. (1993). *Brand equity and advertising: advertising's role in building strong brands*. New Jersey: Lawrence.

Aaker, D. A. (1996). *Building strong brands*. New York: Free Press.

Aaker, J. L. (1995). *Brand personality: conceptualization, measurement and underlying psychological mechanisms,* Doctoral Dissertation, Stanford University.

Aaker, J. L. (1997). Dimensions of brand personality. *Journal of Marketing Research, 34*(August), 347–356.

Aaker, J. L., & Fournier, S. (1995). A brand as a character, a partner and a person: three perspectives on the question of brand personality. In F. R. Kardes, & S. Mita (Eds.), *Advances in Consumer Research* (pp. 391–395). Provo, UT: Association for Consumer Research.

Aaker, J. L., Benet-Martinez, V., & Garlolera, J. (2001). Consumption symbols as carriers of culture: a study of Japanese and Spanish brand personality constructs. *Journal of Personality and Social Psychology, 81*(5), 492–508.

Aiken, L. S., & West, S. G. (1991). *Multiple regression: testing and interpreting interactions*. CA, Beverly Hills: Sage.

Amato, S., Esposito Vinzi, V., & Tenenhaus, M. (2004). A global goodness-of-fit index for PLS structural equation modeling. *Oral communication to PLS Club*, HEC School of Management, France, March 24.

Anderson, J. C., & Gerbing, D. W. (1988). Structural equation modeling in practice: a review and recommended two-step approach. *Psychological Bulletin, 103*(3), 411–423.

Andreou, P. T. (1994). *Brand relationships: a new approach to branding*, PhD Dissertation. University of South Hampton.

Arnett, D. B., Laverie, D. A., & Meiers, A. (2003). Developing parsimonious retailer equity indexes using partial least squares analysis: a method and applications. *Journal of Retailing, 79*, 61–170.

Arora, R. (1982). Validation of an S-O-R model for situation, enduring, and response components of involvement. *Journal of Marketing Research, 19*(November), 505–516.

Ball, A. D., & Tasaki, L. H. (1992). The role and measurement of attachment in consumer behavior. *Journal of Consumer Psychology, 1*(2), 155–172.

Barclay, D., Higgins, C., & Thompson, R. (1995). The partial least squares (PLS) approach to causal modeling: personal computer adoption and use as an illustration. *Technology Studies, 2*(2), 285–309.

Baron, R. M., & Kenny, D A. (1986). The moderator-mediator variable distinction in social psychological research: conceptual, strategic and statistical considerations. *Journal of Personality and Social Psychology, 51*(6), 1173–1182.

Baumgartner, H., & Homburg, C. (1996). Applications of structural equation modeling in marketing and consumer research: a review. *International Journal of Research in Marketing, 13*, 139–161.

Bearden, W. O., Netemeyer, R. G., & Mobley, M. E. (1993). *Handbook of marketing scales: multi-item measures for marketing and consumer behavior research*. Newbury Park, California: Sage.

Bhat, S., & Reddy, S. K. (1998). Symbolic and functional positioning of brands. *Journal of Consumer Marketing, 15*(1), 32–43.

Blackston, M. (1992). Observations: building brand equity by managing the brand's relationships. *Journal of Advertising Research, 32*(May/June), 79–83.

Bollen, K.A., & Ting, K. (2000). A tetrad test for causal indicators. *Psychological Research Methods, 5*, 3–22.

Bontis, N. (1998). Intellectual capital: an exploratory study that develops measures and models. *Management Decision, 36*(2), 63–76.

Boomsma, A. (1983). *On the robustness of LISREL (maximum likelihood estimation) against small sample size and nonnormality*, Doctoral Dissertation. University of Groningen.

Bromley, D. B. (2000). Psychological aspects of corporate identity, image and reputation. *Corporate Reputation Review, 3*(3), 240–252.

Browne, M. W. (1984). Asymptotically distribution-free methods for the analysis of covariance structures. *British Journal of Mathematical and Statistical Psychology, 37*, 62–83.

Byrne, B. M., & Campbell, T. L. (1999). Cross-cultural comparisons and the presumption of equivalent measurement and theoretical structure: a look beneath the surface. *Journal of Cross-Cultural Psychology, 30*(5), 555–574.

Callaghan, W. M., & Wilson, B. J. (1998). The role of the category in brand equity studies: a brand attitudinal segmentation perspective. In *Australia and New Zealand Marketing Educators Conference,* Dunedin, New Zealand.

Callcott, M F., & Lee, W-N. (1994). A content analysis of animation and animated spokes-characters in television commercials. *Journal of Advertising, 23*(4), 1–12.

Capara, G V., Barbaranelli, C., & Gianluigi, G. (1998). Personality as metaphor: extension of the psychlexical hypothesis and the five factor model to brand and product personality description. *European Advances in Consumer Research, 3*, 61–69.

Caprara, G. V., Barbaranelli, C., & Gianluigi, G. (2001). Brand personality: how to make the metaphor fit? *Journal of Economic Psychology, 22*, 377–395.

Carpenter, G. S., Glazer, R., & Nakamoto, K. (1994). Meaningful brands from meaningless differentiation: the dependance on irrelevant attributes. *Journal of Marketing Research, 31*(3), 339–350.

Cassel, C. M., Hackl, P., & Westlund, A. H. (1999). Robustness of partial least-squares method for estimating the latent variable quality structures. *Journal of Applied Statistics, 26*(4), 435–446.

Chen, F., Bollen, K. A., Paxton, P., Curran, P. J., & Kirby, J. B. (2001). Improper solutions in structural equation models: causes, consequences, and strategies. *Sociological Methods and Research, 29*(4), 468–508.

Chin, W. W. (1998) The partial least squares approach to structural equation modelling. In G. A. Marcoulides (Ed.), *Modern methods for business research* (pp. 295–336). Mahwah, NJ: Lawrence Erlbaum.

Chin, W. W. (2002). Multigroup analysis with PLS [WWW document]. URL http://disc-nt.cba.uh.edu/chin/plsfaq/multigroup.htm

Chin, W. W., Marcolin, B. L., & Newsted, P. R. (1996). A partial least squares latent variable modeling approach for measuring interaction effects: results from a monte carlo simulation study and voice mail emotion/adoption study. In J. I. DeGross, S. Jarvenpaa, & A. Srinivasan (Eds.), *Proceedings of the Seventeenth International Conference on Information Systems,* (pp. 21–41), Cleveland, Ohio.

Chin, W. W., Marcolin, B. L., & Newsted, P. R. (2003). A partial least squares latent variable modeling approach for measuring interaction effects: results from a monte carlo simulation study and an electronic-mail emotion/adoption study. *Information Systems Research, 14*(2), 189–217.

Chin, W. W., & Newsted, P. R. (1999). Structural equation modeling analysis with small samples using partial least squares. In R. H. Hoyle (Ed.), *Statistical strategies for small sample research* (pp. 307–341). Thousand Oaks, CA: Sage.

Cohen, J. (1988). *Statistical Power Analysis for the Behavioral Sciences.* Hillsdale, New Jersey.Lawrence Erlbaum.

Cortina, J. M., Chen, G., & Dunlap, W. P. (2001). Testing interaction effects in LISREL: examination and illustration of available procedures. *Organizational Research Methods, 4*(4), 324–360.

Cronbach, L. J. (1951). Coefficient alpha and the internal structure of tests. *Psychmetrika, 16*(3), 297–334.

Davis, K. E., & Todd, M. J. (1982). Friendship and Love Relationships. In K. E. Davis, & T. Mitchell (Ed.), *Advances in descriptive psychology* (pp. 79–122). Greenwich, CT: JAI.

Day, E., Royne S. M., & Camacho, A. (1995). Opportunities for involvement research: a scale-development approach. *Journal of Advertising, 24*(3), 69–75.

Deane, J., Smith, G., & Adams, A. (2003). Sports sponsorship and brand personality- the Ryder Cup Team and IBM. *International Journal of Sports Marketing and Sponsorship, 5*(3), 193–208.

Denham, S. A., Blair, K. A., DeMulder, E., Levitas, J., Sawyer, K., Auerbach-Major, S., & Queenan, P. (2003). Preschool emotional competence: pathway to social competence? *Child Development, 74*(1), 238–256.

Dick, A., & Sauer, P. L., (1992). Using moderator variables in structural equation models. In L. McAlister, & M. L. Rothschild (Eds.), *Advances in Consumer Research* (pp. 637–640). Provo, UT: Association for Consumer Research.

Dillon, W. R., Mulani, N., & Kumar, A. (1987). Offending estimates in covariance structure analysis: comments on the causes and solutions to Heywood cases. *Psychological Bulletin, 101*(1), 126–135.

Efron, B., & Tibshirani, R. J. (1993). *An introduction to the bootstrap.* New York: Chapman and Hall.

Falk, R. F., & Miller, N. B. (1992), *A primer for soft modeling.* Akron, Ohio: The University of Akron Press.

Field, A. (2005). *Discovering statistics using SPSS,* 2nd Edn. London: Sage.

Fornell, C., & Larcker, D. F. (1981). Evaluation structural equation models with unobservable variables and measurement error. *Journal of Marketing Research, 18*(February), 39–50.

Fornell, C., & Bookstein, F. L. (1982) Two structural equation models: LISREL and PLS applied to consumer exit-voice theory. *Journal of Marketing Research, 19*(November), 440–452.

Fornell, C., & Cha, J. (1994). Partial least squares. In R. P. Bagozzi (Ed.), *Advanced methods of marketing research* (pp. 52–78). Oxford: Basil Blackwell.

Fournier, S. (1994). *A consumer-brand relationship framework for strategic brand management,* Doctoral Dissertation. University of Florida.

Fournier, S. (1995). Understanding consumer-brand relationships. *Harvard Business School Working Paper, 96–018*, 1–57.

Fournier, S. (1998). Consumers and their brands: developing relationship theory in consumer research. *Journal of Consumer Research, 24*(March), 343–373.

Fuller, K. (1995). The Levi 501 campaign. *Admap, 349*(March), 20–23.

Gaski, J. (1984). Theory of power and conflict in channels of distribution. *Journal of Marketing, 48*(Summer), 9–29.

Gaski, J. F., & Nevin, J. R. (1985). The differential effect of exercised and unexercised power in a marketing channel. *Journal of Marketing Research, 22*(May), 130–142.

Gefen, D., & Straub D. (1997). Gender difference in the perception and use of e-mail: an extension to the technology acceptance model. *MIS Quarterly, 21*(4), 389–400.

Gerbing, D. W., & Anderson, J. C. (1985). The effect of sampling error and model characteristics on parameter estimation for maximum likelihood confirmatory factor analysis. *Multivariate Behavioral Research, 20*, 255–271.

Goodhue, D., Lewis, W., & Thompson, R. (2006). PLS, small sample size, and statistical power in MIS research. In *Proceedings of the 39*th *Hawaii Conference on Systems Sciences.* IEEE Computer Society, p. 10.

Gordon, M. E., & van der Sprong, F. (1998). Relationship marketing effectiveness: the role of product category. In *6th International Colloquium in Relationship Marketing* (pp. 1–14). Auckland, New Zealand: The University of Auckland.

Grace, D., & O'Cass, A. (2003). Child care services: an exploratory study of choice, switching and search behaviour. *European Journal of Marketing, 31*(1/2), 107–132.

Grace, D., & O'Cass, A. (2005). An examination of the antecedents of repatronage intentions across different retail store formats. *Journal of Retailing and Consumer Services, 12*, 227–243.

Hair, J. F Jnr., Anderson, R. E., Tatham, R. L., & Black, W. C. (1995). *Multivariate data analysis: with Readings.* New Jersey: Prentice Hall.

Hattie, J. (1985). Methodology review: assessing unidimensionality of tests and items. *Applied Psychological Measurement, 9*(2), 139–164.

Hanby, T. (1999). Brands- dead or alive? *Journal of the Market Research Society, 41*(1), 7–18.

Henseler, J., & Fassott, G. (2010). Testing moderating effects in PLS path models: an illustration of available procedures. In V. Esposito Vinzi, W. W. Chin, J. Henseler & H. Wang (Eds.), *Handbook of partial squares* (pp. 715–737), Heidelberg: Springer.

Hess, J. S. (1998). *A multidimensional conceptualization of consumer brand relationships: the differential impact of relationship dimensions on evaluative relationship outcomes*, Doctoral Dissertation. University of Colorado, U.S.A.

Holmes-Smith, P., & Rowe, K. J. (1994). The development and use of congeneric measurement models in school effectiveness research: improving the reliability and validity of composite and latent variables for fitting multilevel and structural equation models. In *International Congress for School Effectiveness and Improvement.* Melbourne: The World Congress Centre.

Homburg, C., & Giering, A. (2001). Personal characteristics as moderators of the relationship between customer satisfaction and loyalty- an empirical analysis. *Psychology and Marketing, 18*(1), 43–66.

Hoyle, R. H. (1995). *Structural equation modeling: concepts, issues and applications.* Thousand Oaks, California: Sage.

Jain, K., & Srinivasan, N. (1990). An empirical assessment of multiple operationalizations of involvement. In M. Goldberg, G. Gorn, & R. Pollay (Eds.), *Advances in consumer research* (pp. 594–602). The Association for Consumer Research: Provo, UT.

Jarvis, C.B., MacKenzie, S.B., & Podsakoff, P. M. (2003). A critical review of construct indicators and measurement model misspecification in marketing and consumer research. *Journal of Consumer Research, 30*(September), 199–218.

Ji, M. F. (2002). Children's relationships with brands: "true love" or "one-night" stand? *Psychology and Marketing, 19*(4), 369–387.

Jöreskog, K. G. (1967). Some contributions to maximum likelihood factor analysis. *Psychometrika, 32*, 443–482.

Jöreskog, K. G., & Yang, F. (1996). Nonlinear structural equation models: the Kenny-Judd model with interaction effects. In G. A. Marcoulides, & R. E. Schmacker (Eds.), *Advanced structural modeling: Issues and techniques* (pp. 57–88). Mahwah, NJ: Lawrence Erlbaum.

Kapferer, J-N., & Laurent, G. (1985). Consumer's Involvement Profile: New Empirical Results. In E. Hirschman, & M Holbrook. (Eds.), *Advances in Consumer Research. Prov, UT. The Association for Consumer Research*, 290–295.

Kapferer, J-N., & Laurent, G. (1986). Consumer involvement profiles: a new practical approach to consumer involvement. *Journal of Advertising Research, 25*(December–January), 49–56.

Khalifa, M., & Cheng, S. K. N. (2002). Adoption of mobile commerce: role of exposure. In *Proceedings of the 35th Hawaii International Conference on Systems Science*, Vol 9. Hawaii: IEEE Computer Society.

Kwong, T. C. H., & Lee, M. K. O. (2002). Behavioral intention model for the exchange mode internet music piracy. In *Proceedings of the 35th Hawaii International Conference on Systems Sciences*, Vol 9. Hawaii: IEEE Computer Society.

Laurent, G., & Kapferer, J-N. (1985). Measuring consumer involvement profiles. *Journal of Marketing Research, 12*(February), 41–53.

Lee, S. Y. (1980). Estimation of covariance structure models with parameters subject to functional restraints. Psychometrika, *45*, 309–324.

Lee, D. Y. (2000). Retail bargaining behaviour of American and Chinese customers. *European Journal of Marketing, 34*(1/2), 190–206.

Limayem, M., Hirt, S. G., & Chin, W. W. (2001). Intention does not always matter: the contingent role of habit on IT usage behavior. In *Global co-operation in the new millennium. The 9th European Conference on Information Systems* (pp. 274–286). Slovenia: Bled.

Lohmöller, J-B. (1989). *Latent variable path modeling with partial least squares.* Heidelberg, Germany: Physica-Verlag.

Low, G. S., & Mohr, J. J. (2001). Factors affecting the use of information in the evaluation of marketing communications productivity. *Journal of the Academy of Marketing Sciences, 29*(1), 70–88.

Malhotra, N. K. (1996). The impact of the academy of marketing science on marketing scholarship: an analysis of the research. *Journal of the Academy of Marketing Science, 24*(4), 291–298.

McCracken, G. (1988). *Culture and consumption.* Bloomington: University Press.

Netemeyer, R. G., Krishnan, B., Pullig, C., Wang, G., Yagci, M., Dean, D., Ricks, J., & Wirth, F. (2004). Developing and validating measures of facets of customer-based brand equity. *Journal of Business Research, 57*(2), 209–224.

Nunnally, J. C., & Bernstein, I. H. (1994). *Psychometric theory.* New York: McGraw-Hill.

O'Cass, A., & Pecotich, A. (2005). The dynamics of voter behavior and influence in electoral markets: a consumer behavior perspective. *Journal of Business Research, 58*, 406–413.

Paulssen, M., & Fournier, S. (2005) Individual difference in consumer relationship – an attachment perspective. In *Proceedings of the 34th European Marketing Academy: Rejuvenating Marketing: Contamination, Innovation, Integration.* Milan, Italy.

Ping, R. A. (1995). A parsimonious estimating technique for interaction and quadratic latent variables. *Journal of Marketing Research, 32*(August), 336–347.

Ping, R. A. (1996). Latent variable interaction and quadratic effect estimation: a two-step technique using structural equation analysis. *Psychological Bulletin, 119*(1), 166–175.

Ping, R. A. (1996). Latent variable regession: a technique for estimating interaction and quadratic coefficients. *Multivariate Behavioral Research, 31*(1), 95–120.

Ping, R. A. (2005). Second-order latent variable interactions, and second-order latent variables. [WWW document: online paper]. URL http://home.att.net/(rpingjr/2NCUTDWN_6.DOC

Podsakoff, P. M., & Organ, D. W. (1986). Self-reports in organizational research: problems and prospects. *Journal of Management, 12*(4), 531–544.

Raykov, T. (2001). On the use and utility of the reliability coefficient in social and behavioral research. *Quality and Quantity, 35*, 253–263.

Reinartz, W., Krafft, M., & Hoyer, W. D. (2003). Measuring the customer relationship management construct and linking it to performance outcomes. *INSEAD Working Paper Series.* 2003/02/MKT:1–51.

Reinartz, W., Krafft, M., & Hoyer, W. D. (2004). The customer relationship management process: its measurement and impact on performance. *Journal of Marketing Research, 41*(3), 293–305.

Richins, M. L., & Bloch, P. H. (1986). After the new wears off: the temporal context of product involvement. *Journal of Consumer Research, 13*(September), 280–285.

Rigdon, E. E., & Ferguson Jr., C. E. (1991). The performance of the polychoric correlation coefficient and selected fitting functions in confirmatory factor analysis. *Journal of Marketing Research, 28*(November), 491–497.

Rigg, C., & Sparrow, J. (1994). Gender, diversity and working styles. *Women in Management Review, 9*(1), 9–16.

Rindskopf, D. (1984). Structural equation models. *Sociological Methods and Research, 13*(1), 109–119.

Sirgy, J. M. (1982) Self-concept in consumer behavior: a critical review. *Journal of Consumer Research*, *9*(December), 287–300.

Steenkamp, J-B. E. M., & van Trijp, H. C. M. (1991). The use of LISREL in validating marketing constructs. *International Journal of Research in Marketing*, *8*, 283–299.

Suh, J-C., & Yi, Y. (2006). When brand attitudes affect the customer satisfaction-loyalty relations: the moderating role of product involvement. *Journal of Consumer Psychology*, *16*(2), 145–155.

Swap, W. C., & Rubin, J. Z. (1983). Measurement of interpersonal orientation. *Journal of Personality & Social Psychology, 44*(1), 208–219.

Tabachnick, B. G., & Fidell, L. S. (1996). *Using multivariate statistics,* 3rd Edn. New York: HarperCollins College Publishers.

Tenenhaus, M., Esposito Vinzi, V., Chatelin, Y-M., & Lauro, C. (2005). PLS path modeling. *Computational Statistics & Data Analysis, 48*(1), 159–205.

Thakor, M. V., & Kohli, C. S. (1996). Brand origin: conceptualization and review. *Journal of Consumer Marketing, 13*(3), 27–42.

Venables, B. T., Rose, G. M., & Gilbert, F. W. (2003). Measuring the brand personality of non-profit organizations. In *Advances in Consumer Research* (pp. 379–380). Duluth, MN: Association for Consumer Research.

Venaik, S. (1999). *A model of global marketing in multinational firms: an empirical investigation,* PhD Dissertation. The Australian Graduate School of Management, Sydney.

Venaik, S., Midgley, D. F., & Devinney, T. M. (2001). Autonomy, networking and interunit learning in a model of MNC subsidiary innovation and performance. *AGSM Working Paper.* (May), 1–47.

Venaik, S., Midgley, D. F., & Devinney, T. M. (2005). Dual paths to performance: the impact of global pressures on MNC subsidiary conduct and performance. *Journal of International Business Studies, 36*(6), 655–678.

Werts, C. E., Linn, R. L., & Jöreskog, K. G. (1974). Intraclass reliability estimates: testing structural assumptions. *Educational and Psychological Measurement, 34*, 25–33.

Wilson, B., Callaghan, W., & Stainforth, G. (2003). An exploratory study of interpersonal orientation: a dimension worthy of investigation in consumer-brand relationship measurement? In *Proceedings of the Fifth Australasian Services Research Workshop* (pp. 74–81). Melbourne, Monash University.

Wold, H. (1980). Model construction by partial least squares. In J. Kmenta, & J. Ramsey (Eds.), *Evaluation of econometric models* (pp. 47–74). London: Academic Press.

Wold, H. (1981). Model construction and evaluation when theoretical knowledge is scarce: theory and application of partial least squares. In J. Kmenta, & J. Ramsey (Eds.), *Evaluation of econometric models* (pp. 47–74). New York: Academic Press.

Yang Jonsson, F. (1998). Modeling interaction and nonlinear effects: a step-by-step Lisrel example. In R. E. Schumacker, & G. A. Marcoulides (Eds.), Interaction and nonlinear effects in structural equation modeling (pp. 17–42). Mahwah, NJ: Lawrence Erlbaum and Associates.

Yang-Wallentin, F., Schmidt, P., Davidov, E., & Bamberg, S. (2004). Is there any interaction effect between intention and perceived behavioural control? *Methods of Psychological Research Online, 8*(2), 127–157.

Zaichkowsky, J. L. (1985). Measuring the involvement construct. *Journal of Consumer Research, 12*(December), 341–352.

Zaichkowsky, J. L. (1994). The personal involvement inventory: reduction, revision, and application to advertising. *Journal of Advertising, 23*(4), 59–70.

Zhang, P., Li, N., & Sun, H. (2006). Affective quality and cognitive absorption: extending technology acceptance research. In *Proceedings of the 39thHawaii International Conference on Systems Sciences.* IEEE Computer Society, Vol 11.

Part III
Tutorials

Chapter 28
How to Write Up and Report PLS Analyses

Wynne W. Chin

Abstract The objective of this paper is to provide a basic framework for researchers interested in reporting the results of their PLS analyses. Since the dominant paradigm in reporting Structural Equation Modeling results is covariance based, this paper begins by providing a discussion of key differences and rationale that researchers can use to support their use of PLS. This is followed by two examples from the discipline of Information Systems. The first consists of constructs with reflective indicators (mode A). This is followed up with a model that includes a construct with formative indicators (mode B).

28.1 Introduction

This intent of this paper is to provide an introduction with corresponding examples to assist social scientists interested on how to write up research that employs PLS path analyses. Due to page limitations, the scope of discussion will be tailored towards survey based studies with specific examples from Information Systems research.

While a number of papers have been written dealing with appropriate reporting of covariance based SEM analyses (CBSEM) (Hoyle and Panter 1995; Steiger 1988, 2001; McDonald and Moon-Ho 2002), this is less so for Partial Least Squares. At first glance, it would seem that a researcher can simply follow the same process employed by covariance based SEM researchers. But, unreflectively following the same procedures may also overemphasize or possibly incorporate aspects that are idiosyncratic to that particular methodology. For example, it can arguably be said that there tends to be more emphasis spent in CBSEM papers on the adequacy of how well proposed models account for all item covariances based on the chi-square statistic and various goodness of fit indices. In contrast, as discussed in more detail in other papers in this handbook, PLS path analysis does not focus on accounting

W.W. Chin
Department of Decision and Information Sciences, Bauer College of Business, University of Houston, TX, USA
e-mail: wchin@uh.edu

V. Esposito Vinzi et al. (eds.), *Handbook of Partial Least Squares*, Springer Handbooks of Computational Statistics, DOI 10.1007/978-3-540-32827-8_29,

for measurement item covariances. Rather, depending on the particular model specified by the researcher, only the variances of dependent variables (item or construct level) variances are considered. Therefore, since the dominant paradigm is CBSEM and most reviewers are trained in the use and reporting (at times unreflectively) of CBSEM models, it would seem appropriate to spend a little upfront time discussing why goodness of fit indices and chi-square statistics are not expected to have as prominent a role in PLS reports. In addition, other key methodological distinctions will be presented. It is hoped at the end of this initial discussion, researchers using PLS will be in better position to provide various reasons justifying both their choice of using PLS and why lack of usage or reporting of goodness of fit measures should not necessarily be viewed as a deficit. Then, we begin with an example of reporting a model with all reflective items. This is followed with one that incorporates formative measures.

28.2 On Using PLS Versus CBSEM

At this point in time, given that most readers and reviewers of research articles are likely to have more experience with CBSEM methods than PLS, it can be argued that researchers employing PLS analysis are obliged to provide some initial discussion as to the rationale for their use of this particular technique. Specifically, it can be viewed as an education process in explaining the underlying "raison d'être" for both CBSEM and PLS. Often, the cookbook like recipe taught to students on CBSEM analysis and reporting are continued in a unreflective manner when it comes to expectations for PLS papers. Rather than being competitive, it can be argued that the use of PLS is often complementary to CBSEM for research endeavors and may potentially be better suited depending on the specific empirical context and objectives. For consideration, some of the key issues and/or justifications used in the past are:

- Degree of Emphasis on Covariance Explanation
- Soft Distributional Assumptions
- Exploratory in Nature
- Modeling Formative Measurement Items
- Higher Order Molar and Molecular Models
- High Model Complexity as Criterion
- Sample Size Requirement
- Accuracy of Parameter Estimation
- Eschewing the "True" Model for Prediction Focus
- Determinate Scores/Indices for Predictive Relevance
- Ease of Model Specification and Model Interpretation

Let's consider each one in detail.

28.3 Degree of Emphasis on Covariance Explanation

More than a decade ago Chin (1998a) noted that there tends to be an immediate reliance on the use of overall model fit (or goodness of fit) indices among CBSEM researchers without consideration of the full suite of information that should also be used to evaluate the adequacy of the model being considered. In fact, there may be a mistaken inference among some people between overall model fit and the specific term of "goodness of model fit" with "goodness of model." As Chin (1998a, pp. xii–xiii) noted:

> A final issue is the over-reliance towards overall model fit (or goodness of fit) indices. "Where is the goodness of fit measures?" has become the 90s mantra for any SEM based study. Yet, it should be clear that the existing goodness of fit measures are related to the ability of the model to account for the sample covariances and therefore assume that all measures are reflective. SEM procedures that have different objective functions and/or allow for formative measures (e.g., PLS) would, by definition, not be able to provide such fit measures. In turn, reviewers and researchers often reject articles using such alternate procedures due to the simple fact that these model fit indices are not available.

In actuality, models with good fit indices may still be considered poor based on other measures such as the R-square and factor loadings. The fit measures only relate to how well the parameter estimates are able to match the sample covariances. They do not relate to how well the latent variables or item measures are predicted. The SEM algorithm takes the specified model as true and attempts to find the best fitting parameter estimates. If, for example, error terms for measures need to be increased in order to match the data variances and covariances, this will occur. Thus, models with low R square and/or low factor loadings can still yield excellent goodness of fit.

Thus, in contrast to the component based algorithm of PLS, CBSEM primarily focuses on selecting appropriate estimates for the structural paths among latent constructs and the corresponding roadmap connecting all item measures. Moreover, the CBSEM algorithm does not follow the PLS approach which explicitly creates constructs scores by weighting sums of items underlying each latent variable. Rather, all latent variables are viewed as intangible and primarily the conduit connecting item measures. Loosely speaking, the CBSEM algorithm attempts to provide estimates for all open structural paths and measurement loadings such that the summation of all pathways connecting any two items result in an implied covariance is as similar to those obtained from the sample data.

For example, in Fig. 28.1, the model specified has one pathway connecting items B1 and E1. The algorithm attempts to provide the best set of numeric estimates (i.e., for b1, p1, p3, p4, and e1) such that the product of those estimates along with construct variances ends up being as similar to the covariance between B1 and E1 obtained from the sample data set. Similarly, the item loadings connecting E1 and E2 would also yield an implied covariance that must be compared with those obtained from the actual data set. Thus, the algorithm seeks to find the "best" set of parameters estimates for a given model such that all the implied covariances matches those observed from the sample data set. As the quote by Chin earlier noted, the

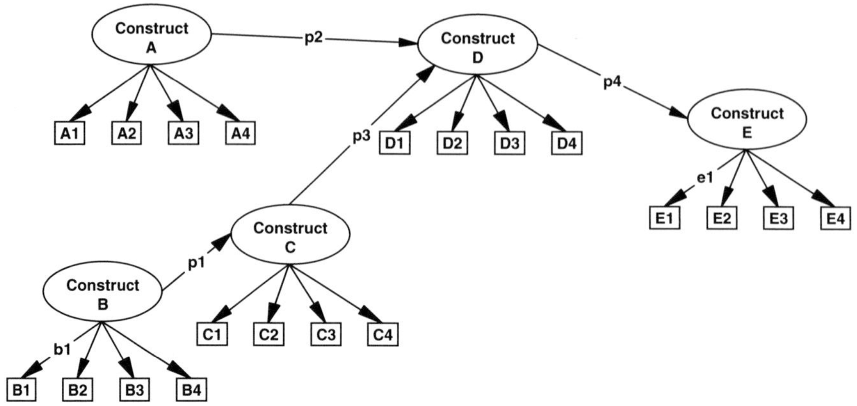

Fig. 28.1 Hypothetical model explaining item covariances

resulting path and loading estimates may end up being relatively low in magnitude and yet provide a good overall fit. Chin (1998a, xii) goes on to say:

> Therefore, pure reliance of model fit follows a Fisherian scheme similar to ANOVA which has been criticized as ignoring effect sizes (e.g., Cohen 1990, p. 1309). Instead, closer attention should be paid to the predictiveness of the model. Are the structural paths and loadings of substantial strength as opposed to just statistically significant? Standardized paths should be around 0.20 and ideally above 0.30 in order to be considered meaningful. Meehl (1990) has argued that anything lower may be due to what he has termed the crud factor where "everything correlate to some extent with everything else" (p. 204) due to "some complex unknown network of genetic and environmental factors" (p. 209). Furthermore, paths of 10, for example, represents at best a 1% explanation of variance. Thus, even if they are "real," are constructs with such paths theoretically interesting?

A few additional points are worth considering regarding the example in Fig. 28.1. The overall model fit involves all item covariances. There is no differentiating between proximity of constructs. As a full information algorithm, CBSEM attempts to reduce the discrepancy between model implied covariances from estimates with those obtained from the data. If the choice of estimates provides a larger overall reduction in discrepancies of implied with observed covariances for construct E items and construct B items than for construct E items and construct D item, then it will do so. Thus, there is no relative importance placed on whether a researcher wishes to explain the covariances of items for neighboring constructs versus those further separated in a nomological network. Accounting for the covariance for items B1 and E1, as an example, is equally as important as for D1 and E1.

Also as a full information approach, model misspecification can have a substantial impact. For example, including one item that does not belong with a particular construct can impact estimates obtained throughout the model. Likewise, path estimates can be quite different if a relevant path is left out (e.g., direct path from construct B to E). It is for these reasons, that CBSEM analysis is not only viewed as confirmatory in nature, but often considered by many as requiring relatively strong theoretical and substantive background knowledge for adequate deployment. Model misspecification with missing structural paths or multidimensional items placed under one construct can ripple through the entire model estimation process. In contrast, PLS estimates are limited to the immediate blocks a particular construct is structurally connected. Item weights and loadings for construct D, for example, are developed based on the inner weight relationships with constructs A, C, and E. Construct B being two links removed is not directly used in the PLS iterative algorithm.

28.4 Soft Distributional Assumptions

Predictor specification forms the basis for PLS modeling. Whereas a covariance-based maximum likelihood (ML) estimation rests on the assumptions of a specific joint multivariate distribution and independence of observations, the PLS approach does not make these hard assumptions. Rather, the PLS technique of model building uses very general, soft distributional assumptions which often led to this approach being termed *soft modeling*. As Lohmöller (1989) noted: "it is not the concepts nor the models nor the estimation techniques which are 'soft,' only the distributional assumptions" (p. 64).

Because PLS makes no distributional assumption other than predictor specification in its procedure for estimating parameters, traditional parametric-based techniques for significance testing/evaluation would not be appropriate. Wold (1980, 1982) argued for tests consistent with the distribution-free/predictive approach of PLS. In other words, rather than based on covariance fit, evaluation of PLS models should apply prediction-oriented measures that are also nonparametric. To that extent, the R-square for dependent LVs, the Stone-Geisser (Stone 1974; Geisser 1975) test for predictive relevance, Fornell and Larcker (1981) average variance extracted measure, and bootcross validation are used to assess predictiveness, while resampling procedures such as jack knifing and bootstrapping are used to examine the stability of estimates.

Chin (1998b, pp. 315–316) also noted that identical distributions are not assumed. Specifically, he said:

> For any two cases say n and n+1, no assumption is made that the residuals ν_n and ν_{n+1} have the same distribution. Nor is independence of cases required because no specification was made regarding the correlation between two different cases (i.e., $\mathrm{Cov}[\nu_n, \nu_{n+1}]$). In general, a sufficient condition for consistency in LS estimates is that as the number of observations go toward infinity, the sum of the correlations between cases must stay below infinity (i.e., $\sum_i |cor(\nu_n, \nu_{n+i})| < Infinity$; Wold 1988, p. 589).

In general, predictor specification could be viewed as a least squares counterpart to the distributional assumptions of ML modeling. At the same time, PLS avoids the assumptions that observations follow a specific distributional pattern and that they must be independently distributed. Therefore, no restrictions are made on the structure of the residual covariances and under PLS modeling the residual variance terms are minimized.

28.5 Exploratory in Nature

As we've noted, CBSEM typically employs a full information maximum likelihood estimation process that yields parameter estimates that are consistent and a chi-square statistic that is correct under the assumption of a "true" model being tested. But also by employing a full information procedure, a poorly developed construct where some of the item measures are weak or inappropriately measuring some other latent construct or a theoretical model with misspecified paths can bias other estimates throughout the proposed model. PLS, being a limited-information, component-based least squares alternative, tends to be less affected. The weights developed for each construct take into account only those neighboring constructs it is structurally connected. Because of this, some researcher often use the argument that they used PLS because both the theoretical knowledge and substantive knowledge for the domain they are studying is limited. As such, some conclude that PLS is primarily appropriate for exploratory studies where theoretical knowledge is relatively scarce and, possibly and inappropriately believe that CBSEM is superior to PLS for establishing theoretical models.

In fact, there are other instances beyond initial exploratory stages that PLS is well suited. It should not be construed that PLS is not appropriate in a confirmatory sense nor in well researched domains. As to be shown as an example, it may be the case that the researcher begins with a well established baseline model where both theory and measures have been rigorously developed. Instead, as an incremental study, it builds on a prior model by developing both new measures and structural paths. Depending how extensive the model is, there may be a desire to use PLS to constrain the new construct and measures to its immediate nomological neighborhood of constructs and avoid possible CBSEM estimation bias that can be affected by minor modeling or item selection errors.

28.6 High Model Complexity as Criterion

The prior discussion leads to a topic that is rarely ever considered: the objectives and requirements of the modeling process. Meehl (1990, p. 114), discussing the concept of *verisimilitude* (i.e., truth-likeness) from philosophy, noted that models are always imperfect and vary in the degree to which they approximate reality in two ways. *Incompleteness* deals with how well the complexities of the real world are represented in the model. *Falseness* examines how well contradictions between the model and the world are represented. Rozeboom (2005) noted that Meehl's notion of *verisimilitude* urged us to recognize that scientific theories are never impeccably veridical in all respects, and practical theory adjudication requires a researcher to ask not whether a model is true but how a model is true and to what degree it is true.

As Chin et al. (2008, p. 294) note:

> Most SEM studies seem to focus on the falsity of a model as opposed to its completeness. In part because of algorithmic constraints, few SEM models are very complex (i.e., have

> a large number of latent variables). Emphasis on model fit tends to restrict researchers to testing relatively elementary models representing either a simplistic theory or a narrow slice of a more complex theoretical domain. As an example, Shah and Goldstein (2006), in their review of 93 SEM-based articles in operations management, found an average of 4.4 latent variables per model with a range of 1–12 latent variables, and between 3 and 80 manifest variables with a mean of 14. MacCallum (2003, p. 118) concluded that "the empirical phenomena that yield the population variances and covariances in Σ are far too complex to be fully captured by a linear common factor model with a relatively small number of common factors."

Blalock (1979, p. 881), in his presidential address concerning measurement and conceptualization problems in sociology three decades earlier, made the case that "reality is sufficiently complex that we will need theories that contain upwards of fifty variables if we wish to disentangle the effects of numerous exogenous and endogenous variables on the diversity of dependent variables that interest us." He later stated (Blalock 1986) that in formulating theories and models, there is a natural incompatibility between generalizability, parsimony, and precision, and that one of these desired characteristics must be sacrificed when conducting research. Blalock, therefore, argued for excluding the criterion of parsimony in order to allow models to describe more diverse settings and populations by replacing "constants" reflecting such settings with explanatory variables.

It is under this backdrop of high complexity that PLS, regardless of whether applied under a strong substantive and theoretical context or limited/exploratory conditions, comes to the fore relative to CBSEM. Due to the algorithmic nature requiring inverting of matrices, users often run into difficulties handling larger models with 50 or more items measures using CBSEM. As the model complexity with associated number of items increase, not only does the chance of obtaining poor model fits increase, but so will the memory limitations in our current computer systems where the model either simply will not run or take an extraordinarily long time.

Thus, the question becomes whether the goal is to explain the covariances of a relatively small set of measured items based on a few underlying latent constructs or to focus on the complex interrelationships among a large set of factors that more closely mirrors the study context. The former may work well in experimental settings, whereas more complex models capturing many factors related to attitudes, opinions, and behaviors over time could be difficult to fully capture using CBSEM. In these instances, component-based methods such as PLS or path analysis may be very useful, especially if one places greater emphasis on the completeness portion of Meehl's notion of verisimilitude.

28.7 Sample Size Requirement

A side benefit of the partial nature of the PLS algorithm is that the sample size requirements when using PLS for complex models are likely much smaller than required for CBSEM (Chin and Newsted 1999). This can be ascertained as a first approximation by determining the specific portion of the model that has the largest number of predictors for a particular dependent variable and then applying Cohen's power tables (1988) relative to the effect sizes one wishes to detect. In other words, the researcher needs to determine which dependent variable (either at the structural level or item measure level) has the highest number of predictors (i.e., arrows directed). Since this represents the largest regression performed during the PLS iterative process, this would be the logical starting point for choosing an adequate sample to insure an adequate level of accuracy and statistical power. Ideally, if one wishes to customize the sample size estimation with specific effect sizes for the structural paths and include a certain amount of measurement error (normal or nonnormal), running a Monte Carlos simulation would be a better approach (e.g., Majchrak et al., 2005).

Figure 28.2 provides an exaggerated hypothetical example where the PLS algorithm has an advantage to that of CBSEM. As depicted, we have a linear sequential process connecting 100 constructs each with 100 reflective indicators. To estimate this within a standard CBSEM software requires initially calculating the covariances among the 10,000 indicators in the model. This represents a lower triangular matrix of 50,005,000 variances and covariances. This matrix size is prohibitive. Current computer memory cannot invert a matrix of this size. But within the PLS framework, we can see that all dependent variables only have one predictor. Therefore, throughout the PLS iterative process, PLS only performs a series of simple OLS regression (i.e., correlations). Thus, depending on the effect sizes for paths and loadings, the case can be made that sample size can be extremely small relative to the complexity of the model. If you were to use an OLS regression rule of 20 cases per dependent variable, this particular model would suggest 20 cases is enough. To play it safe, one might recommend 100 or 200 to improve accuracy. But this amount is several orders of magnitude better that what can be accomplished with CBSEM even if the software memory allows it to be estimated.

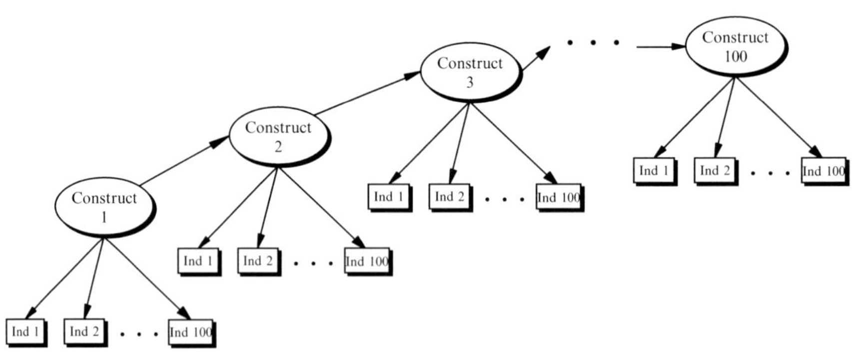

Fig. 28.2 Hypothetical example of a model requiring minimal sample size under PLS relative to CBSEM

28.8 Accuracy of Parameters Estimation

We now come to the case of accuracy of model estimates. Fornell and Bookstein (1982) have shown that PLS often provides component-based loadings and structural paths similar to SEM without requiring the distributional assumptions. Moreover, PLS estimates can be obtained with smaller sample sizes relative to model complexity (as discussed earlier). Yet, critics would often make the case that PLS estimates are not as efficient or potentially even biased relative to those obtained by CBSEM. A key argument for this position is that the case values for the constructs (i.e., PLS estimated scores) are "inconsistent" relative to CBSEM model analysis because PLS components are aggregates of the observed variables and include measurement error (Chin 1998b). This bias tends to manifest itself in somewhat higher estimates for loadings and lower structural path estimates. The estimates will approach the "true" parameter values when both the number of indicators per construct and sample size increase. This limiting case has been termed "consistency at large" (Wold 1982, p. 25).

What if the underlying population model is not covariance-based? Other researchers have suggested that these estimated biases were calculated relative to the covariance-based ML estimation, which presupposes that the underlying model is "true" and the generated data are covariance-based. Schneeweiss (1990, p. 38) noted that the consistency-at-large notion is really a "justification for using PLS as an estimation method to estimate LISREL parameters in cases where the number of manifest variables is large." He argued that PLS can be seen as a consistent estimator of parameters and latent variables as long as we ask the question of which population parameters we are attempting to estimate. If we are estimating the parameters for the population model as defined by PLS, then we have the advantage of "treating PLS as a method to define parameters and latent variables that are useful for describing the relations that may exist between blocks of observable (manifest) variables" (p. 38), even if the data cannot be regarded as stemming from a covariance model. Under these conditions, PLS will estimate model parameters consistently. If, on the other hand, the data are generated from a covariance-based model, PLS will produce inconsistent estimates. To date, papers running Monte Carlo simulation to test PLS estimation have always employed an underlying covariance-based model for data generation. No other underlying latent variable generating model (PLS based or otherwise) have been used.

Therefore, while PLS can be used in a confirmatory sense following a covariance-based orientation, it can also be used for testing the appropriateness of a block of indicators in a predictive sense and for suggesting potential relations among blocks without necessarily making any assumptions regarding which LV model generated the data. As Wold (1980) noted:

> The arrow scheme is usually tentative since the model construction is an evolutionary process. The empirical content of the model is extracted from the data, and the model is improved by interactions through the estimation procedure between the model and the data and the reactions of the researcher. Consequently, the researcher should begin with a generous number of observables-indicators in the various blocks. To use many observables makes

> for rich empirical content of the model and is favorable to the accuracy of the PLS estimation procedure. In the interaction between the data and the original model it will become apparent which indicators are relevant and which should be omitted. (p. 70)

28.9 Formative Measurement Items

A default assumption for CBSEM analysis is that the items or indicators used to measure a LV are *reflective* in nature. Such items are viewed as affected by the same underlying concept (i.e., the LV). Yet a common and serious mistake often committed by researchers is to inadvertently apply *formative* indicators (also known as cause measures) in an SEM analysis. Formative items are multidimensional in nature, but are the most immediate/antecedent items that produce/form/cause the LV to exist in its current state. As an initial conceptual approach for sorting through this, one needs to look at all the items used for a particular construct and determine whether they are tapping into the same underlying issue or factor. In other words, if the underlying construct was to change in magnitude, would all its items change as well? Alternatively, one can do the following thought exercise: Is it necessarily true that if one of the items (assuming all coded in the same direction) were to suddenly change in a particular direction, the others will change in a similar manner? If the answer is no and the items suggest multidimensionality and may, in fact, be formative. If so, the resulting CBSEM estimates would be invalid.

Figure 28.3 provides a graphical representation of these two modes for modeling indicators to latent variables. First introduced by Blalock (1964), formative indicators are defined as measures that form or cause the creation or change in an LV (Chin and Gopal 1995, pp. 58–59; Chin 1998b; Jarvis et al. 2003). Yet, a quarter century later, Cohen et al. (1990) found that this is a common mistake in psychological and sociological journals leading to serious questions concerning the validity of the results and conclusions. Attempts to explicitly model formative indicators in a CBSEM analysis have been shown to lead to identification problems with efforts to work around them generally unsuccessful (MacCallum and Browne 1993).

Since PLS explicitly estimates the outer weights to form construct scores, modeling formative indicators is much less problematic. A construct with formative indicators (whether endogenous or exogenously modeled) must be connected to at least one other construct to yield meaningful information since the multiple regression weights that PLS estimates are intended to overlap with neighboring latent variable blocks. Otherwise, without some structural linkage, the weights would end up being identical. This differs from modeling reflective indicators where the weights are meant to form the single best score to maximally predict its own measures (i.e., the first principal component).

There has been a mistaken assumption by some that all weights estimated by PLS are formative in nature. The likely reason is based on the perspective that the act of performing a weighted summation of items to create a construct score is the same as forming a construct. While this is technically true in the strict sense, the direction of the arrows linking measures to construct nonetheless can have a

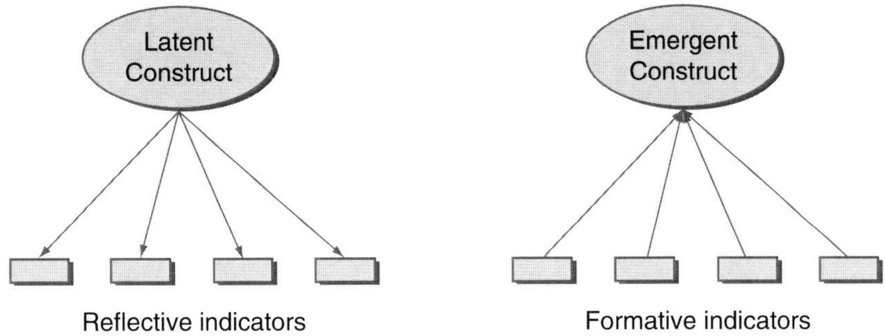

Fig. 28.3 Latent Construct with reflective indicators (Mode A) and emergent construct with formative indicators (Mode B)

dramatic effect on the weights that are produced. In the case of formative indicators, there is no emphasis on predicting its own measures. Rather, the objective is to obtain weights that create the best variate or construct score such that it maximally correlates with the neighboring constructs. Thus, PLS based formative indicators are inwards directed to maximize the structural portion of the model.

28.10 Higher Order Molar and Molecular Construct Scores

Higher order latent variables are often useful if a researcher wishes to model a level of abstraction higher than those first order constructs used in a basic CBSEM and PLS model. Due to the determinate nature of the PLS algorithm that explicitly weights measurement indicators to create construct scores, two types of higher order constructs can be modeled: what Chin and Gopal (1995) termed as molar and molecular higher order constructs. Molecular 2nd order constructs represent a higher level of abstraction with arrows pointing to its respective first order constructs (see Fig. 28.3). Whereas a second order molar model would have the arrow in the opposite direction going from the first order constructs to the higher second order one. In the case of CBSEM, researchers are limited only to second order molecular model. Moreover, an implicit equality constraint is placed among the ratio of the paths between the first and second order LVs (Fig. 28.4).

In the context of PLS, modeling either molecular or molar models is easily accomplished with existing PLS software. According to Chin et al. (1996, appendix A):

> Second order factors can be approximated using various procedures. One of the easiest to implement is the approach of repeated indicators known as the hierarchical component model suggested by Wold (cf. Lohmöller 1989, pp. 130–133). In essence, a second order factor is directly measured by observed variables for all the first order factors. While this approach repeats the number of manifest variables used, the model can be estimated by the standard PLS algorithm. This procedure works best with equal numbers of indicators for each construct (Fig. 28.5).

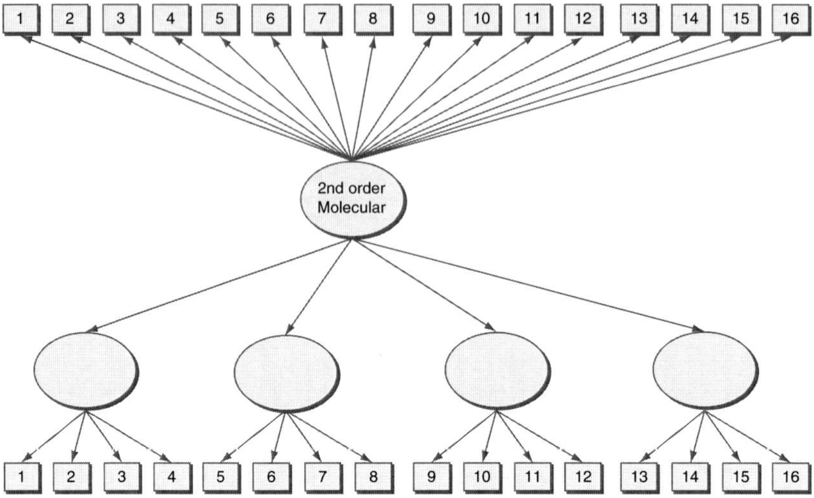

Fig. 28.4 Second order molecular model in PLS

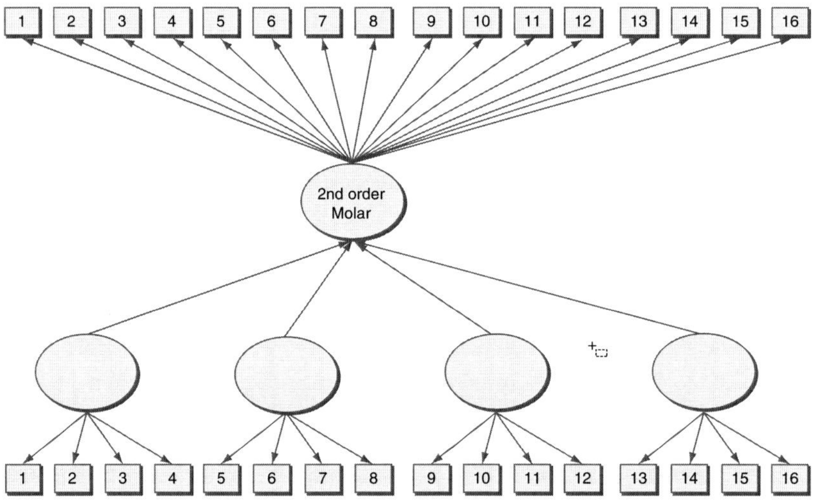

Fig. 28.5 Second order molar model in PLS

When considering such models, Chin (1998a) posed several questions that should be considered. The first is the purpose of these models. Is this second order factor expected to mediate fully the relationship of the first order factors when applied in a theoretical model? To postulate the existence of second order factor that sits in a vacuum holds little value. Rather, it must be related to other factors in a conceptual model. Because a second order factor such as a molecular model is modeled as being at a higher level of abstraction and reflected by first order factors, it needs to be related with other factors that are at a similar level of abstraction independent of whether these other factors are inferred from measured items or other first order factors. Therefore, it is imperative that this be demonstrated by embedding such second order factor models within a nomological network (i.e., used as a consequent and/or predictor of other LVs).

Tests of validity for a second order factor model should, by analogy, follow the same process that is used to examine the validity of first order factors. The first step applies the conceptual thought experiment discussed earlier for formative/reflective items, but at this higher order level. In essence, one asks whether the first order factors actually taps into the same underlying second order LV or are factors that form the second order LV (Chin 1998a).

28.11 Determinate Scores/Indices for Predictive Relevance

While the preceding discussion on higher order constructs indicated how the determinate scores that are provided by the PLS algorithm can provide flexibility in modeling both molar and molecular higher order constructs, another consideration is simply the scores at the first order level. These scores can be scaled in different ways (e.g., normalized or 0 to 100 points). But these scores actually are immediately interpretable from a predictive perspective. CBSEM, in contrast, has an inherent indeterminacy (Steiger 1979).

28.12 Eschewing the "True" Model for Prediction Focus

The benefits of CBSEM are predicated to a large part on the accuracy of the model being tested. Being a full-information procedure, we have noted that model fits and estimates can be influenced by many sources of error, including simply having one or two poor measures that do not belong with the other measures for a particular construct. A model positing that only a single factor can explain the inter-item covariances with items for other downstream constructs in a nomological network is likely incorrect. Subsets of items may be correlated because they are mutually affected by other underlying trait or method factors. Nonlinear relationships may also exist between the construct and its measures. The key question becomes how robust are the estimates that are obtained from models that are imperfect representations of the underlying "real world."

This leads to a recent statement from Cudeck and Henley (2003, p. 378) who question the notion of ever uncovering a "real world" by saying:

> A realistic perspective is that although a healthy skepticism to complex statistical results is appropriate, there are no true models to discover... The purpose of a mathematical model is to summarize data, to formalize the dynamics of a behavioral process, ***and to make predictions***. All of this is scientifically valuable and can be accomplished with a carefully developed model, even though the model is false. [emphasis added]

They go on to say (2003, p. 381):

> Even more tenable is the viewpoint that there is no population operating model of any kind. Researchers advocate a model that hopefully makes some sense of results. There is no pretense that the final product is true. From this perspective, the **purpose of the analysis is to summarize the data, to describe the change process algebraically, to formalize an expert's opinion about development with empirical support, and to make predictions**. These are extremely valuable activities both theoretically and practically. **This process differs from mere armchair theorizing in that a useful model must fit data well and must make sense scientifically**. If the enterprise is successful, the end result is a structure that accounts for data and is consistent with someone's theory of behavior. It actually is not all that easy to reach these goals. When it can be accomplished, the achievement is no less impressive because the model is not the truth. [emphasis added]

Such a viewpoint tends to be more akin to the American philosophical perspective of pragmatism (Menand 2002; Diggins 1994) which holds that both the meaning and the truth of any idea is a function of its practical outcome. Pragmatism according to Charles Sanders Peirce focuses on what the truth of statements (i.e., our analytical claims) means in terms of action (i.e., consequences of truth). Alternatively, William James' view of pragmatism loosely equates truth and usefulness as in "if something is true it is useful, and if it isn't useful, then talking about its truth doesn't make sense."

28.13 Ease of Model Specification and Model Interpretation

Although ease of model specification and reduction in the complexity regarding model identification should not be a primary basis for choosing PLS over CBSEM, it probably is worth considering at the margins. For PLS, as a component based approach with explicit estimation via indicator weights, a researcher only needs to specify the block of indicator representing each construct in question and the structural paths among all constructs. For CBSEM analysis, additional considerations such as model identification, measurement scale adequacy for the discrepancy estimator, setting the metric for each construct, and other SEM constraints need to be addressed.

The results from a PLS analysis can also be arguably said to be easier to assess. As long as the individual has a solid understanding of traditional multiple regression analysis and interpretation, PLS results are similarly interpreted. Rather than determining whether various model fit indices are appropriate, we focus on variance explained (i.e., the predictiveness of the model).

Summarizing, decisions or justification regarding the use of PLS instead of SEM depend in part on whether the researcher

- Places less premium on explaining the covariances of all item measures,
- Avoid negative impact due to errors in modeling or item usage,
- Values soft distributional assumptions,
- Sees the research not simply exploratory in nature, but interactive,

- Has formative measurement items,
- Requires flexibility in modeling higher order Molar and Molecular models,
- Is interested in obtaining determinate scores/indices that are predictive
- Has high model complexity,
- Faces relatively smaller sample size,
- Is less concerned with accuracy of parameter estimation or do not hold the belief in the notion of an underlying covariance based latent variable generating mechanism,
- Wants to shift the perspective of a "True" Model towards a Prediction Focus, and
- Values Ease of Model Specification & Model Interpretation

28.14 Clear Reporting

Before we begin, it is probably useful to reiterate, as in the case of all statistical analyses, that clear communication of the study context can assist both during the review process and in building a cumulative tradition for any discipline. Enough information need to be provided to understand (a) the population from which the data sample was obtained, (b) the distribution of the data to determine the adequacy of the statistical estimation procedure, (c) the conceptual model to determine the appropriateness of the statistical models analyzed, and (d) statistical results to corroborate the subsequent interpretation and conclusions. In addition, the computer program and version number, and specific estimation settings different from the program's default need to be reported.

In general, PLS models tested can easily be described through graphical representation and simple language. The graph needs to provide enough information to the reader regarding (a) the measurement model, which links LVs to its respective block of items and (b) the structural model connecting LVs. Luckily, as a component based approach, recursive models are automatically identified. Thus, in contrast to CBSEM, there is no need to articulate other parameter specifications such as which variances and paths are fixed and which are freed to be estimated.

28.15 Two Step Approach for Reporting Results

After providing the underlying rationale and justification for the use of PLS followed by the proposed model, the next step is to present results. In many respects, this approach is still heavily influenced by CBSEM reporting (especially the measurement portion) since it follows the notion of an underlying covariance based latent variable generating mechanism. Nevertheless, a very common approach is to present results in two phases. The first is to focus on the reliability and validity of the item measures used. The logic is that if you are not confident that the measures are representing the constructs of interest, there is little reason to use them to test the theoretical model in question. But if the measures are shown to be adequate, then the validity and results of the theoretical model (i.e., structural portion) is then presented.

28.16 Model Evaluation: Measurement Model Results

Thus, the first part in evaluating a model is to present what is termed the measurement model results. Here, we focus on the reliability and validity of the measures used to represent each construct. Ideally, this portion provides an evaluation on how accurate (i.e., reliable) the measures are and also their convergent and discriminant validities. One approach to obtain the measurement results is to first draw all possible structural links among the constructs you plan to use and then set the PLS inner weighting option using the factorial scheme. This essentially ignores the directionality of the arrows among constructs and simply performs pair wise correlations to establish inner weights. Alternatively, if you wish to determine the reliability and validity of your measures within the context of your actual structural model, you would report all measurement results, but with the particular structural model you are testing. Ideally, you should do both and compare both measurement results.

There are two sets of information that results from the preceding setup and are generally available from standard PLS software. Each set represent tests of discriminant validity (Chin 1998b). The first group of results is meant to show that a construct is more strongly related its own measures than with any other construct by examining the overlap in variance. Essentially, the argument is that if a specific construct is more correlated with another construct than with its own measures, there is the possibility that the two constructs share the same types of measures and are not conceptually distinct. Alternatively, it indicates that the two sets of items do a poor job of discriminating or differentiating the two underlying concepts you believe exists. To test for this, we compare the square root of the average variance extracted (AVE) with the correlations among constructs.

AVE was originally proposed by Fornell and Larcker (1981). It attempts to measure the amount of variance that an LV component captures from its indicators relative to the amount due to measurement error. AVE is only applicable for mode A (outward-directed) blocks. The AVE is calculated as follows:

$$AVE = \frac{\left(\sum \lambda_i{}^2\right) \operatorname{var} F}{\left(\sum \lambda_i{}^2\right) \operatorname{var} F + \sum \Theta_{ii}}$$

where λ_i, F, and Θ_{ii}, are the factor loading, factor variance, and unique/error variance respectively.

When all the indicators are standardized, this measure would be the same as the average of the communalities in the block. Fornell and Larcker (1981) suggested that this measure can also be interpreted as a measure of reliability for the

LV component score and tends to be more conservative than composite reliability ρ_c. Ideally, AVE should be greater than 0.50 meaning that 50% or more variance of the indicators should be accounted for.

While many researchers have compared the square root of AVE to construct correlations, you can equivalently compare the average variance extracted with the squared correlations among constructs. In either case, it provides a basis to see whether each construct is more highly related to its own measures than with other constructs. Overall, presenting AVE with squared correlations have two advantages. It provides a more intuitive interpretation since it represents the percentage overlap (i.e., shared variance) among constructs and construct to indicators and it is tends to be easier to distinguish the differences. Table 28.1 provides an example with ten constructs.

It is typical to also include the composite reliability measure, ρ_c, for each block of indicators. Composite reliability developed by Werts, Linn, and Jöreskog (1974) is a measure of internal consistency and is calculated as follows:

$$\rho_c = \frac{(\sum \lambda_i)^2 \text{ var } F}{(\sum \lambda_i)^2 \text{ var } F + \sum \Theta_{ii}},$$

where λ_i, F, and Θ_{ii}, are the factor loading, factor variance, and unique/error variance respectively.

In comparison to Cronbach's alpha, this measure does not assume tau equivalency among the measures with its assumption that all indicators are equally weighted. Therefore, while alpha tends to be a lower bound estimate of reliability, ρ_c is a closer approximation under the assumption that the parameter estimates are accurate. Finally, ρ_c like AVE is only applicable for LVs with reflective indicators (i.e., mode A blocks).

Depending on how much more information one wishes to provide, a table such as just presented can be a logical place to also include Cronbach alpha statistics as well as means, standard deviations, number of items and other descriptive statistics related to the construct scores. Alternatively, it can be in a separate table as depicted in Table 28.2.

The second and more detailed set of information examines how each item relates to each construct. Not only should each measure be strongly related to the construct it attempts to reflect, but it should not have a stronger connection with another construct. Otherwise, such a situation would imply that the measure in question is unable to discriminate as to whether it belongs to the construct it was intended to measure or to another (ie., discriminant validity problem). Table 28.3 provides an example from a different data set comparing correlations of each item to its intended construct (i.e., loadings) and to all other constructs (i.e., cross loadings). As Chin (1998b) notes, going down a particular construct column, you should expect to see item loadings to be higher than the cross loadings. Similarly, if you scan across a particular item row, you should expect to see that any item be more strongly related to its construct column than any other construct column. If this is found to be the

Table 28.1 Inter-construct correlations and reliability measures

Composite reliability	Average variance extracted		Intention	ra	eou	cmpt	Image	vis	tr	rd	vlt	Attitude
			Correlations of among Constructs									
0.963	0.947	intention	1.000									
0.962	0.850	ra	0.653	1.000								
0.952	0.817	eou	0.596	0.483	1.000							
0.949	0.908	cmpt	0.660	0.678	0.549	1.000						
0.864	0.758	image	0.212	0.396	0.169	0.284	1.000					
0.852	0.688	vis	0.563	0.395	0.415	0.408	0.178	1.000				
0.812	0.691	tr	0.629	0.380	0.517	0.456	0.123	0.588	1.000			
0.886	0.814	rd	0.586	0.448	0.614	0.475	0.165	0.562	0.549	1.000		
0.843	0.761	vlt	−0.417	−0.403	−0.190	−0.299	−0.164	−0.359	−0.267	−0.257	1.000	
0.949	0.853	attitude	0.659	0.566	0.642	0.646	0.309	0.437	0.498	0.560	−0.233	1.000
		Squared correlations of among Constructs										
Composite Reliability	Average Variance extracted		intention	ra	eou	cmpt	image	vis	tr	rd	vlt	attitude
0.963	0.897	intention	1.000									
0.962	0.722	ra	0.426	1.000								
0.952	0.667	eou	0.355	0.233	1.000							
0.949	0.824	cmpt	0.435	0.459	0.302	1.000						
0.864	0.575	image	0.045	0.157	0.029	0.081	1.000					
0.852	0.473	vis	0.317	0.156	0.172	0.167	0.032	1.000				
0.812	0.478	tr	0.395	0.144	0.267	0.208	0.015	0.345	1.000			
0.886	0.663	rd	0.343	0.201	0.377	0.226	0.027	0.315	0.302	1.000		
0.843	0.579	vlt	0.174	0.163	0.036	0.090	0.027	0.129	0.071	0.066	1.000	
0.949	0.728	attitude	0.434	0.320	0.412	0.417	0.095	0.191	0.248	0.314	0.055	1.000

Table 28.2 Descriptive statistics for each construct

Construct	Number of items	Mean	Standard deviation
Relative Advantage (ra)	11	5.0265	1.2000
Perceived Ease-of-use (eou)	10	5.3914	1.0608
Compatibility (cmpt)	4	5.0820	1.3162
Image	5	3.2760	1.4037
Visibility (vis)	7	5.4762	.9156
Trialability (tr)	5	4.6516	1.3926
Result Demonstrability (rd)	4	5.3915	1.1158
Voluntariness (vlt)	4	3.0837	1.3502
Attitude	7	5.5677	.9071
Intention	3	6.7584	1.6170

Table 28.3 Outer model loadings and cross loadings

	Loadings and cross-loadings for the measurement (outer) model.				
	Useful	Ease of use	Resources	Attitude	Intention
U1	**0.95**	0.40	0.37	0.78	0.48
U2	**0.96**	0.41	0.37	0.77	0.45
U3	**0.95**	0.38	0.35	0.75	0.48
U4	**0.96**	0.39	0.34	0.75	0.41
U5	**0.95**	0.43	0.35	0.78	0.45
U6	**0.96**	0.46	0.39	0.79	0.48
EOU1	0.35	**0.86**	0.53	0.42	0.35
EOU2	0.40	**0.91**	0.44	0.41	0.35
EOU3	0.40	**0.94**	0.46	0.40	0.36
EOU4	0.44	**0.90**	0.43	0.44	0.37
EOU5	0.44	**0.92**	0.50	0.46	0.36
EOU6	0.37	**0.93**	0.44	0.42	0.33
R1	0.42	0.51	**0.90**	0.41	0.42
R2	0.37	0.50	**0.91**	0.38	0.46
R3	0.31	0.46	**0.91**	0.35	0.41
R4	0.28	0.38	**0.90**	0.33	0.44
A1	0.80	0.47	0.39	**0.98**	0.54
A2	0.80	0.44	0.41	**0.99**	0.57
A3	0.78	0.45	0.41	**0.98**	0.58
I1	0.48	0.38	0.46	0.58	**0.97**
I2	0.47	0.37	0.48	0.56	**0.99**
I3	0.47	0.37	0.48	0.56	**0.99**

case, the claim can be made for discriminant validity at the item level. Specifically, we can say that each item loads more highly on their own construct than on other constructs and that all constructs share more variance with their measures than with other constructs.

At this point, it is worth noting that while the discriminant validity based on correlations can be easily determined in our example, this is not necessarily always the case. In situations where the cross loadings seem to be quite close in magnitude to the item loading, it may require an alternative presentation where you square all the loadings and cross loadings. In fact, while the current norm among researchers is to present loadings and cross loadings, we can argue that presenting the square of the loadings and cross loadings is more intuitive. For example, while a standardized loading 0.8 compared to a cross loading of 0.7 may raise concerns among naïve researchers pointing out that there is a 0.1 difference, providing squared results gives a more intuitive interpretation since it represents the percentage overlap between an item and any construct. In our hypothetical example, the item relationship to its own construct has shared variance of 64% (i.e., 0.8*0.8), while that shared with some other item is 49% (i.e., 0.7*.07). If we take a look at item U1 as another example, the shared variance to its own construct is 90% (i.e., 0.95*0.95) whereas it only overlaps most at 61% with attitude (i.e., 0.78*0.78). Since the goal is to have a strong nomological network where constructs at the structural level are closely related, this difference seems reasonable.

In addition to discriminant validity, one also needs to examine convergent validity which is defined as the extent to which blocks of items strongly agree (i.e., converge) in their representation of the underlying construct they were created to measure. In other words, how high are each of the loadings and are they more or less similar? If you have measures that are mixed and have a wide range (e.g., varying from 0.5 to 0.9), this would raise concern about whether your measures are truly a homogenous set that primarily captures the phenomenon of interest. But, with both a higher average loadings and narrower range such as from 0.7 to 0.9 you would have greater confidence that all items help (i.e., converge) in estimating the underlying construct. While there is no set range or minimum, the narrower the range and higher the lowest loading is the more you can assume convergent validity.

28.17 Model Evaluation: Structural Model Results

Having established the appropriateness of the measures, the next step is to provide evidence supporting the theoretical model as exemplified by the structural portion of your model. As discussed earlier, a major emphasis in PLS analysis is on variance explained as well as establishing the significance of all path estimates. Specifically, predictive power of the structural model is assessed by the R^2 values of the endogenous constructs. Similar to its counterparts in OLS regression, PLS R^2 results represent the amount of variance in the construct in question that is explained by the model.

Thus, for a given PLS model, we can start by looking at the R-squares for each dependent LV in the structural model provided by PLS. This is obtained because the case values of the LVs are determined by the weight relations. The corresponding standardized path estimates can also be examined and interpreted in the same manner. Finally, the change in R-squares can be explored to see whether the impact of a particular independent LV on a dependent LV has substantive impact. Specifically, the effect size f^2 can be calculated as:

$$f^2 = \frac{R^2_{included} - R^2_{excluded}}{1 - R^2_{included}} \tag{28.1}$$

where $R^2_{included}$ and $R^2_{excluded}$ are the R-squares provided on the dependent LV when the predictor LV is used or omitted in the structural equation respectively. f^2 of 0.02, 0.15, and 0.35, similar to Cohen (1988) operational definitions for multiple regression, can be viewed as a gauge for whether a predictor LV has a small, medium, or large effect at the structural level.

If you are examining sets of predictors for a dependent construct, where the baseline model is compared with adding two or more additional LVs, you can perform an F test which is calculated as follows:

$$F = \frac{\frac{R_2^2 - R_1^2}{k_2 - k_1}}{\frac{1 - R_2^2}{N - k_2 - 1}}$$

With $k_2 - k_1$, N-$k_2 - 1$ degrees of freedom

where R_1^2 is for the baseline model and R_2^2 is the superset model that includes the additional LVs, k_2 is the number of predictors for the superset model and k_1 is the number of predictors for the baseline, and N is the sample size.

In terms of significance, the conventional wisdom since Chin (1998b) first introduced its use for PLS estimation is to apply bootstrapping. The bootstrap approach represents a nonparametric approach for estimating the precision of the PLS estimates. N samples sets are created in order to obtain N estimates for each parameter in the PLS model. Each sample is obtained by sampling with replacement from the original data set (typically until the number of cases are identical to the original sample set). Various approaches for estimating confidence intervals have been developed (see Efron and Tibshirani 1993, for more details). The simplest is a semi parametric approach that uses the N bootstrap estimates for each parameter of interest to calculate the standard error and associated t test. But both a percentile or BCA approach would be completely distribution free.

An alternative resampling procedure, but less utilized in recent years is the jackknife. In general, the jackknife is another inferential technique that assesses the variability of a statistic by examining the variability of the sample data rather than using parametric assumptions. Developed in the late 1940s and 1950s, the jackknife can be used to provide both estimates and compensate for bias in statistical estimates by developing robust confidence intervals. The general approach, in contrast to bootstrapping, is "delete n cases" where n is typically 1. Parameter estimates are calculated for each instance and the variation in the estimates are analyzed.

The basic steps for performing jackknife on a parameter estimate θ of the population value θ_p (e.g., factor weight or loading, structural path) is as follows:

1. Calculate the parameter using the entire sample data. Let's call this θ.
2. Partition the sample into subsamples according to the deletion number d. The first subsample represents the full sample with the first d cases removed. The second subsample has the next d cases deleted. Thus, a full sample set of 100 cases with a deletion number of 2 results in 50 subsamples where each subsample has 98 cases.
3. For each of the n subsamples (say the ith subsample), calculate the pseudo-jackknife value J_i as follows:

$$J_i = n^* \theta - (n - 1)\theta_i \tag{28.2}$$

4. Calculate the mean of the pseudovalues to yield the jackknife estimate JM of the population parameter θ_p as follows:

$$JM = \frac{\sum J_i}{n} = n^* \theta - (n - 1)^* \frac{\sum \theta_i}{n} \tag{28.3}$$

5. Treat the pseudovalues as approximately independent and identically randomly distributed (Tukey 1958) and calculate the standard deviation (SD) and standard error (SE) as follows:

$$\begin{aligned} SD &= \frac{\sqrt{\sum_i (J_i - JM)^2}}{n - 1} \\ SE &= \frac{SD}{\sqrt{n}} \end{aligned} \tag{28.4}$$

6. The jackknifed t-statistic with n-1 degrees of freedom (where n is the number of subsamples) is used to test the null hypothesis that θ_p is not different from θ_0.

$$t\text{-}statistic = \frac{(JM - \theta_0)}{SE} \tag{28.5}$$

where θ_0 is normally zero.

Although the pseudovalues are asymptotically independent, Gray and Schucany (1972, pp. 138–162) advise adjusting the "t-statistic" to account for possible interdependence. If the intraclass correlation between pseudovalues is r, the t-statistic should be adjusted by multiplying it with the following correction factor:

$$\sqrt{\frac{1 - r}{1 - (n - 1)^* r}} \tag{28.6}$$

Gray and Schucany suggested the use 1/n for r, which results in the correction factor as follows:

$$\sqrt{\frac{n-1}{2n-1}} \tag{28.7}$$

Overall, jackknife estimation tends to take less time for standard error estimation under the joint assumption that the bootstrap procedure utilizes a confidence estimation procedure other than the normal approximation and the number of resamples are larger than those of the jackknife. Conversely, the jackknife is viewed as less efficient than the bootstrap because it can be considered as an approximation to the bootstrap (Efron and Tibshirani 1993, pp. 145–146). In general, both the jackknife and bootstrap standard errors should converge.

With the preceding discussion in mind, Figs. 28.6 and 28.7 drawn from a study by George, Hinson, and Chin (2000) provide an example of presenting some of the structural model results. Figure 28.6 represents a baseline model and Fig. 28.7 incorporates the additional predictor construct labeled attitude. The effect size for attitude, while significant, is considered small at 0.045. But also important is that attitude mediates the influence of the construct Ease of Use. While all other structural paths remain approximately the same both prior and after the inclusion of attitude, Ease of Use changes from having a significant standardized beta of 0.128 to non-significant. This is a sufficient test within the specific context that attitude fully

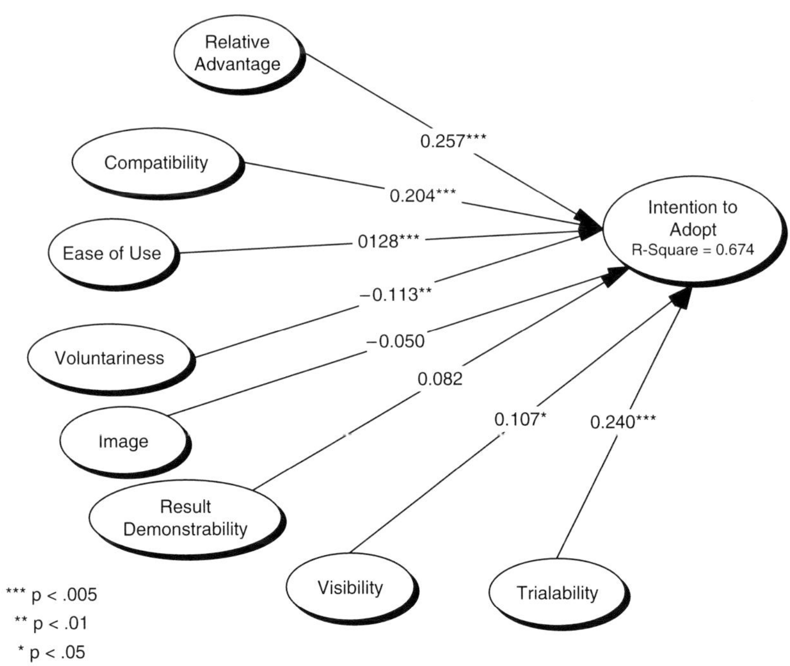

Fig. 28.6 Structural model with path coefficients (without Attitude)

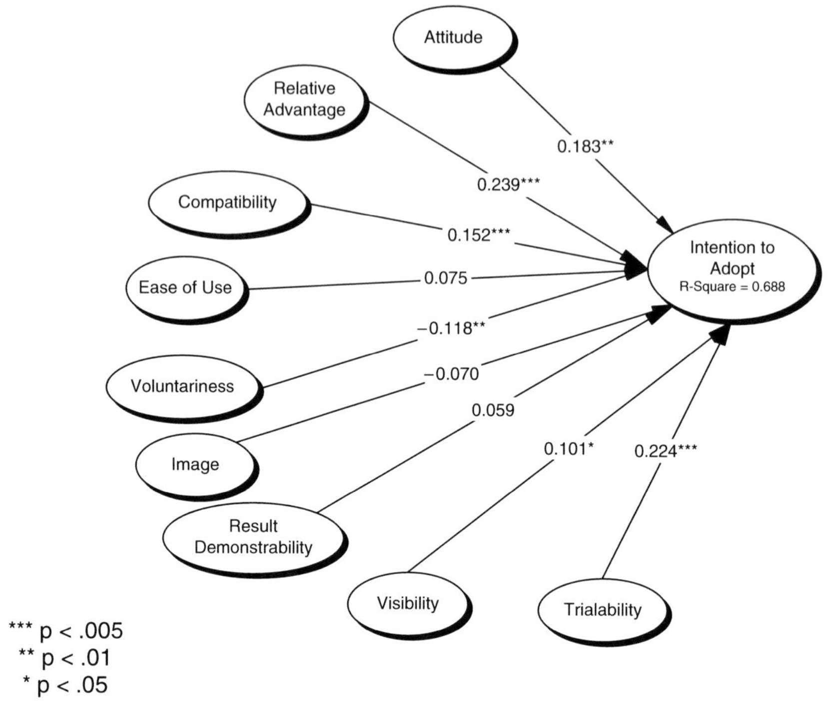

Fig. 28.7 Structural model with path coefficients (including Attitude)

mediates the impact of Ease of Use on Intention. In other words, if the inclusion of a new construct into a model changes the path of an existing construct from significant to non-significant, you have established full mediation for this new construct.

To assess the significance of indirect effects, one needs to explicitly model the two paths both directed in and out of the mediating construct. Figure 28.8 represents a simplistic case of only one indirect path. While the bootstrap results from both Ease of Use to Attitude and Attitude to Intention is shown to be significant, this does not necessarily guarantee the indirect effect of 0.12 (i.e., 0.65*0.183) is significant. It is recommended that for assessing the significance of indirect paths in a PLS structural model, you should simply apply the same bootstrapping procedure as done elsewhere with path analysis.

The two step bootstrapping procedure for testing mediation is as follows:

1. Use the specific model in question with both direct and indirect paths included and perform N bootstrap resampling (e.g., 1,000 resamples) and explicitly calculate the product of direct paths that form the indirect path being assessed.
2. Estimate the significance using either percentile bootstrap or bias corrected corrected bootstrap which has been shown to have the least biased confidence intervals, greatest power to detect nonzero effects and contrasts, and the most accurate overall Type I error (Williams and MacKinnon 2008).

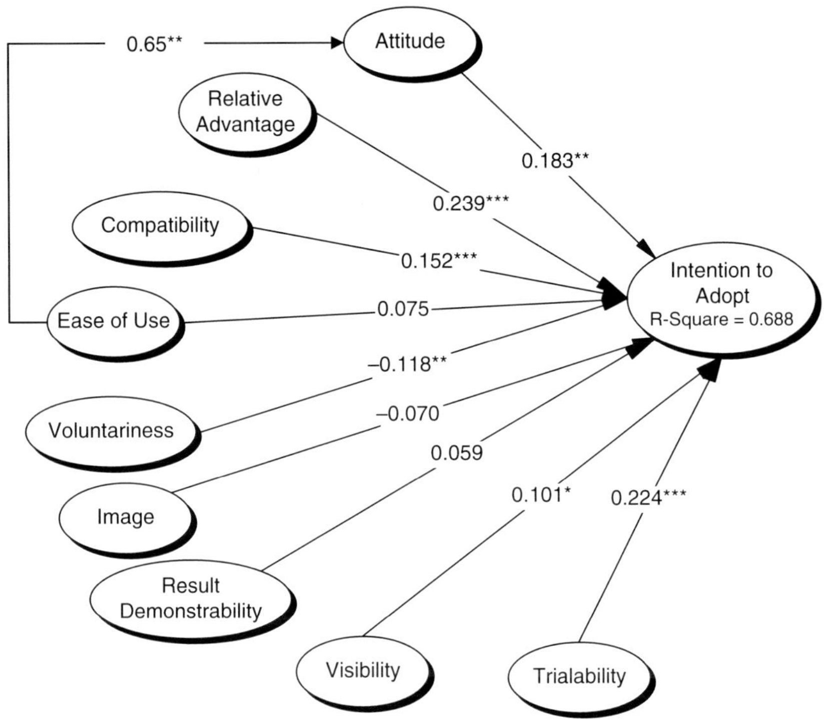

Fig. 28.8 Structural model with indirect path effect through attitude

In the case of our example, using this distribution free procedure, the results did indicate that the indirect path from Ease of Use to Intention through Attitude was significant ($p < 0.005$). Coupled with the fact that the results also showed the direct effect of Ease of Use to Intention as non-significant, we can conclude that Attitude fully mediates the impact of Ease of Use. Hypothetically, if we had found the direct effect from Ease of Use to be smaller, but statistically significant, we would label Attitude as a partial mediator.

28.17.1 Q^2 Predictive Relevance

Besides looking at the magnitude of the R-square as a criterion for predictive relevance, we can also apply the predictive sample reuse technique as developed by Stone (1974) and Geisser (1975). This technique represents a synthesis of cross-validation and function fitting with the perspective that the prediction of observables or potential observables is of much greater relevance than the estimation of what are often artificial construct-parameters (Geisser 1975, p. 320). The sample reuse technique has been argued as fitting the soft modeling approach of PLS like hand in glove (Wold 1982, p. 30).

The PLS adaptation of this approach follows a blindfolding procedure that omits a part of the data for a particular block of indicators during parameter estimations and then attempts to estimate the omitted part using the estimated parameters. Specifically, the blindfolding procedure takes a block of say N cases and K indicators and takes out a portion of the N by K data points. Using an omission distance D, the first point (case 1 indicator 1) is removed and then every other D data point as we move across each column and row is omitted until we reach the end of the data matrix. With the remaining data points, estimates are obtained by treating the missing values via pairwise deletion, mean substitution, or an imputation procedure. The sum of squares of prediction error (E) is calculated when the omitted data points are then predicted. The sum of squares errors using the mean for prediction (O) is also calculated. The omitted data points are returned and we shift over to the next data point in the data matrix (case 1 indicator 2) as the starting point for a new round of omission. A new E and O are calculated. This continues until D sets of Es and Os are obtained. The predictive measure for the block becomes:

$$Q^2 = 1 - \frac{\sum_D E_D}{\sum_D O_D} \tag{28.8}$$

Thus, without any loss of freedom, Q^2 represents a measure of how well-observed values are reconstructed by the model and its parameter estimates. $Q^2 > 0$ implies the model has predictive relevance whereas $Q^2 < 0$ represents a lack of predictive relevance. As in the case of f^2, changes in Q^2 can be used to assess the relative impact of the structural model on the observed measures for each dependent LV:

$$q^2 = \frac{Q^2_{included} - Q^2_{excluded}}{1 - Q^2_{included}} \tag{28.9}$$

Different forms of Q^2 can be obtained depending on the form of prediction. A cross-validated communality Q^2 is obtained if prediction of the data points is made by the underlying latent variable score, whereas a cross-validated redundancy Q^2 is obtained if prediction is made by those LVs that predict the block in question. One would use the cross-validated redundancy measure to examine the predictive relevance of one's theoretical/structural model.

According to Wold (1982, p. 33), the omission distance D should be a prime integer between the number of indicators K and cases N. Furthermore, the choice of the omission distance D need not be large. Experience shows that D from 5 to 10 is feasible as long as N is large.

For the model depicted in Fig. 28.6, we obtained a cross-validated redundancy Q^2 of 0.585 and a cross-validated communality Q^2 of 0.731. In general, a cross-validated redundancy Q^2 above 0.5 is indicative of a predictive model.

Recently, a global criterion of goodness of fit (i.e., GoF index) has been proposed by Tenenhaus et al. (2004). The intent is to account for the PLS model performance at both the measurement and the structural model with a focus on overall prediction performance of the model. The GoF index is obtained as the geometric mean of the

Table 28.4 GoF index results

	GoF	GoF (Bootstrap)	Standard error	Critical ratio (CR)	Lower bound (95%)	Upper bound (95%)
Absolute	0.672	0.678	0.021	31.904	0.631	0.722
Relative	0.909	0.883	0.018	51.678	0.840	0.916
Outer model	0.979	0.974	0.011	89.827	0.950	0.988
Inner model	0.928	0.906	0.015	62.779	0.877	0.939

average communality index and the average R^2 value:

$$GOF = \sqrt{\overline{Communality} * \overline{R^2}}$$

While the utility of this index is likely best applied for models with reflective indicators, the case has been made that there is a natural tradeoff when using formative indicators where the inner model predictiveness is increased as the expense of the outer model. For more detail, please see the Esposito Vinzi, Trinchera, and Amato chapter in this book. Table 28.4 presents the results for the same model depicted in Fig. 28.6 and corroborating the results of the cross-validated redundancy Q^2, the relative GoF was above the 0.90 threshold suggestive of a good model. Finally, the bootstrap cross validation Relative Performance index (see Chin chapter in this book) of 23.12 provides yet another example of how the PLS estimates provide predictive improvement relative to an equally weighted simple summed regression.

Example 2: Application of PLS with a Formative Construct - the Case of Perceived Resources

As a final example of reporting PLS results, we examine a portion of the results produced by Mathieson et al. (2001) that assesses the adequacy of formative measures. In their paper, they took a well established model for predicting individual usage of information technology (IT) and included a new construct called perceived resources.

The baseline model (see Fig. 28.9) depicts an individual's intention to use an IT as predicted by both one's attitude (in an evaluative sense) toward usage (labeled Attitude) and one's cognitive belief that the use of the IT will lead to performance gains (labeled Usefulness). Attitude is seen as partially mediating the impact of Usefulness. In addition, one's belief in the ease of use of the IT is modeled indirectly impacting intention through the attitude and usefulness. Finally Intention is seen as leading to actual System Usage. For completeness, the loadings, weights, composite reliability, AVE, Q^2, and other measures as discussed earlier should be presented. But due to space limitations, we focus primarily on those results used to validate the R measures.

Mathieson et al. (2001) extended the baseline model in order to enhance predictiveness under conditions where the ability to use an IT is not entirely under

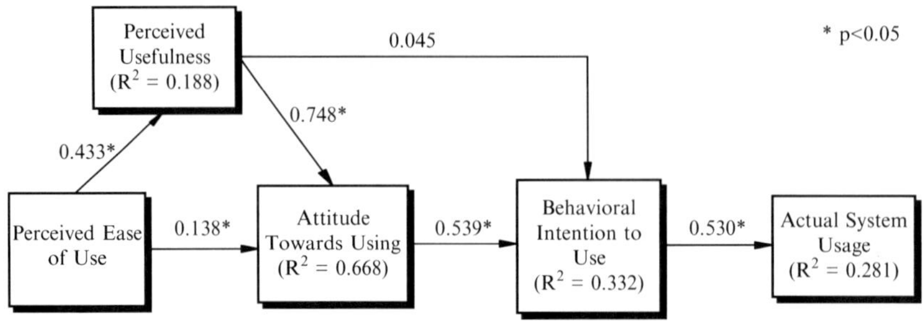

Fig. 28.9 Baseline model

Table 28.5 The perceived resource instrument. (Fully-anchored Likert scales were used. Responses to all items ranged from Extremely likely (7) to Extremely unlikely (1).)

General Items	
1.	I have the resources, opportunities and knowledge I would need to use a database package in my job.
2.	There are no barriers to my using a database package in my job.
3.	I would be able to use a database package in my job if I wanted to.
4.	I have access to the resources I would need to use a database package in my job.
Specific items	
5.	I have access to the hardware and software I would need to use a database package in my job.
6.	I have the knowledge I would need to use a database package in my job.
7.	I would be able to find the time I would need to use a database package in my job.
8.	Financial resources (e.g., to pay for computer time) are not a barrier for me in using a database package in my job.
9.	If I needed someone's help in using a database package in my job, I could get it easily.
10.	I have the documentation (manuals, books etc.) I would need to use a database package in my job.
11.	I have access to the data (on customers, products, etc.) I would need to use a database package in my job.

the volition of the individual. They developed a new construct called perceived resources. Perceived resources (R) is defined as the extent to which an individual believes that he or she has the personal and organizational tools needed to use an IT. Thus, separate from the notion of assessing one's own ability, R attempts to capture how the perception of the presence or absence of resources or opportunities can impact one's attitude and intention toward using an IT.

Two sets of items were developed to measure R (see Table 28.5). One set of four indicators (R1 through R4) consisted of reflective measures that tap into the general feeling of having enough resources, whereas the other set (R5 through R11) attempted to capture a comprehensive set of formative indicators that help create that perception.

For the R formative measures, the previous recommendations of examining AVE and correlations, composite reliability, and loadings versus cross loadings do not apply since formative items are viewed as multidimensional and not similar measures (in a convergent validity sense) reflecting the same underlying construct. Therefore, we must present other results. The first analysis is to compare the two sets of measures via a two-block redundancy model. The redundancy model is specified based on the original design of the questions. In other words, one set was designed with reflective indicators in mind whereas the other was meant to be formative. The extent to which both modes of assessing R are successful can be partly determined by the structural path linking them. In general, we would expect a path of 0.80 or above to be suggestive of securing an adequate (i.e., comprehensive) set of formative measures assuming convergent validity (i.e., adequate loadings) for the reflective set. A path of 0.90 or above would indicate an extremely strong result.

Figure 28.10 provides the results of the redundancy analysis. The path of 0.87 between the two modes of assessing R indicates a strong convergence and implies an adequate coverage of the perceptions in the formative set. The topmost estimate for each measurement path represents the regression estimates. For the formative case, the estimates represent the multiple regression weights as opposed to the component loadings for the reflective case. In turn, the numbers in the parentheses for the formative block represent the component loadings (simple regression between the indicator and the LV component scores). Conversely, for the reflective block, the parentheses represent the weights. Bootstrap resampling was performed to examine the significance of the weights for the formative block and loadings for the reflective block. Overall, the loadings for the reflective set were uniformly high around 0.9 with a composite reliability ρ_c of 0.95 and a AVE of 0.81. Among the formative measures, R5, R6, R7, and R8 were all significant ($p < 0.01$) with weights of 0.59, 0.27, 0.13, and 0.10 respectively. This empirically suggests that the overall impression of available resources for IT usage is primarily formed by access to necessary hardware, software, and knowledge.

The interpretation of LVs with formative indicators in any PLS analysis should be based on the weights. As in the case of a canonical correlation (Harris 1989),

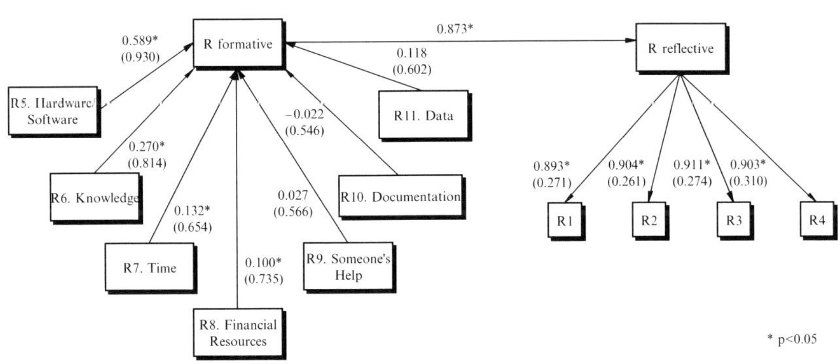

Fig. 28.10 PLS results for a redundancy model. (*indicates significant estimates $p < 0.01$)

the weights provide information as to what the makeup and relative importance are for each indicator in the creation/formation of the component. Because the intraset correlations for each block was never taken into account in the estimation process, use of the loadings would be misleading. Comparison of loadings among indicators within a block of formative indicators would therefore be nonsensical. At best, loadings can be used for identifying which indicator makes the best surrogate for the component score.

The next step is to examine the validity of the R measures as applied into the nomological network of the basic model. R (mode A) is placed as a new predictor. Although R was theoretically developed to predict the Attitude and Intention constructs, structural paths to Usefulness and Ease of Use were also included in an exploratory sense. Results indicate significant paths of 0.22 and 0.51 respectively. The full structural level results are presented in Fig. 28.11. All paths going from R were found to be significant. But in terms of substantive effects, R had more impact on Intention than Attitude. This is determined in terms of changes in both the R-squares and Q^2 as measured by f^2 and q^2. The f^2 for intention and attitude were 0.12 and 0.04. Thus, R has an approximately medium effect on Intention above and beyond the contributions provided by Usefulness and Attitude. The impact on Attitude, on the other hand, was smaller.

In calculating q^2, blindfold analyses were performed with varying omission distances. Specifically, D of 7, 37, and 97 were used. The results were very similar (i.e., to the third decimal point). The cross-validated redundancy Q^2 went from 0.15 to

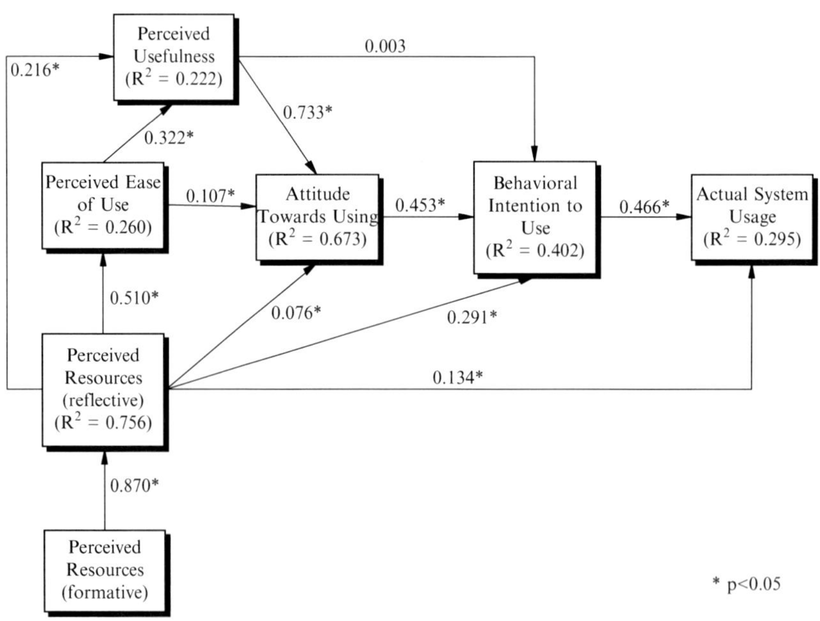

Fig. 28.11 Impact of including perceived resources (reflective measures)

0.26 when we included R implying a q^2 of 0.15 reflecting, again, a medium impact. Similarly, the q^2 for Attitude was small with an effect of 0.026 for Q^2 increasing from 0.60 to 0.61.

The validity of the measurement model was then assessed by examining the loading and cross-loadings (not shown here for space limitation). Because all measures are reflective (i.e., mode A analysis), we can examine the individual loadings for each block of indicators. All standardized loadings should be greater that 0.707. This condition was met in the study. But it should also be noted that this rule of thumb should not be as rigid at early stages of scale development. Loadings of 0.5 or 0.6 may still be acceptable if there exist additional indicators in the block for comparison basis. The composite reliability ρ_c for each construct was also above 0.95.

If items with low loadings in a mode A block are encountered, possible reasons are (a) that the item is simply unreliable, (b) it may be influenced by additional factors such as a method effect or some other concept, or (c) the construct itself is multidimensional in character (thus items where created capturing different issues). For the last situation, one might partition the items into more coherent blocks or simply remove the item. For the first situation, keeping the item will likely still increase predictiveness since the PLS algorithm will still weight it to the extent it helps minimize residual variance as long as other more reliable indicators exist. This, of course, assumes the poor loading is due only to noise. This would not be the case if the indicator cross-loads higher with other LVs. Only in situation (b) would you have to remove it for lack of discriminant validity.

Finally, we can replace the mode A measures with the mode B measures of R. If the mode A measures do approximate R well, the pattern of structural relationships we saw with the reflective measures should also appear. The only difference would be an increase in the magnitude of the paths connected to R because mode B minimizes the residuals at the structural level. The results, as provided in Fig. 28.12, did occur as expected. In particular, we see the R-square increase from 0.40 to 0.44 for Intention, 0.67–0.69 for Attitude, 0.26–0.35 for Ease of Use, and 0.22–0.324 for Usefulness. While not shown here, a similar analysis can be made to assess the impact of R to System Usage. In terms of structural paths to the main endogenous constructs of Attitude, Intention, and System Usage, both sets of measures showed it had the most substantive impact to Intention. In contrast, both yield the same conclusion of having lesser impact on Attitude and System Usage.

As we look at the weights and loadings associated with our model using the formative measures of R (see Table 28.6), we notice that only indicators R6, R7, and R10 have a significant impact with 0.23, 0.56, and 0.41 respectively. Substantively, this would suggest that time is the most important resource, followed by documentation, and then knowledge in forming an overall perception of resources that facilitate or hinder using an IT. We also see a difference in the impact of these measures relative to the earlier Redundancy model. Whereas, in both analyses, knowledge (R6) and Time (R7) were factors influential in forming one's overall perception of resources, the significance of the other factors varied. These results highlight the importance of the nomological context in which measures are used. Whether

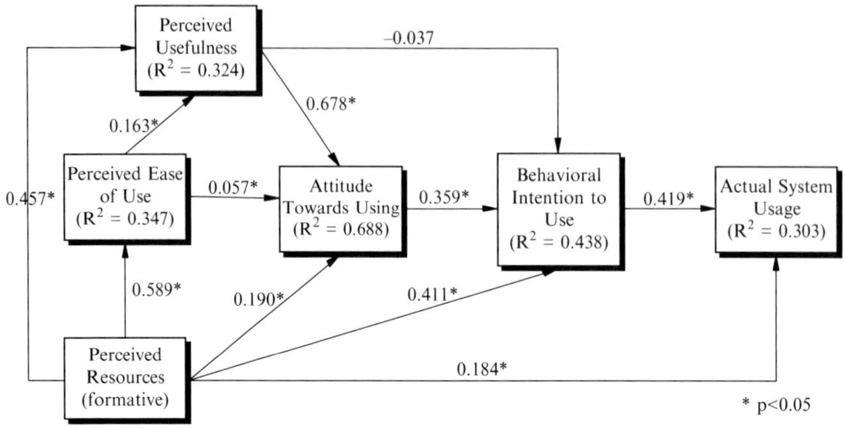

Fig. 28.12 PLS results for the extended model using formative measures

Table 28.6 Outer Model weights and loadings for model using formative R measures (n.s. = nonsignificant)

Indicator	Weight	Loading
U1	0.1782	0.9494
U2	0.1727	0.9591
U3	0.1740	0.9462
U4	0.1665	0.9555
U5	0.1752	0.9541
U6	0.1822	0.9600
EOU1	0.1764	0.8592
EOU2	0.1846	0.8973
EOU3	0.1843	0.9256
EOU4	0.1909	0.8883
EOU5	0.2003	0.9095
EOU6	0.1763	0.9221
A1	0.3364	0.9778
A2	0.3439	0.9863
A3	0.3408	0.9779
I1	0.3360	0.9644
I2	0.3473	0.9822
I3	0.3433	0.9821
TIMES	0.6746	0.8294
LENGTH	0.6026	0.7360
R5. Hardware/Software	−0.0024 (n.s.)	0.5832
R6. Knowledge	0.2284	0.7290
R7. Time	0.5611	0.8549
R8. Financial Resources	−0.1015 (n.s.)	0.5597
R9. Someone's Help	0.0698 (n.s.)	0.5941
R10. Documentation	0.4120	0.7592
R11. Data	0.0874 (n.s.)	0.6847

formative or reflective, loadings and weights may change for a given construct as it is applied in different contexts for which it was originally developed.

In summary, this second example provides both an instance of how PLS can be used in a confirmatory sense and presenting results for validating formative indicators. An existing theoretical model with an established set of measures was used as the basis for further theoretical and measurement development. As depicted in Fig. 28.13, the validation process for the formative items depends on whether you have access to a validated reflective set. Without the reflective measures, there would be less evidence as to whether the researcher was successful in estimating the particular construct. In a situation where a researcher has only formative measures, the predictive capabilities of that block of measures would be the primary bases for validation. If the formative indicators is applied in a theoretical model where a reflective set had been used in the past, a structural pattern comparison can be made. Specifically, we would expect that the structural paths linking the emergent construct with other constructs should follow the same pattern as those estimated in previous studies that applied the latent construct using reflective measures. In our example, both formative and reflective sets of measures were created to estimate the

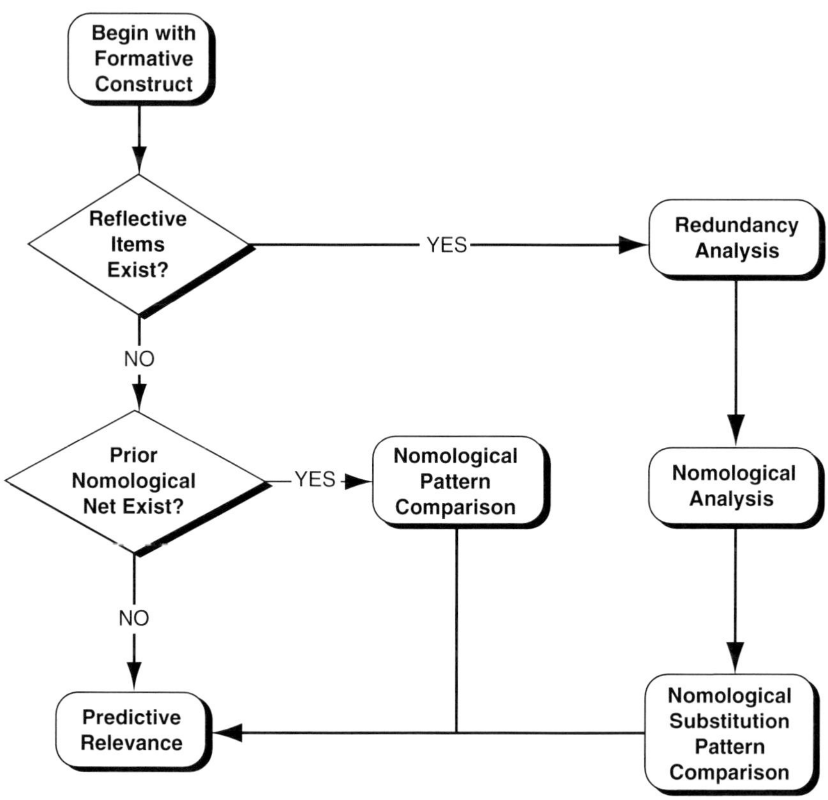

Fig. 28.13 Formative construct validation roadmap

same underlying construct. In so doing, results can be presented to show that both sets converged toward the same LV. The formative set was compared to the reflective set via a redundancy analysis. The reflective set, being new, also had to be validated in the context of the baseline model. This process would follow the same procedure as in our first example. Then a pattern substitutability comparison is made where the structural paths are compared for each set in the proposed nomological network.

References

Blalock, H. M., Jr. (1964). *Causal inferences in nonexperimental research*. Chapel Hill, NC: University of North Carolina Press.

Blalock, H. M., Jr. (1979). The presidential address: measurement and conceptualization problems: the major obstacle to integrating theory and research. *American Sociological Review, 44*, 881–894.

Blalock, H. M., Jr. (1986). Multiple causation, indirect measurement and generalizability in the social sciences. *Synthese, 68*, 13–36.

Chin, W. W. (1998a). Commentary: issues and opinion on structural equation modeling. *MIS Quarterly, 22*, vii–xvi.

Chin, W. W. (1998b). The partial least squares approach for structural equation modeling. In G. A. Marcoulides (Ed.), *Modern methods for business research* (pp. 295–336). Mahwah, NJ: Lawrence Erlbaum Associates.

Chin, W. W., & Gopal, A. (1995). Adoption intention in GSS: Relative importance of beliefs. *The Data Base for Advances in Information Systems, 26*(2&3), 42–64.

Chin, W. W., Marcolin, B. L., & Newsted, P. R. (1996). A partial least squares latent variable modeling approach for measuring interaction effects: Results from a monte carlo simulation study and voice mail emotion/adoption study. In J. I. DeGross, S. Jarvenpaa, & A. Srinivasan (Eds.) *Proceedings of the Seventeenth International Conference on Information Systems* (pp. 21–41).

Chin, W. W., & Newsted, P. R. (1999). Structural equation modeling analysis with small samples using partial least squares. In R. Hoyle (Ed.), *Statistical strategies for small sample research* (pp. 307–341). Thousand Oaks, CA: Sage Publications.

Chin, W. W., Peterson, R. A., Brown, S. P. (2008). Structural equation modeling in marketing: some practical reminders. *Journal of Marketing Theory and Practice, 16*(4), 287–298.

Cudeck, R., & Henley, J. J. (2003). A realistic perspective on pattern representation in growth data: Comment on Bauer and Curran (2003). *Psychological Methods, 8*, 378–383.

George, B., Hinson, S., & Chin, W. W. (2000). *Modeling the Technology Adoption Decision: The Impact and Generalizability of the Perceived Characteristics Of Innovating Inventory On Email Adoption*. Paper presented at the Diffusion Interest Group in Information Technology (DIGIT), – December 10, 2000. Brisbane, Australia: Queensland University of Technology.

Cohen, J. (1988). *Statistical power analysis for the behavioral sciences*. Hillside, NJ: L. Erlbaum Associates.

Cohen, J. (1990). Things I have learned (so far). *American Psychologist, 45*(12), 1304–1312.

Cohen, P., Cohen, J., Teresi, J., Marchi, M., & Velez, C. N. (1990). Problems in the measurement of latent variables in structural equations causal models. *Applied Psychological Measurement, 14*, 183–196.

Diggins, J. P. (1994). *The promise of pragmatism: modernism and the crisis of knowledge and authority*. Chicago: The University of Chicago Press.

Efron, B., & Tibshirani, R. J. (1993). *An introduction to the bootstrap (monographs on statistics and applied probability, #57)*. New York: Chapman & Hall.

Fornell, C., & Bookstein, F. L. (1982). Two structural equation models: LISREL and PLS applied to consumer exit-voice theory. *Journal of Marketing Research, 19*, 440–452.

Fornell, C., & Larcker, D. F. (1981). Evaluating structural equation models with unobserved variables and measurement error. *Journal of Marketing Research, 18*, 39–50.

Harris, R. J. (1989). A canonical cautionary. *Multivariate Behavioral Research, 24*(1), 17–39.

Hoyle, R. H., & Panter, A. T. (1995). Writing about structural equation models. In R. H. Hoyle (Ed.), *Structural equation modeling: concepts, issues, and applications* (pp. 158–176). Thousand Oaks, CA: Sage Publications.

Geisser, S. (1975). The predictive sample reuse method with applications. *Journal of the American Statistical Association, 70*, 320–328.

Gray, H. L., & Schucany, W. R. (1972). *The generalized jackknife statistic*. New York: Marcel Dekker.

Jarvis, C. B., MacKenzie, S. B., & Podsakoff, P. M. (2003). A critical review of construct indicators and measurement model misspecification in marketing and consumer research. *Journal of Consumer Research, 30*, 199–218.

Lohmöller, J.-B. (1989). *Latent variable path modeling with partial least squares*. Heidelberg: Physica-Verlag.

MacCallum, R. C. (2003). Working with imperfect models. *Multivariate Behavioral Research, 38*, 113–139.

MacCallum, R. C., & Browne, M. W. (1993). The use of causal indicators in covariance structure models: Some practical issues. *Psychological Bulletin, 114*(3), 533–554.

Majchrak, A., Beath, C., Lim, R., & Chin, W. W. (2005). Managing client dialogues during information systems design to facilitate client learning. *MIS Quarterly, 29*(4), 653–672.

Mathieson, K., Peacock, E., & Chin, W. W. (2001). Extending the technology acceptance model: The influence of perceived user resources. *The Data Base for Advances in Information Systems, 32*(3), 86–112.

McDonald, R. P., & Moon-Ho R. H. (2002). Principles and practice in reporting structural equation analyses. *Psychological Methods, 7*, 64–82.

Meehl, P. E. (1990). Appraising and amending theories: The strategy of lakatosian defense and two principles that warrant it. *Psychological Inquiry, 1*(2), 108–141.

Menand, L. (2002). *The metaphysical club: A story of ideas in America*. New York: Farrar, Straus and Giroux.

Rozeboom, W. W. (2005). Meehl on metatheory. *Journal of Clinical Psychology, 61*, 1317–1354.

Schneeweiss, H. (1990). Models with latent variables: LISREL versus PLS, Contemporary Mathematics, 112, 33–40.

Shah, R., & Goldstein, S. M. (2006). Use of structural equation modeling in operations management research: looking back and forward. *Journal of Operations Management, 24*, 148–169.

Steiger, J. H. (1979). Factor indeterminacy in the 1930's and the 1970's – some interesting parallels. *Psychometrika, 44*, 157–167.

Steiger, J. H. (1988). Aspects of person-machine communication in structural modeling of correlations and covariances. *Multivariate Behavioral Research, 23*, 281–290.

Steiger, J. H. (2001). Driving fast in reverse, the relationship between software development, theory, and education in structural equation modeling. *Journal of the American Statistical Association, 96*, 331–338.

Stone, M. (1974). Cross-validatory choice and assessment of statistical predictions. *Journal of the Royal Statistical Society, Series B, 36*(2), 111–133.

Tenenhaus, M., Amato, S., & Esposito Vinzi, V. (2004). A global goodness-of-fit index for PLS structural equation modelling, In *Proceedings of the XLII SIS Scientific Meeting* (pp. 739–742). Vol. Contributed Papers, CLEUP, Padova.

Tukey, J. W. (1958). Bias and confidence in not-quite large samples. *Annals of Mathematical Statistics, 29*, 614.

Werts, C. E., Linn, R. L., & Jöreskog, K. G. (1974). Intraclass reliability estimates: Testing structural assumptions. *Educational and Psychological Measurement, 34*, 25–33.

Williams, J., & MacKinnon, D. P. (2008). Resampling and distribution of the product methods for testing indirect effects in complex models. *Structural Equation Modeling, 15*, 23–51.

Wold, H. (1980). Model construction and evaluation when theoretical knowledge is scarce: Theory and application of partial least squares. In J. Kmenta & J. B. Ramsey (Eds.), *Evaluation of econometric models*. New York: Academic Press.

Wold, H. O. A. (1982). Soft modeling: The basic design and some extensions. In K. G. Jöreskog & H. O. A. Wold (Eds.), *Systems under indirect observations: Causality, structure, prediction, Part 2* (pp. 1–54). Amsterdam, The Netherlands: North-Holland.

Wold, H. (1988). Specification, predictor. In S. Kotz & N. L. Johnson (Eds.), Encyclopedia of statistical sciences (Vol. 8), New York: Wiley, 587–599.

Chapter 29
Evaluation of Structural Equation Models Using the Partial Least Squares (PLS) Approach

Oliver Götz, Kerstin Liehr-Gobbers, and Manfred Krafft

Abstract This paper gives a basic comprehension of the partial least squares approach. In this context, the aim of this paper is to develop a guide for the evaluation of structural equation models, using the current statistical methods methodological knowledge by specifically considering the Partial-Least-Squares (PLS) approach's requirements. As an advantage, the PLS method demands significantly fewer requirements compared to that of covariance structure analyses, but nevertheless delivers consistent estimation results. This makes PLS a valuable tool for testing theories. Another asset of the PLS approach is its ability to deal with formative as well as reflective indicators, even within one structural equation model. This indicates that the PLS approach is appropriate for explorative analysis of structural equation models, too, thus offering a significant contribution to theory development. However, little knowledge is available regarding the evaluating of PLS structural equation models. To overcome this research gap a broad and detailed guideline for the assessment of reflective and formative measurement models as well as of the structural model had been developed. Moreover, to illustrate the guideline, a detailed application of the evaluation criteria had been conducted to an empirical model explaining repeat purchasing behaviour.

29.1 Introduction

The analysis of interdependencies across latent variables concerns empirical research done in many areas of economy and social sciences, which has led to a growing interest in the analysis of structural equation models (Baumgartner and

---

O. Götz and M. Krafft
University of Münster, Marketing Centrum Münster,
Am Stadtgraben 13-15, 48143 Münster, Germany
e-mail: o.goetz@uni-muenster.de, mkrafft@uni-muenster.de

K. Liehr-Gobbers
Hering Schuppener Consulting, Kreuzstraße 60, 40210 Düsseldorf, Germany
e-mail: kliehr@heringschuppener.com

V. Esposito Vinzi et al. (eds.), *Handbook of Partial Least Squares*, Springer Handbooks of Computational Statistics, DOI 10.1007/978-3-540-32827-8_30,

Homburg 1996, pp. 140). An indication of this development is the growing number of papers where covariance structure analysis is used in national and international journals (Krafft et al. 2003, pp. 95; Homburg and Baumgartner 1995, p. 1095). This may be one of the most popular methods with which to estimate structural equation models, but it can only be utilized if various requirements concerning data, theory and the operationalization of latent variables are fulfilled. For instance, maximum likelihood estimation, which is frequently used in covariance structure analysis, is only efficient and unbiased when the assumption of multivariate normality is met. Furthermore, a sample size ranging between 150 and 400 is recommended when maximum likelihood estimation is used in a covariance-based analysis (Hair et al. 2006 p. 740–743). However, the predominant utilization of covariance structure analysis to analyze reflective measurement models has led to a remarkable number of incorrect specifications regarding formative measurement models (Fassott 2006, pp. 76–78; Jarvis et al. 2003, pp. 206; Cohen et al. 1990, pp. 184–186; Albers 2010, pp. 411–427). Cohen et al. (1990), for example, examined 15 papers in which covariance-based approaches had been used and showed that a substantial number of latent variables had been inadequately specified by treating formative measurement models as if they were reflective measurement models. Another requirement for covariance structure analysis is to achieve identification within the use of formative measurement models (Jarvis et al. 2003, p. 213). An alternative approach to address these issues is the Partial Least Squares (PLS) approach for the analysis of structural equation models.

After its initial frequent application in the early 1980s, the PLS approach has recently attracted renewed interest from applied researchers. Simultaneously, software packages with which to analyze structural equation models with PLS have become more readily available (LVPLS, PLS-Graph, PLS-GUI, SmartPLS, SPAD PLS). Temme et al. (2010) provides a detailed comparison of current PLS software. Nevertheless, there is still uncertainty regarding appropriate criteria with which to evaluate PLS models that contain both reflective and formative measures.

This incertitude is emphasized by the following example: While Bollen (1989) pointed out that traditional validity assessments and classical test theory do not cover formative indicators, Berscheid et al. (1989) used internal consistency and indicator reliability as criteria for evaluating a formative construct. Although the PLS approach's main advantage lies in the unrestricted coverage of reflective and formative measurement models, few authors have used PLS to analyze formative constructs. They either unquestioningly apply the same evaluation criteria that they used for testing reflective constructs in respect of formative measurement models, or state that the existing criteria cannot be used for formative models without presenting alternative, more appropriate criteria (Fornell et al. 1990, pp. 1252; Bontis 1998, p. 69; Hulland 1999, pp. 199–201; Tan et al. 1999, p. 950; Alpert et al. 2001, pp. 177; Sarkar et al. 2001, pp. 705–710; O'Cass 2002, pp. 69–71; Soo et al. 2002, p. 37).

Given this background, the objective of this paper is to present guidelines for a comprehensive evaluation of structural equation models, including both formative and reflective constructs, and taking the current methodological discussion on PLS into account.

In Sect. 29.2, the evaluation of PLS models' quality deals with the question of the extent to which specified PLS models are deemed appropriate for describing the effects between latent variables. For this evaluation, we propose a two-tiered process (Götz and Liehr-Gobbers 2004). In the first step, the reflective as well as the formative measurement models are evaluated in terms of their overall quality. In the second step, the extent to which the PLS model reproduces the real data structure, i.e., the indicator values, is evaluated. Subsequent to the examination of model fit in respect of measurement models, the structural model's evaluation is presented in detail.

We demonstrate the procedure in more detail in Sect. 29.3 by using an example from a study of customer loyalty behavior. The paper concludes with a summary and outlook in Sect. 29.4.

29.2 Evaluation of the Model Quality

Similar to covariance structure analysis, applying the PLS algorithm requires an extensive model evaluation. Specifically, the extent to which a specified model is appropriate for describing the effects between the constructs under investigation needs to be demonstrated. The evaluation of the model quality follows a multi-level process. After demonstrating how measurement models can be evaluated in PLS, and stressing the diversity of such an evaluation in a comparison of reflective and formative measurement models, the assessment of the structural model will be described in detail.

29.2.1 Evaluation of Measurement Models

The measurement or outer model specifies the relationship between observable variables and the underlying construct. In this context, the search for and investigation of suitable indicators are an important step with regard to the operationalization of such a construct (Churchill 1979, pp. 67).

There are different ways of operationalizing a construct. Subject to the hypothesized effect's direction and the nature of the relationship between latent constructs and their indicators, one can differentiate between reflective and formative indicators. Figure 29.1 clarifies this issue: the arrows either point from the construct to the (reflective) indicators, or in the opposite direction from the (formative) indicators to the shared construct.

A measurement model can either include reflective or formative indicators exclusively, or consist of both – reflective and formative – indicators, depending on the observed construct (Fornell and Bookstein 1982, pp. 292–294). The decision whether a construct should be operationalized with formative and/or reflective indicators should be based on theoretical considerations.

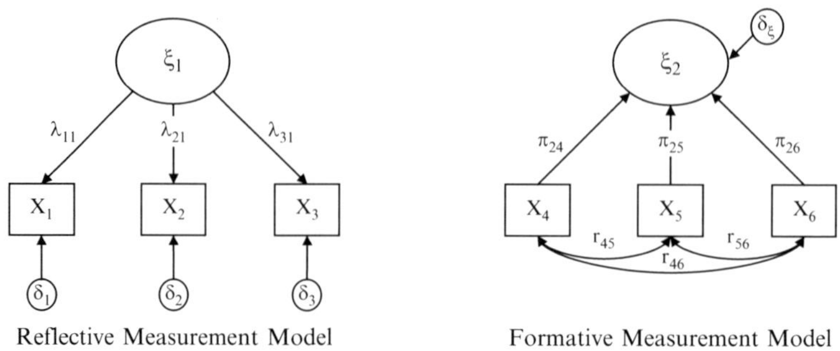

Fig. 29.1 Comparison of reflective and formative measurement models

29.2.1.1 Evaluation of Reflective Measurement Models

When a measurement model is operationalized reflectively, each indicator represents an error-afflicted measurement. This measurement error can be split into a random and systematic part. The random part includes all factors that influence a construct measurement's results unsystematically. The systematic measurement error is, however, not dependent on random measurement errors, but occurs at each repetition and always at the same level (Churchill 1987, pp. 381). A measurement is totally reliable if the random measurement error is zero. A measurement is completely valid if both error parts equal zero (Churchill 1987, p. 382).

Starting with this basic definition, the literature discusses several criteria for validating reflective constructs. The paragraphs that follow describe a detailed procedure for evaluating reflective constructs in respect of the PLS approach.

According to Bagozzi (1979), Churchill (1979) and Peter (1981) four basic evaluation types can be differentiated:

1. *Content validity:* According to Bohrnstedt (1970, p. 92), the content validity reveals to what extent a measurement model's variables belong to the domain of the construct. The principal component analysis is an appropriate method for examining the indicators' underlying factor structure (Vinzi et al. 2003, pp. 5; Bohrnstedt 1970, p. 92). After establishing the construct's indicators in respect of uni-dimensionality, further evaluation is required regarding the reliability and validity.
2. *Indicator reliability:* The indicator reliability specifies which part of an indicator's variance can be explained by the underlying latent variable. A common threshold criterion is that more than 50% of an indicator's variance should be explained by the latent construct. This implies that for loadings λ of the latent constructs on an indicator variable x or y, values larger than 0.7 are acceptable. This threshold value also means that the shared variance between a construct and its indicator is larger than the variance of the measurement error (Vinzi et al. 2003, pp. 5; Bohrnstedt 1970, p. 92). Weak loadings are frequently observed in empirical research, especially when newly developed scales are used (Hulland

1999, p. 198). However, reflective indicators should be eliminated from measurement models if their loadings within the PLS model are smaller than 0.4 (Hulland 1999, p. 198).

3. *Construct reliability:* Although small indicator reliabilities may point to a given indicator's inadequate measurement of a construct, it is usually more important that all the construct's indicators jointly measure the construct adequately (Bagozzi and Baumgartner 1994, p. 402). This can be assessed by means of the construct reliability (Rodgers and Pavlou 2003, pp. 24), which requires indicators assigned to the same construct to reveal a strong mutual association. Subsequently, the composite reliability measure (synonymous with factor reliability, or Jöreskog's rho) can be used to check how well a construct is measured by its assigned indicators. According to Fornell and Larcker (1981, p. 45), composite reliability in reflective measurement models is defined as follows:

$$\textit{Composite reliability}(\rho) = \frac{(\sum_i \lambda_{ij})^2}{(\sum_i \lambda_{ij})^2 + \sum_i \mathrm{var}(\varepsilon_{ij})} \tag{29.1}$$

λ_i indicates the loading of indicator variable i of a latent variable, ε_i indicates the measurement error of indicator variable i, and j represents the flow index across all reflective measurement models. The composite reliability can vary between 0 and 1. Values larger than 0.6 are frequently judged as acceptable (e.g., Bagozzi and Yi 1988, p. 82). The composite reliability is, similar to Cronbach's alpha, a measure of a reflective construct's construct reliability, yet it includes the actual factor loading, whereas the alpha uses equal weighting. Indicators showing weak correlations with the measurement model's remaining indicators have to be eliminated.

In academic publications, the most commonly used reliability coefficient is Cronbach's alpha, which is a generalized measure of a uni-dimensional, multi-item scale's internal consistency (Cronbach 1951; Peterson 1994). This criterion is defined as:

$$\text{Cronbach's alpha} : \alpha = \left(\frac{N}{N-1}\right) * \left(1 - \frac{\sum_{i=1}^{N} \sigma_i^2}{\sigma_t^2}\right) \tag{29.2}$$

Cronbach's alpha quantifies how well a set of indicators measures a uni-dimensional latent construct. If the data have a multidimensional structure, this alpha will usually be low. In this context, N is equivalent to the number of indicators assigned to the factor. σ_i^2 indicates the variance of indicator i. σ_t^2 represents the variance of the sum of all the assigned indicators' scores. A basic assumption is that the average covariance among indicators has to be positive. Therefore one can easily see that Cronbach's alpha varies between 0 and 1. An issue in assessing Cronbach's alpha is that correlations among indicators and scale length are critical, influencing alpha. In addition, sample size has a significant effect on the precision of the estimation of alpha. A common threshold for sufficient values of Cronbach's alpha is 0.6 (Hair et al. 2006, p. 102). Furthermore, Nunnally (1978, pp. 245) provides a short discussion about sufficient thresholds for alpha.

4. *Convergent validity:* In classical test theory convergent validity is based on the correlation between responses obtained by maximally different methods of measuring the same construct (Peter 1981, p. 136). This involves several problems. In addition to the practical problem of developing different methods, a major problem is in selecting "maximally different methods" and avoiding shared method variance. However, some authors may argue that indicators of a reflective construct can be treated as different methods to measure the latent construct. A common measure to examine convergent validity is the average variance extracted (AVE), which is formally defined as follows (Fornell and Larcker 1981, pp. 45):

$$AVE = \frac{\sum_i \lambda_i^2}{\sum_i \lambda_i^2 + \sum_i \text{var}(\varepsilon_i)} \tag{29.3}$$

AVE includes the variance of its indicators captured by the construct relative to the total amount of variance, including the variance due to measurement error. An AVE of less than 0.5 is considered insufficient, as more variance is due to error variance than to indicator variance (Homburg and Giering 1996, p. 12; Rodgers and Pavlou 2003, p. 25).

5. *Discriminant validity:* Besides considering the indicator and construct reliability, a thorough validation procedure also requires the evaluation of a measurement (or structural) model's discriminant validity. Discriminant validity is defined as the dissimilarity in a measurement tool's measurement of different constructs. A necessary condition for discriminant validity is that the shared variance between the latent variable and its indicators should be larger than the variance shared with other latent variables (Hulland 1999, p. 199).

According to Fornell and Larcker (1981, p. 46), discriminant validity is proven if a latent variable's AVE is larger than the common variances (squared correlations) of this latent variable with any other of the model's constructs. After having checked for discriminant validity, the reflective measurement model's validation process has been completed.

With the exception of content validity, all the evaluation criteria for reflective measurement models as discussed above are basic PLS-Graph 3.0 outputs.

29.2.1.2 Evaluation of Formative Measurement Models

In contrast to reflective models, formative measurement models reverse the direction of causality in as far as the indicators form or constitute the latent variable. This causality reversal demands a different interpretation and evaluation of the measurement model. Consequently, the statistical evaluation criteria for reflective measurement models cannot be directly transferred to formative measurement models (Diamantopoulos 1999, pp. 453).

Following the structure of the reflective measurement models' assessment in 2.1.1, a detailed discussion of how to evaluate formative measurement models follows below:

1. *Content validity* Contrary to reflective measurement models, neither the content validity nor the uni-dimensionality criterion can, strictly speaking, be used to assess formative measurement models and/or their quality (Bollen and Lennox, 1991; Cohen et al. 1990; Chin and Gopal, 1995).

Suggested Procedure: In a formative measurement model, content validity should already be ensured when the model is specified (i.e., before the data are collected), because every single indicator measures a specific facet of the latent construct. Omitting an indicator would therefore mean omitting a part of the latent construct. Consequently, all facets of the formative construct should be considered.

The scientific literature has discussed different approaches to determining whether a construct has a more formative or reflective nature. On the one hand, expert judgments are deemed appropriate (Diamantopoulos and Winklhofer 2001, p. 271; Rossiter 2002, p. 306). On the other hand, some researchers provide rules for determining whether a construct is formative or reflective (Chin 1998a, p. 9; Jarvis et al. 2003, p. 203). Nevertheless, the boundaries between formative and reflective specifications are rather fuzzy. Many latent variables measure – depending on the context of the study – as either formative or reflective. This is emphasized in the MIMIC model examples below. As an alternative, Bollen and Ting (2000) propose the tetrad test as a criterion for construct specification, which takes into consideration that formative indicators do not necessarily correlate, whereas reflective indicators do, and is an evaluation of the given data's correlation structure. According to this test, a construct cannot be reflective if there is no or only a little correlation. Gudergan et al. (2003), Venaik et al. (2004) and Albers (2010) present a detailed description of the tetrad test as well as Bucic and Gudergan (2004), who applied the vanishing tetrad test to evaluate formative scales.

In this context, it has to be noted that a construct's reflective specification can be excluded if the tetrad test is "positive." However, if the outcome of the tetrad test is "negative," it does not automatically mean that the specific construct is formative, as a formative measurement model's indicators do not *have to* but *may* correlate. Therefore, ultimately, the tetrad test cannot resolve whether a construct should be specified formatively or reflectively, either.

Nevertheless, if a construct seems to be specified formatively, all facets of the construct have to be included. To support this, a pre-test could be applied to measure the extent of the similarity between the a priori intended and the actually occurring indicator assignment (expert validity). Following this step, the indicators must be assigned to their respective constructs during a pretest. Anderson and Gerbing (1991, p. 734) recommended measures for evaluating the assignment's uniqueness and its relevance in respect of the content, which can be quantified if based on experts' statements.

2. *Indicator reliability:* Contrary to reflective measurement models, the assessment of formative measurement models' reliability makes little sense, as a measurement model's formative indicators do not have to be correlated (Chin 1998b, p. 306).

Suggested Procedure: Instead of checking the indicator reliability, it is more logical to compare each indicator's weights by means of the PLS approach. One could thus determine which indicators contribute most substantially to the construct ("indicator relevance").

Formative constructs' valid indicators can reveal positive, negative or no correlations. Consequently, the different indicators' weights must not be interpreted as factor loadings, but should rather be compared to determine their relative contribution to the relevant construct (Sambamurthy and Chin 1994, pp. 231). Formative indicators' weights are frequently smaller than reflective items' loadings. The PLS approach optimizes the indicators' weights to maximize the explained variance of the dependent variable(s) in the model. Therefore, a formative construct's rather small absolute weights should not be misinterpreted as a poor measurement model (Chin 1998b, p. 307).

While indicators with very small loadings are frequently eliminated within reflective measurement models, this procedure should not be applied in formative measurement models, as theoretical and conceptual considerations have led to indicators being assigned to the construct. Another reason is that a measurement model's formative indicators do not have to be correlated, so that the elimination of an indicator with a small weight could lead the omission of a substantial part of the latent construct (Bollen and Lennox 1991, p. 308; Jarvis et al. 2003, p. 202). However, the elimination of an indicator from a formative measurement model is recommended if substantial multicollinearity occurs. After the formative indicators have been derived from the construct description and have passed a pre-test, they have to be immediately checked for multicollinearity, which indicates the indicators' degree of linear dependency. While reflective items have to be highly correlated due to the model's factor analytic construction, substantial collinearity in formative models can lead to the results, i.e., the parameter estimations, being highly biased. Substantial collinearity within indicators consequently complicates ascertaining the individual indicators' distinct influence on the latent construct (Diamantopoulos and Winklhofer 2001, p. 272). As formative measurement models are based on the principles of multiple regression analysis, the beta-coefficients' standard errors inflate with increasing multicollinearity, and their estimation becomes less reliable (inefficiency of estimates). If perfect multicollinearity is given, the regression analysis cannot be calculated at all (Backhaus et al. 2003, p. 88). Various testing procedures can be applied to reveal collinearity within a model:

An inspection of all indicators' correlation matrix can serve as a first indication of pairwise collinearity. The Variance Inflation Factor (VIF) is a metric for multicollinearity, i.e., collinearity between more than two indicators. The VIF is calculated as the inverse of the tolerance value (Eckey et al. 2001, p. 93; Hair et al. 2006, pp. 227). The term VIF is derived from the fact that its square root is the degree to which the standard error has been increased due to multicollinearity. There is no clear threshold value for multicollinearity. As a rule of thumb, the VIF should not exceed a value of 10, but, in general, the critical value should be defined individually and be based on practical considerations in respect of each analysis. Green et al. (1988, p. 457), for example, argued that no multiple correlation of a regression's variables should exceed the dependent variable's multiple correlation with the indicators.

3. *Construct reliability:* Contrary to the procedure in reflective measurement models, no evaluation is allowed of formative constructs that are based on the internal consistency measure (Hulland 1999, p. 201). Mathieson et al. (1996) formulated this circumstance as follows: "Since the latent variable is viewed as an effect rather than a cause of the item responses, internal consistency is irrelevant." On the one hand, a reason for this can be found in the fact that formative indicators do not have to be highly correlated (Krafft 1999, pp. 124; Rossiter 2002, pp 307). On the other hand, the suggestion to eliminate indicators with rather small weights argues against the application of internal consistency metrics. As stated earlier, indicators of formative measurement models should not be eliminated even if they show small weights, as the operationalized construct's conceptual domain may otherwise not be fully covered.

Suggested Procedure: Reinartz et al. (2004, pp. 298) suggest using external validity as an evaluation criterion for formative measurement models, as it is quite often possible to operationalize a construct formatively as well as reflectively. However, covering a latent construct's entire scope by means of formative indicators is hardly possible. In such cases, reflective indicators can be used to quantify the error terms. This MIMIC (Multiple effect indicators for multiple causes) model, developed by Hauser and Goldberger (1971, pp. 81), provides an opportunity to measure a construct with both formative and reflective indicators. In such a case, reflective indicators serve as formative measurement models' external validation. An example of a construct that can be measured both formatively and reflectively, is "drunkenness," with the construct's formative indicators being the amount of different alcoholic beverages (e.g., beer, wine, champagne) consumed. The reflective operationalization could be accomplished by measuring the blood alcohol level as well as the ability to coordinate, articulate, and concentrate. The operationalization of the construct by means of reflective indicators allows the measurement error to be determined (Chin 1998a, p. 9). In this specific case, a substantial error term could be attributed to the fact that the consumption of alcoholic sweets or medicine could also lead to drunkenness. These indicators were not, however, included in the formative operationalization. Another example is the operationalization of the construct "product quality" (Stone-Romero and Stone 1997). This construct can be operationalized

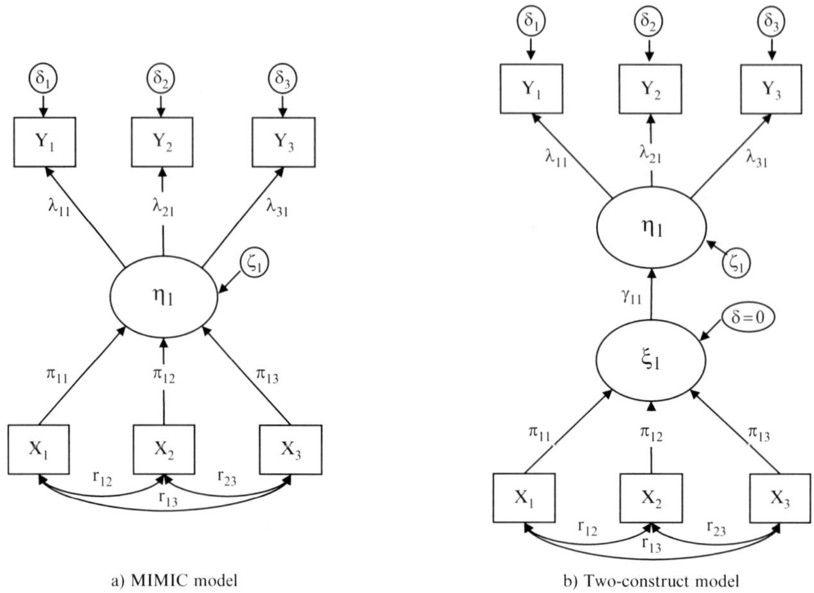

Fig. 29.2 Alternative specifications for a construct that is operationalized formatively as well as reflectively. Source: Adapted from Diamantopoulos and Winklhofer (2001), p. 272

by means of formative indicators such as "appealing design," "high-quality functionality," and "adequate product weight." Product quality can also be measured by means of reflective indicators to determine the formative operationalization's error term. Reflective items could, for example, be "the product is of high quality," "my quality expectations have been met," "I will not complain about the product," and "my quality expectations have been exceeded." Figure 29.2 clarifies the hypothetical interrelation.

Currently, only SPAD PLS supports the specification of variables by means of the MIMIC model. An alternative specification for quantifying the error terms is to use the two-construct model (cp. Fig. 29.2, type b) that integrates an additional "phantom variable" (Rindskopf 1984), which represents the construct's reflective operationalization (Diamantopoulos and Winklhofer 2001, pp. 272–274). In such a case, the two-construct model can also be applied to evaluate the formative measurement model. If a strong and significant association between the latent and the phantom variable is confirmed, external validity is proven. If there are no reflective indicators with which to identify the phantom variable, nomological validity can be investigated by means of the association between the construct operationalized formatively and another latent variable ("dependent variable") within the model (Diamantopoulos and Winklhofer 2001, p. 273). The analyzed structural relationship should have a thorough theoretical base and be empirically supported. For instance, in the example presented above, the relationship between "product quality" and "customer retention" could be used to validate the formative measurement model. If a strong positive relationship is detected between product quality and customer retention, this can be regarded as an indication of the applied measurement models' nomological validity.

4. *Convergent validity:* and d*iscriminant validity:* As formative indicators do not have to be strongly interrelated, the convergent and discriminant validity (according to Fornell and Larcker 1981, p. 46) by no means represents a reasonable criterion for evaluating formative measurement models.

The final evaluation of reflective as well as formative measurement models is carried out in the context of the PLS approach with the help of significance tests, which can be conducted by asymptotic t-statistics generated by resampling techniques (Chin 1998b, pp. 318–320).

For the evaluation of formative measurement models, PLS-Graph 3.0 presents the following evaluation criteria: the weights of each indicator (indicator relevance), the intensity and direction as well as the significance of all relationships between the indicators and constructs. The significance, intensity and direction of inter-construct relationships can be used for the evaluation of external (nomological) validity.

29.2.2 Evaluation of the Structural Equation Model

The structural model covers the relationships among hypothetical constructs. Latent variables that only predict other latent variables are called exogenous variables, while a latent variable that is a dependent variable in at least one causal relationship is called an endogenous variable. The relationships between constructs are also hypothesized in accordance with theoretical and logical reasoning. For PLS, the structural model has to be designed as a causal chain. This model type is known as a recursive type, i.e., there is no loop in the path model.

In contrast to covariance-based approaches, the PLS method does not allow statistical tests to measure the calibrated model's overall goodness, which is mainly due to the assumption of distribution-free variance. Alternatively, non-parametrical tests can be applied to evaluate the structural model's quality. A logical metric for judging the structural (or inner) model is the endogenous variables' determination coefficient (R^2). Similar to a multiple regression's coefficients, the evaluation of the model's quality should also be based on the path coefficients' directions and significance levels (Chin 1998b, p. 316).

The determination coefficient (R^2) reflects the level or share of the latent construct's explained variance and therefore measures the regression function's "goodness of fit" against the empirically obtained manifest items (Backhaus et al. 2003, p. 63). R^2 is a normalized term that can assume values between 0 and 1. According to Backhaus et al. (2003), no generalizable statement can be made about acceptable threshold values of R^2. Whether this determination coefficient is deemed acceptable or not rather depends on the individual study. However, the larger R^2 is, the larger the percentage of variance explained.

The PLS structural model's individual path coefficients represent standardized beta coefficients resulting from the least-squares method or estimation. The goodness of the path coefficients estimated in PLS can be tested by means of asymptotic t-statistics, which are also obtained by resampling methods (Venaik et al. 2001, p. 20). Paths that are insignificant, or show signs contrary to the hypothesized direction, do not support a prior hypothesis, while significant paths showing the hypothesized direction empirically support the proposed causal relationship. The hypotheses are tested by quantifying the structural equation paths' significance with an appropriate resampling method and by examining all the hypothesized relationships' absolute values.

Besides inspecting the R^2 metrics of all endogenous variables, the change in the determination coefficient also shows whether an independent latent variable has a substantial influence on the dependent latent variable. Similarly to traditional partial F-tests, Cohen (1988, p. 410–413) developed the so-called "effect size" f^2. Contrary to the F-test, the effect size f^2 does not refer to the sample at all, but to the basic population of the analysis, therefore no degrees of freedom need be considered. This is justified by the fact that if a variance-based structural equation model "proceeds more naturally with [...] squared correlation values, it is more convenient to work directly with f^2 rather than f." (Cohen 1988, p. 410)

The effect size f^2 is defined as follows:

$$\text{Effect size}: f^2 = \frac{R^2_{incl} - R^2_{excl}}{1 - R^2_{incl}} \tag{29.4}$$

The change in the dependent variable's determination coefficient is calculated by estimating the structural model twice, i.e., once with and once without the independent latent variable (R^2_{incl} and R^2_{excl}). Values for f^2 of 0.02, 0.15, or 0.35 indicate the latent exogenous variable's weak, moderate or substantial influence on the particular latent endogenous variable (Cohen 1988, p. 413; Chin 1998b, p. 316).

The model's predictive validity can be tested by means of the non-parametric Stone–Geisser test (Geisser 1975, p. 320; Stone 1975; Fornell and Cha 1994, pp. 71–73; Chin 1998a, p. 15). This test uses a so-called "blindfolding" procedure, which systematically assumes that a part of the raw data matrix is missing during the parameter estimation. Even with missing values, it is possible to estimate parameter and construct values. For cross-validation purposes, two data sets are needed: one set for the model estimation and the other for determining the full model's predictive validity. The blindfolding procedure removes some data from the sample and treats these data as missing in the estimation. In the next step, the obtained parameter estimates are used to reconstruct the raw data that the blindfolding procedure assumes are missing. Consequently, the blindfolding technique produces general cross-validation metrics as well as the parameter estimates' jackknifing standard deviation.

Similar to the determination coefficient (R^2) in OLS, the Stone–Geisser test criterion Q^2 is interpreted without loss of degrees of freedom. It shows how well the data collected empirically can be reconstructed with the help of the model and the PLS parameters (Fornell and Cha 1994, p. 72).

Formally, the Stone–Geisser test criterion can be displayed as:

$$\text{Stone-Geisser test criterion}: Q_j^2 = 1 - \frac{\sum\limits_k E_{jk}}{\sum\limits_k O_{jk}} \tag{29.5}$$

The predictive errors are calculated as the difference between the true values of the data omitted from the blindfolding procedure and the predicted values, using parameter estimates from the remaining data points. E_{jk} represents the squares of the prediction errors, while O_{jk} represents the squares of the trivial prediction error provided by the mean of the remaining data from the blindfolding procedure. Index j indicates the observed endogenous measurement model, and k represents the index for all indicators of the measurement model. If this test criterion is larger than 0, the model is considered to have predictive validity, otherwise, the model cannot be granted predictive relevance (Fornell and Cha 1994, p. 73; Chin, 1998b). The model can, however, have predictive relevance, if the sum of the remaining residuals from the model estimation is lower than a trivial estimation. For a detailed description of the Stone–Geisser test criterion, see Fornell and Cha (1994, pp. 71–73).

With the exception of effect size f^2, all evaluation criteria discussed in this section are part of the standard output of PLS-Graph 3.0.

The following section illustrates the PLS approach's evaluation procedure and introduces a consumer behavior example to explain repeat purchasing intention by means of customer satisfaction and the homogeneity of the service offering.

29.3 Empirical Application of Model Quality's Evaluation by Means of Partial Least Squares (PLS) Analysis

This section outlines the evaluation of a PLS model's quality to explain repeat purchasing behavior. The developed model examines the relationship between customer satisfaction, homogeneity of the service offering and their effect on customer loyalty behavior. The focus of this chapter is not centered on the substantive model as such, but on the evaluation process when the PLS method is applied.

Customer loyalty has long been the focus of interest. Customer satisfaction is a necessary precondition for customer loyalty, which is in turn a key driver of profit growth and performance (Kotler 1994; Reichheld 1993; Heskett et al. 1997; Reinartz and Kumar 2000). Previous studies have focused on specific determinants of customer loyalty, such as the link between service quality and customer loyalty (Bitner 1990; Boulding et al. 1993; Cronin and Taylor 1992), between product quality and customer loyalty (Woodside and Taylor 1978), customer satisfaction and customer loyalty (Giering 2000), and the effect of available product and service alternatives on customer loyalty (Sriram and Mummalaneni 1990). Various definitions of the term "customer loyalty" have also evolved (Anderson and Sullivan 1993; Olsen 2002; Yi and Jeon 2003). This study, however, focuses specifically on the customer's repurchase intention (Taylor and Baker 1994). Drawing on cognitive dissonance theory

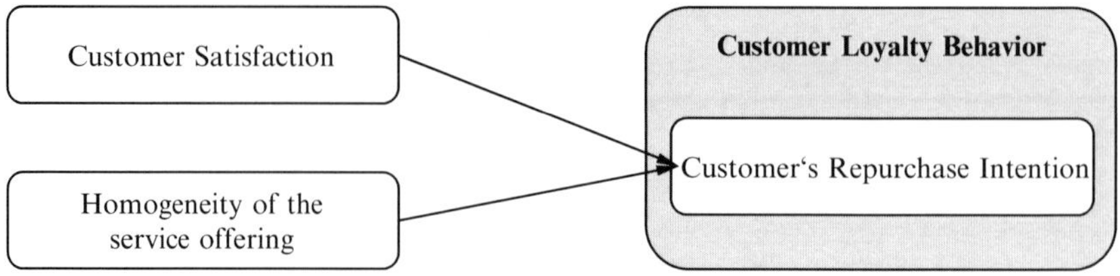

Fig. 29.3 Conceptual model to explain customer loyalty

(Festinger 1978), learning theory and risk theory (Sheth and Parvatiyar 1995), customer satisfaction and homogeneity of the service offering have been identified as constructs that impact customer's repurchase intention. Figure 29.3 outlines the conceptual framework.

Customer satisfaction was measured by means of ten formative items, adapted from Ganesh, Arnold, and Reynolds, which measure different facets of customer satisfaction (e.g., tariff transparency, issuing of an invoice, reliability of the energy supply, the qualification and friendliness of the employees, and additional services) as well as by means of one item that measured the overall customer satisfaction in respect of external validation (Ganesh et al. 2000). Homogeneity of the service offering was measured by two reflective indicators adapted from Burnham et al. (2003). Based on the work of Giering (2000), a three-item scale was constructed in respect of the customer's repurchase intention.

To test our customer loyalty model, data were collected for a German energy supplier by means of a telephone survey. The study focused on a random sample frame of 300 private customers. Altogether, 273 surveys were usable. The model was then analyzed by using SPSS 13 and PLS Graph 3.0 (Chin and Fry 2004).

Before the hypotheses could be investigated, each construct was assessed for reliability and validity. The reflective measures (representing the homogeneity of the service offering and customer repurchase intentions) were deemed satisfactory.

The principal component analysis (PCA) confirmed the uni-dimensionality of both constructs, indicating a high content validity. The homogeneity of the service offering and the customer's repurchase intention's indicators loaded on one principal component each (72.5% and 60.4%). Communalities ranged between 0.60 and 0.83.

The loadings of all the PLS analysis's reflective indicators were examined to assess the indicator reliability. The item loadings ranged between 0.64 and 0.96, in other words, one indicator's explained variance was below the 0.5 level, but still above the 0.4 level. This item was therefore not deleted.

To check how well the reflective constructs are measured by their assigned indicators, internal consistency metrics (e.g., Cronbach's alpha and composite reliability) can be used. Table 29.1 indicates Cronbach's alpha (α), composite reliability (ρ), and average variance explained (AVE). All common thresholds were thus met for construct reliability in accordance with the number of indicators. Discriminant validity was satisfied, with the correlation between the homogeneity of the service offer and customer repurchase intention ($r = -0.213$) being substantially lower than the square root of the average variance extracted (AVE).

Table 29.1 Internal consistency of reflective constructs

Construct	No. of Ind.	Cron. Alpha (α)	Comp Rel (ρ)	AVE
Homogeneity of the service offering	2	0.57	0.79	0.66
Customer's repurchase intention	3	0.80	0.89	0.72

Table 29.2 Goodness of formative measurement model of customer satisfaction

Satisfaction with ...	Indicator	Weight	t-value
transparency of tariff	cs1	0.18	1.69
price-performance ratio	cs2	0.28	2.53
reliability of energy supply	cs3	0.10	1.42
exposure to renewable energy and environmental friendliness	cs4	0.03	0.33
issuing of an invoice	cs5	0.17	1.82
expertise of the employees	cs6	0.23	2.12
information offered (e.g., internet, customer magazine)	cs7	0.25	2.30
additional services (e.g., arrangements, customer card)	cs8	0.04	0.33
friendliness of the employees	cs9	0.13	1.07
reachability in case of problems	cs10	0.12	1.23

The evaluation of reflective measurement models, as mentioned in Sect. 29.2.1.1, is not adequate for formative measurement models. To ensure content validity, a pre-test was applied to ensure a complete definition of customer satisfaction as a formative construct. Experts supported the content validity and approved the a priori assignment of the indicators to constructs.

The weights of the customer satisfaction indicators were obtained through PLS estimation. The indicators' weights and their bootstrap t-statistics (cf. Chin 2010) are presented in Table 29.2. The variables "satisfaction with price-performance ratio" and "information offered" contribute most effectively to customer satisfaction.

The elimination of a formative indicator is, however, only recommended if high multicollinearity occurs. The maximum variance inflation factor (VIF) came to 1.99, which is far below the common cut-off threshold of ten (Hair et al. 2006, p. 230). The average VIF of 1.59 across all indicators also indicates that with respect to the ten indicators, collinearity does not seem to pose a problem.

A two-construct model was applied to test the external validity of the formative measurement model (see Fig. 29.2 in Sect. 29.2.1.2). The postulated strong (0.61) and significant (t-value $=$ 14.37) connection between the formative and reflective measurement model of customer satisfaction confirmed the external validity. The R^2 for the reflective construct (phantom variable) (0.37) indicates that much of the

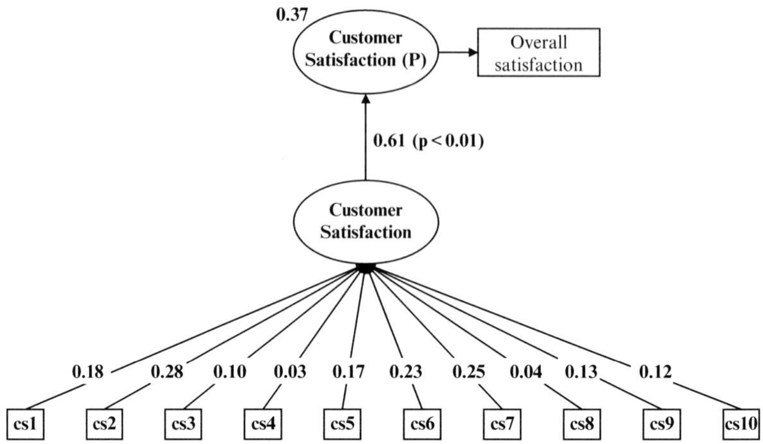

Fig. 29.4 External validation of customer satisfaction

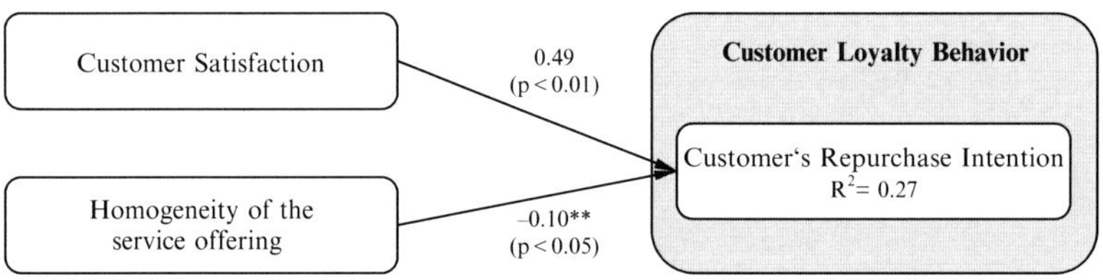

Fig. 29.5 Structural model to explain customer loyalty behavior

variance in "overall satisfaction" could be explained by the formative measurement model. Figure 29.4 illustrates the external validation of customer satisfaction.

The starting point for judging our structural (inner) model is the determination coefficient (R^2). The estimated model fits the survey data well, with an R^2 for customer's repurchase intention equal to 0.27. Thus, our parsimonious model consists of two factors that are strongly associated with customer loyalty behavior. Customer satisfaction (0.49, $p < 0.01$) and homogeneity of the service offering (-0.10, $p < 0.05$) significantly influence repurchase intention. The results of the model are presented in Fig. 29.5.

The change in the determination coefficient shows whether an independent latent variable has a substantial influence on the dependent latent variable. Table 29.3 reports the effect size in respect of the exogenous constructs.

Customer satisfaction seems to be the key explanatory factor in terms of incremental variance explained in the dependent variable.

The predictive relevance of the model was tested by means of the Stone–Geisser test (see Sect. 29.2.2). Considering the Q^2 value of 0.005, which is only slightly above the common threshold (i.e., larger than zero), and Q^2's standard deviation of 0.28, the predictive relevance of the customer loyalty model seems t o be doubtful.

Table 29.3 Relative explanatory power (Effect Size)

Construct	$R^2_{excluded}$	Effect size (f^2)
Homogeneity of the service offering	0.262	0.012
Customer satisfaction	0.048	0.306

29.4 Summary

The aim of this paper has been to develop a guide for the evaluation of structural equation models, using the actual available methodological knowledge by specifically considering the PLS approach's requirements.

It has been established that the PLS method demands significantly fewer requirements compared to that of covariance structure analyses, but nevertheless delivers consistent estimation results. This makes PLS a valuable tool for testing theories. In this context, another advantage of the PLS approach is its ability to deal with formative as well as reflective indicators, even within one structural equation model. This indicates that the PLS approach is appropriate for explorative analyses of structural equation models, too, and thus offers a significant contribution to theory development.

The procedure that the authors have presented of conceptualizing, operationalizing and evaluating structural equation models uses the above-mentioned advantages. The individual steps for the assessment of reflective and formative measurement models as well as structural models have also been described in detail. Moreover, to illustrate the guideline, an empirical application was undertaken of model quality's evaluation to explain repeat purchasing behavior.

In summary, it can be stated that the PLS method expands the spectrum of structural equation model analysis, indicating a future change of focus in empirical research. Further conceptual development and empirical validation of the presented approach should therefore play an important role in future research papers. In addition, future research should also deal with the theory-based derivation of more evaluation criteria and the (further) development of adequate software.

References

Albers, S. (2010). PLS and Success Factor Studies in Marketing. In: V. Esposito Vinzi, W. Chin, J. Hensler, and H. Wold (Eds.,) *Handbook of partial least squares*, (pp.411–427), Heidelberg: Springer.

Alpert, F., Kamins, M., Sakano, T., Onzo, N., & Graham, J. (2001). Retail buyer beliefs, attitude and behavior toward pioneer and me-too follower brands. *International Marketing Review, 18*, (2), 160–187.

Anderson, E. W., & Sullivan, M. W. (1993). The antecedents and consequences of customer satisfaction for firms. *Marketing Science, 12*, pp. 125–143.

Anderson, J. C., & Gerbing, D. W. (1991). Predicting the performance of measures in a confirmatory factor analysis with a pretest assessment of their substantive validities. *Journal of Applied Psychology, 76*, (5), pp. 732–740.

Backhaus, K., Erichson, B., Plinke, W., & Weiber, R. (2003). *Multivariate analysemethoden: eine anwendungsorientierte einführung* (10th ed.). Berlin: Springer.

Bagozzi, R. P. (1979). The role of measurement in theory construction and hypothesis testing: toward a holistic model. *In* O. C. Ferrell, S. W. Brown, C. W. Lamb (Eds.), *Conceptual and theoretical developments in marketing* (pp. 15–32). Chicago: American Markting Association.

Bagozzi, R. P., & Baumgartner, H. (1994). The evaluation of structural equation models and hypothesis testing. *In* R. P. Bagozzi (Ed.), *Principles of marketing research* (pp. 386–422). Cambridge: Blackwell.

Bagozzi, R. P., & Yi, Y. (1988). On the evaluation of structural equation models. *Journal of the Academy of Marketing Science, 16*(1), 74–94.

Baumgartner, H., & Homburg, C. (1996). Applications of structural equation modelling in marketing and consumer research: a review. *International Journal of Research in Marketing, 13*, 139–161.

Berscheid, E.; Snyder, M., & Omoto, A. M. (1989). The relationship closeness inventory: assessing the closenness of interpersonal relationships. *Journal of Personality and Social Psychology, 57*, 792–807.

Bitner, M. J. (1990). Evaluating service encounters: the effects of physical surroundings and employee responses. *Journal of Marketing, 54*(2), 69–82.

Bohrnstedt, G. W. (1970). Reliability and validity assessment in attitude measurement. In G. F. *Summers* (Ed.), *Attitude measurement* (pp. 80–99). London: Rand McNally.

Bollen, K. A. (1989). *Structural equations with latent variables*. New York: Wiley.

Bollen, K. A., & Lennox, R. (1991). Conventional wisdom on measurement: a structural equation perspective. *Psychological Bulletin, 110*(2), 305–314.

Bollen, K. A., & Ting, K. (2000). A tetrad test for causal indicators. *Psychological Methods, 5*, 3–22.

Bontis, N. (1998). Intellectual capital: an exploratory study that develops measures and models. *Management Decision, 36*(2), 63–76.

Boulding, W., Kalra, A., Staelin, R., & Zeithaml, V. A. (1993). A dynamic process model of service quality: from expectations to behavioral intentions. *Journal of Marketing Research, 30*, 7–27.

Bucic, T., & Gudergan, S. P (2004). Formative versus reflective measurement implications for explaining innovation in marketing partnerships. In: Proceedings of the Australia New Zealand Marketing Academy Conference.

Burnham, T. A., Frels, J. K., & Mahajan, V. (2003). Consumer switching costs – A typology, antecedents and consequences. Journal of the Academy of Marketing Science, *31*(2), 109–126.

Chin, W., & Fry, T. A. (2004). *PLS-Graph 3.0 Build 1126*. Houston: Department of Decision and Information Science, University of Houston.

Chin, W. W. (1998a). Issues and opinion on structural equation modeling. *Management Information Systems Quarterly, 22*(1), 7–16.

Chin, W. W. (1998b). The partial least squares approach to structural equation modeling. In G. A. Marcoulides (Ed.), *Modern methods for business research* (pp. 295–336) Mahwah: Lawrence Erlbaum.

Chin, W.W. (2010). Bootstrap Cross-Validation Indices for PLS Path Model Assessment. In: V. Esposito Vinzi, W. Chin, J. Hensler, and H. Wold (Eds.,) *Handbook of partial least squares*, (pp. 83–97), Heidelberg: Springer.

Chin, W. W., & Gopal, A. (1995). Adoption intention in GSS: relative importance of beliefs. *DATA BASE for Advances in Information Systems, 26*(2 & 3), 42–64.

Churchill, G. A. (1979). A paradigm for developing better measures of marketing constructs. *Journal of Marketing, 16*, 64–73.

Churchill, G. A. (1987). *Marketing research: methodological foundations* (4th ed.). Fort Worth: Dryden Press.

Cohen, J. (1988). *Statistical power analysis for the behavioral sciences* (2nd ed.). Hillsdale: Lawrence Erlbaum Associates.

Cohen, P., Cohen, J., Teresi, J., Marchi, M., & Velez, C. N. (1990). Problems in the measurement of latent variables in structural equations causal models. *Applied Psychological Measurement, 14*(2), 183–196.

Cronbach, L. J. (1951). Coefficient alpha and the internal structure of tests. *Psychometrika, 16*, 297–334.

Cronin, J. J. J., & Taylor, P. A. (1992). Measuring service quality. *A Re-examination and Extension, Journal of Marketing, 56*(3), 55–68.

Diamantopoulos, A. (1999). Export performance measurement: reflective versus formative indicators. *International Marketing Review, 16*(6), 444–457.

Diamantopoulos, A., & Winklhofer, H. M. (2001). Index construction with formative indicators: an alternative to scale development. *Journal of Marketing Research, 38*, 269–277.

Eckey, H. F.; Kosfeld, R. Draeger, C. (2001). *Ökonometrie* (2nd ed.), Wiesbaden: Gabler.

Fassott, G. (2006). Operationalisierung latenter variablen in strukturgleichungsmodellen: eine standortbestimmung. *Zeitschrift für betriebswirtschaftliche Forschung, 58*, 67–88.

Festinger, L. (1978). *Theorie der kognitiven dissonanz*. Bern: Verlag Hans Huber.

Fornell, C., & Bookstein, F. L. (1982). A comparative analysis of two structural equation models: lisrel and pls applied to market data. In C. Fornell (Ed.), *A second generation of multivariate analysis* (pp. 289–323). New York, Praeger.

Fornell, C., & Cha, J. (1994). Partial least squares. In R. P. Bagozzi (Ed.), *Advanced methods of marketing research* (pp. 52–78). Cambridge: Blackwell.

Fornell, C., & Larcker, D. F. (1981). Evaluating structural equation models with unobservable variables and measurement errors. *Journal of Marketing Research, 18*, 39–50.

Fornell, C., Lorange, P., & Roos, J. (1990). The cooperative venture formation process: a latent variable structural modeling approach. *Management Science, 36*(10), 1246–1255.

Ganesh, J., Arnold, M. J., & Reynolds, K. E. (2000). Understanding the customer base of service providers: an examination of the difference between switchers and stayers. *Journal of Marketing, 64*(30), 1–78.

Geisser, S. (1975). A predictive approach to the random effect model. *Biometrika, 61*(1), 101–107.

Giering, A. (2000). Der Zusammenhang zwischen Kundenzufriedenheit und Kundenloyalität Eine Untersuchung moderierender Effekte. Wiesbaden: Gabler.

Götz, O., & Liehr-Gobbers, K. (2004). Analyse von Strukturgleichungsmodellen mit Hilfe der Partial-Least-Squares (PLS)-Methode. *Die Betriebswirtschaft (DBW)*, 24(6), 716–740.

Green, P. E., Tull, D. S., & Albaum, G. (1988). *Research for marketing decisions* (5th ed.), Englewood Cliffs: Prentice-Hall.

Gudergan, S. P.; Devinney, T. Ellis, R. S. (2003). *A Competence-innovation framework of non-equity alliance performance, institute for the study of business markets*. University Park: The Pennsylvania State University.

Hair, J. F., Black, W. C., Babin, B. J., Anderson, R. E., Tatham, R. L. (2006). *Multivariate Data Analysis* (6th ed.), Upper Saddle River: Pearson.

Hauser, R. M., & Goldberger, A. S. (1971). The treatment of unobservable variables in path analysis. In H. L. Costner (Ed.), *Sociological methodology* (pp. 81–117). London: Jossey-Bass

Heskett, J. L., Sasser, W. E., & Schlesinger, L. A. (1997). *The service profit chain*. New York: The Free Press.

Homburg, C., & Baumgartner, H. (1995). Die kausalanalyse als instrument der marketingforschung: eine bestandsaufnahme. *Zeitschrift für Betriebswirtschaft, 65*(10), 1091–1108.

Homburg, C., & Giering, A. (1996). Konzeptualisierung und operationalisierung komplexer konstrukte. *Marketing ZFP, 18*(1), 5–24.

Hulland, J. (1999). Use of partial least squares (PLS) in strategic management research: a review of four recent studies. *Strategic Management Journal, 20*(4), 195–204.

Jarvis, C. B., Mackenzie, S. B., & Podsakoff, P. M. (2003). A critical review of construct indicators and measurement model specification in marketing and consumer research. *Journal of Consumer Research, 30*, 199–218.

Kotler, P. (1994). *Marketing management – analysis, planning, implementation and control* (8th ed.), Englewood Cliffs, NJ, Prentice-Hill.

Krafft, M. (1999). An empirical investigation of the antecedents of sales force control systems. *Journal of Marketing, 63*, 120–134.

Krafft, M., Haase, K., & Siegel, A. (2003). Statistisch-ökonometrische BWL-forschung: entwicklung, status quo und perspektiven. In M. Schwaiger; D. Harhoff (Eds.), *Empirie und Betriebswirtschaft* (pp. 83–104) Schäffer-Poeschel, Stuttgart: Entwicklungen und Perspektiven.

Mathieson, K., Peacock, E., Chin, W. W. (1996). *Extending the technology acceptance model: The influence of perceived user resources*. Alberta, Canada: University of Calgary.

Nunnally, J. C. (1978). *Psychometric theory*. (2nd ed.), New York: McGraw-Hill Book Company.

O'Cass, A. (2002). Political advertising believability and information source value during elections. *Journal of Advertising, 31*(1), 63–74.

Olsen, P. O. (2002). Comparative evaluation and the relationship between quality, satisfaction and repurchase loyality. *Journal of Academy of Marketing Science, 30*(3), 240–249.

Peter, J. P. (1981). Construct validity: a review of basic issues and marketing practices. *Journal of Marketing Research, 18*, 133–145.

Peterson, R. A. (1994). A meta-analysis of cronbach's coefficient alpha. *Journal of Consumer Research, 21*, 381–391.

Reichheld, F. F. (1993). Loyalty-based management. *Harvard Business Review, 71*, 64–73.

Reinartz, W., Krafft, M., & Hoyer, W. D. (2004). The customer relationship management process: its measurement and impact on performance. *Journal of Marketing Research, 41*, 293–305.

Reinartz, W. Kumar, V. (2000). On the profitability of long-life customers in a noncontractual setting: an empirical investigation and implications for marketing. *Journal of Marketing, 64*(4), 17–35.

Rindskopf, D. (1984). Using phantom and imaginery latent variables to parameterize constraints in linear structural models. *Psychometrika, 49* 37–47.

Rodgers, W., & Pavlou, P. (2003). Developing a predictive model: a comparative study of the partial least squares vs maximum likelihood techniques. (Working Paper). Graduate School of Management, University of California, Riverside.

Rossiter, J. R. (2002). The C-OAR-SE procedure for scale development in marketing. *International Journal of Research in Marketing, 19*(4) 305–335.

Sambamurthy, V., & Chin, W. W. (1994). The effects of group attitudes towards alternative gdss designs on the decision-making performance of computer-supported groups. *Decision Sciences, 25*(2) 215–242.

Sarkar, M. B., Echambadi, R., Cavusgil, S. T., & Aulakh, P. S. (2001). The influence of complementarity, compatibility, and relationship capital on alliance performance. *Journal of the academy of marketing science, 29*(4), 358–373.

Sheth, J., & Parvatiyar, A. (1995). Relationship marketing in consumer markets: antecedents and consequences. *Journal of the Academy of Marketing Science, 23*(4), 255–271.

Soo, C. W., Midgley, D. F., & Devinney, T. M. (2002). The Process of Knowledge Creation in Organizations. (Working Paper). INSEAD, Fontainebleau.

Sriram, V., Mummalaneni, V. (1990). Determinants of Source Loyalty in Buyer-Sellers Relationships. *Journal of Purchasing and Materials Management, 26*(4), 21–26.

Stone-Romero, E. F., Stone, D. L. (1997). Development of a multidimensional measure of perceived product quality. *Journal of Quality Management, 2*(1), 87–111.

Stone, M. (1975). Cross-validatory choice and assessment of statistical predictions. *Journal of the Royal Statistical Society, 36*(2), 111–133.

Tan, W. G., Chan, T., & Gable, G. G. (1999). A structural model of software maintainer effectiveness. In Proceedings of the 10th Australasian Conference on Information Systems, Wellington, pp. 943–954.

Taylor, P., & Baker, T. (1994). An assessment of the relationship between service quality and customer satisfaction in the formation of consumers' purchase intention. *Journal of Retailing, 70*, 163–178.

Temme, D., Kreis, H., & Hildebrandt, L. (2010). A Comparison of Current PLS Path Modeling Software: Features, Ease-of-Use, and Performance. In: V. Esposito Vinzi, W. Chin, J. Hensler, and H. Wold (Eds.,) *Handbook of partial least squares*, (pp. 741–760), Heidelberg: Springer.

Venaik, S., Midgley, D. F., & Devinney, T. (2004). A new perspective on the integration-responsiveness pressures confronting multinational firms. *Management International Review, 44*(1), 15–48.

Venaik, S., Midgley, D. F., & Devinney, T. M. (2001). Autonomy, Networking and Interunit Learning in a Model of MNC Subsidiary Innovation and Performance. http://www.agsm.edu.au/~timdev/research/VMD.PDF.

Vinzi, V. E., Lauro, C., & Tenenhaus, M. (2003). PLS Path Modeling. (Working Paper). DMS – University of Naples, HEC – School of Management, Jouy-en-Josas.

Woodside, A. G., & Taylor, J. L. (1978). Consumer purchase intention of product quality and national advertising. *Journal of Advertising, 7*(1), 48–51.

Yi, Y., Jeon, H. (2003). Effects of loyality programs on value perception, program loyalty and brand loyalty. *Journal of Academy of Marketing Science, 31*(3), 229–40.

Chapter 30
Testing Moderating Effects in PLS Path Models: An Illustration of Available Procedures

Jörg Henseler and Georg Fassott

Abstract Along with the development of scientific disciplines, namely social sciences, hypothesized relationships become increasingly more complex. Besides the examination of direct effects, researchers are more and more interested in moderating effects. Moderating effects are evoked by variables whose variation influences the strength or the direction of a relationship between an exogenous and an endogenous variable. Investigators using partial least squares path modeling need appropriate means to test their models for such moderating effects. We illustrate the identification and quantification of moderating effects in complex causal structures by means of Partial Least Squares Path Modeling. We also show that group comparisons, i.e. comparisons of model estimates for different groups of observations, represent a special case of moderating effects by having the grouping variable as a categorical moderator variable. We provide profound answers to typical questions related to testing moderating effects within PLS path models:

1. How can a moderating effect be drawn in a PLS path model, taking into account that the available software only permits direct effects?

2. How does the type of measurement model of the independent and the moderator variables influence the detection of moderating effects?

3. Before the model estimation, should the data be prepared in a particular manner? Should the indicators be centered (by having a mean of zero), standardized (by having a mean of zero and a standard deviation of one), or manipulated in any other way?

J. Henseler
Nijmegen School of Management, Radboud University Nijmegen, P.O. Box 9108, 6500 HK Nijmegen, The Netherlands
e-mail: j.henseler@fm.ru.nl

G. Fassott
Department of Marketing, University of Kaiserslautern, Postfach 30 49, 67653 Kaiserslautern, Germany
e-mail: fassott@wiwi.uni-kl.de

V. Esposito Vinzi et al. (eds.), *Handbook of Partial Least Squares*, Springer Handbooks of Computational Statistics, DOI 10.1007/978-3-540-32827-8_31,

4. How can the coefficients of moderating effects be estimated and interpreted? And, finally:

5. How can the significance of moderating effects be determined?

Borrowing from the body of knowledge on modeling interaction effect within multiple regression, we develop a guideline on how to test moderating effects in PLS path models. In particular, we create a graphical representation of the necessary steps to take and decisions to make in the form of a flow chart. Starting with the analysis of the type of data available, via the measurement model specification, the flow chart leads the researcher through the decisions on how to prepare the data and how to model the moderating effect. The flow chart ends with the bootstrapping, as the preferred means to test significance, and the final interpretation of the model outcomes.

30.1 Moderating Effects – An Overview

Along with the development of scientific disciplines, namely social sciences, the complexity of hypothesized relationships has steadily increased (Cortina 1993). As Jaccard and Turrisi (2003) established, there are basically six types of relationships that can occur within causal models: (1) direct effects when an independent variable, X, causes a dependent variable, Y; (2) indirect effects (also called mediating effects) when an independent variable, X, has an impact on a third variable, Z, which then influences the dependent variable, Y; (3) spurious effects when a correlation between two variables stems from a common cause, Z; (4) bidirectional effects when two variables, X and Y, influence each other; (5) unanalyzed effects; and (6) moderating effects (also called interaction effects) when a moderator variable influences the strength of the direct effect between the independent variable, X, and the dependent variable, Y. Figure 30.1 shows the symbolic representations of the different causal relationships.

The detection and estimation of direct effects is a central domain of PLS path modeling and is thus an inherent part of almost all of this volume's contributions. The nature of path modeling particularly supports the examination of indirect effects. A good example of this is the contribution by Helm et al. (2010). Typically, neither spurious effects and unanalyzed effects nor bidirectional effects are accounted for in PLS path models. Moreover, the requirement of recursivity in standard PLS path models (Lohmöller 1989) inhibits investigating bidirectional effects.

Besides the examination of direct effects, researchers are more and more interested in moderating effects. Moderating effects are evoked by variables whose variation influences the strength or the direction of a relationship between an exogenous and an endogenous variable (Baron and Kenny 1986, p. 1174). The causes of moderating effects are called "moderator variables" or just "moderators." Moderator variables can either be metric (e. g., consumer psychological constructs like arousal or intelligence) or categorical (e. g., gender or social class) in nature. Interestingly, group comparisons, i.e. comparisons of model estimates for different

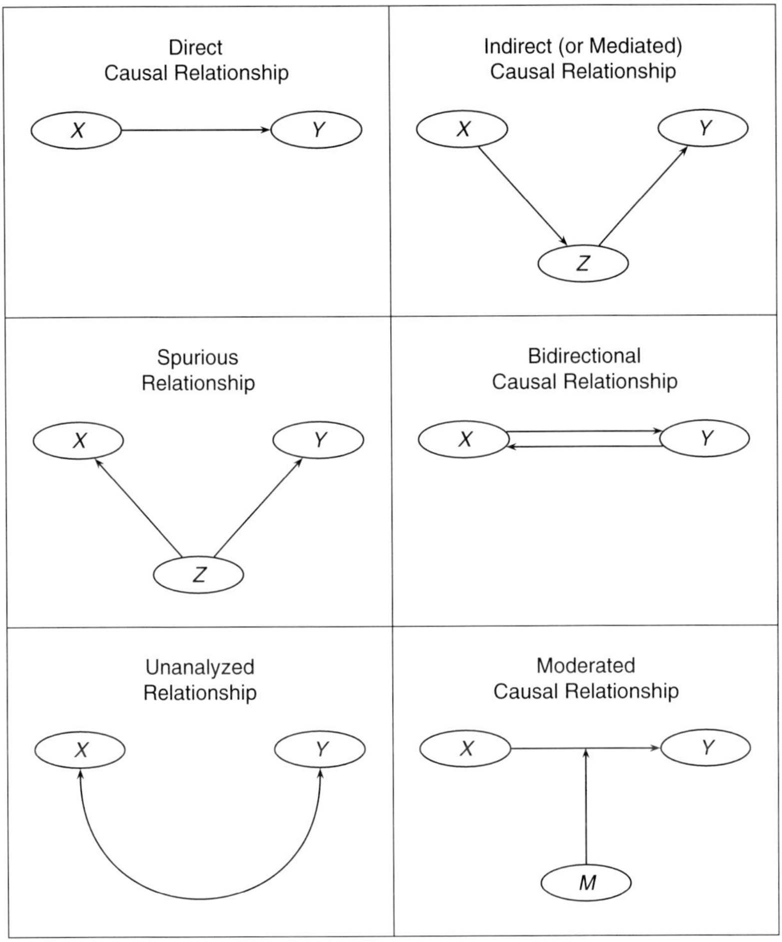

Fig. 30.1 Examples of causal relationships between latent variables (slight variation of Jaccard and Turrisi (2003, p. 2))

groups of observations, can be regarded as a special case of moderating effects. The grouping variable is nothing more than a categorical moderator variable.

One example of the examination of moderating effects is a paper by Homburg and Giering (2001): They find that age and income have significant effects on the strength of the relationship between customer satisfaction and customer loyalty. In that context, age and income serve as moderator variables. Other examples are presented in this volume, particularly in the contributions by Chin and Dibbern (2010), Eberl (2010), Tenenhaus et al. (2010), and Streukens et al. (2010).

In the majority of structural equation models, moderating effects are not taken into account, although in the literature the importance of moderators for the understanding of complex relationships is emphasized repeatedly (c. f. Chin et al.

(2003), p. 193; Homburg and Giering (2001), p. 47). This neglect of moderating effects leads to a lack of relevance: Relationships that hold true regardless of the context factors are often trivial. For example, marketing researchers and practitioners will not be surprised that customer satisfaction and customer loyalty usually correlate positively. However, serious progress could be achieved in scientific knowledge if an answer is found to the question of the circumstances under which this relationship is extremly strong or extremely weak. Research questions of the latter type rely on the identification and quantification of moderating effects.

Once a marketing research project has determined *that* moderating effects should be accounted for, the question arises of *how* this should be realized. One of the first frameworks for identifying moderating effects in marketing research was presented by Sharma et al. (1981). But 20 years later, Irwin and McClelland (2001, p. 101) still state that the proper use of moderated regression models, under which they subsumate OLS regression as well as AN(C)OVA, logistic regression and structural equation modeling, "...is not a minor issue in marketing." This statement certainly holds true for other social sciences, too.

30.2 PLS Path Modeling and Moderating Effects

The purpose of this contribution is to illustrate the identification and quantification of moderating effects in complex causal structures by means of PLS path modeling. The use of PLS path modeling in order to identify and quantify other types of causal relationships is discussed elsewhere in this volume (Esposito Vinzi 2006; Helm et al. 2010).

To date, only a few methodologically oriented articles have been dedicated to the detection of moderating effects in PLS path models, among them Chin et al. (2003) and Eggert et al. (2005). Discussions among researchers, for example in internet forums like www.smartpls.de, show that there is a strong need for clarification of how moderating effects can be integrated into PLS path models. Researchers who want to test moderating effects within PLS path models have to cope with a number of questions:

- How can a moderating effect as depicted in Fig. 30.2 be drawn in a PLS path model, taking into account that the available software only permits direct effects?
- How does the type of outer model of the independent and the moderator variables (outwards directed as in mode A, or inwards directed as in mode B) influence the detection of moderating effects?
- Before the model estimation, should the data be prepared in a particular manner? Should the indicators be centered (by having a mean of zero), standardized (by having a mean of zero and a standard deviation of one), or manipulated in any other way?
- How can the coefficients of moderating effects be estimated and interpreted?
- How can the significance of moderating effects be determined?

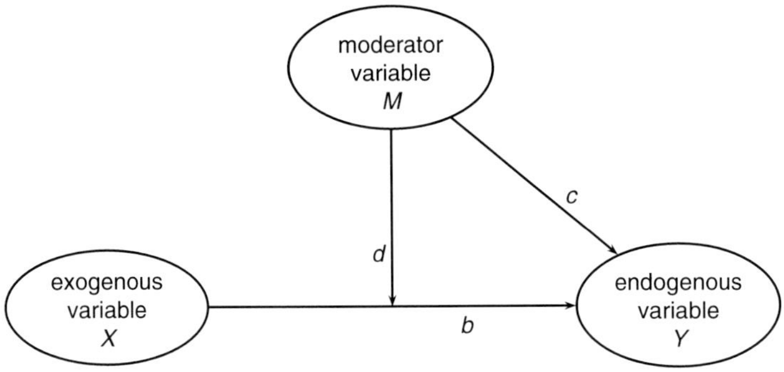

Fig. 30.2 A simple model with a moderating effect (d)

In order to better understand how moderating effects can be estimated and tested within PLS path models, it is useful to first have a look at the nature of PLS path modeling. As Esposito Vinzi et al. (2010) explain, PLS path models are estimated in two steps. Firstly, in an iterative process, latent variable scores are estimated for each latent variable. Secondly, these latent variable scores enter as independent and dependent variables (depending on their position in the path model) into one or more (OLS) regressions. Owing to the nature of the second step, most of the recommendations for testing moderating effects in multiple regression hold for PLS path modeling as well. We can thus rely on the body of research on interaction effects in linear regressions, as it is presented in, for example, Aiken and West (1991) or Jaccard and Turrisi (2003).

The remainder of this contribution is structured as follows: Firstly, we will transfer the body of research on how moderating effects are tested within multiple regression to PLS path modeling. This means that we restrict our perspective to the structural model. Secondly, we will discuss how the measurement or outer model has to be designed in order to support the testing of moderating effects in the structural model. Thirdly, we will compile the raised issues in a guideline on how to test moderating effects in PLS path models.

30.3 Structural Model Considerations

When we speak of moderating effects in the context of PLS path modeling, we always mean a moderated relationship within the structural model. This means that we are interested in the moderating effects of latent variables on the direct relationships between latent variables. Throughout the remainder of this paper, we will use the smallest possible type of structural model as an exemplary model consisting of a dependent, an independent, and a moderator variable. Figure 30.2 shows such a simple model with a moderating effect. The moderating effect (d) is symbolized by an arrow pointing to the direct relationship, (b), which is hypothesized as moderated.

In general, there are two common approaches to estimate moderating effects with regression-like techniques: the product term approach and the group comparison approach. We will present both approaches and discuss their strengths and weaknesses.

30.3.1 Moderating Effects as Product Terms

In order to develop the structural equation for the exemplary model, we first consider only the main effects. The main effects of the two independent variables X and M on the dependent variable Y can be expressed by the following equation:

$$Y = a + b \cdot X + c \cdot M \tag{30.1}$$

Here, a is the intercept, and b and c are the slopes of X and M, respectively. Note that if you once partially derive equation 30.1 with respect to X and M, you receive the change in Y depending on the change in one predictor if the other predictor is held constant. Obviously, these first partial derivatives are b and c, respectively.

In order to include the moderating effect, its nature has to be clear. Keeping in mind Baron and Kenny's (1986, p. 1174) definition of a moderator as a "... variable that affects the direction and/or strength of the relation between an independent or predictor variable and a dependent or criterion variable," the moderating effect can easily be added. The idea of a moderating effect is that the slope of the independent variable is no longer constant, but depends linearly on the level of the moderator. The structural equation of the model depicted in Fig. 30.2, including the moderating effect can thus be mathematically formulated as follows:

$$Y = a + (b + d \cdot M) \cdot X + c \cdot M \tag{30.2}$$

In this equation, the slope of X depends on the level of M. Equation (30.2) can be rearranged to have either of the following two forms:

$$Y = a + b \cdot X + c \cdot M + d \cdot (X \times M) \tag{30.3}$$

$$= (a + c \cdot M) + (b + d \cdot M) \cdot X \tag{30.4}$$

Equation (30.3) forms the basis of the following discussion. This equation also delivers the answer to the question of how moderating effects can be integrated into a PLS path model. The solution is a so-called interaction term $X \times M$, an additional latent variable in the structural model covering the product of the independent and the moderator variable. Figure 30.3 illustrates this approach. At this stage, four comments should be made.

Firstly, in general, the regression parameters a, b, and c in formulae (30.2) and (30.3) will differ from those in formula (30.1). The reason for this is that regression

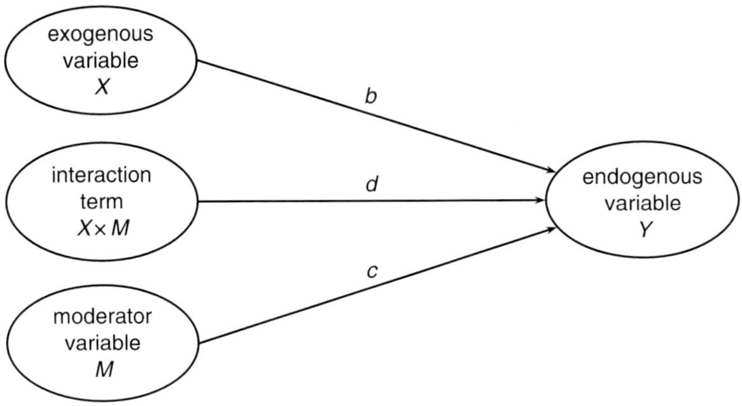

Fig. 30.3 Transcript of the model in Fig. 30.2 for PLS path models

parameters in regression functions with product terms, as in formulae (30.2) and (30.3), no longer represent main effects, but single effects. Single effects mean that they describe the strength of an effect when all the other components of the product term have a value of zero. This idea becomes especially clear when looking at formula (30.4): For a given level of M, Y is expressed by X in the form of a single regression with intercept $a + (c \cdot M)$ and slope $(b + d \cdot M)$.

Secondly, besides controlling for the focal effects b and d, the structural equation (30.3) should control for the direct effect c, too. Only when all the components of a product term are included in the regression model in a direct form, does the product term represent the moderating effect (Irwin and McClelland 2001; Cohen 1978; Cronbach 1987). Sometimes called "reduced" models as in, for example, $Y = a + d \cdot (X \times M)$, are not appropriate to determine moderating effects, because they would overestimate the size of the moderating effect (Carte and Russell 2003).

Thirdly, note that the formula (30.2) is a type of regression formula and therefore requires metric data. If the independent or the moderator variable are categorical with more than two categories ($l > 2$), the respective variable has to be dichotomized as described in Sect. 30.4.3.

Fourthly, the interaction term, i. e. the product of the independent variable X and the moderator variable M, is as such commutative. This fact implies that mathematically it does not matter which variable is the independent and which one the moderator variable. Both the interpretations are equally legitimate.

30.3.2 Determining Moderating Effects Through Group Comparisons

Especially if either of the independent or moderator variable is not continuous, an alternative technique for identifying moderating effects in structural equation modeling is widely suggested. "If one or both of the interacting variables is discrete, or can be made so, researchers can apply a 'multisample' approach, with the interaction effects becoming apparent as differences in parameter estimates when the same model is applied to different but related sets of data" (Rigdon et al. 1998, p. 1).

When the moderator variable is categorical (as, e. g., sex, race, class) it can be used as a grouping variable without further refinement. However, when a metrically scaled variable is used as a grouping variable, it first has to be transformed into a categorical variable. The prevailing technique is dichotomization here, i. e. the moderating variable is divided into two value categories, "high" and "low". There are mainly two ways to dichotomize a latent construct: either using the indicator values or the construct values. If the indicators have an interpretable mean, the following decision rule can be used to determine to which group each observation should belong:

- If all indicator values are above the mean, the grouping value is "high".
- If all indicator values are below the mean, the grouping value is "low".
- Otherwise, the observation should not be assigned to any group.

While this dichotomization rule is in general fine for reflective constructs, it may be problematic with formative constructs: Formative constructs do not necessarily have to correlate with one another. Consequently, many observations may be discarded. In that case, or if the indicators have no interpretable mean, a different decision rule can be applied:

- If the moderator variable's latent variable score of an observation lies within the upper third, the grouping value is set to "high".
- If the moderator variable's latent variable score of an observation lies within the lower third, the grouping value is set to "low".
- Otherwise, the observation is not assigned to any group.

Another popular method is the so-called median split. Observations whose moderator score is above the median, are said to have a high moderator value; observations whose moderator score is below the median, are said to have a low moderator value. The selection of one of the suggested grouping methods for a particular research question is up to the researcher.

Once the observations are grouped, the model with the direct effects is estimated separately for each group of observations. Differences in the model parameters between the different data groups are interpreted as moderating effects. Figure 30.4 depicts the proposed procedure by having a dichotomous (or dichotomized by the researcher) moderator variable. In this example, the direct relationship b between the exogenous latent variable X and the endogenous latent variable Y is compared across G groups. The superscript g symbolizes that all values $X^{(g)}$, $Y^{(g)}$, and $b^{(g)}$ are estimated for every group g ($g = 1, \dots, G$) separately.

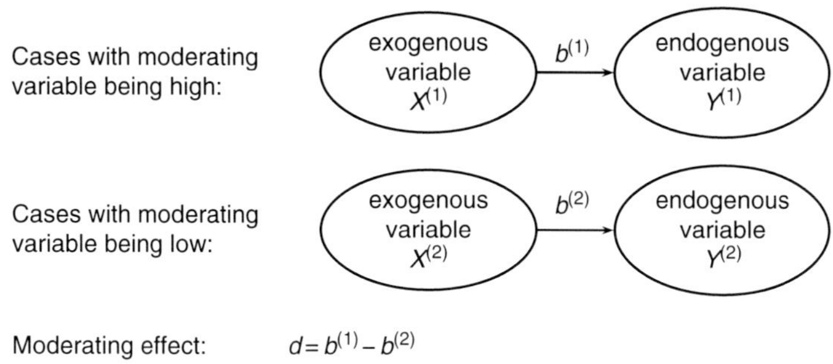

Fig. 30.4 Detecting a moderating effect (d) through group comparisons

30.3.3 Comparison of Both Approaches

The product term approach is a straightforward modelation of a moderating effect if the moderator influences the strength of the moderated direct relationship linearly. As long as the construct measurement is invariant across groups, the product term approach and the group comparison approach lead to the same results.

Obviously, the group comparison is suboptimal for continuous moderating variables: Firstly, due to the dichotomization, a part of the moderator variable's variance is lost for analysis. Secondly, observations that cannot be unambiguously allocated to a single group are ignored for analysis. Thirdly, the assignment of observations to groups is rather arbitrary. Given that the only indication of the moderator effect size is the parameter difference $d = b^{(1)} - b^{(2)}$ as depicted in Fig. 30.4, the arbitrariness of assignation opens the door for manipulation. However, despite these disadvantages, the group comparison approach is quite popular among researchers, probably because of its ease of use.

One could argue that guidelines developed for structural equation modeling do also hold for PLS path modeling in particular. For example, in the context of structural equation modeling, Rigdon et al. (1998, p. 4) regard the product term approach as the "natural" approach when both interacting variables are continuous, whereas the group comparison approach is seen as the logical choice when one or both of the interacting variables are discrete or categorical. Other researchers (e. g., Reinecke (1999)) suggest first conducting group comparisons in order to obtain a deeper insight, and thereafter applying the product term approach.

In this respect, we provide differing advice concerning the selection of approaches to estimate moderating effects within PLS path models: Given that the results of the product term approach are usually equal or superior to those of the group comparison approach, we recommend always using the product term approach. Only if the moderator variable is categorical, or if the researcher wants a quick overview of a possible moderator effect, could the group comparison approach be considered.

30.3.4 Three-Way Interactions

The issues regarding moderating effects discussed so far do not only hold true for simple moderating effects, but also cover cascaded moderating effects. We speak of cascaded moderating effects if the strength of a moderating effect is influenced by another variable, i. e. the moderating effect is again moderated. In the special case of three interacting variables, i. e. the independent variable and two moderator variables, we find a so-called three-way interaction.

Returning to the example of the direct relation between customer satisfaction and customer loyalty, which was moderated by age and income, we could imagine that the moderating effect of age and income is not constant, but is itself influenced by other variables like product category.

We now consider the simplest possible three-way interaction model. It consists of an independent variable X, two moderator variables, M and N, and the dependent variable, Y. A mathematical representation of the three-way interaction is as follows:

$$Y = aX + bM + cN + d(X \times M) + e(X \times N) + f(M \times N) + g(X \times M \times N) \quad (30.5)$$

This formula can be expressed by a PLS path model with three direct effects and four product terms. The path model would thus be comprised of eight latent variables, including the endogenous variable.

As in the case of a simple moderating effect, all components of the product term should also be entered into the regression function explicitly. In particular, besides the three-way interaction term, all single effects and all two-way interaction effects should be included.

30.4 Measurement Model Considerations

How can the measurement models of all the involved variables contribute to facilitating the estimation and testing of moderating effects? The answer to this question will vary depending on the type of measurement model of the independent and the moderator variable. Three types of measurement models can be distinguished in the present context:

- In *formative* measurement models, the latent variable is regarded as a consequence of its respective indicators (Bollen and Lennox 1991). As the latent variable is defined by its indicators, changing indicators alters the meaning of the latent variable (Diamantopoulos and Winklhofer 2001). It is important to recognize that the latent variable values are sensitive to changes in the importance (the weight) of each indicator, because the indicators can measure different attributes and/or different components and thus do not have to be correlated.
- In *reflective* measurement models, indicators are regarded as consequences of the latent variable to which they belong (Jarvis et al. 2003). Having a common

cause, reflective indicators should be highly correlated. The reflective indicators of a latent variable can be used interchangeably and even to a certain extent be discarded.
- The latent variable is, in fact, a dummy or effects-coded variable. In this case, the latent variable and its indicator are one and the same. The latent variable is thus neither the cause nor the consequence of its indicator.

It is noteworthly that the distinction between the first two types is based solely on the direction of causality. It is not necessarily linked to the choice of the statistical measurement model, i.e. the selection of Mode A, Mode B, or any other mode.

In this contribution, only non-hierarchical measurement models will be discussed. However, the generalization of the techniques (described below) to hierarchical measurement models (as, e. g., assigned to second-order constructs) is straightforward. Examples for the estimation of moderating effects between second-order constructs can be found in the contributions of Streukens et al. (2010) and Wilson (2010).

30.4.1 Moderating Effects with Reflective Constructs: The Product Indicator Approach

In order to model moderating effects of latent variables in structural equation models, Kenny and Judd (1984) proposed building product terms between the indicators of the latent independent variable and the indicators of the latent moderator variable. These product terms serve as indicators of the interaction term in the structural model. Chin et al. (1996, 2003) were the first to transfer this approach to PLS path modeling. They suggest building the products of each indicator of the independent latent variable with each indicator of the moderator variable. These product indicators become the indicators of the latent interaction term. If the independent latent variable has I indicators and the latent moderator variable has J indicators, then the latent interaction variable will have $I \cdot J$ product indicators. Figure 30.5 shows a simple example of the product indicator approach.

One question which is particularly raised in structural equation modeling is whether really all possible indicator products should be built and combined in the interaction term. Jöreskog and Wang (1996) show that already one product indicator is sufficient to estimate the moderating effect. Jonsson (1998) uses several but not all product terms in order to obtain a better estimate of the interaction term's standard error. However, this coincides with a stronger bias of the estimates (Jonsson 1998). In PLS path modeling, statistical inferences are usually based on bootstrap outcomes of the parameter estimates. As it is the variation of parameter estimates across bootstrap samples that determines the range of the confidence interval of a parameter, the correct estimation of the interaction term's path coefficient should be prioritized against the estimation of its standard error. As a conclusion, the approach by Chin et al. (2003) is most promising.

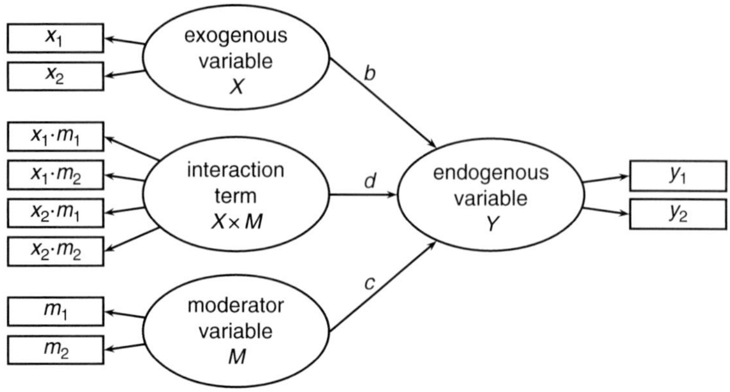

Fig. 30.5 Creating the interaction term with product indicators

30.4.2 Moderating Effects with at Least One Formative Construct: The Two-Stage Approach

If the exogenous variable and/or the moderator variable are formative, the pair-wise multiplication of indicators is not feasible. "Since formative indicators are not assumed to reflect the same underlying construct (i. e., can be independent of one another and measuring different factors), the product indicators between two sets of formative indicators will not necessarily tap into the same underlying interaction effect" (Chin et al. 2003, Appendix D).

Instead of using the product indicators approach, we recommend a two-stage PLS approach for estimating moderating effects when formative constructs are involved. We thereby make use of PLS path modeling's advantage of explicitly estimating latent variable scores. The two stages are built up as follows:

Stage 1: In the first stage, the main effect PLS path model is run in order to obtain estimates for the latent variable scores. The latent variable scores are calculated and saved for further analysis.

Stage 2: In the second stage, the interaction term $X \times M$ is built up as the element-wise product of the latent variable scores of X and M. This interaction term as well as the latent variable scores of X and M are used as independent variables in a multiple linear regression on the latent variable scores of Y.

Figure 30.6 illustrates the two-stage approach. Whilst in the first stage the latent variable scores are estimated, these are used in the second stage to determine the coefficients of the regression function in the form of formula (30.3).

The second stage can be realized by multiple linear regression or be implemented within PLS path modeling by means of single indicator measurement models. However, using only a single indicator measurement for the interaction term and estimating the formative measurement models again is not recommendable, because

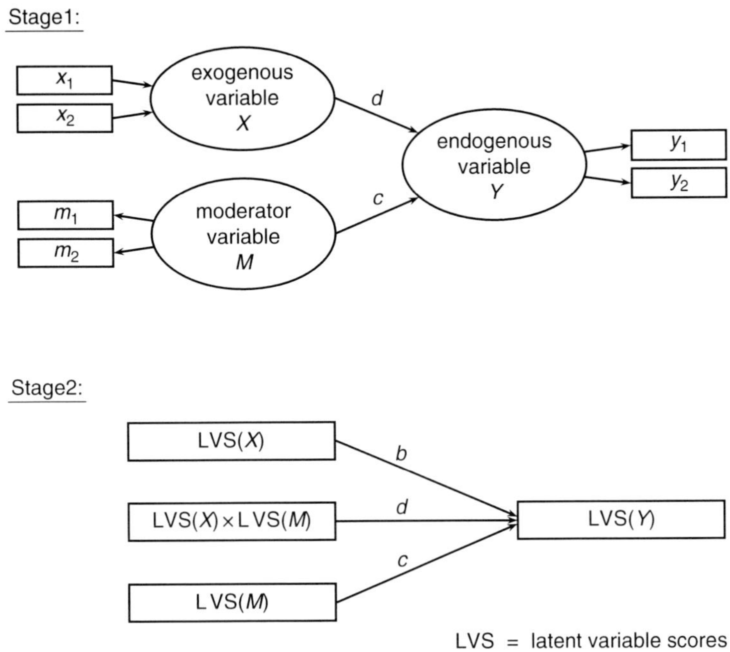

Fig. 30.6 A two-stage PLS approach to model interaction effects with formative constructs involved

the latent variable scores of X and M may alter significantly. Such an alteration can have two negative consequences. Firstly, due to the formative measurement, the construct meaning might change in comparison to the main effects model. Secondly, the element-wise product of the newly estimated latent variable scores of X and M no longer equals the previously calculated interaction term.

An important characteristic of measurement models in PLS path modeling is that latent variables with only one indicator are set equal to this indicator, no matter which type of measurement model is chosen. If all formative interacting variables are measured by single indicators, the researcher can choose either the product indicator approach or the two-stage approach.

30.4.3 Moderating Effects with Categorical Variables

The third possible type of measurement model encompasses the case when the latent variable is a categorical variable described by one indicator. Actually, from a psychometric point of view, since the PLS path modeling algorithm sets the latent variable equal to its single indicator, it is questionable whether the variable is really "latent". However, for simplicity's sake, we accept that the latent variable just equals its respective manifest variable.

As PLS path modeling is based on ordinary least squares regressions, only interval scaled variables can be analyzed. Therefore, categorical variables with more than two categories have to be transformed into sets of dichotomous variables. Several methods for this are suggested in literature; however, we want to focus on the two most widely used: dummy coding, as shown in Table 30.1, and unweighted effects coding, as visualized in Table 30.2.

The most often used coding scheme for dichotomizations of categorical variables is certainly dummification, i. e. a categorical variable with γ categories is dissolved into $\gamma - 1$ distinct 0/1-coded variables. One category is arbitrarily designated as the reference category. The dummy variables indicate the assigned category by a value of one.

If both the independent variable and the moderator variable are categorical, Aiken and West (1991) recommend using a different coding scheme: unweighted effects codes. Table 30.2 shows the corresponding coding scheme for a categorical variable with three categories. The advantage of unweighted effects codes is that they produce ANOVA-like results, i e. unweighted effects codes focus on the explanation of group differences.

Figure 30.7 illustrates how a moderating effect evoked by a categorical moderator variable M with three categories can be drawn and estimated through PLS path modeling. Obviously, the procedure differs from the product indicators approach for reflective measurement models only in as much as categorical variables with more than two categories have to be converted into dichotomous variables; consequently, additional interaction terms have to be eventually considered.

Table 30.1 Three potential dummy variable coding schemes for a categorical variable with three categories

Original variable M		Comparison group					
		Category 1		Category 2		Category 3	
		M_1	M_2	M_1	M_2	M_1	M_2
	1	0	0	1	0	1	0
Categories	2	1	0	0	0	0	1
	3	0	1	0	1	0	0

Table 30.2 Three potential coding schemes for variables with unweighted effects codes for a categorical variable with three categories

Original variable M		Comparison group					
		Category 1		Category 2		Category 3	
		M_1	M_2	M_1	M_2	M_1	M_2
	1	−1	−1	1	0	1	0
Categories	2	1	0	−1	−1	0	1
	3	0	1	0	1	−1	−1

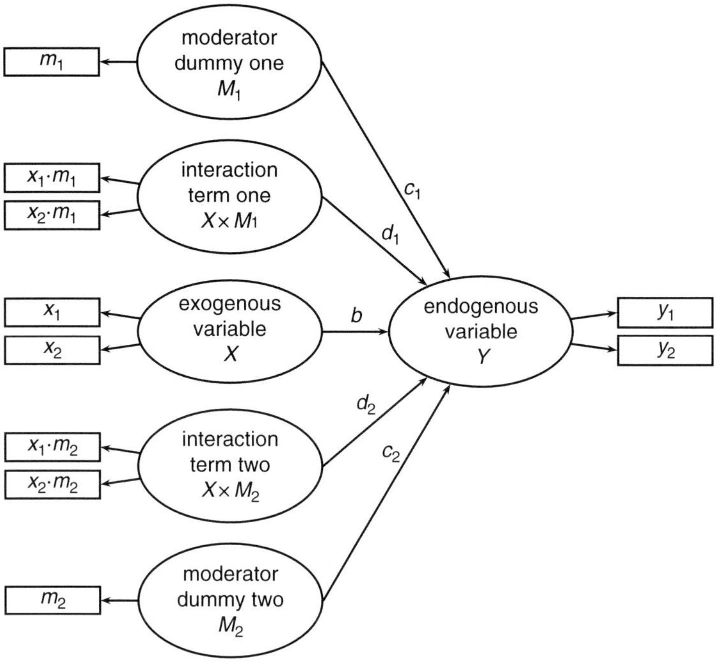

Fig. 30.7 Moderator variable as a categorical variable with three categories

It is noteworthy that as M_1 and M_2 belong to one variable M, it does not make sense to look for three-way interactions of the kind $X \times M_1 \times M_2$. Moreover, it is generally not possible to disentangle a binary variable's effect when, besides the single effect, also higher order effects (e. g., M_1^2 or $M_1 \times M_2$) are modelled.

30.4.4 The Scaling of the Interacting Variables

Another important question is whether the manifest variables should be standardized or not. From additive regression models it is known that linearly rescaling a variable – this explicitly comprises standardization – does not influence either the statistical test or the interpretation of a regression parameter. However, this experience is not transferable to regressions containing product terms.

Aiken and West (1991) regard standardized regression coefficients as particularly problematic in interaction models, because z-scores of products in general do not equal products of z-scores. In other words: In general, a standardized interaction term does not equal the product of its standardized factors. It can be concluded that in order to get interpretable coefficients for moderating effects, the interaction term itself must specifically not be standardized.

Jaccard and Turrisi (2003, p. 68) deliver a smooth illustration of how standardized model parameters can lead to a false conclusion:

> "As one illustration of the limitations of standardized coefficients, consider a simple bivariate regression where we regress a measure of income onto the number of years of education in order to determine the 'value' of a year of education. The analysis is conducted in two differen ethnic groups, African Americans and European Americans. Suppose that the analysis yielded identical standardized regression coefficients in the two groups, indicating that for every 1 *SD* that education changes, income is predicted to change 0.50 *SD*. One might conclude from this that the 'value' of education is the same in the two groups. Suppose that the standard deviation for education is 3.0 in both groups but that for income it is 15,000 for European Americans and 6,000 for African Americans. Such a state of affairs yields unstandardized coefficients of 2,500 European Americans and 1,000 for African Americans. Whereas for European Americans an additional year of education is predicted to be worth $2,500, for African Americans, it is worth only $1,000. There is a clear disparity between the groups that is not reflected in the standardized analysis."

Obviously, the false conclusion might be drawn when firstly, the differences the standard deviation are not taken into account and, secondly, the standard deviation can be meaningfully interpreted. The researcher must thus be aware of potential misinterpretations.

On the one hand, centering and especially standardization of indicators are associated with problems concerning the interpretation of model outcomes. On the other hand, centering also has its specific merits, as we will show.

Multicollinearity is a well-known problem that arises in the context of modeling moderating effects through multiplicative terms. This multicollinearity can lead to serious computational problems (Cohen 1978; Pedhazur 1982). However, in contrast to multicollinearity between two theoretically different variables, which has to be acknowledged by the researcher, "[t]he multicollinearity in the context of regression with higher order terms [which comprise interaction terms; the authors] is due to scaling, and can be greatly lessened by centering variables" (Aiken and West 1991, p. 35). This is also clearly discussed by Marquardt (1980), Smith and Sasaki (1979), and Tate (1984).

A second reason to center the indicators of the independent and the moderator variable lies in the interpretation of the single effects. In Fig. 30.2, the coefficient b represents the slope of the regression of X on Y when M has a value of zero. If zero were not an existing value on the scale of M, the reference point would not be a particularly sensible choice. Centering is an appropriate means of shifting the reference point to the mean and facilitating the interpretation of the parameters (Aiken and West 1991; Finney et al. 1984).

In PLS path models, the latent variable scores are calculated as linear combinations of the corresponding indicators. This characteristic is common to all measurement models, whether mode A or mode B (see vinzi et al. 2010), mode C (see Tenenhaus et al. 2005), mode PLS or mode PCA (as implemented in SPAD PLS, see Temme and Kreis 2010). If one requires a latent variable to be centered, this can thus easily be accomplished by centering all its indicators. However, by means of this approach it is in general not possible to obtain standardized latent variables (except for a single indicator measurement when the indicator is standardized). In order to get standardized latent variables, the standardization has to take place within the PLS estimation algorithm. As this built-in functionality can not distinguish between exogenous and latent moderator variables, also the moderator variable would be standardized. Every interaction term's path coefficient then has to be manually corrected, taking into account the original variance of the interaction term.

Whilst centering is advantageous for metric independent and moderator variables, we have neglected the scaling of the dependent latent variable so far. Is it recommendable to center or standardize the dependent latent variable and, thus, its indicators? Aiken and West (1991, p. 35) point out that "[c]hanging the scaling of the criterion by additive constants has no effect on regression coefficients in equations containing interactions. By leaving the criterion in its original (usually uncentered form), predicted scores conveniently are in the original scale of the criterion. There is typically no reason to center the criterion Y when centering predictors." However, when the researcher is only interested in the relative impact of the predictors and uses standardized indicators for X and M, also the indicators of the dependent variable Y should be standardized in order to get standardized path coefficients.

30.4.5 Software Support

Most PLS software packages offer several possibilities to standardize the manifest variables (Temme et al. 2010). Many of them follow Lohmöller's suggestion for a standardization parameter, the so-called METRIC (Tenenhaus et al. 2005) (see Table 30.3); others allow for a global or project-specific standardization option. Obviously, none of the METRICs is designed to center manifest variables but at the same time leave unchanged their variance. Moreover, it is not possible to standardize or center only some latent variables (or their indicators), while others, in particular the interaction term and eventually the endogenous variable, maintain their original scale. We therefore recommend to always use METRIC=4 if interaction effects are tested. In order to get centered manifest variables, the respective mean should be manually

Table 30.3 Lohmöller's METRIC (Tenenhaus et al. 2005, p. 170)

Variable scales are comparable	Means are interpretable	Variance reflects the importance of a variable	Mean	Variance	Rescaling	METRIC
No			0	1	No	1
Yes	No	No	0	1	Yes	2
Yes	Yes	No	Original	1	Yes	3
Yes	Yes	Yes	Original	Original		4

subtracted from each manifest variable before the manifest variable is integrated into a PLS path model. If standardization of indicators is desired, the indicators have to be divided by their respective standard deviations after centering. Researchers must be cautious not to use in-built standardization mechanisms. Even if the latent independent and moderator variable are standardized, their product must not be a posteriori standardized. Otherwise, the path coefficient of the interaction term is not interpretable.

30.5 Interpreting Moderating Effects

In order to analyze moderating effects, the direct relations of the exogenous and the moderator variable (effects b and c in Fig. 30.3) as well as the relation of the interaction term (effect d in Fig. 30.3) with the endogenous variable Y are examined. The hypothesis on the moderating effect is supported if the path coefficient d is significant – regardless of the values of b and c (Baron and Kenny 1986, p. 1174). Firstly, it has to be determined whether the moderating effects really exist in the population to which the researcher wants to generalize the research results, i.e. a test has to be done whether the path coefficient capturing the moderating effect differs significantly from zero. Secondly, the strength of the identified moderating effect has to be assessed.

30.5.1 Determining the Significance of Moderating Effects

As PLS path modeling does not rely on distributional assumptions, direct inference statistical tests of the model fit and the model parameters are not available. As a solution to this, bootstrapping is recommended (Chin 2010). Bootstrapping is a non-parametric technique for estimating standard errors of the model parameters (Efron and Tibshirani 1993). The quotient of (1) a model parameter and (2) its standard deviation is asymptotically Student t distributed. The significance of model parameters and, in particular, the coefficient of the interaction term, can be determined by means of respective tables.

In the case of group comparisons, the researcher is interested in whether certain path coefficients differ across groups. The Chow test is a parametric test that gives evidence of this (Chow 1978). This issue was discussed in marketing by Bass and Wittink (1975). Nowadays, the procedure is well described in econometrics primers (c.f. Gujarati 2003). The input necessary for a Chow test can be obtained by overall and group-wise bootstrap analyses. However, in the context of PLS path modeling, the Chow test's use of distributional assumptions is regarded as suboptimal. Alternatively, non-parametric approaches can be used. In their contribution, Chin and Dibbern (2010) present a permutation-based approach that provides the possibility to test for different path coefficients among groups. As their approach is distribution-free, it should be the first choice.

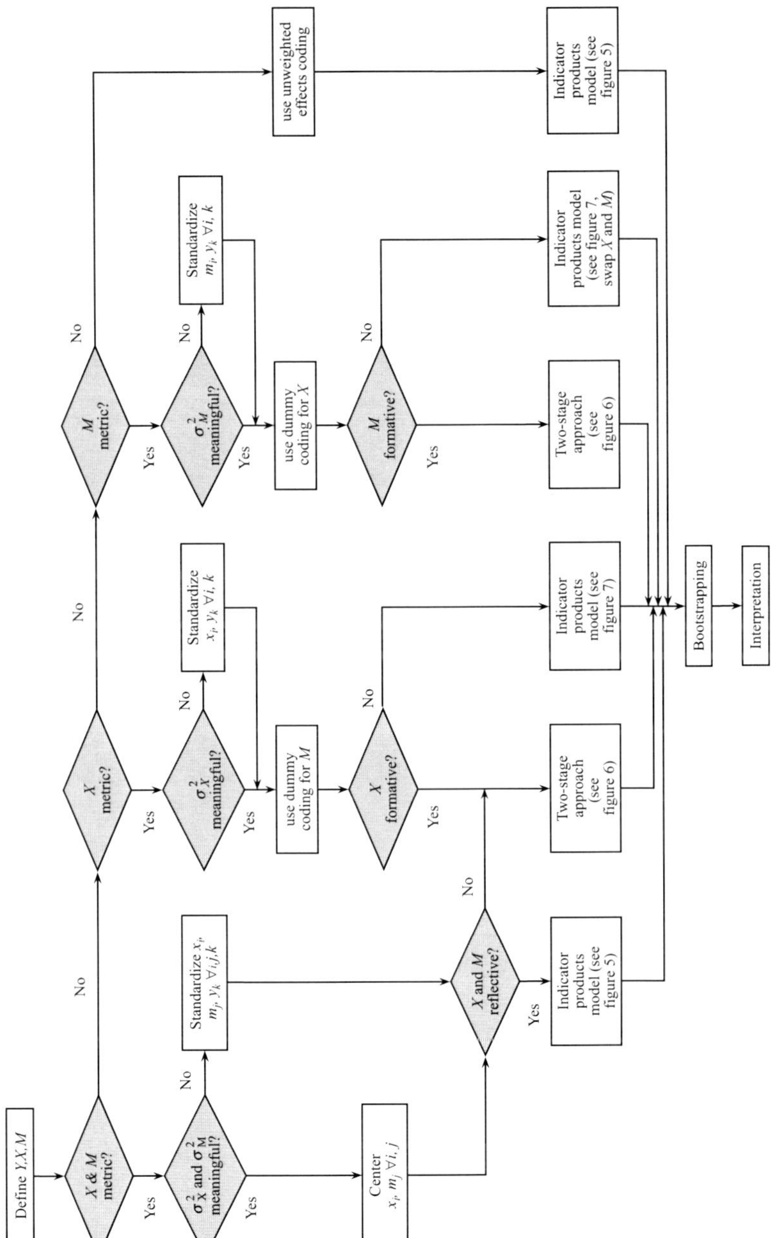

Fig. 30.8 Framework for determining moderating effects in PLS path models

30.5.2 Determining the Strength of Moderating Effects

The estimated path coefficient b describes the exogenous variable's influence on the endogenous variable when the moderator variable is zero (i. e., its mean). The path coefficient d of the interaction term indicates to which extent the exogenous variable's influence on the endogenous variable changes depending on the moderating variable. In the case of standardized variables, the following interpretation is possible: If the moderator variable is one, i. e. one standard deviation higher than its mean, the exogenous variable's influence on the endogenous variable is $b + d$ (in the nomenclature of Fig. 30.3).

Further, the moderating effect can be assessed by comparing the proportion of variance explained (as expressed by the determination coefficient R^2) of the main effect model (i. e. the model without moderating effect) with the R^2 of the full model (i. e. the model including the moderating effect). This idea also underlies the effect size. Drawing on Cohen (1988, p. 410–414), we suggest calculating the effect size f^2 with the following formula:

$$f^2 = \frac{R^2_{\text{model with moderator}} - R^2_{\text{model without moderator}}}{1 - R^2_{\text{model with moderator}}} \tag{30.6}$$

Moderating effects with effect sizes f^2 of 0.02 may be regarded as weak, effect sizes from 0.15 as moderate, and effect sizes above 0.35 as strong. Chin et al. (2003) state that a low effect size (f^2) does not necessarily imply that the underlying moderator effect is negligible: "Even a small interaction effect can be meaningful under extreme moderating conditions, if the resulting beta changes are meaningful, then it is important to take these conditions into account" (Chin et al. 2003, p. 211).

30.6 A Framework for Determining Moderating Effects in PLS Path Models

Within this contribution, several procedures were presented to estimate and test moderating effects by means of PLS path modeling. The selection of procedures should be based on the model specification as well as the type of data available. Instead of a verbal summary of the discussed issues, we give a graphical representation in the form of a flow chart (see Fig. 30.8). Starting with the analysis of the type of data available, via the measurement model specification, this flow chart leads the researcher through the decisions on how to prepare the data and how to model the moderating effect. The flow chart ends with bootstrapping, as the preferred means to test significance, and the final interpretation of the model outcomes.

References

Aiken, L. S., & West, S. G. (1991). *Multiple regression: testing and interpreting interactions*. Newbury Park: Sage.

Baron, R. M., & Kenny, D. A. (1986). The moderator-mediator variable distinction in social psychological research: conceptual, strategic, and statistical considerations. *Journal of Personality and Social Psychology, 51*(1), 1173–1182.

Bass, F., & Wittink, D. (1975). Pooling issues and methods in regression analysis with examples in marketing research. *Journal of Marketing Research, 12*(4), 414–425.

Bollen, K. A., & Lennox, R. (1991). Conventional wisdom on measurement: A structural equation perspective. *Psychological Bulletin, 110*(2), 305–314.

Carte, T. A., & Russell, C. J. (2003). In pursuit of moderation: Nine common errors and their solutions. *MIS Quarterly, 27*(3), 479–501.

Chin, W., & Dibbern, J. (2010). A permutation based procedure for multi-group PLS analysis: results of tests of differences on simulated data and a cross cultural analysis of the sourcing of information system services between germany and the USA. In: V. Esposito Vinzi, W. Chin, J. Hensler, and H. Wold (Eds.,) *Handbook of partial least squares*, Heidelberg: Springer.

Chin, W. W. (2010). Bootstrap cross-validation indices for PLS path model assessment. In V. Esposito Vinzi, W. W. Chin, J. Henseler, & H. Wang (Eds.), *Handbook of Partial Least Squares*. Heidelberg: Springer.

Chin, W. W., Marcolin, B. L., & Newsted, P. N. (2003). A partial least squares latent variable modeling approach for measuring interaction effects: results from a monte carlo simulation study and an electronic-mail emotion/adoption study. *Information Systems Research, 14*(2), 189–217.

Chin, W. W., Marcolin, B. L., & Newsted, P. R. (1996). A partial least squares latent variable modeling approach for measuring interaction effects. Results from a monte carlo simulation study and voice mail Emotion/Adoption study. In J. I. DeGross, S. Jarvenpaa, & A. Srinivasan (Eds.), *Proceedings of the Seventeenth International Conference on Information Systems*, pp. 21–41, Cleveland, OH.

Chow, G. (1978). Test of equality between sets of coefficients in two linear regressions. *Econometrika, 28*, 591–605.

Cohen, J. (1978). Partialed products *Are* interactions; partialed powers *Are* curve components. *Psychological Bulletin, 85*(4), 858–866.

Cohen, J. (1988). *Statistical power analysis for the behavioral sciences*. Hillsdale, NJ: Lawrence Erlbaum.

Cortina, J. M. (1993). Interaction, nonlinearity, and multicollinearity: Implications for multiple regression. *Journal of Management, 19*, 915–922.

Cronbach, L. J. (1987). Statistical tests for moderator variables: Flaws in analysis recently proposed. *Psychological Bulletin, 102*(3), 414–417.

Diamantopoulos, A., & Winklhofer, H. M. (2001). Index construction with formative indicators: an alternative to scale development. *Journal of Marketing Research*, 38(2), 269–277.

Eberl, M. (2010). An application of PLS in multi-group analysis: the need for differentiated corporate-level marketing in the mobile communications industry. In: V. Esposito Vinzi, W. Chin, J. Hensler, and H. Wold (Eds.,) *Handbook of partial least squares*, Heidelberg: Springer.

Efron, B., & Tibshirani, R. J. (1993). *An introduction to the bootstrap*. New York: Chapman & Hall.

Eggert, A., Fassott, G., & Helm, S. (2005). Identifizierung und quantifizierung mediierender und moderierender effekte in komplexen kausalstrukturen. In F. Bliemel, A. Eggert, G. Fassott, & J. Henseler (Eds.), *Handbuch PLS-Pfadmodellierung. Methode, Anwendung, Praxisbeispiele*. Stuttgart: Schäffer-Poeschel.

Esposito Vinzi, V. (2010). PLS Path Modeling: From Foundations to Recent Developments and Open Issues for Model Assessment and Improvement. In V. Esposito Vinzi, W. W. Chin, J. Henseler, & H. Wang (Eds.), *Handbook of Partial Least Squares*. Heidelberg: Springer.

Finney, J. W., Mitchell, R. E., Cronkite, R. C., & Moos, R. H. (1984). Methodological issues in estimating main and interactive effects: Examples from coping/social support and stress field. *Journal of Health and Social Behavior, 25*, 85–98.

Guinot, C., Mauger, E., Malvy, D., Latreille, J., Ambroisine, L., Ezzedine, K., Galan, P., Hercberg, S., & Tenenhaus, M. (2010). Use of ULS-SEM and PLS-SEM to Measure a Group Effect in a Regression Model Relating Two Blocks of Binary Variables. In V. Esposito Vinzi, W. W. Chin, J. Henseler, & H. Wang (Eds.), *Handbook of Partial Least Squares*. Heidelberg: Springer.

Gujarati, D. (2003). *Basic econometrics*, 4th Edn. New York: McGraw-Hill.

Helm, S., Eggert, A., & Garnefeld, I. (2010). Modelling the impact of corporate reputation on customer satisfaction and loyalty using PLS. In V. Esposito Vinzi, W. W. Chin, J. Henseler, & H. Wang (Eds.), *Handbook of Partial Least Squares*. Heidelberg: Springer.

Homburg, C., & Giering, A. (2001). Personal characteristics as moderators of the relationship between customer satisfaction and loyalty: An empirical analysis. *Psychology & Marketing, 18*(1), 43–66.

Irwin, J. R., & McClelland, G. H. (2001). Misleading heuristics and moderated multiple regression models. *Journal of Marketing Research, 38*(1), 100–109.

Jaccard, J., & Turrisi, R. (2003). *Interaction effects in multiple regression*, 2nd Edn. Thousand Oaks: Sage.

Jarvis, C. B., MacKenzie, S. B., & Podsakoff, P. M. (2003). A critical review of construct indicators and measurement model misspecification in marketing and consumer research. *Journal of Consumer Research, 30*(2), 199–218.

Jonsson, F. Y. (1998). Nonlinear structural equation models: The Kenny–Judd model with interaction effects. In R. E. Schumacker, & G. A. Marcoulides (Eds.), *Interaction and nonlinear effects in structural equation modeling* (pp. 17–42). Mahwah, NJ: Lawrence Erlbaum.

Jöreskog, K. G., & Wang, F. (1996). Nonlinear structural equation models: The Kenny–Judd model with interaction effects. In G. A. Marcoulides, & R. E. Schumacker (Eds.), *Advanced structural equation modeling: issues and techniques* (pp. 57–88). Mahwah, NJ: Lawrence Erlbaum.

Kenny, D. A., & Judd, C. M. (1984). Estimating the nonlinear and interactive effects of latent variables. *Psychological Bulletin, 96*, 201–210.

Lohmöller, J.-B. (1989). *Latent variable path modeling with partial least squares*. Heidelberg: Physica.

Marquardt, D. W. (1980). You should standardize the predictor variables in your regression models. *Journal of the American Statistical Association, 75*, 87–91.

Pedhazur, E. J. (1982). *Multiple regression in behavioral research*. New York: Holt, Rinehart & Winston.

Reinecke, J. (1999). Interaktionseffekte in Strukturgleichungsmodellen mit der Theorie des geplanten Verhaltens. Multiple Gruppenvergleiche und Produktterme mit latenten Variablen. *ZUMA-Nachrichten, 23*(45), 88–114.

Rigdon, E. E., Schumacker, R. E., & Wothke, W. (1998). A comparative review of interaction and nonlinear modeling. In R. E. Schumacker, & G. A. Marcoulides (Eds.), *Interaction and nonlinear effects in structural equation modeling*, (pp. 1–16). Mahwah, NJ: Lawrence Erlbaum.

Sharma, S., Durand, R. M., & Gur-Arie, O. (1981). Identification and analysis of moderator variables. *Journal of Marketing Research, 18*(3), 291–300.

Smith, K. W., & Sasaki, M. S. (1979). Decreasing multicollinearity: a method for models with multiplicative functions. *Sociological Methods and Research, 8*, 35–56.

Streukens, S., Wetzels, M., Daryanto, A., & de Ruyter, K. (2010). Analyzing Factorial Data Using PLS: Application in an Online Complaining Context. In V. Esposito Vinzi, W. W. Chin, J. Henseler, & H. Wang (Eds.), *Handbook of Partial Least Squares*. Heidelberg: Springer.

Tate, R. L. (1984). Limitations of centering for interactive models. *Sociological Methods and Research, 13*, 251–271.

Temme, C., & Kreis, H. (2010). A Comparison of Current PLS Path Modeling Software: Features, Ease-of-Use, and Performance. In V. Esposito Vinzi, W. W. Chin, J. Henseler, & H. Wang (Eds.), *Handbook of Partial Least Squares*. Heidelberg: Springer.

Tenenhaus, M., Vinzi, V. E., Chatelin, Y.-M., & Lauro, C. (2005). PLS path modeling. *Computational Statistics & Data Analysis, 48*(1), 159–205.

Chapter 31
A Comparison of Current PLS Path Modeling Software: Features, Ease-of-Use, and Performance

Dirk Temme, Henning Kreis, and Lutz Hildebrandt

Abstract After years of stagnancy, PLS path modeling has recently attracted renewed interest from applied researchers in marketing. At the same time, the availability of software alternatives to Lohmöller's LVPLS package has considerably increased (PLS-Graph, PLS-GUI, SPAD-PLS, SmartPLS). To help the user to make an informed decision, the existing programs are reviewed with regard to requirements, methodological options, and ease-of-use; their strengths and weaknesses are identified. Furthermore, estimation results for different simulated data sets, each focusing on a specific issue (sign changes and bootstrapping, missing data, and multi-collinearity), are compared.

31.1 Introduction

When it comes to modeling relationships between latent variables, mainly two different methodological approaches can be distinguished: Covariance structure analysis on the one hand and PLS path modeling (not to be confused with PLS regression) on the other. Although both methods emerged roughly at the same time, their development took a rather diverse course. Since the introduction of the first LISREL version in the early 1970s, the software available for covariance structure analysis has experienced substantial progress with respect to ease-of-use and methodological capabilities. Graphical interfaces in programs like AMOS or LISREL have freed

D. Temme
Chair of Retailing and Service Management, Schumpeter School of Business and Economics, Bergische Universität Wuppertal, Gaußstr. 20, 42097 Wuppertal, Germany
e-mail: Temme@wiwi.uni-wuppertal.de

H. Kreis
Marketing-Department, Freie Universität Berlin, School of Business and Economics, Otto-von-Simson-Str. 19, 14195 Berlin, Germany
e-mail: henning.kreis@fu-berlin.de

Lutz Hildebrandt
Institute of Marketing, Humboldt University Berlin, Unter den Linden 6, 10099 Berlin, Germany
e-mail: hildebr@wiwi.hu-berlin.de

V. Esposito Vinzi et al. (eds.), *Handbook of Partial Least Squares*, Springer Handbooks of Computational Statistics, DOI 10.1007/978-3-540-32827-8_32,

the user from having to specify his/her model in matrix or equation form. Simultaneously, estimation methods for non-normal/categorical data as well as multi-level, multi-group, and finite mixture models have emerged, thus offering a wide range of possible applications. Meanwhile, covariance structure analysis is arguably one of the most popular methods used in the social sciences (e.g., marketing).

In contrast, PLS path modeling has, until recently, rarely been applied in marketing although its basic algorithms were developed in the 1970s and the first software packages were publicly available in the 1980s (*LVPLS* (Lohmöller 1984), PLSPath (Sellin 1989). The rather limited use of PLS path modeling in the last decades can be explained to a considerable degree by the lack of progress regarding the software's ease-of-use and methodological options. Recently, however, this situation has changed tremendously. Currently, researchers can choose between several alternative software solutions (*PLS-GUI*, *VisualPLS*, *PLS-Graph*, *SmartPLS*, *SPAD-PLS*) which provide a clear improvement especially in terms of user-friendliness. Furthermore, growing need in modeling so-called formative constructs, particularly in marketing and management/organizational research (e.g., Diamantopoulos and Winklhofer 2001; Jarvis et al. 2003; MacKenzie et al. 2005), has stimulated great interest in applying the PLS path modeling approach. Although models with formative constructs can, in principle, also be estimated within covariance structure analysis (e.g., MIMIC models), doing so causes specific identification problems which are not an issue in PLS (e.g., MacCallum and Browne 1993).

Against the background of a growing number of PLS software packages and an increasing differentiation in the programs' capabilities, a comprehensive review would help researchers to decide on the specific PLS program to be used in their studies. To the best of our knowledge, no such review of PLS path modeling software currently exists. In order to close this gap, we aim at providing an informative software overview by identifying specific strengths and weaknesses of the relevant programs. In the remaining part of the article, we offer a brief description of each software package; in addition, screenshots will give an impression of how analyses are set up in the different programs. Subsequently, the software is assessed with respect to the following criteria: requirements, methodological options, and ease-of-use. Next, estimation results for different simulated data sets, each focusing on a specific issue, are compared. Finally, the main conclusions of the study are discussed.

31.2 PLS Path Modeling Software

Besides *LVPLS*, the software overview includes several more recent software packages for PLS path modeling: *PLS-GUI*, *VisualPLS*, *PLS-Graph*, *SPAD-PLS*, and *SmartPLS*. Following the description of *LVPLS*, we will discuss *PLS-GUI* and *VisualPLS* which are basically graphical interfaces to *LVPLS*. Finally, the remaining programs are characterized. In contrast to the former software, these programs are more or less self-contained implementations of the algorithms developed by Wold (1982, 1985) and Lohmöller (1987). This review includes those program versions available to the authors as of August 2006. It should be noted, that all programs

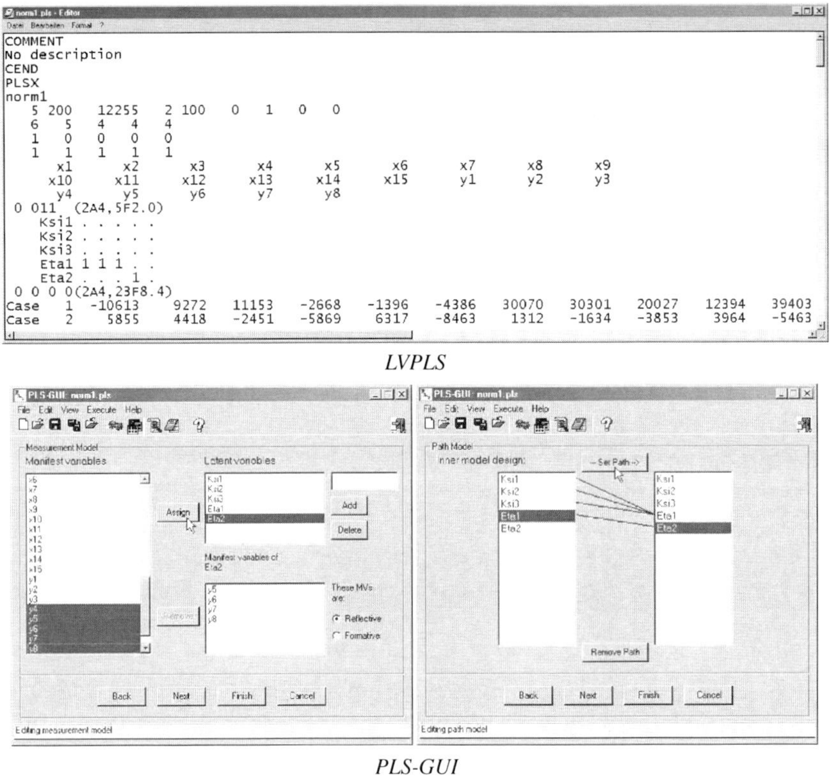

Fig. 31.1 Specification of Path Models in PLS Software: *LVPLS*, *PLS-GUI*

(except *LVPLS*) are constantly under development and can therefore be expected to offer additional features in the future.

LVPLS: The DOS-based program *LVPLS 1.8* (Lohmöller 1987) includes two different modules for estimating path models. Whereas LVPLSC analyzes the covariance matrix of the observed variables, the LVPLSX module is able to process raw data. In order to specify the input file an external editor is necessary. The input specification requires that the program parameters are defined at specific positions in the file – a format which resembles punchcards (see upper panel in Fig. 31.1). Results are reported in a plain text file. The program offers blindfolding and jackknifing as resampling methods in case raw data has been analyzed. When analyzing covariance/correlation matrices, resampling techniques cannot be applied.

PLS-GUI: The Windows-based *PLS-GUI* (Li 2005) provides a graphical interface for *LVPLS* which supports both the analysis of raw data (LVPLSX) as well as covariance information (LVPLSC). To specify a model, the user is led through a stepwise procedure which offers a menu at each step (see lower panel in Fig. 31.1). Additional options (e.g., weighting schemes, missing data code) are to be chosen in a separate window. The program finally creates an input file which is processed by the executable file pls.exe of *LVPLS*. If required, the input file can be modified by the user. The output is the same as for *LVPLS*. The current version offers a bootstrap option as an additional feature not provided by *LVPLS*.

VisualPLS: VisualPLS (Fu 2006a) is a graphical user interface for *LVPLS* running in the Windows environment which enables the analysis of raw data only. The path model is specified by drawing the latent variables and by assigning the indicators in a pop-up window (see upper panel in Fig. 31.2). Based on the graphical model, the program produces a separate *LVPLS* input file, which is run by LVPLSX (pls.exe). Different formats of input data are supported. The results are offered as *LVPLS* output (plain text file) as well as in HTML/Excel format. In addition, a path model showing the estimated parameters is displayed. Beyond blindfolding and jacknifing, bootstrapping has been integrated. Special support for specifying moderating effects and second order factors is offered.

PLS-Graph: PLS-Graph (Chin 2003) is a Windows-based program which uses modified routines of *LVPLS*, but only processes raw data (LVPLSX). In order to specify the model, a graphical interface can be used which provides some tools for drawing a path diagram (see lower panel in Fig. 31.2). Different options (e. g., weighting scheme, resampling method) can be chosen from a menu. Although the generated input file is a text file, it can only be processed by *PLS-Graph*, but not by *LVPLS*. Estimation results are presented in ASCII format as well as in a graphical path model; resampling methods include blindfolding, jackknifing, and bootstrapping.

SPAD-PLS: This program is part of the comprehensive data analysis software SPAD (running under Windows) which is offered by the French company Test&Go. *SPAD-PLS* (Test&Go 2006) does not process covariance information but needs raw data instead. Models can be specified with a menu or graphically in a Java applet; the remaining settings may be adjusted in additional menu windows (see upper panel in Fig. 31.3). Different options for handling missing data (but see section 31.3.2.1) and multi-collinearity are provided. Results are reported both as a path diagram and as text or Excel file; blindfolding, jackknifing, and bootstrapping (including confidence intervals) are available. In the non-graphical manual mode transformations of latent variables (squares, cross-products) can be specified.

SmartPLS: Since *SmartPLS* (Ringle et al. 2005) is Java-based, it is independent from the user's operating system. Again, only raw data can be analyzed. The model is specified by drawing the structural model for the latent variables and by assigning the indicators to the latent variables via "drag & drop" (see lower panel in Fig. 31.3). The output is provided in HTML, Excel or Latex format, as well as a parameterized path model. Bootstrapping and blindfolding are the resampling methods available. Like in *VisualPLS*, the specification of interaction effects is supported. A special feature of *SmartPLS* is the finite mixture routine (FIMIX)

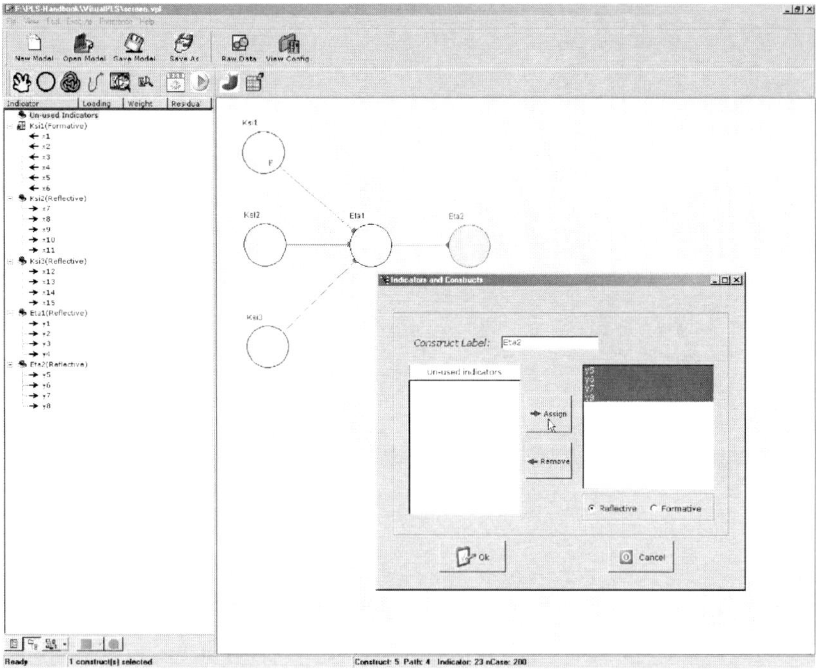

VisualPLS

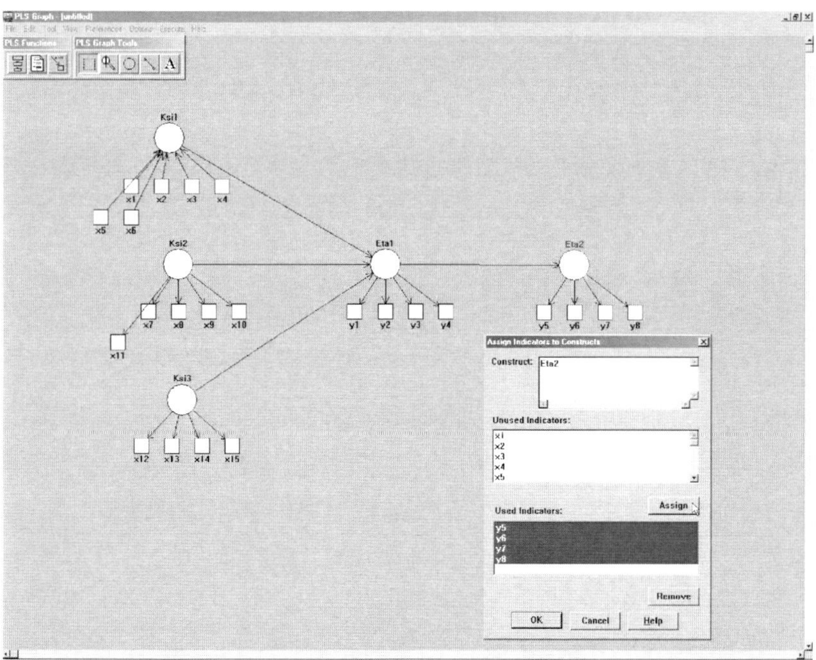

PLS-Graph

Fig. 31.2 Specification of Path Models in PLS Software: *VisualPLS*, *PLS-Graph*

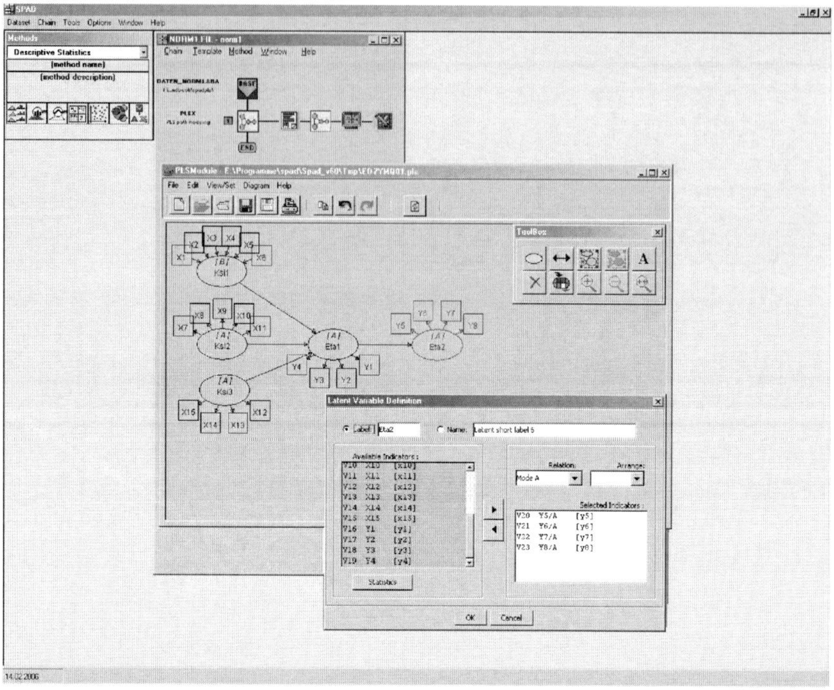

SPAD-PLS

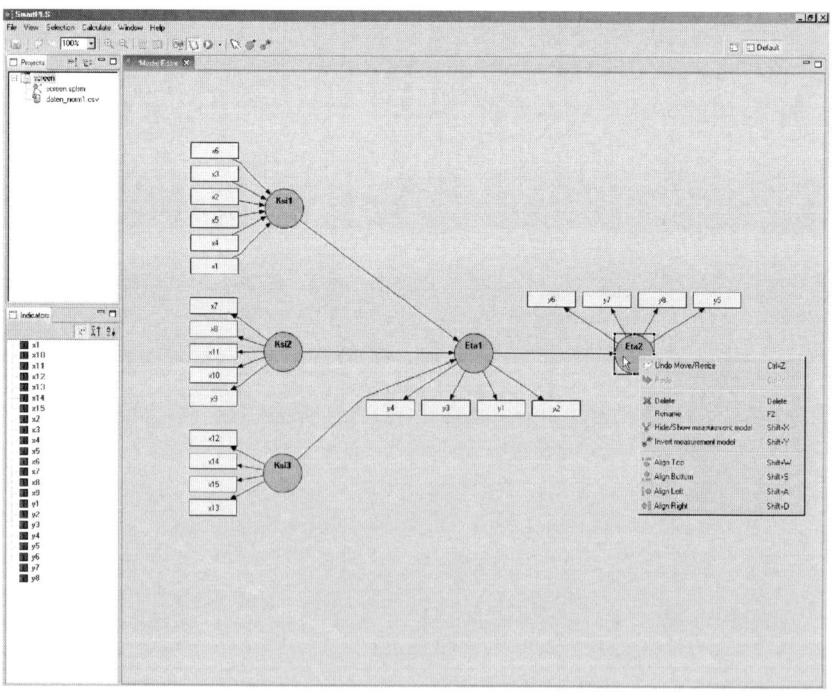

SmartPLS

Fig. 31.3 Specification of Path Models in PLS Software: *SPAD-PLS*, *SmartPLS* (see Chap. 9). Such an option might be of interest if unobserved heterogeneity is expected in the data (McLachlan and Peel 2000).

31.3 Comparison and Recommendations

In order to support the user in making an informed decision about the software to be used in his/her study, programs are compared with respect to several features which can be subsumed under the following headings: requirements (e. g., operating system, data), methodological options (e. g., weighting scheme, resampling methods), and ease-of-use (e.g., specification, output format). In addition, we also point to some issues the researcher should pay attention to when using a specific program. The main properties of the programs are summarized in Tables 31.1, 31.2, and 31.3.

Table 31.1 Overview of PLS Path Modeling Software (Part 1). Details represent the stage of development as of August 2006

<table>
<tr><th colspan="2">Features</th><th>LVPLS 1.8
Lohmöller (1987)</th><th>PLS-GUI 2.0.1
Li (2005)</th></tr>
<tr><td rowspan="5">Requirements</td><td>Operating system</td><td>DOS</td><td>Windows</td></tr>
<tr><td>Data</td><td colspan="2">Raw data / covariance matrix</td></tr>
<tr><td>Scale level</td><td colspan="2">Metric / binary exogenous variables</td></tr>
<tr><td>Definition of missing values (MV)</td><td>Individual definition of MV for each variable</td><td>Common definition of MV for all variables</td></tr>
<tr><td>Data format</td><td>.inp (ASCII)</td><td>.dat (ASCII)</td></tr>
<tr><td rowspan="5">Methodology</td><td>Data metric</td><td colspan="2">• Mean=0, Var=1
• Mean=0, Var=1, rescal.
• Mean=1, rescal.
• Original</td></tr>
<tr><td>Missing data treatment</td><td colspan="2">Fixed (pairwise elimination and/or imputation of means (see Section 31.3.2.1))</td></tr>
<tr><td>Weighting scheme</td><td colspan="2">Factor-, centroid-, or path weighting</td></tr>
<tr><td>Resampling</td><td>• Blindfolding
• Jackknifing</td><td>• Blindfolding
• Jackknifing
• Bootstrapping</td></tr>
<tr><td>Cross-validation</td><td>• CV-redundancy
• CV-communality</td><td>Not available</td></tr>
<tr><td rowspan="6">Ease-of-use</td><td>Specification</td><td>Text editor</td><td>Quasi graphically</td></tr>
<tr><td>Output</td><td colspan="2">ASCII</td></tr>
<tr><td>Graphical output</td><td colspan="2">Not available</td></tr>
<tr><td>Documentation</td><td>Lohmöller (1984)</td><td>Li (2003)</td></tr>
<tr><td>Internet</td><td>not available</td><td>http://dmsweb.moore.sc.edu/yuanli/pls-gui/</td></tr>
<tr><td>Availability</td><td colspan="2">Freeware</td></tr>
</table>

Table 31.2 Overview of PLS Path Modeling Software (Part 2). Details represent the stage of development as of August 2006

	Features	*VisualPLS 1.04* Fu (2006a)	*PLS-Graph 3.00* Chin (2003)
Requirements	Operating system	Windows	
	Data	Raw data	
	Scale level	Metric / binary exogenous variables	
	Definition of missing values (MV)	Common definition of MV for each variable	
	Data format	.dat (ASCII), .csv	.raw (ASCII)
Methodology	Data metric	• Mean=0, Var=1 • Mean=0, Var=1, rescal. • Mean=1, rescal. • Original	
	Missing data treatment	Fixed (pairwise elimination and/or imputation of means (see Section 31.3.2.1))	
	Weighting scheme	Factor-, centroid-, or path weighting	
	Resampling	Blindfolding, jackknifing, and bootstrapping	
	Cross-validation	CV-redundancy and CV-communality	
	Special features	Interaction- and 2nd-order factor model support	Individual and construct level sign correction for bootstrapping
Ease-of-use	Specification	Graphically	
	Output	ASCII, Excel, HTML	ASCII
	Graphical output	Path diagram	
	Documentation	Fu (2006b)	Chin (2001)
	Internet	http://www2.kuas.edu.tw/prof/fred/vpls/index.html	http://www.cba.uh.edu/plsgraph/
	Availability	Freeware	

31.3.1 Requirements

Comparing the software with respect to their system requirements reveals that users of UNIX/LINUX or Mac systems have to use the platform-independent *SmartPLS* program. Further requirements concern the analyzed data. All programs at present expect that the indicators of the latent variables are continuous, or – for instance in the case of rating scales with 5 or more answer categories – approximate a continuous scale. In addition, binary exogenous variables can be included in the analysis. If only covariance matrices are available as data input, the choice is currently restricted to *LVPLS* or *PLS-GUI*. Except for *LVPLS*, all programs require a common definition of missing values for all variables (e.g., −999). In general, all programs are able to process ASCII data although some software requires a conversion into specific data formats (e.g., .sba in *SPAD-PLS*). *SPAD-PLS* also supports data formats of common software packages like SPSS and SAS which are converted in a data editor or exchange module.

Table 31.3 Overview of PLS Path Modeling Software (Part 3). Details represent the stage of development as of August 2006

	Features	*SPAD-PLS* Test&Go (2006)	*SmartPLS 2.0 M3* Ringle et al. (2005)
Requirements	Operating system	Windows	Independent (Java)
	Data	Raw data	
	Scale level	Metric / binary exogenous variables	
	Definition of missing values (MV)	Common definition of MV for each variable	
	Data format	.sba (ASCII, SPSS, SAS)	.txt (ASCII), .csv
Methodology	Data metric	• Mean=0, Var=1 • Mean=0, Var=1, rescal. • Mean=1, rescal. • Original	
	Missing data treatment	Pairwise elimination or imputation of means, NIPALS/EM*	Casewise elimination or imputation of means
	Weighting scheme	Factor-, centroid-, or path weighting	
	Resampling	• Blindfolding • Jackknifing • Bootstrapping	• Blindfolding • Bootstrapping
	Cross-validation	• CV-redundancy • CV-communality	• CV-redundancy • CV-communality
	Special features	PLS regression for weights and path coefficients; confidence intervals for jacknifing and bootstrapping; contribution to R^2; check of unidimensionality of latent variables (eigenvalues)	Finite-mixture PLS; Interaction model support; Cronbach's alpha
Ease-of-use	Specification	Graphically	
	Output	ASCII, Excel	HTML, Latex, Excel
	Graphical output	Path diagram	
	Documentation	Vinzi et al. (2004)	Hansmann and Ringle (2004)
	Internet	http://www.testandgo.com	http://www.smartpls.de
	Availability	Test&Go	Freeware

*Not implemented in the test version used for this review

31.3.2 Methodological Options

31.3.2.1 Missing Data

Data sets where at least some values of their variables are missing are ubiquitious in empirical research. In order to deal with missing data, several alternative approaches have been proposed (e.g., Little and Rubin 2002). *LVPLS* offers a specific treatment in the case of missing data which combines mean value imputation and pairwise deletion in the course of the estimation (Lohmöller 1984; for a more comprehensive description see Tenenhaus et al. 2005). This missing data treatment is also provided by the graphical interfaces (*PLS-GUI*, *VisualPLS*) as well as by *PLS-Graph* and *SPAD-PLS*. In contrast, *SmartPLS* offers two options equivalent to some data preprocessing which either substitute the mean over all available cases of a variable for the missing values or which delete those cases with missing data (casewise deletion). Since casewise deletion throws away a lot of useful information and thus leads to lower efficiency, this procedure is not to be recommended. Even the other traditional methods of dealing with missing data (i.e., pairwise deletion, mean imputation) have several shortcomings such as computing covariances (mode B) based on different sample sizes and biased parameter estimates (Allison 2002; Haitovsky 1968), for example. Meanwhile, more advanced data imputation methods are announced for the next release of *SPAD-PLS* which will include an EM algorithm as well as the NIPALS approach.

31.3.2.2 Multi-collinearity

Multi-collinearity can be a problem both for the estimation of indicator weights in the case of formative constructs (mode B) and for the estimation of the relationships among latent variables. Possible means to detect severe multi-collinearity with respect to formative indicators are inspecting the correlation matrix, calculating the variance inflation factors, or examining the condition index (see, for example, Chap. 30). *SPAD-PLS* at present is the only program which addresses the problem of multi-collinearity by providing a PLS regression routine for estimating weights (Mode PLS) and path coefficients (PLS regression instead of OLS regression). PLS regression searches for a set of components which decompose the vector y of the endogenous variable and the matrix X of explanatory variables in such a way that the explained covariance between y and X is maximized (for details on PLS regression see Chap. 16).

31.3.2.3 Resampling Methods

Since one of the appealing features of PLS path modeling is the fact that it does not rest on any distributional assumptions, significance levels for the parameter estimates which are based on normal theory are, strictly speaking, not suitable. Therefore, information about the variability of the parameter estimates and hence their significance has to be generated by means of resampling procedures. Whereas *LVPS* only offers blindfolding and jacknifing, all recent software packages include a bootstrap option. In order to assess the quality of the estimated model, several criteria for model validation have been proposed in the literature (for a discussion see, for example, Tenenhaus et al. (2005) as well as Chaps. 3 and 30). To calculate cross-validation indices, blindfolding is necessary and now offered by all programs. Except for *PLS-GUI*, cross-validated communality and redundancy measures are also provided in the programs' output by request.

In order to derive valid standard errors or t-values, applying bootstrapping is superior to the other two resampling methods (see Chap. 3). Therefore, in the following we will focus on the former. The bootstrap procedure approximates the sampling distribution of an estimator by resampling with replacement from the original sample. An important issue is that in PLS the signs of the latent variables are indetermined. Since arbitrary sign changes in the parameter estimates of the various bootstrap samples can increase their standard error to a substantial degree, procedures have been developed to correct for sign reversals (for a more comprehensive discussion of this issue see Chap. 3). Here, both *PLS-Graph* and *SmartPLS* allow the user to choose between two correction procedures: In the first option (individual sign changes), the sign of each individual outer weight is made equal to the corresponding sign in the original sample. Because this procedure does not check for the overall coherence of the model as would be done if mental "reverse coding" (Chin 2000) were performed, this option should be used with special care. The second option (construct level changes) compares the loadings for each latent variable with the original loadings and reverses the sign of the weights if the absolute value of the summed difference between the original and the bootstrap loadings is greater than the absolute value of the sum of the original loadings and the bootstrap loadings (Tenenhaus et al. 2005). However, both procedures do not guarantee that sign changes are properly handled. The graphical interfaces *PLS-GUI* and *VisualPLS* only offer construct level correction. Since *SPAD-PLS* uses the elements of the first eigenvector of a principal components analysis with predominantly positive signs, sign control aligns the signs in the bootstrap samples to those of the original sample.

Another possibility to gauge the significance of the PLS estimates is to calculate the confidence intervals from the bootstrap samples. So far this option using the percentile method is only implemented in *SPAD-PLS*.

31.3.2.4 Other Features

With respect to the inner weights, all programs offer the weighting schemes for estimating the inner model (centroid-, factor-, and path weighting) already available in *LVPLS*. A topic of special interest is the use of different sets of starting values for determining the outer weights. The starting values can have an impact on the sign of the estimated weights or factor loadings and therefore also on the path coefficients (see the simulation results for data set 1 in Section 31.4). Although this is not a statistical issue, it is important for the interpretation of the estimation results. None of the programs currently allow users to specify their own set of starting values. For those programs with fixed starting values (*LVPLS*, *PLS-GUI*, *VisualPLS*, *PLS-Graph*, and *SmartPLS*), rearranging the order of indicators in a single block is the only means of exerting an influence on the sign (Tenenhaus et al. 2005). In *SPAD-PLS*, starting values are flexible insofar as the elements of the first eigenvector of a principal components analysis (PCA) with predominantly positive signs are used (Tenenhaus et al. 2005). *SPAD-PLS* provides normalized weights if all outer weights are positive as well as latent variable scores in the original metric.

31.3.3 Ease-of-Use

Compared to *LVPLS*, all recent PLS software is considerably more user-friendly. This is especially true for programs where the user can specify the model graphically and where the program displays a parameterized path diagram as output (*VisualPLS*, *PLS-Graph*, *SPAD-PLS*, and *SmartPLS*). Particularly mentionable are the following features: In *PLS-Graph*, *SPAD-PLS*, and *SmartPLS*, it is easy to change the data set without having to specify the model again. Additionally, it is possible to save the complete analysis (including data set, model, and results) into a single project file.

VisualPLS and *SmartPLS* both give assistance in constructing product indicators for path models with interaction effects. The user can choose between mean centering and standardizing the corresponding manifest variables. Whereas *VisualPLS* only calculates product terms and includes them as new variables, *SmartPLS* directly adds the latent interaction term with its measures to the graphical path model. The program even modifies the indicator product terms automatically if the measurement models of the latent predictor/moderator variables are changed. In the case of reflective indicators, the interaction module is a convenient feature. However, the option should not be used in the case of formative constructs (for a discussion of estimating interaction effects in PLS path modeling see Chin et al. (2003) as well as Chaps. 27 and 32).

Most programs provide rich tool boxes which help to improve the layout of the path diagrams (color, size, text etc.). This especially applies to *SPAD-PLS* and *SmartPLS*. Even though graphical tools are not available under *PLS-GUI*, model specification is nevertheless fairly easy. For all programs, user manuals document the application with example data. Additional information on *SPAD-PLS*, but also on PLS path modeling in general, can be accessed on the website www.esisproject.com of the European Satisfaction Index System (ESIS). Overall, PLS path analyses can be performed after a few initial practice sessions with all of the recent software.

31.4 Comparison Based on Simulated Data

Since all programs for PLS path modeling more or less use the same basic algorithms, estimation results should not differ for data sets without any "problematic" characteristics. In order to provoke distinct results, we therefore created three different data sets, each focusing on a specific issue: First, we demonstrate that programs can produce different solutions with respect to the parameter signs under certain conditions. Second, parameter estimates differ across programs if missing data are present. Third, we focus on the case that latent exogenous variables show a substantial degree of multi-collinearity. Since the programs' capabilities to cope with the data characteristics described above in part differ, the three simulated data sets have been analyzed with *PLS-GUI*, *VisualPLS*, *PLS-Graph*, *SPAD-PLS*, and *SmartPLS*. All data simulations have been performed in the statistical environment R.

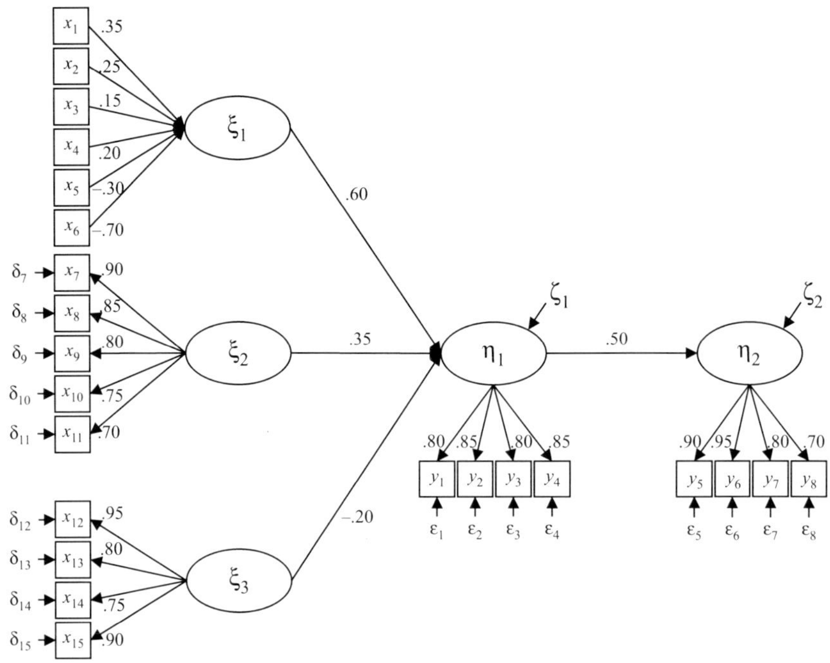

Fig. 31.4 Path Model Used for Simulating Data Set 1 (all variables are standardized)

31.4.1 Data Set 1 – Sign Changes and Bootstrapping

The first data set ($N = 200$) has been generated according to the parameterized path model in Fig. 31.4. Because specification issues with respect to the measurement models for latent variables (reflective versus formative models) have recently been discussed rather intensively in the marketing research literature (e. g., Diamantopoulos and Winklhofer 2001; Jarvis et al. 2003), we specify two different kinds of measurement models for the exogenous latent variables (ξ_1: formative/mode B, ξ_2 and ξ_3: reflective/mode A). Since formative indicators do not necessarily imply a specific pattern of correlations among them (Nunnally and Bernstein 1994), a negative influence of the manifest variables x_5 and x_6 on the latent variable ξ_1 has been specified. For both endogenous variables η_1 and η_2, only reflective measurement models are supposed (mode A).

Comparing the results for the programs used in our study reveals the following: Absolute parameter values are almost identical across the programs. For specific relations, however, the signs differ across the software packages (as reported in Table 31.4). Whereas *PLS-GUI*, *VisualPLS* and *SPAD-PLS* reproduce the signs of the population values used for the simulation, *PLSGraph* and *SmartPLS* generate opposite signs for the weights of the indicators x_1 to x_6. As a consequence, the estimated effect of the exogenous latent variable ξ_1 on the latent endogenous variable η_1

Table 31.4 Comparisons of the results for Data Set 1 – Estimates and signs for selected parameters

			Programs		
			All	*PLS-GUI/ VisualPLS/ SPAD-PLS*	*PLS-Graph/ SmartPLS*
			Absolute values	*Signs*	
Measurement model ξ_1	*Weights*	x_1	0.073	+	−
		x_2	0.339	+	−
		x_3	0.225	+	−
		x_4	0.335	+	−
		x_5	-0.345	−	+
		x_6	-0.822	−	+
Structural model	*Path coeff.*	$\xi_1 \rightarrow \eta_1$	0.447	+	−
		$\xi_2 \rightarrow \eta_1$	0.323	+	+
		$\xi_3 \rightarrow \eta_1$	-0.180		−
		$\eta_1 \rightarrow \eta_2$	0.482	+	+

differs likewise. This finding can be explained by different sets of starting values. Whereas *LVPLS* uses the sequence $1, 1, \ldots, -1$ as starting values for each block, *SmartPLS*, for example, uses the value 1 for all weights of a block. By performing mental "reverse coding" (Chin 2000), the different solutions can be aligned. Thus, from a statistical point of view, sign changes across the programs are not an issue, but applied researchers should be sensitized to think thoroughly about the expected signs of the relationships between the manifest and latent variables as well as the effects between the latent variables. A peculiar finding emerges for *VisualPLS*: The signs for the weights of the formative construct ξ_1 and the path coefficient for its effect on η_1 in *LVPLS* (see Table 31.4) are reversed in the displayed path model. Since the reversed signs are used in the bootstrap procedure, they are reported in Table 31.5.

As discussed above, arbitrary sign changes can have a severe influence on the bootstrap results if not properly controlled for. Therefore, 500 bootstrap samples (each with $N = 200$) have been analyzed with the various programs. Construct level sign change is applied since for ξ_1 the signs of the weights differ within the block. The results with respect to the bootstrap means/standard errors and the t-ratios are reported in Table 31.5. There are substantial differences in the time needed for the different programs to produce the bootstrap results for our sample. *SPAD-PLS* is by far the fastest software (the run took less than 5 s), followed by *SmartPLS* and *PLSGraph* (about 30 s). Both graphical interfaces for *LVPLS*, i. e. *PLS-GUI* and *VisualPLS* were rather slow in providing the bootstrap estimates (about 1 min and 40 s).

Whereas the graphical interfaces for *LVPLS* as well as *PLS-Graph* and *Smart-PLS* produce similar results, both the bootstrap means/standard errors and the t-ratios of *SPAD-PLS* in part differ considerably. For example, for the path of ξ_1 to η_1 the t-ratio is less than half the ratio which results from the other programs; the same applies to the weight for x_6. These differences might be explained by the

Table 31.5 Comparisons of the Bootstrap Results for Data Set 1 – selected parameters

			Original estimate	*Bootstrap*														
				means					*standard errors*					*t-ratios*				
				PLS-GUI	*Visual PLS*	*PLS-Graph*	*SPAD-PLS*	*Smart PLS*	*PLS-GUI*	*Visual PLS*	*PLS-Graph*	*SPAD-PLS*	*Smart PLS*	*PLS-GUI*	*Visual PLS*	*PLS-Graph*	*SPAD-PLS*	*Smart PLS*
Measurement model ξ_1	*Weights*	x_1	0.073	0.070	-0.127	-0.056	0.078	-0.073	0.139	0.092	0.144	0.139	0.138	0.525	-0.790	-0.511	0.527	-0.528
		x_2	0.339	0.325	-0.321	-0.336	0.371	-0.318	0.114	0.123	0.113	0.150	0.119	2.974	-2.763	-3.011	2.260	-2.843
		x_3	0.225	0.225	-0.222	-0.239	0.242	-0.219	0.125	0.110	0.116	0.135	0.120	1.800	-2.047	-1.941	1.679	-1.887
		x_4	0.335	0.314	-0.321	-0.305	0.358	-0.303	0.125	0.121	0.139	0.132	0.132	2.680	-2.761	-2.426	2.536	-2.544
		x_5	-0.345	-0.334	0.317	0.314	-0.366	0.333	0.137	0.134	0.131	0.152	0.138	-2.518	2.577	2.638	-2.271	2.501
		x_6	-0.822	-0.784	0.794	0.785	-0.881	0.790	0.091	0.086	0.090	0.203	0.092	-9.033	9.575	9.182	-4.048	8.921
Structural model	*Path coeff.*	$\xi_1 \rightarrow \eta_1$	0.447	0.459	-0.460	-0.459	0.457	-0.460	0.049	0.050	0.048	0.108	0.052	9.122	-8.932	-9.228	4.130	8.666
		$\xi_2 \rightarrow \eta_1$	0.323	0.321	0.318	0.322	0.323	0.326	0.051	0.055	0.052	0.045	0.054	6.333	5.870	6.169	7.205	5.950
		$\xi_3 \rightarrow \eta_1$	-0.180	-0.184	-0.182	-0.190	-0.191	-0.181	0.053	0.048	0.056	0.069	0.052	-3.396	-3.779	-3.217	-2.617	3.447
		$\eta_1 \rightarrow \eta_2$	0.482	0.485	0.482	0.482	0.476	0.486	0.051	0.053	0.053	0.040	0.054	9.451	9.042	9.135	12.103	8.897

idiosyncratic way *SPAD-PLS* determines the starting values for each block and the corresponding sign control (Tenenhaus et al. 2005, p. 184). However, *SPAD-PLS* does not consistently produce the lowest t-ratios.

31.4.2 Data Set 2 – Missing Data

In order to compare the results of the different programs in the case of missing data, a very simple model is used for data simulation in which a formative construct only influences one latent variable measured by reflective indicators. Since in *LVPLS* missing data treatment depends on whether data is missing on a whole block or just on some (but not all) manifest variables, two different missing data schemes are applied. For the formative construct it is assumed that values are missing for only some manifest variables. In contrast, for the reflective endogenous latent variable, missing data are produced such that values are absent for all of its indicators. Missing data (about 10 % for each variable) have been generated completely at random (MCAR).

Since the programs in part offer different options in the case of missing data, some discrepancies in the results are expected. However, at least the graphical interfaces for *LVPLS* (*PLS-GUI*, *VisualPLS*) should produce the same results as Lohmöller's program. The actual results nevertheless show unexpected differences. Obviously, these differences are caused by an incorrect setup of the input file to *LVPLS*. In both interfaces one is allowed to specify one specific value for missing data (e.g., -1) which is then used to add a "missing data case" to the data. In the input file, this value should exactly correspond to the missing data values in the raw data. In *PLS-GUI* and *VisualPLS*, however, this code is transferred in a way which does not correspond to the Fortran format specified for reading the data. If the raw data has decimal places, this means that the missing data code differs from the missing values contained in the data. For example, in the following input file generated by *PLS-GUI* the variables have four decimal places. For the given Fortran format, a value of -1 (for missing data) is written as -10000 whereas the missing data case includes a -1 instead.

```
COMMENT
LVPLS input file generated by PLS-GUI 2.0.1
CEND
PLSX
Missing Data - Demonstration
   2-201   12255   2 100   0   1   0   0
   6   4
   1   0
   1   1
      x1       x2       x3       x4       x5       x6       y1       y2       y3
      y4
 0 011   (2A4,2F2.0)
     Ksi . .
     Eta 1 .
 0 0 0 0(2A4,10F8.4)
MISSING       -1       -1       -1  ...       -1       -1       -1       -1
Case    1   -1376   -10000     4377  ...   -3641     4779    -4490    -4477
```

```
Case   2   -3732  -12294  -10674   ...    -8361   13691   -5158   -1034
Case   3    6682   12600   -3648   ...    -2730   -9584   -2378   -3407
Case   4    3872  -11195   13664   ...    -1358   10213   -5583    8641
Case   5    7190  -15879  -10000   ...   -10000  -10000  -10000  -10000
Case   6  -17634  -10000   19101   ...   -15699   -8155   -3705  -14140
...
Case 199    4730  -17174    8129   ...    10000   19261   12698    9710
Case 200  -14363  -10805   -1490   ...    -3009    2840    4169    2750
STOP
```

Correcting the wrong coding in the missing data case of the input file for *LVPLS* produced by the interfaces and running it with the executable PLS file leads to the correct results. In *PLS-Graph* and *SPAD-PLS*, the missing data procedure is implemented correctly. The more advanced methods to deal with missing data (EM algorithm, NIPALS) announced for *SPAD-PLS* were not available in the test version for this review. Since *SmartPLS* only allows for mean imputation or casewise deletion, different results emerge compared to the other programs.

31.4.3 Data Set 3 – Multi-collinearity

Multi-collinearity can be a problem for the estimation of the relationships within (formative) measurement models as well as the effects among the latent variables.

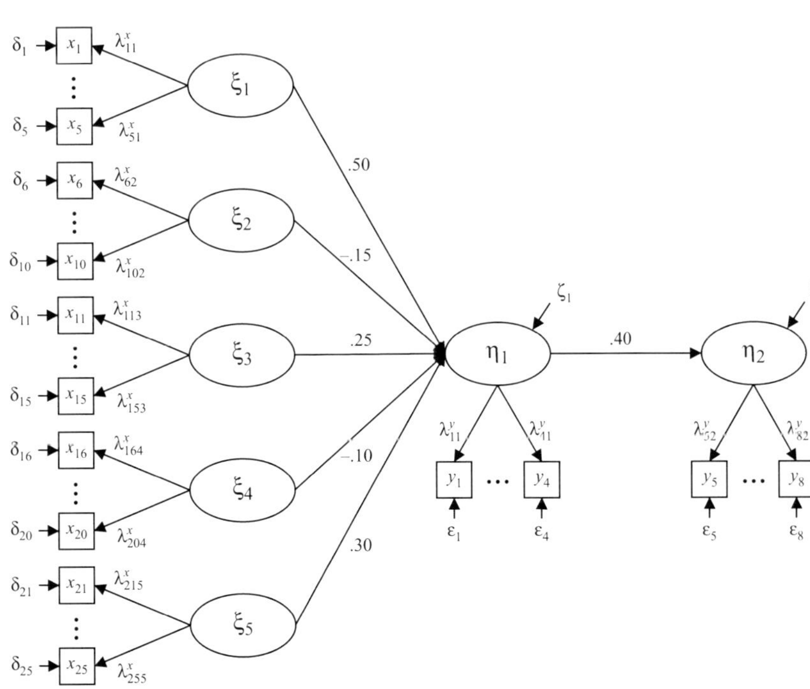

Fig. 31.5 Path Model Used for Simulating Data Set 3 (all variables are standardized)

Table 31.6 Comparisons of the Results for Data Set 3 – OLS versus PLS Regression Path Coefficients

			True values	*OLS regression*	*PLS regression*
Structural Model	*Path coefficients*	$\xi_1 \rightarrow \eta_1$	0.50	-0.003	0.029
		$\xi_2 \rightarrow \eta_1$	-0.15	0.382	0.378
		$\xi_3 \rightarrow \eta_1$	0.25	0.156	0.114
		$\xi_4 \rightarrow \eta_1$	0.10	0.288	0.298
		$\xi_5 \rightarrow \eta_1$	0.30	0.004	0.008
		$\eta_1 \rightarrow \eta_2$	0.40	0.441	0.441

So far *SPAD-PLS* is the only program which takes this problem into account by offering an option to use PLS regression in the estimation of the outer weights and the path coefficients. Here we only focus on the problem of multicollinearity at the latent construct level. We therefore compare the results of *SPAD-PLS* with PLS regression in the case of multi-collinearity with the results of the remaining programs based on common OLS regressions. A data set ($N = 100$) with five correlated exogenous latent variables has been created (see the model in Fig. 31.5).

The resulting variance inflation factors (VIF) for these constructs are between VIF = 16 and VIF = 38. According to general rules of thumb (e.g., Kutner et al. 2004), values above VIF = 10 allude to a potentially severe problem of multi-collinearity. The results reported in Table 31.6 show very similar estimates for the path coefficients both under OLS and PLS regression. In addition, the highest contribution to R^2 is determined for those two exogenous variables which have the smallest "true" effect size (ξ_2 and ξ_4). Given the great discrepancies between the "true" values and the estimated coefficients, *SPAD-PLS* does not really seem to cure the problem of multi-collinearity, at least in our data set.

31.5 Conclusion

In this review on PLS path modeling programs, *LVPLS* and the more recent software packages (*PLS-GUI*, *VisualPLS*, *PLSGraph*, *SPAD-PLS*, and *SmartPLS*) have been characterized and compared with each other. A special emphasis has been placed on the criteria ease-of-use and methodological options. Whereas specifying path models in *LVPLS* is rather inconvenient, all recent programs have made a huge step with respect to ease-of-use, reaching now the same level as the software used in covariance structure analysis. Individual strengths in user-friendliness have been identified, such as supporting the estimation of interaction effects (*VisualPLS* and *SmartPLS*) and helpful export options (*SPAD-PLS* and *SmartPLS*).

One main methodological improvement is the bootstrap procedure for assessing the significance of parameter estimates, which is now implemented in all software packages and supplements the blindfolding and jacknifing resampling routines of *LVPLS*. A specific strength of *SPAD-PLS* is the estimation of bootstrap confidence intervals for the parameters. Model validation is another important aspect; although some measures like the goodness-of-fit index (Tenenhaus et al. 2005) have been discussed in the literature, so far only the blindfolding cross-validation indices (cv-redundancy and cv-communality) are offered. The performance of the different programs has also been tested on data sets with missing data and multi-collinearity. Here, both *PLS-GUI* and *Visual-PLS* provide an incorrect missing data code for the *LVPLS* input file. A major improvement in dealing with missing data is expected for the next release of *SPAD-PLS*.

Multi-collinearity is a problem both for the estimation of weights in the case of formative constructs and the estimation path coefficients. To cure this problem, *SPAD-PLS* has implemented a PLS regression routine. In our study, results for simulated data, however, are very similar to those resulting from OLS regression. This issue should be the subject of a comprehensive Monte Carlo study.

Overall, there is considerable demand for implementations of the various methodological advances documented, for example, in this volume.

References

Allison, P. D. (2002). *Missing data.* Thousand Oaks, CA: Sage.

Chin, W. W. (1993–2003). *PLS Graph – Version 3.0.* Soft Modeling Inc.

Chin, W. W. (2000). *Frequently asked questions – Partial Least Squares & PLS-Graph.* http://disc-nt.cba.uh.edu/chin/plsfac.htm.

Chin, W. W. (2001). *PLS-Graph User's Guide Version 3.0.* Houston, Texas: C. T. Bauer College of Business, University of Houston.

Chin, W. W., Marcolin, B. L., & Newsted, P. N. (2003). A partial least squares latent variable modeling approach for measuring interaction effects: Results from a monte carlo simulation study and an electronic-mail emotion/adoption study. *Information Systems Research, 14*(2), 189–217.

Diamantopoulos, A., & Winklhofer, H. (2001). Index construction with formative indicators: an alternative to scale development. *Journal of Marketing Research, 38*(2), 269–277.

Fu, J.-R. (2006a). *VisualPLS – Partial Least Square (PLS) Regression – An Enhanced GUI for Lvpls (PLS 1.8 PC) Version 1.04.* Taiwan, ROC: National Kaohsiung University of Applied Sciences.

Fu, J.-R. (2006b). *VisualPLS – Partial Least Square (PLS) Regression – An Enhanced GUI for Lvpls (PLS 1.8 PC) Version 1.04.* http://www2.kuas.edu.tw/prof/fred/vpls/index.html.

Haitovsky, Y. (1968). Missing data in regression analysis. *Journal of the Royal Statistical Society, 30*, 67–82.

Hansmann, K.-W., & Ringle, C. M. (2004). *SmartPLS Benutzerhandbuch.* Hamburg: Universität Hamburg.

Jarvis, C. B., MacKenzie, S. B., & Podsakoff, P. M. (2003). A critical review of construct indicators and measurement model misspecification in marketing and consumer research. *Journal of Consumer Research, 30*, 199–218.

Kutner, M. H., Nachtsheim, C. J., Neter, J., & Li, W. (2004). *Applied linear statistical models*, 5th Edn. Homewood, IL: McGraw-Hill/Irwin.

Li, Y. (2003). *PLS-GUI User Manual – A graphic user interface for LVPLS (PLS-PC 1.8) – Version 1.0.* Columbia, SC: University of South Carolina.

Li, Y. (2005). *PLS-GUI – Graphic user interface for partial least squares (PLS-PC 1.8) – Version 2.0.1 beta.* Columbia, SC: University of South Carolina.

Little, R. J. A., & Rubin, D. B. (2002). *Statistical analysis with missing data*, 2nd Edn. Hoboken, NJ: Wiley.

Lohmöller, J.-B. (1984). *LVPLS Program Manual - Version 1.6.* Köln: Zentralarchiv für Empirische Sozialforschung, Universität zu Köln.

Lohmöller, J.-B. (1987). *PLS-PC: Latent variables path analysis with Partial Least Squares - Version 1.8 for PCs under MS-DOS.*

MacCallum, R. C., & Browne, M. W. (1993). The use of causal indicators in covariance structure models: Some practical issues. *Psychologicval Bulletin, 114*, 533–541.

MacKenzie, S. B., Podsakoff, P. M., & Jarvis, C. B. (2005). The problem of measurement model misspecification in behavioral and organizational research and some recommended solutions. *Journal of Applied Psychology, 90*(4), 710–730.

McLachlan, J., & Peel, D. (2000). *Finite mixture models.* New York: Wiley.

Nunnally, J. C., & Bernstein, I. H. (1994). *Psychometric theory*, 3rd Edn. New York: McGraw-Hill.

Ringle, C. M., Wende, S., & Will, A. (2005). *SmartPLS – Version 2.0.* Hamburg: Universität Hamburg.

Sellin, N. (1989). *PLSPATH - Version 3.01. Application Manual.* Hamburg: Universität Hamburg.

Tenenhaus, M., Vinzi, V. E., Chatelin, Y.-M., & Lauro, C. (2005). PLS path modeling. *Computational Statistics and Data Analysis, 48*(1), 159–205.

Test&Go. (2006). *Spad Version 6.0.0.* Paris, France.

Vinzi, V.E., Tenenhaus, M., & Chatelin, Y. (2004). ESIS-PLS software. In *Demonstration at Club PLS 2004 Meeting*, Jouy-en-Josas, France.

Wold, H. (1982). Soft modeling: The basic design and some extensions. In Jöreskog, K. G., & Wold, H., (Eds.), *Systems under indirect observation. Part II* (pp. 1–54). Amsterdam: North-Holland.

Wold, H. (1985). Partial least squares. In Kotz, S., & Johnson, N. L., (Eds.), *Encyclopaedia of statistical sciences*, Vol 6 (pp. 581–591). New York: Wiley.

Chapter 32
Introduction to SIMCA-P and Its Application

Zaibin Wu, Dapeng Li, Jie Meng, and Huiwen Wang

Abstract SIMCA-P is a kind of user-friendly software developed by Umetrics, which is mainly used for the methods of principle component analysis (PCA) and partial least square (PLS) regression. This paper introduces the main glossaries, analysis cycle and basic operations in SIMCA-P via a practical example. In the application section, this paper adopts SIMCA-P to estimate the PLS model with qualitative variables in independent variables set and applies it in the stand storm prevention in Beijing. Furthermore, this paper demonstrates the advantage of lowering the wind erosion by Conservation Tillage method and shows that Conservation Tillage is worth promotion in Beijing sand storm prevention.

32.1 Introduction to SIMCA-P

32.1.1 About SIMCA-P Software

SIMCA-P is developed by Umetrics, which is mainly used for the methods of principle component analysis (PCA) and partial least square (PLS) regression. It is a kind of user-friendly software based on Windows: the operations of models in SIMCA-P are very convenient to handle and the results can be easily illustrated by plots and lists, which present the explanations of the models in kinds of forms. At present,

Z. Wu and H. Wang
School of Economics and Management, Beihang University, 37 Xueyuan Road, Haidian District, Beijing 100083, China
e-mail: binship@126.com, wanghw@vip.sina.com

D. Li
Agricultural Bank of China, Beijing 100036, China
e-mail: zh.ldp@intl.abocn.com

J. Meng
School of Statistics, Central University of Finance and Economics, Beijing 100081, China
e-mail: mengjie517@126.com

V. Esposito Vinzi et al. (eds.), *Handbook of Partial Least Squares*, Springer Handbooks of Computational Statistics, DOI 10.1007/978-3-540-32827-8_33,

SIMCA-P has been a standard tool in PLS regression analysis for researchers in many fields of science and technology.

32.1.2 Some Glossaries in SIMCA-P

There are several special glossaries in SIMCA-P system, which would help us to gain a mastery of the software.

(1) Project. SIMCA-P is organized into projects. A project is a folder including the models with the relative statistics and results.
(2) List. All the data are listed in tables in the system. The first row with the variable names is marked as the Primary variable ID. The first column is marked as identification numbers.
(3) Dataset. The set of processing data is known as a Dataset. A project may contain several datasets.
(4) Model. Models are mathematical representations of your process and are developed using the data specified in the workset and with a specified model type.
(5) Workset. A workset is the set of data processed by the current active model. A workset can contain all the data, or be a subset of the primary data, with a particular treatment of the variables, such as role (predictor variables X, or responses Y),scaling, transformation, lagging, etc.
(6) Block. A block is a combination of the variables with same role. For example, the Y block in a PLS model refers to all the dependent variables (responses).
(7) Class. The observations of a dataset can be spitted into different set for different purposes, known as class.

32.1.3 The Analysis Cycle

It is convenient to do PCA or PLS estimation with SIMCA-P. Users can get the analysis results after several steps in accordance with the principles of PCA or PLS methods. The analysis cycles can be summarized as follows.

(1) Start a project. Users should import the primary data form file or databases to create a new project.
(2) Preprocess the data. View or modify a SIMCA-P data set. For example, it is easy to generate new variables as functions of existing ones or from model results. Users can do similar operations to preprocess the data.
(3) Prepare the workset. The default workset is the whole data set with all variables as X at the project start, and the default model (unfitted) is a principal component model of X. Users should change the role of the variables to fit other models.
(4) Fit the models. After all the preparing procedures, users can do the estimation.
(5) Detect the outliers. Display the score scatter plot to show the possible presence of outliers, groups, and other patterns in the data. Users should exclude the outliers from the workset and go back to step (4) to fit a new model.

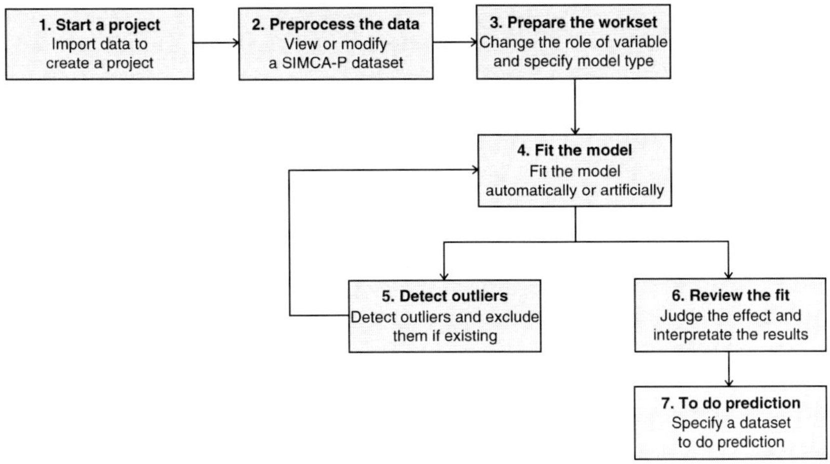

Fig. 32.1 Road map to SIMCA-P

(6) Review the fit. After a fit, the whole spectrum of plots and lists are available for model interpretation. Users should judge the effect of the fitted model and decide whether to do prediction.
(7) To do prediction. Build the prediction set from the primary or any secondary data sets to do prediction.

The above steps can be shown as the above road map (Fig. 32.1).

32.2 The Basic Operations of SIMCA-P

The example below will illustrate the main operations of SIMCA-P. The data in this example describes the relationship between body condition and sports grade of people. The predictors reflect one's body condition including avoirdupois, cummerbund and pulse. The responses are three grades of physical exercise including chin-up, curl and high jump. 20 persons have been selected. Table 32.1 shows the original data set (Jone Neter, used by Tenenhaus 1998).

32.2.1 The Main Window of SIMCA-P

Double click on the SIMCA-P icon on the desktop, the main window opens and displays as Fig. 32.2. (Note: Before this operation, the project with the above data in Table 32.1 has been created. Otherwise the active model status window will not display.)

The main window includes the following parts.

(1) The command menu bar. The name and the folder of a project will be shown in the title. The menu in the bar includes the entire functions menu.

Table 32.1 Observed data of body condition and sports grade

No	avoirdupois	cummerbund	pulse	chin-up	curl	high jump
1	191	36	50	5	162	60
2	189	37	52	2	110	60
3	193	38	58	12	101	101
4	162	35	62	12	105	37
5	189	35	46	13	155	58
6	182	32	56	4	101	42
7	211	38	56	8	101	38
8	167	34	60	6	125	40
9	176	31	74	15	200	40
10	154	33	56	17	251	250
11	169	34	50	17	120	38
12	166	33	52	13	210	115
13	154	34	64	14	215	105
14	247	46	50	1	50	50
15	193	36	46	6	70	31
16	202	37	62	12	210	120
17	176	37	54	4	60	25
18	157	32	52	11	230	80
19	156	33	54	15	225	73
20	138	33	68	2	110	43

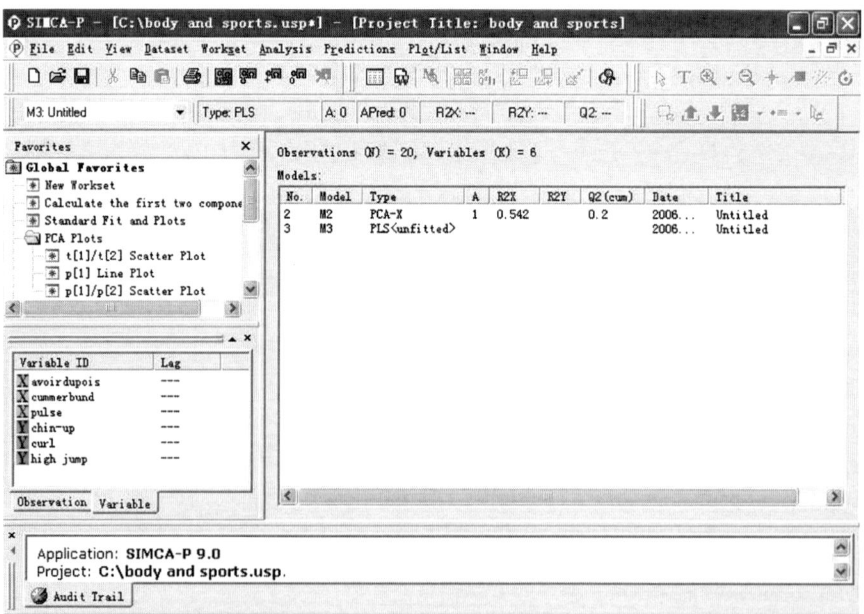

Fig. 32.2 Main window of SIMCA-P

(2) Standard and shortcut bar. These shortcut buttons are for activating command menus and plots. Pressing a button will perform a certain task.
(3) Plot and maker bar. Use the buttons in Plot toolbar to insert labels or text in plot, enlarge and read positions in graphs, get information about observations or variables, show a regression line in scatter plots or rotate 3D graphs. The main function of Maker toolbar is to exclude or include the observations or variables in the active model and create a new model.
(4) The Favorites window. The Favorites window contains commands and plots, which are marked with different symbols. Double click on a symbol will execute a command or open a specified plot.
(5) The Workset bar. The bar displays the variables and observations in the workset and their status.
(6) The active model status window. The window shows the information about all the models, such as model name, type, number of components, etc.
(7) The Audit Trail window. The log events are shown in this window.

32.2.2 The Operations of SIMCA-P

The important operations are as followings.

(1) Import the data and create a project

Data can only be imported from file or data bases, but not by keyboard. The system supports more than 10 types of files, such as txt, xls, mat, etc. Select File|New|Get data from file, a standard dialog box opens to enter the file type, name and source address of the data file to be imported (Fig. 32.3).

After importing the data file, the first page of the import wizard opens (Fig. 32.4). The row with the variable names is by default marked as the Primary variable ID and colored in dark green. You can select any other row as the Primary variable ID. If the Primary variable ID has not been specified, SIMCA-P creates the Primary variable ID as Var_1, Var_2, etc. The column with observation numbers or names is colored in dark yellow. Select any desired column as Primary Observation ID. Data are colored in white and text are colored in blue.

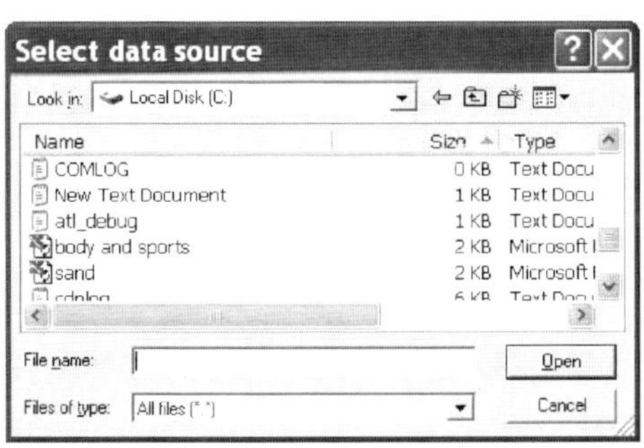

Fig. 32.3 Import the data file

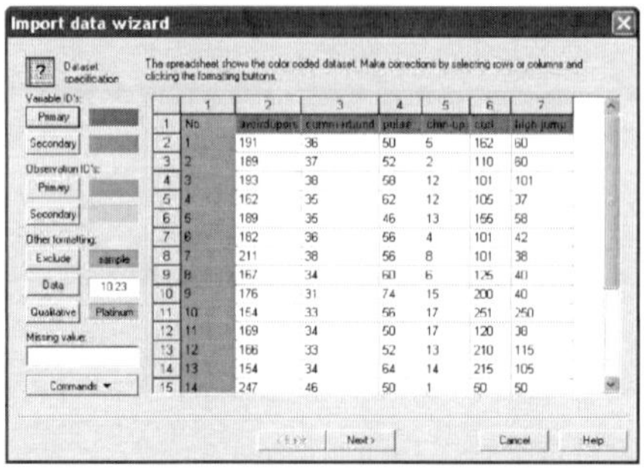

Fig. 32.4 Import data wizard

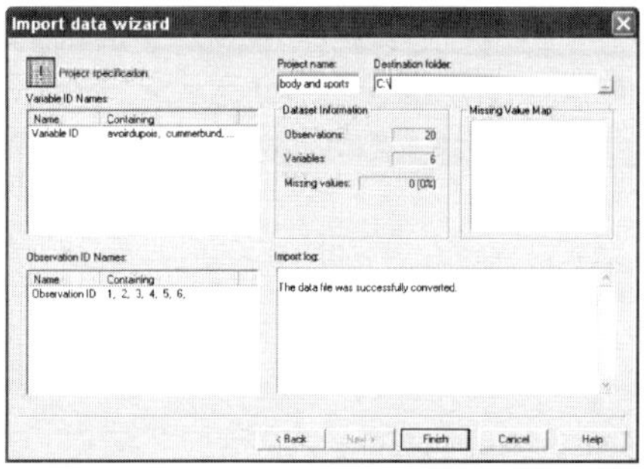

Fig. 32.5 Specification of the project

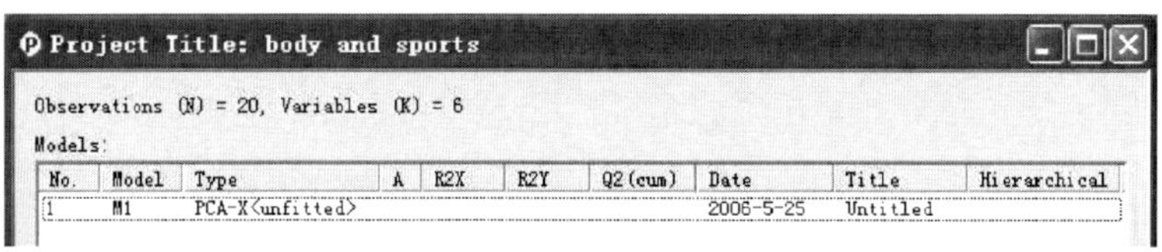

Fig. 32.6 The active model status window

In the Import data wizard, you can do other operations by pressing the buttons on the left window.

Click on Next, the project specification page of the import wizard displays (Fig. 32.5). Users should specify the project name and the folder to save the work file. The file type is usp. The window still displays other information about the data set.

Click on Finish and the data set is imported. A project has been created (Fig. 32.6). The default workset is the whole data set with all variables as X and scaled to unit variance. The associated model is PCA.

(2) Explore the data

Before fitting the model, you should understand the data comprehensively. Select Dataset|Quick info|Variables/Observations, a window opens with the name of the

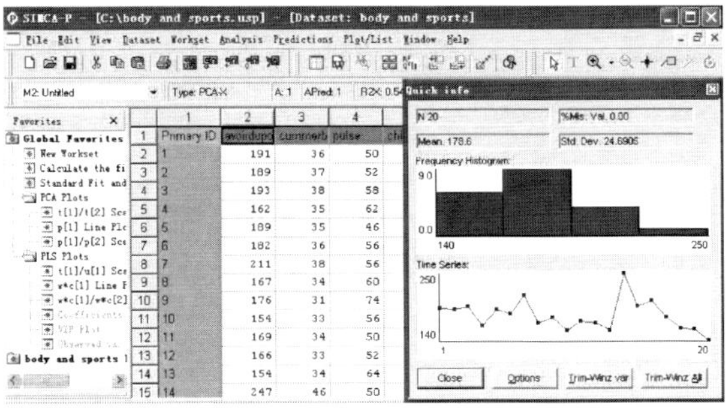

Fig. 32.7 Usual statistics and dataset

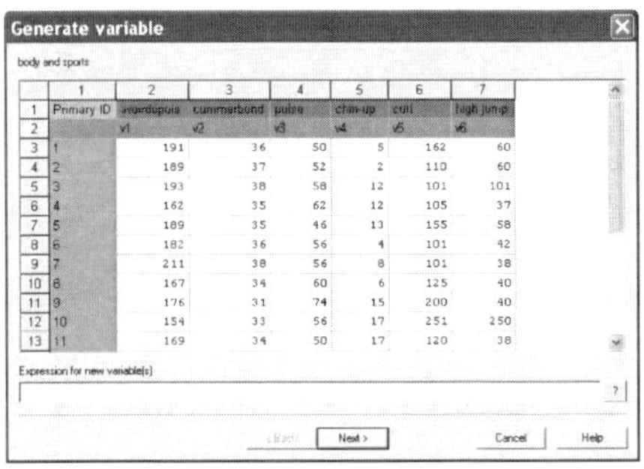

Fig. 32.8 Generate variables

variable/observation and default options (Fig. 32.7). The usual statistics are showed in the window, such as number, missing values, mean, etc.

In some case, a new variable should be generated from raw data. Select Dataset| Generate Variable, SIMCA-P opens the wizard window displaying the active data set in a spreadsheet (Fig. 32.8).

Enter the expression defining the new variable and click on Next. SIMCA-P displays the new variable, with its formula, statistics and Quick info plots (Fig. 32.9).

Click finish, and the new variable is added at the end of the active dataset.

(3) Create the workset and set model options

After the primary data set is loaded, all the variables are selected as X variables (predictors). The active model type is PCX (Principle Component Analysis of the X variables).

In order to change the role of the variables or observations, you can select Workset|Edit to open the Overview page of the workset dialog with the current observations and variables and their attributes (Fig. 32.10).

The workset is organized into pages. Select the desired page to change the attributes of the observations or variables. In order to do PLS estimation, you can select Workset|Edit to change roles of variables by marking the variable y and

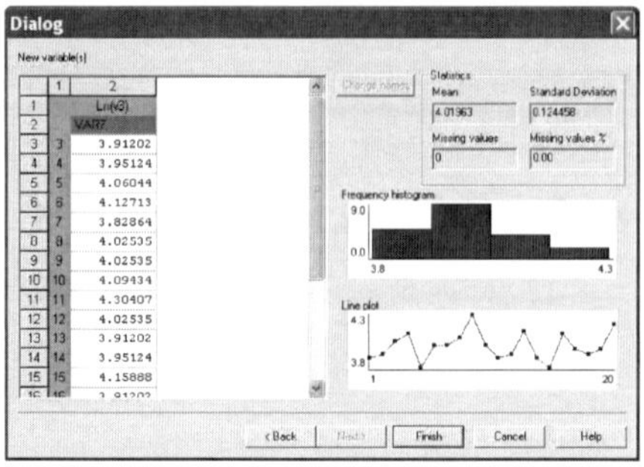

Fig. 32.9 Info of new variable

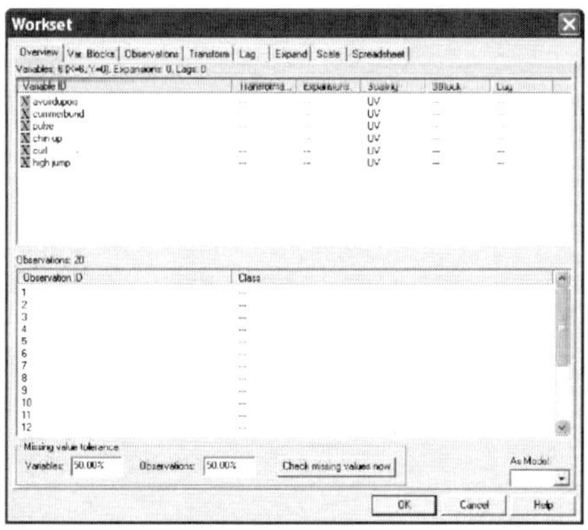

Fig. 32.10 Overview page of the workset dialog

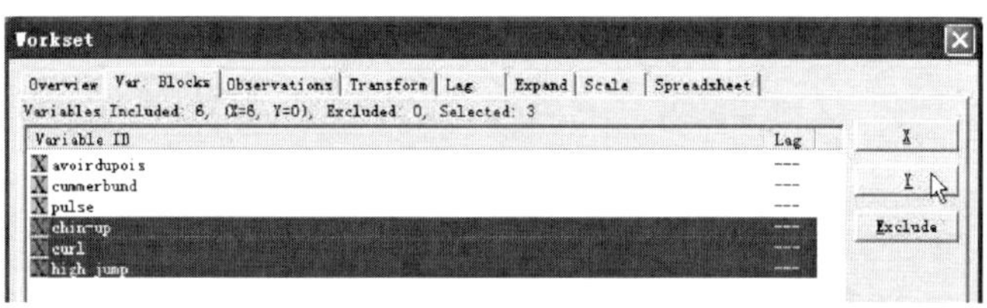

Fig. 32.11 Specify variables as responses

clicking on the desired button Y (Fig. 32.11). You can also set class of observations in the Observations page.

After the above procedures, you can select Workset|Options to set the options of the current active model (Fig. 32.12).

(4) Fit the model

You can select Analysis menu to fit the model. The model is by default non hierarchical base model. The model type is decided according to the role of variables, including PCA on X-block, PCA on Y-block, etc. The methods of fit include autofit,

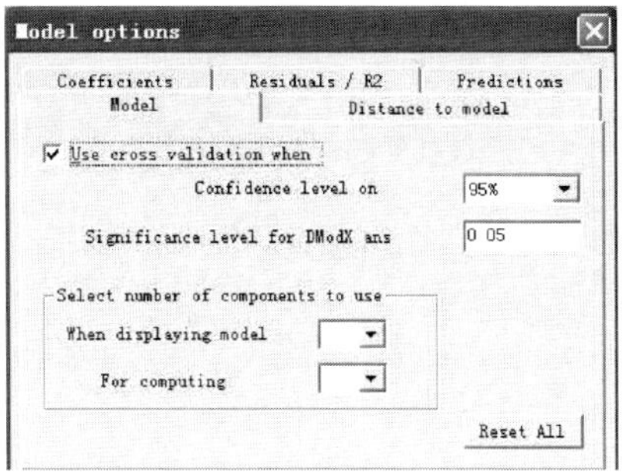

Fig. 32.12 Model options

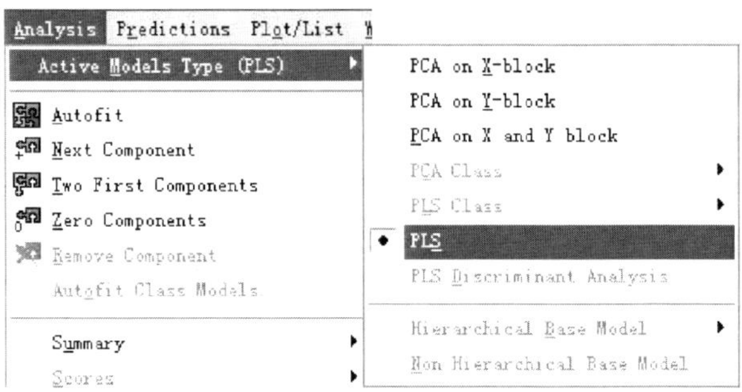

Fig. 32.13 Fit the model

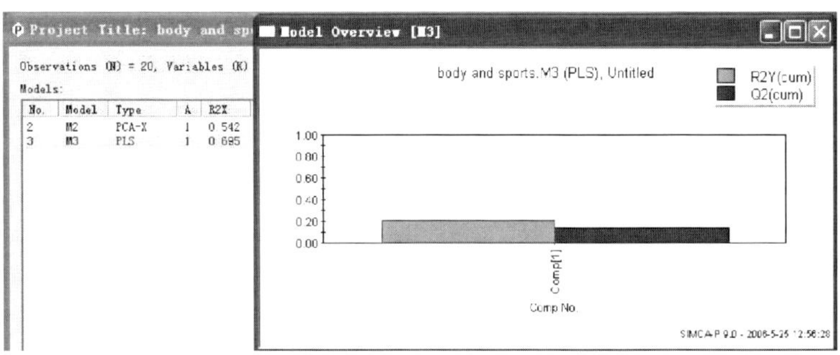

Fig. 32.14 Model overview

next component, 2 first components, next component, zero component, remove component and autofit class models (Fig. 32.13).

Select Analysis|Autofit, SIMCA-P extracts as many components as considered significant. When you fit a model, a plot window opens and displays the cumulative R2 and Q2 for the X(PCA) or Y(PLS) matrix (Fig. 32.14).

After fitting a model, you can mark the model and click on Active Model Type|Hierarchical Base Model and select scores, residuals, or both as variables in

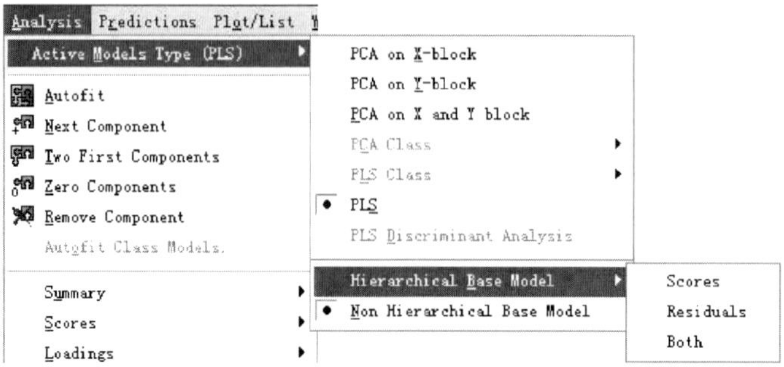

Fig. 32.15 Hieraarchical base model

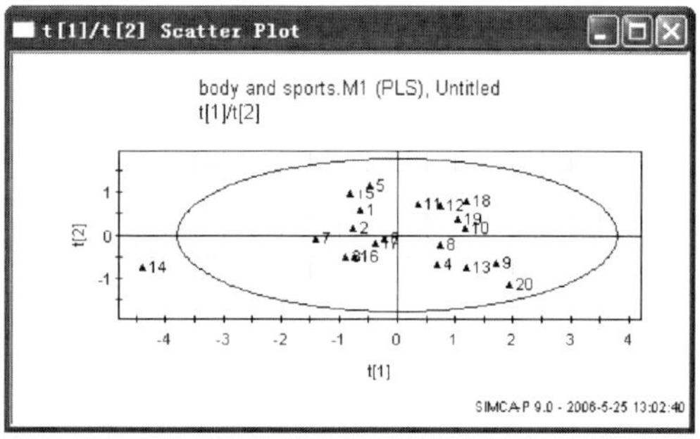

Fig. 32.16 t[1]/t[2] scatter plot

another model (Fig. 32.15). The scores, residuals, or both would be added to the workset to be used as variables in another model.

(5) Detect the outliers

Double click t[1]/t[2] Scatter Plot in Favorite window to display score scatter plot after fitting (Fig. 32.16). These plots show the possible presence of outliers, groups, and other patterns in the data. In order to illustrate this plot, we extract two components.

In Fig. 32.16, observation 14 is outside the 95% confidence region of the model. This means observation 14 is an outlier. In order to eliminate the effect of observation 14, you should exclude this observation from the workset. Press Mark item button in Marker toolbar and mark observation 14, and then press the red arrow button. Resultingly, observation 14 is excluded from workset and a new unfitted model is created without observation 14 (Fig. 32.17).

(6) Review the results

You can select Analysis to plot or list some statistics, including scores, loading, coefficients, etc. For example, you can select Analysis|Summary|List to show the individual cumulative R2 and Q2 for each Y variable (Fig. 32.18).

In fact, you can double click a ceratin symbol in Favorites window to execute a command. The Favorites bar is similar to a customized Navigation Bar. It contains commands and plots. They are marked with different symbols for specified

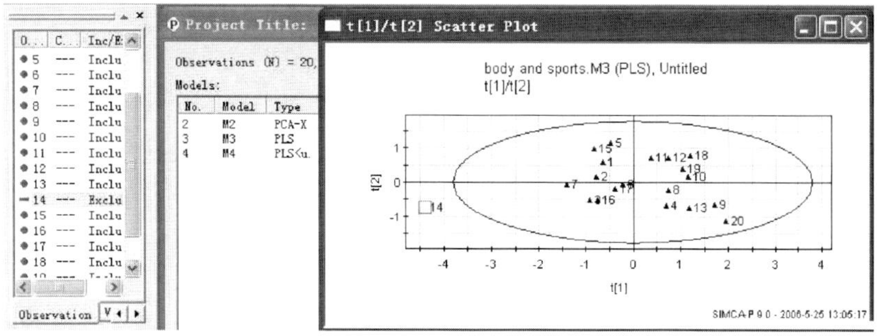

Fig. 32.17 Detect the outliers

	1	2	3	4	5	6	7
1	Var ID (Primary)		R2VY	R2VY(cum)	Q2VY	Q2 limit	Q2VY(cum)
2	Total	Comp 1	0.209447	0.209447	0.135405	0.05	0.135405
3		Comp 2	0.0294908	0.238938	-0.0752204	0.05	0.0703701
4							
5	chin-up	Comp 1	0.236348	0.236348	0.155521	0.05	0.155521
6		Comp 2	0.0495692	0.285917	-0.122236	0.05	0.0710727
7							
8	curl	Comp 1	0.350593	0.350593	0.267608	0.05	0.267608
9		Comp 2	0.037041	0.387634	-0.0758256	0.05	0.212074
10							
11	high jump	Comp 1	0.0413997	0.0413997	-0.0169132	0.05	-0.0169132
12		Comp 2	0.00186238	0.0432621	-0.0373566	0.05	-0.0549015
13							

Fig. 32.18 Summary of the results

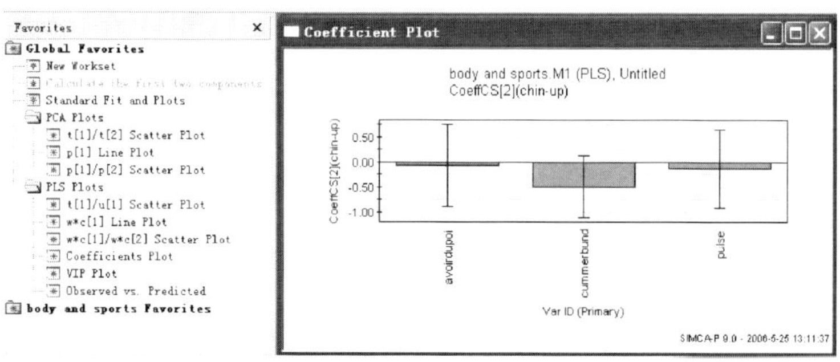

Fig. 32.19 Coefficients Plot

plots/lists and for a command (works on the active model). For example, you can double click on the Coefficients Plot in Favorites window to show the coefficients plot (Fig. 32.19).

Besides the results included in Analysis menu and Favorites window, you can also select Plot/List menu to plot or list all the results. The Plot/List menus allow you to plot and list input data such as observations and variable values, compute elements such as scaling weights, variable variances, etc., as well as results such as loadings, scores, predictions, etc., of all the fitted models.

(7) To do Prediction

After fitting, the workset is by default specified as prediction set. If you want to build a prediction set by combining observations from different data sets, or removes observations from the prediction set, select Predictions|Specify Predictions Set|Specify.

The observations are displayed in the left window (Fig. 32.20). Select the ones you want in the prediction set and move them to the right window.

After specifying the prediction set, you can select Predictions menu to obtain the prediction information about the current model. For example, you can select Distance to Model|Y Block|Line Plot to display this plot (Fig. 32.21).

The residual standard deviation of an observation in the Y space is proportional to the observation distance to the hyper plane of the PLS model in the corresponding space. SIMCA-P computes the observation distances to the PLS model in the Y

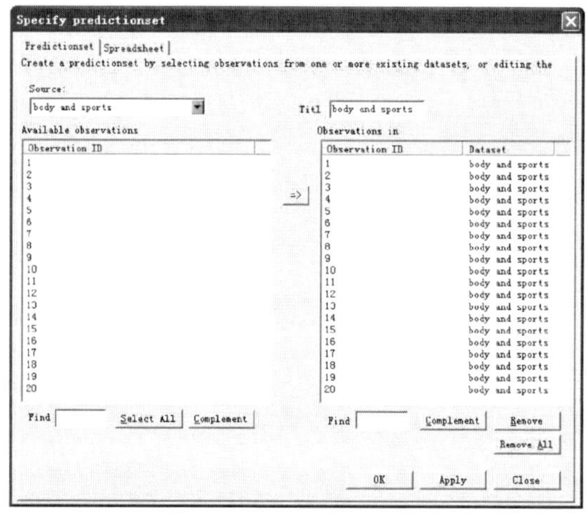

Fig. 32.20 Specify predictions set

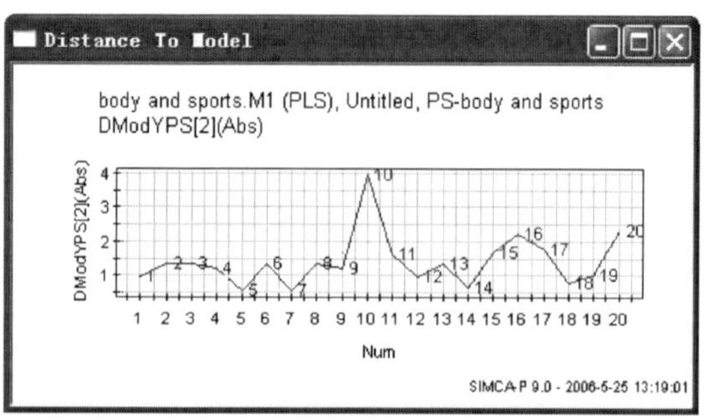

Fig. 32.21 Distance to model

space (DModY) and displays them as line plots. A large DModY value indicates that the observation is an outlier in the Y space. By default, these distances are computed after all extracted components.

32.3 Application

In this section, we provide an application of SIMCA-P. Sandstorms have been a big barrier against the development of the world, which results in an annual global loss of 48 billion USD, including 6.5 billion USD in China. In recent years, sand storms in Beijing have caused many serious problems. Investigations showed that about 70 percent of the sand in these storms are generated by wind erosion of dry, fallow farmland around the city. Consequently, the study on wind erosion of soil becomes very important in sand storm prevention (Shen et al. 2000; Li and Gao 2001; Gao 2002; Zang 2003).

In this research, the Water Content in Soil (x_1), Soil Particle Size(x_2),the Rate of Straw Mulching (x_3) and the Type of Farmland is defined as four independent variables (IVs). The Type of Farmland is a qualitative variable (QV) consisting of the following four categories: sand farmland, traditional tillage farmland, grass farmland and Conservation Tillage farmland. These categories are regarded as different types of farmland. To establish a regression model of Wind Erosion Rate (y)with the above four IVs, the Type of Farmland should be transformed into four dummy variables (DVs), D_1, D_2, D_3, D_4. Table 32.2 shows the original data set. The sample size is 16.

Based on the data in Table 32.2, the regression model can be written as follows:

$$y = u + \beta_1 x_1 + \beta_2 x_2 + \beta_3 x_3 + \alpha_1 D_1 + \alpha_2 D_2 + \alpha_3 D_3 + \alpha_4 D_4 \qquad (32.1)$$

Table 32.2 Wind erosion rate and IVs

No	y	x_1	x_2	x_3	D_1	D_2	D_3	D_4
1	11.674	3.623	0.651	12.4	1	0	0	0
2	13.812	3.623	0.651	12.4	1	0	0	0
3	15.260	3.623	0.651	12.4	1	0	0	0
4	12.160	3.623	0.651	12.4	1	0	0	0
5	6.021	6.291	0.266	13.8	0	1	0	0
6	8.598	6.291	0.266	13.8	0	1	0	0
7	10.395	6.291	0.266	13.8	0	1	0	0
8	7.331	6.291	0.266	13.8	0	1	0	0
9	3.689	10.210	0.337	45.4	0	0	1	0
10	5.339	10.210	0.337	45.4	0	0	1	0
11	5.971	10.210	0.337	45.4	0	0	1	0
12	4.893	10.210	0.337	45.4	0	0	1	0
13	2.768	8.883	0.339	58.5	0	0	0	1
14	4.167	8.883	0.339	58.5	0	0	0	1
15	4.357	8.883	0.339	58.5	0	0	0	1
16	4.111	8.883	0.339	58.5	0	0	0	1

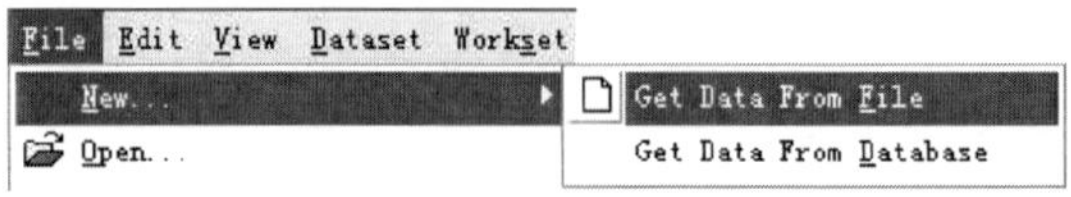

Fig. 32.22 Get the data from file

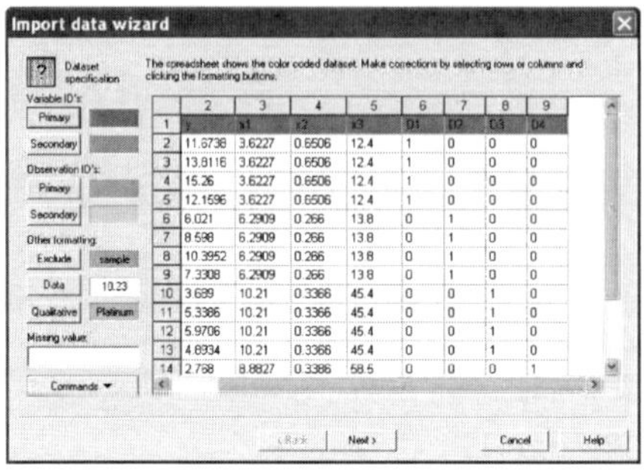

Fig. 32.23 Import data

It is clear that the following equation always exists in the model (1):

$$D_1 + D_2 + D_3 + D_4 = 1 \tag{32.2}$$

The above results show that there is full multicollinearity between the IVs. We have used SAS 8.0 to obtain the estimation. The system provides the following notes: the model is not of full rank; the least-squares solutions for the parameters are not unique; some statistics will be misleading. A reported DF of 0 or B means that the estimate is biased. Therefore, OLS method is invalid in this case study.

Consequently, we adopted PLS to establish the regression model, which was executed by SIMCA-P 9.0.

At the beginning, select File|New|Get data from file to import the primary dataset and create a new project (Fig. 32.22).

The raw data was stored in c:\ and the name of the source file is sand.dif. After the file was selected, the window of Import data wizard was displayed (Fig. 32.23).

The first row with the variable names is by default marked as the Primary variable ID. The first column is by default marked as identification numbers.Click on Next in Import data wizard when finished.

After importing the data, we should specify the project name and select the destination folder to save the project (Fig. 32.24).

After the above procedures, a project has been created. By default all variables are selected as X. The active model type is PCX (Fig. 32.25).

In order to adopt PLS estimation, we can select Workset|Edit to change all the options of the default model in Workset Window. Variables are displayed with their roles (X, Y or excluded (–)). To change roles, mark the variable y and click on the desired button Y (Fig. 32.26).

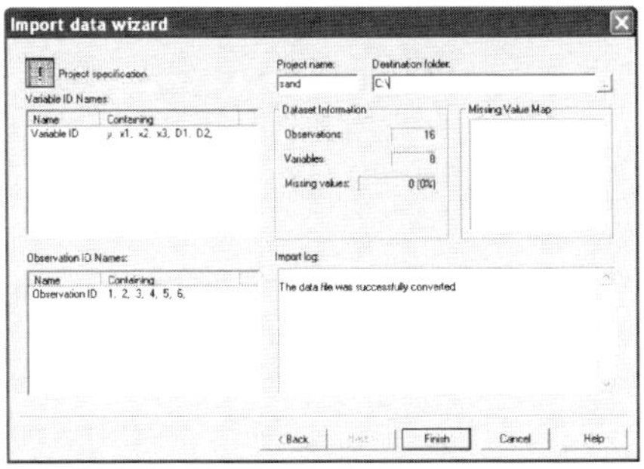

Fig. 32.24 Specification of the project

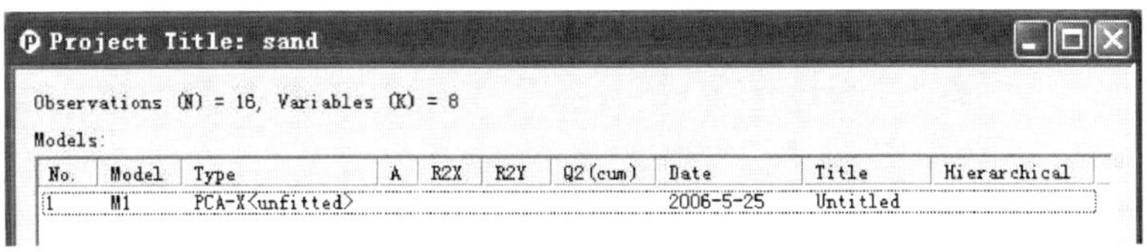

Fig. 32.25 The default model

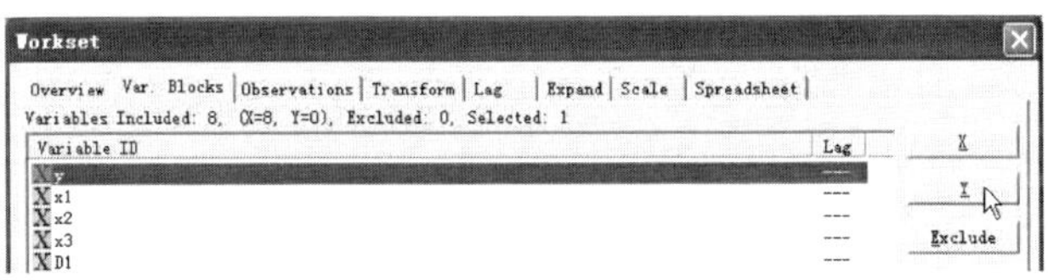

Fig. 32.26 Change the role of variable y

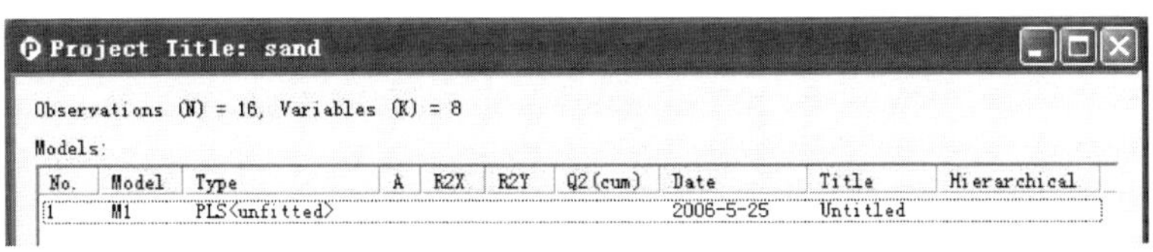

Fig. 32.27 Change the role of variable y

When we exit the Workset window, the model type has been changed from PCX to PLS. This model is unfitted and is the active model (Fig. 32.27).

SIMCA-P extracts one component according to the cross validation rules after selecting Analysis|Autofit. The right plot displays the cumulative R2 and Q2 for the Y (PLS) matrix after the extracted component (Fig. 32.28).

Double click t[1]/u[1] Scatter Plot in Favorites window to display the t_1/u_1 plot (Fig. 32.29).

The plot indicates a good fit corresponding to the small scatter around the straight line. It proves that there is a strong linear correlation between Wind Erosion Rate and its IVs. So the linear regression is fundamentally established.

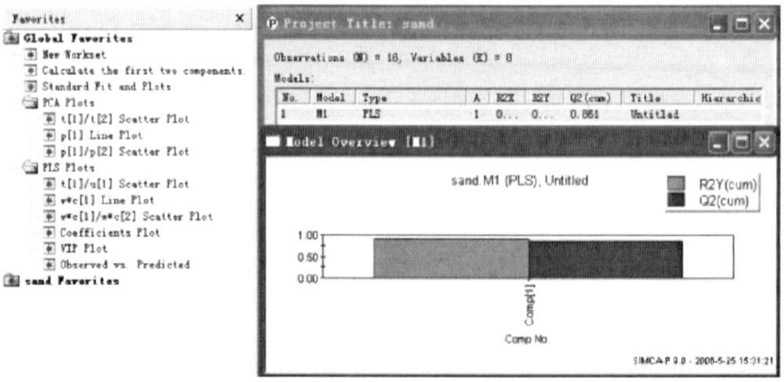

Fig. 32.28 The PLS model

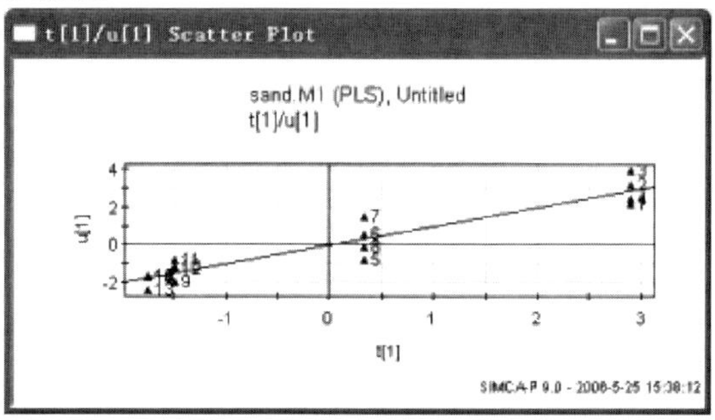

Fig. 32.29 t[1]/u[1] scatter plot

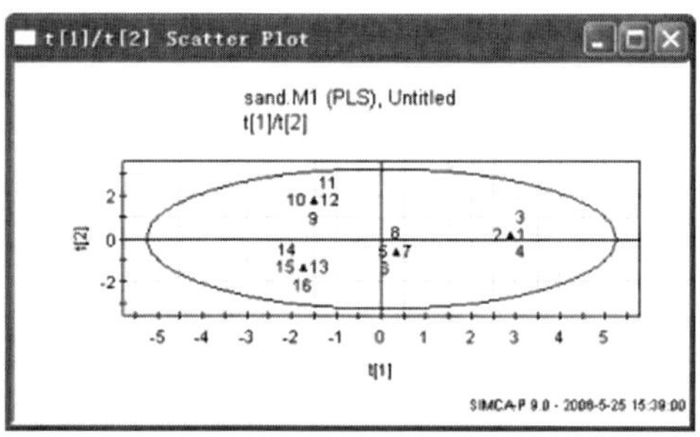

Fig. 32.30 t[1]/t[2] scatter plot

For a better illustration of the regression results, we extracted two components. The cumulative Q2 for the extracted components is 0.848 and it can explain 72.6% variation of IVs and 90.2% variation of y.

Double click t[1]/t[2] Scatter Plot in Favorites window to display a two-dimensional score plot (Fig. 32.30).

SIMCA-P draws the confidence ellipse based on Hotelling T2. Observations situated outside the ellipse are outliers. Figure 32.30 shows no outliers.

Double click Observed vs. Predicted in Favorites window to shows the observed values vs. the fitted or predicted values (Fig. 32.31).

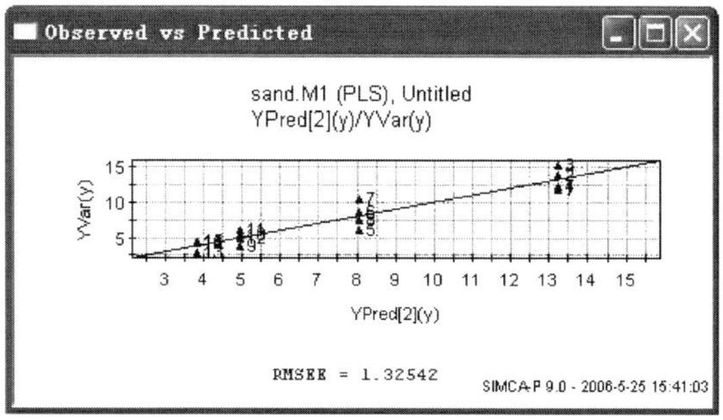

Fig. 32.31 Observed vs. Predicted values

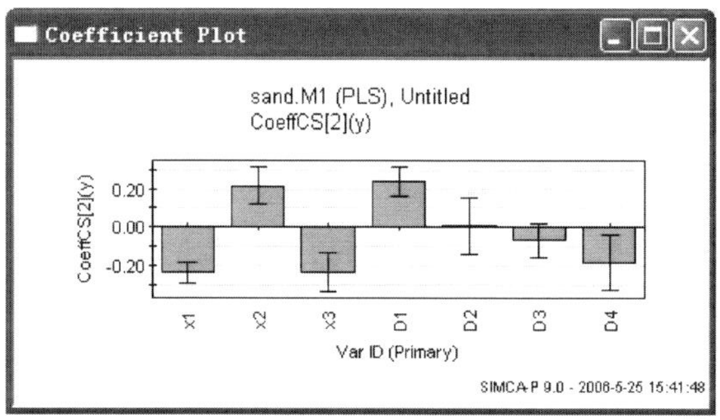

Fig. 32.32 Standardized regression coefficients

Figure 32.31 demonstrates that the estimation by PLS is effective. The model consisting of original variables is estimated as follows:

$$y = 9.36 - 0.36x_1 + 5.54x_2 - 0.03x_3 + 2.11D_1 + 0.10D_2 - 0.59D_3 - 1.62D_4 \quad (32.3)$$

Double click Coefficients Plot in Favorites window to show standardized regression coefficients of the model (Fig. 32.32).

According to Fig. 32.32, the larger the size of soil particle, the more serious the wind erosion. Furthermore, because Soil Water Content and Straw Mulching Rate are negatively correlated with Wind Erosion Rate, these are beneficial in easing the soil wind erosion problem by adding Soil Water Content and increasing the Straw Mulching Rate. Considering the different kinds of farmland, we conclude that the Conservation Tillage farmland has the lowest wind erosion rate, while the Sand farmland has the highest.

Double click w*c[1]/w*c[2] Scatter Plot in Favorites window to show both the X-weights (w or w*) and Y-weights (c) and thereby the correlation structure between X and Y (Fig. 32.33).

Since the Conservation Tillage method cultivates the farmland in a shallow way and leaves the crop residues on the land surface as much as possible, it is the most effective way to prevent wind erosion. Additionally, it can increase land coverage

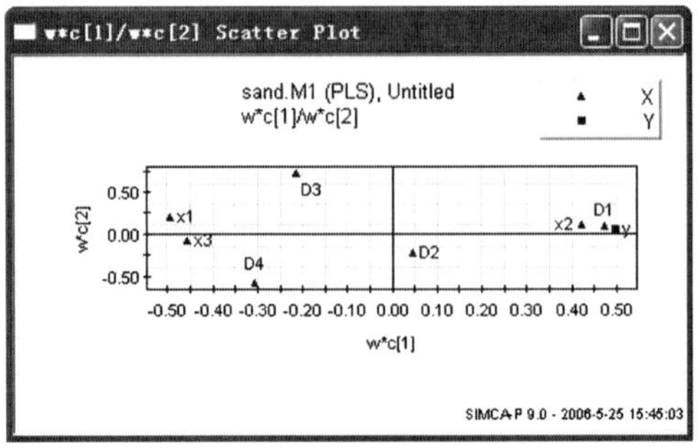

Fig. 32.33 Loading plots of IVs

rate, prevent water and soil loss and enlarge the level of production. Therefore, it is worthwhile promoting the Conservation Tillage method both for the prevention of sand storms in the Beijing area and for agricultural production.

32.4 Conclusion

This paper has introduced fitting modes by employing SIMCA-P. It is obvious that SIMCA-P is an effective tool to conduct multivariate data analysis. In the part of empirical research, the results show that, compared with OLS, PLS is preferable in dealing with QVs. In the investigation, the PLS model not only illustrated the factors of soil wind erosion, which conformed fairly well with reality, but also demonstrated that Conservation Tillage method is the most effective way to ease soil wind erosion. The results provide valuable information for Beijing sand storm prevention.

References

UMETRI AB: SIMCA-P for Windows: Maltivariate Modeling, Analysis and SPC of Process Data. UMETRI AB (1996)

Tenenhaus, M.: La Rgression PLS: Thorie et Pratique. TECHNIP, Paris (1998)

Shen, Y.C., Yang, Q.Y., Jing, K, Xu, J.X.: Sand-storm and Dust-storm in China and Prevention and Control. Journal of Arid Land Resources and Environment, **14(3)**, 11–14 (2000)

Li, L.J., Gao, Q.X.: Source Analysis of Beijing Sand-Dust in 2000, Research of Environmental Sciences, **14(2)**, 1–4 (2001)

Gao, Q.X.: The Dust Weather of Beijing and Its Impact, China Environmental Science, **22(5)**, 468–471 (2002)

Zang, Y.: Experimental Study on Soil Erosion by Wind under Conservation Tillage, **19(2)**, 56–60 (2003)

Chapter 33
Interpretation of the Preferences of Automotive Customers Applied to Air Conditioning Supports by Combining GPA and PLS Regression

Laure Nokels, Thierry Fahmy, and Sébastien Crochemore

Abstract A change in the behavior of the automotive customers has been noticed throughout the last years. Customers feel a renewed interest in the intangible assets of perceived quality and comfort of environment. A concrete case of study has been set up to analyze the preferences for 15 air conditioning supports. Descriptive data obtained by flash profiling with five experts on the photographs of 15 air conditioning supports are treated by Generalized Procrustes Analysis (GPA). The preferences of 61 customers are then explained by Partial Least Squares (PLS) regression applied to the factors selected from the GPA. The results provided by the XLSTAT GPA and PLS regression functions help to quickly identify the items that have a positive or negative impact on the customers' preferences, and to define products that fit the customers' expectations.

33.1 Introduction

Due to an overall standardization of technical performances of cars, the automotive industry concentrates on ergonomics, safety, design and sensations. In order to create attractive passenger cells and to answer the customers' expectations, *sensory analysis* is integrated into the vehicles' design.

Sensory analysis allows to measure customers' preferences, and to describe products on the basis of human perceptions, such as the sight, the touch or the odor (Crochemore and Nesa 2004).

A study on air conditioning supports is conducted in order to define, explain and anticipate the future customers' behaviors.

L. Nokels and S. Crochemore
Materials Engineering Department, Technocentre Renault, 1, avenue du Golf, API TCR LAB 2 52, 78 288 Guyancourt Cedex, France
e-mail: laure.nokels@renault.com, sebastien.crochemore@renault.com

T. Fahmy
Addinsoft, 40 rue Damrémont, 75018 Paris, France
e-mail: fahmy@xlstat.com

V. Esposito Vinzi et al. (eds.), *Handbook of Partial Least Squares*, Springer Handbooks of Computational Statistics, DOI 10.1007/978-3-540-32827-8_34,

33.2 Sensory Methodology

Flash profiling is a method that makes it possible to quickly obtain the descriptions given by a group of judges for a series of products (Dairou and Sieffermann 2002).

In this study, five sensory experts used their own terms to describe the 15 air conditioning supports on the basis of one photograph for each support. A total of 99 attributes have been used by the experts, with, for each expert, a minimum of 4 and a maximum of 27. Based on the attributes they chose, the experts ranked the photographs.

During a second session, 61 customers gave for each air conditioning photography a preference rating on a continuous scale (from 0 “Don’t like it” to 10 “Like it very much”).

33.3 Statistical Methodology

33.3.1 Generalized Procrustes Analysis

A *Generalized Procrustes Analysis* (GPA) was used in this study on the data obtained from the experts, in order to homogenize and relate the various attributes, and to obtain after a Principal Components Analysis (PCA) a reduced number of factors summarizing them.

GPA is a multivariate data analysis method (Gower 1975) that allows to apply repeatedly until convergence, three types of transformations (translation, rotation, rescaling) to a series of m configurations of n points described in a p-dimensional space. The transformations are performed in order to reduce the sum of the Euclidean distances between the m configurations and the *consensus configuration*, which is the mean of the m configurations. After these transformation steps, a Principal Component Analysis (PCA) is often performed on the consensus configuration in order to reduce the number of dimensions while concentrating the variability on the first axes. The PCA transformation is then applied to the m configurations.

Once the GPA algorithm has converged, we obtain three types of results:

- Results that allow evaluating how well the algorithm performed.
- Results on the consensus configuration.
- Results on the m individual configurations.

GPA has been very early applied to the field of sensory data analysis (Harries and MacFie 1976), where it allows to solve several problems that may arise and make the analysis difficult:

- Experts might use the rating scales in different ways, either because they tend to give higher or lower ratings in average, or because the range of the scale they use is different.

- Experts might use different descriptors to rate the similar characteristics of the products.
- Experts might use several descriptors that could be summarized by a single concept.

The scaling issue is solved by the translation and rescaling steps of the GPA. The problem of the conceptual similarity between different descriptors can be solved by the rotation steps. Last, the reduction of the number of dimensions can be solved by the PCA step.

As is often the case with methods aimed at decreasing the number of dimensions, the key question is how many dimensions should be kept. The scree plot of the PCA allows to visually decide how many dimensions should be taken into account, but it is very empirical and not very reliable after a GPA, as the consensus configuration is an average of the m configurations.

A *permutation test* has been developed by Wu et al. (2002) in order to determine how many dimensions are significant to describe the consensus configuration. The Wu test is based on an F statistic that measures how much of the variability of the m initial configurations is represented by a given factor of the consensus configuration. Permuting many times the initial configurations makes it possible to obtain a distribution for the F statistic. If the observed F statistic is greater than the value that corresponds to the selected confidence interval, then the dimension should be taken into account.

Prior to this test, another permutation test had been developed by Wakeling et al. (1992) to test whether the GPA impact on the variance reduction is significant or not. This test is based on the Rc coefficient that corresponds to the ratio of the variance of the consensus configuration and the variance before GPA is computed for each permutation. The rows of each configuration are randomly permuted in order to obtain a distribution of Rc. The Rc coefficient for the original data is computed and the corresponding quantile in the distribution obtained from the permutations is determined, and compared to a significance level.

33.3.2 Partial Least Square Regression

A Partial Least Square (PLS) regression is run to explain the standardized customers' preferences using as explanatory variables the factors obtained from the GPA/PCA steps.

In sensory data analysis it is very common that the number of observations (in our case observations are products) is much lower than the number of explanatory variables. The latter often correspond to experts' ratings, or physico-chemical descriptors of the products.

While applying classical linear regression is not possible in such situations, PLS regression offers a very interesting framework, as it allows to automatically decrease the number of dimensions by taking into account the covariance structure of both the explanatory and the dependent variables.

Furthermore, there are often many dependent variables (the consumers' preferences). As in our case we will be decreasing the number of explanatory variables using the GPA, the PLS regression is used here more because it allows to treat simultaneously many dependent variables, trying to find a common structure within these variables.

However, that advantage of the PLS regression, that you can model several dependent variables using the same set of explanatory variables, must be considered with caution. If your population is too heterogeneous, and if the information corresponding to the heterogeneity is not carried by the explanatory variables, the quality of the PLS models might be penalized. This problem was pinpointed by Tenenhaus et al. (2005). To improve the quality of the PLS models, *hierarchical clustering* can be used to cluster the consumers using the t components of the PLS regression on the whole population. The use of the t components allows considering the *heterogeneity* at the model level, and not at the variable level which would be of little effect.

Some authors make a distinction between the PLS-1 regression where only one dependent variable is being modeled and the PLS-2 regression where two or more variables can be modeled. As the PLS-2 algorithm can be applied to the case where there is only one dependent variable, we will consider later that a PLS regression is a PLS-2 regression.

To analyze the results of the PLS models we have extensively used here the outputs of the PLS regression that include several graphical representations that facilitate the interpretation.

33.4 Application to Air Conditioning Supports

33.4.1 Generalized Procrustes Analysis

A Generalized Procrustes Analysis (GPA) is performed on the descriptive data with the GPA function of the XLSTAT software (Addinsoft 2006). The GPA function is available in the XLSTAT-MX and XLSTAT-ADA modules, which are respectively dedicated to sensory data analysis and multiple tables data analysis techniques.

When flash profiling is used to describe products, the number of dimensions that are spontaneously chosen might vary from one expert to another, leading to very heterogeneous configurations in terms of dimensionality. In our case study, from 4 to 27 descriptors have been quoted depending on the expert.

XLSTAT requires that all configurations have the same number of dimensions. This issue is easily solved by replacing the missing dimensions by columns of zeroes, which has no influence on the results.

When one applies GPA on free profiles data, the consensus configuration is usually not easy to interpret, as it is likely that each dimension of the consensus configuration is a mixture of heterogeneous concepts. In order to facilitate the

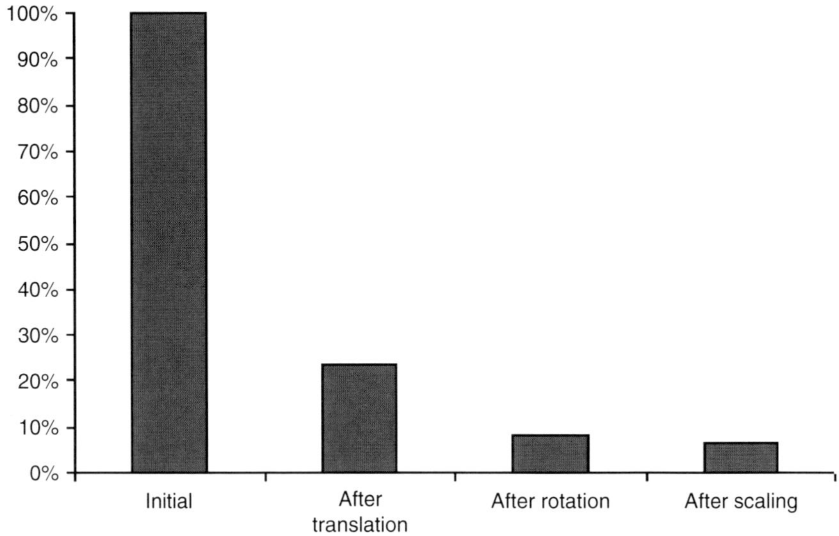

Fig. 33.1 Percentage of residual variance after each GPA transformation

interpretation of the results, we first analyzed the terms used by the experts, in order to position identical or close terms in the same column of the configurations: This allowed to reduce the number of dimensions from the 99 different terms that were used by the experts, down to 44 descriptors.

The GPA was highly efficient as it decreased the residual sum of squares down to 6% of the initial value, as shown in Fig. 33.1.

We can see that most of the variance reduction is provided by the translation. This indicates that the five experts tend to use differently the rating scale in terms of position. The scaling effect is low. This shows that the experts use similar ranges, in terms of width, of the rating scale.

GPA allows to analyze the results it provides using two point of views: the objects, here the air conditioning supports, and the configurations, here the experts. This is well illustrated by the analysis of the residuals of the objects and configurations.

The residuals for an object correspond to the sum of squared distances between the object as described in the various configurations (after the GPA) and the object in the consensus configuration, for all the dimensions kept after the PCA. The residuals for a configuration correspond to the sum of squared distances between the configuration described by the various objects and the consensus configuration, for all the dimensions kept after the PCA. Figure 33.2 shows the residuals for each air conditioning support.

We can see on Fig. 33.2 than even after the GPA, there is quite heterogeneity among the objects, the residuals being sometimes more than three time higher for some objects.

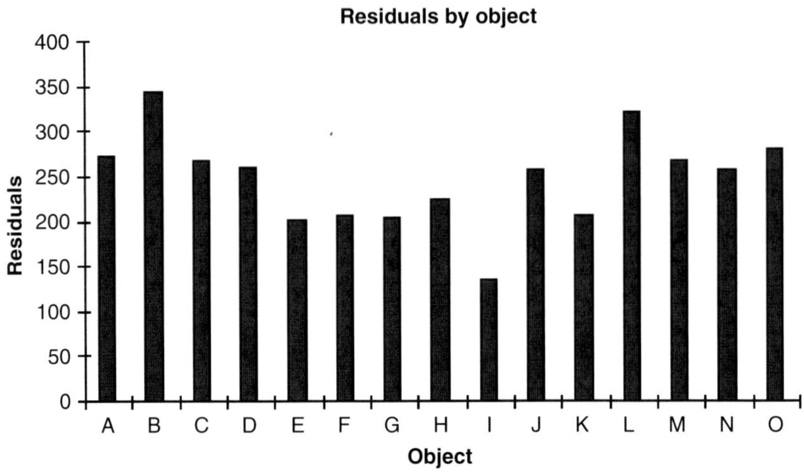

Fig. 33.2 Residuals by object

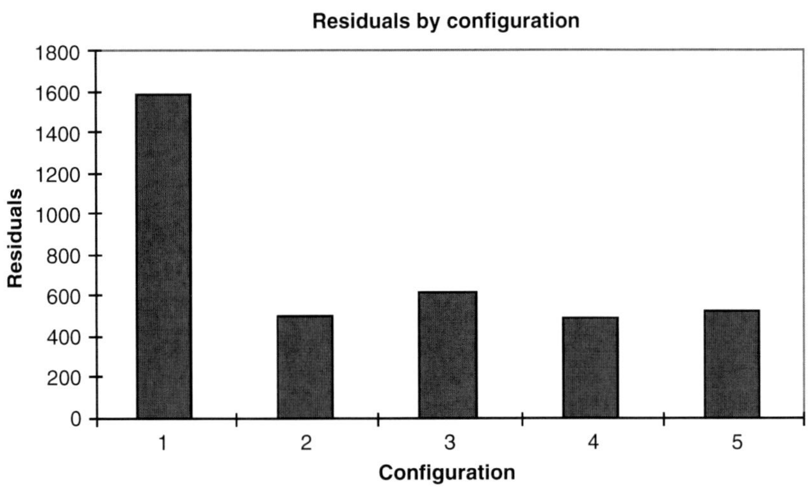

Fig. 33.3 Residuals by configuration

The residuals by configuration (Fig. 33.3) indicate that the first configuration is significantly more different from the consensus configuration than the four other ones.

However, this does not tell us whether we can considerer that the GPA was efficient in reaching a consensus or not. In order to verify this assumption, we use the consensus test developed by Wakeling et al. (1992).

On Fig. 33.4 we see that the Rc coefficient obtained after the GPA is far above those obtained after 5,000 permutations. However, we can also see there is still some

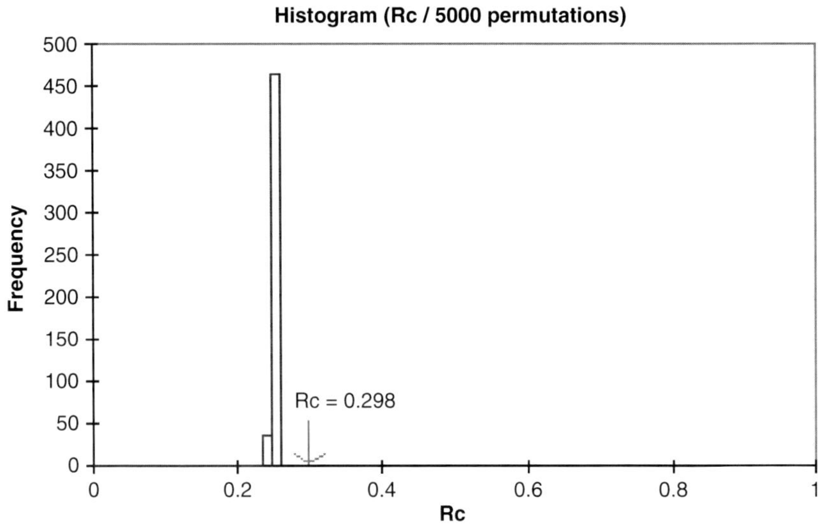

Fig. 33.4 Histogram of the Rc coefficient, and observed Rc coefficient

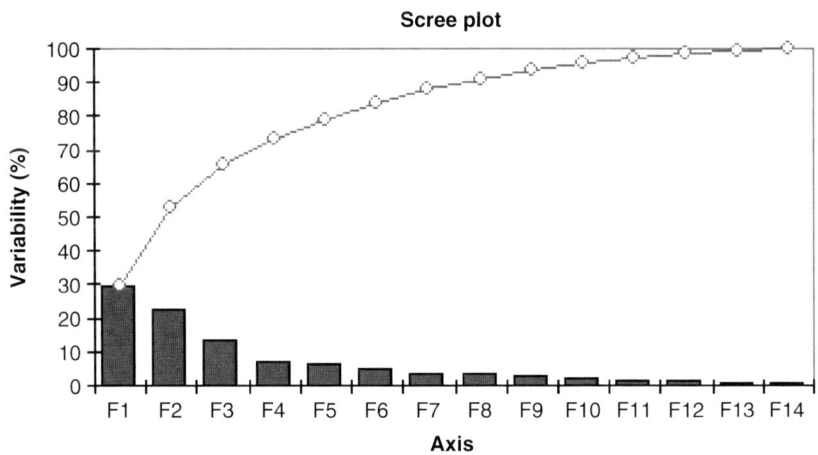

Fig. 33.5 Percentage of variability carried by the PCA axes

disagreement between the experts, as the value of Rc would be one in the case of a perfect consensus between the experts.

The permutation test, run to determine the number of dimensions that should be retained after the PCA step of the GPA, allows us to conclude that four dimensions should be enough to describe the consensus configuration. The analysis of the eigenvalues, which corresponding scree plot is displayed in Fig. 33.5, shows that the first four factors bring 72% of the total variability.

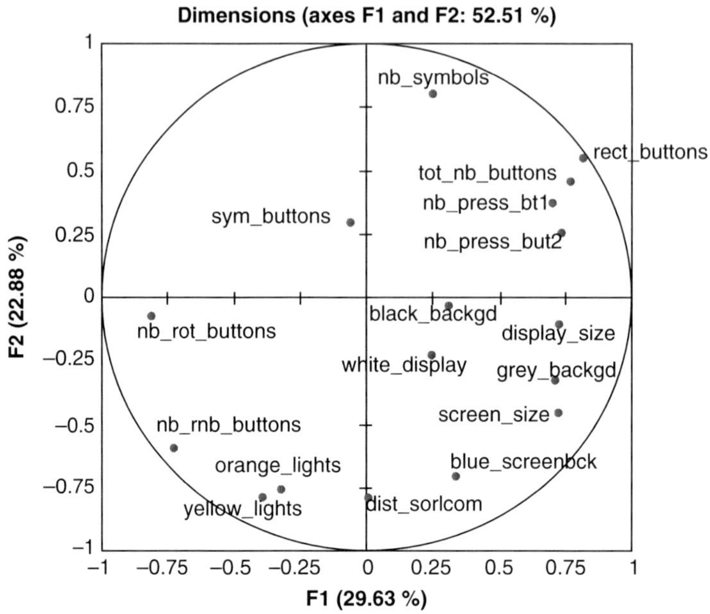

Fig. 33.6 Correlation circle (factors 1 and 2)

On Fig. 33.6 only the dimensions that have a correlation greater than 0.7 or lower than −0.7 with the four selected factors have been displayed, i.e. 17 dimensions.

The first axis obtained from the GPA/PCA is positively correlated with "*display size*", "*rectangular buttons*" and "*number of press buttons1*", but negatively correlated to "*number of rotative buttons*". The descriptors "*distance screen to commands*", "*orange lights*" and "*yellow lights*" are correlated with the second axis, and negatively correlated to "*number of symbols*".

On Fig. 33.7, we see that the third axis is strongly positively correlated with the "*black screen background*". The fourth axis is positively correlated with "*symmetric buttons*".

33.4.2 Partial Least Square Regression

Before modeling them, the preference scores *(ps)* corresponding to each customer have first been standardized in order to avoid location and scale effects (Howell 1995); for each consumer j we compute the mean $\mu(j)$ and the standard deviation $\sigma(j)$. The standardized preference score for an air conditioning support i is computed as follows:

$$\mathrm{ps_{st}(i,j)} = \frac{\mathrm{ps(i,j)} - \mu(\mathrm{j})}{\sigma(\mathrm{j})}.$$

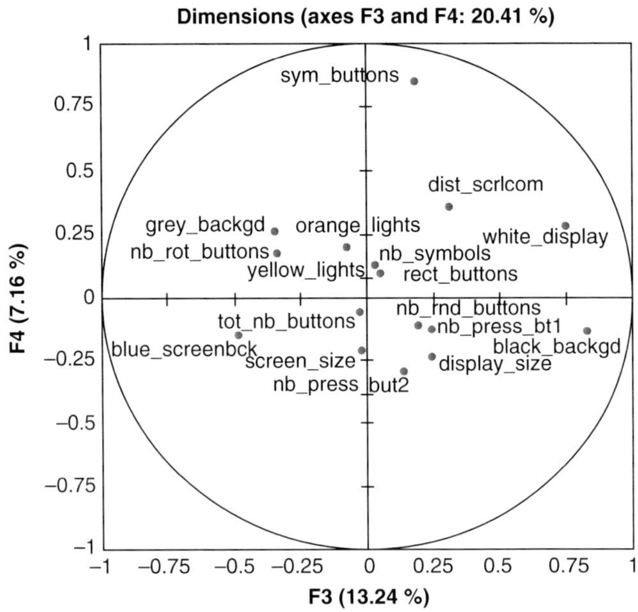

Fig. 33.7 Correlation circle (factors 1 and 2)

Table 33.1 Quality of fit indexes for the PLS regression

Index	Comp1	Comp2
Q^2 cum	−0.010	−0.022
R^2Y cum	0.170	0.288
R^2X cum	0.250	0.500

The standardized preferences of the 61 consumers are modeled using partial least squares (PLS-2) regression (Wold 1995), with as explanatory variables, the four factors obtained from the GPA/PCA step.

The regression is performed using the PLS regression function of XLSTAT (Addinsoft 2006), while taking into account the methodology presented by Tenenhaus et al. (2005). The first two components are chosen to display the results.

The cumulative Q^2 amounts to only −0.022 with two components and becomes even worse with more. We considered that the poor fit of the model was due to the heterogeneity of the customers' preferences (Table 33.1).

In order to remove some of the heterogeneity, the customers are grouped using the Agglomerative Hierarchical Clustering (AHC) function provided by XLSTAT (Addinsoft 2006). The AHC is run on the four *t* components, using the Euclidean distance and the Ward method.

Four groups have been generated (Fig. 33.8). The four groups include respectively 13, 8, 17, and 23 and consumers.

A PLS regression is then run on each group to explain the preferences.

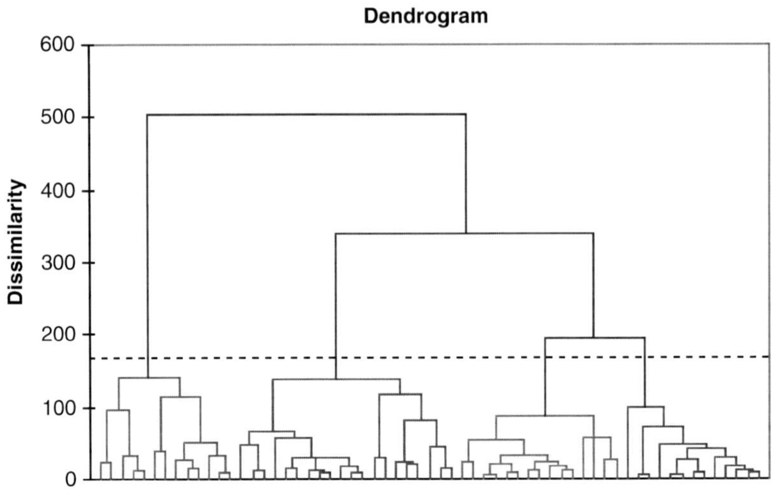

Fig. 33.8 Agglomerative Hierarchical Clustering

Table 33.2 Quality of fit indexes for the PLS regression on group 1

Index	Comp1	Comp2
Q^2 cum	0.267	0.216
R^2Y cum	0.392	0.436
R^2X cum	0.250	0.500

33.4.2.1 PLS Regression on the Data of the First Group

The quality of this PLS regression for this group is much better than it was for the whole population: using the first two components the cumulative R^2Y reaches 0.436 and the cumulative Q^2 equals 0.216 (Table 33.2).

According to the correlation map (Fig. 33.9), this group shows a preference for the air conditioning photographs F and L, but it clearly rejects B, C, D, and E. We see that the customers' preferences of this group are positively correlated with the first factor of the GPA. Using that information we deduce that consumers from group 1 prefer air conditioning supports with rectangular buttons, a large display, and buttons to press rather than buttons to rotate.

The analysis of the Variable Importance for Projection (VIP) coefficients confirms that the F1 factor is the most important in the models (Fig. 33.10).

Although we are not interested in using the models in this type of applications, it can be of interest to have a look at the PLS models. For example, for the consumer J21, we have a fine model with an R^2 of 0.6. The standardized coefficients of the model are computed in order to allow to determine if they are significant or not, and so that we can compare the coefficients (in linear regression these coefficients are often referred to as beta coefficients). The confidence intervals are estimated

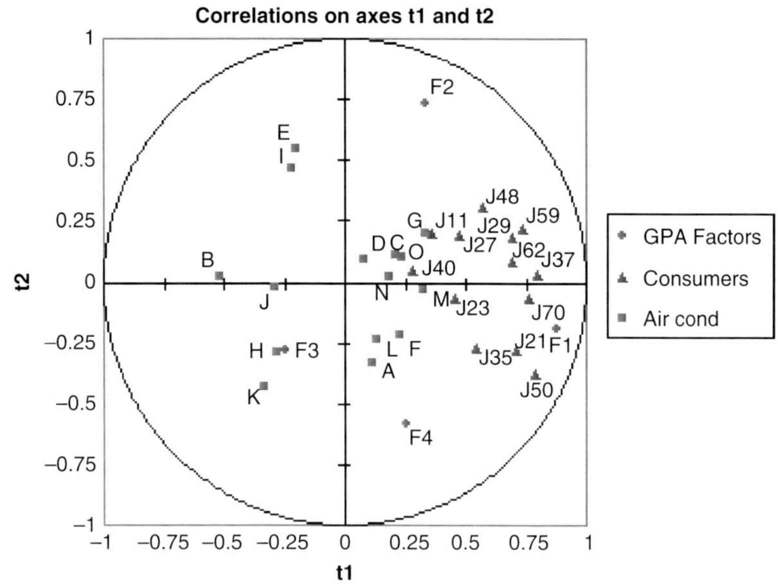

Fig. 33.9 Correlation between the t_1 and t_2 components with the air conditioning systems (■), the consumers (▲) and the GPA factors (•) (group 1)

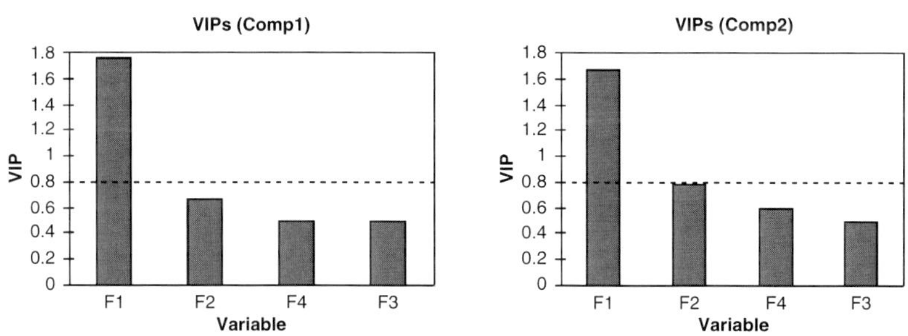

Fig. 33.10 Variable importance for the projection (group 1)

using a jackknife method. Figure 33.11 allows to confirm that only the factor F1 is significant in the model.

33.4.2.2 PLS Regression on the Data of the Second Group

The quality of the PLS regression for this group is poor as shown below (Table 33.3). This means that either the necessary information to explain the preferences is not provided by the four GP4 factors, or that a linear model is not suited for that group.

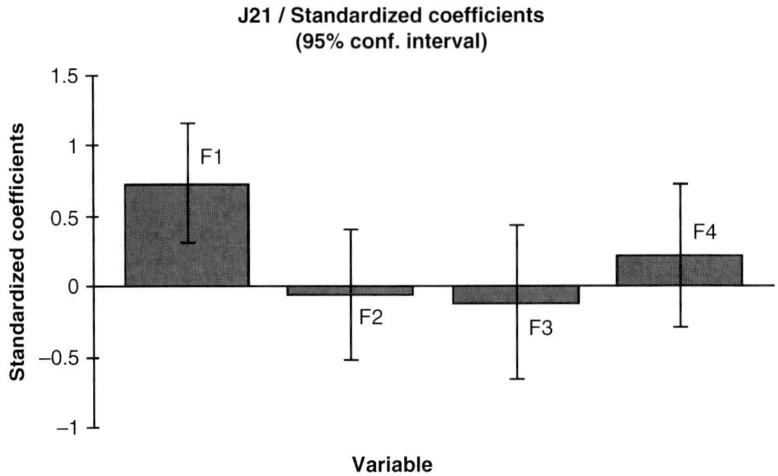

Fig. 33.11 Standardized coefficients and 95% confidence intervals (group 1)

Table 33.3 Quality of fit indexes for the PLS regression on group 2

Index	Comp1	Comp2
Q^2 cum	−0.071	−0.171
R^2Y cum	0.173	0.283
R^2X cum	0.250	0.500

Table 33.4 Quality of fit indexes for the PLS regression on group 3

Index	Comp1	Comp2
Q^2 cum	0.196	0.166
R^2Y cum	0.384	0.455
R^2X cum	0.250	0.500

33.4.2.3 PLS Regression on the Data of the Third Group

The quality of this PLS regression is almost as good as for the first group: using the first two components the cumulative R^2Y reaches 0.455 and the cumulative Q^2 equals 0.166 (Table 33.4).

According to the correlation map (Fig. 33.12), this group shows a preference for the air conditioning photographs F and L, but it clearly rejects B, C, D, and E. We see that the customers' preferences of this group are positively correlated with the fourth factor of the GPA/PCA, and negatively correlated to the second and third factors.

The analysis of the Variable Importance for Projection (VIP) coefficients indicates that the factor 2 is the most important in the models (Fig. 33.13).

Using what we established earlier for the second factor of the GPA, we understand that this group prefers the systems with symmetric buttons with more symbols, and with no yellow or orange lights.

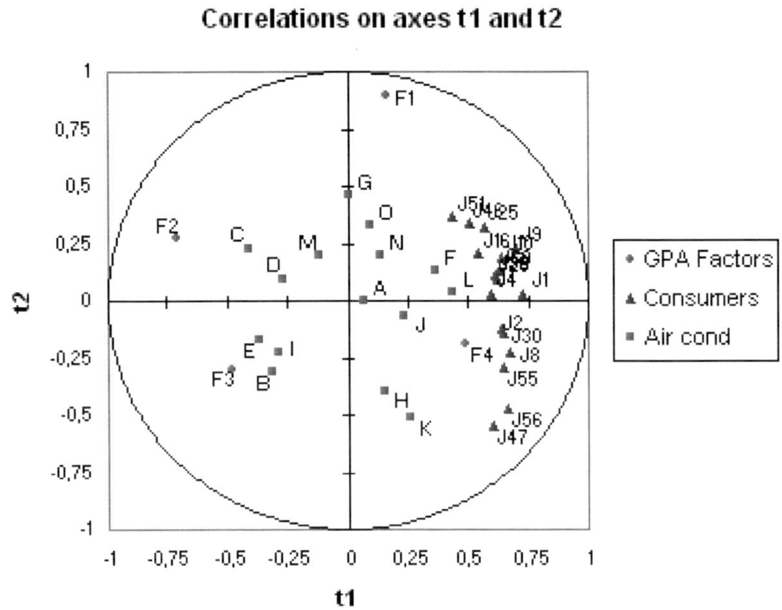

Fig. 33.12 Correlation between the t_1 and t_2 components with the air conditioning systems (■), the consumers (▲) and the GPA factors (•) (group 3).

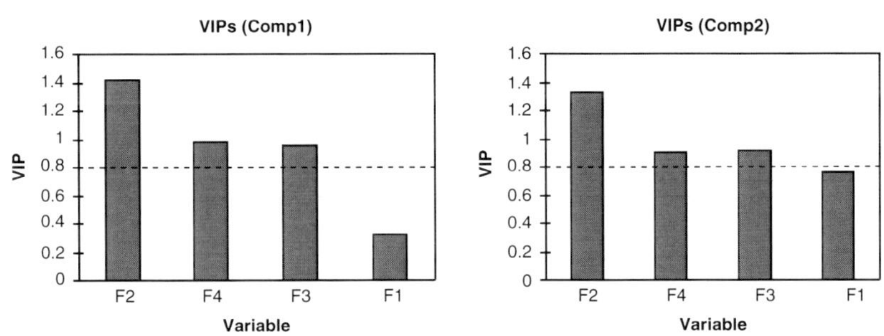

Fig. 33.13 Variable importance for the projection (group 3)

33.4.2.4 PLS Regression on the Data of the Fourth Group

The quality of this PLS regression is poor for this group: the cumulative R^2Y equals only 0.282 using two components and the cumulative Q^2 equals −0.065 (Table 33.5). With three components, the cumulative Q^2 decreases to −0.202.

Table 33.5 Quality of fit indexes for the PLS regression on group 3

Index	Comp1	Comp2
Q^2 cum	−0.066	−0.065
R^2Y cum	0.184	0.282
R^2X cum	0.250	0.500

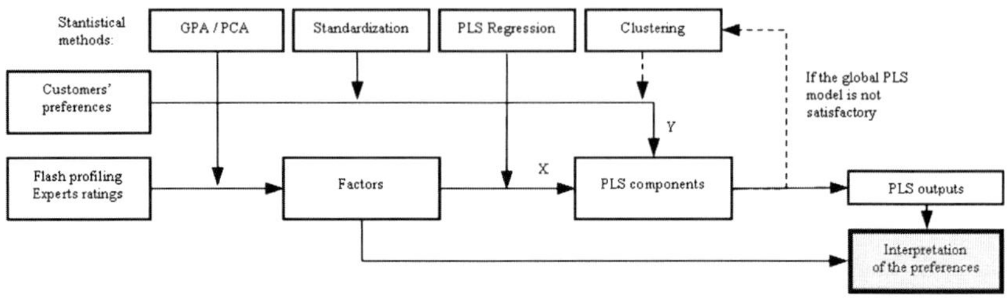

Fig. 33.14 Modeling customers' preferences using GPA/PCA and PLS regression

33.4.3 Conclusion

We developed a methodology that allows to sequentially apply GPA, PCA and PLS regression to preference data in order to explain the customers' preferences on the basis of sensory descriptors. A clustering step was necessary in order to reduce the heterogeneity of the preference data and to obtain models that can be interpreted.

We believe this approach should be more and more used in the coming years especially by the growing community of sensory analysts. The methodology used in this study is summarized on Fig. 33.14.

Future developments should allow to automatically translate the results obtained after the GPA and PLS steps in the original space of experts terms. As we are here dealing with linear combines, displaying the PLS models in the original space would make a lot of sense and would facilitate the interpretation of results.

References

Addinsoft. (2006). XLSTAT, a data analysis and statistical solution for Microsoft Excel. Paris, France.

Crochemore, S., & Nesa, D. (2004). The use of the sensory analysis to the perception of materials: How to objectivize tactile. First sino-french workshop on sensory evaluation and intelligent techniques, dec. 17–19 2003, Shanghai, R.P. Chine.

Dairou, V., & Sieffermann, J. M. (2002). A comparison of 14 jams characterized by conventional profile and a quick original method, the flash profile. *Journal of Food Science, 67*(2), 826–834.

Gower, J. C. (1975). Generalized procrustes analysis. *Psychometrika, 40*(1), 33–51.

Harries, J. M., & MacFie, H. J. H. (1976). The use of a rotational fitting technique in the interpretation of sensory scores for different characteristics. *Journal of Food Technology*, 11, 449–456.

Howell, D. C. (1995). Fundamental statistics for behavioural science, 3rd edn. Belmont: Duxbury.

Tenenhaus M., Pagès J., Ambroisine, L., & Guinot, C. (2005). PLS methodology to study relationships between hedonic judgements and product characterisitcs. *Food Quality and Preference, 16*(4), 315–325.

Wakeling, I. N., Raats, M. M., & MacFie, H. J. H. (1992). A new significance test for consensus in generalized Procrustes analysis. *Journal of Sensory Studies, 7*, 91–96.

Wold, S. (1995). PLS for multivariate linear modelling. In: H. van de Waterbeemd (Ed.), QSAR: Chemometric methods in molecular design, Vol 2 (pp. 195–218). Weinheim, Germany: Wiley-VCH.

Wu, W., Gyo, Q., de Jong, S., & Massart, D. L. (2002). Randomisation test for the number of dimensions of the group average space in eneralised Procrustes analysis. *Food Quality and Preference, 13*, 191–200.

Index

MIX
Papier aus verantwortungsvollen Quellen
Paper from responsible sources
FSC® C105338

Printed by Books on Demand, Germany